Cognitive Technologies

For further volumes:
http://www.springer.com/series/5216

Paolo Petta · Catherine Pelachaud · Roddy Cowie
Editors

Emotion-Oriented Systems

The Humaine Handbook

With 104 Figures and 35 Tables

Springer

Editors
Prof. Paolo Petta
Österreichische Studiengesellschaft für Kybernetik
Austrian Research Institute for Artificial Intelligence (OFAI)
Freyung 6/6
1010 Vienna
Austria
paolo.petta@ofai.at

Prof. Catherine Pelachaud
CNRS-LTCI, TELECOM ParisTech
37–39 rue Dareau
75014 Paris
France
catherine.pelachaud@telecom-paristech.fr

Prof. Roddy Cowie
Queen's University Belfast
Department of Psychology
Room 01.531, David Keir Building
Belfast BT7 1NN
United Kingdom
roddy.cowie@qub.ac.uk

Managing Editors
Prof. Dov M. Gabbay
Augustus De Morgan Professor of Logic
Department of Computer Science
King's College London
Strand, London WC2R 2LS, UK

Prof. Dr. Jörg Siekmann
Forschungsbereich Deduktions- und Multiagentensysteme, DFKI
Stuhlsatzenweg 3, Geb. 43
66123 Saarbrücken, Germany

Cognitive Technologies ISSN 1611-2482
ISBN 978-3-642-15183-5 e-ISBN 978-3-642-15184-2
DOI 10.1007/978-3-642-15184-2
Springer Heidelberg Dordrecht London New York

Library of Congress Control Number: 2010938791

ACM Computing Classification (1998): I.2, H.1, H.5

Cover design: KuenkelLopka GmbH, Heidelberg

Printed on acid-free paper

Springer is part of Springer Science+Business Media (www.springer.com)

Contents

Part I "Theories and Models" of Emotion

Editorial: "Theories and Models" of Emotion 3
Etienne B. Roesch

Emotion: Concepts and Definitions 9
Roddy Cowie, Naomi Sussman, and Aaron Ben-Ze'ev

Emotions in Social Interactions: Unfolding Emotional Experience . . . 31
Claudia Marinetti, Penny Moore, Pablo Lucas, and Brian Parkinson

Biological and Computational Constraints to Psychological Modelling of Emotion 47
Etienne B. Roesch, Nienke Korsten, Nickolaos F. Fragopanagos, John G. Taylor, Didier Grandjean, and David Sander

Part II Signals to Signs

Editorial: "Signals to Signs" – Feature Extraction, Recognition, and Multimodal Fusion 65
Kostas Karpouzis

The Automatic Recognition of Emotions in Speech 71
Anton Batliner, Björn Schuller, Dino Seppi, Stefan Steidl, Laurence Devillers, Laurence Vidrascu, Thurid Vogt, Vered Aharonson, and Noam Amir

Image and Video Processing for Affective Applications 101
Maja Pantic and George Caridakis

Multimodal Emotion Recognition from Low-Level Cues 115
Maja Pantic, George Caridakis, Elisabeth André, Jonghwa Kim, Kostas Karpouzis, and Stefanos Kollias

Physiological Signals and Their Use in Augmenting Emotion Recognition for Human–Machine Interaction 133
R. Benjamin Knapp, Jonghwa Kim, and Elisabeth André

Part III Data and Databases

Editorial: "Data and Databases" 163
Ellen Douglas-Cowie

Principles and History 167
Roddy Cowie, Ellen Douglas-Cowie, Ian Sneddon, Anton Batliner, and Catherine Pelachaud

Issues in Data Collection 197
Roddy Cowie, Ellen Douglas-Cowie, Margaret McRorie, Ian Sneddon, Laurence Devillers, and Noam Amir

Issues in Data Labelling 213
Roddy Cowie, Cate Cox, Jean-Claude Martin, Anton Batliner, Dirk Heylen, and Kostas Karpouzis

The HUMAINE Database 243
Ellen Douglas-Cowie, Cate Cox, Jean-Claude Martin, Laurence Devillers, Roddy Cowie, Ian Sneddon, Margaret McRorie, Catherine Pelachaud, Christopher Peters, Orla Lowry, Anton Batliner, and Florian Hönig

Part IV Emotion in Interaction

Editorial: "Emotion in Interaction" 287
Christopher Peters

Fundamentals of Agent Perception and Attention Modelling 293
Christopher Peters, Ginevra Castellano, Matthias Rehm, Elisabeth André, Amaryllis Raouzaiou, Kostas Rapantzikos, Kostas Karpouzis, Gaultiero Volpe, Antonio Camurri, and Asimina Vasalou

Generating Listening Behaviour 321
Dirk Heylen, Elisabetta Bevacqua, Catherine Pelachaud, Isabella Poggi, Jonathan Gratch, and Marc Schröder

Coordinating the Generation of Signs in Multiple Modalities in an Affective Agent 349
Jean-Claude Martin, Laurence Devillers, Amaryllis Raouzaiou, George Caridakis, Zsófia Ruttkay, Catherine Pelachaud, Maurizio Mancini, Radek Niewiadomski, Hannes Pirker, Brigitte Krenn, Isabella Poggi, Emanuela Magno Caldognetto, Federica Cavicchio, Giorgio Merola, Alejandra García Rojas, Frédéric Vexo, Daniel Thalmann, Arjan Egges, and Nadia Magnenat-Thalmann

Representing Emotions and Related States in Technological Systems . . 369
Marc Schröder, Hannes Pirker, Myriam Lamolle, Felix Burkhardt, Christian Peter, and Enrico Zovato

Embodied Conversational Characters: Representation Formats for Multimodal Communicative Behaviours 389
Brigitte Krenn, Catherine Pelachaud, Hannes Pirker, and Christopher Peters

Part V Emotion in Cognition and Action

Overview of Emotion in Cognition and Action 419
Lola Cañamero

A Bottom-Up Investigation of Emotional Modulation in Competitive Scenarios 427
Lola Cañamero and Orlando Avila-García

Novelty Processing and Emotion: Conceptual Developments, Empirical Findings and Virtual Environments 441
Didier Grandjean and Christopher Peters

Cognitive Evaluations and Intuitive Appraisals: Can Emotion Models Handle Them Both? 459
Fiorella de Rosis, Cristiano Castelfranchi, Peter Goldie, and Valeria Carofiglio

Anticipation and Emotion 483
Cristiano Castelfranchi and Maria Miceli

Socially Situated Affective Systems 501
Sabine Payr and Peter Wallis

Part VI Persuasion and Communication

Editorial: "Persuasion and Communication" 523
Massimo Zancanaro

Emotion in Persuasion from a Persuader's Perspective: A True Marriage Between Cognition and Affect 527
Maria Miceli, Fiorella de Rosis, and Isabella Poggi

Approaches to Verbal Persuasion in Intelligent User Interfaces 559
Marco Guerini, Oliviero Stock, Massimo Zancanaro, Daniel J. O'Keefe, Irene Mazzotta, Fiorella de Rosis, Isabella Poggi, Meiyii Y. Lim, and Ruth Aylett

Non-verbal Persuasion and Communication in an Affective Agent . . . 585
Elisabeth André, Elisabetta Bevacqua, Dirk Heylen, Radoslaw Niewiadomski, Catherine Pelachaud, Christopher Peters, Isabella Poggi, and Matthias Rehm

Computational Humour 609
Carlo Strapparava, Oliviero Stock, and Rada Mihalcea

Part VII Usability

Editorial: "Usability" 637
Jarmo Laaksolahti and Kia Höök

The Design and Evaluation Process 641
Joseph Jofish Kaye, Jarmo Laaksolahti, Kia Höök, and Katherine Isbister

Understanding Users and Their Situation 657
Ylva Fernaeus, Katherine Isbister, Kia Höök, Jarmo Laaksolahti, and Petra Sundström

Generating Ideas and Building Prototypes 671
Katherine Isbister, Kia Höök, Petra Sundström, and Jarmo Laaksolahti

Evaluation of Affective Interactive Applications 687
Kia Höök, Katherine Isbister, Steve Westerman, Peter Gardner, Ed Sutherland, Asimina Vasalou, Petra Sundström, Joseph Jofish Kaye, and Jarmo Laaksolahti

Part VIII Ethics and Good Practice

Editorial: 'Ethics and Good Practice' – Computers and Forbidden Places: Where Machines May and May Not Go 707
Roddy Cowie

Principalism: A Method for the Ethics of Emotion-Oriented Machines . 713
Sabine Döring, Peter Goldie, and Sheelagh McGuinness

The Ethical Distinctiveness of Emotion-Oriented Technology: Four Long-Term Issues 725
Peter Goldie, Sabine Döring, and Roddy Cowie

Emotion-Oriented Systems and the Autonomy of Persons 735
Holger Baumann and Sabine Döring

Ethics in Emotion-Oriented Systems: The Challenges for an Ethics Committee 753
Ian Sneddon, Peter Goldie, and Paolo Petta

Glossary 769

Index 781

Contributors

Vered Aharonson Tel Aviv Academic College of Engineering, Tel Aviv, Israel, vered@afeka.ac.il

Noam Amir Department of Communication Disorders, Tel Aviv University, Tel Aviv, Israel, noama@post.tau.ac.il

Elisabeth André University of Augsburg, Augsburg, Germany, Andre@informatik.uni-augsburg.de

Orlando Avila-García Open Canarias S.L., Santa Cruz de Tenerife, Spain, newoavila@hotmail.com

Ruth Aylett Heriot-Watt University, Edinburgh, UK, ruth@macs.hw.ac.uk

Anton Batliner Lehrstuhl für Mustererkennung, Friedrich-Alexander-Universität Erlangen, Erlangen, Germany, batliner@informatik.uni-erlangen.de

Holger Baumann Universitärer Forschungsschwerpunkt Ethik, Universität Zürich, Zürich, Switzerland, baumann@ethik.uzh.ch

Aaron Ben-Ze'ev Department of Philosophy, University of Haifa, Haifa, Israel, benzeev@research.haifa.ac.il

Elisabetta Bevacqua University of Paris 8, now at CNRS, Telecom-ParisTech, Paris, France, elisabetta.bevacqua@telecom-paristech.fr, e.bevacqua@iut.univ-paris8.fr

Felix Burkhardt Deutsche Telekom Laboratories, Berlin, Germany, felix.burkhardt@t-systems.com

Antonio Camurri University of Genova, Genoa, Italy, antonio.camurri@unige.it

Lola Cañamero School of Computer Science, University of Hertfordshire, College Lane, Hatfield, Herts, UK, L.Canamero@herts.ac.uk

George Caridakis Image, Video and Multimedia Systems Lab, National Technical University of Athens, Athens, Greece, GCari@image.ece.ntua.gr

Valeria Carofiglio Department of Informatics, University of Bari, Bari, Italy, Carofiglio@di.uniba.it

Cristiano Castelfranchi Institute of Cognitive Sciences and Technologies, National Research Council, Rome, Italy; Department of Communication Sciences, University of Siena, Siena, Italy, cristiano.castelfranchi@istc.cnr.it

Ginevra Castellano Queen Mary University of London, London, UK, ginevra@dcs.qmul.ac.uk

Federica Cavicchio CIMeC Università di Trento, Trento, Italy, federica.cavicchio@unitn.it

Roddy Cowie Department of Psychology, Queen's University Belfast, Belfast, Northern Ireland, UK, roddy.cowie@qub.ac.uk

Cate Cox Department of Psychology, Queen's University Belfast, Belfast, Northern Ireland, UK, C.Cox@qub.ac.uk

Laurence Devillers LIMSI-CNRS, Orsay, France, Devil@limsi.fr

Sabine Döring Department of Philosophy, University of Tübingen, Tübingen, Germany, mail@sabinedoering.de

Ellen Douglas-Cowie Department of Psychology, Queen's University Belfast, Belfast, Northern Ireland, UK, E.Douglas-Cowie@qub.ac.uk

Arjan Egges Universiteit Utrecht, Utrecht, The Netherlands, egges@cs.uu.nl

Ylva Fernaeus SICS, Stockholm, Sweden, Ylva@dsv.su.se

Nickolaos F. Fragopanagos Department of Mathematics, Kings College, London, UK, nikofrago@gmail.com

Alejandra García Rojas Ecole Polytechnique Fédérale de Lausanne, Lausanne, Switzerland, alejandra.garciarojas@epfl.ch

Peter Gardner Institute of Psychological Sciences, University of Leeds, Leeds, UK, P.H.Gardner@leeds.ac.uk

Peter Goldie Department of Philosophy, University of Manchester, Manchester, UK, peter.goldie@manchester.ac.uk

Didier Grandjean Neuroscience of Emotion and Affective Dynamics Laboratory, Department of Psychology, University of Geneva, Geneva, Switzerland; Swiss Centre for Affective Sciences, Geneva, Switzerland, Didier.Grandjean@unige.ch

Jonathan Gratch University of Southern California, Marina del Rey, CA, USA, gratch@ict.usc.edu

Marco Guerini Fondazione Bruno Kessler-Irst, Povo, Trento, Italy, guerini@fbk.eu

Dirk Heylen Faculty of Electrical Engineering, Mathematics and Computer Science, University of Twente, Enschede, The Netherlands, d.k.j.Heylen@ewi.utwente.nl

Florian Hönig Lehrstuhl für Mustererkennung, Friedrich-Alexander-Universität Erlangen, Erlangen, Germany, Hoenig@informatik.uni-erlangen.de

Kia Höök Department of Computer and Systems Sciences, Stockholm University/KTH, Kista, Sweden, Kia@dsv.su.se

Katherine Isbister Center for Computer Games Research, IT University of Copenhagen, Copenhagen, Denmark, KIsbister@itu.dk

Kostas Karpouzis Image, Video and Multimedia Systems Lab, Institute of Communications and Computer Systems, National Technical University of Athens, Athens, Greece, kkarpou@image.ece.ntua.gr

Joseph Jofish Kaye Nokia Research Center, Palo Alto, CA, USA, jofish.kaye@nokia.com

Jonghwa Kim University of Augsburg, Augsburg, Germany, Kim@informatik.uni-augsburg.de

R. Benjamin Knapp School of Music and Sonic Arts, Queen's University Belfast, Belfast, Northern Ireland, UK, B.Knapp@qub.ac.uk

Stefanos Kollias Image, Video and Multimedia Systems Lab, Institute of Communications and Computer Systems, National Technical University of Athens, Athens, Greece, Stefanos@cs.ntua.gr

Nienke Korsten Department of Mathematics, Kings College, London, UK, nienke.korsten@kcl.ac.uk

Brigitte Krenn Austrian Research Institute for Artificial Intelligence, Vienna, Austria, brigitte.krenn@ofai.at

Jarmo Laaksolahti Department of Computer and Systems Sciences, Stockholm University/KTH, Kista, Sweden, jarmo@sics.se

Myriam Lamolle Université Paris VIII, Paris, France, m.lamolle@iut.univ-paris8.fr

Meiyii Y. Lim Heriot-Watt University, Edinburgh, UK, myl@macs.hw.ac.uk

Orla Lowry Department of Psychology, Queen's University Belfast, Belfast, Northern Ireland, UK, E.Douglas-Cowie@qub.ac.uk

Pablo Lucas University of Bath, England, P.Lucas@bath.ac.uk

Nadia Magnenat-Thalmann University of Geneva, Geneva, Switzerland, thalmann@miralab.unige.ch

Emanuela Magno Caldognetto Institute of Cognitive Sciences and Technologies, Rome, Italy, emanuela.magno@pd.istc.cnr.it

Maurizio Mancini InfoMus Lab, Università di Genova, Genoa, Italy, maurizio@infomus.org

Claudia Marinetti Department of Psychology, Katholieke Universiteit Leuven, Leuven, Belgium, claudia.marinetti@chch.oxon.org

Jean-Claude Martin Computer Sciences Laboratory for Mechanics and Engineering Sciences (LIMSI), Paris, France, Martin@limsi.fr

Irene Mazzotta University of Bari, Bari, Italy, mazzotta@di.uniba.it

Sheelagh McGuinness Centre for Professional Ethics, Keele University, Staffordshire, UK, S.McGuinness@peak.keele.ac.uk

Margaret McRorie Department of Psychology, Queen's University Belfast, Belfast, Northern Ireland, UK, M.McRorie@qub.ac.uk

Giorgio Merola Università Roma Tre, Rome, Italy, merogio@hotmail.com

Maria Miceli Institute of Cognitive Sciences and Technologies, CNR, Rome, Italy, maria.miceli@istc.cnr.it

Rada Mihalcea Department of Computer Science, University of North Texas, Denton, TX, USA, rada@cs.unt.edu

Penny Moore Department of Psychology, Oxford University, Oxford, UK, penny.moore@psy.ox.ac.uk

Radek Niewiadomski Telecom ParisTech, Paris, France, niewiado@telecom-paristech.fr

Radoslaw Niewiadomski CNRS, Telecom-ParisTech, Paris, France, radoslaw.niewiadomski@telecom-paristech.fr

Daniel J. O'Keefe Northwestern University, Evanston, IL, USA, d-okeefe@northwestern.edu

Maja Pantic Department of Computing, Imperial College, London, UK; Faculty of Electrical Engineering, Mathematics and Computer Science, University of Twente, Enschede, The Netherlands, M.Pantic@imperial.ac.uk

Brian Parkinson Department of Psychology, Oxford University, Oxford, UK, brian.parkinson@psy.ox.ac.uk

Sabine Payr Austrian Research Institute for Artificial Intelligence, Vienna, Austria, Sabine.Payr@ofai.at

Catherine Pelachaud CNRS-LTCI, TELECOM ParisTech, Paris, France, Catherine.Pelachaud@telecom-paristech.fr

Christian Peter Fraunhofer IGD, Rostock, Germany, christian.peter@igd-r.fraunhofer.de

Christopher Peters Coventry University, Coventry, UK, Christopher.Peters@coventry.ac.uk

Paolo Petta Österreichische Studiengesellschaft für Kybernetik, Austrian Research Institute for Artificial Intelligence, Vienna, Austria, paolo.petta@ofai.at

Hannes Pirker Austrian Research Institute for Artificial Intelligence, Vienna, Austria, hannes.pirker@ofai.at

Isabella Poggi University of Rome 3, Rome, Italy, Poggi@uniroma3.it

Amaryllis Raouzaiou National Technical University of Athens, Athens, Greece, araouz@image.ntua.gr

Kostas Rapantzikos National Technical University of Athens, Athens, Greece, rap@image.ntua.gr

Matthias Rehm University of Augsburg, Augsburg, Germany, matthias.rehm@informatik.uni-augsburg.de

Etienne B. Roesch Centre for Integrative Neuroscience and Neurodynamics, University of Reading, Reading, UK, contact@etienneroes.ch

Zsófia Ruttkay University of Twente, Enschede, The Netherlands, zsofi@cs.utwente.nl

David Sander Department of Psychology, University of Geneva, Geneva, Switzerland; Swiss Centre for Affective Sciences, Geneva, Switzerland, david.sander@unige.ch

Marc Schröder Deutsches Forschungsinstitut für Künstliche Intelligenz, Saarbrücken, Germany, schroed@dfki.de

Björn Schuller Institute for Human-Machine Communication, Technische Universität München, Munich, Germany, schuller@tum.de

Dino Seppi Fondazione Bruno Kessler-Irst, Trento, Italy, seppi@fbk.eu

Ian Sneddon Department of Psychology, Queen's University Belfast, Belfast, Northern Ireland, UK, I.Sneddon@qub.ac.uk

Stefan Steidl Lehrstuhl für Mustererkennung, Friedrich-Alexander-Universität Erlangen, Erlangen, Germany, steidl@informatik.uni-erlangen.de

Oliviero Stock Fondazione Bruno Kessler-Irst, Povo, Trento, Italy, stock@fbk.eu

Carlo Strapparava Fondazione Bruno Kessler-Irst, Povo, Trento, Italy, strappa@fbk.eu

Petra Sundström Department of Computer and Systems Sciences, Stockholm University/KTH, Kista, Sweden, Petra@dsv.su.se

Naomi Sussman Department of Philosophy, University of Haifa, Haifa, Israel, snaomi@research.haifa.ac.il

Ed Sutherland Institute of Psychological Sciences, University of Leeds, Leeds, UK, E.J.Sutherland@leeds.ac.uk

John G. Taylor Department of Mathematics, Kings College, London, UK, john.g.taylor@kcl.ac.uk

Daniel Thalmann Ecole Polytechnique Fédérale de Lausanne, Lausanne, Switzerland, daniel.thalmann@epfl.ch

Asimina Vasalou University of Bath, Bath, UK, MinaV@luminainteractive.com

Frédéric Vexo Ecole Polytechnique Fédérale de Lausanne, Lausanne, Switzerland, frederic.vexo@epfl.ch

Laurence Vidrascu Iminent SA, Paris, France, laurence.vidrascu@free.fr

Thurid Vogt Multimedia Concepts and their Applications, University of Augsburg, Augsburg, Germany, thurid.vogt@informatik.uni-augsburg.de

Gaultiero Volpe University of Genova, Genoa, Italy, gualtiero.volpe@unige.it

Peter Wallis University of Sheffield, Sheffield, UK, P.Wallis@dcs.shef.ac.uk

Steve Westerman Institute of Psychological Sciences, University of Leeds, Leeds, UK, S.J.Westerman@leeds.ac.uk

Massimo Zancanaro Fondazione Bruno Kessler-Irst, Povo, Trento, Italy, Zancana@fbk.eu

Enrico Zovato Loquendo S.p.A., Torino, Italy, enrico.zovato@loquendo.com

Introduction: History, HUMAINE and This Handbook

Emotion pervades human life in general and human communication in particular. The point has been made often and elegantly, for instance, in the opening of de Sousa's piece on emotion in the Stanford Encyclopaedia of Philosophy:

> No aspect of our mental life is more important to the quality and meaning of our existence than emotions. They are what make life worth living, or sometimes ending. (de Sousa, 2010).

The pervasiveness of emotion sets information technology a challenge. Traditionally, it has focused on allowing people to accomplish a practical task in the most efficient way and has set emotion to one side – or expected users to. That is acceptable when the technology is a small part of life; but the more technology becomes interwoven through the fabric of life, the more unsatisfactory it becomes to expect that people will suspend their emotional nature and habits when they are interacting with the technology. The problem is particularly acute for groups who find it difficult to adopt the required unemotional stance (for instance, because of age or cognitive limitations); or in situations that push people away from that stance (for instance, because they are challenging or relaxed); or in activities where emotional factors play a central role (for instance, learning or modifying entrenched habits).

The problem began to attract attention during the 1990s, for instance, with work on the synthesis of voices that convey at least some emotional colouring rather than a remorseless monotone (Murray and Arnott, 1993). The research came to general notice with Picard's groundbreaking book '*Affective Computing*' (1997). Around the same time, the European Commission funded a series of projects on emotion and computing led by John Taylor and Stefanos Kollias (Cowie et al., 2001). The European work led to a major project called HUMAINE. The name reflected the spirit of the enterprise, which was to 'humanise' computing. It was from that project that this handbook grew.

HUMAINE included distinguished figures from a wide range of academic backgrounds, some of whom have gone on to publish collections that deal in-depth with the issues that arise from the standpoint of their own discipline. For example, Peter Goldie (2010) edited the *Oxford Handbook of Philosophy of Emotion*, and Klaus Scherer's team brought together research on the psychological underpinnings of the area in their 'Blueprint' for an affectively competent agent (Scherer et al., 2010). HUMAINE members were also deeply involved in conferences whose proceedings are an essential resource – *Affective Computing and Intelligent Interaction* in 2007 and 2009 (Paiva et al., 2007; Cohn et al., 2009) and the Royal Society's workshop on the Computation of Emotion in Man and Machines (Robinson and al Kaliouby, 2009). However, it was always part of the intention that the members of HUMAINE, along with key members of the wider community, would produce a book that offered newcomers to the area an academically sound introduction to the whole range of disciplines that are involved – technical, empirical, and conceptual. This is that book.

The fundamental aim of the book is to provide readers with a broad overview of the areas that should be familiar to any team that wants to do credible work on affective/emotion-oriented computing. That is easy to say, but it is actually not straightforward to form a cohesive map of the areas that are relevant. One of the tasks that HUMAINE faced was to form a working understanding of the material that needed to be addressed and the way it could be parcelled out. Eventually eight working parts were identified. The chapters are grouped into those eight parts. The next part of this introduction summarises them in turn.

It is worth neutralising a point of terminology at this stage. Some descriptions of the part use the term 'emotion'; others use 'affect'. 'Emotion' is used in this editorial because it is generally understood. Debates about the advantages of different terminologies are left to the substantial chapters.

It is also worth commenting on critiques of what those who write them call 'affective computing'. Some of these have received a considerable amount of publicity (e.g. Gaver, 2009; Turkle, 2010). Certainly some visions of affective computing do fully deserve criticism. They may fail to register the first point that was made in this Introduction, that emotion is something that pervades life rather than a discrete phenomenon. They may fail to acknowledge the complexity of even the most basic kinds of emotional competence. They may fail to acknowledge that engaging seriously with emotion requires engagement with several established disciplines. They may imply that they are creating machines with the ability to feel as humans do, rather than simply to take account of the way humans signal various kinds of feelings.

The spirit of this collection is that if research does any of these things, then it deserves criticism. On the other hand, the criticism ought to rebound on writers who criticise the research area as a whole without having taken the trouble to understand how issues like these are actually understood and handled within it. Readers are invited to think through the relationship between criticisms and the research effort that they see reflected in these chapters.

Part I: Theories and Models

Designing systems that can engage with emotion depends on having an appropriate understanding of emotion. The outstanding problem in this area is the temptation to overestimate the approaches to emotion that come easily to hand. On the one side, 'folk psychology' gives people a comfortable feeling that they understand emotion simply because it is part and parcel of their everyday life. Sadly, what folk psychology provides is a massively simplified image of an immensely diverse and complex set of phenomena. Research that accepts it uncritically can all too easily be diverted into dead ends, such as studying stereotyped renditions of smiles and angry voices, on the false assumption that they are an adequate representation of the way emotion is expressed in reality. Many more misdirections wait – such as false assumptions about the effect of cheerful voices or alignment between parties in a conversation. On the other side, the shelves of academic libraries are well stocked with theories that offer admirably clear prescriptions. It is natural to assume that picking them up and implementing them is the scientifically correct thing to do. Alas, clear theories in this area are generally opening bids in what the parties know will be a long negotiation. They are extremely useful points of reference, sometimes even when they are thoroughly discredited; but they are not like the theory of evolution, or the theory of relativity, of the theory of plate tectonics.

In that spirit, the chapters in Part I set out to give a broad, tolerant overview of ideas that are known to be useful. The chapters approach from three perspectives – broadly cognitive, broadly social, and broadly computational neuroscience. It is well known that differences arise when people take these different perspectives. The chapters do not attempt to hide that, but neither do they seek out confrontational positions. The aim is that a reader who has absorbed the material in all three will have a mature idea of the resources and the difficulties that typify the area.

Part II: Signals to Signs

The second part is one that engineers are drawn to, because it seems to be a straightforward application of standard techniques – particularly techniques from signal processing. There are well-developed technologies for face recognition, voice recognition, automatic speech recognition, and so on. Emotion recognition seems to offer an opportunity to apply those in a different domain. Up to a point that is true, but there is now a solid body of research on the specific problem of emotion recognition, and it has identified a great many wheels that need not be reinvented (and shown that some of them come off).

The single most important point to make about the material in this section is that it treats emotion detection as a multimodal task. There are separate chapters on recognition from speech and visual cues, because there are quite highly developed technologies there. There is also a chapter on inputs that are often seen as a royal road to emotion recognition, that is, sensors that detect the bodily changes associated with strong emotional reaction – heart rate, skin conductance, and so on. All of those sources do yield information, but critically, all of them are limited. Speech carries limited amounts of information about how positive or negative the speaker is, and it tends to be intermittent. When it is taking place, it seriously complicates extraction of information from the face. That information in turn is limited outside social situations; and in social situations, it is not necessarily sincere. Bodily changes do not have those limitations, but they are very hard to distinguish from effects of non-emotional physical or mental activity. In recognition of those lessons, the section explicitly includes a chapter on multimodal recognition.

Part III: Data and Databases

Many of the technologies in the area depend on substantial collections of recordings, associated with 'labels' that describe their emotional content or emotion-related features. That is linked to the caution about theory that was expressed above. Systems that engage competently with emotion need a depth of information about detail that traditional theory does not provide; and direct recordings of the phenomena are the only obvious source.

From that point of view, a suitable collection of recordings, furnished with suitable labels, serves much the same kind of function as a theory of human behaviour, except that it covers details that theories omit. If that is so, then building suitable databases is a way of doing basic science in this area. The chapters in this section are written in that spirit. They look at the development of that viewpoint; then at the issues of making suitable collections and developing suitable descriptive schemes; and then they describe a prototype for the kind of database that seems to be needed.

Part IV: Emotion in Interaction

The original intention was that this part would be a mirror image of Part II. One dealt with the transition from pixels (or other signal elements) to high-level descriptions: the other would deal with the transition from high-level descriptions to pixels (or other signal elements). Experience showed that it was wrong to expect that kind of symmetry. The areas that were difficult on the upward path (from pixels to high level descriptors) were not particularly interesting on the downward path; but it was a major challenge for research concerned with the downward path to recreate patterns that were there at ground level in the upward path.

As a result, the chapters in this part highlight problems at a very different level. The chapters in Part I: Theories and Models indicate that emotion has intimate links to attention (a person who is frightened of the knife does not gaze casually at the couples on the beach). How can we generate an agent that shows emotionally appropriate attention patterns? Two people who are in emotional rapport will co-ordinate their behaviour in particular ways. How can we generate an agent that displays the relevant kinds of co-ordination? Behind those, how can the relevant flows of events be represented?

Part V: Emotion in Cognition and Action

The emphasis in the previous sections has been on behavioural patterns – facial and vocal – that are relatively specific to emotion. However, everybody knows that emotion can be signalled by the way a person goes about making a meal or shopping. That is because emotion has pervasive effects on different aspects of cognition and action, such as the way people perceive, move, evaluate, think, and make choices. The chapters in Part V are concerned with modeling those effects. Progress at that level would pay multiple dividends: ability to simulate convincingly emotional behaviour; ability to recognise signs of emotion that are not contained in specific signals; and ability to anticipate how a person in a particular emotional state might react to a particular 'move' that the agent considered making.

Simulating the processes involved in the emotional modulation of our cognition and action is a huge task, and at present no single approach comes close to solving it. Hence the chapters reflect three main, and in our view complementary, approaches. On the one side, 'embodied' approaches use models based on biological processes to capture a style of fast, automatic processing that is characteristic of the emotions more closely related to survival. On the other side, 'rational' approaches use symbolic representations to model key mental processes. Some of these are grounded in cognitive science and deal with processes such as appraisal and anticipation. Perhaps less obviously, others draw on ideas from sociology and address the way emotion functions in interactions and relationships.

Part VI: Persuasion and Communication

It is a feature of emotion that a great deal of the communication associated with it is nonverbal. That does not mean that a balanced view of emotion can ignore verbal communication. On the contrary, there are specific and important types of communication that depend on words and emotion working in tandem. The chapters in this section concentrate on two of those.

The first, which is more extensively covered, is persuasion. Recognising that emotion is fundamental to persuasion is hardly a new insight. One of the most influential early discussions of emotion is Aristotle's. The book where it occurs is the *Rhetoric*, which is about swaying people's minds to a particular viewpoint – in other words, persuasion. It takes on a very different colour in this context, though. It is a deep challenge to integrate the logical formalisms that are usually used to construct arguments with representations that express emotional values. The chapters cover the ground systematically, looking first at the part emotion plays in persuasion; then at verbal aspects of persuasion; then at the nonverbal elements that need to accompany them to convey that the agent is trustworthy, credible, and so on.

Last but not least, the section considers humour. It is very understandable that people tend not to take humour seriously. However, humour plays an enormous part in creating an appropriate emotional climate. Anyone who doubts it only needs to think how difficult it is for politicians to win elections if they totally lack a sense of humour. It is unlikely that a humourless artificial agent would be much more popular – particularly if our efforts to lighten the tone were met invariably with a brick wall.

Part VII: Usability

There is no point in building systems that simulate human emotional behaviour exquisitely if people hate them (at least, no practical point). The chapters in this section focus attention on the problem of gauging what kind of system people will actually welcome into their lives. That calls for a shift of perspective which is more challenging than one might imagine.

The section highlights an analysis of human beings that is quite different from the kind of analysis that was introduced at the beginning of the book. It considers people as actors in particular social settings, who are drawn to artefacts that enhance the life experiences that are open to them in those settings. In that context, it is natural to view emotion as a feature of the way that people relate to things and other people in their environment, rather than as a (largely) inherited set of behaviours and dispositions. The editors regard the difference between the two not as a conflict, but as a shift of focus (wide angle to zoom, so to speak): not everyone agrees.

A second key point is that questions about people's response should not be held back until the product is complete. Instead, they need to be integrated into the design process. That calls for methods of accessing people's emotional responses that can realistically be integrated into the design cycle and that connect with designers' thought processes. Some of them will raise eyebrows among people with a formal scientific training. However, design is not a formal science. It is a practical culture and one that it is extremely important for the scientific discipline to engage with.

On the other hand, the section also covers more recognisably scientific evaluation methods. Few, if any other sources, present as comprehensive a review of them.

Part VIII: Ethics and Good Practice

Research on human beings takes it for granted that there are ethical questions it has to address. It should be no surprise that when technology moves into similar areas, it has to face similar questions. After all, the aim is to lower barriers that currently divide humans from machines. The aim of lowering the barriers is to benefit human beings. However, it is in the nature of lowering barriers that what flows across them will not necessarily be beneficial.

Subtle minds have been thinking for at least 3,000 years about the ethics of interactions between human and human. By comparison, there has been hardly any time to think about interactions between humans and machines with human-like capacities. Arguably there has been no time at all. In the domain of emotion, the machines that we actually have may well be less competent than lizards, and it is hard to doubt that they are less competent than crows. That should be borne in mind when voices elsewhere make pronouncements based on machines in science fiction movies. The voices in this collection try to deal with the ethical issues raised by technologies that an informed person might realistically imagine.

At a conceptual level, the chapters consider three types of issue that the area needs to be aware of. The first is identifying a suitable basis for ethical judgements. The second is understanding the basis for moral concerns, whether or not they are justified. The third is recognising the human characteristics that are most deeply involved in the ethical concerns.

Finally, the section looks at the mainstay of practical responses, that is, the ethics committee. They are the instrument that related areas have developed to manage their ethical challenges. Experience in those areas offers a resource that is well worth assimilating.

The outline of the chapters underlines the range of disciplines that the area involves. The fact that it is so interdisciplinary has always meant that incomers brought different perspectives, paradigm assumptions, and values to it. As a result, HUMAINE meetings rarely passed without one party transgressing across lines that another regarded as fundamental. It would be wrong to pretend that the tensions were all resolved or are resolved now. There are areas where the contributors do not agree and will not pretend to. The collection tries to represent them fairly, not to present an illusion of unanimity. The fact that there are disagreements is a true reflection of the fact that the challenge being addressed is large and very difficult. It is right and proper that different groups should have different ideas about the best way to address it.

Nevertheless, it is a hallmark of the teams who took part in HUMAINE that they were thoroughly exposed to others who had different backgrounds and priorities and acquired the ability to disagree constructively and with mutual respect. It is still a shock to encounter teams who did not have that experience and find it difficult to register that their own perspective may not hold all the answers.

Perhaps the single most positive outcome of the book would be to widen access to that feature of HUMAINE. Reading articles written by people with diverse backgrounds is not the same as meeting them, and talking to them, and coming to appreciate where they are coming from. However, it can go part way. The enterprise of humanising computing is too large and too diverse to work without establishing that kind of human underpinning: and that, of course, depends on engaging not only intellectually but emotionally.

Paolo Petta
Austrian Research Institute for Artificial Intelligence, Vienna, Austria, paolo.petta@ofai.at

Catherine Pelachaud
CNRS – LTCI, TELECOM ParisTech, Paris, France, catherine.pelachaud@telecom-paristech.fr

Roddy Cowie
Department of Psychology, Queen's University Belfast, Belfast, Northern Ireland, UK, roddy.cowie@qub.ac.uk

References

Cohn J, Nijholt A, Pantic M (eds) (2009) International conference on affective computing & intelligent interaction. IEEE Computer Society Press, Los Alamitos, Amsterdam

Cowie R, Douglas-Cowie E, Tsapatsoulis N, Votsis G, Kollias S, Fellenz W, Taylor JG (2001) Emotion recognition in human-computer interaction. IEEE Signal Process Mag 18:32–80

Gaver W (2009) Designing for emotion (among other things). Philos Trans R Soc B 364(1535):3597–3604

Goldie P (ed) (2010) The oxford handbook of philosophy of emotion. Oxford University Press, Oxford

Murray I, Arnott JL (1993) Toward the simulation of emotion in synthetic speech. J Acoust Soc Am 93:1097–1108

Paiva A, Prada R, Picard R (eds) (2007) Proceedings of the 2nd international conference on Affective Computing and Intelligent Interaction, Lisbon, Portugal, 12–14 Sept 2007. Lecture Notes in Computer Science, vol 4738. Springer, Berlin

Picard RW (1997) Affective computing. The MIT Press, Cambridge, MA

Robinson P al Kaliouby R (ed) (2009) Computation of emotions in man and machines. Philos Trans R Soc B 364(1535):3439–3604

Scherer K, Banziger T, Roesch EB (2010) A blueprint for affective computing: a sourcebook. Oxford University Press, Oxford

de Sousa R (2010) Emotion. The Stanford Encyclopaedia of Philosophy. http://plato.stanford.edu/entries/emotion/. Accessed on 2 November 2010

Turkle S (2010) In good company? On the threshold of robotic companions. In: Wilks Y (ed) Close engagements with artificial companions. John Benjamins, Amsterdam/Philadelphia, PA, pp 3–10

Part I
"Theories and Models" of Emotion

Editorial: "Theories and Models" of Emotion

Etienne B. Roesch

Abstract Of all work-packages of the HUMAINE Network of Excellence, WP3 "Theories and Models" was probably the most heterogeneous group that one can think of. It was composed of researchers from psychology, cognitive neuroscience, philosophy, ethology and a wide range of disciplines in computer science including neural networks, artificial intelligence and signal processing. We aimed at describing, informing and advising other work-packages of relevant emotion theories and models, as well as scientifically studying emotion, taking advantage of all resources available in HUMAINE.

1 Scope of WP3 "Theories and Models"

One of the main challenges faced by whoever seeks to scientifically address emotion – or simply to use emotion theories in a particular (engineering) framework – is to comprehend the variety of emotion theories available. Emotion theories differ greatly with respect to the components they represent and the levels of processing they address. This rather disparate landscape may yield confusion, as not all theories are good for all purposes. Throughout HUMAINE, we used one representation of this landscape in the form of Table 1. This table, taken from Scherer and Peper (2001), represents major theoretical threads and attempts to describe the boundaries for each theory. HUMAINE members were thus able to identify the particular component and the particular level that was most suited for their needs. Most chapters of this book will go back to the theory and model of their choice, and describe how it was applied in their area of expertise; the hope is that reproducing this table here will help the reader to grasp how theories and models relate to each other.

Another big challenge faced by HUMAINE members was the identification of the *transition rules* that allow to pass from one level of conceptualisation to

E.B. Roesch (✉)
Centre for Integrative Neuroscience and Neurodynamics, University of Reading, Reading, UK
e-mail: contact@etienneroes.ch

P. Petta et al. (eds.), *Emotion-Oriented Systems*, Cognitive Technologies,
DOI 10.1007/978-3-642-15184-2_1, © Springer-Verlag Berlin Heidelberg 2011

Table 1 Representation of the landscape of emotion theories (Scherer and Peper, 2001)

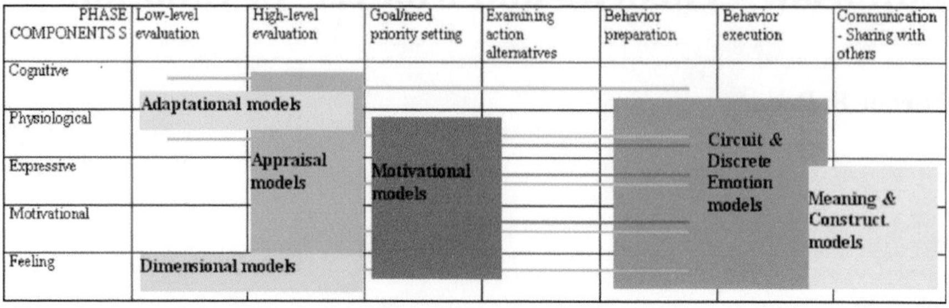

Fig. 1 Relations between the levels of representations addressed by emotion theories

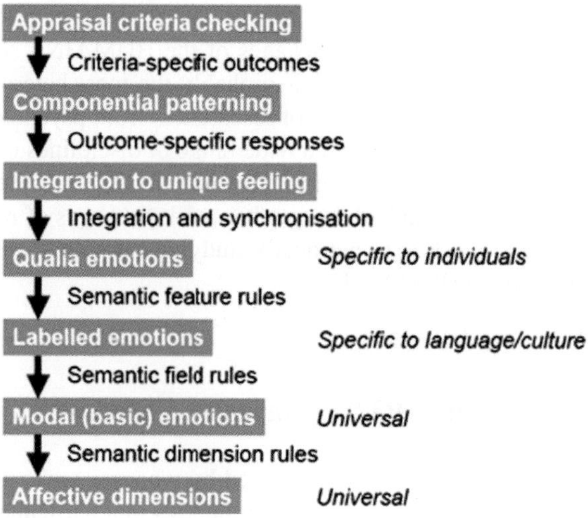

the next. Figure 1 illustrates this exercise and introduces some ways to integrate different approaches. For instance, passing from basic emotion theories to a dimensional perspective may be achieved through semantic dimension rules that describe the relationship between the features of modal emotions and the dimensions of affect. The GRID study, for instance, is one attempt to define these transition rules (see Sect. 3.3).

2 Outreach Efforts

Throughout HUMAINE, we produced several documents aiming at introducing the network to emotion theories. Each partner had particular goals to achieve, and thus needed answers to particular questions. One of our first goals was thus to establish a dialogue between theory-oriented groups and engineering-oriented groups to identify areas of common interest and means to bridge the gaps. In this section, we briefly describe some of these outreach efforts. Most of the documents resulting from this dialogue are reproduced on the HUMAINE portal.

2.1 The Proceedings of the First HUMAINE Workshop in Geneva

WP3 organised the first HUMAINE workshop. It was held in Geneva, from June 17 to 19, 2004. The aim of the workshop was explicitly to bridge the gaps between the many disciplines and individual expertise of HUMAINE, by attempting to create a cohesive affective science research community. This meeting paved the ground for the interactions and collaborations that followed, as reflected in the content of this book. The main achievement was the identification of ways to achieve interdisciplinarity by establishing a common language. The proceedings of the workshop can be found online at this address: http://emotion-research.net/projects/humaine/ws/wp3/ (last visit: November 14, 2010).

2.2 Definition of Concepts

One of the first deliverables we produced reviewed some of the most important concepts in the scientific study of emotions and provided pointers to the relevant literature. Significant scientific contributions were extended in journal articles (e.g. Scherer, 2005, and ensuing responses in *Social Science Information*). A whole chapter in this section of the book is dedicated to this topic. More information can be found at http://emotion-research.net/projects/humaine/deliverables/D3c.pdf (last visit: November 14, 2010).

2.3 The GRID Study

One of the main outcome of WP3 has been the GRID instrument, developed to address the semantic profiles of emotion words in different languages (Roesch et al., 2006; Scherer, 2005; Fontaine et al., 2007. This instrument comprehensively gathers 144 features, representing the 6 components explicitly assumed by most current emotion theorists as centrally relevant to the domain of emotion. Applied to three different cultures, we found robust evidence for at least four dimensions to represent the semantic content of emotion words: in order of importance, evaluation–pleasantness, potency–control, activation–arousal and unpredictability (of the occurring event). Whereas the first three dimensions resemble the space suggested half a century ago (Wundt, 1905), the fourth dimension, unpredictability, is not reported in most studies. This latter dimension reflects the urgent reaction to novel stimuli and unfamiliar situations. Of most interest, it renders an explicit continuum differentiating the semantic spaces of surprise, fear and anxiety. More information about the GRID project can be found at http://www.iccra.net/grid-project (last visit: November 14, 2010).

2.4 *Blueprint for Affective Computing: A Source Book (Oxford University Press, Series in Affective Science)*

One of the main outreach activity of WP3 is the publication of a textbook for all those interested in computational models of emotion based on the state of the art in current scientific investigation of affect in emotion psychology and affective neuroscience (Scherer et al., 2010). Its aim is to present systematic theoretical conceptualisations of the processes underlying emotional reactions and their implications for various fields in applied affective computing. It provides an accessible description of the structures, functions and mechanisms underlying emotional reactions, including processes involved in the elicitation of emotional responses, as well as expressive and physiological responses generated during emotional responses, and interpersonal perception of emotional responses.

3 Conclusion

This section of the book gathers chapters representative of the heterogeneity found in WP3. In this respect, WP3 was probably the most extreme case of inter-disciplinary collaborations.

The chapter by Cowie, Sussman and Ben-Ze'ev reviews the concepts used in the scientific study of emotions. In what constitutes a genuine *tour de force*, they identify the commonalities and differences between theories and point at the ambiguities that may confuse someone new to the field. They spell out the features that are proposed by emotion theorists as central in the definition of emotion and attempt to wrap it all together.

The chapter by Marinetti, Moore, Lucas and Parkinson explains the role of emotions in social life. Emotions undeniably play a crucial role in the way social relationships develop and, they argue, cannot be reduced to static entities. Instead, they propose a framework within which can be explained both explicit and implicit emotional reactions in terms of dynamic interactions involving all the components of emotion.

The chapter by Korsten, Roesch, Fragopanagos, Taylor, Grandjean and Sander addresses emotion through the lens of computational neuroscience. They review the current state of the field and spell out the conflicts that arise from the literature, and means to address them.

References

Fontaine JRJ, Scherer KR, Roesch EB, Ellsworth PC (2007) The world of emotions is not two-dimensional. Psycho Sci 18(12):1050–1057

Roesch EB, Fontaine JRJ, Scherer KR (2006) The world of emotions is two-dimensional. or is it? In *HUMAINE Summer School*, Genoa, Italy, 2006. http://emotion-research.net/ws/summerschool3/RoeschFontaineScherer-06-Genova.pdf. Last accessed on 14 November 2010

Scherer KR (2005) What are emotions? How can they be measured? Soc Sci Inf 44(4):695–729
Scherer KR, Peper M (2001) Psychological theories of emotion and neuropsychological research. In: Gainotti G, (ed). Handbook of Neuropsychology, vol 5, 2nd ed. Elsevier, Amsterdam pp 17–49
Scherer KR, Baenziger T, Roesch EB (eds) (2010) A blueprint for affective computing. Oxford University Press, Oxford, UK
Wundt W (1905) Grundzüge der physiologischen Psychologie [Fundamentals of physiological Psychology], 5th edn. Engelmann, Leipzig

Emotion: Concepts and Definitions

Roddy Cowie, Naomi Sussman, and Aaron Ben-Ze'ev

Abstract This chapter deals with the task of defining and describing emotion. What do people mean when they identify emotion as a key domain for computing? How are "emotions" related to, and differentiated from, other affective phenomena? The chapter considers the definitions of emotions (and other affective states) formulated by scientists and those that are implicit in everyday language. Empirical results regarding the conception of emotion in everyday life (e.g. frequency of emotional reports in different contexts) are presented and discussed. The focus is always on the way conceptual and terminological issues impact affective computing.

1 Introduction

This chapter is about using words to describe phenomena and situations involving emotion and/or affect. The background is acute awareness, arising from a decade's experience in technologies dealing with emotion and/or affect, that this is an area where words have a double-edged quality. They have central roles to play – in communication both between people and between people and machines and in helping researchers to order their thoughts; but they may also set traps. One of the main goals of the chapter is to help people in the area to recognise some of those traps and to deal with them.

An obvious kind of trap involves restriction. Relying on a very limited set of verbal resources can push people towards a seriously oversimplified conception of the area. A subtler kind of trap involves ambiguity. The word 'emotion' itself illustrates that kind of trap. Philosophers have written that emotion pervades human life (Stocker and Hegeman 1996), and in one sense of the word, it seems obviously true. That is presumably why so many people feel that it is important to take account of

R. Cowie (✉)
Department of Psychology, Queen's University Belfast, Belfast, Northern Ireland, UK
e-mail: roddy.cowie@qub.ac.uk

P. Petta et al. (eds.), *Emotion-Oriented Systems*, Cognitive Technologies,
DOI 10.1007/978-3-642-15184-2_2,

emotion when they are designing artefacts, including computer systems, for humans. And yet, people also tend to agree that the word emotion 'strictly' refers to a very specific kind of state. States of that kind occur briefly and occasionally (perhaps once a day) in everyday life. It is a common kind of ambiguity: the word has two senses, one broad and one narrow, as do 'cat', 'boat', 'Kleenex', and a great many other words. Nevertheless, the consequences can be serious. Enthusiasts who argue that technology should engage with emotion are likely to be talking about something that pervades human life. Sceptics are more likely to be questioning the case for engagement with something that occurs briefly and occasionally. That kind of mutual noncomprehension leads all too easily into deadlocks that have the potential to do real damage to the field.

The need to communicate with a range of outsiders constrains the way these problems can be handled. In some areas, experts can eliminate problems rooted in semantics by agreeing among themselves to use words in specialised ways. But in this area, experts need to be able to give a fair picture of what they can do, or aspire to, to outside colleagues, funding bodies, institutions, interested firms, and others; and their systems need to be understood by, and perhaps to communicate with, 'naïve users'. As a result, the only obvious way forward for the foreseeable future is for experts to develop a sophisticated understanding of the language in general use and the substantial issues that lie behind it. The approach in this chapter is to point people towards that kind of sophistication.

2 Plato's Middle Ground

A key first step is to stand back from words and consider what it is that people who are interested in emotion and computing want to engage with. After that, one can ask how well the words that come to hand express the underlying interest.

Writers have often imagined a being who is as intelligent as we are, or more so, but whose mind can only process information in a strictly rational way. Probably the most famous example is Star Trek's Mr Data. Many people who work in emotion-oriented computing seem to be motivated by a sense that a being like Mr Data would lack something, an ingredient X, that is central to being human, and that technology could and should engage more systematically with that ingredient X.

Star Trek did not invent the idea of such an ingredient. On the contrary, the Star Trek character captures the imagination because he reflects a widespread intuition. Influential versions were articulated by Augustine and, before him, by Plato.

Plato's version (*The Republic*, Book IV) is a degree subtler than Star Trek's. It proposes a three-part division of the mind. As in Star Trek, it was accepted that reason was something distinct. The main debate was whether the rest of mind involved one category or two. Plato argued for two. At the lowest level were pure appetites, simple and amoral, which reason either controls or is controlled by. Between appetite and reason was spirit – exemplified by anger – which is inherently attuned to social and moral issues and capable of allying with reason. On the whole, modern research also separates off the phenomena that Plato would have called appetitive. They tend to be called 'drives' in modern parlance. There are exceptions, such as Rolls (1999), who takes hunger as an archetypal emotion; but that is not the norm.

Despite differences of detail, Plato and Mr Data reflect the same broad kind of intuition: that phenomena like anger are instances of something that plays a very large part in making human life what it is. 'Plato's middle ground' seems an apt enough phrase to describe the domain without making specific commitments about its boundaries and contents. The point of introducing the phrase is that it seems to express what people in emotion-oriented/affective computing feel intuitively technology should engage with.

Unfortunately, translating the intuition into well-defined words and concepts is fraught with difficulty. Many words are naturally associated with the elusive category. They include emotion, feeling, expression, passion, and affect. All of them pose the same kind of problem. In the right circumstances, they can be used to designate something like Plato's middle ground. However, each of them also has at least one other sense, corresponding to a specific part of the domain. That creates an immense potential for confusion both between parties and within a single person's thought.

2.1 Common Terms and Their Ambiguities

When non-experts want to describe the domain as a whole, their first choice tends to be 'emotion' and its cognate forms ('emotional', 'emotive', etc). Negative forms in particular fit the role quite well. To say that someone is unemotional or emotionless conveys that factors which affect most people most of the time are not operating. In terms of Plato's picture, the middle ground has shrunk to nothing, leaving the field (perhaps disturbingly) to reason and amoral appetite. Positive forms are more problematic, though.

A study for HUMAINE demonstrated the difference experimentally using video recordings (provided by a TV company) of people dealing with challenges in a novel outdoor environment. The company's psychologist selected about 5 h of material that she regarded as representative of the types of experience found in the whole. Four raters watched the tapes and indicated moment by moment which of three categories best described their impression of the person being recorded – experiencing emotion in the full sense of the word; unemotional; or in an intermediate state involving elements of emotion, but not emotion in the full sense of the word. Ratings divided as follows. States perceived as unemotional made up 7% of the total. States perceived as emotional in the strong sense made up 14%. The remaining 79% was perceived as intermediate, with elements of emotionality, but not emotion in the strong sense.

Studies like this can only give ball park estimates, but those are all that matters here. It seems clear that one sense of the word emotion refers to something that makes up a small part of human life – perhaps about a sixth even in challenging circumstances. But the word has another sense, which refers to something much less sharply defined, and much commoner. That sense comes to the fore when the negative form, 'unemotional', is used. There is an intriguing variant of the same point in Augustine (1984): he claimed that complete absence of emotion (*apatheia*) did not belong in this life (City of God, XIV9).

There is not the same direct evidence for other words, but related patterns seem to hold. When the heroine in a romantic novel sobs that the hero is utterly devoid of feeling, the void that she has in mind probably corresponds quite well to Plato's middle ground. However, 'feeling' in its more precise sense conveys a domain that is different from Plato's in important ways. First, it refers strictly to phenomena that are subjective and part of consciousness. Second, it very definitely includes phenomena that are much more basic than the ones Plato had in mind, as in 'I have no feeling in this leg'.

'Expression' is included here mainly because respected figures argue that what is currently described as research on emotion should talk about expression instead (Campbell, 2003). However, 'expression' in the precise sense misses the mark in much the same way as 'feeling', but in the opposite direction. First, it refers strictly to phenomena that are objective and observable. What lies beneath (such as feelings) is strictly no part of the domain – even if it is being expressed, but especially if it is not. Second, in the strict sense, even some objective signs are excluded – a racing heart, rash behaviour at the steering wheel, selective attention, and so on.

The term 'passion' is included for a similar reason. At one stage, it was the word that philosophers most often used to describe phenomena that we would call emotional. In the present era, though, it tends to have a narrower sense, referring to states where feeling overwhelms reason. Some of Plato's middle ground is like that, but not very much.

'Affect' is a word that deserves special attention, because it is much used in the area, and it has a very curious semantic profile. It is rarely used in everyday discourse. Insofar as it has a generally accepted meaning, it signifies something akin to emotion, but broader in some sense. Experts have taken it up and given it a great variety of more precise senses, often grounded in a theory which implies that emotional and emotion-related phenomena divide naturally in particular ways.

The term was taken up early in the history of psychology. William James noted that German writers used it to refer to "a general seizure of excitement (. . .) which is what I have all along meant by an emotion" (1920, p. 358). Freud gave it a specific sense in the context of his psychoanalytic theory, and that sense gained some currency (Rapaport, 1953). Medicine also adopted the term, but with quite inconsistent usages. For instance, Dark defines it explicitly as an inward state: 'The feeling-tone accompaniment of an idea or mental representation. It is the most direct psychic derivative of instinct and the psychic representative of the various bodily changes by means of which instincts manifest themselves'. In contrast, Abess defines it as 'observable behavior that represents the expression of a subjectively experienced feeling state (emotion)'.

In the mid-twentieth century, psychologists like Hilgard (1980) and Sylvan Tomkins took up a philosophical tradition of using 'affect' as a name for a division based on Augustine's. Tomkins described affect as the person's 'heart, his feelings, his affects' (1964, p vii). His usage was very broad indeed. He explicitly included overt signs as part of affect and used the term to cover not only the standard emotions but also arousal, hunger, pain, commitment, and various other states that would not generally be called emotional, some of which Plato would clearly have excluded from his middle category.

Another set of usages has gained currency more recently. Panksepp (2003) describes affect as 'the feelings associated with emotional processes', and Russell (2003) describes it as similar to 'what is commonly called a feeling'. Both tie affect specifically to feeling. However, they also make it clear that their sense of the word has another level. They use 'affect' to describe states with a dual character, which involves both experience and physiology. For instance, Russell (op. cit.) describes core affect as 'a neurophysiological state that is consciously accessible as a simple, nonreflective feeling'. When the activity of the neural systems involved is reflected in consciousness, we experience it as feelings; but the systems can be active without being reflected as feelings.

The appeal to physiology has hidden subtleties. There is a recurrent suggestion that a person's affective state could be established by monitoring the relevant neural events, using current or near-future technology. In contrast, nobody expects to be able to decipher a person's beliefs by similar means in the foreseeable future – presumably because correspondences between beliefs and neural events are felt to be much more intricate. A much grosser correspondence seems to be assumed in the case of affect as understood by Panksepp, Russell, and others. That may hold for some aspects of anger, for instance, but correlates of the fact that anger is directed towards a particular person or thing, and is felt to be morally right, seem likely to be as subtle as the correlates of belief.

Research that describes itself as 'affective computing' seems on the whole to lean towards a use of the term that is broadly similar to Panksepp's and Russell's. It is likely (though not certain) to be particularly concerned with states that might be identified by monitoring some relevant neural events. If so, it represents a particular way of approaching Plato's middle ground, guided by a particular scientific model of what gives that ground its character. That is why this chapter does not use the term 'affective computing' to describe the whole enterprise of trying to engage computing with Plato's middle ground. It would be like using the term 'Catholic' to describe all mainstream Christian denominations – defensible in principle, but likely to be confusing in practice.

The particular ambiguities that have been discussed involve principles that have quite wide-ranging effects. There is a well-known phrase *pars pro toto*, meaning 'the part stands for the whole'. It captures a feature of discourse in the area, which is that at least in casual conversation or writing, terms like 'feeling' or 'expression', or simply examples like anger, can stand well enough for the whole of Plato's middle ground. In formal contexts, though, an opposite principle seems to apply: *pars invadet totum*, the part usurps the whole. An investigator may begin using a term like ‘expressiveness’ intending to refer to Plato’s middle ground or at least most of it – *pars pro toto*. But as a research effort develops, the narrow sense tends to shoulder others aside, so that it becomes difficult to justify considering anything that is not expression in the narrow sense and mandatory to consider aspects of expression that are not particularly relevant to the original conception – *pars invadet totum*.

The way words move between broad and narrow senses causes a multitude of difficulties; and yet it is not something that can be eradicated from language, not least because it is useful to non-experts. However, there are strategies that make it easier to avoid the most negative consequences.

2.2 Systematising Vocabulary/Emotion Terms

A time-honoured strategy for dealing with ambiguity is to set key terms in phrases that direct people reasonably reliably to one sense rather than the other. A key attraction of the strategy is that it gives people access to multiple ways of bounding and dividing the domains associated with Plato’s middle ground – divisions which everyday language implies, but does not make it easy to separate cleanly. Various options of that kind have been explored within HUMAINE.

‘Pervasive emotion’ emerges as a reasonably satisfactory way to refer to whatever is present in most of life, but absent when people are emotionless (which the data given earlier suggest happens rather rarely). The term is adapted from Stocker and Hegeman (1996). It is the single most convenient description of a domain roughly coextensive with Plato’s middle ground. It is surprisingly difficult to find a term that expresses the narrower sense satisfactorily. ‘Emergent emotion’ poses fewer problems than other options that have been considered. It reflects a widely held interpretation of this kind of state, which is that it involves multiple elements coming together to form a distinctive *Gestalt* (Scherer and others describe the effect as synchronisation) which either dominates the way a person acts and thinks or needs to be held in check by a deliberate effort. To complete the set, ‘emotional life’ has been used to refer to the sum total of the states, processes, experiences, and actions that are substantially influenced by pervasive emotion and therefore distinguish human life as it normally is from the life of a being who is always and completely emotionless.

Both feelings (which are internal) and expression (which is public) are key elements of emotional life. There are feelings that we do not consider emotional (such as pain). It is not self-evident why people distinguish them from emotional feelings. Plato suggests a distinction based on different relationships to moral and intellectual systems; Ortony and his collaborators propose that the hallmark of emotional feelings is an element of positive or negative evaluation (Ortony and Turner 1990). ‘Passion’ is used technically by philosophers to distinguish emotional phenomena that carry a distinctive kind of compulsion. In the context of computing, the natural default is to assume that ‘affect’ refers to phenomena, including feelings, which (*ex hypothesi*) arise rather directly from the activity of particular neural systems. More specific senses can be identified by a standard phrase (e.g. 'core affect') or the name of someone who uses the sense (affect in Panksepp's sense or Freud's sense, etc.).

The point of systematising vocabulary like this is to help people to avoid traps. There are various kinds of traps that may be easier to identify and avoid once the distinctions are registered. An obvious example is setting out to take account of what intuitively seem to be pervasive human characteristics; using the term emotion to describe them (as is natural) and being drawn (by the principle of *pars invadet totum*) into research on the rare phenomenon of emergent emotion (for instance, collecting databases full of examples of emergent emotion or building agents that simulate it). Similarly, if the term used to describe the pervasive characteristics is 'affect', it is natural to be drawn towards a search for correlates of hypothetical biological processes and to lose sight of the cognitive and moral aspects that mark off the pervasive phenomenon from phenomena like thirst. Choosing to study feeling (which is private in the narrow sense) makes it easy to lose sight of the interpersonal aspects of the domain; choosing to study expression carries the temptation to gloss over hard questions about what lies behind the expression and what is read into it; and so on. Many of these carry with them the trap of conveying to the general public that one is going to do much more than one actually is – a trap that can be deadly in the long run for the reputation of disciplines that fall into it.

Traps like these are nobody's fault. They exist because emotional life is a huge, complicated domain, and people are continually looking for ways to make their dealings with it manageable. Proposed solutions that ignore those realities are unlikely to work.

2.3 Coda

An obvious question should be faced before the end of this section. Plato has provided the framework for the discussion: but why should anyone take his framework seriously?

Tripartite divisions have been surprisingly widely accepted. A division derived from Plato via Augustine, using the terms cognition, conation, and affect, remains widely used (Hilgard 1980). Ortony and his group (Norman et al. 2003) propose another related division (using the terms reflective, affective, and reflexive). However, the root answer is that what matters here is not whether Plato's analysis is correct, but whether it reflects ideas that people bring to the field, and thereby helps them to clarify their thinking. A decade in the field suggests that people who enter it often do have something like Plato's middle ground in mind. Other schemes involving slightly more or slightly different subdivisions may capture that intuitive sense marginally better, but the differences are not particularly important in this context.

There may be people whose motivation is different – whose aim is simply to study a domain that is fully and accurately captured by the strict sense of one of the other terms – emergent emotion, or affect in the sense of Panksepp or Russell, or passion. They may be quite clear in themselves about what they want to study. But if they want to convey it to others, even they may benefit from having some sense of what another person might mistakenly assume – *pars pro toto* – that they were studying.

On the other hand, people who are drawn to something like Plato's middle ground intuit that what they are trying to engage with makes up such a large part of human life that its importance for technology hardly needs arguing. Evidence like the TV study cited above indicates that their intuition is right, however tricky it may be to put it into words.

3 Describing Fragments of Emotional Life

This section shifts focus from the macro-task of naming a large domain to the micro-task of describing individual parts. It tries to draw together the main descriptive resources – both terms and concepts – that are needed to convey what is happening in a particular situation where emotion is a key factor. Throughout the section, 'emotion' is used in the sense of 'pervasive emotion' (*pars pro toto*) unless the context indicates otherwise.

Between them, philosophy and psychology provide a very rich set of descriptive resources. It is easy to underestimate the resource, because key ideas are often associated with (apparently) conflicting theories, as if they were alternatives. Broadly speaking, conflicts tend to involve claims that a particular set of concepts captures the central essence of emotion. It is not clear how much technology needs to be concerned with claims about essence. If those debates are set aside, contributions by a range of theorists can be seen as acute descriptions of factors that may or may not define the essence of emotion, but that are certainly relevant to describing what one can expect to find happening when emotion is present.

The sheer number of factors that has been identified is a key point in itself. It is a sharp reminder that like it or not, the domain of emotional life is massively complex; and that there are good reasons to be wary of any model that appears to reduce it to a few simple concepts. Issues are grouped under a few headings here for convenience. Different headings could certainly be chosen, but these provide a structure and progression that seem useful.

3.1 Units

Emotional life is typically divided into two major types of part, illustrated by William James's urgent reaction to a bear in the woods, on the one hand, and by Lord Jim's lifelong shame at jumping ship as a young man, on the other. The term 'emotional episode' will be used for a case like James's, where the person's mental state changes briefly but deeply, and 'established emotion' to describe units like Lord Jim's shame, which are quite likely to last for a lifetime. An established emotion is likely to underlie many emotional episodes, but it tends to be dispositional most of the time (Goldie, 2000).

These large-scale units in turn are individuated by features of various different types. Some are components (that is to say, processes or structures in their own right); others are attributes (that is to say, properties of processes or structures). The rest of this section is concerned with setting out the key types of component and property that give an individual emotional episode or established emotion its character.

3.2 Dimensions

It is often natural and useful to describe emotion in terms of a few dimensions (it may or may not be theoretically profound). Three dimensions are very standard. *Valence* describes the value (positive or negative) of the feelings involved. *Activation* describes the strength of the individual's disposition to act. *Potency* describes the individual's sense that he/she has the power to deal with relevant events.

Many other dimensions have been proposed, of which two will be mentioned here. Unpredictability is included because it emerges as a key factor in a particularly well-constructed study (Roesch et al., 2006). *Engagement* (as opposed to detachment) is rarely mentioned in the psychological literature, though Ortony (2002) used the term 'caring' to express what seems to be a related concept. However, it does concern technologists working on 'presence' in virtual reality (i.e. the sense of being materially engaged with the virtual surroundings, rather than essentially distanced from them). In that context, it is widely assumed that full emotional responsiveness to virtual surroundings implies, and depends on, engagement with them (Huang and Alessi, 1999).

3.3 Feeling

Distinctive kinds of feeling are among the obvious hallmarks of an emotional episode. They are notoriously difficult to describe. Two main ideas about description are standard. One involves dimensions. Russell and his collaborators refer to the characteristic feeling element of emotion as 'core affect', implying that it bears a special relationship to neurophysiology. They suggest that it is characterised primarily by valence, secondarily by activation (Russell and Feldman Barett, 1999). On the other hand, William James (1884) proposed that the feeling element of emotion consisted of awareness of somatic changes (in heart rate, breathing, etc.) associated with the emotion. His idea seems to be partly true: injuries that prevent detection of somatic changes do alter the quality of emotional feelings, but they do not eliminate them.

Most investigators accept that emotion can exist without feeling, most obviously because the other hallmarks of an emotional episode can exist without conscious emotional feelings. (Note that in James's example of the bear, fear reactions precede fear feeling.) That is one of the arguments for defining emotion in terms of the activity of systems with a particular link to feeling rather than feelings per se. According to authors like Panksepp (2003), the work of these systems need not impinge on awareness; but when it does, it has a characteristic quality, which we convey by saying that it is felt rather than analytic. *Ex hypothesi*, these systems generate evaluations that are felt rather than calculated, inclinations to act that are felt rather than deliberately decided, and so on. It would help to explain the lasting appeal of James' proposal if they had close biological links to the systems that generate visceral feelings. Its strength would then come from an intuitive sense that the same kinds of system were in play in emotional and visceral feelings – that the two were cut from the same cloth, so to speak.

3.4 Appraisal

One of the hallmarks of an emotional episode is nicely captured by Ben-Ze'ev's phrase 'partial perception' (2000): it involves a selective grasp of a situation, which highlights what is relevant to the 'weal or woe' (Arnold 1960) of key players. The best known developments of that idea propose relationships between emotion categories and value-oriented 'appraisals' of the situation.

There are many specific descriptions of appraisal. A well-developed example is due to Scherer's group (Sander et al., 2005). It describes a sequence of 'stimulus evaluation checks' which makes sense logically and fits data collected by the group. It proposes that the onset of emergent emotion involves a series of checks, in the following sequence:

- Relevance (including sub-checks for novelty, intrinsic pleasantness, and relevance to the subject's goals and needs),
- Implications (including sub-checks for causal attribution, outcome probability, discrepancy from expectations, goal conduciveness, and urgency),
- Coping potential (including sub-checks for the controllability of the event and the subject's power to affect its course and/or to adjust to its consequences), and
- Normative significance (including sub-checks concerned with the way outcomes relate to one's own values and to society's).

One of the analyses that has had most impact in technology, the framework proposed by Ortony et al. (1988), is also rooted in appraisal theory. Their analysis in turn exists in both an extended and a reduced version. It is beyond the scope of this chapter to evaluate different forms of appraisal theory.

3.5 Emotional Colouring

It is natural to picture the result of appraisal checks as a representation in which descriptors specifying emotion-related qualities are attached to significant things and relationships in the relevant situation. The idea can be expressed by saying that the representation is emotionally coloured – with the colours indicating whether key features of the surroundings are pleasant, conducive to the person's goals, within the person's power to control, morally acceptable, and so on.

The metaphor extends to the general meaning of concepts as well as impressions of individual situations. A large body of work by Osgood and his collaborators (Osgood, 1957) showed that the emotional colouring of concepts (their term was 'feeling tone') could be summarised reasonably well in terms of three dimensions (evaluation, potency, and activity). It would be interesting to revisit the area in the light of the richer descriptive systems that appraisal theorists have developed since.

3.6 Action Tendency

There is a long-standing recognition that emotion tends to close the gap between having an impression of the situation and acting on the impression. For instance, Aristotle cited the impulses to/for revenge as a defining feature of anger. Frijda (1987) reintroduced a related concept in the modern era and his term 'action tendencies' is widely used. He argued that tendencies to act in particular (biologically significant) ways were integral to emotion and were central to distinguishing among emotions with a direct biological significance – tendency to approach is the kernel of desire, tendency to avoid is the kernel of fear, tendency to reject is the kernel of disgust, and so on.

There is another sense in which emotion is bound up with instigating action, which may or may not be fundamentally separate. It involves motivation. It is not clear how tight the connection is. Motivation is often linked more to Plato's lowest category of appetite (hunger, thirst, pain). However, emotional colouring is obviously a factor in motivation (you will do more for someone you like or fear); and emotional episodes certainly motivate or demotivate. Sylvan Tomkins (1991) proposed that emotions act as amplifiers that modulate basic drives. It is not unlike a metaphor that Plato used: he imagined emotion as the good horse that responded directly to the driver (rationality), drawing its less co-operative companion (appetite) along with it.

3.7 Expression

Actions with a communicative element are among the most characteristic components of emotional episodes – smiling, weeping, screaming, and so on. Theorists from different traditions have understood these in substantially different ways.

Simplifying grossly, accounts that appeal to evolution have tended to assume that expressions of emotion are produced by innate mechanisms which automatically generate external signs of significant internal states, with socially defined operations (display rules) capable of concealing or mimicking the innate patterns (though not usually perfectly). In contrast, social psychologists argue that the patterns are fundamentally communicative: smiles are directed to people, not automatic externalisations of an inner state. The two lead to different research strategies. For instance, evolutionists assume that expression which is uncontaminated by display rules has a privileged status and should be sought out. For social psychologists, it is a fiction that draws people into studying situations that are ecologically unrepresentative (misguided in rather the same way as trying to study sitting behaviour without the complicating factor of a chair).

3.8 Emotional Modes of Action and Cognition

Emotion affects not only what people do, but also the way they do it (of course, the line is often blurred). Some of the effects flow from underlying shifts in the way people perceive and think under the influence of emotion. There are well-documented examples at many levels of cognition.

A practically important example of effects on attention is called 'weapon focus' – exclusive concentration on a single, focal detail of a scene (the gun) to the exclusion of other features which are actually important (the gunman's face). The effect seems to be due to the perceptual processes that evoke emotional responses (Laney et al., 2004). There is a substantial literature on the way anxiety affects perceptual and related process – attentional control, depth of processing, and speed of processing. Eysenck's work on anxiety illustrates a well-developed analysis of the issues surrounding these effects (1997, 2007). Positive conditions tend to generate extensive and well-organised memories, and positive affect promotes their recall later. It also fosters flexible and creative thought, can speed decision making, and affects risk taking, not necessarily in obvious ways (Isen, 1998). Negative moods tend to increase people's impression of the effort that a task requires (Gendolla and Krüsken, 2002). Marked emotion tends to reduce coherent verbal communication (Cowie and Cornelius, 2003).

These effects are practically important for emotion-oriented computing. Consider, for example, the implications of ability to recognise when emotion is impairing a driver's perception of risk, or a pupil's ability to learn, or a manager's ability to communicate clearly, or a worker's readiness to sustain effort, and so on.

3.9 Connectedness

Usually (perhaps always) describing an emotional episode depends on referring beyond the person who has an emotion experience to various significant objects and significant others. That is already implicit in several of the points above, but deserves to be drawn out. An appraisal is an appraisal *of* people, events, or things; and expressions of emotion tend to be directed *to* particular people in the context of an audience (couples know how dramatically the sudden appearance of a child or an in-law can affect the expression of various emotions).

Philosophers use the term intentionality to express the fact that emotions are *about* something, which is called the object of the emotion (as against the subject who experiences it). 'Connectedness' is broader and aims to cover both that kind of linkage and the linkages involving others who are involved in an emotional episode, but not the object of the emotion.

The list of connections above is far from exhaustive. An emotional episode may be about one thing (mortality, for instance); prompted by another (a poetry reading, perhaps); with causal roots in events long past (such as a bereavement). Shame before an audience which approves of an action may arise because another audience (present in the mind rather than in reality) would not approve. There is no obvious end to the permutations. In addition, mixed emotions often involve connections with multiple different events, bearing different emotional colourings – gladness that a gap in life is to be filled, sadness about the loss that created the gap, concern that it might still not work out, and so forth.

These issues have points in common with claims about 'groundedness'. Advocates of 'groundedness' argue that certain kinds of representation can only emerge by a causal process from the reality that the agent is part of. However, it seems fair to say that their concern is with a different level of connection, defined by the fact that a history of causal interactions has moulded the symbolic medium in which connections, with current or past events, are expressed. It is not clear what hinges on the existence of such a level, and the two concepts are probably best kept apart.

3.10 Impressions of Emotion

Emotional episodes typically involve more than one person; and when they do, understanding how signs of emotion are registered is as essential as understanding how they are emitted. Scherer (2003) recently reasserted the point in a neat form using a lens model adapted from Brunswik. There are several ways of conceptualising registration.

Detection paradigms consider whether an objectively verifiable state is identified correctly. That approach is clearly appropriate in certain application areas – emotional intelligence tests, for instance, and lie detection (where emotion-related signs are assumed to be pivotal).

Experiential paradigms are broadly comparable to certain areas of psychophysics, where it is accepted that subjective experience may have its own dimensions. In a parallel way, it makes sense to consider whether (for instance) dimensional descriptions capture the way we perceive other people's emotions under certain circumstances, whether or not they capture the intrinsic nature of our own and so on.

Control paradigms consider how variables affect behaviour rather than experience. The two can be very different. For instance, it is well known in perception that the behaviour of visually guided grasping is not affected by variables that distort conscious reports of size and distance. There is evidence that in a parallel way, responses to others' emotion may reflect variables that are not reflected in conscious impressions of the other.

Timing needs to be considered alongside these distinctions. Taking time to identify a single static state is not the same as registering in real time how a person's emotional balance and focus is shifting, which is what people have to do when they participate in an emotionally coloured interaction. Perceived flow of emotion seems an apt term for the kind of impression that underpins real-time interaction.

Subjects perceive (or fail to perceive) their own emotions as well as other people's. That would seem to involve forming explicit representations of changing flows and pressures that are at work in their own heads (so to speak). Helping people to perceive their own emotions is one of the application areas that is regularly considered for emotion-oriented computing.

3.11 Category Labels

This section has listed various kinds of resources that are relevant to describing a particular fragment of emotional life. It has deliberately left until last the resource that people typically consider first, that is, words like 'mood', 'anger'. The reason is that people find it very easy to think of emotional life as a collection of events that correspond reasonably closely to salient category labels. It is a prime case of *pars invadet toto* – attention is pulled onto the special cases which are close to category archetypes, leaving the mass of everyday phenomena that are far from the archetypes sidelined.

Databases that use naturalistic material highlight the issue. Research teams repeatedly observe that what they find often is not particularly well described by any standard category label and often seems to involve multiple categories (Douglas-Cowie et al., 2003, 2005). Even when category labels do fit, they do not in themselves provide information that is crucial to understanding the events (for example, triumph at a football match is likely to differ in a great many ways, visible and invisible, from triumph in a court room). The point is not that category labels have no part to play. On the contrary, they are considered at length in the next section. The point is that they are one kind of resource among many.

A significant divergence from that view should be acknowledged. Many theorists argue that a few qualitatively distinct neural systems give rise to the whole of emotional life, and that the most important category terms are linked very directly to those systems. The term 'basic emotion' is used in various ways, but one of them is to describe states that are hypothesised to relate simply and directly to one of these systems.

It is not obvious how important that idea is in practice for emotion-oriented computing. It may be that a few systems underlie the complex emotions involved in attending the funeral of a former colleague who had been able but quarrelsome; but it does not necessarily do much more to illuminate the emotions than knowing about retinal receptor types does to illuminate the complex visual experience of watching faces at the funeral.

In that situation, technologists may make different choices. Some will judge that it is worth pursuing the idea that there are a few basic systems underlying emotional life. Others will choose to treat category labels as simply one resource among many to be used in engaging with an inherently complex set of phenomena. Discovering the relative success of the two approaches in computing may contribute a good deal to resolving the theoretical disputes within psychology: it is hard to see how anything but a computational approach can establish how much of the complexity of real emotional life each can accommodate.

The kind of framework that has been set out in this section is intimately related to the development of databases. Many of the ideas were prompted by the surprising difficulty of describing the phenomena that were observed in naturalistic recordings. In turn, the descriptive system in the HUMAINE database is a practical simplification of the ideas presented in this section.

4 Classifying Emotion-Related States

One of the most natural ways of thinking about emotional life is in terms of something like a taxonomy of states. If a standard emotion word such as 'fear' refers to a species of individual states, then 'emergent emotion' refers to a genus, and 'emotional life' refers to a family of states. The idea is not as straightforward as it might look, but it does provide a useful framework in which to set out the various kinds of description that deal with states related in some way to pervasive emotion.

4.1 Mindsets and Personal States

People are sceptical, and reasonably so, when emotion and related terms are used as a catchall for everything that is conveyed by non-verbal communication. Clearly states which can reasonably be thought of as related to emotion are only part of larger domains. A natural name for a domain one step larger is 'mindsets'. Included at that level might be social states (dominance, deference, and so on) and cognitive states (confusion, interest, etc). 'Personal states' is a natural term for a broader domain still, including, for instance, states of health and well-being (ill, vigorous, and so on) as well as mindsets.

It is a useful approximation to say that emotion-related states are a particular kind of mindset. That allows (for instance) people who are working on classification of states observed in meetings to say that they are interested in states which are mindsets, but not (in the main) emotional. It is approximate rather than exact because an actual person (unlike Mr Data) is unlikely to be in a state that has no emotional elements. The division is between descriptors which do and do not refer to emotion-related factors in the person's state, not between states that do and do not have emotional elements.

To complicate matters still further, a large proportion of terms that are not explicitly emotional nevertheless include emotion as a likely factor. For example, courage seems to be essentially a behaviourally defined state; and yet its meaning has intimate links to emotional factors (such as controlling fear). Is courage, then, to be considered an emotion-related state? Lazarus (1999) illustrates a slightly different kind of connection when he observes that it is unrealistic to discuss stress without reference to emotion, even though they can be separated conceptually.

Questions like these probably have to be handled pragmatically and in context. So if one wants to model a brave man's behaviour, it may be enough to consider courage as a matter of risk assessment; but instilling courage probably depends on engaging with its emotional aspects.

4.2 Generic Emotion-Related States

On the taxonomic metaphor, emotion-related states make up a family and can be divided into genera – obvious examples being moods and emergent emotion. One might assume that there would be well-accepted ways of subdividing emotion-related states into genera and that they would be backed up by data on prevalence. It is clearly of interest to emotion-oriented computing to know which kinds of states are actually common, since (other things being equal) it makes sense to orient systems to common phenomena rather than rare ones.

In fact, there have been few systematic attempts to provide a set of categories that between them cover the whole domain of emotion-related states. A table due to Scherer, given below, provided a starting point for research on the topic in HUMAINE. The research extended the table in two stages: the first a priori and the second empirical.

The a priori stage started by considering features that distinguish emotion-related states from other mindsets. The most obvious of those are dimensions of emotion that were listed in the last section: valence, activation, potency, and engagement. It makes sense that the genera might include states that are distinguished by an unusual level of one factor, but not the others. Mood falls neatly into that framework, since it is often described as a state distinguished mainly by valence. Intuitively, there do seem to be states that correspond to the other dimensions in similar ways, involving heightened or lowered sense of potency or control, activity, and engagement or seriousness about events at the focus of attention.

Similarly, the states listed in Scherer's table are either stable or follow a relatively set trajectory. One type of state that changes more freely with time has already been

Table 1 Genera of emotion-related states (after Scherer et al., 2004, p 11)

Design features	Emergent emotions	Interpersonal stances	Moods	Attitudes	Affective dispositions
Impact on behaviour	++	+		+	+
Intensity	++	+	+	+	
Rapidity of change	++	++	+		
Brevity	++	+	+		
Event focus	++	+			
Appraisal elicitation	++				
Synchronisation	++				

mentioned, that is, established emotion, which is usually latent, but occasionally translates into an emotional episode. On a shorter time scale, there can be oscillation between a sustained mood-like state and outbursts of emergent emotion. Recognising transition as an issue suggests that transitional states are sometimes sustained: people can simmer on the edge of anger, but not quite succumb. The category 'emotionless' was also added, for obvious reasons.

Empirical studies followed up those ideas. Naïve participants were given lists of genera based on the reasoning outlined above. For each category, they were asked to assess whether they had experienced episodes that fitted the description; and if so, to give brief accounts of them. Most accounts fitted the categories and confirmed that they were recognisable, but some described experiences that were not well captured by the a priori framework. That led to additions in two areas. Stances towards things and situations were added to stances towards people; and a category involving more enduring orientations to people (described as bonds) was also added. At the same time, the term 'attitude' was abandoned, because participants clearly used it to mean something quite different from what was intended. It is noticeable that the term has also acquired quite different meanings in different academic literatures, taking on one technical sense in a literature derived from Ajzen asnd his colleagues (1988) and another in linguistics (O'Connor, 1973), which has sometimes been equated with Scherer's term 'interpersonal stance' (Wichmann, 2002). Like 'affect', 'attitude' seems to have a facility for picking up multiple meanings; and it needs to be used cautiously for that reason.

The revised list was used in an 'ambulatory study' (Wilhelm et al., 2004). Ten participants were given a protocol in which each generic category was named, described, and illustrated with an example given by a participant in the previous study. Each of them was then contacted by phone 50 times at random times over a period of weeks. They responded by identifying the generic descriptor that best reflected their state at the time. The main results are summarised below:

Established emotion	0.9%
Emergent emotion (suppressed)	1.7%
Emergent emotion (full-blown)	1.5%
Mood/emergent emotion oscillation	1.5%
Mood	36.1%
Stance towards object/situation	25.6%
Interpersonal stances	2.4%
Interpersonal bonds	4.1%
Altered state of arousal	21.9%
Altered state of control	3.9%
Altered state of seriousness	0.4%
Emotionless	0.0%
None of the above	0.0%

These data reinforce points made earlier through the TV study: people are rarely emotionless and not often in a state of full-blown emotion. They also show that between the two is a variety of states which are very common – some of which have established names, but by no means all. Hence if emotion-oriented computing wants to address the emotion-related states which occur commonly, it cannot let itself be guided exclusively by the labels that are available in everyday language.

The work described here clearly needs to be developed. But it provides at least a preliminary overview of the states that make up Plato's middle ground and some protection against being drawn (*pars invadet toto*) into studying emotion in a sense that accounts for a very small proportion of emotional life.

4.3 Specific Emotion Terms

Words like anger, joy – what is being called the species level here – clearly have a kind of priority in the domain of emotion. In Rosch's terms, they appear to be the basic level in this domain – as 'dog' or 'cat' do in the domain of animals. It is natural to assume that they are essentially the names of states and more specifically of affective states. A good deal of research has gone into showing why that kind of assumption is at best a first approximation and can be quite seriously inappropriate for emotion-oriented computing in particular.

There are sources which offer to analyse words at this level as co-ordinates in a space with a low number of dimensions. Whissell's dictionary of affect is a particularly thorough example of that approach, and it is reasonable to interpret the co-ordinates as a description of affect. That kind of analysis is often useful, but it is very far from sufficient.

A first kind of problem concerns variation in the scope of words. Words like anger and love can famously refer to many different states – hot and cold anger, sexual and nurturant love, and so on (Russell and Fehr, 1994; Sternberg, 1988). They can also refer to one-off emergent episodes (anger at a rude shop assistant) or established emotion (anger at UK policy in Iraq). Linked to that, there is evidence that there are material differences between the anger evoked by things that are physically present and the anger evoked by remembering past events (Stemmler et al., 2001). Observing naturalistic data also underlines the frequency of states which do not exactly fit any category, either because they are intermediate or because they are blends (using that term to mean that two or more emotions seem to be coexisting – happiness at one aspect of the current situation, sadness at another). If the word 'anger' describes a species, it is a species much more variable than (for instance) dogs.

A deeper problem is that specific emotion words do not simply refer to inner affective states (however varied). Their meaning is bound up with a variety of complex judgments, many related to the fact that words in everyday language must allow people other than the subject to apply them (otherwise they could not form part of a common language). Because of that, words that involve inner states must be rather complicated instruments whose rules of use are logically bound to both intra- and interpersonal elements.

The point is taken up in an article by Sabini and Silver (2005) and a reply by Cowie (2005). Sabini and Silver argue that terms like jealously and anger, shame and embarrassment may refer to the same affective state. The difference lies in factors surrounding that affective kernel, some internal to the subject experiencing the emotion, some external. The reply draws out the implications for the task of assigning everyday emotion-related words as humans do and identifies eight types of consideration that appear relevant to the task. Clearly one type of consideration involves the internal feeling state. But assigning emotion terms also depends on the objective events which prompt an emotion (if a successful person has genuinely insulted an unsuccessful one, the latter's emotional response would be called anger; if the latter has perceived an insult in quite innocent behaviour, it would be called envy). It depends too on evaluation of the person's character (the word 'shame' is applied when a person accepts that negatively evaluated actions reflect a genuine deficiency in themselves, 'embarrassment' when he or she does not). Similar points can be made about the other types of consideration: these involve the person's appraisal of circumstances; the involuntary signs that he or she gives; his or her choice of action; the manner in which the action is undertaken; and the observer's evaluation of action.

Obviously these ideas are closely related to what has already been said about describing fragments of emotional life. What they add is that assigning standard emotion words is not a simpler task that can be dissociated from the complexities described there. On the contrary, the conditions for using specific emotion terms are bound up with the overall complexity of emotional life.

Related but distinctive problems arise over the right to use particular words. A parent may describe a child as sulking. A machine which attached the same label to the child would be presuming a right to make moral judgments which the recipient might dispute and might expect to be smashed. In general, it is not at all obvious what rights people might attribute to computers; and that means there are open questions about the emotion words they could or should use.

These considerations are not abstruse entertainment for the philosophically minded. If emotion-oriented systems are to use emotion terms appropriately, they need to use them in accordance with the complex criteria that actually govern their use in natural languages. That means recognising that although it is a convenient approximation to think of specific emotion words as labels for species of states, it is nevertheless an approximation.

5 Where Have All the Theories Gone?

This review has said very little about theorists and their positions per se, though it has drawn ideas from them liberally. This section touches briefly on major theoretical positions for the sake of reference.

Relationships between emotion and the body have been a recurring issue. Descartes depicted passion as the mind being disrupted by turbulence in the body. William James identified emotional feelings with awareness of bodily changes. His son-in-law and antagonist, Walter Cannon, identified them with changes in the midbrain. Late twentieth century research has used brain imaging to identify brain centres associated with emotion, led by figures like Damasio, LeDoux, and Davidson.

Darwin can be seen as a branch from that line. His view of emotion was biological, but emphasised the evolutionary constraints that made particular behavioural patterns adaptive, in humans and animals. Frijda (1987) developed a distinctive evolutionary approach which emphasised the action tendencies associated with emotions.

An influential synthesis of these approaches, most strongly associated with Ekman, uses the concept of basic emotions. By that he means that emotions are of several discrete types, each with a cluster of characteristics, laid down by evolution and rooted in discrete physiological systems. A great variety of approaches are broadly similar, for instance, those of Plutchik (1980), Cosmides and Tooby (2000), and many others. Some of these emphasise the discreteness of the hypothetical affective systems, others their ability to interact with cognition (e.g. Panksepp, 2003).

Cognitive approaches emphasise the integral part that cognitive processes, and particularly appraisal, play in emotion. They are usually traced to Arnold (1960), with influential formulations due to Lazarus, Le Doux, Ortony and his colleagues, and Scherer. Strong versions of cognitivism regard emotions as informationally encapsulated brain processes (LeDoux, 2000), whose feeling component is relatively unimportant. Weaker versions see emotion as essentially an amalgam of cognition and motivation (Lazarus, 1999).

In contrast to evolutionists, social constructivists emphasise the role of culture in giving emotions their meaning and coherence (e.g. Averill, 1980; Harre, 1986). The position held by Russell (2003) can be regarded as a distinctive variant, which considers affect an underlying biological substrate that enters into a variety of processes shaped by social and other factors. Other social theorists have argued that emotion is fundamentally an attribute of interactions between people rather than of individuals (Parkinson, 1995).

Even this short summary gives some sense of the variety of theories that carry weight in the field. That reflects the fact that nobody has yet identified a single, unifying kernel round which all that is known about emotion can be organised in a completely coherent, satisfying way. However, it does not reflect a field in utter turmoil. A large body of knowledge exists, and the bulk of the chapter has tried to reflect it. It remains a challenge to find a truly satisfying way of organising it. The scale of the challenge should not be underestimated.

6 Conclusion

From top to bottom, emotion language is more complex than it looks. That gives rise to traps when people forge ahead relying on a model which is or seems appealingly simple, but which in fact conceals both the complexity of language and the complexity of the real phenomena involved.

The strategy of this chapter has been to alert people to the complexities at both levels. No doubt many readers will find that thoroughly unsatisfying and look for articles that offer more concrete or elegant prescriptions. It may be that when they have worked their way through a sufficient number of traps, they will come back.

References

Ajzen I (1988) Attitudes, personality, and behavior. Open University Press, Milton Keynes

Arnold MB (1960) Emotion and personality. Columbia University Press, New York, NY

Augustine (1984) The city of God (trans: Bettenson H.). Penguin Classics, London

Averill JR (1980) A constructivist view of emotion. In: Kellerman H, Plutchik, R. (eds.) Emotion:theory, research and experience, Vol. 1. NY, pp 305337 Academic, New York, NY

Ben Ze'ev A (2000) 'I only have eyes for you': the partiality of positive emotions. J, Theory Soc Behav, 30(3): 341–351

Campbell N (2003) Databases of expressive speech. In: Proceedings of Oriental COCOSDA Workshop, Singapore, October 2003

Cosmides L, Tooby J (2000) Evolutionary psychology and the emotions. In: Lewis M. Haviland-Jones J.M. (eds). Handbook of Emotions, 2nd edn. Guilford, New York, NY

Cowie R (2005) What are people doing when they assign everyday emotion terms? Psychol Inq. 16(1): 11–18

Cowie R, Cornelius R (2003) Describing the emotional states that are expressed in speech. Speech Commun 40: 5–32

Douglas-Cowie E, Campbell N, Cowie R, Roach P (2003) Emotional speech: Towards a new generation of databases. Speech Commun 40: 33–60

Douglas-Cowie E, Devillers L, Martin J, Cowie R, Savvidou S, Abrilian S, Cox C (2005) Multimodal databases of everyday emotion: facing up to complexity. In: Proceedings of Interspeech 2005, Lisbon, Portugal, 2005, pp 813–816

Eysenck MW (1997) Anxiety and cognition: a unified theory. Psychology Press, Hove, East Sussex, UK

Eysenck MW, Derakshan N, Santos R, Calvo MG (2007) Anxiety and cognitive performance: attentional control theory. Emotion, 7(2): 336–353

Frijda NH (1987) The emotions. Studies in emotion and social interaction (Ser.). Cambridge University Press, Cambridge

Gendolla GHE, Krüsken J (2002) Mood, task demand, and effort-related cardiovascular response. Cog, Emo, 16: 577–603

Goldie P (2000) The emotions: a philosophical exploration. Clarendon Press, Oxford

Harré R (1986) The Social construction of emotions. Blackwell, Oxford

Hilgard ER (1980) The trilogy of mind: cognition, affection and conation. J Hist Behav Sci 16: 107–117

Huang MP, Alessi NE (1999) Presence as an emotional experience. In: Westwood JD, Hoffman HM, Robb RA, Stredney D, (eds.) Medicine meets virtual reality: the convergence of physical and informational technologies options for a new era in healthcare. IOS Press, Amsterdam, pp 148–153

Isen AM (1998) On the relationship between affect and creative problem solving. In: Russ S. (ed.) Affect, creative experience, and psychological adjustment. Braun-Brumfield, Ann Arbor, MI, pp 3–17

James W (1984) What is an emotion? Mind, 9: 188–205

James W, Perry RB (1920) Collected essays and reviews. Longmans, Green, New York, NY

Laney C, Campbell HV, Heuer F, Reisberg D (2004) Memory for thematically arousing events. Mem, Cogn 32: 1149–1159

Lazarus R (1999) The cognition-emotion debate: A bit of history. In: Dalgleish T, Power M. (eds.) Handbook of cognition and emotion, Wiley, Hoboken, NJ, pp 3–19
LeDoux JE (2000) Emotion circuits in the brain. Ann. Rev. Psychol 23: 155–184
Norman DA, Ortony A, Russell DM (2003) Affect and machine design: lessons for the development of autonomous machines. IBM Syst J 42: 38–44
O'Connor GF, Arnold JD (1973) Intonation in colloquial English 2nd edn. Longman, London
Ortony A (2002) On making believable emotional agents believable. In: Trappl R, Petta P, Payr S. (eds.) Emotions in humans and artifacts. MIT Press, Cambridge, MA, p 189
Ortony A, Turner TJ (1990) What's basic about basic emotions? Psychol Rev, 97: 315–331
Ortony A, Clore GL, Collins A (1988) The cognitive structure of emotions. Cambridge University Press, Cambridge
Osgood CE (1957) The measurement of meaning. University of Illinois Press, Urbana, IL
Panksepp J (2003) At the interface of the affective, behavioral, and cognitive neurosciences: decoding the emotional feelings of the brain. Brain Cogn. 52: 4–14
Parkinson B (1995) Ideas and realities of emotion. Routledge, London
Plato (2003) The Republic (trans: Lee HDP), 2nd edn. Penguin Books, London
Plutchik R (1980) Emotion: a psychoevolutionary synthesis. Harper & Row, New York, NY
Rapaport D (1953) On the psycho-analytic theory of affects. Intl. J Psycho-Anal 34: 177–198
Roesch EB, Fontaine JRJ, Scherer KR (2006) The world of emotions is two-dimensional.. or is it? In: HUMAINE Summer School, Genoa, Italy URL http://emotion-research.net/ws/summerschool3/RoeschFontaineScherer-06-Genova.pdf. Accessed on 11 October 2010
Rolls ET (1999) The brain and emotion. Oxford University Press, Oxford
Russell JA (2003) Core affect and the psychologcial construction of emotion. Psychol Rev, 110(1): 145–172
Russell JA, Fehr B (1994) Fuzzy concepts in a fuzzy hierarchy: Varieties of anger. J, Pers, Soc, Psychol, 67: 186–205
Russell JA, Feldman Barett L (1999) Core affect, prototypical emotional episodes, and other things called emotion: dissecting the elephant. J Pers Soc Psychol, 76(5): 805–819
Sabini J, Silver M (2005) Why emotion names and experiences don't neatly pair. Psychol Inq. 16: 11–48
Sander D, Grandjean D, Scherer KR (2005) A systems approach to appraisal mechanisms in emotion. Neural Netw 18(4): 317–352
Scherer KR (2003) Vocal communication of emotion: a review of research paradigms. Speech Commun 40: 227–256
Scherer KR, Roesch EB, Bänziger T (2004) Preliminary plans for exemplars: Theory humaine deliverable D3c. Technical report, University of Geneva, 2004. URL http://emotion-research.net/projects/humaine/deliverables/D3c.pdf. Accessed on 11 October 2010
Stemmler G, Heldmann M, Pauls C, Scherer T (2001) Constraints for emotion specificity in fear and anger: the context counts. Psychophysiology, 69: 275–291
Sternberg RJ (1988) Triangulating love. In: Barnes ML, (ed.) The psychology of love. Yale University Press, New Haven, CT
Stocker M, Hegeman E (1996) Valuing emotions. Cambridge University Press, Cambridge
Tomkins SS (1964) Affect, imagery, consciousness vol. I. Springer, New York, NY
Tomkins SS (1991) Affect, imagery, consciousness vol. III. The negative affects: anger and fear. Springer, New York, NY
Wichmann A (2002) Attitudinal intonation and the inferential process. In: Proceedings of Speech Prosody 2002 Aix-en-Provence, France, April 11–13, 2002. http://www.isca-speech.org/archive/sp2002/sp02_011.pdf. Accessed on 10 November 2010
Wilhelm P, Schoebi D, Perrez M (2004) Frequency estimates of emotions in everyday life from a diary methodś perspective: a comment on scherer et al.ś survey-study "emotions in everyday life". Soc Sci Inf 43(4): 647–665

Emotions in Social Interactions: Unfolding Emotional Experience

Claudia Marinetti, Penny Moore, Pablo Lucas, and Brian Parkinson

Abstract This chapter takes into account the role of emotions in social interactions, both face-to-face and video-mediated. Emotions are conceptualised as ongoing processes rooted in dynamic social contexts, which can shape both implicit and explicit emotional responses. Emotion interactions are therefore considered as continuously developing, thanks to the relationship between interactants and between them and the surrounding environment. Theories of emotions as non-static phenomena are illustrated before presenting a review of literature on regulating processes in emotional interactions. Finally, based on the theoretical framework described in this chapter, comparisons between an emotionally competent human and an emotionally competent artificial agent are drawn.

1 Introduction

In the unremitting complexity of social life, emotions play a key role in defining and regulating our relationships with others and, more generally, with the environment surrounding us. Our emotional reactions to other people influence how those others react to us and, to a certain extent, how future encounters will develop. At the same time, our own emotional behaviour is shaped by others' thoughts and deeds. Although emotions undeniably have personal and subjective aspects, they are usually experienced in a social context and acquire their significance in relation to this context (see also the "Socially Situated Affective Systems" chapter in Part V).

1.1 Emotions as Communication Tools

Social processes affect the way in which our emotions unfold at a number of different levels. At the cultural level, what we define as an emotion, and what kind

---

C. Marinetti (✉)
Department of Psychology, Katholieke Universiteit Leuven, Leuven, Belgium
e-mail: claudia.marinetti@chch.oxon.org

P. Petta et al. (eds.), *Emotion-Oriented Systems*, Cognitive Technologies,
DOI 10.1007/978-3-642-15184-2_3,

of emotion is appropriate in a specific situation, depends on the set of established values, norms, and customs within one's own society. Narrowing the focus of our analysis, at the group level our emotions towards other individuals and other groups are affected by our belonging to an in-group with whose members we share a common social identity. Finally, at the interpersonal level, other individuals' emotions affect our emotions and our emotions affect them in an ongoing process of mutual exchange (Parkinson et al., 2005).

Emotion and its components (e.g. appraisals, bodily changes, action tendencies, expression, regulation, and subjective feeling), therefore, play an important role in social interactions, social comparison, and social influence processes. For example, Harré and Gillett (1994) emphasise the communicative function of emotion feelings and displays in the events of everyday life by treating them as psychologically equivalent to verbal statements. From this point of view, emotional expression cannot be reduced to a simple direct manifestation of an internal emotional state, but has to be considered as emerging from an interactive context. There is, for instance, evidence that facial displays are affected by both emotional and social factors (Parkinson, 2005). Indeed, several authors have argued that facial displays can serve both emotion-expressive and social communicative functions (e.g. Cacioppo et al., 1992; Hess et al., 1995). Regardless of whether we consider facial displays as depending directly on emotions or on social motives, it is clear that facial behaviour has a great impact in social interactions. Effective communication, for instance, requires not only that we properly interpret other people's expressions, but also that we correctly assess the extent to which others can read our expressions (Kenny and De Paulo, 1993). Indeed, our own misappraisal of the emotional expression that we have shown to another during an interaction could have important consequences for our interpretation of that person's responses to us, and thus for the course of the interaction itself (Muttiallu, The effects of display rules on the "illusion of transparency": moderating factors of partner identity and positivity of emotion, "unpublished").

Successful emotional communication, though, is not restricted to top-down processes like the ones just illustrated, but necessitates an ongoing reciprocal adjustment between interactants that can often happen at an implicit level. That is, emotional responses are also constructed in a bottom-up way during an interaction in which we do not always register the emotional significance of specific indicators of emotion.

1.2 Explicit and Implicit Responses to Emotions

Some authors have focused on the existence of two modes of interpersonal response to emotions: explicit and implicit (Hsee et al., 1992; Parkinson and Lea, in press). At the explicit level, people seem to gain knowledge about one another's emotional states by consciously processing relevant information. Thus, verbal and non-verbal reactions during an interaction are categorised as signs, signals, or symptoms of emotion. Furthermore, the nature of our response during an emotional interaction is influenced by the cultural norms and scripts invoked by attributing emotion to another on the basis of perceived responses. On the other hand, people often react to aspects of another person's emotion at an implicit level without registering the emotional meaning of response components. This is the case, for example, for mimicking others' expressions on-line (Hatfield et al., 1994) and for automatically adjusting our own rhythms of movement to match those of our interaction partners (Bernieri et al., 1988). However, the distinction between implicit and explicit processes does not necessarily imply that they have to be considered as separate and independent systems. Indeed, most of our everyday interactions involve a complicated combination of implicit and explicit responses to emotions. Each interactant can explicitly transmit emotional information or implicitly convey it, while the other person's reaction can be at the same level or at a different one (e.g. an implicit "symptom" of one person's emotion may be registered explicitly by the other, and vice versa).

Therefore, it seems possible that we can regulate our emotional reactions both deliberately and automatically, sharing others' emotional states or adopting complementary or contrasting ones, sometimes without ever registering the emotion that they are experiencing. Automatic on-line mutual adjustments during an emotional interaction are a vital part of how we respond. From this point of view, any intelligent system programmed to respond only to the explicit, categorical meaning of emotion signals as indications of the emotion itself is likely to miss out on potentially important parts of the usual interpersonal process. Indeed, it is recognised that the existence of multiple communication channels is critical. Early embodied conversational agents (ECAs) tried to convey emotion using analyses of static faces showing full-blown emotions. The results were recognisable, but perplexing and somehow forced and artificial (see Schröder and Cowie, 2006). So, users are unlikely to perceive an embodied agent that works on the basis of emotion decoding as a properly responsive interactant. They may not explicitly know what is missing, but some of the sense of engagement will be lacking.

In the following section we will describe Scherer's (2001) multicomponent model of emotions and Fogel's (1993; Fogel et al., 1992) dynamic systems approach to emotions and communication as examples of how emotions may develop during interactions. We will then illustrate some specific effects emerging during emotional encounters and regulating our responses to such interactions. Finally, we will make a comparison between humans and artificial agents in emotional interactions.

2 Theoretical Explanations of Emotions as Ongoing Processes

2.1 Multicomponent Theory of Emotion

Extending the traditional stimulus–appraisal–response model, Scherer (2001) emphasises the multicomponent nature of emotion and proposes a dynamic sequential model in which emotion is considered as an evolved, continually developing mechanism, allowing for flexibility and adaptation to the changing environment. Its primary adaptive role is to focus attention onto situations or events of importance to an organism's well-being.

The emotional episode itself is viewed as a sequence of synchronised changes in several organismic subsystems following the identification of a significant event. A series of stimulus evaluation checks (SECs) provide the information necessary for an adaptive response. A novelty check scans the environment for changes; if and only if a change is detected, further checks evaluate the nature of the event. Such checks include assessing whether the stimulus should be avoided or approached, whether it is consistent with current goals, the coping potential needed to deal with the stimulus, and whether any action taken would conform to social conventions or expectations.

According to this model, different emotional states (e.g. anger, fear, or joy) emerge from the cumulative process of changes in the various subsystems. The emotional process is thus seen as a continually fluctuating pattern of change brought about by the particular pattern of outputs from SECs. Consequently, emotional states are the result of multiple components each serving particular adaptive functions. Table 1 shows the relationship between the various subsystems and the components, which serve the adaptive functions of emotion. In contrast to theories emphasising the cognitive processing as antecedents of emotion, Scherer stresses that many of the appraisal mechanisms thought to be responsible for various emotional states may rely on automatic processing conducted by hard-wired or innate mechanisms.

An emotional episode is conceptualised as starting with the evaluation of an event as significant to the organism. All aspects of the appraisal process are highly interrelated, with each SEC depending on the result of prior checks; the process is seen as a sequence in which the SECs are performed in a fixed order. For example, it seems logical to assume that the nature of the stimulus needs to be determined before an

Table 1 Relationships between organismic subsystems and the functions and components of emotion (after Scherer, 2001)

Emotion function	Organismic subsystem and major substrata	Emotion component
Evaluation of objects and events	Information processing (CNS)	Cognitive component (appraisal)
System regulation	Support (CNS, NES, ANS)	Neurophysiological component (bodily symptoms)
Preparation and direction of action	Executive (CNS)	Motivational component (action tendencies)
Communication of reaction and behavioural intention	Action (SNS)	Motor expression (facial and vocal expression)
Monitoring of internal state and organism–environment interaction	Monitor (CNS)	Subjective feeling component (emotional experience)

CNS = central nervous system; NES = neuro-endocrine system; ANS = automatic nervous system; SNS = somatic nervous system assessment of coping potential can be made. However, the ensuing appraisal process is not a one-shot matter but an evolving cycle triggering reappraisals which are fed back into the system continually throughout the emotional episode. Like in most appraisal theories, the emotional episode is seen as depending on the evaluation of a stimulus and requires some form of transformation or meaning extraction to be performed. Intuitively one can imagine how emotional episodes may arise in such a way, especially when the event encountered has adaptive significance. However, it may also be possible for subjective feelings to arise when the triggering event is less easy to identify (e.g. generalised anxiety).

One advantage of the multicomponent account is that it does not overestimate the need for cognitive involvement in emotional causation and acknowledges that emotions can be generated with little cognitive effort. Similarly, Parkinson (2001) shares the view that appraisal is an unfolding emotional process in which emotional meaning could be reached in an implicit way. Both accounts agree that low-level processes may be responsible for eliciting emotion and allow for minimal cognitive involvement in emotional causation.

While Scherer's account focuses on appraisal processes within the intrapsychic arena, however, Parkinson argues that appraisal may be interpersonally distributed, with each party contributing pieces of explicit and implicit information towards the ongoing social dialogue (see also Lewis, 1993). Rather than viewing emotional processes as occurring solely within the individual, the responsibility of appraisal is taken away from the separate resources of each individual and is shared throughout the interaction. No person needs to register emotional significance of what is happening because the ultimate meaning is dictated by, and emerges from, the social process itself. Holding an interpersonally distributed emotional representation further reduces the need for cognitive assessments to be performed by a particular individual. This account introduces the possibility of emotional meaning being achieved through a shared process of appraisal in which the overall emotional impact depends on the particular characteristics of the ongoing social interaction.

While the multicomponent approach emphasises the evaluations that occur within an individual in response to a significant event, it does acknowledge that situation and contextual factors affect the evaluation process. Scherer states that the cultural and social context, the nature of the situation, and in particular the nature of the interpersonal relationships within that situation will all influence the appraisal process. Like Parkinson, Scherer shares the view that emotions are not static manifestations. Scherer stresses the dynamic nature of emotion as a process of the continually changing states of its components.

The theories presented in this section agree that attention to the ongoing nature of emotional episodes and accommodation of implicit mechanisms that feed into these processes are both important features of any workable conceptualisation of emotion processes. Both Scherer and Parkinson share the view that emotional episodes are not static manifestations resulting from high-level cognitive calculations, but are ongoing, evolving multicomponent processes. They also suggest that any intelligent system capable of interacting emotionally with humans would ideally need to understand the multicomponent, implicit/explicit nature of emotional elicitation and experience. It would also need to be sensitive to, and capable of being regulated by, the resulting, ongoing emotional interaction itself. In summary, the multicomponent description of emotional episodes suggests that it would be computationally challenging to implement virtual software agents capable of managing human-like social interaction.

2.2 Social Process Theory of Emotion and Continuous Process Model of Communication

Another model that takes the active nature of emotions into account is Fogel's (1993; see also Fogel et al., 1992) dynamic systems approach to emotions and communication. Fogel presents a theoretical framework in which emotion is regarded not only as a multicomponent construct, but also as an ongoing process whose boundaries are not rigidly defined as single, discrete occurrences. In fact, the capacity of particular events to trigger specific emotions is said to vary in accordance with the dynamics of the emotion in relation to social context; for example, an action (e.g. bursting into a laugh) may elicit pleasant emotions on one occasion but not on another, depending on the appropriateness of the action to the situation in question (e.g. a dinner vs. a funeral), the interactants' prior emotional state (e.g. happiness vs. anger or sadness), and the structure of the physical environment (e.g. a noisy crowd vs. a silent assembly).

The starting point of Fogel's social process theory of emotion is to consider emotions not as states but as self-organising dynamic processes closely tied to the course of an individual's activity in a social and physical context. From this point of view, the subcomponents of the unfolding emotional process create a self-organising system by interacting with, influencing, and imposing constraints on each other. However, none of the single subcomponents determine the others in an a priori predictable way, nor are any of these subcomponents individually sufficient to cause emotion, or fully delineated and shaped in the absence of interaction. Moreover, integration and coordination of the emotional components do not necessarily depend on an internal system, but can instead arise from the co-regulated interchange between persons and persons and environment (bottom-up, rather than top-down control processes). Such co-regulated transaction, in turn, can happen at both an explicit and an implicit level.

Fogel (1993) explains co-regulation as "a social process by which individuals dynamically alter their actions with respect to the ongoing and anticipated actions of their partners" (p. 34). In this sense, the outcomes of co-regulated interactions are not results of either an explicit plan or scheme inside one individual, or of an exchange of messages produced by discrete communication signals, but instead emerge from the dynamics of the interaction and the constraints on the communication system.

Unlike discrete state communication systems models, Fogel's theory does not allocate the distinct roles of sender and receiver to interactants engaged in information exchange. Moreover, considering communication as a dynamic process, it is hard to isolate specific signals carrying definite messages. In everyday communication, the same information can be understood in different ways by distinct interactants and can also have different meanings to the same person in different situations.

In emotional communication, in particular, interactants are constantly exchanging explicit and implicit information, modifying each other's emotional responses as they occur. Teasing a friend, for instance, may have different outcomes on our emotions depending on our friend's reactions. If the reaction is laughter, our enjoyment may be increased, but if the response is an expression of disapproval or sadness, our initial amusement may turn to guilt. More generally, during ordinary social interactions one person continuously gives feedback to another, who in turn reacts adjusting his or her own flow of actions, thus creating a consensual, negotiated interactional frame. Emotions, therefore, can hardly be defined as discrete units within the individual; they develop in a shared social context and as such are continuous and never static.

3 Regulating Processes in a Dynamic Emotional World

The flow of interpersonal exchange during social interaction is often regulated by both explicit and implicit processes. Correspondingly, our dynamic emotional experience can be simultaneously influenced by both top-down and bottom-up systems. In this section, we focus on synchrony, dissynchrony, mimicry, and emotional contagion as examples of such phenomena, before using the example of video-mediated interactions to illustrate how the different components of affect communication can interact.

3.1 Synchrony and Dissynchrony

Condon and Ogston (1966) first introduced the concept of behavioural entrainment, or synchrony, to describe an individual's adjustment of behaviour to coordinate with the rhythms of his or her interactants. Analysing patterns of change within and between the behaviour of individuals, these authors found that in normal behavioural patterns harmonious configurations of change between body movements and speech arise, both at an intra-individual (self-synchrony) and at an interactional (interactional synchrony) level.

Bernieri et al. (1988) characterised interactional synchrony as involving (a) direct imitation or mirroring of others' movements, affects, and attitudes; (b) congruence between behavioural cycles of two or more people; and (c) perception of a new meaningful, coordinated whole event created by the unification of concurrent behavioural factors. Although this synchronisation of behaviour is observable in principle, it is not usually attended to. One can perceive another person's engagement or disengagement without explicitly knowing which aspects of his or her behaviour triggered such awareness. Several authors (e.g. Coy, 2001; Wallbott, 1995) consider synchrony, along with motor mimicry and emotional contagion, as a mechanism that helps people to reach interpersonal mutuality in an immediate and unconscious way.

Despite the emphasis that is usually put on implicit regulation of behavioural synchronisation in social interactions, some authors have called attention to the role of more overt acts of information transmission. In a study on the development of mutuality and the subsequent achievement of intersubjectivity, Tronick et al. (1977) suggest that interactants may modify their affective and attentional displays to match or clash with those of their interaction partners, in order to communicate more or less desire to be involved in the specific social interaction. In this case, synchrony would be a way to communicate interest and approval (see also Kendon, 1970), while dissynchrony, the opposite effect, would be a means of interrupting or modifying the current interaction (cf. Tiedens and Fragale, 2003). Support for this communicative interpretation of synchrony and dissynchrony comes from a study by Bernieri and colleagues (1988). These authors point out that low levels of rated synchrony between interactants do not always reflect an absence of synchrony (i.e. asynchrony), but may instead depend upon affects and actions being out of phase as a result of one or more interaction partner's deliberate mismatching.

Behavioural entrainment, therefore, can be seen as influenced by both automatic, unconscious processes and explicit communicative actions deliberately intended to modify the emotional interaction. While communicating, individuals can instinctively adjust to each other's movements and affect reaching a high degree of engagement; yet, during the same interaction, one of the partners may intentionally try to redefine or recalibrate the frame of the communication by varying the degree of synchrony with the other. Again, this intentional regulation of entrainment might be registered implicitly or explicitly by the other interactant, leading to different consequences on the unfolding emotional interaction.

3.2 *Imitation, Mimicry, and the Chameleon Effect*

While synchrony involves interactants jointly constructing a coordinated pattern of behaviours using a kind of bodily dialogue, mimicry refers more specifically to the direct imitation of another's behaviour. At times, we might mimic someone on purpose to ingratiate ourselves to them. For example, training courses for salespeople (and in relationship skills more generally) often explicitly encourage the use of mimicry and imitation as tools for creating smoother interactions and for enhancing interpersonal impressions. Further, imitation may be used to draw other people's attention to specific features of their behaviour, thereby recalibrating the ongoing interaction.

However, imitation effects often seem to be unintentional. Chartrand and Bargh (1999) refer to the non-conscious mimicry of facial expressions, speech patterns, postures, mannerism, and other behaviours of one's interaction partner as the "chameleon effect". These authors propose that perception and interpretation of another person's behaviour automatically activate corresponding behavioural representations in the self, which in turn increase our own tendency to behave in a congruent manner (perception–behaviour link). Thus, mimicry can be the involuntary behavioural consequence of perceiving another's behaviour; moreover, perception of a similar behaviour by the other strengthens the interaction, creating feelings of empathy. In an interesting study, Bailenson and Yee (2005) report unconscious mimicry and a subsequent increase in rapport even when people interact with an embodied artificial intelligence agent. Mimicking agents were viewed by participants as more persuasive and pleasant than non-mimicking ones, even though participants apparently failed to notice mimicry at an explicit level.

Investigators have also studied the influence of mood and emotions on mimicry. For example, Van Baaren et al. (2006) found that good moods lead people to greater automatic imitation of other people's behaviour than bad moods. The authors account for these findings by reference to the contrasting informational implications of positive and negative affect (e.g. Schwarz, 1990) as safety and danger signals, respectively. Thus, good moods lead people to process information in a more holistic and spontaneous way whereas bad moods lead people to process information more analytically (e.g. Mackie and Worth, 1991; Schwarz and Bless, 1991). Similarly, good moods may lead people to be less reflective and more spontaneous in regulating their behaviour leaving them more susceptible to automatic influences.

Niedenthal et al. (2001), on the other hand, explored the role of mimicry in understanding facial behaviour in emotional interactions. Their work has found that individuals in a particular emotional state detect changes in another's emotion expression better if this expression is congruent with their own emotion than if it is incongruent. This seems to be due to the fact that people mimic emotion-congruent expressions more easily than emotion-incongruent ones. As a result, individuals detect changes in an emotion-congruent expression because these changes produce a noticeable alteration in their own facial behaviour.

Finally, mimicry, as well as synchrony, seems to have an impact on the development of shared emotions during interactions, a phenomenon known as emotional contagion. This phenomenon will be discussed in the next section.

3.3 Emotional Contagion

The pervasive automatic tendency towards mimicry and behavioural coordination may generate emotional episodes by inducing a corresponding emotional state in the mimicker. Hatfield and colleagues (1994) coined the term primitive emotional contagion to describe the process whereby individuals catch the emotions of those around them. While emotional contagion is sometimes thought to involve conscious perceptions and evaluations, Hsee et al. (1992) argue that "generally the process by which people feel others' emotional states is fairly non-conscious primitive, and automatic" (p. 2). During the course of social interaction the natural tendency to mimic others results in a synchrony of facial, postural, and vocal expressions.

The synchrony of affective behaviours manifests in a shared emotional state among the interactants. Specifically, subjective emotional states are affected moment to moment by the activation and/or feedback from such mimicry. Emotional states may be influenced by either central nervous system commands which direct mimicry, afferent feedback from facial, verbal, and postural movements, or conscious attribution of affect based on self-perception of expressive behaviour. Emotional contagion is hence conceptualised as a two-stage process in which mimicry and the activation of feedback from mimicry result in a corresponding subjective emotional state among the interactants. As other theorists have also argued (e.g. Öhman, 1988; Posner and Snyder, 1975; Shiffrin and Schneider, 1977), Hatfield et al. (1994) propose that much of the processing of emotional information occurs outside of conversant awareness; the theory draws on the subtlety with which this implicit information affects behaviour. From this perspective, subjective emotional experience is heavily influenced by the subconscious monitoring and reaction to implicit emotional information presented throughout the social dialogue. While Hatfield and colleagues maintain that the emotional contagion effect is predominantly the result of automatic responses occurring outside awareness, it is likely that explicit processes also regulate contagion mechanisms.

Presumably emotional contagion is a reciprocal process in which one person's emotions affect another's and vice versa. If contagion were to operate at an entirely unconscious and automatic level during social interaction, one would expect to see emotions rapidly intensifying between people and potentially escalating to extreme levels. For example one person's fear would induce a corresponding emotional state in the other, whose reaction would in turn influence the first person's, and so on; in this scenario it would not take too long for a state of emotional hysteria to develop. While Hatfield et al. (1994) cite several examples throughout history where this has in fact happened (e.g. the dancing manias of the Middle Ages, the great fear of 1789, and the New York City riots of 1863), instances of hysterical contagion of this kind are relatively few and far between. It is likely that conscious or unconscious control processes often override any natural tendency to mimic and regulate contagion.

Rather than being purely a primitive reflex-type reaction, it seems highly possible that mimicry is sometimes employed as an adaptive communication tool and as such is influenced by the social context. For example, the nature of the relationship between interacting individuals can either facilitate or thwart mimicry. Hatfield et al. (1994) argued that liking and closeness encourage mimicry, although there is no clear evidence concerning whether emotional contagion results from liking, or liking increases mimicry which in turn causes emotional convergence (or both). However, Bucy and Bradley (2004) have shown that counter-mimicry and counter-empathetic emotional responses are evoked by emotional expressions deemed inappropriate to the situation. This would suggest that the mechanisms underlying mimicry are subject to some form of assessment in relation to the social context.

Perhaps the least convincing aspect of the emotional contagion hypothesis is the postulated role of autonomic feedback from facial/postural movements in the overall emotional experience. Available evidence in fact suggests that effect sizes from facial feedback are likely to be small (e.g. Tourangeau and Ellsworth, 1979). It is questionable whether sensory feedback is in fact the main mechanism by which emotional convergence occurs. It is more likely that individuals make appraisals of other people's emotional reactions in order to make sense of the situation and ultimately to decide how they will respond emotionally (e.g. social referencing, Sorce et al., 1985). Thus, emotions are partly determined by interpersonal rather than internally generated feedback. In sum the mechanisms postulated are likely to involve the interaction of both explicit and implicit processes which create and regulate contagion effects.

3.4 Video-Mediated Interactions

Before introducing a general comparison between emotional-competent humans and emotional-competent artificial agents in the next section, we want to consider video-mediated communication as a specific kind of non-face-to-face interaction between humans which can show the implications of explicit and implicit signals, and lack of them, during emotional encounters (see also Parkinson, 2008).

In a study on the impact of audio-visual technology on informal communication in workplaces, Heath and Luff (1992) pointed out how some features of such technology may transform the impact of visual and vocal conduct and introduce asymmetries in interpersonal interactions. In particular, face-to-face interactants may respond automatically to cues arising from gaze direction or physical gestures that are only registered in peripheral vision, whereas users communicating via video-conferencing may completely fail to pick up these signals. This might be due, for example, to a two-dimensional vs. three-dimensional representation of faces, to the physical arrangement of screens and cameras, or to the size and flatness of the monitor. In any case, the result is an unbalanced coordination of the interaction with a restricted possibility for regulation of the communication, even when explicit attempts to adjust the conversation are made by one of the interactants.

Parkinson and Lea (in press) specifically investigated the impact of the constraints associated to video-mediated communication on transmission and regulation of emotions. Introducing transmission delays in a video-conferencing system, the authors showed that such gaps in the flow of the conversation can interfere with the establishment of rapport between individuals. The reason for this might be found in a lack of temporal synchrony or of temporal complementarity (i.e. limitation in immediate feedback).

Limitations in communication technology such as videos and cameras can therefore decrease shared positive affect and may even lead to the development of negative emotions such as frustration between interactants. As illustrated in the next section, this kind of reaction can also apply when humans are interacting with artificial agents whose emotional interactional skills are often limited to recognition of full-blown, explicitly categorised emotions.

4 Comparison Between Humans and Artificial Agents in Emotional Interactions

The required level of sophistication of an agent's affective architecture depends directly on the purpose of its design. That is, not every application needs to provide an elaborate veridical simulation of the emotional aspects of human behaviour. However, if software entities such as ECAs are intended to permit face-to-face interaction with human beings, the underlying components and processes of their affective architecture are likely to require sufficiently elaborated behavioural and expressive mechanisms to allow believable engagement with humans (Pelachaud, 2006). Despite recent significant developments, most affective ECAs can only occasionally fulfil their purposes without inciting abusive behaviour or negative emotions from users. The breakdown of user empathy towards an artificial entity caused by the lack of human-like appearance, emotional expression, and motion displays during an interaction can contribute both to user dissatisfaction and reluctance to engage in further communicative attempts (Angeli and Carpenter, 2005).

In the context of human–computer (HC) interaction, agent technologies clearly require additional development, especially in terms of better integrating cognitive and emotional features that can appropriately resemble at least a minimum set of adaptive features found in human social behaviour (Ventura et al., 2005). These include, for example, implicit and explicit regulatory responses as described earlier in this chapter together with ECA design recommendations discussed in chapters "Coordinating the Generation of Signs in Multiple Modalities in an Affective Agent" and "Generating Listening Behaviour" of Part IV. Good practice obviously requires that the escalation of user frustration while interacting with artificial entities is either actively avoided or quickly corrected. However, to date, most of the available agent affective architectures provide at best poor explicit and efficient means to automatically deal with or learn from these situations (Barkhuysen et al., 2005). If emotional interactions with artificial agents are to be truly comparable to those with humans, it is necessary to overcome various computational challenges that currently undermine positive perceptions of the technologies used to compose artificial social entities. In order to enable more successful synthesis of agent behaviour, research endeavours will probably involve not only the enhancement of architectural processes and data structures, but also more realistic adaptation of emotional facial expressions and other multimodal communication capabilities such as the capability to process natural language via sound or text interfaces. Examples include agents that simultaneously mimic human facial expressions accompanied with gestures (Caridakis et al., 2007), and guidance protocols for collection of multimodal emotional data between human–human interactions (Zara et al., 2007).

Increasing the anthropomorphism of the agent interface is known to affect the user's experience and its evaluation. This is mainly due to the resulting higher user expectations regarding agent believability and possible intelligent traits (Fong et al., 2002). Numerous examples could be cited of how subtle visual or behavioural imperfections can transform positive perceptions of artificial entities into generally uncomfortable experiences that are sustained by cognitive dissonance (Masahiro, 2005). For example users sometimes report unsettling experiences whilst interacting with real android-like robots, usually triggered by incongruous behaviour. On the other hand, if an entity is clearly unrecognisable as having human-like behaviour or expressivity, even simpler traits resembling acceptable cognitive-affective responses are more likely to be noticed and may stimulate empathy and social engagement from the users. Animated cartoon characters are probably the best example of the latter observation. Therefore, pragmatic understanding of the impact of graphical agents and their internal processing structures are directly relevant to the design of ECAs. Nevertheless, the precise way of addressing such issues is not always completely clear, as it depends heavily on the application purpose (and also sometimes on good understanding of the cultural context in question).

Several different artificial intelligence approaches are currently available to model affective agents, but most usually include agent architectures focused only on symbolic (logic-based) or sub-symbolic (connectionist) data representations and processing mechanisms. Each has its specific strengths and, although a limited number of hybrid systems exist, there is no consensus on how to integrate such benefits in a single computational system (Schröder and Cowie, 2006). A number of authors have proposed similar component requirements for affective architectures in order to create entities capable of successful human–agent emotional interactions. Examples include appraisal and interpretation mechanisms influenced by action tendencies, management of emotional personality profiles for processing contextual human information, believable interaction capabilities, and supervised or unsupervised machine-learning techniques that can take advantage of past interaction experiences (Payr, 2001). To facilitate design, it is necessary to take into account the expected functionalities of an agent and the ways in which the user interface might assist the interaction based on their context-dependent requirements.

The ability to communicate emotions is often regarded as crucial for the usability of socially interactive agents such as affective ECAs and certain types of robots, because users will probably prefer, and be more accepting of, entities that can provide a sense of comfort and usefulness during their interactive sessions (Masahiro, 2005). Emotions need not be visible in order for a system to fulfil its designated purposes; for example, an entertaining entity will probably have a different design and different operational principles than a software entity intended to improve decision-making. In this sense, clear specification and analysis of relevant user-centred metrics can help to understand the impact of different systems and facilitate their subsequent comparison. These metrics may include, for instance, rating criteria for classifying interactions that are aimed at conveying emotions in real time during the generation of speech and facial expressions, attention and appraisal skills, degrees of freedom for adapting body postures (e.g. gazes and gestures), social competence (e.g. conformity to social norms), and suggestive use of colours on the interface available to the human user. Conversational agents processing text for assisting the accomplishment of specific tasks pose different design and implementation issues than those that, for instance, would sense physical proximity in game-like educational software. Social coping mechanisms are vital as data collected during social interactions are often incomplete or ambiguous, and the handling of such events can influence user perceptions of how much their task was ameliorated by the agent's performance. Recommendations to ECA designers depend heavily on which HC interfaces are available to implement architectures that will interact with users or help them to execute their tasks. It is important to take the extent to which users will rely on ready-to-use interactive features into account (i.e. the relevance of text, sound or image expressive modalities). Whilst user interface preferences vary according to individual criteria, design effectiveness can improve substantially, simply by focusing on concrete contributions to the user experience with regard to what the system ought to facilitate.

5 Conclusion

In our everyday face-to-face encounters, as well as in any video-mediated conversations or dealings with artificial agents, the way we respond to each other, whether consciously or unconsciously, is essential to the creation of understanding and rapport, which in turn can increase our enthusiasm for continuing the interaction. Throughout this chapter we have illustrated how emotions cannot be reduced to fixed entities, and how emotional encounters are based on ongoing processes that are typically embedded in a dynamic social context. Moreover, we have seen how both explicit responses to emotions (i.e. conscious processing and categorisation of emotional information) and implicit ones (i.e. registration of emotion response components) can influence the unfolding interpersonal interactions. In order to better understand emotional communication and to develop agents that can naturally interact with humans, we conclude that attention should be focused on the development of dynamic models of social interaction that do not consider emotions as static units but as multicomponent constructs that are continuously regulated by the interaction itself and by the social and cultural context in which it unfolds.

References

Angeli DA, Carpenter R (2005) Stupid computer! Abuse and social identities. In: Proceedings of Abuse: The dark side of human-computer interaction, An INTERACT 2005 workshop. http://www.agentabuse.org/deangeli.pdf. Accessed 27 February 2008

Bailenson JN, Yee N (2005) Digital chameleons: automatic assimilation of nonverbal gestures in immersive virtual environments. Psych Sci 16(10):814–819

Barkhuysen P, Krahmer EJ, Swerts M (2005) Problem detection in human-machine interactions based on facial expressions of users. Speech Commun 45:343–359

Bernieri F, Reznick JS, Rosenthal R (1988) Synchrony, pseudosynchrony, and dissynchrony: Measuring the entrainment process in mother-infant interactions. J Pers Soc Psych 54(2): 243–253

Bucy EP, Bradley MM (2004) Presidential expressions and viewer emotion: counterempathic responses to televised leader displays. Soc Sci Info 43:59–94

Cacioppo JT, Bush LK, Tassinary LG (1992) Micro expressive facial actions as a function of affective stimuli: Replication and extension. Pers Soc Psych Bull 18:515–526

Caridakis AG, Bevacqua Raouzaiou E, Mancini M, Karpouzis K, Malatesta L, Pelachaud C (2007) Virtual agent multimodal mimicry of humans. Lang Res Eval 41:367–388
Chartrand TL, Bargh JA (1999) The chameleon effect: The perception-behavior link and social interaction. J Pers Soc Psych 76(6):893–910
Condon WS, Ogston WD (1966) Sound film analysis of normal and pathological behavior patterns. J Nerv Ment Dis 143(4):338–347
Coy KR (2001) Rhythm within collaboration: A qualitative study of male and female dyads. Dissertation Abst Int Sect B Sci Eng 61:6691
Fogel A (1993) Developing through relationships: origins of communication, self, and culture. University of Chicago Press, Chicago
Fogel A, Nwokah E, Dedo JY, Messinger K, Dickson KL, Matusov E, Holt SA (1992) Social process theory of emotion: a dynamic systems approach. Soc Devt 1:122–142
Fong TW, Nourbakhsh I, Dautenhahn K (2002) A survey of socially interactive robots: Concepts, design, and applications. Tech report CMU-RI-TR-02-29, Robotics Institute, Carnegie Mellon University. infoscience.epfl.ch/record/30017/files/CMU-RI-TR-02-29.pdf. Accessed 27 February 2008
Harré R, Gillett G (1994) The discursive mind. Sage, Thousand Oaks, CA
Hatfield E, Cacioppo JT, Rapson RL (1994) Emotional contagion. Cambridge University Press, New York, NY
Heath C, Luff P (1992) Media space and communicative asymmetries: preliminary observations of video-mediated interaction. Hum-Comput Int 7:315–346
Hess U, Banse R, Kappas A (1995) The intensity of facial expression is determined by underlying affective state and social situation. J Pers Soc Psych 69:280–288
Hsee CK, Hatfield E, Chemtob C (1992) Assessments of the emotional states of others: Conscious judgments versus emotional contagion. J Soc Clin Psych 11(2):119–128
Kendon A (1970) Movement coordination in social interaction: Some examples described. Acta Psychol 32:100–125
Kenny DA, De Paulo BM (1993) Do people know how others view them? an empirical and theoretical account. Psych Bull 114(1):145–161
Lewis M (1993) Self-conscious emotions: Embarrassment, pride, shame, and guilt. In: Lewis M, Haviland JM. (eds) Handbook of emotions. Guilford, New York, NY, pp 563–573
Mackie DM, Worth LT (1991) Feeling good, but not thinking straight: The impact of positive mood on persuasion. In: Forgas. JP (ed) Emotion and social judgments. Pergamon Press, Oxford, pp 201–219
Masahiro M (2005) On the Uncanny Valley. In: Proceedings of the Humanoids-2005 workshop: views of the Uncanny Valley, 2005. http://www.theuncannyvalley.org/. Accessed 27 February 2008
Niedenthal PM, Brauer M, Halberstadt JB, Innes-Ker AH (2001) When did her smile drop? Facial mimicry and the influences of emotional state on the detection of change in emotional expression. Cog Emo 15(6):853–864
Öhman A (1988) Nonconscious control of automatic responses: A role for Pavlovian conditioning? Bio Psych 27:113–135
Parkinson B (2001) Putting appraisal in context. In: Scherer KR, Schorr A, Johnstone T, (eds) Appraisal processes in emotion: Theory, methods, research. Oxford University Press, London, pp 173–186
Parkinson B (2005) Do facial movements express emotions or communicate motives? Pers Soc Psych Rev 9(4):278–311
Parkinson B (2008) Emotions in direct and remote social interaction: getting through the spaces between us. Comp Human Behav 24:1510–1529
Parkinson B, Lea MF. Video-linking emotions. In: Kappas A, Kramer N. (eds) Face-to-face communication over the internet: issues, research, challenges. Cambridge University Press, Cambridge, (in press)

Parkinson B, Fischer AH, Manstead ASR (2001) Emotion in social relations: cultural, group, and interpersonal processes. Psychology Press, New York, NY

Payr S (2001) The virtual other: aspects of social interaction with synthetic characters. Appl Artif Intell 15:493–519

Pelachaud C (2006) Humaine D6f exemplar: how to build an affective interactive ECA, rev. on February 2007. Emotion-research.net/deliverables/D6f_v1.5 Feb 2007.pdf. Accessed 27 February 2008

Posner MI, Snyder CRR (1975) Attention and cognitive control. In: Solso RL, (ed), Information processing and cognition: the Loyola Symposium. Erlbaum, Hillsdale, NJ, pp 55–87

Scherer KR (2001) Appraisal considered as a process of multilevel sequential checking. In: Scherer KR, Schorr A, Johnstone T, (eds) Appraisal processes in emotion: theory, methods, research. Oxford University Press, Oxford, pp 92–120

Schröder M, Cowie R (2006) Developing a consistent view on emotion-oriented computing. In: Renals S, Bengio S (eds) Machine learning for multimodal interaction: LNCS 3869. Springer, Berlin, pp 194–205

Schwarz N (1990) Feelings as information: informational and motivational functions of affective states. In: Higgins ET, Sorrentino RM (eds) Handbook of motivation and cognition, vol 2. Guilford Press, New York, pp 527–561

Schwarz N, Bless H (1991) Happy and mindless, but sad and smart? The impact of affective states on analytic reasoning. In: Forgas JP. (ed) Emotion and social judgments. Pergamon, Oxford, pp 55–71

Shiffrin RM, Schneider W (1977) Controlled and automatic human information processing: Perceptual learning automatic attention and a general theory. Psych Rev 84:127–190

Sorce JF, Emde RN, Campos J, Klinnert MD (1985) Maternal emotional signaling: Its effect on the visual cliff behavior of 1-year-olds. Devt Psych 21(1):195–200

Tiedens LZ, Fragale A R (2003) Power moves: Complementarity in dominant and submissive nonverbal behavior. J Pers Soc Psych 84:558–568

Tourangeau R, Ellsworth PC (1979) The role of facial response in the experience of emotion. J Pers Soc Psych 37:1519–1531

Tronick ED, Als H, Brazelton TB (1977) Mutuality in mother-infant interaction. J Comm 27: 74–79

Van Baaren RB, Fockenberg DA, Holland RW, Janssen L, van Knippenberg A (2006) The moody chameleon: the effect of mood on non-conscious mimicry. Soc Cogn 24(4):426–437

Ventura J, Ventura M, Olabe JC (2005) Embodied conversational agents: developing usable agents. Proceedings of IEEE Southeast Conference, Fort Lauderdale, Florida doi: 10.1109/SECON.2005.1423322

Wallbott HG (1995) Congruence, contagion, and motor mimicry: Mutualities in nonverbal exchange. In: Markova I, Graumann CF, Foppa K. (eds) Mutualities in dialogue. Cambridge University Press, New York, NY, pp 82–98

Zara A, Maffiolo V, Martin J-C, Devillers L (2007) Collection and annotation of a corpus of human-human multimodal interactions: emotion and others anthropomorphic characteristics. In: Paiva A, Prada R, Picard RW. (eds) 2nd international conference on affective computing and intelligent interaction. Springer, New York, NY

Biological and Computational Constraints to Psychological Modelling of Emotion

Etienne B. Roesch, Nienke Korsten, Nickolaos F. Fragopanagos, John G. Taylor, Didier Grandjean, and David Sander

Abstract Emotion modelling comes in many guises, from strictly behaviourist black box models to intricately detailed computational architectures. In this chapter, an investigation of the conflicts and common grounds between the different approaches is presented, touching on psychological theories, neuroscience, and artificial neural networks (ANN). Also included is an overview of state-of-the-art ANN, simulating several emotion-related processes in the human brain.

1 From Psychology to Cognitive Neuroscience, to Modelling, and Back

The middle 1960s have been the cradle of a *cognitive revolution* (Gardner, 1985). The explicit goal of this movement was to depart from behaviourism, which dominated most of the twentieth century, by posing a new paradigm. Behaviourism, the "science of the black box", believed that the purpose of psychology was to study and predict behaviour. Little importance, if at all, was put on the underpinnings of the behaviour. Unlike behaviourists, the proponents of the cognitive revolution wanted to figure out how the mind, and ultimately the brain, processes information. First chronicler of this revolution, Gardner (1985) depicted this new scientific landscape as groups of disciplinary researchers attempting to bridge the gaps between their disciplinary expertise (Fig. 1, panel a). His conclusion at the time was that there was no consensual research paradigm, no common set of hypotheses, and no consensual methodology. Things have evolved since then and, even though the utopia of witnessing truly interdisciplinary approaches to the mind is never fully met, strong ties between disciplines are being built.

E.B. Roesch (✉)
Centre for Integrative Neuroscience and Neurodynamics, University of Reading, Reading, UK
e-mail: contact@etienneroes.ch

P. Petta et al. (eds.), *Emotion-Oriented Systems*, Cognitive Technologies,
DOI 10.1007/978-3-642-15184-2_4,

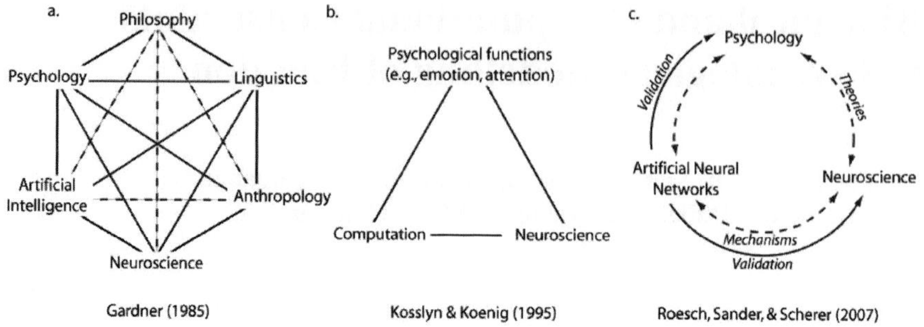

Fig. 1 (**a**) Connections among the cognitive sciences, after Gardner (1985). *Unbroken lines* = strong interdisciplinary ties. *Broken lines* = weak interdisciplinary ties. (**b**) The *triangle* of the cognitive neuroscience, adapted from Kosslyn and Koenig (1992). (**c**) Proposal for an implementation of interdisciplinary dynamics between psychology, neuroscience, and artificial neural networks modelling (Roesch et al., 2007)

Proof is the perspective taken by Kosslyn and Koenig (1992) who formalised the cognitive neuroscience triangle (Fig. 1, panel b), in which the three complementary objects of research (behaviours, computations, and brain substrates) are mutually constraining each other. In particular, a general challenge for cognitive neuroscientists, which is considered in the present chapter, is to investigate psychological functioning in a biologically plausible way that can produce computational models – with both the biological and the computational levels constraining psychological models of cognition (Roesch et al. 2007). These constraints can be expressed in the form of the goodness of fit between the models proposed by each discipline (Fig. 1, panel c).

The goal of cognitive neuroscience is very ambitious. It is nothing less than "to map the information-processing structure of the human mind and to discover how this computational organisation is implemented in the physical organization of the brain" (Tooby and Cosmides, 2000). By being as explicit as possible, researchers can account for healthy and pathological functioning and design new research in such a way as to address new predictions. This approach is often referred to as the "boxes-and-arrows" approach, because of the graphical representation researchers use to describe the topology of cognitive systems and the interactions and dynamics between their constituents. Boxes represent the computations that are taking place. They take inputs, and deliver outputs, and interact in fixed manners. Arrows represent the mechanisms of interaction, the dynamics of the information flow (Marr, 1982).

Sander and Koenig (2002) developed structural and functional principles for such a description applied to emotional processing. The constraints they extracted review the basics of emotion literature and contain criteria that all models of emotions should be able to account for. Functional principles encompass criteria related to the evaluation of the stimuli, the expression of the emotion, and the subjective experience of the organism. These are complemented with computational criteria describing the types of input, and output, that would be relevant to emotion models. They also emphasise the need for biological plausibility. The mind being intrinsically tied to the cerebral substrate, biological plausibility constraints posit that, in order to adequately represent cognitive functioning, models should be as close as possible to what is known of the nervous system (Kosslyn and Koenig, 1992). However, some argue that, in a first stage of the modelling process, biological implausibility can be an asset to understand and characterise the problem, examine and evaluate the information flow using a representation that is sometimes easier to grasp (Dror and Gallogly, 1999).

Extending on Sander and Koenig's approach to the modelling of emotion, one can emphasise a few points and make the following recommendations to both theoretical and applied research groups.

Recommendation 1 – Formal Description of Endogenous and Exogenous Structures

A model accounts for a specific system (endogenous structure), which only makes sense in a specific context (exogenous structure). Therefore, we argue that models should be embedded in a context as detailed as possible. The main purpose of this principle is to ease the comparison first between models and second with experimental data.

Recommendation 2 – Formal Description of Inputs and Outputs

Inputs can be of several types, coming from the environment of the organism (e.g. visual, tactile) or from internal stimulation (e.g. somatic information). Similarly, outputs can feed forward into deeper processes or backward into the system. This information being at the core of the model, researchers need to provide a detailed description of each arrow they draw.

Recommendation 3 – Formal Description of Transfer Functions

Models describe the interaction of several processes. Each process computes the information it receives and transmits the processed information in an output further. What the process does is not necessarily clear, and the dynamics involved can be very dependent on the interpretation of the researcher. We therefore argue that great effort should be invested into the description of the processes, in relation to the input it receives and the output it transmits.

Recommendation 4 – Formal Description of the Information Flow

The information transmitted may change and evolve. Therefore, we argue that researchers should describe the sequence of actions that occur in the system as precisely as possible. The time course of the information in the model is of critical importance to compare it with experimental data.

Recommendation 5 – Formal Description of State Transitions

At any point in time, a model is in an observable state, expressing a particular mode of behaviour of the system and can only be in that particular state. Consequently, the model could potentially be described by more than one state. We therefore argue that researchers should describe the states the model can be in, as well as the possible transitions that can follow. This recommendation could help the comparison with behavioural and neuroimaging data.

Critically, adopting a cognitive neuroscience approach to emotion may help to distinguish between alternative theories of emotional processing, by providing new constraints that are brain based as much as computation based. Until now, two major classes of psychological theories of emotion have dominated research in cognitive neuroscience of emotion, with an emphasis on brain-based evidence: (1) basic emotions theories and (2) dimensional theories (see also the chapter "Emotion: Concepts and Definitions" by Cowie et al., this volume):

(1) The dominant view of emotion during the last century, and which is probably still the most influential in current emotion research, is represented by the so-called basic emotions models (Ekman, 2003; Izard, 2007). These models convey the notion that there is a finite number of separate emotions, shaped by evolutionary pressure, that can be combined to form blended and more complex emotions. Whereas researchers do not agree on the exact number of such emotions, they all agree that "the various classes of emotion are mediated by separate neural systems" (LeDoux, 1996). Most of the recent cognitive neuroscience research on emotion has attempted to identify specific brain regions implementing these distinct basic emotions (Ekman, 1999) such as fear, disgust, anger, sadness, and happiness. Indeed, a large corpus of data suggests that signals of fear and disgust are processed by distinct neural substrates (see Calder et al., 2001). Functional imaging of the normal human brain (e.g. Phillips et al., 1997) and behavioural investigations of brain-damaged patients (e.g. Calder et al., 2000) revealed a crucial involvement of insula and basal ganglia in processing disgust signals. On the other hand, animal research (e.g. LeDoux, 1996), behavioural studies of brain-damaged patients (e.g. Adolphs et al., 1994), and functional imaging in healthy people (e.g. Morris et al., 1996) suggested that the amygdala is a key structure for responding to fear-related stimuli (see also Öhman and Mineka, 2001). Mineka and Öhman (2002) even proposed that "the amygdala seems to be the central brain area dedicated to the fear module" (see also Öhman and Mineka, 2001). More tentative evidence suggests a similar segregation of processes related to anger, sadness, and happiness, particularly during recognition of facial expression (e.g. Blair et al., 1999). On the basis of neuropsychological dissociations between fear, disgust, and anger, Calder et al. (2001, 2004) encouraged neuropsychologists to adopt the basic emotions framework in order to understand and dissect the emotion system.

(2) From another perspective, all emotions are considered to be represented in a common multidimensional space. For example, Wundt (1905) proposed that the nature of each emotion category is defined by its position within three orthogonal dimensions: pleasantness–unpleasantness, rest–activation, and relaxation–attention. It has been argued that emotional response and stimulus evaluation might

primarily be characterised by two dimensions: valence (negative–positive) and intensity (low–high) (see also Anderson and Sobel, 2003; Hamann, 2003). Using functional neuroimaging, Anderson et al. (2003) and Anderson and Sobel (2003) found that amygdala activation correlated with the intensity but not the valence of odours, whereas distinct regions of orbitofrontal cortex were associated with valence independent of intensity. Similarly, Small et al. (2003) dissociated regions responding to taste intensity and taste valence: structures such as the middle insula and amygdala coded for intensity irrespective of valence, whereas other structures such as the orbitofrontal cortex showed valence-specific responses. However, adopting a theoretically based approach, recent results show that four dimensions (valence, potency, arousal, and unpredictability) are needed to satisfactorily represent similarities and differences in the meaning of emotion words, Fontaine et al. (2007) questioning the ability of two-dimensional models to fully represent emotional systems.

However, instead of adopting either the discrete emotions or the dimensional views, some researchers have proposed an alternative approach, by parsing emotions into distinct subcomponents at the process level and determining dynamic interactions between these processes. If Damasio (1994, 1999) distinguished neural systems involved in emotion from those involved in feelings, Panksepp (1998) proposed four primitive systems (seeking, fear, rage, and panic systems) combined with special-purpose socio-emotional systems (for sexual lust, maternal care, and roughhousing play). With an emphasis on psychopathology, Gray (1994) distinguished three types of behaviour (fight, active avoidance, and behavioural inhibition), each mediated by different neural circuits. Similarly, Davidson (1995) proposed differential systems in the two cerebral hemispheres underlying approach-related emotions and withdrawal-related emotions.

These diverse approaches illustrate the lack of a consensual definition of emotion (see Kleinginna and Kleinginna, 1981). With the aim to elaborate on a working definition, Scherer (1984, 2001) integrated different theoretical elements and proposed a multicomponent approach to emotions (see also the chapter "Emotion: Concepts and Definitions" by Cowie et al., this volume). This definition is part of the theoretical framework laid by appraisal theorists of emotion, which represent an attempt to parse the underlying functional mechanisms in terms compatible with brain-based and computation-based evidence.

A key aspect of appraisal is that the specificity and the differentiation of an emotion may depend on a multifactorial evaluation of the meaning and consequences of an event, given the individual's goals, needs, and values, as well as the current context. To our knowledge, until recently, neither basic emotion theorists nor dimensional theorists have been centrally concerned with the nature of processes that may not only detect relevant events, but also evaluate their consequence and meaning in a context-dependent manner. By contrast, such evaluation is central to componential appraisal theories of emotion (Ellsworth, 1991; Ellsworth and Scherer, 2003; Frijda, 1986; Lazarus, 1999; Roseman and Smith, 2001; Smith and Lazarus, 1993; Scherer, 2001). Such an approach conceptualises the behavioural meaning of an event for the individual (and thus the resulting emotion) on the basis of multiple complementary criteria including novelty, agreeableness, goal conduciveness, coping potential, and norm compatibility (Scherer, 2001). Therefore, a cognitive neuroscience account of appraisal processes in emotion may offer new avenues of investigation and possibly account for results that otherwise remain difficult to explain (Sander et al., 2005). Among the neural networks involved in emotional processing, a few critical structures have been intensely investigated, but their respective roles remain difficult to link with theories of emotion. These structures include the amygdala, ventral striatum, dorsolateral prefrontal cortex, superior temporal sulcus, somatosensory-related cortices, orbitofrontal cortex, medial prefrontal cortex, fusiform gyrus, cerebellum, and anterior cingulate (Adolphs et al., 2002; Adolphs, 2003; Damasio, 1998; Davidson and Irwin, 1999; Pessoa, 2008; Rolls, 1999). As already mentioned, it has been proposed that the insula is particularly involved in processing "disgust" (Calder et al., 2001; Phillips et al., 1997). However, the insula was also found to be activated during the experience of sadness (George et al., 1996), fear conditioning (e.g. Büchel et al., 1999), and processing of fearful faces (Anderson et al., 2003), challenging this "basic emotion" approach. Moreover, Morris et al. (1998) showed that the anterior insula was responsive to increasing intensity of fear in faces, and Phelps et al. (2001) proposed that the insular cortex might "be involved in conveying a cortical representation of fear to the amygdala". Therefore, it appears that the insula, as a structure, might not be uniquely involved in disgust-related mechanisms. In particular, a specific account of the function of the amygdala, derived from appraisal research, can help to constrain and inform models of emotion. Contrary to the assumption that the amygdala is central to a "fear module" only (Öhman and Mineka, 2001), in accord with a discrete emotion model, some brain imaging studies suggest that this structure contributes to the processing of a much wider range of negative affective stimuli (for a review, see Sander et al., 2003). As the amygdala seems also involved in the processing of positive events, it was suggested that it modulates arousal, independently of the valence of the elicitor (e.g. Anderson and Sobel, 2003) – potentially supporting dimensional theories of emotion. However, it has been shown that equally intense stimuli differentially activate the dorsal amygdala (e.g. Whalen et al., 2001), and that arousal ratings in a patient with an amygdala lesion are impaired for negative, but not positive, emotions (Adolphs 1999). These results seem to contradict the view that the amygdala codes for arousal irrespective of valence. From another perspective, it has been proposed that the computational profile of the human amygdala might meet the core appraisal concept of relevance detection (see Sander et al., 2003), a view that integrates several findings on the amygdala and suggests that it may be central in processing self-relevant information. In general terms, an event is relevant for an organism if it can influence (positively or negatively) the attainment of his or her goals, the satisfaction of his or her needs, the maintenance of his or her own well-being, and the well-being of his or her species. Evaluation of relevance may then elicit the corresponding affective responses and behaviours. These responses may include enhanced sensory analysis and enhanced encoding into memory, as well as autonomic, motor, and cognitive effects. A review of the literature is consistent with this idea that amygdala processes do not respond just to the intrinsic valence or arousal level of a stimulus, but to the subjectively appraised relevance (see Sander et al., 2003, 2005). As an example, one can mention the study by Winston et al. (2005) that demonstrated that the amygdala exhibits an intensity-by-valence interaction in olfactory processing. These authors were able to show that the amygdala responds differentially to high (vs. low)-intensity odours for pleasant and unpleasant smells (as previously reported by Anderson and Sobel, 2003 and Small et al. 2003) but, critically, not for neutral smells. This recent study therefore concluded that the amygdala codes neither for intensity nor for valence per se, but for an integration of these dimensions.

In conclusion, a critical function of the cognitive neuroscience of emotion is to design new experiments specifically testing for critical predictions of these current major psychological theories. In particular, determining whether the amygdala is primarily involved in (1) fear or negative valence, (2), intensity or arousal coding, and (3) relevance detection would, respectively, provide support to (1) discrete emotions theories, (2) dimensional theories, or (3) appraisal theories. Although advanced neuroscience approaches have helped to start testing these current major psychological models, computational approaches will certainly bring to the new field of affective neuroscience critical concepts and methods for testing the respective predictions.

2 Biologically Plausible Artificial Neural Networks to Test Models of Emotion

In addition to neural network modelling on a functional or psychological basis as described above, some researchers have attempted to approach implementable emotion modelling from a neural perspective. The aim of this research is to implement a mechanism with known neural correlates, in order to draw a direct analogy to neural mechanisms. Thus, the ultimate goal is not necessarily to create a network architecture that could have practical applications, but rather to increase our understanding of the workings of the brain by combining the knowledge that we already have in implementable and testable mechanisms. Not only does implementation allow us to grasp the workings of mechanisms that are otherwise too complex to reason through, but it also allows us to assess if non-implemented models are realistic. In this section, we will show how this way of modelling can and has been used to investigate emotional processes.

The architecture of individual neurons or nodes in biologically realistic models is based on the behaviour of neurons or groups of neurons in the brain. A typical example of a single neuron-like node is a leaky integrate and fire neuron that integrates its inputs and fires when a certain threshold is reached. This produces a spiking response, consisting of short bursts of activation. Larger groups of neurons can be represented by a more complex input–output function or architecture, producing a graded response which can be seen as an average of a group of spiking neurons, with overall more gradual changes in the output pattern. Pioneers in the area of biologically realistic neural modelling are Hodgkin and Huxley (1952), who were the first to describe biological neuron data (of the squid giant axon) in mathematical terms in the early 1950s. Their mathematical neuron descriptions were used in many models, which were then again tested experimentally, creating a scientific dialogue between experimental and modelling research. More than half a century later, models of various neural processes that are in one way or another related to emotion, such as conditioning (Grossberg, 1971; Suri and Schultz, 1999; Sutton and Barto, 1981) and attention (Servan-Schreiber et al., 1998; Sperling et al., 2001), have been created, but we are only just beginning to hypothesise about the mechanisms that underlie truly emotional processes.

The following summary of the neural structures involved in affective processes is very short and therefore incomplete and coarse. However, it is necessary to give an introduction to the neural correlates presented in the neural models described later, but it is beyond the scope of this chapter to give a more lengthy overview, hence this brief summary. For a more extensive review, see Lane et al., (2000).

Amygdala – As described in the previous section, the amygdala has always been an important structure in neural emotion research and is proven to have a large role in emotional processing, by assigning some kind of value to the perceived objects of the environment.

Orbitofrontal cortex (OFC) – This area is known to be active in (emotional) face recognition (Adolphs et al., 2003). It also has a role in conditioning, its activation being related to the value of the expected punishment or reward (Rolls, 2004; Schoenbaum and Roesch, 2005).

Anterior cingulate cortex (ACC) – The ACC is involved in decision making and premotor functions. It is important for the production of emotional responses, i.e. arousal (Critchley, 2005), as well as for self-regulation of affect (Phan et al., 2005). It also has a strong role in pain perception (Sewards and Sewards, 2002).

Insula – This region is activated during recognition and production of disgust, but (as described above) has also been found to respond to sadness, fear, and reward and has recently also been implicated in the hedonic (feeling) aspect of pain (Sewards and Sewards, 2002).

Nucleus accumbens (Nacc) – Part of the ventral striatum, this area is involved in conditioning processes and the anticipation of rewards and punishments (Knutson et al., 2005).

Thalamus – This structure relays sensory information to the rest of the brain. Of particular interest in our endeavour, it provides a fast, subcortical pathway to the amygdala.

Ventral tegmental area (VTA) – A prediction error signal is generated in this area, which is positive when an unpredicted reward is received and negative when a predicted reward is not received. This signal is governed by the neurotransmitter dopamine.

Based mainly on LeDoux fear conditioning research (LeDoux, 1996), Armony et al. (1995, 1997) developed a range of models specific to fear processing

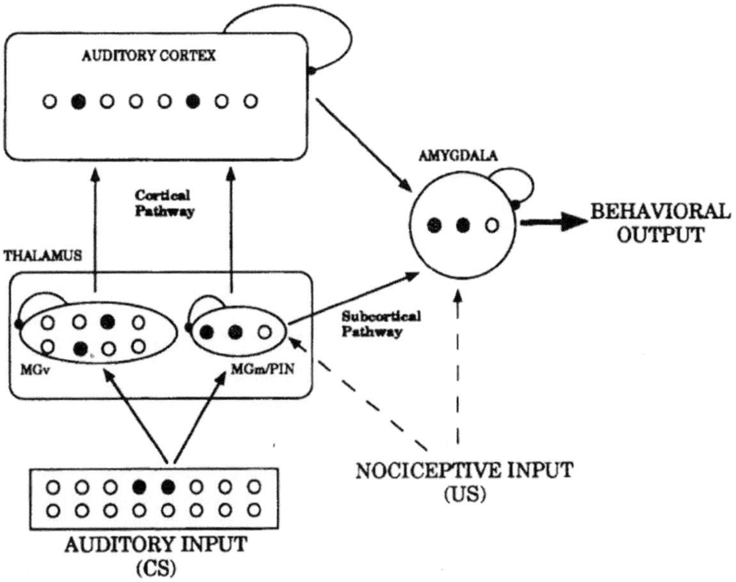

Fig. 2 Diagram of the network modelled by Armony et al. (1995, 1997). Brain structures are composed of individual artificial neurons. Patterns of activation are represented with *gray shading* (*black bullets* = maximum activation; *white bullets* = zero activation). Structures are self-inhibitory. Feedforward connections are excitatory, and learning occurs following Hebbian rules

and fear learning, employing cortical and thalamic modulation of the amygdala, which produces behavioural output, reproducing many experimental results in great detail (Fig. 2).

This has provided support for the "basic emotions" view of the amygdala as a fear module. However, as described earlier in this chapter, there is evidence that the amygdala also has a role in other emotional processes, which is not at all incompatible with its large role in fear processing. It only means that the role of the amygdala is not restricted to fear.

Another recent example of a simulation of an emotional process is that of Wagar and Thagard (2004), who propose the nucleus accumbens (NAcc) to be a central structure in the integration of cognitive and emotional neural signals for decision making. Modules representing the ventromedial prefrontal cortex (VMPFC), amygdala, hippocampus, and ventral tegmental area (VTA) are also included in this model. VMPFC activation represents a prediction of the outcome of a given response. The amygdala is activated only if this response is emotionally laden, and hippocampal activation simply represents the current context (see Fig. 3 for a schematic outline of connections between these structures in the model). In this model, the NAcc is continuously deactivated by the VTA. This means that more than one input is needed to elicit an output from this structure. So when both the hippocampus and VMPFC are active, which means the VMPFC prediction applies to the current context, a response is elicited. A response can also be elicited when

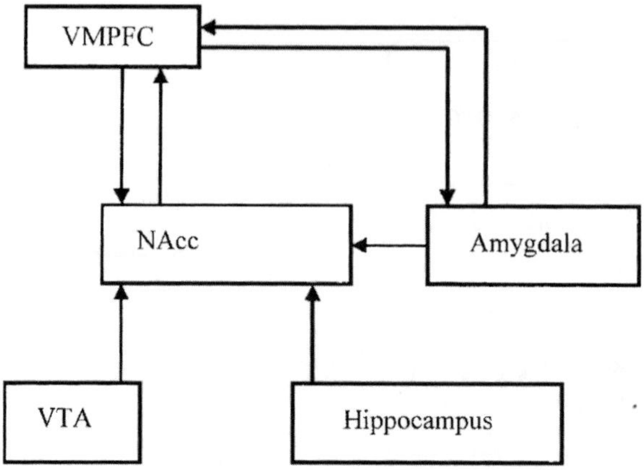

Fig. 3 Outline of GAGE model of cognitive-emotional integration (Wagar and Thagard, 2004). VMPFC = ventromedial prefrontal cortex, NAcc = nucleus accumbens, VTA = ventral tegmental area. The VMPFC receives sensory input. The amygdala processes contextual somatic states, and the joint action of the VTA and the hippocampus acts as a gating mechanism over the prefrontal cortex throughput in NAcc neurons

VMPFC and amygdala are simultaneously active, but because these activations tend to be short as they respond to nonpermanent stimuli (unlike the context), there is a very short time frame in which this coactivation can produce a response.

Implementation of this model with spiking neurons performing the IOWA gambling task produces NAcc output data that are similar to behavioural data in normal subjects and to data in VMPFC-lesioned patients when the VMPFC module is damaged. This provides a potential mechanism for partly emotion-based decision making, giving a computational account of somatic marker theory (Damasio, 1994) through reciprocal connections between VMPFC and amygdala. In this theory, emotion influences decision making through somatic markers (a kind of emotional memory) associated with the stimulus, stored in the VMPFC. This account is not compatible with basic emotion theories, since amygdala activation represents more than one possible emotional marker. In contrast, neither discrete emotion theories nor an appraisal-based approach is directly contradicted by this model.

Taylor and Fragopanagos (2005) have applied the COrollary Discharge of Attention Movement (CODAM) model (Taylor and Rogers, 2002; Korsten et al., 2006) to emotional phenomena. Originally developed to model consciousness and attention from an engineering control approach, CODAM has proven to be applicable in the modelling of processes with an emotion–attentional overlap, such as the emotional–attentional blink, whereby emotion has an influence on ongoing attentional processing. Figure 4 shows an outline of the CODAM model of the emotional–attentional blink, in which an amygdala module is added to the original CODAM attentional blink model (Fragopanagos et al., 2005) to simulate emotional influences on attentional processing.

In this model, an attentional control signal is produced by the Inverse Model Controller (IMC) module (analogous to parietal cortical areas), which influences

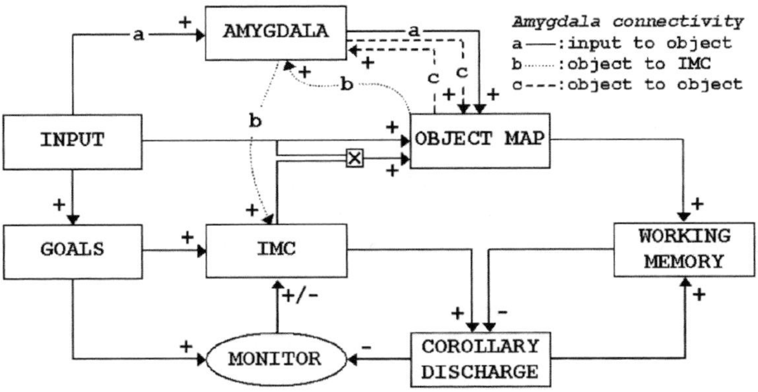

Fig. 4 CODAM model of the emotional–attentional blink

working memory (located in parietal cortex as well) through modulation of the object map (primary and associative cortices). The corollary discharge and monitor module (cingulate cortex) provide error correction to IMC and working memory, partly through comparisons to goals (prefrontal cortex). See the original article for a more detailed description (Fragopanagos et al., 2005). The paradigm simulated with this model is that of Anderson and Phelps (2001), in which the second target in the attentional blink is charged with an emotional load. If this emotional load is sufficient, it cancels the attentional blink that normally renders the second target imperceptible in normal subjects, but not in patients suffering from a lesion of the amygdala. The addition of amygdala modulation in the attentional blink model, and subsequent simulation of amygdala lesions by disabling the amygdala module, produced simulated ERP data that minutely resembled experimental ERP data from both normal subjects and amygdala patients, as well as fMRI data from normal subjects (including amygdala activations). This provides direct support for the theory that the amygdala is the structure creating emotional interference in this task and that it does so through amplification, as in the model.

Based on similar principles, the same authors have also simulated other paradigms where amygdala activation has been connected to emotional influences on attention (Taylor and Fragopanagos, 2005), such as the Pessoa et al. (2002) face/bars divided attention paradigm. The fact that this framework is suitable to model a whole range of emotion–attention paradigms provides support not only for the theory that the amygdala is the core structure in the emotional modulation of attentional processes, but also for the proposed amplification mechanism. Thus, we do gain knowledge not only on the 'where' of this emotional process (amygdala) but also on the 'how'.

The above and numerous other studies have modelled the impact of emotion on attention. However, an important question, of value in getting a better understanding of emotion, is as to if and by how much the reverse occurs. In other words, of the impact of attention on emotion. A series of recent studies (Raymond et al., 2003) have shown that selective attention can influence the emotional value of both selected and ignored items when a complex scene is being attended to. Specifically, ignored items (distracters) were consistently rated less positively in emotional evaluations, following attention selection, relative to (typically) simultaneously presented items (targets). The resulting effect was termed 'Distracter Devaluation' (DD), being measured by how much reduction of the value of a distracter occurs when it has just been inhibited in the above search paradigm. Furthermore, a known electrophysiological index of attention selectivity (the so-called N2pc, measuring the degree to which attention has been selectively focused in one hemisphere or another and at about 200–300 ms post-stimulus) was shown to correlate with the magnitude of the observed distracter devaluation.

A neural model was developed by Fragopanagos et al. (2009) to account for these findings by means of a plausible mechanism linking attention processes to emotional evaluations. This mechanism relies on the transformation of attention inhibition of the distracter into a reduction of the value of that distracter (expected to be coded in the orbitofrontal cortex). The model is successful in reproducing the existing behavioural results as well as the observed link between the magnitude of the N2pc (arising in the attention control system) and the magnitude of DD. Moreover, the model proposes a series of testable hypotheses that call for further experimental investigation. In particular it provides a hypothesis as to how inhibitory features arising from prefrontal cortex can be transferred not only as part of the feedback attention control system but simultaneously as part of the rapid manipulation of a value reward system of the emotional limbic system in the brain. As such it begins to cover a different range of time intervals than that associated with the slower conditioned learning of dopamine-assisted predictions of reward much studied in conditioned learning in animals. These models are, again, not compatible with a basic emotion approach, but could fit in a dimensional framework or an appraisal-based approach with the amygdala as relevance detector.

More recent work has focused on the emergence of emotions per se: Korsten et al. (2007) have simulated different representations of the value of self-esteem and connected this valuation to the emergence of anger and, when disrupted, depression. In this model, the module responsible for the comparison between these different self-esteem values is analogous to the cingulate cortex, whereas the values presented to the cingulate cortex module for comparison are thought to be analogous to the prefrontal and orbitofrontal cortices. Expanding and generalising on this initial model, current work focuses on the development of a neural model in which we suggest that emotions arise from comparisons/conflict between three instances of a particular perceived value (e.g. food reward): the 'actual' value (the amount of food reward possessed at the moment), corresponding to activation of the amygdala, the 'expected' value (the amount of food reward expected to be possessed), activating the orbitofrontal cortex, and the 'normal' value (the amount of food reward that we would normally, on average, expect to be possessing), analogous to activation of prefrontal cortex. When a discrepancy between two of these instances is large enough (we are receiving a larger reward than we were expecting), an emotional response emerges (joy, surprise). Through comparisons of these different value instances, it is possible to differentiate between the basic emotions of joy, anger, hope, fear, sadness, disappointment, and relief. Ongoing work focuses on the implementation of this framework.

These three approaches present various detailed theories of the mechanisms underlying emotional processes and processes influenced by emotion. Not every detail of these models has as yet been verified by experimental research, which is why they could provide good guidance as to which aspects of emotional processing could be suitable for neuroscientific research.

3 Conclusion

By providing both a functional account of the processes involved in emotional processing and a description of the underlying cerebral substrate, psychologists and cognitive neuroscientists explain normal and pathological functioning. Several traditions compete, however, and new approaches need to be developed to disentangle their respective predictions. We showed that biologically plausible artificial neural networks can be used in this enterprise by formalising both the structure and the dynamics of the networks involved in the processing of emotional information. Results of such simulations go in line with cognitive theories of emotion, distributing the processing of information over several nodes of complex networks. New predictions can be extracted from these artificial models, and future work will need to involve both experimenters and modellers.

References

Adolphs R (1999) Social cognition and the human brain. Trends Cogn Sci 3(12):469–479

Adolphs R (2003) Cognitive neuroscience of human social behavior. Nat Rev Neurosci, 4:165–178

Adolphs R, Tranel D, Damasio H, Damasio AR (1994) Impaired recognition of emotion in facial expressions following bilateral damage to the human amygdala. Nature 372(6507):669–672

Adolphs R, Baron-Cohen S, Tranel D (2002) Impaired recognition of social emotions following amygdala damage. J Cogn Neurosci 14(8):1264–1274

Adolphs R, Tranel D, Damasio AR (2003) Dissociable neural systems for recognizing emotions. Brain Cogn 52(1):61–69

Anderson AK, Phelps EA (2001) Lesions of the human amygdala impair enhanced perception of emotionally salient events. Nature 411(6835):305–309

Anderson AK, Sobel N (2003) Dissociating intensity from valence as sensory inputs to emotion. Neuron 39(4):581–583

Anderson AK, Christoff K, Panitz D, De Rosa E, Gabrieli JDE (2003) Neural correlates of the automatic processing of threat facial signals. J Neurosci 23(13):5627–5633

Armony JL, Servan-Schreiber D, Cohen JD, LeDoux JE (1995) An anatomically constrained neural network model of fear conditioning. Behav Neurosci 109(2):246–257

Armony JL, Servan-Schreiber D, Cohen JD, LeDoux JE (1997) Computational modeling of emotion: explorations through the anatomy and physiology of fear conditioning. Trends Cogn Sci 1(1):28–34

Blair RJ, Morris JS, Perrett D, Perrett DI, Dolan RJ (1999) Dissociable neural responses to facial expressions of sadness and anger. Brain 122(5):883–893

Büchel C, Dolan R, Armony JL, Friston KJ (1999) Amygdala-hippocampal involvement in human aversive trace conditioning revealed through event-related functional magnetic resonance imaging. J Neurosci 19(24):10869–10876
Calder AJ, Keane J, Manes F, Antoun N, Young AW (2000) Impaired recognition and experience of disgust following brain injury. Nat Neurosci 3(11):1077–1078
Calder AJ, Lawrence AD, Young AW (2001) Neuropsychology of fear and loathing. Nat Rev Neurosci 2(5):352–363
Calder AJ, Keane J, Lawrence AD, Manes F (2004) Impaired recognition of anger following damage to the ventral striatum. Brain 127(9):1958–1969
Critchley HD (2005) Neural mechanisms of autonomic, affective, and cognitive integration. J Comp Neurol 493(1):154–166
Damasio AR (1994) Descartes's error: emotion, reason, and the human brain. Putnam, New York, NY
Damasio AR (1998) Emotion in the perspective of an integrated nervous system. Brain Res Rev 26(2/3):83–86
Damasio AR (1999) The feeling of what happens: body and emotion in the making of consciousness, 1st edn, Harcourt Brace, New York, NY.
Davidson RJ (1995) Cerebral asymmetry, emotion, and affective style. In: Davidson RJ, Hugdahl K (eds), Brain asymmetry. MIT Press, Cambridge, MA, pp 361–387
Davidson RJ, Irwin W (1999) The functional neuroanatomy of emotion and affective style. Trends Cogn Sci 3(1):11–21
Dror IE, Gallogly DP (1999) Computational analyses in cognitive neuroscience: in defense of biological implausibility. Psychon Bull Rev 6(2):173–182
Ekman P (1999) Basic emotions. In: Dalgleish T, Power MJ. (eds) Handbook of cognition and emotion. Wiley, Chichester
Ekman P (2003) Emotions revealed: recognizing faces and feelings to improve communication and emotional life 1st edn. Weidenfeld & Nicolson, London
Ellsworth PC (1991) Some implications of cognitive appraisal theories of emotion. In: Strongman KT. (ed) International review of studies on emotion. Wiley, River street Hoboken, NJ, pp 143–161
Ellsworth PC, Scherer KR (2003) Appraisal processes in emotion. In: Davdison RJ, Scherer KR, Hill Goldsmith H. (eds), Handbook of affective sciences. Oxford University Press, Oxford, pp 572–595
Fontaine JRJ, Scherer KR, Roesch EB, Ellsworth PC (2007) The world of emotions is not two-dimensional. Psychol Sci 18(12):1050–1057
Fragopanagos N, Kockelkoren S, Taylor JG (2005) A neurodynamic model of the attentional blink. Cogn Brain Res 24(3):568–586
Fragopanagos N, Cristescu T, Goolsby BA, Kiss M, Eimer M, Nobre AC, Raymond JE, Shapiro KL, Taylor JG (2009) Modelling distractor devaluation (DD) and its neurophysiological correlates. Neuropsychologia 47:2354–2366
Frijda NH (1986) The emotions, volume Studies in emotion and social interaction. Cambridge University Press Editions de la Maison des sciences de l'homme, Cambridge
Gardner H (1985) The mind's new science: a history of the cognitive revolution. Basic Books, New York, NY
George MS, Ketter TA, Parekh PI, Herscovitch P, Post RM (1996) Gender differences in regional cerebral blood flow during transient self-induced sadness or happiness. Biol Psychiatry 40:859–871
Gray JA (1994) Three fundamental emotion systems. In: Davdison RJ. (ed) The nature of emotion. Oxford University Press, Oxford
Grossberg S (1971) On the dynamics of operant conditioning. J Theor Biol 33(2):225–255
Hamann S (2003) Nosing in on the emotional brain. Nat Neurosci 6(2):106–108
Hodgkin AL, Huxley AF (1952) A quantitative description of membrane current and its application to conduction and excitation in nerve. J Physiol 117(4):500–544

Izard CE (2007) Basic emotions, natural kinds, emotion schemas, and a new paradigm. Perspect Psychol Sci 2:260–280

Kleinginna PR, Kleinginna AM (1981) A categorized list of emotion definitions, with suggestions for a consensual definition. Motiv Emot 5(4):345–359

Knutson B, Taylor J, Kaufman M, Peterson, R Glover G (2005) Distributed neural representation of expected value. J Neurosci 25(19):4806–4812

Korsten N, Fragopanagos N, Hartley M, Taylor N, Taylor JG (2006) Attention as a controller. Neural Netw 19(9):1408–1421

Korsten N, Fragopanagos N, Taylor JG (2007) Neural substructures for appraisal in emotion: Self-esteem and depression. In: Marques de Sá Joaquim (ed) Proceedings of Internatinal Conference on Artificial Neural Networks – LNCS (ICANN), vol 2. Springer, New York, NY, pp 850–858

Kosslyn SM, Koenig O (1992) Wet mind: the new cognitive neuroscience, 1st edn. Free Press, New York, NY

Lane RD, Nadel L, Ahern G (2000) Cognitive neuroscience of emotion. Oxford University Press, New York, NY

Lazarus RS (1999) Stress and emotion: a new synthesis. Springer, New York, NY

LeDoux JE (1996) The emotional brain: the mysterious underpinnings of emotional life. Simon & Schuster, New York, NY

Marr D (1982) Vision : a computational investigation into the human representation and processing of visual information. Walter H Freeman, San Francisco, CA

Mineka S, Öhman A (2002) Phobias and preparedness: the selective, automatic, and encapsulated nature of fear. Biol Psychiatry 52(10):927–937

Morris JS, Frith CD, Perrett DI, Rowland D, Young AW, Calder AJ, Dolan. RJ (1996) A differential neural response in the human amygdala to fearful and happy facial expressions. Nature 383(6603):812–815

Morris JS, Friston KJ, Buchel C, Frith CD, Young AW, Calder AJ, Dolan RJ (1998) A neuromodulatory role for the human amygdala in processing emotional facial expressions. Brain 121(1):47–57

Öhman A, Mineka S (2001) Fears, phobias, and preparedness: toward an evolved module of fear and fear learning. Psychol Rev 108(3):483–522

Panksepp J (1998) Affective neuroscience: the foundations of human and animal emotions. Oxford University Press, New York, NY

Pessoa L (2008) On the relationship between emotion and cognition. Nat Rev Neurosci, 9(2):148–158

Pessoa L, McKenna M, Gutierrez E, Ungerleider LG (2002) Neural processing of emotional faces requires attention. Proc Natl Acad Sci USA 99(17):11458–11463

Phan KL, Fitzgerald DA, Nathan PJ, Moore GJ, Uhde TW, Tancer ME (2005) Neural substrates for voluntary suppression of negative affect: a functional magnetic resonance imaging study. Biol Psychiatry 57(3):210–219

Phelps EA, O'Connor KJ, Gatenby JC, Gore JC, Grillon C, Davis M (2001) Activation of the left amygdala to a cognitive representation of fear. Nat Neurosci 4(4):437–441

Phillips ML, Young AW, Senior C, Brammer M, Andrew C, Calder AJ, Bullmore ET, Perrett DI, Rowland D, Williams SC, Gray JA, David AS (1997) A specific neural substrate for perceiving facial expressions of disgust. Nature 389(6650):495–498

Raymond JE, Fenske MJ, Tavassoli NT (2003) Selective attention determines emotional responses to novel visual stimuli. Psychol Sci 14(6):537–542

Roesch EB, Sander D, Scherer KR (2007) The link between temporal attention and emotion: a playground for psychology, neuroscience, and plausible artificial neural networks. In: Marques de Sá Joaquim, (ed) Proceedings of international conference on artificial neural networks – LNCS (ICANN), vol 2. Springer, New York, NY, pp 859–868

Rolls ET (1999) The brain and emotion. Oxford University Press, Oxford, England

Rolls ET (2004) The functions of the orbitofrontal cortex. Brain Cogn 55(1):11–29

Roseman IJ, Smith CA (2001) Appraisal theory: overview, assumptions, varieties, controversies. In: Scherer KR, Schorr A, Johnstone T. (eds) Appraisal processes in emotion : theory, methods, research Oxford University Press, New York, NY, pp 3–19
Sander D, Koenig O (2002) No inferiority complex in the study of emotion complexity: a cognitive neuroscience computational architecture of emotion. Cogn Sci Q 2:249–272
Sander D, Grafman J, Zalla T (2003) The human amygdala: An evolved system for relevance detection. Rev Neurosci 14:303–316
Sander D, Grandjean D, Scherer KR (2005) A systems approach to appraisal mechanisms in emotion. Neural Netw 18(4):317–352
Scherer KR (1984) On the nature and function of emotion: a component process approach. In: Scherer KR, Ekman P. (eds) Approaches to emotion Erlbaum, Hillsdale, NJ, pp 293–317
Scherer KR (2001) Appraisal considered as a process of multilevel sequential checking. In: Scherer KR, Schorr A, Johnstone T. (eds) Appraisal processes in emotion: Theory, methods, research Oxford University Press, New York, NY, pp 192–120
Schoenbaum G, Roesch M (2005) Orbitofrontal cortex, associative learning, and expectancies. Neuron 47(5):633–636
Servan-Schreiber D, Bruno RM, Carter CS, Cohen JD (1998) Dopamine and the mechanisms of cognition: part I. a neural network model predicting dopamine effects on selective attention. Biol Psychiatry 43(10):713–722
Sewards TV, Sewards. MA (2002) The medial pain system: neural representations of the motivational aspect of pain. Brain Res Bull 59(3):163–180
Small DM, Gregory MD, Mak YE, Gitelman D, Mesulam MM, Parrish T (2003) Dissociation of neural representation of intensity and affective valuation in human gustation. Neuron 39(4):701–711
Smith CA, Lazarus RS (1993) Appraisal components, core relational themes, and the emotions. Cogn Emot 7:233–269
Sperling G, Reeves A, Blaser E, Lu Z-L, Weichselgartner E (2001) Two computational models of attention. In: Braun J, Koch C, Davis JL. (eds) Visual attention and cortical circuits. MIT Press, Cambridge, MA, pp 177–214
Strongman KT (1996) The psychology of emotion: theories of emotion in perspective 4th edn. Wiley, Chichester
Suri RE, Schultz W (1999) A neural network model with dopamine-like reinforcement signal that learns a spatial delayed response task. Neurosci 91(3):871–890
Sutton RS, Barto AG (1981) Toward a modern theory of adaptive networks: expectation and prediction. Psychol Rev 88(2):135–170
Taylor JG, Fragopanagos N (2005) The interaction of attention and emotion. Neural Netw 18(4):353–369
Taylor JG, Rogers M (2002) A control model of the movement of attention. Neural Netw 15(3):309–326
Tooby J, Cosmides L (2000) Toward mapping the evolved functional organization of mind and brain. In: Gazzaniga MS. (ed) The new cognitive neurosciences . MIT Press, Cambridge, MA, pp 1167–1178
Wagar BM, Thagard P (2004) Spiking phineas gage: a neurocomputational theory of cognitive-affective integration in decision making. Psychol Rev 111(1):67–79
Whalen PJ, Shin LM, McInerney SC, Fischer H, Wright CI, Rauch SL (2001) A functional mri study of human amygdala responses to facial expressions of fear versus anger. Emotion 1(1):70–83
Winston JS, Gottfried JA, Kilner JM, Dolan RJ (2005) Integrated neural representations of odor intensity and affective valence in human amygdala. J Neurosci 25(39):8903–8907
Wundt W (1905) Grundzüge der physiologischen Psychologie [Fundamentals of physiological psychology], 5th edn. Engelmann, Leipzig

Part II
Signals to Signs

Editorial: "Signals to Signs" – Feature Extraction, Recognition, and Multimodal Fusion

Kostas Karpouzis

Abstract Processing of recorded or real-time signals, feature extraction, and recognition are concepts of utmost important to an affect–aware and capable system, since they offer the opportunity to machines to benefit from modeling human behavior based on theory and interpret it based on observation. This chapter discusses feature extraction and recognition based on unimodal features in the case of speech, facial expressions and gestures, and physiological signals and elaborates on attention, fusion, dynamics, and adaptation in different multimodal cases.

Signal processing, feature extraction, and recognition are integral parts of an affect–aware system. The central role of these processes is illustrated in the "map of the thematic areas involved in emotion-oriented computing" included in the "Start here!" section of the HUMAINE portal (Fig. 1)

Here, emotion detection as a whole is strongly connected to "raw empirical data," represented by the "Databases" chapters of this handbook, "usability and evaluation" and "synthesis of basic signs," and also has strong links to "theory of emotional processes." With respect to databases, this Handbook Area discusses the algorithms used to extract features from individual modalities from natural, naturalistic, and acted audiovisual data, the approaches used to provide automatic annotation of unimodal and multimodal data, taking into account different emotion representations as described in the Theory chapters, and the fallback approaches which can be used when the unconstrained nature of these data hampers extraction of detailed features (this also involves usability concepts). In addition to this, studies correlating manual annotation to automatic classification of expressivity have been performed in order to investigate the extent to which the latter can introduce a pre-processing step to the annotation of large audiovisual databases.

Regarding synthesis and embodied conversational agents (ECAs), this chapter discusses how low-level features (e.g., raising eyebrows or hand movements) can

K. Karpouzis (✉)
Image, Video and Multimedia Systems Lab, Institute of Communications and Computer Systems, National Technical University of Athens, Athens, Greece
e-mail: kkarpou@image.ece.ntua.gr

P. Petta et al. (eds.), *Emotion-Oriented Systems*, Cognitive Technologies,
DOI 10.1007/978-3-642-15184-2_5,

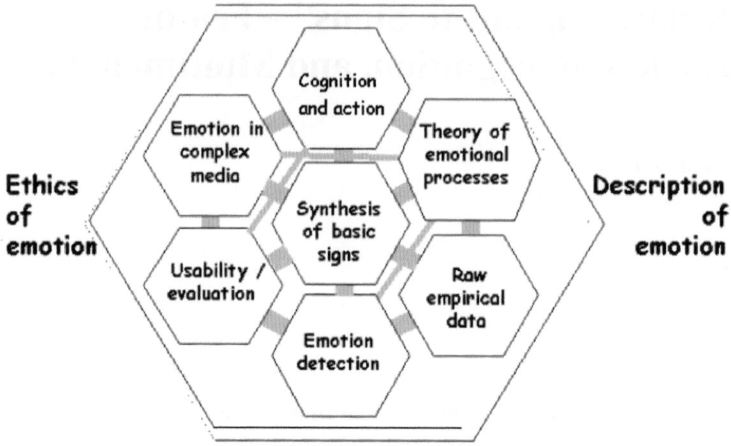

Fig. 1 Map of the thematic areas involved in emotion-oriented computing (from http://emotion-research.net/aboutHUMAINE/start-here, visited 2010-05-31)

be connected to higher level concepts (facial expressions or semantic gestures) using emotion representation theories, how ideas from image processing and scene analysis can be utilized in virtual environments, supplying ECAs with attention capabilities, and how real-time feature extraction from facial expressions and hand gestures can be used to render feedback-capable ECAs.

1 Multimodality and Interaction

The misconceptions related to multimodality in emotion recognition and interaction were discussed in great detail by Oviatt (1999). However, a number of comparative studies (Castellano et al., 2008; Gunes and Piccardi, 2009) illustrate that taking into account multiple channels, either in terms of features or in terms of unimodal decisions or labels, does benefit recognition rates and robustness. Systems can integrate signals at the feature level (Rogozan, 1999) or, after coming up with a class decision at the feature level of each modality, by merging decisions at a semantic level (late identification, Rogozan, 1999; Teissier et al., 1999), possibly taking into account any confidence measures provided by each modality or, generally, a mixture of experts mechanism. Cognitive modeling and experiments indicate that this kind of fusion may happen at feature level (Onat et al., 2007) but discussion regarding the semantics and robustness of each approach is still open in this area.

The inherent multimodality of human interaction can also be exploited in terms of complementing information across channels as well; consider, for instance, that human speech is bimodal in nature (Chen and Rao, 1998). Speech that is perceived by a person depends not only on acoustic cues but also on visual cues such as lip movements or facial expressions. This combination of auditory and visual speech recognition is more accurate than auditory only or visual only, since use of multiple sources generally enhances speech perception and understanding. Consequently, there has been a large amount of research on incorporating bimodality of speech into human–computer interaction interfaces. Lip sync is one of the research topics in this area (Zoric and Pandzic, 2006).

For interactive applications it is necessary to perform lip sync in real time, which is a particular challenge not only because of computational load but also because the low delay requirement reduces the audio frame available for analysis to a minimum. Speech sound is produced by vibration of the vocal cords and then it is additionally modeled by vocal tract (Lewis and Parke, 1986). A phoneme, defined as basic unit of acoustic speech, is determined by vocal tract, while intonation characteristics (pitch, amplitude, voiced/whispered quality) are dependent on the sound source. Lip synchronization is the determination of the motion of the mouth and tongue during speech (McAllister et al., 1997). To make lip synchronization possible, position of the mouth and tongue must be related to characteristics of the speech signal. Positions of the mouth and tongue are functions of the phoneme and are independent of intonation characteristics of speech.

There are many sounds that are visually ambiguous when pronounced. Therefore, there is a many-to-one mapping between phonemes and visemes, where viseme is a visual representation of phoneme (Pandžić and Forchheimer, 2002).

The process of automatic lip sync, as shown in Fig. 2, consists of two main parts (Zoric, 2003). The first one, audio to visual mapping, is a key issue in bimodal speech processing. In this first phase, speech is analyzed and classified into viseme categories. In the second part, calculated visemes are used for animation of virtual character's face. Audio to visual (AV) mapping can be solved on several different levels, depending on the speech analysis that is being used. In Zoric (2003), speech is classified into viseme classes by neural networks and GA is used for obtaining the optimal neural network topology. By introducing segmentation of the speech directly into viseme classes instead of phoneme classes, computation overhead is reduced, since only visemes are used for facial animation. Automatic design of neural networks with genetic algorithms saves much time in the training process.

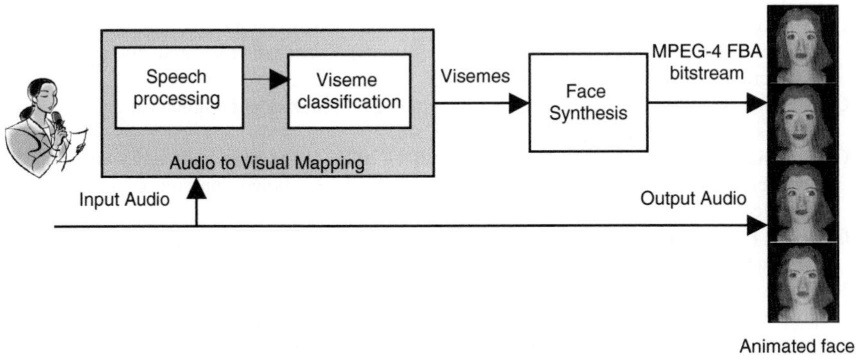

Fig. 2 Schematic view of lip sync system

2 From Emotions to Social Signals and Contexts of Interaction

Features extracted from human faces and bodies, utterances, and physiological signals may also be exploited to provide cues not necessarily related to emotion, but also in an emerging field termed "social signal processing." Here, low-level features, such as prominent points around the eyes, may be used to estimate the user's eye gaze direction, instead of going directly to a high-level concept, such as emotions and from that, the user's degree of engagement to an interacting party or machine (Vertegaal et al., 2001). This effectively demonstrates the strong relation of choosing class names for a machine learning algorithm to the underlying theoretical concepts; detected features from a human face may correspond to wide open eyes, but what *that* fact means is still a question to be answered.

Another open research question related to understanding features and adapting classifiers is the choice of context. Instead of providing "all-weather" feature extraction and recognition systems, the current (and more interesting, user-wise) trend is to exploit what knowledge is available, related to who the user is, where the interaction takes place, and in what application context, in order to fine-tune a general purpose algorithm and choose the relevant, prominent features to track (Karpouzis and Kollias, 2007). In this framework, signal processing techniques may be used in association with knowledge technology concepts to provide features for cognitive structures, effectively bridging diverse disciplines into one, user-centered loop. In addition to this, information from available modalities can also be used to fortify the result in one particular unimodal recognition; in Morency et al. (2007), authors investigate the presence of lexical hints related to posing a question (e.g., subject–verb reversal) in relation to detecting particular body gestures (a head nod), while in Christoudias et al. (2006), features from strong, successful classifications from the audio channel are used to train a visual recognition algorithm in an unsupervised manner.

References

Castellano G, Kessous L, Caridakis G (2008) Emotion recognition through multiple modalities: face, body gesture, speech. In: Peter C, Beale R (eds) Affect and emotion in human–computer interaction. Springer, Heidelberg, pp 92–103

Chen T, Rao R (1998 May) Audio-visual integration in multimodal communication. Proc IEEE Spec Issue Multimedia Signal Process 86:837–852

Christoudias M, Saenko K, Morency L-P, Darrell T (2006) Co-adaptation of audio-visual speech and gesture classifiers. In: Proceedings of the 8th international conference on multimodal interfaces, Banff, AB, Canada

Gunes H, Piccardi M (2009) Automatic temporal segment detection and affect recognition from face and body display. IEEE Trans Syst Man Cybern B 39(1):64–84

Karpouzis K, Kollias S (2007) Multimodality, universals, natural interaction, Humaine plenary presentation. http://tinyurl.com/humaine-context. Accessed 6 Jul 2009

Lewis JP, Parke FI (1986) Automated lip-synch and speech synthesis for character animation. SIGCHI Bull 17(May):143–147. doi: 10.1145/30851.30874 http://doi.acm.org/10.1145/30851.30874

McAllister DF, Rodman RD, Bitzer DL, Freeman AS (1997) Lip synchronization for animation. In: Proceedings of SIGGRAPH 97, Los Angeles, CA

Morency L, Sidner C, Lee C, Darrell T (2007) Head gestures for perceptual interfaces: the role of context in improving recognition. Artif Intell 171(8–9):568–585

Onat S, Libertus K, Koenig P (2007) Integrating audiovisual information for the control of overt attention. J Vis 7(10):1–16

Oviatt S (1999) Ten myths of multimodal interaction. Commun ACM 42(11):74–81

Pandzic IS, Forchheimer R (eds) (2002) MPEG-4 facial animation – the standard, implementation and applications. Wiley, New York, NY. ISBN 0-470-84465-5

Rogozan A (1999) Discriminative learning of visual data for audiovisual speech recognition. Int J Artif Intell Tools 8:43–52

Teissier P, Robert-Ribes J, Schwartz JL (1999) Comparing models for audiovisual fusion in a noisy-vowel recognition task. IEEE Trans Speech Audio Process 7:629–642

Vertegaal R, Slagter R, van der Veer G, Nijholt A (2001) Eye gaze patterns in conversations: there is more to conversational agents than meets the eyes. In: Proceedings of the SIGCHI conference on human factors in computing systems, Seattle, Washington, DC, pp 301–308

Vinciarelli A, Pantic M, Bourlard H (2009 November) Social signal processing: survey of an emerging domain. Image Vis Comput 27(12):1743–1759

Zoric G (2003) Real-time animation driven by human voice. In: Proceedings of ConTEL, Zagreb

Zoric G, Pandzic I (2006) Real-time language independent lip synchronization method using a genetic algorithm. Signal Process 86(12):3644–3656

The Automatic Recognition of Emotions in Speech

Anton Batliner, Björn Schuller, Dino Seppi, Stefan Steidl, Laurence Devillers, Laurence Vidrascu, Thurid Vogt, Vered Aharonson, and Noam Amir

Abstract In this chapter, we focus on the automatic recognition of emotional states using acoustic and linguistic parameters as features and classifiers as tools to predict the 'correct' emotional states. We first sketch history and state of the art in this field; then we describe the process of 'corpus engineering', i.e. the design and the recording of databases, the annotation of emotional states, and further processing such as manual or automatic segmentation. Next, we present an overview of acoustic and linguistic features that are extracted automatically or manually. In the section on classifiers, we deal with topics such as the curse of dimensionality and the sparse data problem, classifiers, and evaluation. At the end of each section, we point out important aspects that should be taken into account for the planning or the assessment of studies. The subject area of this chapter is not emotions in some narrow sense but in a wider sense encompassing emotion-related states such as moods, attitudes, or interpersonal stances as well. We do not aim at an in-depth treatise of some specific aspects or algorithms but at an overview of approaches and strategies that have been used or should be used.

1 Introduction

The study of speech and emotion can be traced back to the first decades of the last century, cf. Scripture (1921), Skinner (1935), and Fairbanks and Pronovost (1939). Whereas such studies were not very frequent during the following decades – one of the exceptions being Williams and Stevens (1972) – the topic began to attract researchers more and more during the eighties. Until the nineties most of these studies could be subsumed under the heading 'basic research in psychology and phonetics/linguistics'; an overview is given, for example, in Scherer (2003). In the nineties, the automatic processing of speech started to address topics beyond pure word recognition. First, higher linguistic levels, for instance, dialogue acts, and

A. Batliner (✉)
Lehrstuhl für Mustererkennung, Friedrich-Alexander-Universität Erlangen, Erlangen, Germany
e-mail: batliner@informatik.uni-erlangen.de

P. Petta et al. (eds.), *Emotion-Oriented Systems*, Cognitive Technologies,
DOI 10.1007/978-3-642-15184-2_6, © Springer-Verlag Berlin Heidelberg 2011

then topics beyond pure information transmission, that is, paralinguistic phenomena, e.g. emotions and attitudes conveyed via the speech channel, were addressed in studies such as Dellaert et al. (1996). At that time, however, almost all data used were 'prompted' and acted, cf. below, modelling the prototypical 'big' *n* emotions, *n* being a figure greater or equal 2 and up to 4, 6, or even more classes. Maybe the first paper dealing with 'natural(istic)' speech and emotions was Slaney and McRoberts (1998). At the turn of the century, researchers began to use non-acted databases from, generally speaking, interactions of humans with information offices/systems, i.e. human–human or human–machine interaction – the role of the machine sometimes played by a human Wizard-of-Oz (WoZ) – such as appointment scheduling or call centre dialogues, cf. Batliner et al. (2000a), Lee et al. (2001), and Ang et al. (2002).

Nowadays, it is widely acknowledged that acted data cannot model naturalistic data sufficiently, as demonstrated by Batliner et al. (2000a) and Wilting et al. (2006), especially because the emotions produced that way are too pronounced and will rather seldom be encountered as such in more realistic data. Thus a (direct) transfer from acted data onto data encountered in realistic applications is not feasible. Acted data are still used to a large extent, e.g. in Ververidis and Kotropoulos (2006); possible applications can be found within the entertainment business, e.g. for data mining in movie archives or for computer games. The main reason for this preponderance, however, is simply that non-acted data are still sparse and most often not available freely. In this chapter, we will concentrate on the genuine approach of automatically recognising/classifying emotional user states signalled in naturalistic (spontaneous) speech. We will deal with acted speech only in order to illustrate specific approaches or methodologies. Nonetheless, the basic requirements of automatic processing are the same for both acted and naturalistic data: large enough size of the database, balanced distribution of classes, large number of speakers, recording quality, class assignment as unequivocal as possible, etc. However, using realistic data requires us to face some more challenges: sparse and very unbalanced data, less pronounced emotions, and definitely the need to explicitly annotate the data, assigning emotion classes. Moreover, the data should be representative for the envisioned application; actually, this is the most important requirement: if we are interested in emotional film scenes, film actors as speakers are adequate – but not necessarily speakers prompted for emotions in the laboratory.

In the field of emotion in speech, two lines of research came together with their own standards and methods which have not converged yet: basic (psychological, clinic, phonetic) research, dealing mostly with acted data, and applied engineering – so far, too often dealing with acted data as well. Naïve conceptualisations of the respective other line of research should be replaced by a mutual understanding of innate constraints and benefits. However, it is beneficial to conceive the study and especially the automatic processing of non-acted, non-prompted emotional states as a topic sui generis.

2 Corpus Engineering

We conceive the term 'corpus engineering' as encompassing all the steps necessary before feature extraction and automatic classification can take place:

1. the design of an application-oriented scenario
2. the recruiting of the necessary personnel such as subjects, supervisors (Wizard-of-Oz), and the experimental setting or the real-life scenario
3. the recordings and – if necessary – subsequent transfer onto storage media with/without resampling of the audio signal
4. the transliteration, i.e. the orthographic transcription of the data, sometimes including the annotation of extra- or non-linguistic events such as breathing or noise
5. the definition and extraction of appropriate units of analysis such as words, chunks, turns, dialogue moves with appropriate criteria (intuitive or based on prosodic, linguistic, or pragmatic criteria)
6. the annotation of emotional states, possibly with subsequent mapping onto fewer main classes
7. evaluating the quality of these annotations by applying some measures of correlation/correspondence
8. some other pre-processing steps like manual processing or correction of automatically processed feature values
9. defining and applying exchange formats

We will sketch (1), (2), and (4) skipping the technical aspects of (3), mention (8)–(9), and concentrate on (5)–(7).

2.1 Databases

A common breakdown of emotion databases is the one into acted/non-acted, induced, and naturalistic databases, cf. Douglas-Cowie et al. (2007). This is a gross taxonomy which does not yet capture pertinent differences: the settings, i.e. the scenarios, are defined and created by the researcher; the outcome is the data that we have to deal with. Here we want to tell apart acted/non-acted and prompted/non-prompted (Schiel, 1999) settings: if the subject acts, he/she is doing as if they were in this specific situation – no matter whether it is about being emotional or not. If emotions are prompted themselves, the subjects have been told that they should produce specific emotions. The subjects can be volunteering or recorded in real-life situations. Inducing emotions means to arrange situations where the subjects are more likely to produce the desired emotional states. Strictly speaking, all these different conditions do not tell us whether our subjects will produce 'natural', realistic emotion-related states or not. It is just more likely that the outcome, i.e. the emotional database, is less natural if acted; induced data, for instance, can be more or less spontaneous, or fully spontaneous. All these differences can be evaluated by applying a perceptive evaluation – either with naïve listeners in a perception experiment or with a more intuitive assessment.

This is a representative but not necessarily exhaustive list of scenarios where non-acted, non-prompted data have been collected, recorded, and used for the automatic classification of emotions in speech in the last decade: mother–child interaction (Slaney and McRoberts, 1998), human–robot interaction (Batliner et al., 2008b), tutoring dialogues (Ai et al., 2006), stress detection in a driving scenario (Fernandez and Picard, 2003), human–human (multi-party) conversation and interaction (Neiberg et al., 2006; Grimm et al. 2008; Schuller et al., 2009a), interaction human-information kiosk (Batliner et al., 2003b), appointment scheduling dialogues (Batliner et al., 2000a, 2003a), surgeons' speech during operations (Schuller et al., 2008), call centre applications using volunteering or real users, WoZ or real systems (Lee et al. 2001; Ang et al. 2002; Batliner et al. 2004; Steidl et al. 2004; Devillers et al. 2005). Some more references to databases mostly with acted data can be found in Cowie et al. (2005). Multi-modal databases and approaches are dealt with in Zeng et al. (2009) with the focus on other modalities, and in Cowie et al., this volume.

2.2 Annotations

Annotations can be automatic or manual or both (first automatic and then edited manually). The first annotation pass is normally the transliteration of what has been said. Even if automatic speech recognition (ASR) can be applied, a manual editing of its results is mandatory if correct transliterations are aimed at. Transliteration conventions are either implicit or following standards put forth, e.g. by LDC (http://www.ldc.upenn.edu/), cf. Devillers et al. (2005), or within the Verbmobil project in Burger et al. (2000). Apart from the 'normal' linguistic events, i.e. the words produced by the speakers, several other para-/extra-linguistic (breathing, sighing, laughter) or non-linguistic (technical noise) events can be annotated. Moreover, there are specific conventions for the annotation of typical spontaneous phenomena such as hesitations, filled or unfilled pauses, false starts, repetitions.

The next step should be to define the units of emotion annotation; these, in turn, are constitutive for the units of analysis used in the classification phase. So far, this has been done mostly on a trivial or on an intuitive basis: the unit is given trivially if simply utterances/dialogue moves/turns are taken – this can be an easy endeavour in a dialogue where the partners alternate as speakers/listeners. If the turns are longer, however, chances are that there is not one and the same emotion throughout this turn. This is of course descriptively less adequate and diminishes the discriminative power of automatic classification. Sometimes, longer turns are segmented on an intuitive notion (de Rosis et al., 2007; Devillers et al. 2005) of prosodic, syntactic, or pragmatic segmentation. In Batliner et al. (2003a) an objective approach towards defining units based on syntactic–prosodic segmentation has been put forth. Another possibility is to segment automatically at prosodic boundaries, using either only pause information or more complex information on intonational/prosodic units. Although there is a high correspondence between such prosodic units and higher syntactic/pragmatic units as shown in Batliner et al. (1998), it is not perfect and thus sub-optimal if it comes to the processing of emotion recognition in a full end-to-end system (Batliner et al., 2000b) because there will be the additional task to time-align the syntactically/semantically 'blind' prosodic units with the units processed by the higher module.

The impact of choosing the appropriate unit of analysis has been underestimated so far. However, the most important initial step is, of course, to find the adequate (number of) emotion labels. To start with, this can be done top-down or data driven: in the first case, the basis is normally a catalogue of theoretically derived or empirically obtained categories, cf. the terms used by Devillers et al. (2005) or the scheme proposed by Craggs and Wood (2004). Theoretically derived dimensional terms can be more or less elaborated (Russel, 1997). In the data-driven approach that has often been employed by more 'application-minded' studies, cf. below, only those categories are used that can be observed (often enough) in the data and are, at the same time, relevant for the intended applications.

The biggest issue in this phase concerns the two questions 'What to annotate' and 'How to annotate'. In the case of naturalistic data, a catalogue of prototypical (basic) emotion categories or dimensions falls short of the phenomena one can find; and what cannot be found cannot be annotated. Of course, different granularities can be chosen for a first annotation pass. In the short history of annotating naturalistic databases, the first studies were normally restricted to modelling a mapping onto a two-way distinction *negative* (encompassing user states such as anger, annoyance, or frustration) vs. the complement, i.e. *neutral*, even if at the beginning, more classes were annotated such as in Ang et al. (2002) neutral, annoyed, frustrated, tired, amused, other, not applicable. The minor reason for this mapping onto negative valence vs. neutral/positive valence was that in the intended application, it is most important to detect 'trouble in communication'. The major reason is simply that for statistical modelling, enough items per class are needed. The default, 'neutral', unmarked state dominates and accounts for up to >90% of the cases. The situation has not changed much recently, cf. Devillers et al. (2005). Neiberg et al. (2006) model, label, and recognize a three-way distinction neutral, emphatic, and negative for one database (voice-controlled telephone service) and for another (multi-party meetings), a three-way emotional valence negative, neutral, and positive. Ai et al. (2006) use a three-way distinction for student emotion in spoken tutoring dialogs: mixed/uncertain, certain, and neutral. Devillers et al. (2005) established an annotation scheme with the possibility to have a mixture of emotions (two labels per segment) and to use a coarse level (8 classes) and a fine-grained level (20 classes) plus neutral for annotation; a coarse label is, for example, anger with the fine-grained sub-classes anger, annoyance, impatience, cold anger, and hot anger. Mower et al. (2009) elaborate on prototypical/consensus vs. non-prototypical/no consensus for the following labels: angry, happy, sad, neutral, frustrated, or excited (audiovisual data, 10 actors). In some few studies, up to seven different emotional user states are classified as in Batliner et al. (2003b, 2008b); however, this 7-class problem cannot be used for real applications because classification performance is simply too low.

There are basically two different strategies answering the question 'How to annotate': we can start with a detailed catalogue of labels and reduce them in a more or less systematic manner to fewer labels to be used in annotation – those that really denote states that can be observed in the data – and to an even smaller set of labels to be used in automatic classification. The catalogue can be obtained from other basic studies or be based on free annotation, cf. Devillers et al. (2005). Alternatively, we can skip this step and establish in a data-driven way a set of labels suited for the intended application; for instance, in a call centre application, we might only want to find out whether the user is getting angry/annoyed, etc., i.e. whether something is going wrong. This would be a task-dependent emotion annotation with the goal of emotion detection in a real system. In the studies conducted so far, the set of labels chosen was mostly intended to be suited for the data, although aiming at the general issue of emotional behaviour annotation. However, emotional states that cannot be observed often enough were skipped in an earlier or later stage of the annotation process. Moreover, there is a certain trade-off between the number of the labellers, their expertise, and the effort to be spent: from methodological reasons, it might be desirable to employ something like >10 naïve labellers or >5 expert labellers to annotate on a fine-grained scale; by that, any 'central tendency' is not corrupted even if one expert or two naïve labellers might go astray. To follow this rule of thumb is, however, almost never feasible. Normally, more than one labellers are employed. This makes it possible to establish measures of agreement, cf. below, and to establish different levels of agreement: apart from the method to allow each labeller to give more than one label per unit, cf. the major and minor label in Devillers et al. (2005), for more labellers, either a correspondence or a majority decision can be defined or a soft vector with percentages can be created (Steidl et al., 2005; Devillers et al., 2005). Further, continuous labelling can be performed over time and space by dimensions as arousal or valence (Cowie et al., 2000). In any case, standards as, e.g. EmotionML in Schröder et al. (2007) can be of use, which allows for all of these labellings. For some scenarios, there can be some 'external ground truth', e.g. the intensity of stress-inducing tasks, a worse performance of the system, physiological measures as indicators of stress (levels). Such an external evidence can be taken as either means for assigning labels or later on as additional feature in the classification phase.

There are two classic criteria for assessing the quality of such labels: validity and reliability. Ecological validity is most important but not easy to measure; thus normally, reliability measures are aimed at such as measures of correlation, correspondence, (weighted) kappa, or (weighted) alpha (Fleiss et al., 1969; Rosenberg and Binkowski, 2004). The use of 'quantised' score ranges, based on such measures, e.g., for kappa, < 0.2 'bad', between 0.2 and 0.4 'moderate', between 0.4, and 0.6 'good', between 0.6 and 0.8 'very good', > 0.8 'excellent' (there are other scalings), seems to be a convenient way of assessing the quality of annotations. As far as we can see, however, it has almost never been used for any decision to be made – for some reasons: a lower kappa score can – apart from being caused by deficiencies in the very score itself – mean that inter- or intra-rater reliability is low because of spurious factors or because there simply are different – and valid – criteria and thresholds for annotation, and/or simply that the task is difficult, etc. Too high scores can be rather suspicious because it can be doubted that they can be obtained when dealing with naturalistic data. Moreover, the ultimate measure (of validity) is on the one hand the performance of the classifier – which itself can be compared with the performance of the annotators by using measures such as proposed in Steidl et al. (2005), and on the other hand, the impact on the users of such systems, cf. Sect. 5.

2.3 Further Processing

The ultimate goal in ASR is fully automatic processing although important steps such as building a lexicon or transliterating the training data are still mostly done manually. Matters are different in the research of emotion in speech: here it is not yet considered to be very important whether processing is manual or not; thus we often observe a mixture of manual and automatic processing. A typical approach is, e.g., to extract acoustic features automatically and linguistic features such as non-verbals or part-of-speech classes semi-automatically or fully based on manual processing. Sometimes, automatically extracted acoustic features are corrected manually, cf. Batliner et al. (2007b) where the manual correction of word segmentation and pitch values is described. Segmentation of higher units into lower ones can be 'blind', i.e. automatic, e.g. by defining fixed length segments or by partitioning each turn into a fixed number of segments, or it can be 'intelligent', e.g. by segmenting into words or other smaller units using other higher level information. A 'blind' segmentation is normally automatic, an 'intelligent' one so far mostly manual. The choice of segmentation strategies is of course conditioned by the type of data used and by the effort needed: turns produced by one speaker taking part in a bi-directional dialogue can be segmented by hand, whereas the effort needed for a more fine-grained (word- or syllable-based) segmentation is considerably higher.

A last and decisive step is the selection of units out of the whole database for feature extraction and classification. Two easy and automatic strategies are almost never employed: simply using all the data or using a randomly chosen sub-sample. This is due to the sparse data problem: the overwhelming majority of the cases belong to the 'uninteresting' default class neutral, cf. Sect. 4.1. Non-neutral cases can often not unequivocally be attributed to one of the 'interesting' classes because they are mixed; often, more prototypical cases are chosen. This is permissible – after all, we can imagine an application looking only for very pronounced cases – but the selection criteria have to be documented clearly: simply to select more prototypical cases by sharpening the threshold criterion can yield a marked performance improvement, cf. Batliner et al. (2005) and Seppi et al. (2008b).

It should be mandatory for writing a paper on recognising emotions in speech, and it is advisable for readers of such papers, to point out explicitly and to find out the strategies used at different stages: what is automatic, what manual, which criteria were intuitive, which objective, and which criteria for selecting the final sample were applied. Intuitive and/or selection criteria as such should not necessarily be forbidden, if stated explicitly. They simply introduce some fuzziness at a certain stage of processing. Their impact on the final results – and it is mostly recognition performance that is remembered by the readers of such studies – can be decisive or small. It would be good practice if the authors themselves pointed out the presumable impact.

3 Features

Feature extraction is a crucial phase in automated emotion recognition. As yet there has not been a large-scale, comprehensive comparison of different feature types; as for preliminary efforts in this direction, cf. Batliner et al. (2006b) and Schuller et al. (2007a). Presenting a comprehensive overview of feature types and feature extraction methods requires some kind of division of features into classes, though there is more than one way to do so. We will present several – alternative and complementing – approaches to grouping features. The most basic distinction to be made is between acoustic vs. linguistic features, as extraction methods for these two types are extremely different. Their relative contribution can also vary greatly, depending on the database being analysed: for acted data, based on scripted speech, linguistic features are normally of no value – apart from some specific applications such as data mining in movie archives. On the other hand, as we come closer to spontaneous real-life speech, these features can gain considerably in importance. Acoustic features are the more 'classic' features which have been in use since the inception of studies in this field, though researchers are far from agreeing which are most important, or whether this can even be determined. In the following subsections we discuss these two feature types separately.

3.1 Acoustic Features

Segmental features are mainly short-term spectra and derived features: MFCC (Lee et al., 2004), LPC, PLP (Perceptual Linear Prediction), etc. (Hermansky, 1990), and Wavelets (Fernandez and Picard, 2003 and Schuller et al., 2007a), TEO (Teager Energy operator) (Fernandez and Picard, 2003 and Zhou et al., 2001), LFPC, LPPC (Nwe et al., 2003). MFCCs are classically used for ASR, normally for modelling segments such as phones and, by that, words. In emotion recognition, they are rather used for modelling longer units of analysis such as utterances/turns, dialogue moves. To this aim, the features are extracted frame-wise and combined by appropriate measures such as averaging or by resorting to dynamic classification such as hidden Markov models. Although originally intended to model segments, these features have been used successfully for supra-segmental units.

Supra-segmental features model the classic prosodic types: pitch, intensity, duration, then voice quality, and long-term spectra. Prosodic features involve two steps: extracting raw prosodic *basic* features, then calculating *structured* features based on this data (Kießling, 1997; Hess et al., 1996). The raw prosodic data are the F0 contour, the intensity contour, and durational data on different levels (lengths of chunks, words, voiced segments, syllables, phonemes). Various errors can creep into the calculations at this stage. The second step involves extracting structured features from the basic prosodic features using various statistics such as mean, standard deviation, percentiles, ranges, peaks, slopes, regressions. Voice quality is a complicated issue in itself, since there are many different measures of voice quality (e.g. Lugger et al., 2006), mostly clinical in origin, though once again standardisation in this area is lacking. Other, less well-known voice quality features were intended towards normal speech from the outset, e.g. those modelling 'irregular phonation', cf. Batliner et al. (2007a). There are several survey papers on prosodic features in automatic speech processing such as Hess et al. (1996) and Nöth et al. (2002) and on their use in emotion modelling, cf. Frick (1985), Scherer et al. (2003), and Johnstone and Scherer (2000).

Features can be low level or high level, i.e. statistic features or those based on pitch models such as MoMel (Hirst et al., 2000) or the Fujisaki model (Fujisaki, 1992). Features can be represented by raw values, i.e. they can be non-perceptual or they can be based on perception models. Normalisation and standardisation of pitch range, pitch mean, speech tempo, etc. are used for modelling perception as well and for making successive measurements coherent with respect to a common scale.

Using another terminology, we can speak about *Low Level Descriptors* (LLDs), i.e. basic measures of feature types, and *functionals* such as mean, percentiles. LLDs account for base contours that usually are extracted by processing a fixed number of samples contained in a sliding window. For example, pitch attributes derive from the F0 contour. Subsequently to the LLD extraction, a number of operators and functionals are applied to obtain a certain feature vector out of each contour. Functionals provide a normalisation over time: base contours associated to words have different lengths, depending on the duration of the words and on the magnitude of the window step; with the usage of functionals, we obtain one feature vector per word, with a constant number of elements.

To reduce the influence of noise and to model temporal variations of LLDs, base contours are usually filtered, and first- and second-order derivatives are extracted. These functionals that can be applied to raw contours range from simple statistics to curve fitting methods or even methods based on perceptual criteria. The most popular statistical functionals cover the first four moments (mean, standard deviation, skewness, and kurtosis). Other functionals are positions of extremes values within a certain temporal context, quartiles, amplitude ranges, zero-crossing rates, roll-on/-off, on-/off-set, and higher level analysis. Curve fitting methods produce regression coefficients, such as slope of polynomial regressions, and regression errors (such as the mean square error between the regression and the original contour). Maybe the most comprehensive list of functionals is given in Schuller et al. (2007a) and Eyben et al. (2009).

We now characterise shortly the different types of acoustic features:

- *Duration* features model temporal aspects; the basic unit is milliseconds (ms) for the 'raw' values. Different types of normalisation can be applied. Note that relative positions on the time axis of base contours like energy and pitch such as maxima or on-/off-set positions do not strictly represent energy and pitch but duration – simply because they are measured in milliseconds and because they are often highly correlated with duration features (Batliner et al., 2001). In other words, duration attributes can be distinguished according to their extraction nature: those that represent temporal aspects of other acoustic base contours, and those that exclusively represent the parameter 'duration' of higher phonological units, like phonemes, syllables, words, pauses, utterances. Duration values are usually correlated with the linguistic features described below: for instance, function words are shorter on average, content words are longer. These two main word classes are not equally distributed across emotion types; this information can be used for classification, no matter whether it is encoded in linguistic or acoustic (i.e. duration) features.
- *Energy (intensity)* features usually model the loudness of a sound as perceived by the human ear, based on the amplitude in different intervals; different types of normalisation are applied. Energy features can model intervals or characterising points. As the intensity of a stimulus increases, the hearing sensation grows logarithmically (decibel scale). It is further well known that sound perception also depends on the spectral distribution and on its duration too. The loudness contour is the sequence of short-term loudness values extracted on a frame base. So-called energy features are finally obtained from the loudness contour by applying functionals.
- The basics of *pitch* extraction have largely remained the same; nearly all Pitch Detection Algorithms (PDAs) are built using frame-based analysis: the speech signal is broken into overlapping frames and a pitch value is inferred from each segment by either autocorrelation (Rabiner, 1977) in its manifold variants and derivatives. Often, the LPC residual or a low-pass filtered version is used over the original signal. Other approaches use the cepstral representation (Noll, 1967) or exploit harmonic information by spectral compression. However, also PDAs in the time domain exist that have the advantage of being able to detect changes per fundamental period, though generally being less reliable. The acoustic equivalent to the perceptual unit pitch is measured in Hertz and often made perceptually more adequate, e.g. by logarithmic/semitone transformation. Intervals, characteristic points, or contours are often modelled.
- The *spectrum* is characterised by formants (spectral maxima) modelling spoken content, especially the lower ones. Higher formants also represent speaker characteristics. Each one is fully represented by position, amplitude, and bandwidth. The estimation of formant frequencies and bandwidths can be based on Linear Prediction Coding (LPC) (Makhoul, 1975) or on cepstral analysis (Davis and Mermelstein, 1980). LPC enables one to model the human vocal tract. Once the spectral envelope is estimated by using the LPC method, a number of spectral features can be computed such as formant band-energies, roll-off, centroid, and

flux. Furthermore, the long-term average spectrum over a unit can be employed: this averages out formant information, giving general spectral trends.

- The *cepstrum*, i.e. the inverse or secondary spectral transform of the logarithm of the spectrum (Bogert et al., 1963), emphasises changes or periodicity in the spectrum, while being relatively robust against noise. Its basic unit is quefrency which is related to frequency. Mel-Frequency-Cepstral-Coefficients (MFCCs) – as homomorphic transform with equidistant band-pass filters on the Mel-scale – tend to strongly depend on the spoken content. Yet, they have been proven beneficial in practically any speech processing task. PLP coefficients (Hermansky, 1990) and the MFCCs are extremely similar, as they both correspond to a short-term spectrum smoothing – the former through an autoregressive model, the latter trough the cepstrum – and to an approximation of the auditory system by filter bank-based methods. At the same time, PLP coefficients are also an improvement of LPC by using the perceptually based Bark filter bank.
- *Voice quality* features model jitter, shimmer, and further micro-prosodic events. Noise to harmonic ratio (NHR) or harmonic-to-noise ratio (HNR) is another measure of the quality of the speech signal. Although they depend in part on other LLDs such as pitch (jitter) and energy (shimmer), they reflect peculiar voice quality properties such as breathiness or harshness. Therefore, they are usually dealt with within a separate feature class. Some of these have several variants and even when their definitions are agreed upon, different software can give different values, due, for example, to differences in pitch extraction methods.
- *Wavelets* give a short-term multi-resolution analysis of time, energy, and frequencies in a speech signal (Daubechies, 1990). Compared to similar parametric representations such as MFCCs, they are superior in the modelling of temporal aspects.
- *Non-linguistic Vocalisations* identify non verbal phenomena such as breathing and laughter. Automatic detection of disfluencies and non-verbals normally requires that the vocabulary used by the ASR engine includes both these entities. Thus they could be subsumed under linguistic features as well.

Other acoustic features that have been used or can be used are TRAPs (Hermansky and Sharma, 1998) or Teager operator (especially for stress detection) (Zhou et al., 2001). The standard acoustic feature types used in many emotion classification studies might be – probably in this order of frequency but not necessarily of importance – pitch, energy, spectrum, cepstrum, voice quality, duration. Traditionally, pitch has been conceived as being most important – this is not backed up by empirical results; note that the reason might not be extraction errors, cf. Batliner et al. (2007b).

3.2 Linguistic Features

Spoken or written text also carries information about the underlying affective state (Arunachalam et al., 2001). This is usually reflected in the usage of certain words or grammatical alterations – which means in turn, in the usage of specific higher semantic and pragmatic entities. A number of approaches exist for this analysis: keyword spotting (Elliott, 1992; Cowie et al., 1999), rule-based modelling (Litman and Forbes, 2003), semantic trees (Zhe and Boucouvalas, 2002), latent semantic analysis (Goertzel et al., 2000), transformation-based learning (Wu et al., 2005), world-knowledge-modelling (Liu et al., 2003), key-phrase spotting (Schuller et al., 2004), string kernels (Schuller et al., 2009b), and Bayesian networks (Breese and Ball, 1998). Context/pragmatic information has been modelled as well, e.g. type of system prompt (Steidl et al., 2004), dialogue acts (Litman and Forbes, 2003, Batliner et al., 2003a), or system and user performance (Ai et al., 2006). Two methods seem to be predominant, presumably because they are shallow representations of linguistic knowledge and have already been frequently employed in automatic speech processing: *(class-based) N-grams* (Polzin and Waibel 2000; Ang et al., 2002; Lee et al., 2002; Devillers et al., 2003) and *Bag-of-Words* (*vector space modelling*), cf. Schuller et al. (2005) and Batliner et al. (2006b); these will be dealt with in the following.

A first step will always be the pre-processing of the text. This seems an easy task for written text, yet, soft string matching (e.g. by Levenshtein Distance) is reported to be advantageous to overcome misspelling, or spelling variations, dialects, etc. Considering analysis from spoken text, only few results for emotion recognition rely on ASR output (Schuller et al., 2005, 2009b) rather than on manual annotation of the data (Batliner et al., 2006b). This comes, as ASR of emotional speech itself is a challenge (Athanaselis et al., 2005; Schuller et al., 2006a, 2007b, 2009b) and might be error prone.

Second, an inventory of term entities, known as vocabulary, needs to be constructed which initially consists of all different words observed in the training corpus; this usually amounts to several thousands. (Note that for instance the balanced affective wordlist (Siegle, 1995) consists of only roughly 300 words.) Eventually, the vocabulary has to be reduced somehow, by stopping or by stemming.

Stopping resembles elimination of irrelevant words. The traditional approach to stopping is an expert-based list of words as function words. Yet, even for an expert it seems hard to judge which words can be of importance in view of the affective context. Data-driven approaches such as salience or information gain-based reduction (see below) are popular. The easiest, yet often effective way, is also stopping by the general minimum frequency of occurrence within a training corpus.

Stemming stands for clustering of morphological variants, i.e. flexions (e.g. by declination or conjugation), of a word by its stem into a *lexeme*. This reduces the number of entries in the vocabulary while at the same time providing more training instances per class. Thereby also words that were not seen in the training can be mapped upon lexemes, for instance, by simple (character) N-gram stemming, cf. below, or by (Iterated) Lovins or Porter stemmers that base on suffix lists and rules for their application (Lovins, 1968; Porter, 1980). A very compact approach to stemming is the use of so-called part-of-speech (POS) classes, such as nouns, verbs, adjectives, particles (Batliner et al., 2006b). Also *sememes*, i.e. semantic units represented by lexemes, can be clustered into higher semantic concepts such as generally positive or negative terms (Batliner et al., 2006b). In addition, non-linguistic vocalisations like sighs and yawns (Russell et al., 2003), laughs (Campbell et al., 2005; Truong and van Leeuwen, 2005), cries (Pal et al., 2006), and coughs (Matos et al., 2006) can easily be integrated into the vocabulary (Batliner et al., 2006b; Schuller et al., 2006b, 2009a).

N-grams and *class-based N-grams* are commonly used for general language modelling. Thereby the posterior probability of a (class of a) word is given by its predecessors from left to right within a sequence of words. For emotion recognition, the probability of each emotion is determined per N-gram of an utterance. Following Zipf's principle of least effort stating that irrelevant function words occur very frequently opposing terms of interest, the number of considered words is reduced to N in order to prevent over-modelling. In addition, word class-based N-grams can be used as well, to better cope with data sparseness.

Nonetheless, in emotion recognition, mostly uni-grams (N=1) have been applied so far (Lee et al., 2002; Devillers et al., 2003), besides bi-grams (N=2) and trigrams (N=3) (Ang et al., 2002). The actual emotion is calculated by the posterior probability of the emotion given the actual word(s) by maximum likelihood or a-posteriori estimation.

Bag-of-Words is a well-known numerical representation form of text in automatic document categorisation (Joachims, 1998). It has been successfully ported to recognise sentiments (Pang et al., 2002) or emotion (Schuller et al., 2005, 2006b). Thereby each word in the vocabulary adds a dimension to a linguistic vector representing the term frequency within the actual utterance. Note that easily, very large feature spaces may occur, which usually require stopping and stemming. The logarithm of frequency is often used; this value is further better normalised by the length of the utterance and by the overall (log)frequency within the training corpus. Also, it is possible not to refer to words, but sequences of them, i.e. Bags-of-N-grams, to overcome the lack of word order modelling (Schuller et al., 2009b).

Note that most vector elements will resemble zero, as feature vectors are constructed for short utterances rather than for longer texts, as in document retrieval, and only few words of the vocabulary will be seen. Support vector machines (cf. below) show high performance for this task. The possibility of early fusion with acoustic features helped make this technique very popular (Schuller et al., 2006b; Batliner et al., 2006b).

The preponderance of acoustics in emotion modelling so far is conditioned by the traditional focus on segmentally identical, acted utterances. For naturalistic data, both acoustic and linguistic features should be employed, both for a deeper understanding and a better classification performance. Basic feature extraction and subsequent computation of structured features employing (combinations of) functionals will certainly be the subject of much research in the future, examined in different contexts. We are far from knowing which feature (type) models best which emotional states in which context. Thus we have to resort to the general advise to use a representative set of features of different types rather than only one type of feature.

4 Classification

The data-driven way to evaluate extracted features and classification performance is to rely on machine learning and/or pattern recognition techniques: we let the machine find and learn regularities in the data. In the past decades, a prolific amount of methods has emerged for automatic modelling and extraction of informative patterns of the data. The number of successive refinements and slight variations of each machine learning algorithm is even bigger. One challenge to address in emotion classification is how to prune into this depth of options and find a good method for this specific task.

Emotion recognition from speech has to deal with noisy, redundant, and correlated features. Furthermore speech feature vectors are often complex and large, contaminated with interferences, background noise, and overlapping signals; this is especially true for naturalistic emotional speech. Thus different studies have shown that the same feature vector can yield very different classification results using different algorithms.

4.1 The Curse of Dimensionality and the Sparse Data Problem

Realistic emotional speech databases are characterised by the following problems: (1) small number of patterns, (2) potentially high number of features, and (3) skewed classes. Typically such databases comprise some hundreds of labelled utterances, while the features for classifying them can be chosen within a high-dimensional space, usually up to some hundreds as well. As the amount of available data is usually fixed, any increase in the feature space rapidly (exponentially in the number of features) leads to regions of the feature space where data are very sparse. This problem is known as 'curse of dimensionality' (Bellman, 1961), and it affects classifiers that divide the feature space into cells. A good rule of thumb requires that the number of patterns should never be lower than twice the number of features. Although some classifiers implicitly and successfully cope with the curse of dimensionality, pre-processing methods such as 'feature selection' and 'feature reduction' are generally applied to the input space. A favourable by-product of reducing the feature space is the reduction of the computational burden and implementation complexity while training the classifier. Both should not be underestimated: the former may lead to no solution at all (in reasonable time), the latter can yield wrong results due to numerical instabilities and overflows. Furthermore, feature reduction and selection methods selectively proceed to discard correlated and non-relevant features, resulting in higher reliability of the results.

Feature reduction consists in the mapping of the input space onto a less-dimensional one, while keeping as much information as possible. Common reduction techniques used in the field of emotion recognition are principal component analysis (PCA), linear discriminant analysis (LDA) and more sophisticated derivations like heteroscedastic discriminant analysis (Ayadi et al., 2007) and independent component analysis (ICA). PCA is the feature transformation that minimises the sum of square error (Jolliffe, 2002). Furthermore, the base of the new space is orthonormal, which means that PCA de-correlates the original features: new features are constructed as linear superpositions so that the first one explains the largest amount of total variance of the data while each subsequent component explains the largest amount of the remaining variance while remaining uncorrelated with previously constructed features. The use of PCA requires the guess of the dimensionality of the target space. This can be done by the Kaiser–Guttman test, Log-Eigenvalue (LEV) diagram, Cattell's scree test (broken stick model), cross-validation, etc.

While PCA is an unsupervised feature reduction method (and thus maybe sub-optimal for specific problems), LDA is a supervised feature reduction method which searches for the linear transformation that maximises the ratio of the determinants of the between-class covariance matrix and the within-class covariance matrix (Fukunaga, 1990). LDA is less used as feature reduction, but it is widely adopted for direct classification (Lee and Narayanan, 2005; Kwon et al., 2003; Batliner et al., 2000a). Finally ICA (Hyvärinen et al., 2001) is the transformation that maps the feature space into an orthogonal space; furthermore, the target features are independent. Both theoretical and practical assumptions must hold, like the non-Gaussianity of the input features and the low dimensionality of the transformed space. There are already some studies adopting ICA (Rahurkar and Hansen, 2003), where both the input space and the output space are kept small.

Feature reduction is not appropriate for feature mining, as the original features are not retained after the transformation. *Feature selection* denotes a set of techniques that remove features which are irrelevant for modelling. This is a combinatorial optimisation problem: the feature space is traversed and at each step of the search, a different feature combination is evaluated. Evaluation is usually done following two possible strategies: the closed-loop 'wrapper' method, which trains and re-evaluates a given classifier at each search step using accuracy as objective function and the open-loop 'filter' method, which maximises simpler objective functions. While a wrapper can consist of any classifier, filter objective functions are usually measures such as information gain ratio (Witten and Frank, 2005) or inter-feature and feature-class correlation (Hall, 1998). As an exhaustive search through all possible feature combinations is unfeasible, faster but sub-optimal search functions are chosen. Most popular thereby is hill-climbing search or random injection as within random or genetic search. Typical hill-climbing procedures are sequential forward (SFS) and backward (SBS) selection by adding (deleting) at each search step the feature reporting the best performance according to the chosen wrapper or filter. SFS and SBS are commonly used (Lee et al., 2001; Lee and Narayanan, 2005; Kwon et al., 2003). Sequential floating forward selection (SFFS) (Pudil et al., 1994; Jain and Zongker, 1997) is an improved SFS method in the sense that at each step, previously selected features are considered for being discarded from the optimal group (SBS steps) to overcome nesting effects. Experiments show SFFS to dominate over other methods (Jain and Zongker, 1997). Note that a good feature selection should de-correlate the feature space to optimise a set of features as opposed to sheer ranking of features. This is in particular the case for wrapper search, which at the same time usually demands considerably higher computational effort. Some studies combine feature selection with feature generation to find better representations and combinations of features by simple mathematical operations such as addition, multiplication, or reciprocal value of features (Batliner et al., 2006b). Also, balancing of the training instances with respect to instances per emotion class may be done before feature selection if these are highly skewed (Schuller et al., 2009c).

With the growing interest in spontaneous data, class skewness or the 'sparse data' problem in the output space came to the fore: many classes are characterised by few observations only. Normally, most cases belong to the neutral class. The skewness of the output space can be addressed by considering proper class weights, by resampling, i.e. (random) up- or down-sampling, or by introducing main classes (clustering similar classes under the same hat). The most frequent couples of main classes are 'neutral vs. non-neutral' and 'positive vs. negative' emotions modelling the 'valence' dimension, where neutral generally encompasses the absence of any emotion while 'positive' emotions span from neutral to happiness.

4.2 Classifiers

A number of reasons speak for considering diverse classifiers for different tasks: mostly high recognition rates (e.g. ability to solve non-linear problems, learn discriminatively, adapt online, generalise, tolerate high dimensionality), adequate modelling (static or dynamic, data- or knowledge-based, model or instance-based, handling of missing feature values and uncertainty, training stability), efficiency and economical factors (real-time capability, low computational cost for training and recognition, low memory requirement, need of only few exemplary instances, easy implementation), and optimal integration in a system context, e.g. (class-wise) provision of confidences or handling of input confidence. These considerations, and the simple availability of implementations in toolboxes such as WEKA (Witten and Frank, 2005) or HTK (Young et al., 2006), led to a considerable bandwidth of variants being used in the recognition of emotion from speech.

Very popular classifiers for emotion recognition are linear discriminant classifiers (LDCs) (Fukunaga, 1990) and k-nearest neighbour (kNN) classifiers (Cover and Hart, 1967): their implementation is easy, the time needed for training is short, unbalanced classes can be handled, and the sensitivity to lack of data in general is small. kNN is a lookup method: the training data are simply stored ('lazy' or instance-based learning, as opposed to model building classifiers) and each new pattern is assigned by averaging its nearest neighbour classes. They are widely used (Dellaert et al., 1996; Petrushin, 1999), with good results for non-acted emotional speech as well (Lee and Narayanan, 2005; Shami and Verhelst, 2007). LDC – as a natural extension of LDA, see Fukunaga (1990) – is basically a classifier with straight-line decision surfaces (hyperplanes). LDA is one possible method of estimating LDC hyperplane parameters by maximisation of class separability (see above). They have often been used (Lee and Narayanan, 2005; Rahurkar and

Hansen, 2003; Kwon et al., 2003; Litman and Forbes, 2003; Batliner et al., 2000a), with a competitive performance (Batliner et al., 2006b) in spite of some limitations: the data should be linearly separable and the method is sensitive to outliers. A natural extension of LDCs is support vector machines (SVM): if the input data have previously implicitly undergone a non-linear transformation, which may have increased or decreased the number of features, and if the linear classifier obeys a maximum-margin fitting criterion, then we obtain an SVM (Vapnik, 1995). SVM provide very good generalisation properties (McGilloway et al., 2000; Lee et al., 2002; Chuang and Wu, 2004; You et al., 2006; Morrison et al., 2007); thus, they became increasingly popular. Note, however, that their performance is not always (way) better than the one obtained by using alternative classifiers (Meyer et al., 2002).

Among the most used non-linear discriminative classifiers are artificial neural networks (ANNs) and decision trees. Feedforward ANNs, also known as multi-layered perceptrons, are equivalent to fitting pre-defined non-linear functions to some given data. Decision surfaces might become very complex and depend on the topology of the network (number of neurons), on the learning algorithm (usually a derivation of the well-known backpropagation algorithm (Rumelhart et al., 1986), and on the activity rules (how the input patterns and the ANN weights are combined to obtain a decision output class). ANNs are therefore not robust to over-fitting and require greater amounts of data to be trained on. Therefore, ANNs are rarely used for acted data (Petrushin, 1999; Martinez and Cruz, 2005) and even less for non-acted, but cf. (Batliner et al., 2000a, 2006b). Recurrent networks can further be complemented by long short-term Memory to integrate emotional context (Wöllmer et al., 2008). Although they are also characterised by the property of handling non-linearly separable data, decision trees are less of a 'black box' compared to SVMs or neural networks, since they are based on simple recursive splits of the data. These splits (yes/no questions usually ranked by information gain) are very readable, especially if the tree has been adequately pruned, i.e. cutoff according to the ranking. Popular decision tree algorithms are C4.5 (Quinlan, 1993) and CART (Breiman et al., 1984). Note, however, that accuracy degrades in case of irrelevant features or noisy patterns. A solution is random forests (RF) (Breiman, 2001), an ensemble of trees each one accounting for a random subset of the input features and learned on variants of the training set by sampling with replacement. They are practically insensitive to the curse of dimensionality (Schuller et al., 2007a).

Apart from the already named kNN, which can be seen as a very basic statistical classifier, one also basic representative of this group is the Naive Bayes classifier (Langley et al., 1992; Good, 1965). It is robust with respect to irrelevant features but its performance may degrade quickly if correlated – even relevant – features are added. Less 'naïve' are Gaussian mixture models (GMM) that employ a number of multivariate Gaussians to model the original densities in the feature space. However, this of course also requires more training data.

Dynamic classifiers like hidden Markov models (HMM), dynamic Bayesian networks (DBN), or simple dynamic time warp (DTW) implicitly warp observed feature sequences over time. No further processing of the raw feature contours on a per-frame basis as pitch or energy is needed (like the application of functionals, to obtain the same number of features for different lengths of units such as turns or words). Among dynamic classifiers, apparently only HMM were studied yet, probably mostly because of the presence of well-elaborated tools such as HTK. For acted emotion there are numerous references as given by ten Bosch (2003); Schuller et al. (2003), and Zeng et al. (2009); for non-acted emotion, fewer are known (Kwon et al., 2003; Wagner et al., 2007; Vlasenko et al., 2007b; Schuller et al., 2009c). The performance of static modelling is usually not reached (ten Bosch, 2003; Schuller et al., 2003), as emotion apparently is better modelled on a timescale above frame level; note that a combination of static features such as minimum, maximum, onset, offset, duration, regression implicitly shape contour dynamics as well. Still, when the spoken content is fixed, the combination of static and dynamic processing may help improve overall accuracy (Vlasenko et al., 2007a). However, it is not clear whether emotion can be satisfyingly modelled using the simplifying Markov assumption that underlies HMM modelling (ten Bosch, 2003).

Ensembles of classifiers (Schuller et al., 2005) combine their individual strengths, and overcome training instability deriving from the sparseness of data. In the highly popular *Bagging* (Breiman, 1996) method, several instances of the same classifier are trained on sub-samples of the data set, usually of the same size, obtained by sampling with replacement. The final decision is then made by majority voting. *Boosting* decides by weighted majority voting after iteratively assigning (high) weights for hardly separable instances throughout learning. Next, *MultiBoosting* combines bias and variance reduction of these two methods by their sequential application. Most powerful, however, is the combination of diverse classifiers by either simple *Voting* (Morrison and Silva, 2007) or introduction of a meta-classifier that learns 'which classifier to trust when' and is trained only on the output of 'base-level' classifiers, known as *Stacking* (Wolpert, 1992). If confidences are provided on lower level, one speaks of *StackingC*. Still, the gain over single strong classifiers as SVM may not justify the extra computational need.

A possibility to use static classifiers for frame-level feature processing is further given by multi-instance learning techniques, where a time series of unknown length is handled as one by SVM or similar techniques (Shami and Verhelst, 2007; Schuller and Rigoll, 2009).

Regression – that is mapping on a continuum rather than on discrete classes – is also used in emotion recognition to handle the dimensional approach. Usually each axis, such as arousal, valence, or dominance, is thereby taken care of by one regression model such as support vector regression (Grimm et al., 2007) or less complex solutions such as multiple linear regression.

Features belonging to different types, e.g. acoustic and linguistic features, can be combined in *early fusion* within the same classifier or the class assignment with or without confidence measures obtained with different classifiers using different features can be combined in *late fusion*, cf. the ROVER approach (Fiscus, 1997) used in Batliner et al. (2006b).

4.3 Evaluation

To assess the performance of a classifier, we have to split the data into train and test. The easiest approach is a percentage split. However, data in emotion recognition are usually sparse, as mentioned. Therefore, it seems desirable to test on all instances: the training set is thereby usually kept as large as possible, the limit being a single pattern at a time for testing; this is repeated j times changing the tested pattern each time. Such a high number of trainings can be infeasible. Splitting the data into $j = 10$ parts, training on 9 parts, and testing on the remaining data is a good, popular compromise, called j-fold cross validation, cf. Salzberg (1997). Throughout partitioning of the data the distribution among classes should be kept, known as stratification. However, the partitioning is usually not explicitly stated, thus not easily allowing for comparative studies. Also, if it is not speaker independent, recognition performance will be too optimistic. Both these downsides can be overcome by leave-one-speaker-out, meaning training with all but one speaker in each cycle (Steidl, 2009), or leave a known group of speakers out to spare computational effort.

Most of the studies report performance measures expressed by accuracy, i.e. the recognition rate (RR), also known as weighted average (WA): the number of correctly recognised patterns divided by the total number of patterns. Given the skewness of spontaneous emotional databases, this is rarely appropriate. A possibility is to measure both, Precision (P, the number of true positives over all positive patterns) and Recall (R, the number of true positives over the number of all reference patterns). When there are more than two classes, it is useful to give a P- and an R-value for each class separately. In this sense R of a class corresponds to the RR of this class. As a general measure over the entire data is useful, we can introduce the mean of the accuracies (RR) over all classes, i.e. the class-wisely averaged classification rate (CL), also known as unweighted average (UA) recall in contrast to weighted average (WA) recall resembling RR (Schuller et al., 2009c). Note that RR and CL for a balanced multi-class recognition problem are always identical; the more the class distribution is unbalanced, the higher the difference between RR and CL. To have a unique measure of the goodness of classification – for comparison aims – the F-measure can be used; it is the harmonic mean of P and R. A similar score can be obtained by averaging UA and WA. The receiver operating characteristic (ROC) curve is independent of the data distribution but has the disadvantage that curves are not easy to compare. It is the plot of R over 1-Specificity (S, the false negative over all negatives). ROC curves are constructed by modifying a threshold during the training of the classifier. Different thresholds correspond to different performance of the classifier (in terms of Recall and Specificity) and thus to different points on the ROC curve, cf. Steidl et al., (2009).

The complete source of information is the confusion matrix. The figures described above all derive from it and try to highlight or smooth some aspects, especially for multiple classes when it might be difficult to interpret or during the training of a classifier when optimisation is achievable only w.r.t. few or one single parameter such as accuracy or F-measure.

Studies eventually end up with the conclusion that a specific classifier is better than another one – which is a conclusion that must not be generalised. Most of the time no significance of the differences is reported. Actually, there are some reasons to handle significance tests with care, for general reasons (Nickerson, 2000) and because of repeated measurements: the more experiments we do on a certain data set, the more probable it is that we accidentally run into some significant results. Significance thresholds should be augmented whenever we increment the number of experiments; however, this is not done very often. The Bonferroni adjustment is a possible choice of a correction factor. For a cookbook on multi-experiment studies, see Salzberg (1997). There are some drawbacks of the Bonferroni correction as it is usually too conservative; these are outlined in Pernegger (1998).

Also, when doing comparative evaluations, everything that is done to modify or prepare the classifier must be done in advance before looking at the test data (Salzberg, 1997). To our knowledge, only few studies in emotion recognition clearly explain what – if any – part of the data have been used for parameter tuning: they describe how the data have been divided into test and training but nothing is said about held-out data for classifier tuning, i.e. a development set; this should be part of future investigations.

Finding, fine-tuning, and evaluating classifiers is a broad topic in its own; although there might be preferences to use one or the other approach in specific fields – such as emotion recognition, it generally suffers from too many degrees of freedom: a strict comparison across studies is practically never possible. Statements such as 'it has been proved that classifier X is superior to classifier Y', should never be generalised. Often it only means that there has been more fine-tuning for X than for Y. In the long run, it might turn out that specific models and classifiers based on them are – on the average – better suited for emotion recognition. However, searching for an optimal classifier alone will not be a panacea; it will not improve unsatisfying recognition rates to such an extent that the intended application will be successful. Anyway, it should be mandatory to document the steps explicitly, e.g. whether a cross-validation has been done speaker-independently or in a speaker-dependent way. This statement holds similarly for comparison across whole studies: what never should be done is simply to compare recognition rates between two studies. Such performance depends crucially on too many factors which have not been standardised yet.

5 Applications

Apart from some 'offline' applications such as data mining in movie archives or screening call-centre agents as for their behaviour against customers, the ultimate goal of the whole endeavour described in this chapter is employing classified emotional user states in an end-to-end system; by end-to-end system we mean 'spontaneous speech, produced by human users as input – generated system reaction such as synthesised speech, produced by the system, as output, and vice versa'.

Several systems have been envisaged so far (Batliner et al., 2006a); example applications are depicted in Burkhardt et al., (2009) and Vogt et al. (2009). The contribution of automatic classification is rather straightforward: each speech unit such as words/chunks/turns/dialogue moves is attributed one out of a rather reduced set of emotion labels, maybe with some probability or confidence measure. This attribution can be correct or wrong – basically the same way as human beings can be right or wrong or disagreeing when estimating the emotions of other human beings. In both cases, some cost function has to be established – is it costly, or does it not matter at all, whether I attribute the wrong emotion or the right one? But it is not only an erroneous classification of emotion which can cause erroneous results: ASR is not perfect. We do not fully know yet whether emotional speech causes more speech recognition errors because it is more difficult than 'normal' speech, or because we simply do not have enough data of this variety to train an ASR engine successfully (Athanaselis et al., 2005; Schuller et al., 2007b, 2009b). In real-life settings, chances are that a worse signal-to-noise ratio will deteriorate ASR and by that, emotion classification; especially using linguistic features might not yield good recognition performance. If ASR is erroneous, this will result in erroneous words and erroneous segmentation, so both acoustic and linguistic features might be computed in a sub-optimal way, resulting in lower classification performance. The impact of erroneous extraction might not be too high if acoustic features are used, cf. Schuller et al., (2007b), but might be problematic if only linguistic information is exploited (Seppi et al., 2008a). Moreover, erroneous ASR is of course not really helpful for processing the user's semantic/pragmatic intentions within the whole system.

ASR normally aims at speaker-independent modelling and recognition; this is state of the art in our field as well. Speaker-dependent processing yields better recognition performance; we want to point out that even if speaker independency is, of course, the ultimate goal, we can imagine applications where speaker-dependent modelling is possible and makes sense. This will always be the case when the speaker can be identified and is a frequent user of the system.

The exchange format with other modules within a full end-to-end system is nowadays normally some XML dialect, cf. Schröder et al. in this volume. However, we do not know yet of any system where really speech and not written language has been used as input into such a representation and subsequent use within a full system – apart from the SmartKom system (Streit et al., 2006) where an implementation of the OCC model (Ortony et al., 1988) had to be restricted to some few so-called use cases. It could be shown that the module was functional on a principled basis in the whole end-to-end system; however, it has to await much more testing and more robust recognition modules to be functional in any practical application.

In this section we want to point out that even if we solved somehow the problems we addressed in this chapter, this is not the end of the story because most of the time, we will have to use ASR output within a 'real system' – and this output inevitably can be erroneous which in turn can cause erroneous processing of not only emotion attributions.

6 Concluding Remarks

In this chapter, we gave an overview of the state of the art in the automatic recognition of real life, natural emotional user states, pointing out problems, pitfalls, and to-do's and not-to-do's. We deliberately refrained from comparing classification performance across studies in terms of recognition rates – this cannot be done in a serious way and would be misleading. We dealt with the full sequence of processing, from conceptualisation to recognition rates, although mostly not in an in-depth manner. We hope to have introduced almost all of the pertinent topics; the references can be used for more detailed information.

As for the future of our topic, the pivotal desideratum is databases; a comparable albeit way easier problem that somehow has been 'solved' – i.e. a satisfying recognition performance has been obtained – in recent time is the performance of automatic dictation systems. Here, the breakthrough came with the use of training material larger by some order of magnitude. However, already the basic unit is not comparable: whereas there can be a fair agreement on what a word is and which word has been produced, there is neither full agreement on what an emotion is nor on the way how to obtain the ground truth, i.e. the types and tokens we want to recognise. Moreover, the creation of databases is expensive, and progress will be slow. Even if the field is emerging – which can be seen from the growing number of contributions to conferences and journal papers – the methodological problem is that practically always, results cannot be compared across studies because too many factors are not kept constant. A few studies have begun to address different databases using the same approaches, cf. Shami and Verhelst (2007) and Batliner et al. (2008a). Initiatives such as CEICES, cf. Batliner et al. (2006b), combining thoroughly annotated data with the fusion of a plethora of different feature types, generated at different sites, might be one way of establishing 'islands of standardisation', i.e. making comparisons across classifiers and features easier and more reliable. The Interspeech 2009 Emotion Challenge, cf. Schuller et al. (2009c), has been the first attempt towards strict comparability and reproducibility of emotion classification results. However, further steps in this direction will be needed to provide comparability among researchers for a multiplicity of remaining challenges.

References

Ai H, Litman D, Forbes-Riley K, Rotaru M, Tetreault J, Purandare A (2006) Using system and user performance features to improve emotion detection in spoken tutoring dialogs. In: Proceedings of the Interspeech, Pittsburgh, PA, September 17–21, pp 797–800

Ang J, Dhillon R, Shriberg E, Stolcke A (2002) Prosody-based automatic detection of annoyance and frustration in human-computer dialog. In: Proceedings of the Interspeech, Denver, September 16–20, pp 2037–2040

Arunachalam S, Gould D, Anderson E, Byrd D, Narayanan S (2001) Politeness and frustration language in child-machine interactions. In: Proceedings of the Eurospeech, Aalborg, September 3–7, pp 2675–2678

Athanaselis T, Bakamidis S, Dologlu I, Cowie R, Douglas-Cowie E, Cox C (2005) ASR for emotional speech: clarifying the issues and enhancing performance. Neural Netw. 18:437–444

Ayadi MMHE, Kamel MS, Karray F (2007) Speech emotion recognition using gaussian mixture vector autoregressive models. In: Proceedings of ICASSP, Honolulu, April 15–20, pp 957–960

Batliner A, Kompe R, Kießling A, Mast M, Niemann H, Nöth E (1998) M = Syntax + Prosody: a syntactic–prosodic labelling scheme for large spontaneous speech databases. Speech Communi 25(4):193–222

Batliner A, Fischer K, Huber R, Spilker J, Nöth E (2000a) Desperately Seeking Emotions: Actors, Wizards, and Human Beings. In: Proceedings of the ISCA workshop on speech and emotion, Newcastle, Northern Ireland, September 5–7, pp 195–200

Batliner A, Huber R, Niemann H, Nöth E, Spilker J, Fischer K (2000b) The recognition of emotion. In: Wahlster W. (ed) Verbmobil: Foundations of speech-to-speech translations. Springer, Berlin, pp 122–130.

Batliner A, Buckow J, Huber R, Warnke V, Nöth E, Niemann H (2001) Boiling down prosody for the classification of boundaries and accents in German and English. In: Proceedings of the Eurospeech, Aalborg, September 3–7, pp 2781–2784

Batliner A, Fischer K, Huber R, Spilker J, Nöth E (2003a) How to find trouble in communication. Speech Commun, 40:117–143

Batliner A, Zeissler V, Frank C, Adelhardt J, Shi RP, Nöth E (2003b) We are not amused - but how do you know? User states in a multi-modal dialogue system. In: Proceedings of the Interspeech, Geneva, September 1–4, pp 733–736

Batliner A, Hacker C, Steidl S, Nöth E, Haas J (2004) From emotion to interaction: lessons from real human-machine-dialogues. In: Affective dialogue systems, proceedings of a tutorial and research workshop, Kloster Irsee, June 14–16, pp 1–12

Batliner A, Steidl S, Hacker C, Nöth E, Niemann H (2005) Tales of tuning – prototyping for automatic classification of emotional user states. In: Proceedings of the Interspeech, Lisbon, September 4–8, pp 489–492

Batliner A, Burkhardt F, van Ballegooy M, Nöth E (2006a) A taxonomy of applications that utilize emotional awareness. In: Proceedings of IS-LTC 2006, Ljubliana, October 9–10, pp 246–250

Batliner A, Steidl S, Schuller B, Seppi D, Laskowski K, Vogt T, Devillers L, Vidrascu L, Amir N, Kessous L, Aharonson V (2006b) Combining efforts for improving automatic classification of emotional user states. In: Proceedings of IS-LTC 2006, Ljubliana, October 9–10, pp 240–245

Batliner A, Steidl S, Nöth E (2007a) Laryngealizations and Emotions: How Many Babushkas? In: Proceedings of the international workshop on paralinguistic speech – between models and data (ParaLing'07), Saarbrücken, August 3, pp 17–22

Batliner A, Steidl S, Schuller B, Seppi D, Vogt T, Devillers L, Vidrascu L, Amir N, Kessous L, Aharonson V (2007b) The impact of F0 extraction errors on the classification of prominence and emotion. In: Proceedings of the ICPhS, Saarbrücken, August 6–10, pp 2201–2204

Batliner A, Schuller B, Schaeffler S, Steidl S (2008a) Mothers, adults, children, pets — towards the acoustics of intimacy. In: Proceedings of the ICASSP 2008, Las Vegas, NV, March 30–April 04, pp 4497–4500

Batliner A, Steidl S, Hacker C, Nöth E (2008b) Private emotions vs. social interaction — a data-driven approach towards analysing emotions in speech. User Model User-Adap Interact 18:175–206

Bellman R (1961) Adaptive control processes. Princeton University Press, Princeton, NJ

Bogert B, Healy M, Tukey J (1963) The quefrency analysis of time series for echoes: cepstrum, pseudo-autocovariance, cross-cepstrum and saphe cracking. In: Rosenblatt M. (ed) Symposium on time series analysis. Wiley, New York, NY, pp 209–243

Breese J, Ball G (1998) Modeling emotional state and personality for conversational agents. Technical Report MS-TR-98-41, Microsoft

Breiman L (1996) Bagging predictors. Mach Learn 26:123–140

Breiman L (2001) Random forests. Mach Learn 45:5–32

Breiman L, Friedman JH, Olshen RA, Stone CJ (1984) Classification and regression trees. Wadsworth and Brooks, Pacific Grove, CA

Burger S, Weilhammer K, Schiel F, Tillman HG (2000) Verbmobil data collection and annotation. In: Wahlster W. (ed) Verbmobil: foundations of speech-to-speech translations. Springer, Berlin, pp 537–549

Burkhardt F, van Ballegooy M, Engelbrecht K-P, Polzehl T, Stegmann J (2009) Emotion detection in dialog systems: applications, strategies and challenges. In: Proceedings of the ACII, Amsterdam, September 10–12, pp 684–689

Campbell N, Kashioka H, Ohara R (2005) No laughing matter. In: Proceedings of the Interspeech, Lisbon, September 12–14, pp 465–468

Chuang Z-J, Wu C-H (2004) Emotion recognition using acoustic features and textual content. In: Proceedings of ICME, Taipei, June 27–30, pp 53–56

Cover T, Hart P (1967) Nearest neighbor pattern classification. IEEE Trans Info Theoy 13:21–27

Cowie R, Douglas-Cowie E, Apolloni B, Taylor J, Romano A, Fellenz W (1999) What a neural net needs to know about emotion words. In: Mastorakis N (ed), Computational intelligence and applications. World Scientific Engineering Society Press, pp 109–114

Cowie R, Douglas-Cowie E, Savvidou S, McMahon E, Sawey M, Schröder M (2000) Feeltrace: an instrument for recording perceived emotion in real time. In: Proceedings of the ISCA Workshop on Speech and Emotion, Newcastle, Northern Ireland, September 5–7, pp 19–24

Cowie R, Douglas-Cowie E, Cox C (2005) Beyond emotion archetypes: databases for emotion modelling using neural networks. Neural Netw 18:371–388

Craggs R, Wood MM (2004) A categorical annotation scheme for emotion in the linguistic content of dialogue. In: Affective dialogue systems, proceedings of a tutorial and research workshop, Kloster Irsee, June 14–16, pp 89–100

Daubechies I (1990) The wavelet transform, time–frequency localization and signal analysis. TransIT 36(5):961–1005

Davis S, Mermelstein P (1980) Comparison of parametric representations for monosyllabic word recognition in continuously spoken sentences. IEEE Trans Acoust Speech Signal Process 29:917–919

Dellaert F, Polzin T, Waibel A (1996) Recognizing emotion in speech. In: Proceedings of the ICSLP, Philadelphia, PA, October 3–6, pp 1970–1973

Devillers L, Vasilescu I, Lamel L (2003) Emotion detection in task-oriented spoken dialogs. In: Proceedings of ICME 2003, IEEE, multimedia human-machine interface and interaction, Baltimore, MD, July 6–9, pp 549–552

Devillers L, Vidrascu L, Lamel L (2005) Challenges in real-life emotion annotation and machine learning based detection. Neural Netw, 18:407–422

Douglas-Cowie E, Cowie R, Sneddon I, Cox C, Lowry O, McRorie M, Martin J-C, Devillers L, Abrilan S, Batliner A, Amir N, Karpousis K (2007) The HUMAINE database: addressing the collection and annotation of naturalistic and induced emotional data. In: Paiva A, Prada R, Picard RW, (eds), Affective computing and intelligent interaction. Springer, Berlin, pp 488–500

Elliott C (1992) The affective reasoner: a process model of emotions in a multi-agent system. Ph.D. thesis, Dissertation, Northwestern University

Eyben F, Wöllmer M, Schuller B (2009) openear - introducing the munich open-source emotion and affect recognition toolkit. In: Proceedings of the ACII, Amsterdam, September 10–12, pp 576–581

Fairbanks G, Pronovost W (1939) An experimental study of the pitch characteristics of the voice during the expression of emotion. Speech Monogr, 6:87–104

Fernandez R, Picard RW (2003) Modeling drivers' speech under stress. Speech Commun 40: 145–159

Fiscus J (1997) A post-processing system to yield reduced word error rates: recognizer output voting error reduction (ROVER). In: Proceedings of the ASRU, Santa Barbara, CA, December 14–17, pp 347–352

Fleiss J, Cohen J, Everitt B (1969) Large sample standard errors of kappa and weighted kappa. Psychol Bull 72(5):323–327

Frick R (1985) Communicating emotion: the role of prosodic features. Psychol Bull 97:412–429
Fujisaki H (1992) Modelling the process of fundamental frequency contour generation. In: Tohkura Y, Vatikiotis-Bateson E, Sagisasaka Y, (eds), Speech perception, production and linguistic structure. IOS Press, Amsterdam, pp 313–328
Fukunaga K (1990) Introduction to statistical pattern recognition. Academic Press, London
Goertzel B, Silverman K, Hartley C, Bugaj S, Ross M (2000) The baby webmind project. In: Proceedings of the annual conference of the society for the study of artificial intelligence and the simulation of behaviour (AISB), Birmingham, April 17–20
Good I (1965) The estimation of probabilities: an essay on modern bayesian methods. MIT Press, Cambridge, MA
Grimm M, Kroschel K, Harris H, Nass C, Schuller B, Rigoll G, Moosmayr T (2007) On the necessity and feasibility of detecting a driver's emotional state while driving. In: Paiva A, Prada R, Picard RW, (eds), Affective computing and intelligent interaction. Springer, Berlin, pp 126–138
Grimm M, Kroschel K, Narayanan S (2008) The vera am mittag german audio-visual emotional speech database. In: Proceedings of the IEEE international conference on multimedia and expo (ICME), Hannover, Germany, June 23–26, pp 865–868
Hall MA (1998) Correlation-based feature selection for machine learning. Ph.D. thesis, Department of Computer Science, Waikato University, Hamilton, NZ
Hermansky H (1990) Perceptual linear predictive (plp) analysis for speech. J Acoust Soc Am (JASA), 87:1738–1752
Hermansky H, Sharma S (1998) Traps - classifiers of temporal patterns. In: Proceedings of the ICSLP, Sydney, November 30–December 04, pp 1003–1006
Hess W, Batliner A, Kießling A, Kompe R, Nöth E, Petzold A, Reyelt M, Strom V (1996) Prosodic modules for speech recognition and understanding in verbmobil. In: Sagisaka Y, Campell N, Higuchi N, (eds), Computing prosody. Approaches to a computational analysis and modelling of the prosody of spontaneous speech. Springer, New York, NY, pp 363–383
Hirst D, Cristo AD, Espesser R (2000) Levels of representation and levels of analysis for intonation. In: Horne M, (ed), Prosody : theory and experiment Kluwer, Dordrecht, pp 51–87
Hyvärinen A, Karhunen J, Oja E (2001) Independent component analysis. Wiley, New York, NY
Jain A, Zongker D (1997) Feature selection: evaluation, application and small sample performance. PAMI 19(2):153–158
Joachims T (1998) Text categorization with support vector machines: learning with many relevant features. In: Nédellec C, Rouveirol C, (eds), Proceedings of ECML-98, 10th European conference on machine learning. Springer, Heidelberg, pp 137–142
Johnstone T, Scherer KR (2000) Vocal communication of emotion. In: Lewis M, Haviland-Jones JM, (eds), Handbook of emotions, chapter 14. 2nd edn. Guilford Press, London
Jolliffe IT (2002) Principal component analysis. Springer, Berlin
Kießling A (1997) Extraktion und Klassifikation prosodischer Merkmale in der automatischen Sprachverarbeitung. Berichte aus der Informatik. Shaker, Aachen
Kwon O-W, Chan K, Hao J, Lee T-W (2003) Emotion recognition by speech signals. In: Proceedings of the Interspeech, Geneva, September 1–4, pp 125–128
Langley P, Iba W, Thompson K (1992) An analysis of Bayesian classifiers. In: Proceedings of the national conference on articial intelligence, San Jose, CA, pp 223–228
Lee C, Narayanan S, Pieraccini R (2001) Recognition of negative emotions from the speech signal. In: Proceedings of the ASRU, Madonna di Campiglio, December 9–13, no pagination
Lee CM, Narayanan SS (2005) Toward detecting emotions in spoken dialogs. IEEE Trans Speech Audio Process 13(2):293–303
Lee CM, Narayanan SS, Pieraccini R (2002) Combining acoustic and language information for emotion recognition. In: Proceedings of the Interspeech, Denver, September 16–20, pp 873–376
Lee CM, Yildirim S, Bulut M, Kazemzadeh A, Busso C, Deng Z, Lee S, Narayanan SS (2004) Emotion recognition based on phoneme classes. In: Proceedings of the Interspeech, Jeju Island, Korea, October 4–8, pp 889–892

Litman D, Forbes K (2003) Recognizing emotions from student speech in tutoring dialogues. In: Proceedings of the ASRU, Virgin Island, November 30–December 3, pp 25–30

Liu H, Liebermann H, Selker T (2003) A model of textual affect sensing using real-world knowledge. In: Proceedings of the 7th International conference on intelligent user interfaces (IUI 2003), Miami, Florida, USA, January 12–15, pp 125–132

Lovins JB (1968) Development of a stemming algorithm. Mech Transl Comput Linguist 11:22–31

Lugger M, Yang B, Wokurek W (2006) Robust estimation of voice quality parameters under real world disturbances. In: Proceedings of the ICASSP, Toulouse, May 15–19, pp 1097–1100, 2006

Makhoul J (1975) Linear prediction: a tutorial review. Proc IEEE 63:561–580

Martinez CA, Cruz A (2005) Emotion recognition in non-structured utterances for human-robot interaction. In: IEEE international workshop on robot and human interactive communication, August 13–15, pp 19–23, 2005

Matos S, Birring S, Pavord I, Evans D (2006) Detection of cough signals in continuous audio recordings using hidden Markov models. IEEE Trans Biomed Eng pp 1078–108

McGilloway S, Cowie R, Douglas-Cowie E, Gielen S, Westerdijk M, Stroeve S (2000) Approaching automatic recognition of emotion from voice: A rough benchmark. In: Proceedings of the ISCA workshop on speech and emotion, Newcastle, Northern Ireland, September 5–7, pp 207–212

Meyer D, Leisch F, Hornik K (2002) Benchmarking support vector machines. Report series no. 78, SFB Adaptive informations systems and management in economics and management science, Wien, Austria, 19 pp

Morrison D, Silva LCD (2007) Voting ensembles for spoken affect classification. J Netw Comput Appl 30:1356–1365

Morrison D, Wang R, Xu W, Silva LCD (2007) Incremental learning for spoken affect classification and its application in call-centres. Int J Intell Syst Tech: Appl 2:242–254

Mower E, Metallinou A, Lee C-C, Kazemzadeh A, Busso C, Lee S, Narayanan S (2009) Interpreting ambiguous emotional expressions. In: Proceedings of the ACII, Amsterdam, pp 662–669

Neiberg D, Elenius K, Laskowski K (2006) Emotion Recognition in Spontaneous Speech Using GMMs. In: Proceedings of the Interspeech, Pittsburgh, PA, September 17–21, pp 809–812

Nickerson RS (2000) Null hypothesis significance testing: a review of an old and continuing controversy. Psychol Methods 5:241–301

Noll AM (1967) Cepstrum pitch determination. J Acoust Soc Am (JASA), 14:293–309

Nöth E, Batliner A, Warnke V, Haas J, Boros M, Buckow J, Huber R, Gallwitz F, Nutt M, Niemann H (2002) On the use of prosody in automatic dialogue understanding. Speech Commun, 36:(1–2), pp 45–62

Nwe T, Foo S, Silva LD (2003) Speech emotion recognition using hidden Markov models. Speech Commun 41:603–623

Ortony A, Clore GL, Collins A (1988) The cognitive structure of emotions. Cambridge University Press, Cambridge

Pal P, Iyer A, Yantorno R (2006) Emotion detection from infant facial expressions and cries. In: Proceedings of ICASSP, Toulouse, May 15–19, pp 809–812

Pang B, Lee L, Vaithyanathan S (2002) Thumbs up? sentiment classification using machine learning techniques. In: Proceedings of the 2002 conference on empirical methods in natural language processing (EMNLP), Philadelphia, PA, July 6–7, pp 79–86

Pernegger T.V (1998) What's wrong with Bonferroni adjustment. Br Med J, 316:1236–1238

Petrushin V (1999) Emotion in speech: recognition and application to call centers. In: Proceedings of artificial neural networks in engineering (ANNIE '99), St. Louis, MO, November 7–10, pp 7–10

Polzin TS, Waibel A (2000) Emotion-sensitive human-computer interfaces. In: Proceedings of the ISCA workshop on speech and emotion, Newcastle, Northern Ireland, September 5–7, pp 201–206

Porter M (1980) An algorithm for suffix stripping. Program 14(3):130–137

Pudil P, Novovicova J, Kittler J (1994) Floating search methods in feature selection. Pattern Recogn Lett 15:1119–1125

Quinlan JR (1993) C4.5: Programs for machine learning. Morgan Kaufmann, San Francisco, CA

Rabiner LR (1977) On the use of autocorrelation analysis for pitch detection. IEEE Trans Acoust Speech Signal Process 25:24–33

Rahurkar MA, Hansen JHL (2003) Towards affect recognition: an ICA approach. In: Proceedings of 4th international symposium on independent component analysis and blind signal separation (ICA2003), Nara, April 1–4, pp 1017–1022

Rosenberg A, Binkowski E (2004) Augmenting the kappa statistic to determine interannotator reliability for multiply labeled data points. In: Dumais DMS, Roukos S, (eds), HLT-NAACL 2004: short papers. Association for Computational Linguistics, Boston, MA, pp 77–80

de Rosis F, Batliner A, Novielli N, Steidl S (2007) 'You are Sooo Cool, Valentina!' Recognizing social attitude in speech-based dialogues with an ECA. In: Paiva A, Prada R, Picard RW, (eds), Affective computing and intelligent interaction, Springer, Berlin, pp 179–190

Rumelhart D, Hinton G, Williams R (1986) Learning internal representations by error propagation. In: Rumelhart D, McClelland L, the PDP Research Group, (eds), Parallel distributed processes: exploration in the microstructure of cognition, vol 1. MIT Press, Cambridge, MA, pp 318–362

Russel JA (1997) How shall an emotion be called? In: Plutchik R, Conte HR (eds), Circumplex models of personality and emotions, chapter 9. American Psychological Association, Washington, DC, pp 205–220

Russell J, Bachorowski J, Fernandez-Dols J (2003) Facial and vocal expressions of emotion. Ann Rev Psychol 54:329–349

Salzberg S (1997) On comparing classifiers: pitfalls to avoid and a recommended approach. Data Min Knowl Discov, **1**(3), 317–328

Scherer KR (2003) Vocal communication of emotion: a review of research paradigms. Speech Commun 40:227–256

Scherer KR, Johnstone T, Klasmeyer G (2003) Vocal expression of emotion. In: Davidson RJ, Scherer KR, Goldsmith HH, (eds), Handbook of affective sciences, chapter 23. Oxford University Press, Oxford NY, pp 433–456

Schiel F (1999) Automatic phonetic transcription of non-prompted speech. In: Proceedings of the ICPhS, San Francisco, CA, August 1–7, pp 607–610

Schröder M, Devillers L, Karpouzis K, Martin J-C, Pelachaud C, Peter C, Pirker H, Schuller B, Tao J, Wilson I (2007) What should a generic emotion markup language be able to represent? In: Paiva A, Prada R, Picard RW, (eds), Affective computing and intelligent interaction. Springer, Berlin, pp 440–451

Schuller B, Rigoll G (2009) Recognising interest in conversational speech – comparing bag of frames and supra-segmental features. In: Proceedings of the Interspeech, Brighton, UK, September 6–10, pp 1999–2002

Schuller B, Rigoll G, Lang M (2003) Hidden Markov model-based speech emotion recognition. In: Proceedings of the ICASSP, Hong Kong, April 6–10, pp 1–4

Schuller B, Rigoll G, Lang M (2004) Speech emotion recognition combining acoustic features and linguistic information in a hybrid support vector machine-belief network architecture. In: Proceedings of the ICASSP, Montreal, QC, Canada, May 17–21, pp 577–580

Schuller B, Müller R, Lang M, Rigoll G (2005) Speaker independent emotion recognition by early fusion of acoustic and linguistic features within ensemble. In: Proceedings of the Interspeech, Lisbon, September 4–8, pp 805–808

Schuller B, Stadermann J, Rigoll G (2006a) Affect-robust speech recognition by dynamic emotional adaptation. In: Proceedings of speech prosody 2006, Dresden, May 2–5, no pagination

Schuller B, Köhler N, Müller R, Rigoll G (2006b) Recognition of interest in human conversational speech. In: Proceedings of the Interspeech, Pittsburgh, PA, September 17–21, pp 793–796

Schuller B, Batliner A, Seppi D, Steidl S, Vogt T, Wagner J, Devillers L, Vidrascu L, Amir N, Kessous L, Aharonson V (2007a) The relevance of feature type for the automatic classification of emotional user states: low level descriptors and functionals. In: Proceedings of the Interspeech, Antwerp, Belgium, August 27–31, pp 2253–2256

Schuller B, Seppi D, Batliner A, Meier A, Steidl S (2007b) Towards more reality in the recognition of emotional speech. In: Proceedings of the ICASSP, Honolulu, April 15–20, pp 941–944

Schuller B, Rigoll G, Can S, Feussner H (2008) Emotion sensitive speech control for human-robot interaction in minimal invasive surgery. In: Proceedings of the 17th International Symposium on robot and human interactive communication, RO-MAN 2008, Munich, Germany, August 1–3, pp 453–458

Schuller B, Müller R, Eyben F, Gast J, Hörnler B, Wöllmer M, Rigoll G, Höthker A, Konosu H (2009a) Being bored? Recognising natural interest by extensive audiovisual integration for real-life application. Image Vis Comput J, Special Issue on Vis Multimodal Anal Hum Spontaneous Behav 27:1760–1774

Schuller B, Batliner A, Steidl S, Seppi D (2009b) Emotion recognition from speech: putting ASR in the loop. In: Proceedings of ICASSP, Taipei, Taiwan. IEEE, April 19–24, pp 4585–4588

Schuller B, Steidl S, Batliner A (2009c) The INTERSPEECH 2009 emotion challenge. In: Proceedings of the Interspeech, Brighton, September 6–10, pp 312–315

Scripture E (1921) A study of emotions by speech transcription. Vox 31:179–183

Seppi D, Gerosa M, Schuller B, Batliner A, Steidl S (2008a) Detecting problems in spoken child-computer interaction. In: Proceedings of the 1st workshop on child, computer and interaction, Chania, Greece, October 23, no pagination

Seppi D, Batliner A, Schuller B, Steidl S, Vogt T, Wagner J, Devillers L, Vidrascu L, Amir N, Aharonson V (2008b) Patterns, prototypes, performance: classifying emotional user states. In: Proceedings of the Interspeech, Brisbane, September 22–26, pp 601–604

Shami M, Verhelst W (2007) Automatic classification of expressiveness in speech: a multi-corpus study. In: Müller C, (ed), Speaker classification II (Lecture notes in computer science / artificial intelligence) vol 4441. Springer, Heidelberg, pp 43–56

Siegle G (1995) The balanced affective word list project. http://www.sci.sdsu.edu/CAL/wordlist/ (accessed October 17, 2010)

Skinner E (1935) A calibrated recording and analysis of the pitch, force, and quality of vocal tones expressing happiness and sadness. Speech Monogr 2:81–137

Slaney M, McRoberts G (1998) Baby Ears: A Recognition System for Affective Vocalizations. In: Proceedings of the ICASSP, Seattle, WA, pp 985–988

Steidl S (2009) Automatic classification of emotion-related user states in spontaneous children's speech. Berlin. PhD thesis, Logos Verlag

Steidl S, Ruff C, Batliner A, Nöth E, Haas J (2004) Looking at the last two turns, I'd say this dialogue is doomed — measuring dialogue success. In: Sojka P, Kopeček I, Pala K, (eds), Text, speech and dialogue, 7th international conference, TSD 2004. Berlin, Heidelberg, pp 629–636

Steidl S, Levit M, Batliner A, Nöth E, Niemann H (2005) "Of all things the measure is man": automatic classification of emotions and inter-labeler consistency. In: Proceedings of ICASSP, Philadelphia, PA, May 12–15, pp 317–320

Steidl S, Schuller B, Batliner A, Seppi D (2009) The hinterland of emotions: facing the open-microphone challenge. In: Proceedings of ACII, Amsterdam, September 10–12, pp 690–697

Streit M, Batliner A, Portele T (2006) Emotions analysis and emotion-handling subdialogues. In: Wahlster W, (ed), SmartKom: foundations of multimodal dialogue systems. Springer, Berlin, pp 317–332

ten Bosch L (2003) Emotions, speech and the ASR framework. Speech Commun 40(1–2):213–225

Truong K, van Leeuwen D (2005) Automatic detection of laughter. In: Proceedings of the interspeech, Lisbon, Portugal, September 4–8, pp 485–488

Vapnik V (1995) The nature of statistical learning theory. Springer, Berlin

Ververidis D, Kotropoulos C (2006) Fast sequential floating forward selection applied to emotional speech features estimated on DES and SUSAS data collection. In: Proceedings of european signal processing Conference (EUSIPCO 2006), Florence, September 4–8, no pagination

Vlasenko B, Schuller B, Wendemuth A, Rigoll G (2007a) Combining frame and turn-level information for robust recognition of emotions within speech. In: Proceedings of Interspeech, Antwerp, Belgium, August 27–31, pp 2249–2252

Vlasenko B, Schuller B, Wendemuth A, Rigoll G (2007b) Frame vs. turn-level: emotion recognition from speech considering static and dynamic processing. In: Paiva A, Prada R, Picard RW, (eds), Affective computing and intelligent interaction. Springer, Berlin, pp 139–147

Vogt T, André E, Wagner J, Gilroy S, Charles F, Cavazza M (2009) Real-time vocal emotion recognition in artistic installations and interactive storytelling: experiences and lessons learnt from CALLAS and IRIS. In: Proceedings of the ACII, Amsterdam, September 10–12, pp 670–677

Wagner J, Vogt T, André (2007) A systematic comparison of different HMM designs for emotion recognition from acted and spontaneous speech. In: Paiva A, Prada R, Picard RW, (eds), Affective computing and intelligent interaction. Springer, Berlin, pp 114–125

Williams C, Stevens K (1972) Emotions and speech: some acoustic correlates. J Acoust Soc Am (JASA) 52:1238–1250

Wilting J, Krahmer E, Swerts M (2006) Real vs. acted emotional speech. In: Proceedings of Interspeech, Pittsburgh, PA, September 17–21, pp 805–808

Witten IH, Frank E (2005) Data mining: practical machine learning tools and techniques, 2nd Edn. Morgan Kaufmann, San Francisco, CA

Wöllmer M, Eyben F, Reiter S, Schuller B, Cox C, Douglas-Cowie E, Cowie R (2008) Abandoning emotion classes – towards continuous emotion recognition with modelling of long-range dependencies. In: Proceedings of Interspeech, Brisbane, September 22–26, pp 597–600

Wolpert D (1992) Stacked generalization. Neural Netw 5:241–259

Wu T, Khan F, Fisher T, Shuler L, Pottenger W (2005) Posting act tagging using transformation-based learning. In: Lin TY, Ohsuga S, Liau C-J, Hu X, Tsumoto S, (eds), Foundations of data mining and knowledge discovery. Springer, Berlin, pp 319–331

You M, Chen C, Bu J, Liu J, Tao J (2006) Emotion recognition from noisy speech. In: Proceedings of ICME, Toronto, ON, July 9–12, pp 1653–1656

Young S, Evermann G, Gales M, Hain T, Kershaw D, Liu X, Moore G, Odell J, Ollason D, Povey D, Valtchev V, Woodland P (2006) The HTK book. Cambridge University Engineering Department, for htk version 3.4 edition

Zeng Z, Pantic M, Roisman GI, Huang TS (2009) A Survey of affect recognition methods: audio, visual, and spontaneous expressions. IEEE Trans Pattern Anal Mach Intell 31(1):39–58

Zhe X, Boucouvalas A (2002) Text-to-emotion engine for real time internet communication. In: Proceedings of the international symposium on communication systems, networks, and DSPs. Staffordshire University, Stoke-on-Trent, July 15–17, pp 164–168

Zhou G, Hansen JHL, Kaiser J.F (2001) Nonlinear feature based classification of speech under stress. IEEE Trans Speech Audio Process 9:201–216

Image and Video Processing for Affective Applications

Maja Pantic and George Caridakis

Abstract Recent advances in the research area of affective computing have broadened the range of application areas of its findings, and additionally, as the state of the art advances in affective computing, other related research areas (computer vision, pattern recognition, etc.) discover new challenges that are related to image and video processing related to the task of automatic affective analysis. Although humans cope, relatively easily, with the task of perceiving facial expressions, gestural expressivity, and other visual cues involved in expressing emotion the automatic counterpart of the task is far from trivial. This chapter summarizes current research efforts in solving these problems and enumerates the scientific and engineering issues that arise in meeting these challenges toward emotion-aware systems.

1 The Problem Domain

Because of its practical importance and the theoretical interest of cognitive and medical scientists (Ekman et al., 2002; Pantic, 2005; Chang et al., 2006), machine analysis of facial expressions attracted the interest of many researchers. For exhaustive surveys of the related work, readers are referred to Samal and Iyengar (1992) for an overview of early works, Tian et al. (2005) and Pantic and Bartlett (2007) for surveys of techniques for detecting facial muscle actions, and Pantic and Rothkrantz (2000, 2000) for surveys of facial affect recognition methods. However, although humans detect and analyze faces and facial expressions in a scene with little or no effort, development of an automated system that accomplishes this task is rather difficult.

M. Pantic (✉)
Department of Computing, Imperial College, London, UK; Faculty of Electrical Engineering, Mathematics and Computer Science, University of Twente, Enschede, The Netherlands
e-mail: M.Pantic@imperial.ac.uk

P. Petta et al. (eds.), *Emotion-Oriented Systems*, Cognitive Technologies,
DOI 10.1007/978-3-642-15184-2_7, © Springer-Verlag Berlin Heidelberg 2011

1.1 Level of Description: Action Units and Emotions

Two main streams in the current research on automatic analysis of facial expressions consider facial affect (emotion) detection and facial muscle action (action unit) detection. These two streams stem directly from two major approaches to facial expression measurement in psychological research (Cohen, 2006): message and sign judgment. The aim of message judgment is to *infer* what underlies a displayed facial expression, such as affect or personality, while the aim of sign judgment is to *describe* the "surface" of the shown behavior, such as facial movement or facial component shape. Thus, a brow furrow can be judged as "anger" (Ekman, 2003; Kapoor et al., 2003) in a message-judgment and as a facial movement that lowers and pulls the eyebrows closer together in a sign-judgment approach. While message judgment is all about interpretation, sign judgment attempts to be objective, leaving inference about the conveyed message to higher order decision making.

FACS (Ekman and Friesen, 1969, 1978) provides an objective and comprehensive language for describing facial expressions and relating them back to what is known about their meaning from the behavioral science literature. Because it is comprehensive, FACS also allows for the discovery of new patterns related to emotional or situational states. For example, what are the facial behaviors associated with driver fatigue? What are the facial behaviors associated with states that are critical for automated tutoring systems, such as interest, boredom, confusion, or comprehension? Research based upon FACS has also shown that facial actions can show differences between those telling the truth and lying at a much higher accuracy level than naive subjects making subjective judgments of the same faces (Cohn and Schmidt, 2004; Fasel et al., 2004).

It is not surprising, therefore, that automatic Action Units (AU) coding in face images and face image sequences attracted the interest of computer vision researchers. Historically, the first attempts to encode AUs in images of faces in an automatic way were reported by Bartlett et al. (2006), Lien et al. (1998), and Pantic et al. (1998). These three research groups are still the forerunners in this research field. The focus of the research efforts in the field was first on automatic recognition of AUs in either static face images or face image sequences picturing facial expressions produced on command. Several promising prototype systems were reported that can recognize deliberately produced AUs in either (near-) frontal-view face images (Anderson and McOwan, 2006; Samal and Iyengar, 1992; Pantic and Rothkrantz, 2003) or profile-view face images (Pantic and Rothkrantz, 2003; Pantic and Patras, 2005). These systems employ different approaches including expert rules and machine learning methods such as neural networks and use either feature-based image representations (i.e., use geometric features like facial points, see Sect. 2.3) or appearance-based image representations (i.e., use texture of the facial skin including wrinkles and furrows, see Sect. 2.3).

One of the main criticisms that these works received from both cognitive and computer scientists is that the methods are not applicable in real-life situations, where subtle changes in facial expression typify the displayed facial behavior rather than the exaggerated changes that typify posed expressions. Hence, the focus of the research in the field started to shift to automatic AU recognition in spontaneous facial expressions (produced in a reflex-like manner). Several works have recently emerged on machine analysis of AUs in spontaneous facial expression data (e.g., Cohn, 2006; Bartlett et al., 1999; Valstar and Pantic, 2006). These methods employ probabilistic, statistical, and ensemble learning techniques, which seem to be particularly suitable for automatic AU recognition from face image sequences (see, e.g., Tian et al., 2001; Lien et al., 1998).

1.2 Facial Expression Configuration and Dynamics

When it comes to research on automatic AU coding, automatic recognition of facial expression configuration (in terms of AUs constituting the observed expression) has been the main focus of the research efforts in the field. However, both the configuration and the dynamics of facial expressions (i.e., the timing and the duration of various AUs) are important for interpretation of human facial behavior. The body of research in cognitive sciences, which argues that the dynamics of facial expressions are crucial for the interpretation of the observed behavior, is ever growing (Ekman et al., 1993; Lee and Kim, 1999). Facial expression temporal dynamics are essential for categorization of complex psychological states like various types of pain and mood; they represent a critical factor for interpretation of social behaviors like social inhibition, embarrassment, amusement, and shame, and they are a key parameter in differentiation between posed and spontaneous facial displays (Ekman et al., 1993). For instance, spontaneous smiles are smaller in amplitude, longer in total duration, and slower in onset and offset time than posed smiles (e.g., a polite smile) (Cohn and Schmidt, 2004). Another study showed that spontaneous smiles, in contrast to posed smiles, can have multiple apexes (multiple rises of the mouth corners – AU12) and are accompanied by other AUs that appear either simultaneously with AU12 or follow AU12 within 1 s (Cohn et al., 2004). Similarly, it has been shown that the differences between spontaneous and deliberately displayed brow actions (AU1, AU2, AU4) are in the duration and the speed of onset and offset of the actions and in the order and the timing of actions' occurrences (Valstar and Pantic, 2006).

In spite of these findings, the vast majority of the past work in the field does not take dynamics of facial expressions into account when analyzing shown facial behavior. Some of the past work in the field has used aspects of temporal dynamics of facial expression such as the speed of a facial point displacement or the persistence of facial parameters over time (e.g., Lien et al., 1998). However, only three recent studies analyze explicitly the temporal dynamics of facial expressions. These studies explore automatic segmentation of AU activation into temporal segments (neutral, onset, apex, offset) in frontal- (Pantic and Bartlett, 2007; Tian et al., 2005) and profile-view (Pantic and Patras, 2005) face videos.

1.3 Facial Expression Intensity and Context Dependency

Facial expressions can vary in intensity. By intensity we mean the relative degree of change in facial expression as compared to a relaxed, neutral facial expression.

It has been experimentally shown that the expression-decoding accuracy and the perceived intensity of the underlying affective state vary linearly with the physical intensity of the facial display (Gu and Ji, 2004). Hence, explicit analysis of expression intensity variation is very important for accurate expression interpretation and is also essential to the ability to distinguish between spontaneous and posed facial behavior discussed in the previous sections. While FACS provides a 5-point intensity scale to describe AU intensity variation and enable manual quantification of AU intensity (Ekman and Friesen, 1978), fully automated methods that accomplish this task are yet to be developed. However, first steps toward this goal have been made. Automatic coding of intensity variation was explicitly compared to manual coding in Bartlett et al. (1999). They found that the distance to the separating hyperplane in their learned classifiers correlated significantly with the intensity scores provided by expert FACS coders.

Rapid facial signals do not usually convey exclusively one type of messages. For instance, squinted eyes may be interpreted as sensitivity of the eyes to bright light if this action is a reflex (a manipulator), as an expression of disliking if this action has been displayed when seeing someone passing by (affective cue), or as an illustrator of friendly anger on friendly teasing if this action has been posed (in contrast to being unintentionally displayed) during a chat with a friend, to mention just a few possibilities. As already mentioned in Sect. 1.3, to interpret an observed facial expression, it is important to know the context in which the observed expression has been displayed – where the expresser is (outside, inside, in the car, in the kitchen, etc.), what his or her current task is, are other people involved, and who the expresser is. Knowing the expresser is particularly important as individuals often have characteristic facial expressions and may differ in the way certain states (other than the basic emotions) are expressed. Since the problem of context-sensing is extremely difficult to solve (if possible at all) for a general case, pragmatic approaches (e.g., activity/application- and user-centered approach) should be taken when learning the grammar of human facial behavior (Pantic et al., 1998; Pantic and Patras, 2006). However, except for a few works on user-profiled interpretation of facial expressions like those of Fasel et al. (2004) and Pantic and Rothkrantz (2004a), virtually all existing automated facial expression analyzers are context insensitive.

1.4 Facial Expression Databases

To develop and evaluate facial behavior analyzers capable of dealing with different dimensions of the problem space as defined above, large collections of training and test data are needed (Pantic and Rothkrantz, 2000; Tian et al., 2001).

A complete overview of existing, publicly available data sets that can be used in research on automatic facial expression analysis is given by Pantic and Bartlett (2007). We will provide here a description of two relevant facial expression databases: the Cohn–Kanade database (Juslin and Scherer, 2005), which is the most widely used database in research on automated facial expression analysis, and the MMI facial expression database (Pantic et al., 2005a; Pantic, 2006), which represents the most comprehensive, online reference set of face images and videos of both deliberate and spontaneously displayed facial expressions.

2 The State of the Art

Although humans detect and analyze faces and facial expressions in a scene with little or no effort, development of an automated system that accomplishes this task is rather difficult. There are several related problems (Pantic et al., 2006). The first is to find faces in the scene independent of clutter, occlusions, and variations in head pose and lighting conditions. Then, geometric facial features such as facial salient points (e.g., the mouth corners) or parameters of an appearance-based facial model (e.g., parameters of a fitted active appearance model) should be extracted from the regions of the scene that contain faces. The system should perform this accurately, in a fully automatic manner and preferably in real time. Eventually, the extracted facial information should be interpreted in terms of facial signals (winks, blinks, smiles, affective states, cognitive states, moods) in a context-dependent (personalized, task-, situation-, and application-dependent) manner. This section summarizes current research efforts in solving these problems and enumerates the scientific and engineering issues that arise in meeting these challenges.

2.1 Face Detection

The first step in facial information processing is face detection, i.e., identification of all regions in the scene that contain a human face. The problem of *finding faces* should be solved regardless of clutter, occlusions, and variations in head pose and lighting conditions. The presence of non-rigid movements due to facial expression and a high degree of variability in facial size, color, and texture make this problem even more difficult. Numerous techniques have been developed for face detection in still images (Wierzbicka, 1993; Larsen and Diener, 1992). Arguably the most commonly employed face detector in automatic facial expression analysis is the real-time face detector proposed by Viola and Jones (2004).

2.2 Facial Feature Extraction

After the presence of a face has been detected in the observed scene, the next step is to extract the information about the displayed facial signals. The problem of *facial feature extraction* from regions in the scene that contain a human face may be divided into at least three dimensions (Pantic et al., 2006):

(a) Is temporal information used?
(b) Are the features holistic (spanning the whole face) or analytic (spanning subparts of the face)?
(c) Are the features view- or volume-based (2D/3D)?

Most of the existing facial expression analyzers are directed toward 2D spatiotemporal facial feature extraction. The usually extracted facial features are either *geometric features* such as the shapes of the facial components (eyes, mouth, etc.) and the locations of facial fiducial points (corners of the eyes, mouth, etc.) or *appearance features* representing the texture of the facial skin including wrinkles, bulges, and furrows. Typical examples of geometric feature based methods are those of Gokturk et al. (2002), who used 19-point face mesh; Chang et al. (2006), who used a shape model defined by 58 facial landmarks; and Pantic et al. (Pantic and Rothkrantz, 2003; Pantic and Bartlett, 2007; Pantic and Patras, 2005; Tian et al., 2005), who used a set of facial characteristic points visible in either frontal or profile view of the face. Typical examples of *hybrid*, geometric and appearance feature based, methods are those of Tian et al. (2005), who used shape-based models of eyes, eyebrows, and mouth and transient features like crows-feet wrinkles and nasolabial furrow, and of Zhang and Ji (2005), who used 26 facial points around the eyes, eyebrows, and mouth and the same transient features as Tian et al. (2005). Typical examples of appearance feature based methods are those of Bartlett et al. (1999), Anderson and McOwan (2006) and Lien et al. (1998), who used Gabor wavelets; Anderson and McOwen (2006), who used a holistic, monochrome, spatial-ratio face template; and Valstar et al. (2006), who used temporal templates.

To illustrate geometric facial feature detection and tracking, the methods developed by Vukadinovic and Pantic (2005) for automatic point localization and by Patras and Pantic (2004) for facial point tracking will be shortly explained.

It has been reported that methods based on geometric features are often outperformed by those based on appearance features using, e.g., Gabor wavelets or eigenfaces (Anderson and McOwan, 2006). Certainly, this may depend on the classification method and/or machine learning approach which takes the features as input. Recent studies like that of Pantic and Patras (2005) and Valstar and Pantic (2006) show that in some cases geometric features can outperform appearance-based ones. Yet, it seems that using both geometric and appearance features might be the best choice in the case of certain facial expressions (Pantic and Patras, 2005).

2.3 Facial Muscle Action Coding

As already mentioned in Sect. 2.1, two main streams in the current research on automatic analysis of facial expressions consider facial affect (emotion) detection and facial muscle action detection such as the AUs defined in FACS (Ekman and Friesen, 1969, 1978). Although FACS provides a good foundation for AU coding of face images by human observers, achieving AU recognition by a computer is not an easy task. A problematic issue is that AUs can occur in more than 7,000 different complex combinations, causing bulges (e.g., by the tongue pushed under one of the lips) and various in- and out-of-image plane movements of permanent facial features (e.g., jetted jaw) that are difficult to detect in 2D face images. Historically, the first attempts to encode AUs in images of faces in an automatic way were reported by Bartlett et al. (2006), Lien et al. (1998), and Pantic et al. (1998). These three research groups are still the forerunners in this research field.

Pantic and her colleagues reported on multiple efforts aimed at automating the analysis of facial expressions in terms of facial muscle actions that constitute the expressions. The majority of this previous work concerns geometric feature based methods for automatic FACS coding of face images. Early work was aimed at AU coding in static face images (Pantic and Rothkrantz, 2003) while more recent work addressed the problem of automatic AU coding in face video (Pantic and Bartlett, 2007; Tian et al., 2005; Pantic and Patras, 2005; Valstar and Pantic, 2006). Based upon the tracked movements of facial characteristic points, as discussed in Sect. 2.3, Pantic and her colleagues mainly experimented with rule-based (Pantic and Bartlett, 2007; Pantic and Patras, 2005) and support vector machine based methods (Tian et al., 2005; Valstar and Pantic, 2006), for recognition of AUs in either near frontal-view or near profile-view face image sequences. As already mentioned in Sect. 2.2, automatic recognition of facial expression configuration (in terms of AUs constituting the observed expression) has been the main focus of the research efforts in the field. In contrast to the methods developed elsewhere, which thus focus onto the problem of spatial modeling of facial expressions, the methods proposed by Pantic and her colleagues address the problem of temporal modeling of facial expressions as well. In other words, these methods are very suitable for encoding temporal activation patterns (onset → apex → offset) of AUs shown in an input face video. This is of importance for there is now a growing body of psychological research that argues that temporal dynamics of facial behavior (i.e., the timing and the duration of facial activity) is a critical factor for the interpretation of the observed behavior (see Sect. 2.2). Black and Yacoob (1997) presented the earliest attempt to automatically segment prototypic facial expressions of emotions into onset, apex, and offset components. To the best of our knowledge, the only systems to date for explicit recognition of temporal segments of AUs are the ones by Pantic and colleagues (Pantic and Bartlett, 2007; Tian et al., 2005; Pantic and Patras, 2005; Valstar and Pantic, 2006). A short explanation of the methods will follow.

Appearance-based approaches to AU recognition such as the ones by Kapoor et al. (2003), Valstar et al. (2006), and Bartlett and colleagues (e.g., Anderson and McOwan, 2006; Lien et al., 1998; Bartlett et al., 1999) differ from those of Pantic and colleagues (e.g., Pantic and Rothkrantz, 2003; Pantic and Bartlett, 2007) and Tian et al. (2005), in that learning the appearance of any AU is based on a set of labeled training data. Hence the limiting factor in appearance-based machine learning approaches is having enough of various labeled examples for a robust system. Previous explorations of this idea showed that, given accurate 3D alignment, at least 50 examples are needed for moderate performance (in the 80% range), and over 200 examples are needed to achieve high precision (Pantic, 2006). An example of appearance-based approaches to AU recognition is the system of Bartlett et al.

3 Facial Affect Recognition

To interpret someone's behavioral cues, including emotional states, people rely mainly on shown facial expressions (Ambadar et al., 2005; Kapoor et al., 2003), and it is not surprising, therefore, that the majority of efforts in affective computing concern automatic analysis of facial displays. For exhaustive surveys of studies on machine analysis of facial affect, readers are referred to Pantic et al. (2006), Pantic and Rothkrantz (2000), Tian et al. (2001) and Pantic (2006). These surveys indicate that the capabilities of currently existing facial affect analyzers are rather limited. More specifically, current facial affect analyzers

- handle only a small set of volitionally displayed prototypic facial expressions of six basic emotions,
- do not perform a context-sensitive analysis (either user-, or environment-, or task-dependent analysis) of the observed facial behavior,
- do not analyze extracted facial expression information on different time scales (i.e., short pre-segmented videos are only handled) – consequently, inferences about the expressed mood and attitude (larger time scales) cannot be made, and
- adopt strong assumptions (i.e., the systems can handle only portraits or nearly frontal views of faces with no facial hair or glasses, recorded under constant illumination and displaying exaggerated prototypic expressions of emotions).

Automatic detection of the six basic emotions under these assumptions, that is, in posed, controlled displays, can be done with reasonably high accuracy. However, detecting these facial expressions in the less constrained environments of real applications is a much more challenging problem which is just beginning to be explored. There have been just a few such tentative efforts aimed at detection of cognitive and psychological states like interest (Ekman and Rosenberg, 2005), pain (Bartlett et al., 1999), and fatigue (Goleman, 1995). An example is the pain detector of Bartlett et al.

Also an attempt to discern spontaneous from volitionally displayed facial behavior has been reported (Valstar and Pantic, 2006). Description of the method will be provided.

3.1 Machine Analysis of Facial Expressions: Challenges

Automating the analysis of facial expressions is important to realize more natural, context-sensitive (e.g., affective) human–computer interaction, to advance studies on human emotion and affective computing, and to boost numerous applications in fields as diverse as security, medicine, and education. Although most of the facial expression analyzers developed so far target human facial affect analysis and attempt to recognize a small set of prototypic emotional facial expressions like happiness and anger (Pantic et al., 1998; Pantic, 2006), some progress has been made in addressing a number of other scientific challenges that are considered essential for realization of machine understanding of human facial behavior. First of all, the research on automatic detection of facial muscle actions, which produce facial expressions, witnessed a significant progress in the past years. A number of promising prototype systems have been proposed recently that can recognize up to

27 AUs (from a total of 44 AUs) in either (near-) frontal-view or profile-view face image sequences (Tian et al., 2001; Pantic, 2006). Further, although the vast majority of the past work in the field does not make an effort to explicitly analyze the properties of facial expression temporal dynamics, a few approaches to automatic segmentation of AU activation into temporal segments (neutral, onset, apex, offset) have been recently proposed (e.g., Pantic and Bartlett, 2007; Pantic and Patras, 2005; Tian et al., 2005). Also, even though most of the past work on automatic facial expression analysis is aimed at the analysis of posed (deliberately displayed) facial expressions, a few efforts were recently reported on machine analysis of spontaneous facial expressions (e.g., Cohn, 2006; Bartlett et al., 1999; Valstar and Pantic, 2006). In addition, exceptions from the overall state of the art in the field include a few works toward detection of attitudinal and non-basic affective states such as interest (Ekman and Rosenberg, 2005), pain (Bartlett et al., 1999), and fatigue (Goleman, 1995); a few works on context-sensitive (user-profiled) interpretation of facial expressions (El Kaliouby and Robinson, 2004; Pantic and Rothkrantz, 2004); and an attempt to explicitly discern in an automatic way spontaneous from volitionally displayed facial behavior (Valstar and Pantic, 2006). However, many research questions raised in Sect. 2.2 remain unanswered and a lot of research has yet to be done.

When it comes to automatic AU detection, existing methods do not yet recognize the full range of facial behavior (i.e., all 44 AUs defined in FACS).

Existing methods for machine analysis of facial expressions discussed throughout this chapter assume that the input data are near frontal- or profile-view face image sequences showing facial displays that always begin with a neutral state. In reality, such assumption cannot be made.

If we consider the state of the art in face detection and facial feature localization and tracking, noisy and partial data should be expected. A facial expression analyzer should be able to deal with these imperfect data and to generate its conclusion so that the certainty associated with it varies with the certainty of face and facial point localization and tracking data.

Another related issue that should be addressed is how to include information about the context (environment, user, user's task) in which the observed expressive behavior was displayed so that a context-sensitive analysis of facial behavior can be achieved.

4 Machine Analysis of Body Gestures

Gesture and sign language recognition has gathered abundant attention in the recent years and the research area has developed an adequate relevant literature. Several approaches have been proposed and tested on a variety of data sets. An extensive review of these techniques is presented in Ong and Ranganath (2005) and Ying and Huang (2001). The first focuses mainly on sign language recognition and classification issues while examining closely hand localization and tracking, and various feature extraction related to automatic analysis of manual signing. In addition to the previous, they examine the linguistic aspect of sign language and non-manual signals and how they would be incorporated in the sign language recognition chain. On the other hand, Wu and Huang (Ying and Huang, 2001) delve more into publications related to hand modeling (shape analysis, kinematics chain and dynamics) and computer vision and pattern recognition issues associated to hand localization and extracting features from image sequences.

One of the most common approaches is to extract features from the input signal and use these features as input to a fine-tuned HMM. Perrin et al. (2004) track finger gestures using laser light and compute three features based on the finger's coordinates. Starner et al. (1998) present two systems based on head- and desk-mounted cameras feeding HMMs with uniform architecture. For each case two experiments were performed by varying the feature set used for classifying the gestures. Vogler and Metaxas (1998) developed a system based on parallel HMMs to recognize American SL gestures with a 3D camera system.

Also variations of the previous group have been widely presented. Hossain and Jenkin (2005) present two variations of HMM, implicit and explicit temporal information encoded, in order to recognize a single gesture type (hand-raise). Lee et al. (1998) propose another HMM variation enhanced by a gesture spotting network for calculating the likelihood threshold for "pick a winner" situations. Lee implemented the PowerGesture which recognizes 10 PowerPoint continuous commands with an average detection rate of 98% and recognition rate of 93%. Ozer et al. (2005) utilize one HMM per articulated body part. Their main focus is on the real-time aspect of the overall system: their image processing modules feed a graph matching module; this forms the input of the system. Wilson and Bobick (1999) introduce parametric HMMs to cope with gesture variations.

Other approaches have adopted other machine learning and artificial intelligence techniques. Juang and Ku (2005) present FTRFN (fuzzified TSK-type recurrent fuzzy network) and they test the proposed system on 10 trajectories achieving an average of 92% recognition rate. Mu-Chun (2000) presents a fuzzy rule based on hyperrectangular composite neural networks (HRCNNs) for selecting models. Hong et al. (2000) perform 2D gesture recognition using manually constructed finite state machines and have promising results reaching an average of 92% on two gesture data sets, one based on hand gestures and one based on mouse gestures, although the data sets seem quite small (three and four classes, respectively). Wong and Cipolla (2006) achieved 80–93% recognition rate over nine quite elementary gestures by adopting a sparse Bayesian classifier. As inputs to the classifier they utilized motion gradient orientation features over continuous video streams. Yang et al. (2002) base their algorithm on motion trajectories or ensembles of them which feed a time delay neural network. Huang et al. (1998) present an isolated gesture recognition system, which uses as input the monocular sequence of dynamic single-handed gesturing by dynamic time warping.

Finally, there have been some approaches combining more than one technique. Mantyla et al. (2000) present a system for static gestures recognition using a self-organizing mapping scheme of Kohonen while a hidden Markov model is used for recognizing dynamic gestures. The input of the two systems was acquired by acceleration sensors attached to a mobile device. These two systems (SOM and HMM) are not combined in any way but each one is utilized in different gesture types. Black and Jepson (1998) present an extension of the "condensation" algorithm in which gestures are modeled as temporal trajectories of the velocity of the tracked hands. Fang et al. (2001) present an additional layer enhancing the HMM architecture with SOFM and improving their recognition rate by 5%. In a more recent work, the same group of researchers introduced a fuzzy decision tree in an attempt to reduce the search space of recognized classes without loss of accuracy.

References

Ambadar Z, Schooler J, Cohn JF (2005) Deciphering the enigmatic face: the importance of facial dynamics in interpreting subtle facial expressions. Psychol Sci 16(5):403–410

Ambady N, Rosenthal R (1992) Thin slices of expressive behavior as predictors of interpersonal consequences: a meta-analysis. Psychol Bull 111(2):256–274

Anderson K, McOwan PW (2006) A real-time automated system for recognition of human facial expressions. IEEE Trans Syst Man Cybern B 36(1):96–105

Bartlett MS, Hager JC, Ekman P, Sejnowski TJ (1999) Measuring facial expressions by computer image analysis. Psychophysiology 36(2):253–263

Bartlett MS, Littlewort G, Frank MG, Lainscsek C, Fasel I, Movellan J (2006) Fully automatic facial action recognition in spontaneous behavior. In: Seventh IEEE international conference on automatic face and gesture recognition (FG 2006), April 10–12, Southampton, UK, pp 223–230

Bartlett MS, Viola PA, Sejnowski TJ, Golomb BA, Larsen J, Hager JC, Ekman P (1996) Classifying facial actions. Adv Neural Inf Process Syst 8:823–829

Black MJ, Jepson AD (1998) Recognizing temporal trajectories using the condensation algorithm. In: Proceedings of the 3rd international conference on Face and Gesture Recognition FG, IEEE Computer Society, Washington, DC

Black M, Yacoob Y (1997) Recognizing facial expressions in image sequences using local parameterized models of image motion. Comput Vis 25(1):23–48

Bobick AF, Wilson AD (1997) ªA state-based approach to the representation and recognition of gesture. IEEE Trans Pattern Anal Mach Intell 19(12):1325–1337

Chang Y, Hu C, Feris R, Turk M (2006) Manifold based analysis of facial expression. J Image Vis Comput 24(6):605–614

Cohen MM (2006) Perspectives on the face. Oxford University Press, Oxford, UK

Cohn JF (2006) Foundations of human computing: facial expression and emotion. In: Proceedings of the ACM international conference on Multimodal Interfaces, Banff, Canada, November 2–4, pp 233–238

Cohn JF, Reed LI, Ambadar Z, Xiao J, Moriyama T (2004) Automatic analysis and recognition of brow actions in spontaneous facial behavior. In: Proceedings of the IEEE international conference on systems, man and cybernetics, The Hague, Netherlands, October 10–13, pp 610–616

Cohn JF, Schmidt KL (2004) The timing of facial motion in posed and spontaneous smiles. J Wavelets Multiresolution Inf Process 2(2):121–132

Ekman P (2003) Darwin, deception, and facial expression. Ann N Y Acad Sci 1000:205–221

Ekman P, Friesen WF (1969) The repertoire of nonverbal behavioral categories – origins, usage, and coding. Semiotica 1:49–98

Ekman P, Friesen WF (1978) Facial action coding system. Consulting Psychologist Press, Palo Alto, CA

Ekman P, Friesen WF, Hager JC (2002) Facial action coding system. A Human Face, Salt Lake City

Ekman P, Huang TS, Sejnowski TJ Hager JC (eds) (1993) NSF understanding the face.A Human Face Store, Salt Lake City, (see Library)

Ekman P, Rosenberg EL, (eds) (2005) What the face reveals: basic and applied studies of spontaneous expression using the FACS. Oxford University Press, Oxford, UK

El Kaliouby R, Robinson P (2004) Real-time inference of complex mental states from facial expressions and head gestures. Proc Int Conf Comput Vis Pattern Recogn 3:154

Fang G, Gao W, Ma J (2001) Signer-independent sign language recognition based on SOFM/HMM, Recognition, Analysis, and Tracking of Faces and Gestures in Real-Time Systems, 2001. In: Proceedings of the IEEE ICCV Workshop on RATFG-RTS'01, Vancouver, Canada, July 13, pp 90–95

Fasel B, Monay F, Gatica-Perez D (2004) Latent semantic analysis of facial action codes for automatic facial expression recognition. In: Proceedings of the 6th ACM SIGMM international workshop on Multimedia Information Retrieval, New York, NY, October 10–16, pp 181–188

Fridlund AJ (1997) The new ethology of human facial expression. In: Russell JA, Fernandez-Dols JM (eds) The psychology of facial expression. Cambridge University Press, Cambridge, MA, pp 103–129

Gokturk SB, Bouguet JY, Tomasi C, Girod B (2002) Model-based face tracking for view independent facial expression recognition. In: 5th IEEE international conference on automatic face and gesture recognition (FGR 2002), Washington, DC, May 20–21, pp 272–278

Goleman D (1995) Emotional intelligence. Bantam Books, New York, NY

Gu H, Ji Q (2004) An automated face reader for fatigue detection. In: Sixth IEEE international conference on automatic face and gesture recognition (FGR 2004), IEEE Computer Society, Seoul, Korea, May 17–19, pp 111–116

Hong P, Turk M, Huang TS (2000) Gesture modeling and recognition using finite state machines. In: Proceedings of the 4th IEEE international conference and Gesture Recognition, Mar 2000, Grenoble

Hossain M, Jenkin M (2005) Recognizing hand-raising gestures using HMM. In: Proceedings of the 2nd Canadian conference on Computer and Robot Vision (CRV′05) – vol 00 (9–11 May 2005). CRV, IEEE Computer Society, Washington, DC, pp 405–412

Huang Y, Zhu Y, Xu G, Zhang H (1998) Spatial-temporal features by image registration and warping for dynamic gesture recognition. In: IEEE international conference on Systems, Man, and Cybernetics, vol 5, 11–14 Oct 1998, San Diego, pp 4498–4503

Juang CF, Ku K-C (2005 Aug) A recurrent fuzzy network for fuzzy temporal sequence processing and gesture recognition. IEEE Trans Syst Man Cybern B 35(4):646–658

Juslin PN, Scherer KR (2005) Vocal expression of affect. In: Harrigan J, Rosenthal R, Scherer K (eds) The new handbook of methods in nonverbal behavior research. Oxford University Press, Oxford

Kanade T, Cohn JF, Tian Y (2000) Comprehensive database for facial expression analysis. In: 4th IEEE international conference on automatic face and gesture recognition (FGR 2000), IEEE Comptuer Society, Grenoble, France, March 26–30, pp 46–53

Kapoor A, Qi Y, Picard RW (2003) Fully automatic upper facial action recognition. In: Proceedings of the IEEE international workshop on Analysis and Modeling of Faces and Gestures, Nice, France, pp 195–202

Larsen RJ, Diener E (1992) Promises and problems with the circumplex model of emotion. Emotion 13:25–59. In: Clark MS (ed) Review of personality and social psychology. Sage, Newbury Park, CA

Lee C, Ghyme S, Park C, Wohn K (1998) The control of avatar motion using hand gesture. In: Proceedings of the ACM symposium on Virtual Reality Software and Technology (Taipei, Taiwan, 2–5 Nov 1998). VRST ′98, ACM Press, New York, NY, pp 59–65

Lee H-K, Kim JH (1999) ªAn HMM-based threshold model approach for gesture recognitionº. IEEE Trans Pattern Anal Mach Intell 21(10):961–973

Li SZ, Jain AK, (eds) (2005) Handbook of face recognition. Springer, New York, NY

Lien JJJ, Kanade T, Cohn JF, Li CC (1998) Subtly different facial expression recognition and expression intensity estimation. In: Proceedings of the IEEE international conference on Computer Vision and Pattern Recognition, Research Triangle Park, North Carolina, pp 853–859

Mantyla VM, Mantyjarvi J, Seppanen T, Tuulari E (2000) Hand gesture recognition of a mobile device user. In: IEEE international conference on Multimedia and Expo (ICME 2000), New York, NY, vol 1, pp 281–284

Mu-Chun S (2000) A fuzzy rule-based approach to spatio-temporal hand gesture recognition. IEEE Trans Syst Man Cybern C Appl Rev 30(2):276–281

Ong SCW, Ranganath S (2005) Automatic sign language analysis: a survey and the future beyond lexical meaning. IEEE Trans Pattern Anal Mach Intell 27(6):873–891

Ozer IB, Lu T, Wolf W (2005) Design of a real-time gesture recognition system: high performance through algorithms and software. Signal Process Mag IEEE 22(3):57–64

Pantic M (2005) Affective computing. In: Pagani M (ed) Encyclopedia of multimedia technology and networking, vol 1. Idea Group Reference, Hershy, PA, pp 8–14

Pantic M (2006) Face for ambient interface. Lect Notes Artif Intell 3864:35–66

Pantic M, Bartlett MS (2007) Machine analysis of facial expressions. In: Kurihara K (ed) Face recognition. Advanced Robotic Systems, Vienna, Austria

Pantic M, Patras I (2005) Detecting facial actions and their temporal segments in nearly frontal-view face image sequences. In: Proceedings of the IEEE international conference on systems, Man and Cybernetics, Salerno, Italy, pp 3358–3363

Pantic M, Patras I (2006) Dynamics of facial expression: recognition of facial actions and their temporal segments from face profile image sequences. IEEE Trans Syst Man Cybern B 36(2):433–449

Pantic M, Pentland A, Nijholt A, Huang TS (2006) Human Computing and machine understanding of human behaviour: a Survey. In: Proceedings of the ACM international conference Multimodal Interfaces, Banff, Canada, pp 239–248

Pantic M, Rothkrantz LJM (2000) Automatic analysis of facial expressions – the state of the art. IEEE Trans Pattern Anal Mach Intell 22(12):1424–1445

Pantic M, Rothkrantz LJM (2003) Toward an affect-sensitive multimodal human-computer interaction. Proc IEEE 91(9):1370–1390

Pantic M, Rothkrantz LJM (2004a) Facial action recognition for facial expression analysis from static face images. IEEE Trans Syst Man Cybern B 34(3):1449–1461

Pantic M, Rothkrantz LJM (2004b) Case-based reasoning for user-profiled recognition of emotions from face images. In: Proceedings of the 2004 IEEE international conference on multimedia & expo, ICME 2004, IEEE, Taipei, Taiwan, June 27–30, pp 391–394

Pantic M, Rothkrantz LJM, Koppelaar H (1998) Automation of non-verbal communication of facial expressions. In: Proceedings of the conference on Euromedia '98, De Montfort University, Leicester, UK, January 5–7, pp 86–93

Pantic M, Sebe N, Cohn JF, Huang TS (2005) Affective multimodal human-computer interaction. In: Proceedings of the ACM international conference on Multimedia, Hilton, Singapore, pp 669–676

Pantic M, Valstar MF, Rademaker R, Maat L (2005) Web-based database for facial expression analysis. In: Proceedings of the IEEE international conference on Multimedia and Expo, pp 317–321. www.mmifacedb.com

Patras I, Pantic M (2004) Particle filtering with factorized likelihoods for tracking facial features. In: Sixth IEEE international conference on automatic face and gesture recognition (FGR 2004), IEEE Computer Society, Seoul, Korea, May 17–19, pp 97–102

Perrin S, Cassinelli A, Ishikawa M (2004) Gesture recognition using laser-based tracking system. In: Sixth IEEE international conference on automatic face and gesture recognition (FGR 2004), IEEE Computer Society, Seoul, Korea, May 17–19, pp 541–546

Russell JA (1994) Is there universal recognition of emotion from facial expression? Psychol Bull 115(1):102–141

Samal A, Iyengar PA (1992) Automatic recognition and analysis of human faces and facial expressions: A survey. Pattern Recogn 25(1):65–77

Starner T, Weaver J, Pentland A (1998) Real-time American sign language recognition using desk and wearable computer-based video. IEEE Trans Pattern Anal Mach Intell 20(12):1371–1375

Tian YL, Kanade T, Cohn JF (2001) Recognizing action units for facial expression analysis. IEEE Trans Pattern Anal Mach Intell 23(2):97–115

Tian YL, Kanade T, Cohn JF (2005) Facial expression analysis. In: Li SZ, Jain AK (eds) Handbook of face recognition. Springer, New York, NY, pp 247–276

Valstar MF, Pantic M (2006) Fully automatic facial action unit detection and temporal analysis. Proc IEEE Int Conf Comput Vis Pattern Recogn 3:149

Valstar MF, Pantic M, Ambadar Z, Cohn JF (2006) Spontaneous vs. posed facial behavior: automatic analysis of brow actions. In: Proceedings of the ACM international conference on Multimodal Interfaces, Banff, Alberta, Canada, pp 162–170

Valstar MF, Pantic M, Patras I (2004) Motion history for facial action detection from face video. Proc IEEE Int Conf Syst Man Cybern 1:635–640

Viola P, Jones M (2004) Robust real-time face detection. J Comput Vis 57(2):137–154

Vogler C, Metaxas D (1998) ªASL recognition based on a coupling between HMMs and 3D Motion Analysis.º In: Proceedings of the 6th IEEE international conference on Computer Vision (ICCV-98), Narosa Publishing House, Bombay, India, January 4–7, pp 363–369

Vukadinovic D, Pantic M (2005) Fully automatic facial feature point detection using Gabor feature based boosted classifiers. In: Proceedings of the IEEE international conference on Systems, Man and Cybernetics, The Big Island, Hawaii, pp 1692–1698

Wierzbicka A (1993) Reading human faces. Pragmat Cogn 1(1):1–23

Wilson I, Bobick A (1999) Parametric hidden Markov models for gesture recognition. IEEE Trans Pattern Anal Mach Intell 21(9):884–900

Wong S, Cipolla R (2006) Continuous gesture recognition using a sparse Bayesian classifier. In: Proceedings of the 18th international conference on Pattern Recognition – vol 01, 2006. ICPR, IEEE Computer Society, Washington, DC, pp 1084–1087

Yang MH, Ahuja N, Tabb M (2002) Extraction of 2d motion trajectories and its application to hand gesture recognition. IEEE Trans Pattern Anal Mach Intell 24(8):1061–1074

Yang MH, Kriegman DJ, Ahuja N (2002) Detecting faces in images: a survey. IEEE Trans Pattern Anal Mach Intell 24(1):34–58

Ying W, Huang TS (2001) Hand modeling, analysis and recognition. Signal Process Mag IEEE 18(3):51–60

Zhang Y, Ji Q (2005) Active and dynamic information fusion for facial expression understanding from image sequence. IEEE Trans Pattern Anal Mach Intell 27(5):699–714

Multimodal Emotion Recognition from Low-Level Cues

Maja Pantic, George Caridakis, Elisabeth André, Jonghwa Kim, Kostas Karpouzis, and Stefanos Kollias

Abstract Emotional intelligence is an indispensable facet of human intelligence and one of the most important factors for a successful social life. Endowing machines with this kind of intelligence towards affective human–machine interaction, however, is not an easy task. It becomes more complex with the fact that human beings use several modalities jointly to interpret affective states, since emotion affects almost all modes – audio-visual (facial expression, voice, gesture, posture, etc.), physiological (respiration, skin temperature, etc.), and contextual (goal, preference, environment, social situation, etc.) states. Compared to common unimodal approaches, many specific problems arise from the case of multimodal emotion recognition, especially concerning fusion architecture of the multimodal information. In this chapter, we firstly give a short review for the problems and then present research results of various multimodal architectures based on combined analysis of facial expression, speech, and physiological signals. Lastly we introduce designing of an adaptive neural network classifier that is capable of deciding the necessity of adaptation process in respect of environmental changes.

1 Human Affect Sensing: The Problem Domain

The ability to detect and understand affective states and other social signals of someone with whom we are communicating is the core of social and emotional intelligence. This kind of intelligence is a facet of human intelligence that has been argued to be indispensable and even the most important for a successful social life (Goleman, 1995). When it comes to computers, however, they are socially ignorant (Pelachaud et al., 2002). Current computing technology does not account for the fact that human–human communication is always socially situated and that

M. Pantic (✉)
Department of Computing, Imperial College, London, UK; Faculty of Electrical Engineering, Mathematics and Computer Science, University of Twente, Enschede, The Netherlands
e-mail: M.Pantic@imperial.ac.uk

P. Petta et al. (eds.), *Emotion-Oriented Systems*, Cognitive Technologies,
DOI 10.1007/978-3-642-15184-2_8,

discussions are not just facts but part of a larger social interplay. Not all computers will need social and emotional intelligence and none will need all of the related skills humans have. Yet, human–machine interactive systems capable of sensing stress, inattention, confusion, and heedfulness and capable of adapting and responding to these affective states of users are likely to be perceived as more natural, efficacious, and trustworthy (Picard, 1997; Picard, 2003; Pantic, 2005). For example, in education, pupils' affective signals inform the teacher of the need to adjust the instructional message. Successful human teachers acknowledge this and work with it; digital conversational embodied agents must begin to do the same by employing tools that can accurately sense and interpret affective signals and social context of the pupil, learn successful context-dependent social behaviour, and use a proper affective presentation language (e.g. Pelachaud et al., 2002) to drive the animation of the agent. Automatic recognition of human affective states is also important for video surveillance. Automatic assessment of boredom, inattention, and stress would be highly valuable in situations in which firm attention to a crucial but perhaps tedious task is essential (Pantic, 2005; Pantic et al., 2005). Examples include air traffic control, nuclear power plant surveillance, and operating a motor vehicle. An automated tool could provide prompts for better performance informed by assessment of the user's affective state. Other domain areas in which machine tools for analysis of human affective behaviour could expand and enhance scientific understanding and practical applications include specialized areas in professional and scientific sectors (Ekman et al., 1993). In the security sector, affective behavioural cues play a crucial role in establishing or detracting from credibility. In the medical sector, affective behavioural cues are a direct means to identify when specific mental processes are occurring. Machine analysis of human affective states could be of considerable value in these situations in which only informal, subjective interpretations are now used. It would also facilitate research in areas such as behavioural science (in studies on emotion and cognition), anthropology (in studies on cross-cultural perception and production of affective states), neurology (in studies on dependence between emotion dysfunction or impairment and brain lesions), and psychiatry (in studies on schizophrenia and mood disorders) in which reliability, sensitivity, and precision of measurement of affective behaviour are persisting problems.

While all agree that machine sensing and interpretation of human affective information would be widely beneficial, addressing these problems is not an easy task. The main problem areas can be defined as follows:

- *What is an affective state?* This question is related to psychological issues pertaining to the nature of affective states and the best way to represent them.
- *Which human communicative signals convey information about affective state?* This issue shapes the choice of different modalities to be integrated into an automatic analyzer of human affective states.
- *How are various kinds of evidence to be combined to optimize inferences about affective states?* This question is related to how best to integrate information across modalities for emotion recognition.

In this section, we briefly discuss each of these problem areas in the field. The rest of the chapter is dedicated to a specific domain within the second problem area – sensing and processing visual cues of human affective displays.

1.1 What Is an Affective State?

Traditionally, the terms "affect" and "emotion" have been used synonymously. Following Darwin, discrete emotion theorists propose the existence of six or more basic emotions that are universally displayed and recognized (Ekman and Friesen, 1969; Keltner and Ekman, 2000). These include happiness, anger, sadness, surprise, disgust, and fear. Data from both Western and traditional societies suggest that non-verbal communicative signals (especially facial and vocal expression) involved in these basic emotions are displayed and recognized cross-culturally. In opposition to this view, Russell (1994) among others argues that emotion is best characterized in terms of a small number of latent dimensions, rather than in terms of a small number of discrete emotion categories. Russell proposes bipolar dimensions of arousal and valence (pleasant versus unpleasant). Watson and Tellegen propose unipolar dimensions of positive and negative affect, while Watson and Clark proposed a hierarchical model that integrates discrete emotions and dimensional views (Larsen and Diener, 1992; Watson et al., 1995a, 1995b). Social constructivists argue that emotions are socially constructed ways of interpreting and responding to particular classes of situations. They argue further that emotion is culturally constructed and no universals exist. From their perspective, subjective experience and whether or not emotion is better conceptualized categorically or dimensionally is culture specific. Then there is lack of consensus on how affective displays should be labelled. For example, Fridlund argues that human facial expressions should not be labelled in terms of emotions but in terms of behavioural ecology interpretations, which explain the influence a certain expression has in a particular context (Fridlund, 1997). Thus, an "angry" face should not be interpreted as *anger* but as *back-off-or-I-will-attack*. Yet, people still tend to use *anger* as the interpretation rather than *readiness-to-attack* interpretation. Another issue is that of culture dependency; the comprehension of a given emotion label and the expression of the related emotion seem to be culture dependent (Matsumoto, 1990; Watson et al., 1995a). In summary, previous research literature pertaining to the nature and suitable representation of affective states provides no firm conclusions that could be safely presumed and adopted in studies on machine analysis of human affective states and affective computing. Also, not only discrete emotional states like surprise or anger are of importance for the realization of proactive human–machine interactive systems, but also sensing and responding to behavioural cues identifying attitudinal states like interest and boredom, to those underlying moods, and to those disclosing social signalling like empathy and antipathy are essential (Pantic et al., 2006). Hence, in contrast to traditional approach, we treat affective states as being correlated not only to discrete emotions but to other, aforementioned social signals as well. Furthermore, since it is not certain that each of us will express a particular affective state by modulating the same communicative signals in the same way nor is it certain that a particular modulation of interactive cues will be interpreted always in the same way independently of the situation and the observer, we advocate that pragmatic choices (e.g. application- and user-profiled choices) must be made regarding the selection of affective states to be recognized by an automatic analyzer of human affective feedback (Pantic and Rothkrantz, 2003; Pantic et al., 2005, 2006).

1.2 Which Human Behavioural Cues Convey Information About Affective State?

Affective arousal modulates all human communicative signals (Ekman and Friesen, 1969). However, the visual channel carrying facial expressions and body gestures seems to be most important in the human judgment of behavioural cues (Ambady and Rosenthal, 1992). Human judges seem to be most accurate in their judgment when they are able to observe the face and the body. Ratings that were based on the face and the body were 35% more accurate than the ratings that were based on the face alone. Yet, ratings that were based on the face alone were 30% more accurate than ratings that were based on the body alone and 35% more accurate than ratings that were based on the tone of voice alone (Ambady and Rosenthal, 1992). These findings indicate that to interpret someone's behavioural cues, people rely on shown facial expressions and to a lesser degree on shown body gestures and vocal expressions. However, although basic researchers have been unable to identify a set of voice cues that reliably discriminate among emotions, listeners seem to be accurate in decoding emotions from voice cues (Juslin and Scherer, 2005). Thus, automated human affect analyzers should at least include facial expression modality and preferably they should also include (one or both) modalities for perceiving body gestures and tone of the voice. Finally, while too much information from different channels seem to be confusing to human judges, resulting in less accurate judgments of shown behaviour when three or more observation channels are available (e.g. face, body, and speech) (Ambady and Rosenthal, 1992), combining those multiple modalities (including speech and physiology) may prove appropriate for realization of automatic human affect analysis.

1.3 How Are Various Kinds of Evidence to Be Combined to Optimize Inferences About Affective States?

Humans simultaneously employ the tightly coupled modalities of sight, sound, and touch. As a result, analysis of the perceived information is highly robust and flexible. Thus, in order to accomplish a multimodal analysis of human behavioural signals acquired by multiple sensors, which resembles human processing of such information, input signals should not be considered mutually independent and should not be combined only at the end of the intended analysis as the majority of current studies do. The input data should be processed in a joint feature space and according to a context-dependent model (Pantic and Rothkrantz, 2003). The latter refers to the fact that one must know the context in which the observed behavioural signals have been displayed (who the expresser is, what his or her current environment and task are, when and why did he or she display the observed behavioural signals) in order to interpret the perceived multi-sensory information correctly (Pantic et al., 2006).

2 Classification and Fusion Approaches

2.1 Short-Term, Low-Level Multimodal Fusion

The term multimodal has been used in many contexts and across several disciplines. In the context of emotion recognition, a multimodal system is simply one that responds to inputs in more than one modality or communication channel (e.g. face, gesture, and speech prosody in our case, writing, body posture, linguistic content, and others) (Kim and André, 2006; Pantic, 2005). Jaimes and Sebe use a human-centred approach in this definition; by modality we mean mode of communication according to human senses or type of computer input devices. In terms of human senses, the categories are sight, touch, hearing, smell, and taste. In terms of computer input devices, we have modalities that are equivalent to human senses: cameras (sight), haptic sensors (touch), microphones (hearing), olfactory (smell), and even taste (Taylor and Fragopanagos, 2005). In addition, however, there are input devices that do not map directly to human senses: keyboard, mouse, writing tablet, motion input (e.g. the device itself is moved for interaction), and many others.

Various multimodal fusion techniques are possible (Zeng et al., 2009). Feature-level fusion can be performed by merging extracted features from each modality into one cumulative structure and feeding them to a single classifier, generally based on multiple hidden Markov models or neural networks. In this framework, correlation between modalities can be taken into account during classifier learning. In general, feature fusion is more appropriate for closely coupled and synchronized modalities, such as speech and lip movements, but tends not to generalize very well if modalities differ substantially in the temporal characteristics of their features, as is the case between speech and facial expression or gesture inputs. Moreover, due to the high dimensionality of input features, large amounts of data must be collected and labelled for training purposes.

Taylor and Fragopanagos describe a neural network architecture in Taylor and Fragopanagos (2004, 2005) in which features, from various modalities, that correlate with the user's emotional state are fed to a hidden layer, representing the emotional content of the input message. The output is a label of this state. Attention acts as a feedback modulation onto the feature inputs, so as to amplify or inhibit the various feature inputs, as they are or are not useful for the emotional state detection. The basic architecture is thus based on a feedforward neural network, but with the addition of a feedback layer (IMC in Fig. 1 below), modulating the activity in the inputs to the hidden layer.

Results have been presented for the success levels of the trained neural system based on a multimodal database, including time series streams of text (from

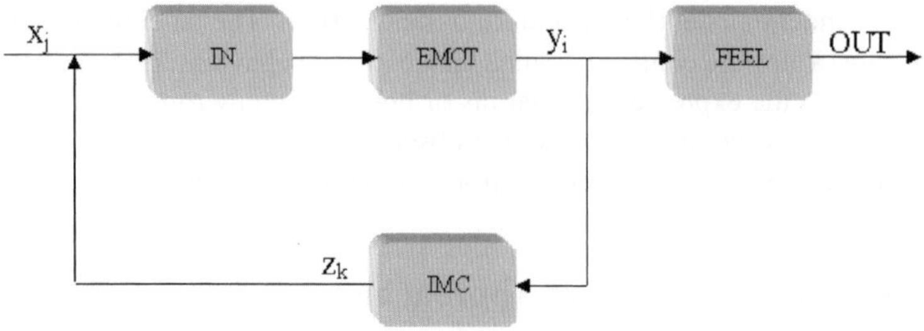

Fig. 1 Information flow in a multimodal emotion recognizer. IMC, inverse model controller; EMOT, hidden layer emotional state; FEEL, output state emotion classifier

an emotional dictionary), prosodic features (as determined by a prosodic speech feature extraction), and facial features (facial animation parameters). The obtained results are different for different viewers who helped to annotate the data sets. These results show high success levels on certain viewers while lower (but still good) levels on other ones. In particular, very high success was obtained using only prediction of activation values for one user who seemed to use mainly facial cues, whilst a similar, but slightly lower success level was obtained on an annotator who used predominantly prosodic cues.

Other two annotators appeared to use cues from all modalities, and for them, the success levels were still good but not so outstanding. This leads to the need for a further study to follow-up the spread of such cue extraction across the populace, since if this is an important component, then it would be important to know how broad is this spread, as well as to develop ways to handle such a spread (such as having a battery of networks, each trained on the appropriate subset of cues). It is, thus evident that adaptation to specific users and contexts is a crucial aspect in this type of fusion.

Decision-level fusion caters for integrating asynchronous but temporally correlated modalities. Here, each modality is first classified independently and the final classification is based on fusion of the outputs of the different modalities. Designing optimal strategies for decision-level fusion is still an open research issue. Various approaches have been proposed, e.g. sum rule, product rule, using weights, max/min/median rule, and majority vote. As a general rule, semantic fusion builds on individual recognizers, followed by an integration process; individual recognisers can be trained using unimodal data, which are easier to collect.

3 Cases of Multimodal Analysis

3.1 Recognition from Speech and Video Features

Visual sources can provide significant information about human communication activities. In particular, lip movement captured by stationary and steerable cameras can verify or detect that a particular person is speaking and help improve speech recognition accuracy. The proposed approach is similar to human lip reading and consists of adding features like lip motion and other visual speech cues as additional inputs for recognition. This process is known as speech reading (Luettin et al., 1996; Potamianos et al., 2003), where most audio-visual speech recognition approaches consider the visual channel as a parallel and symmetric information source to the acoustic channel, resulting in the visual speech information being captured explicitly through the joint training of audio-visual phonetic models. As a result, in order to build a high-performance recognition system, large collections of audio-visual speech data are required. An alternative to the fusion approach is to use the visual and acoustic information in an asymmetric manner, where the tight coupling between auditory and visual speech in the signal domain is exploited and the visual cues used to help separate the speech of the target speaker from background speech and other acoustic events. Note that in this approach the visual channel is considered only up to the signal processing stage, and only the separated acoustic source is passed on to the statistical modelling level. In essence, the visual speech information here is used implicitly through the audio channel enhancement. This approach permits flexible and scalable deployment of audio-visual speech technology.

In the case of multimodal natural interaction (Caridakis et al., 2006), authors used earlier recordings during the FP5 IST Ermis project (FP5 IST ERMIS, 2007), where emotion induction was performed using the SAL approach. This material was labelled using FeelTrace (Cowie et al., 2000) by four labellers. The activation valence coordinates from the four labellers were initially clustered into quadrants and were then statistically processed so that a majority of decision could be obtained about the unique emotion describing the given moment. The corpus under investigation was segmented into 1,000 tunes of varying length. For every tune, the facial feature input vector consisted of the FAPs produced by the processing of the frames of the tune, while the acoustic input vector consisted of only one value per SBPF (segment-based prosodic feature) per tune. The fusion was performed on a frame basis, meaning that the values of the SBPFs were repeated for every frame of the tune. This approach was preferred because it preserved the maximum of the available information since SBPFs are meaningful only for a certain time period and cannot be calculated per frame.

In order to model the dynamic nature of facial expressivity, authors employed RNNs (recurrent neural networks – Fig. 2), where past inputs influence the processing of future inputs (Elman, 1990). RNNs possess the nice feature of modelling explicitly time and memory, catering for the fact that emotional states are not fluctuating strongly, given a short period of time. Additionally, they can model emotional transitions and not only static emotional representations, providing a solution for diverse feature variation and not merely for neutral to expressive and back to neutral, as would be the case for HMMs. The implementation of a RNN was based on an Elman network, with four output classes (three for the possible emotion quadrants, since the data for the positive/passive quadrant was negligible, and one for neutral affective state) resulting in a data set consisting of around 10,000 records. To cater for the fact that facial features are calculated per frame while speech prosody features are constant per tune, authors maintain the conventional input neurons met in

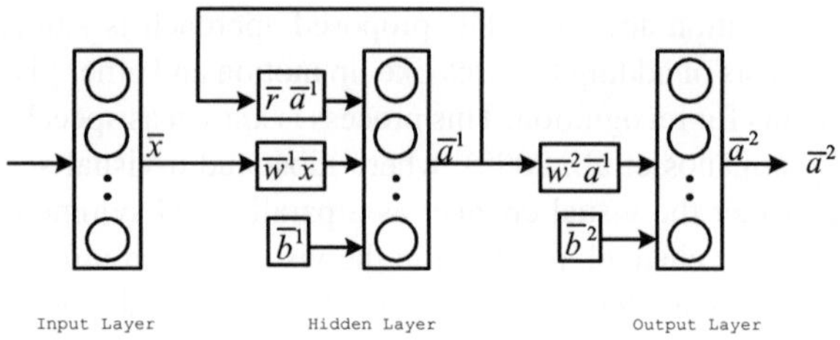

Fig. 2 Structure and functionality of a recursive neural network

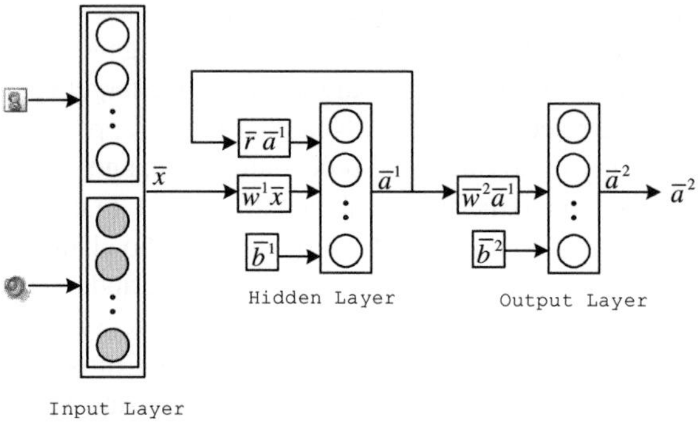

Fig. 3 Modified version of the Elman network

all neural networks, while for the auditory modality features, they use static value neurons (modified version shown in Fig. 3). The classification efficiency, for facial only and audio only, was measured at 67 and 73%, respectively, but combining the two modalities resulted in a recognition rate of 79%. This fact illustrates the ability of the proposed method to take advantage of multimodal information and the related analysis. After further processing by removing very short tunes (less than 10 frames or half a second), recognition rates in the naturalistic database rise to 98.55%.

3.2 Recognition from Physiological and Speech Features

Kim and Andre in Kim et al. (2005) and Kim and André (2006) studied various methods for fusing physiological and voice data at the feature level and the decision level, as well as a hybrid integration scheme. The results of the integrated recognition approach were then compared with the individual recognition results from each modality using the multimodal corpus of speech and physiological data we recorded within the FP6 NoE Humaine for three subjects. After synchronized segmentation of bimodal signals, we obtained a total of 138 features, 77 features from the five-channel biosignals (EMG, BVP, SC, RSP, Temp), and 61 features from the speech segments.

Feature-level fusion is performed by merging the calculated features from each modality into one cumulative structure, selecting the relevant features, and feeding

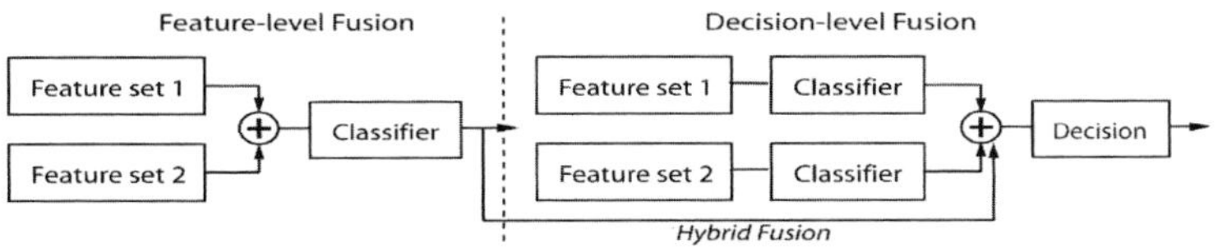

Fig. 4 Feature, decision and hybrid fusion of affective cues

them to a classifier. Decision-level fusion caters for integrating asynchronous but temporally correlated modalities (Fig. 4). Each modality is first classified independently by the classifier, and the final decision is obtained by fusing the output from the modality-specific classification processes. Three criteria, maximum, average, and product, were applied to evaluate the posterior probabilities of the unimodal classifiers at the decision stage. As a further variation of decision-level fusion, we employed a new hybrid scheme of the two fusion methods in which the output of feature-level fusion is also fed as an auxiliary input to the decision-level fusion stage. In Table 1 the best results are summarized that we achieved by the classification schemes described above. We classified the bimodal data subject

Table 1 Recognition results in rates (1.0 = 100% accuracy) achieved by using SBS, LDA, and leave-one-out cross-validation

System	High/pos	High/neg	Low/neg	Low/pos	*Average*
Subject A					
Biosignal	0.95	0.92	0.86	0.85	0.90
Speech signal	0.64	0.75	0.67	0.78	0.71
Feature fusion	0.91	0.92	1.00	0.85	**0.92**
Decision fusion	0.64	0.54	0.76	0.67	0.65
Hybrid fusion	0.86	0.54	0.57	0.59	0.64
Subject B					
Biosignal	0.50	0.79	0.71	0.45	0.61
Speech signal	0.76	0.56	0.74	0.72	0.70
Feature fusion	0.71	0.56	0.94	0.79	**0.75**
Decision fusion	0.59	0.68	0.82	0.69	0.70
Hybrid fusion	0.65	0.64	0.82	0.83	0.73
Subject C					
Biosignal	0.52	0.79	0.70	0.52	0.63
Speech signal	0.55	0.77	0.66	0.71	0.67
Feature fusion	0.50	0.67	0.84	0.74	**0.69**
Decision fusion	0.32	0.77	0.74	0.64	0.62
Hybrid fusion	0.40	0.73	0.86	0.71	0.68
All: subject independent					
Biosignal	0.43	0.53	0.54	0.52	0.51
Speech signal	0.40	0.53	0.70	0.53	0.54
Feature fusion	0.46	0.57	0.63	0.56	**0.55**
Decision fusion	0.34	0.50	0.70	0.54	0.52
Hybrid fusion	0.41	0.51	0.70	0.55	0.54

dependently (subjects A, B, and C) and subject independently (All) since this gave us a deeper insight on what terms the multimodal systems could improve the results of unimodal emotion recognition. We performed both feature-level fusion and decision-level fusion using LDA (linear discriminant analysis) in combination with SBS (sequential backward searching).

The results show that the performance of the unimodal systems varies not only from subject to subject but also for the single modalities. During our experiment, we could observe individual differences in the physiological and vocal expressions of the three test subjects. As shown in Table 1, the emotions of user A were more accurately recognized by using biosignals (90%) than by his voice (71%), whereas it was inverse for users B and C (70 and 67% for voice and 61 and 63% for biosignals). In particular for subject A, the difference between the accuracies of the two modalities is sizable. However, no suggestively dominant modality could be observed in the results of subject-dependent classification in general, which may be used as a decision criterion in the decision-level fusion process to improve the recognition accuracy. Overall, we obtained the best results for feature-level fusion. For instance, we got an acceptable recognition accuracy of 92% for subject A when using feature-level fusion which considerably went down, however, when using decision-level or hybrid fusion. Generally, feature-level fusion is more appropriate for combining modalities with analogous characteristics. As the data for subject A show, a high accuracy obtained for one modality may be declined by a relatively low accuracy from another modality when fusing data at the decision level. This observation indicates the limitations of the decision-level fusion scheme we used, which is based on a pure arithmetic evaluation of the posterior probabilities at the decision stage rather than a parametric assessment process (Fig. 5).

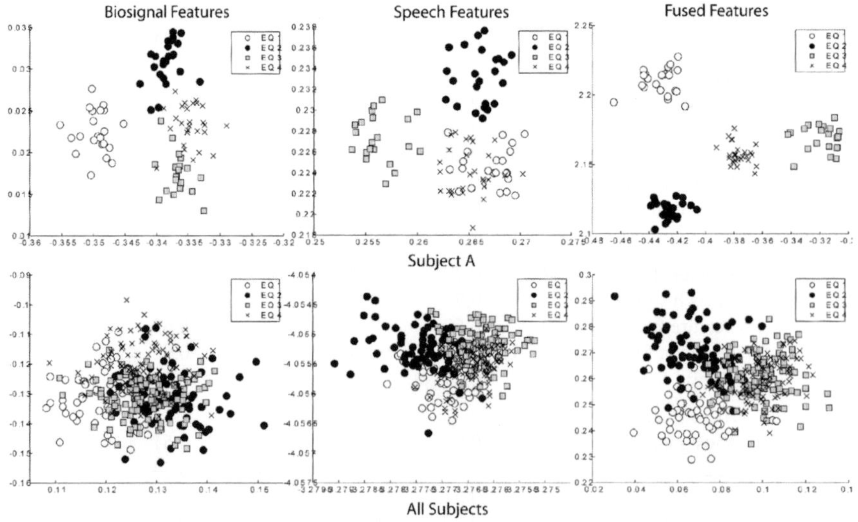

Fig. 5 Fisher projection examples of the bimodal features

3.3 Recognition from Facial Expressions and Hand Gesturing

In Karpouzis et al. (2004) and Balomenos et al. (2006), gestures are utilized to support the outcome of the facial expression analysis subsystem, since in most cases they are too ambiguous to indicate a particular emotion. However, in a given context of interaction, some gestures are obviously associated with a particular expression – e.g. hand clapping of high frequency expresses joy, satisfaction – while others can provide indications for the kind of the emotion expressed by the user. In particular, quantitative features derived from hand tracking, like speed and amplitude of motion, fortify the position of an observed emotion; for example, satisfaction turns to joy or even to exhilaration, as the speed and amplitude of clapping increases.

Given a particular context of interaction, gesture classes corresponding to the same emotional state are combined in a "logical OR" form. Table 2 shows that a particular gesture may correspond to more than one gesture classes carrying different affective meaning. For example, if the examined gesture is clapping, detection of high frequency indicates joy, but a clapping of low frequency may express irony and can reinforce a possible detection of the facial expression disgust.

Although face is the main 'demonstrator' of user's emotion (Ekman and Friesen, 1975), the recognition of the accompanying gesture increases the confidence of the result of facial expression subsystem. Further research is necessary to be carried out in order to define how powerful the influence of a gesture in the recognition of an emotion actually is. It would also be helpful to define which, face or gesture, is more useful for a specific application and change the impact of each subsystem on the final result.

In the current implementation the two subsystems are combined as a weighted sum. Let b_k be the degree of belief that the observed sequence presents the kth emotional state, obtained from the facial expression analysis subsystem, and EI_k be the corresponding emotional state indicator, obtained from the affective gesture analysis subsystem, then the overall degree of belief d_k is given by

$$d_k = w_1 \cdot b_k + w_2 \cdot EI_k$$

Table 2 Correlation between gestures and emotional states

Emotion	Gesture class
Joy	Hand clapping – high frequency
Sadness	Hands over the head – posture
Anger	Lift of the hand – high speed
	Italianate gestures
Fear	Hands over the head – gesture
	Italianate gestures
Disgust	Lift of the hand – low speed
	Hand clapping – low frequency
Surprise	Hands over the head – gesture

where the weights w_1 and w_2 are used to account for the reliability of the two subsystems as far as the emotional state estimation is concerned. In their implementation, the authors used w_1=0.75 and w_2=0.25. These values enables the affective gesture analysis subsystem to be important in cases where the facial expression analysis subsystem produces ambiguous results while at the same time leaves the latter subsystem to be the main contributing part in the overall decision system.

4 The Need for Adaptivity

In Caridakis et al. (2008), Caridakis builds on Balomenos et al. (2006) and Karpouzis et al. (2004), to provide adaptivity characteristics for decision- and feature-level fusion. In this case, a fuzzy logic-based system was derived, based on the formulation shown in Fig. 6.

While the multimodal system outperforms both unimodal (face and gesture) ones, it is clear that the ability of the system to adapt to the specific characteristics and user/situation contexts of the interaction is crucial. This approach still fails to model the interplay between the different modalities, a fact which one can exploit to fortify the results obtained from an individual modality (e.g. correlation between visemes and phonemes) or resolve uncertainty in cases where one or more modalities are not dependable (e.g. speech analysis in the presence of noise can be assisted by visually extracting visemes and mapping them to possible phonemes). The latter approach is termed dominant modality recoding model. Nevertheless, identification

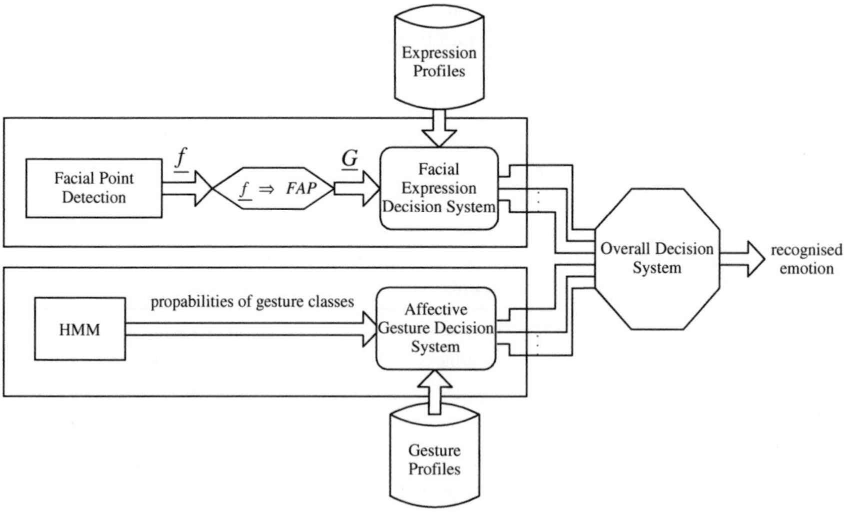

Fig. 6 Overall architecture of the multimodal emotion recognition process in Caridakis et al. (2008)

of dominant modalities is another open issue, which could be resolved if (performance) confidence levels could be estimated in each unimodal case and used thereafter.

4.1 Detecting the Need for Adaptation

The purpose of this mechanism is to detect when the output of the neural network classifier is not appropriate and consequently to activate the adaptation algorithm at those time instances when a change of the environment occurs.

Let us index images or video frames in time, denoting by $\boldsymbol{x}(k, N)$, the feature vector of the kth image or image frame, following the image at which the adaptation of the Nth network occurred. Index k is therefore reset each time adaptation takes place, with $\boldsymbol{x}(0, N)$ corresponding to the feature vector of the image, where the nth adaptation of the network was accomplished. Adaptation of the network classifier is accomplished at time instances where its performance deteriorates, i.e. the current network output deviates from the desired one. Let us recall that vector $\boldsymbol{c}$ expresses the difference between the desired and the actual network outputs based on weights $\boldsymbol{w}_b$ and is applied to the current data set $\boldsymbol{S_c}$. As a result, if the norm of vector $\boldsymbol{c}$ increases, network performance deviates from the desired one and adaptation should be applied. On the contrary, if vector $\boldsymbol{c}$ takes small values, then no adaptation is required. In the following we denote this vector as $\boldsymbol{c}(k, N)$ depending upon feature vector $\boldsymbol{x}(k, N)$.

Let us assume that the Nth adaptation phase of the network classifier has been completed. If the classifier is then applied to all instances $\boldsymbol{x}(0, N)$, including the ones used for adaptation, it is expected to provide classification results of good quality. The difference between the output of the adapted network and that produced by the initially trained classifier at feature vector $\boldsymbol{x}(0, N)$ constitutes an estimate of the level of improvement that can be achieved by the adaptation procedure. Let us denote this difference by $e(0, N)$ and let $e(k, N)$ denote the difference between the corresponding classification outputs when the two networks are applied to the feature set of the kth image or image frame (or speech segment) following the Nth network adaptation phase. It is anticipated that the level of improvement expressed by $e(k, N)$ will be close to that of $e(0, N)$ as long as the classification results are good. This will occur when input images are similar to the ones used during the adaptation phase. An error $e(k, N)$, which is quite different from $e(0, N)$, is generally due to a change of the environment. Thus, the quantity $\alpha(k, N) = |e(k, N) - e(0, N)|$ can be used for detecting the change of the environment or equivalently the time instances where adaptation should occur. Thus, no adaptation is needed if $\alpha(k, N) > T$, where T is a threshold which expresses the max tolerance, beyond which adaptation is required for improving the network performance. In case of adaptation, index k is reset to zero, while index N is incremented by one.

Such an approach detects with high accuracy the adaptation time instances in cases of both abrupt and gradual changes of the operational environment since the comparison is performed between the current error difference $e(k, N)$ and the one obtained right after adaptation, i.e. $e(0, N)$. In an abrupt operational change, error $e(k, N)$ will not be close to $e(0, N)$; consequently, $\alpha(k, N)$ exceeds threshold T and adaptation is activated. In the case of a gradual change, error $e(k, N)$ will gradually deviate from $e(0, N)$ so that the quantity $\alpha(k, N)$ gradually increases and adaptation is activated at the frame where $\alpha(k, N)>T$.

Network adaptation can be instantaneously executed each time the system is put in operation by the user. Thus, the quantity $\alpha(0, 0)$ initially exceeds threshold T and adaptation is forced to take place.

4.2 The Adaptive Neural Network Architecture

Let us assume that we seek to classify, to one of, say, p available emotion classes ω, each input vector $\boldsymbol{x}_i$ containing the features extracted from the input signal. A neural network produces a p-dimensional output vector $\boldsymbol{y}(\boldsymbol{x}_i)$

$$\underline{y}(\underline{x}_i) = \left[p^i_{\omega_1} \; p^i_{\omega_2} \cdots p^i_{\omega_p} \right]^{\mathrm{T}} \tag{1}$$

where $p^i_{\omega_j}$ denotes the probability that the ith input belongs to the jth class.

Let us first consider that the neural network has been initially trained to perform the classification task using a specific training set, say,

$$S_b = \left\{ \left(\underline{x}'_1, \underline{d}'_1\right), \ldots, \left(\underline{x}'_{m_b}, \underline{d}'_{m_b}\right) \right\}$$

where vectors $\underline{x}'_i$ and $\underline{d}'_i$ with $i = 1, 2, \cdots, m_b$ denote the ith input training vector and the corresponding desired output vector consisting of p elements, respectively.

Then, let $\boldsymbol{y}(\boldsymbol{x}_i)$ denote the network output when applied to a new set of inputs, and let us consider the ith input outside the training set, possibly corresponding to a new user or to a change of the environmental conditions. Based on the above described discussion, slightly different network weights should probably be estimated in such cases, through a network adaptation procedure.

Let $\boldsymbol{w}_b$ include all weights of the network before adaptation, and $\boldsymbol{w}_\alpha$ the new weight vector which is obtained after adaptation is performed. To perform the adaptation, a training set $\boldsymbol{S}_c$ has to be extracted from the current operational situation composed of (one or more), say, $\boldsymbol{m}_c$ inputs:

$$S_c = \left\{ \left(\underline{x}_1, \underline{d}_1\right), \ldots, \left(\underline{x}_{m_c}, \underline{d}_{m_c}\right) \right\}$$

where $\boldsymbol{x}_i$ and $\boldsymbol{d}_i$ with $i = 1, 2, \ldots, \boldsymbol{m}_c$ similarly correspond to the ith input and desired output data used for adaptation, respectively. The adaptation algorithm that is activated, whenever such a need is detected, computes the new network weights $\boldsymbol{w}_\alpha$, minimizing the following error criteria with respect to weights:

$$E_a = E_{c,a} + \eta E_{f,a}, E_{c,a} = \frac{1}{2}\sum_{i=1}^{m_c} \left\| \underline{z}_a(\underline{x}_i) - \underline{d}_i \right\|_2, E_{f,a} = \frac{1}{2}\sum_{i=1}^{m_b} \left\| \underline{z}_a(\underline{x}'_i) - \underline{d}'_i \right\|_2 \quad (2)$$

where $E_{c,\alpha}$ is the error performed over training set S_c ("current" knowledge), $E_{f,\alpha}$ is the corresponding error over training set S_b ("former" knowledge); $z_\alpha(\boldsymbol{x}_i)$ and $z_\alpha(\boldsymbol{x}'_i)$ are the outputs of the adapted network, corresponding to input vectors $\boldsymbol{x}_i$ and $\boldsymbol{x}'_i$, respectively, of the network consisting of weights $\boldsymbol{w}_\alpha$. Similarly $z_b(\boldsymbol{x}_i)$ would represent the output of the network consisting of weights $\boldsymbol{w}_b$ when accepting vector $\boldsymbol{x}_i$ at its input; when adapting the network for the first time $z_b(\boldsymbol{x}_i)$ is identical to $\boldsymbol{y}(\boldsymbol{x}_i)$. Parameter η is a weighting factor accounting for the significance of the current training set compared to the former one and $||.||_2$ denotes the L_2 norm.

The goal of the training procedure is to minimize the above equation and estimate the new network weights $\boldsymbol{w}_\alpha$, i.e. $\boldsymbol{w}^0{}_\alpha$ and $\boldsymbol{w}_\alpha$, respectively. Let us first assume that a small perturbation of the network weights (before adaptation) $\boldsymbol{w}_b$ is enough to achieve good classification performance. Then

$$\underline{w}_a = \underline{w}_b + \Delta \underline{w}$$

where $\Delta \boldsymbol{w}$ are small increments. This assumption leads to an analytical and tractable solution for estimating $\boldsymbol{w}_\alpha$, since it permits linearization of the non-linear activation function of the neuron, using a first-order Taylor series expansion.

Equation (2) indicates that the new network weights are estimated taking into account both the current and the previous network knowledge. To stress, however, the importance of current training data, one can replace the first term by the constraint that the actual network outputs are equal to the desired ones, that is

$$z_a(\underline{x}_i) = d_i \quad i = 1, ..., m_c, \quad \text{for all data in } S_c \quad (3)$$

This equation indicates that the first term of (2), corresponding to error $E_{c,\alpha}$, takes values close to 0, after estimating the new network weights. Through linearization, solution of (3) with respect to the weight increments is equivalent to a set of linear equations

$$\underline{c} = \mathbf{A} \cdot \Delta \underline{w}$$

where vector $\boldsymbol{c}$ and matrix $\mathbf{A}$ are appropriately expressed in terms of the previous network weights. In particular

$$\underline{c} = \left[z_a(\underline{x}_1) \cdots z_a(\underline{x}_{m_c})\right]^{\mathrm{T}} - \left[z_b(\underline{x}_1) \cdots z_b(\underline{x}_{m_c})\right]^{\mathrm{T}}$$

expressing the difference between network outputs after and before adapting to all input vectors in $\boldsymbol{S}_c$. $\boldsymbol{c}$ can be written as

$$\underline{c} = \left[d_1 \cdots d_{m_c}\right]^{\mathrm{T}} - \left[z_b(\underline{x}_1) \cdots z_b(\underline{x}_{m_c})\right]^{\mathrm{T}} \quad (4)$$

Equation (4) is valid only when weight increments $\Delta \boldsymbol{w}$ are small quantities. It can be shown (Doulamis et al., 2000) that given a tolerated error value, proper bounds θ and φ can be computed for the weight increments and input vector $\boldsymbol{x}_i$ in $\boldsymbol{S}_c$.

Let us assume that the network weights before adaptation, i.e. $\boldsymbol{w}_b$, have been estimated as an optimal solution over data of set $\boldsymbol{S}_b$. Furthermore, the weights after adaptation are considered to provide a minimal error over all data of the current set $\boldsymbol{S}_c$. Thus, minimization of the second term of (2), which expresses the effect of the new network weights over data set $\boldsymbol{S}_b$, can be considered as minimization of the absolute difference of the error over data in $\boldsymbol{S}_b$ with respect to the previous and the current network weights. This means that the weight increments are minimally modified, resulting in the following error criterion:

$$E_S = \left\| E_{f,a} - E_{f,b} \right\|_2$$

with $E_{f,b}$ defined similarly to $E_{f,a}$, with z_α replaced by z_b in (2) (Park et al., 1991) shows that the above equation takes the form of

$$E_S = \frac{1}{2}(\Delta \underline{w})^{\mathrm{T}} \cdot \mathbf{K}^{\mathrm{T}} \cdot \mathbf{K} \cdot \Delta \underline{w} \tag{5}$$

where the elements of matrix $\mathbf{K}$ are expressed in terms of the previous network weights $\boldsymbol{w}_b$ and the training data in $\boldsymbol{S}_b$. The error function defined by (5) is convex since it is of squared form. The constraints include linear equalities and inequalities. Thus, the solution should satisfy the constraints and minimize the error function in (5). The gradient projection method is adopted to estimate the weight increments.

Each time the decision mechanism ascertains that adaptation is required, a new training set $\boldsymbol{S}_c$ is created, which represents the current condition. Then, new network weights are estimated, taking into account both the current information (data in $\boldsymbol{S}_c$) and the former knowledge (data in $\boldsymbol{S}_b$). Since the set $\boldsymbol{S}_c$ has been optimized over the current condition, it cannot be considered suitable for following or future states of the environment. This is due to the fact that data obtained from future states of the environment may be in conflict with data obtained from the current one. On the contrary, it is assumed that the training set $\boldsymbol{S}_b$, which is in general based on extensive experimentation, is able to roughly approximate the desired network performance at any state of the environment. Consequently, in every network adaptation phase, a new training set $\boldsymbol{S}_c$ is created and the previous one is discarded, while new weights are estimated based on the current set $\boldsymbol{S}_c$ and the old one $\boldsymbol{S}_b$, which remains constant throughout network operation.

References

Ambady N, Rosenthal R (1992) Thin slices of expressive behavior as predictors of interpersonal consequences: a meta-analysis. Psychol Bull 111(2):256–274

Balomenos T, Raouzaiou A, Ioannou S, Drosopoulos A, Karpouzis K, Kollias S (2006) Emotion analysis in man–machine interaction systems. In: Bengio S, Bourlard H (eds) Machine learning

for multimodal interaction. Lecture notes in computer science, vol 3361. Springer, Berlin, pp 318–328
Caridakis G, Karpouzis K, Kollias S (2008) User and context adaptive neural networks for emotion recognition. Neurocomputing, Elsevier, 71(13–15):2553–2562
Caridakis G, Malatesta L, Kessous L, Amir N, Raouzaiou A, Karpouzis K (2006) Modeling naturalistic affective states via facial and vocal expressions recognition. In: International conference on multimodal interfaces (ICMI'06), Banff, AB, 2–4 Nov 2006
Cowie R, Douglas-Cowie E, Savvidou S, McMahon E, Sawey M, Schröder M (2000) FEELTRACE: an instrument for recording perceived emotion in real time. In: ISCA workshop on speech and emotion, Northern Ireland, pp 19–24
Doulamis N, Doulamis A, Kollias S (2000) On-line retrainable neural networks: improving performance of neural networks in image analysis problems. IEEE Trans Neural Netw 11(1):1–20
Ekman P, Friesen WF (1969) The repertoire of nonverbal behavioral categories – origins, usage, and coding. Semiotica 1:49–98
Ekman P, Friesen W (1975) Unmasking the face. Prentice-Hall, Englewood Cliffs, NJ<loc>
Ekman P, Huang TS, Sejnowski TJ, Hager JC (eds) (1993) NSF understanding the face. A Human Face eStore, Salt Lake City (see Library)
Elman JL (1990) Finding structure in time. Cogn Sci 14:179–211
FP5 IST ERMIS (2007) http://www.image.ntua.gr/ermis. Accessed 30 Oct 2007
Fridlund AJ (1997) The new ethology of human facial expression. In: Russell JA, Fernandez-Dols JM (eds) The psychology of facial expression. Cambridge University Press, Cambridge, MA, pp 103–129
Goleman D (1995) Emotional intelligence. Bantam Books, New York, NY
Juslin PN, Scherer KR (2005) Vocal expression of affect. In: Harrigan J, Rosenthal R, Scherer K (eds) The new handbook of methods in nonverbal behavior research. Oxford University Press, Oxford
Karpouzis K, Raouzaiou A, Drosopoulos A, Ioannou S, Balomenos T, Tsapatsoulis N, Kollias S (2004) Facial expression and gesture analysis for emotionally-rich man–machine interaction. In: Sarris N, Strintzis M (eds) 3D modeling and animation: synthesis and analysis techniques. Idea Group, Hershey, PA, pp 175–200
Keltner D, Ekman P (2000) Facial expression of emotion. In: Lewis M, Haviland-Jones JM (eds) Handbook of emotions. Guilford Press, New York, NY, pp 236–249
Kim J, André E (2006) Emotion recognition using physiological and speech signal in short-term observation. In: Perception and interactive technologies, LNAI 4201. Springer, Berlin, Heidelberg, pp 53–64
Kim J, André E, Rehm M, Vogt T, Wagner J (2005) Integrating information from speech and physiological signals to achieve emotional sensitivity. In: Proceedings of the 9th European conference on speech communication and technology, Lisbon, Portugal
Larsen RJ, Diener E (1992) Promises and problems with the circumplex model of emotion. In: Clark MS, (ed) Review of personality and social psychology, vol 13. Sage, Newbury Park, CA, pp 25–59
Luettin J, Thacker N, Beet S (1996) Active shape models for visual speech feature extraction. In: Storck DG, Hennecke ME (eds) Speechreading by humans and machines. Springer, Berlin, pp 383–390
Matsumoto D (1990) Cultural similarities and differences in display rules. Motiv Emot 14:195–214
Pantic M (2005) Affective computing. In: Pagani M (ed) Encyclopedia of multimedia technology and networking, vol 1. Idea Group Reference, Hershy, PA, pp 8–14
Pantic M, Pentland A, Nijholt A, Huang TS (2006) Human computing and machine understanding of human behaviour: a survey. In: Proceedings of the ACM international conference on multimodal interfaces, Banff, Alberta, Canada, pp 239–248
Pantic M, Rothkrantz LJM (2003) Toward an affect-sensitive multimodal human–computer interaction. Proc IEEE 91(9):1370–1390

Pantic M, Sebe N, Cohn JF, Huang TS (2005) Affective multimodal human–computer interaction. In: Proceedings of the 13th annual ACM international conference on Multimedia, pp 669–676

Park D, EL-Sharkawi MA, Marks RJ II (1991) An adaptively trained neural network. IEEE Trans Neural Netw 2:334–345

Pelachaud C, Carofiglio V, De Carolis B, de Rosis F, Poggi I (2002) Embodied contextual agent in information delivering application. In: Proceedings of the international conference on autonomous agents and multi-agent systems. Bologna, Italy

Picard RW (1997) Affective computing. The MIT Press, Cambridge, MA

Picard RW (2003) Affective computing: challenges. Int J Human–Comput Stud 59(1–2):55–64

Potamianos G, Neti C, Gravier G, Garg A (2003 Sept) Automatic recognition of audio-visual speech: recent progress and challenges. Proc IEEE 91(9):1306–1326

Russell JA (1994) Is there universal recognition of emotion from facial expression? Psychol Bull 115(1):102–141

Taylor J, Fragopanagos N (2004) Modelling human attention and emotions. Proc 2004 IEEE Int Joint Conf Neural Netw 1:501–506.

Taylor J, Fragopanagos N (2005) The interaction of attention and emotion. Neural Netw 18(4): 353–369

Watson D, Weber K, Assenheimer JS, Clark LA, Strauss ME, McCormick RA (1995a) Testing a tripartite model: I. Evaluating the convergent and discriminant validity of anxiety and depression symptom scales. J Abnorm Psychol 104:3–14

Watson D, Clark LA, Weber K, Smith-Assenheimer J, Strauss ME, McCormick RA (1995b) Testing a tripartite model: II. Exploring the symptom structure of anxiety and depression in student, adult, and patient samples. J Abnorm Psychol 104:15–25

Zeng Z, Pantic M, Roisman G, Huang T (2009) A survey of affect recognition methods: Audio, visual, and spontaneous expressions. IEEE Trans Pattern Anal Mach Intell 31(1)

Physiological Signals and Their Use in Augmenting Emotion Recognition for Human–Machine Interaction

R. Benjamin Knapp, Jonghwa Kim, and Elisabeth André

Abstract In this chapter we introduce the concept of using physiological signals as an indicator of emotional state. We review the ambulatory techniques for physiological measurement of the autonomic and central nervous system as they might be used in human–machine interaction. A brief history of using human physiology in HCI leads to a discussion of the state of the art of multimodal pattern recognition of physiological signals. The overarching question of whether results obtained in a laboratory can be applied to ecological HCI remains unanswered.

1 Introduction

The three indicators of emotional state currently used in emotion research have been "evaluative reports, overt actions, and physiological responses" (Bradley and Lang, 2007). Evaluative reports (e.g., questionnaires and verbal anecdotes) and overt actions (e.g., facial gestures, vocal utterances, and body position) have been discussed in other chapters within this handbook. Informed by many diverse fields of science including neuro-physiology, psychophysiology, and human–computer interaction (HCI), we will explore this chapter incorporating the use of physiological signals in emotion recognition and the use of this information as part of human–machine interaction.

Emotions are neuro-physiological processes (e.g., Cacioppo and Gardner, 1999), and attempting to decipher emotional state in the absence of direct measures of physiological changes is indeed ignoring a wealth of relevant and sometimes vital information. In spite of this, the focus of much of the recent research on using emotion as a component of HCI has been on facial, gestural, and speech recognition. (In the major conference on affective computing in 2007, papers using physiological

---

R.B. Knapp (✉)
School of Music and Sonic Arts, Queen's University Belfast, Belfast, Northern Ireland, UK
e-mail: B.Knapp@qub.ac.uk

P. Petta et al. (eds.), *Emotion-Oriented Systems*, Cognitive Technologies,
DOI 10.1007/978-3-642-15184-2_9, © Springer-Verlag Berlin Heidelberg 2011

measures to detect affect were outnumbered – in a ratio of about 4:1 – by papers trying to detect affect using facial expression, body gestures, and speech analysis.) One argument for this research bias is based on the incorrect assumption that humans cannot detect physiological changes unless they are revealed by overt actions. However, as has been stated many times, "it is a mistake to think of physiology as something that people do not naturally recognize. A stranger shaking your hand can feel its clamminess (related to skin conductivity); a friend leaning next to you may sense your heart pounding; students can hear changes in a professor's respiration that give clues to stress; ultimately, it is muscle tension in the face that gives rise to facial expressions" (Picard et al., 2001). It should also be noted that revealing changes that are not detectable by human contact indeed might be the most important contribution of physiological measures to emotion recognition (Cowie et al., 2001; Hudlicka, 2003).

In addition to arguments based on the neuro-physiological underpinnings of emotion and the recognition of emotion based on physiological state in everyday interaction, there is yet another strong reason for studying physiological signals when analyzing emotional state. As has been pointed out in publications too numerous to cite, each and every single modality used for emotion recognition has its limitations. Four examples among the different measurement modalities (including physiological measurement) are as follows:

(1) Self-reporting of emotional state can be erroneous or incomplete due to various psychological factors.
(2) Speech and facial patterns are strongly linked to social interaction, and so they tend to provide very little information in the case of a person who is not interacting socially.
(3) Facial or body gestures can be muted or altered in an attempt to conceal emotional state.
(4) Physiological response correlates of emotion (even within one individual) are highly context dependent (see Bradley and Lang, 2007 for an overview).

Combining multiple modalities, where any single modality can be "noisy" or ambiguous (including physiological signals), has the potential to show significant improvement in recognizing patterns that may provide clues to emotional state (e.g., Kim and André, 2006).

1.1 A Note on the Name "Affective Computing" and Physiological Measurement

Physiological measurement has an interesting connection with discipline names. Many psychologists use the term "affect" to describe a global phenomenon with rather direct links to physiology. For instance, Russell's group uses the term "core affect" to describe a primitive sense of positiveness or negativeness; energy or sloth that they hypothesize underpins emotion and other complex phenomena. They also propose that it has links to certain relatively gross kinds of brain activity (Russell, 2003, see also chapter "Editorial: 'Theories and Models' of Emotion"). It is natural to think that what body sensors measure may be linked to affect in that sense rather than the subtler phenomena that philosophers call emotions. Hence some researchers who focus on physiological measurement as part of human–computer interaction also find it natural to call their work "affective computing." However, as has been mentioned and will be discussed further in this chapter, there are indeed unequivocal correlations between physiological changes and specific emotional states and using the term "affective" might be inappropriately limiting.

2 Definitions of "HCI-Appropriate," Emotion-Related Physiological Signals

Before discussing the history and usage of human physiology as part of a repertoire of emotion measurement, it would make sense to define the physiological signals to be used and briefly explain how they are currently measured. There are a great many textbooks and articles that define these signals and explore their usage in fields ranging from medicine and psychology to bioengineering and instrumentation. Two classic books are *The Handbook of Psychophysiology* (Cacioppo et al., 2007), which defines each physiological signal and summarizes its history and usage within the broad field of psychophysiology, and *Medical Instrumentation, Application and Design* (Webster, 1998), a classic engineering handbook that reviews physiological signals in the context of the medical instrumentation needed to capture each signal.

Among the many methods for measuring changes in human physiology, this section will focus on briefly defining those physiological parameters or signals which are currently being used in the fields of psychophysiology and emotion *and* are measurable in a relatively unobtrusive way (and thus could be envisaged to be part of a human–machine interface). It should be emphasized that many physiological signals have only recently met these criteria due to advances in measurement technologies. Thus, one can expect that the introduction of new technologies will serve to expand this list over the coming years (or months!). A good review (although not entirely complete) of so-called ambulatory monitoring systems, systems that can be used in mobile environments and are relatively unobtrusive, can be found in Ebner-Priemer and Kubiak (2007). In defining these key physiological signals, we will categorize them into physiological signals that originate from the autonomic and somatic components of the peripheral nervous system and physiological signals that originate from the central nervous system (see Fig. 1 for a taxonomy of the nervous system).

2.1 Physiological Data Acquisition

In collecting data from physiological sensors, the signals must ultimately be captured by a data acquisition system connected to a digital system ranging from an

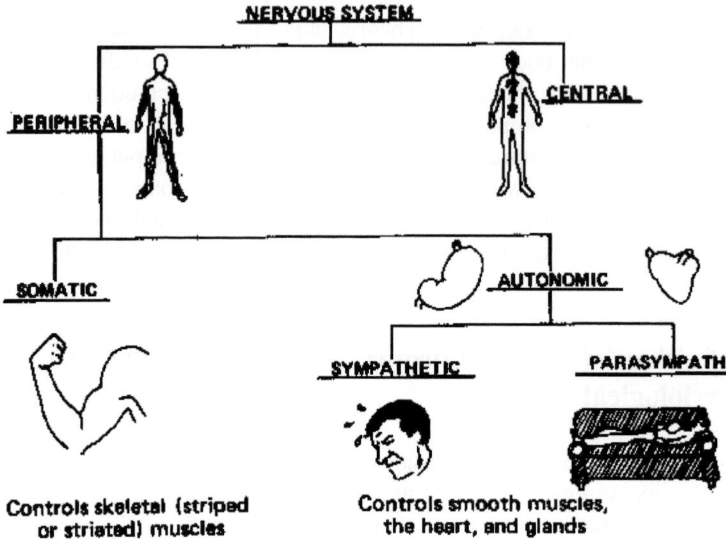

Fig. 1 Taxonomy of the nervous system (from online.sfsu.edu/~psych200/unit5/u5m5.gif)

embedded processor to a super-computer. The choice of available data acquisition systems is vast. Some of the important factors in narrowing this range of choices include the following:

- *Signal isolation*: In order to reduce the chance of electrical shock, *the data acquisition system must electrically isolate the individual from "wall" or "mains" power as well as from the computer itself.* This can be achieved by using a battery-powered wireless data acquisition system or by using a system specifically designed for physiological data acquisition that already incorporates optical or magnetic isolation.
- *Obtrusiveness*: There is a considerable trade-off in the number and type of physiological signals chosen and the obtrusiveness of the measurement. At the limits, either not enough data is being measured or the individual cannot participate in the interaction in any way that is not interfered with by the measurement equipment. While this is one of the most difficult choices in using physiological measures for emotion measurement, there is little to no literature on the impact of the ecology of the measurement on the emotional estimation "accuracy."

It should be pointed out that while so-called wireless data acquisition systems are usually superior to wired systems in terms of their obtrusiveness, they might indeed include many wires connecting the physiological sensors to the wireless transmitter. There are very few commercially available systems that combine the sensor with the transmitter to eliminate all wires and are most commonly found in the consumer sports arena from companies such as Polar, Suunto, or Nike.

- *Sampling rate and resolution*: These will be based on the bandwidth and dynamic range of the physiological signal acquired. The range of sampling rates is considerably reduced for wireless data acquisition systems.
- *Synchronization of multiple data streams*: In many cases, more than one channel of physiological data is analyzed simultaneously (or combined with video or audio streams). The data acquisition system – both hardware and software – must have the capability of synchronizing multiple streams of data with multiple sampling rates.
- *Shielding and differential amplification*: The amplitude of the voltage of physiological signals can be as small as a fraction of a microvolt. In order to avoid contamination from other signals, so-called noise signals, electrical shielding of the signal wires from sensor to data acquisition system is extremely important. Also, data acquisition systems using a technique known as differential amplification, amplifying only the difference between two sensor signals, should be used if possible. By amplifying only the difference between two sensor signals, any noise that is common to both sensors will be considerably reduced.

2.2 Autonomic and Somatic Nervous System

The somatic component of the peripheral nervous system is concerned with sensing information that happens outside the body and is responsible for the voluntary control of our skeletal muscles to interact with this external environment. Signals that measure this voluntary control of the muscles are measuring aspects of the somatic nervous system. The autonomic (ANS) component of the peripheral nervous system is responsible for sensing what happens within the body and regulating involuntary responses including those of the heart and smooth muscles (muscles that control such things as the constriction of the blood vessels, the respiratory tract, and the gastrointestinal tract). There are two components of the ANS, the parasympathetic and the sympathetic (see Figs. 2 and 3). The parasympathetic component is responsible for slowing the heart rate and relaxing the smooth muscles. The sympathetic component of the ANS is responsible for the opposite, i.e., raising the heart rate and constricting the blood vessels which causes, among other effects, an increase in blood pressure. It is also responsible for changes in skin conductivity. The sympathetic response is slower and longer lasting than the parasympathetic response and is associated with the so-called flight or fight reaction. Physiological signals that measure the involuntary responses of the peripheral nervous system are measuring aspects of the autonomic nervous system.

2.2.1 Electrodermal System

Measurement of the electrodermal activity or EDA is one of the most frequently used techniques to capture the affective state of users, especially for exploring

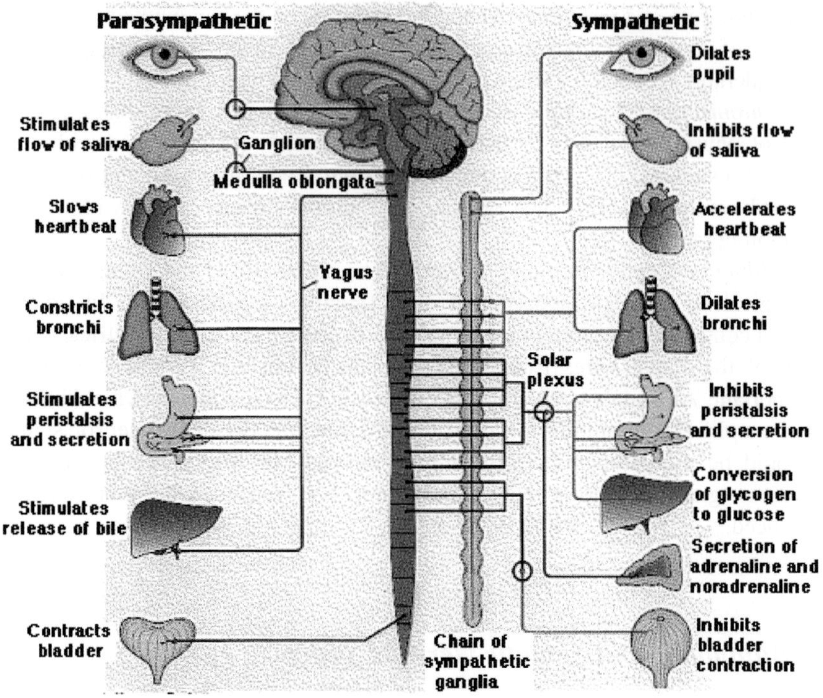

Fig. 2 Autonomic nervous system (from http://www.wickersham.us/anne/images/autonomic.gif)

attention and varying arousal in emotion. It is controlled by the sympathetic nervous system. EDA sensors (electrodes) measure the ability of the skin to conduct electricity. A small fixed voltage is applied to the skin through the electrodes and the skin's current conduction or resistance is measured (this is preferred over applying a current and measuring the voltage produced). The value of this conductivity is usually in the range of 2–20 μS (500 kΩ–50 kΩ). The skin conductivity consists of two separate components. There is a slow-moving tonic component or skin conductance level (SCL) that indicates a general activity of the perspiratory glands from temperature or other influences and a faster phasic component or skin conductance response (SCR – also known as galvanic skin response or GSR) that is influenced by emotions and the level of arousal. For example, when a subject is startled or experiences anxiety, there will be an increase in the skin conductance due to increased quantity of sweat in the sweat ducts of the glands.

EDA is most significant on the palm of the hands and the bottom of the feet. The two most common measurement techniques are thus

1. to place an electrode on each of two fingers (usually the thumb and the index finger) and
2. to place two electrodes across the palm.

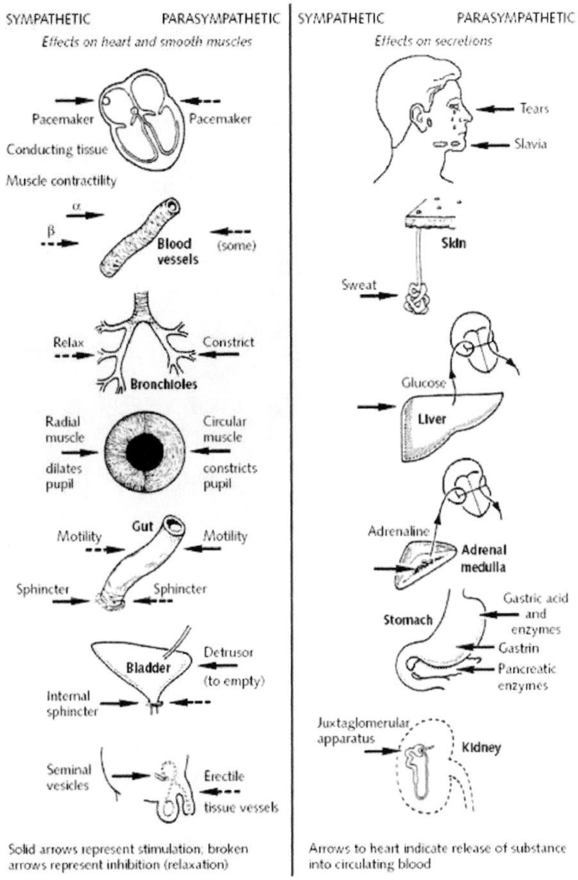

Fig. 3 Autonomic nervous system (after Jennett (1989) Human Physiology)

For the most accurate recordings, the electrodes consist of silver–silver chloride cup electrodes. However, in using the measurement of EDA as part of an HCI system, the EDA is commonly measured using standard conductive metal plates (Fig. 4).

Changes in physical activity, environmental conditions such as temperature and humidity, movement of the electrodes on the skin, and changes in pressure on the electrodes all serve to confound the measurement of EDA and must be mitigated. Physical activity and ambient conditions primarily affect the SCL and should be measured using other modalities (motion sensors, temperature and humidity sensors, etc.) in an attempt to limit the artifact.

In addition to the amplitude of the SCR and SCL, temporal parameters of the SCR such as latency, rise and decay time are all important features of the EDA used in determining attention and emotional state.

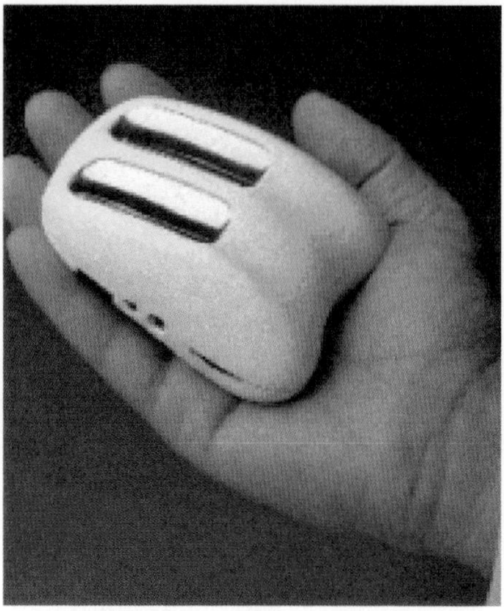

Fig. 4 Mouse used to measure EDA (from Thought Technology)

2.2.2 Cardiovascular System

Heart Rate and Heart Rate Variability

Another important physiological correlate of emotion is the frequency or the period of contraction of the heart muscle. As shown in Fig. 3, the heart rate (HR) is controlled by both the parasympathetic response (decreasing HR) and sympathetic response (increasing HR). The heart rate or period can be derived from many measurement techniques as will be discussed below. The higher frequency changes (0.15–0.4 Hz) are heavily influenced by breathing [respiratory sinus arrhythmia (RSA)], especially with younger and more physically fit individuals. The lower frequency changes (0.05–0.15 Hz) are not influenced by the RSA and can reveal other aspects of the ANS.

***Electrocardiography (ECG)*:** Electrocardiography measures electrical changes associated with the muscular contraction of the heart. More specifically, the ECG results from the sino-atrial (SA) node and atrio-ventricular (AV) node of the heart electrically activating the first of two small heart chambers, the atria, and then the two larger heart chambers, the ventricles. Particularly, the contraction of the ventricles produces the specific waveform known as the QRS complex as shown in Fig. 5. The heart rate is most commonly derived from the ECG by measuring the time between R components of the QRS complex – the so-called R–R interval.

The ECG is measured at the body surface across the axis of the heart. Electrodes are placed on the skin; they transduce the electric field caused by the previously mentioned electrical activity of the heart to electron flow in the measurement lead. This signal is then high-pass filtered to remove long-term offset and amplified by a factor of around 1,000. In medicine, the ECG is most commonly measured using

Fig. 5 ECG wave

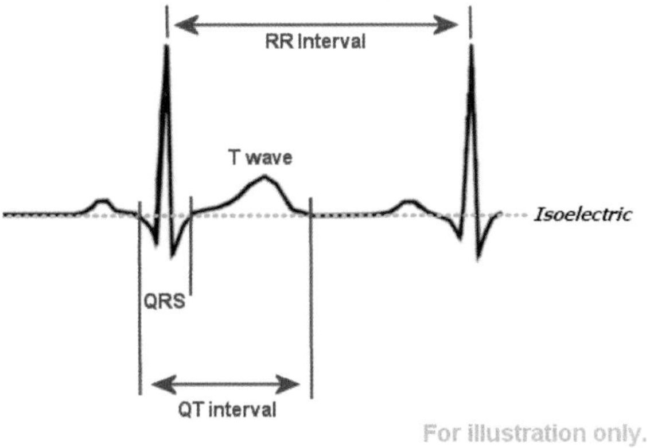

Fig. 6 Watch with wireless ECG-derived heart rate monitor (www.suunto.com)

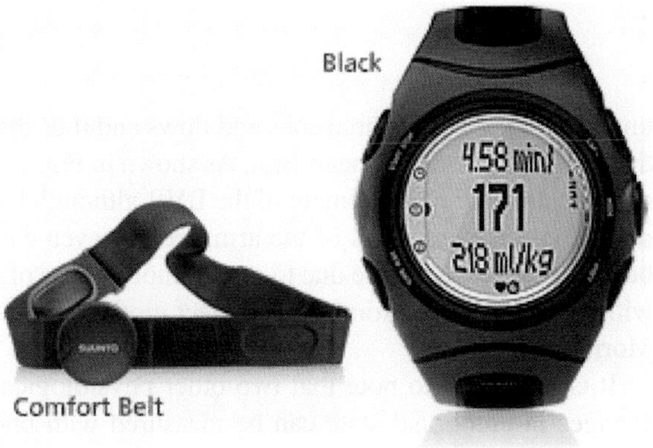

a standard 12-lead (electrode) configuration; however, when only the heart rate is being measured, this can be reduced as low as a 2-lead configuration as shown in Fig. 6. Indeed, as is done with exercise equipment, heart rate measurements can even be recorded simply by measuring the voltage potential from hand to hand (although not nearly as accurate as chest leads).

In addition to measuring heart rate, one other aspect of the ECG wave that is important to emotion research is the amplitude of the T wave shown in Fig. 5. This is because the T wave amplitude has been found to be an indicator of activity of the sympathetic nervous system.

Photoplethysmography and Blood Volume Pulse (BVP): An alternative to measuring heart rate directly from the electrical activity of the heart is to measure the pulsation of blood flow through the vascular system. The most common technique for achieving this is to shine light from an infra-red LED into the skin and measure either the amount of light transmission through the skin or the amount of light reflection from the skin (or both). As the heart beats, the perfusion of the blood vessels

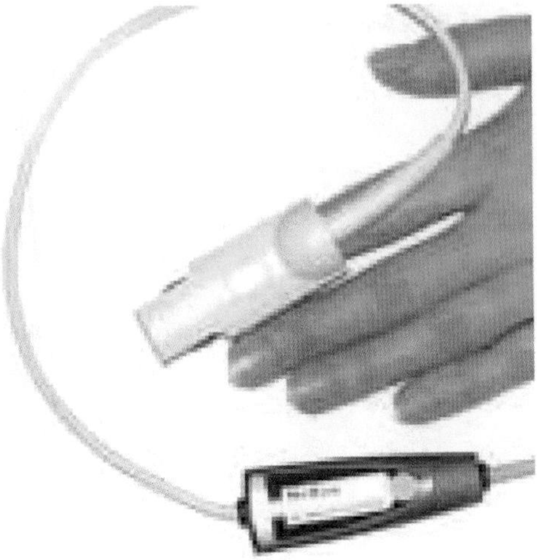

Fig. 7 Photoplethysmography (from http://www.medis-de.com)

underneath the light source ebbs and flows and thus the absorption characteristics of the light pulses with the heart beat. As shown in Fig. 7, the finger tip is the most common location for measurement of the BVP, although it can also be measured in other places on the extremities of the arm, feet, or even earlobe. Photoplethysmography becomes quite inaccurate due to even minor motion of the body and is often coupled with the use of accelerometers to detect motion and to attempt to compensate (e.g., Morris et al., 2008).

It is important to note that two other physiological parameters associated with changes in emotional state can be measured with photoplethysmography: peripheral blood perfusion (e.g., Kunzman and Gruhn 2005) and blood oxygen saturation (SpO_2) levels (e.g., Karekla et al., 2004). The effect of the change in magnitude of the BVP because of changes in peripheral blood perfusion can also cause the measurement of HR and HRV to be somewhat less accurate due to missed beats.

Impedance plethysmography works similar to photoplethysmography except that instead of measuring the change in blood perfusion by measuring the change in light reflectance/transmission, impedance plethysmography measures the change in the skin's capability to conduct electricity. This is most often measured by applying a small AC current across the chest and measuring the change in voltage. This technique has enabled the creation of clothing that can measure HR and HRV.

In addition to modulation of the heart rate, emotional changes can influence the contractility of the blood vessels. This change can affect blood pressure and perfusion as well as skin temperature.

As mentioned previously, peripheral perfusion can be quite ergonomically measured by photoplethysmography. Just as simply, skin temperature can be measured by placement of a thermistor on the skin surface (although this is confounded considerably by ambient temperature). Ecological measurement of blood pressure,

Fig. 8 Portapress ambulatory blood pressure monitor (www.finepress.com)

however, has still not been achieved. This is because the most common (and most accurate) non-invasive techniques for measuring blood pressure require that a cuff is placed on the arm and is inflated to cut off arterial blood flow. The acoustical changes that appear as this flow is returning to normal [first the appearance of a heart rhythm (*k*) sound and then the disappearance of the *k* sound] as the cuff is deflated indicate the systolic and diastolic blood pressure, respectively. This inflation and deflation of the cuff make it difficult to measure continuous blood pressure changes. Although expensive and large, one of the few ambulatory devices available is the Portapress monitor shown in Fig. 8.

It is argued that pulse wave velocity, the measurement of the time it takes for a pulse to move down the arm, is proportional to changes in blood pressure. See Harata et al. (2006) for a review. The pulse arrival time can be measured at two locations using photoplethysmography and then the pulse transit time (and thus pulse wave velocity) can be calculated. An example of a current attempt at using the ecological measurement of pulse wave velocity for emotion tracking can be seen in Fig. 9. There is relatively general agreement that this measurement cannot be used for measuring the absolute value of blood pressure due to the affects of the contractility of the arm's vascular system.

2.2.3 Respiratory System

The respiratory system is one of several systems of the body that are under both autonomic and voluntary (somatic nervous system) control. Thus respiration rate and depth as an indicator of emotional state must be used with caution and always viewed in context. In controlled environments, it has been found that variation of respiration rate generally decreases with relaxation. Startle events and tense situations may result in momentary respiration cessation and negative emotions generally cause irregularities in the respiration pattern.

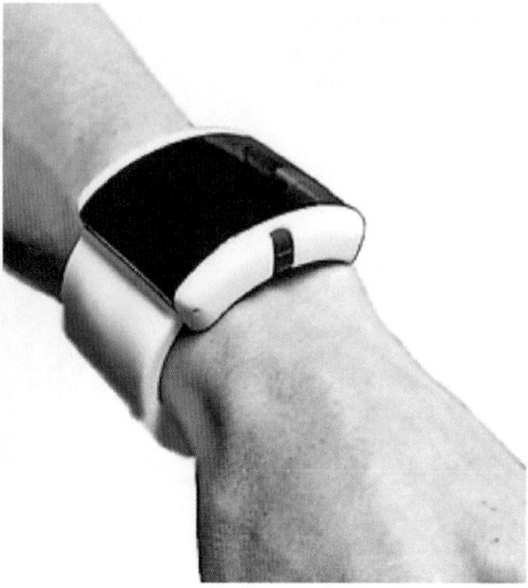

Fig. 9 Pulse wave velocity measurement of blood pressure (www.exmocare.com)

The respiration signal (breathing rate and intensity) is commonly acquired by using a strain gauge or a piezo sensor embedded in an elastic band worn around the chest. The sensor measures the expansion and contraction of the band, which is proportional to the respiration rate. Two bands are often used, one placed on the upper chest and another around the lower abdomen, to measure the depth of inspiration and exhalation.

Another technique for measuring respiration is impedance plethysmography – using the fact that the impedance of the chest cavity varies with respiration. This is highly advantageous because both heart rate and respiration can be determined from one chest strap or article of clothing.

One other ecological technique of measuring respiration rate is to analyze the respiratory sinus arrhythmia (RSA) from the ECG waveform (as discussed above). As with impedance plethysmography, both heart rate and respiration rate can be determined using a single chest band. Although not nearly as accurate as piezo or strain-gauge bands (due to baseline variation and noise in the ECG signal), with improving signal processing techniques, the so-called ECG-derived respiration (EDR) rate has become an increasingly common method of determining respiration rate (e.g., Yeon et al., 2007).

Blood oxygen saturation levels (SpO_2) as well as the pCO_2 levels can be used as another parameter in quantifying emotional response. The cause of changes in the oxygenation of the blood is multi-factorial. That is, it is not simply a function of respiration or cardiovascular activity or any one other system, but a combination of many systems. Photoplethysmography, as discussed previously, is the most ecological technique for measuring blood oxygenation.

2.2.4 Visible (Overt) Effects of Autonomic Physiological Changes: Tears, Eye Blinks, Pupil Dilation, and "Goose Bumps"

There are several overt changes that can directly indicate activity of the autonomic nervous system. While this chapter focuses on physiological changes, since these overt changes can be a direct (uncognitively mediated) function of the underlying physiology, they are worth mentioning. Overt properties of the eye that fall into this category include tears (tear volume and ocular hydration), eye blinks, and pupil dilation, which can be measured with various visual recognition systems that can co-exist with HCI. All have been associated with various changes in emotional state. Visual recognition systems have also been used in attempting to quantify "goose bumps" or "goose flesh" or the pilomotor reflex. The anecdotal reporting of the pilomotor reflex is commonly mentioned in descriptions of emotional response.

2.2.5 Muscle Activity

As has been discussed in other chapters of this book, it is well known that overt facial gestures are a well-studied indicator of emotional state. However, before visible movement occurs on the face, activation of the underlying musculature must occur. Indeed, there are many circumstances where there are measureable changes in the activation of the facial muscles and no visible facial gesture. This can be due to "rapid, suppressed, or aborted" expressions (Cacioppo et al., 1992) or due to the actual attachment of the muscular structure to the skin. Thus, the ability to measure muscular activation in the face is another point in which physiological measurement can supplement measurement of overt changes. Changes in muscular tension in other areas of the body such as the arms may also indicate an overall level of stress or, as with the face, indicate the presence of overt gestures that cannot be viewed with visual observation.

Measurement of muscle tension is commonly achieved using surface electromyography (sEMG or just EMG). Surface electromyography measures muscle activity by detecting the electrical potential that occurs on the skin when a muscle is flexed. This electrical potential is created by motor neurons depolarizing or "firing" causing the muscle fibers to contract. The rate (frequency) of depolarization is proportional to the amount of contraction (until the individual motor neuron begins to saturate). At the same time, more and larger motor neurons are recruited and begin to fire simultaneously. Thus, as shown in Fig. 10, an increase in muscle contraction is seen as an increase in amplitude of the EMG as well as a modulation of the frequency spectrum. The structure of motor neurons and the muscle fibers they innervate is called a motor unit and the potential measured by the surface EMG is commonly referred to as a motor unit action potential or MUAP. As with measurement of the ECG, EMG measurement involves the use of electrodes which are placed on the skin surface (Fig. 11).

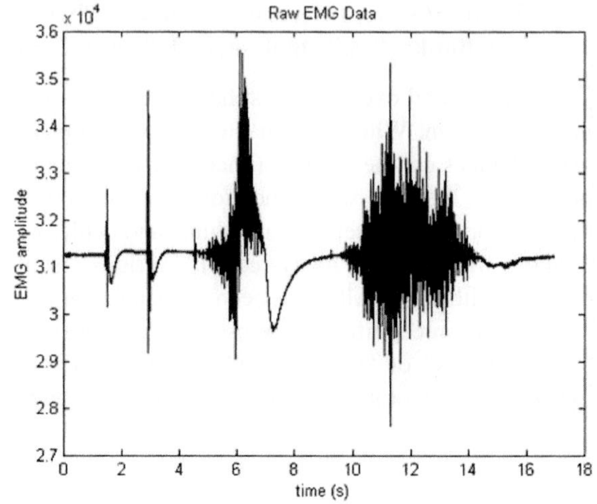

Fig. 10 EMG waveform. Note the four increasingly large contraction events indicated by an increase in amplitude and spectral complexity of the EMG

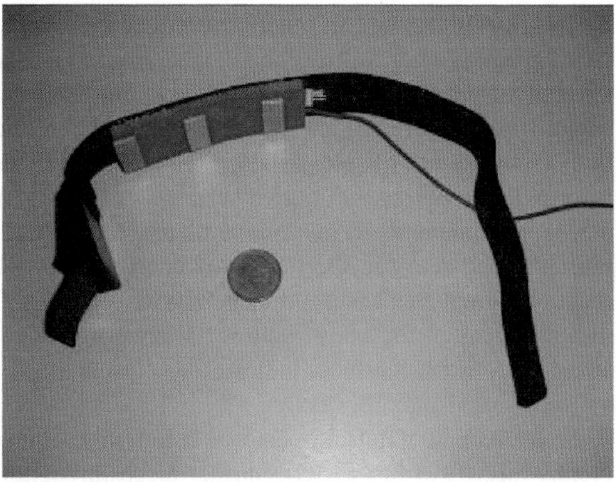

Fig. 11 An ecological EMG band for wearing on arm or leg (www.infusionsystems.com)

2.3 Central Nervous System

The central nervous system (CNS) is composed of the brain and the spinal cord and is responsible for processing information and controlling the activity of the peripheral nervous system. Over the past decade, functional imaging of the brain (imaging of the dynamic function rather than the static condition of the brain) during emotion-inducing activities has been yielding an increasing body of knowledge of how and where emotions are processed in the brain (Bradley and Lang, 2007). The capability to compare and correlate CNS activity as measured with functional imaging with the activities of the ANS as measured using the techniques discussed previously has the potential to reveal mapping functions between

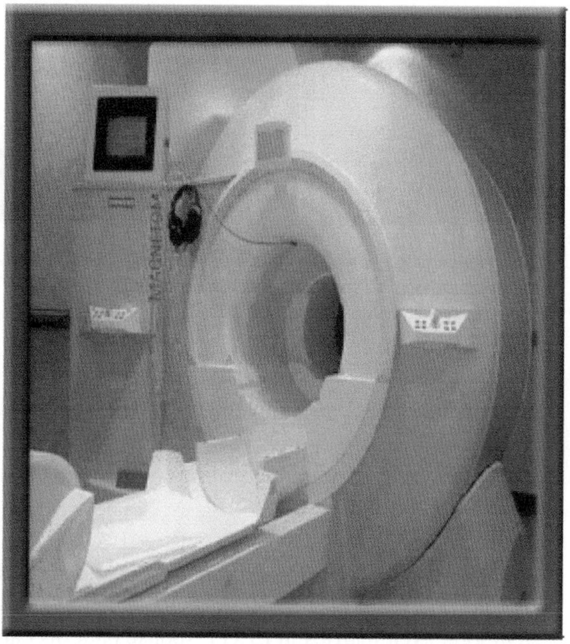

Fig. 12 fMRI machine

emotional stimulation and physiological response. Unfortunately, because of the enormous size, cost, and operational/environmental constraints, the most important functional imaging techniques such as positron emission tomography (PET) and functional magnetic resonance imaging (fMRI) are unlikely to be used as part of a human–machine interaction scenario for several years to come (see Fig. 12).

Electroencephalography (EEG) measures the electrical activity of the brain as it appears at the surface of the skull. As the millions of neurons within the brain "fire," the electric field generated by the electrochemical process can be measured using surface recording electrodes that function in a similar manner as the electrodes used for EMG and ECG. However, unlike ECG signals which are typically measured using up to 12 electrodes, EEG recordings can use anywhere from 3 to 256 (and even more!) electrodes (see Fig. 13). The choice of electrode quantity will depend on the desired number of locations on the head to be measured and the desired spatial resolution within any given location. The measurement of the EEG is also considerably more difficult than either the EMG or the ECG because many of the important features of the EEG signal are more than 1,000 times smaller than the EMG or ECG signal. Measurement of signals below 1 μV is common and can require that more attention is paid to cleaning the electrode site and applying electrolyte. Additionally, considerable signal averaging (comb filtering) and spectral and spatial filtering are required to remove non-EEG signals, "noise," from the EEG signal. These "noise"

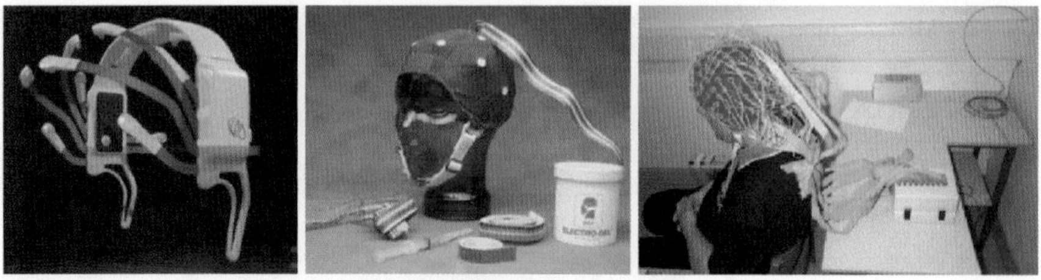

Fig. 13 Three examples of the range of electrode quantity found in EEG interfaces. From *left* to *right*: the emotivEpoc (www.emotiv.com), a standard 10–20 EEG array (www.biopac.com), and a 256 lead array (www.biosemi.com)

signals include EMG and movements of the eyes causing baseline potential shifts (EOG).

There are three general areas of research into the use of EEG measurement as an indicator of emotion:

1. *Spatial location and distribution of EEG signals*: While emotions are not located in any one particular location of the brain (Phan et al., 2004), amplitude and temporal asymmetries in EEG response patterns can yield insight into emotional state (e.g., Costa et al., 2006).
2. *The temporal response patterns of the EEG to stimuli – evoked response potentials (ERPs)*: The pattern of the time, location, and amplitude of the EEG response to auditory, visual, or emotional imagery can yield clues to emotional state and processing. While the ERP is studied at locations across the skull, the time response is limited to a narrow range from 100 mS (fast) to 1,000 mS (slow). This is one of the largest areas of research in the physiological correlates of emotional state – especially in the area of the correlation of ERPs and emotionally evocative visual stimuli (e.g., Holmes et al., 2003; Schupp et al., 2003).
3. *The frequency structure of the EEG signal – "brain waves"*: Synchronization of neural activity underneath the recording electrodes gives rise to oscillatory behavior in a collection of frequency bands. These band ranges include the following:

   - Delta band (1–4 Hz)
   - Theta band (4–8 Hz)
   - Alpha band (8–13 Hz)
   - Beta band (13–30 Hz)
   - Gamma band (36–44 Hz)

Correlations between the presence, timing, and location of these frequencies on the skull have been found to be related to several aspects of attention, vigilance, and emotional state (e.g., Sebastiani et al., 2003).

3 A Very Brief History of Physiological Measurement and Emotion

The history of using physiological signals as a means for recognizing emotion is a relatively short one. The use of this information as a means for augmenting human–machine interaction has only just begun.

3.1 Measuring the Correlations Between Physiology and Emotion

History is worth knowing in the field of emotion and physiology, because it is an area where popular conceptions of science are deeply attached to old ideas. William James (1890) captured the public imagination when he claimed (in 1884) that the essence of emotion was awareness of visceral changes that occur in response to extreme situations (such as meeting a bear in the woods) (Cannon, 1927). If that were so, then measuring the visceral changes directly would allow artificial systems to detect a person's emotions as well as the person him- or herself – or possibly better. In reality, even contemporary experts who are sympathetic to James accept that his idea captures only part of the truth. Injuries that prevent visceral feedback do not nullify emotion; visceral states are not as sharply distinguished as emotions are; experienced emotion can be changed by manipulating cognitive state, but not visceral states; and so on.

Not long after James released his idea, prominent figures (including Jung) took up the idea that skin conductivity could reveal otherwise hidden psychological events. Lie detection quickly became a high-profile application for it and related techniques. The idea captured the public imagination to the extent that employers were held liable for failing to use lie detection before hiring employees who went on to commit crimes. However, by 1959, it was becoming clear that standard forms of lie detection had questionable scientific validity (Lykken, 1998). It was not the physiological measures that were the focus of the uncertainty, but the methodology of the protocol and specifically the questions that were being asked. In studying emotion and, most importantly, physiology and emotion, it was becoming clear that context and induction techniques were critical.

Over the ensuing decades, psychophysiologists have built up an enormous body of evidence on relationships between physiological measures and emotion. Two good summaries are in Cacioppo et al. (2000) and Bradley and Lang (2007). This research has shown that if other variables are meticulously controlled, there are associations between emotional states and physiological variables. The phrase "meticulous control" is key. Physiological signals are subject to multiple influences. They are affected not only by emotion but also by almost any kind of effort, mental or physical. Traditional experiments dealt with these problems by creating situations that prevented irrelevant variables from intruding. Even with meticulous control, recent literature surveys such as found in Kreibig et al. (2007) show that correlations between physiological changes and changes in emotional state are not always consistent.

3.2 Physiology and Human–Machine Interaction

The question then becomes, "If, to find correlates of emotional states, physiological signals must be measured in meticulously controlled environments, how then can they be used as part of an ecological human–computer interaction paradigm?" To see how researchers have answered this, we must examine what has occurred in three separate fields.

3.2.1 Physiological Control of Music

The first endeavors to use physiological signals to control machines occurred well before mice and GUIs and even PCs existed. One of the most interesting examples, as shown in Fig. 14, was Alvin Lucier's piece in 1965 entitled *Music for Solo Performer*. In this performance he sonified the alpha activity of his EEG. While the idea was to listen to changes of cognitive state, i.e., there was no direct intention of quantifying emotional state, the synchronization of alpha activity is proportional to relaxation and so there was probably more than a little self-induction of a low activation emotion.

In 1978, Dick Raaijmaker used EMG, EDA, ECG, and acoustic measurement of respiration to sonify the level of stress of an individual dismounting a bicycle over the course of 30 min. While a large component of the changes in physiological state was caused by physical exertion, changes in emotional state throughout the course of the piece clearly influenced the sound.

The increasing use of physiological state (and by possible unintentional consequence, emotional state) as a tool for artistic expression led to the development by Knapp and Lusted in 1987 of the BioMuse (Knapp and Lusted, 1990). This was one of the first commercially available systems which enabled musical performers to use physiological state (in this case EMG, ECG, and EEG) to control consumer electronic musical instruments and introduced the capability to control the newly ubiquitous PC.

Fig. 14 From Alvin Lucier's *Music for Solo Performer* – controlling sound with EEG

3.2.2 Physiological Control of Computers (Without Emotional Assessment)

At this time, research on using physiological signals to augment human–computer interaction became established with work on EMG (Putnam and Knapp, 1993), EOG (LaCourse and Hludik, 1990), and the new field of brain–computer interfaces (BCIs) (Wolpaw et al., 1998) using EEG. Much of this work was targeted at improving interaction for those with disabilities (Lusted and Knapp, 1996). For example, ERPs from the EEG were combined with EOG to augment mouse control (Patmore and Knapp, 1995).

Research on physiologically augmented interfaces has continued to expand into many arenas including assistive living and computer gaming. From these investigations, new, more ecological interfaces are being created (Knapp and Lusted, 2005) from which forms the foundation for applying physiological measurement of emotion to HCI.

3.2.3 Physiological Control of Computers with Emotional State Assessment

Combining the research on physiological interfaces with the ongoing research on psychophysiology and emotion, Picard and her colleagues at the MIT Media Lab began developing what she termed "affective interfaces" and "affective clothing" (Picard and Healey, 1997). Interfaces ranging from jewelry to gloves (see Fig. 15) were being used to investigate whether correlates of emotional state could be found with ecological interfaces.

Many new physiologically based systems are currently being created in research centers and from commercial enterprises. Some of these can be seen in Figs. 4, 11,

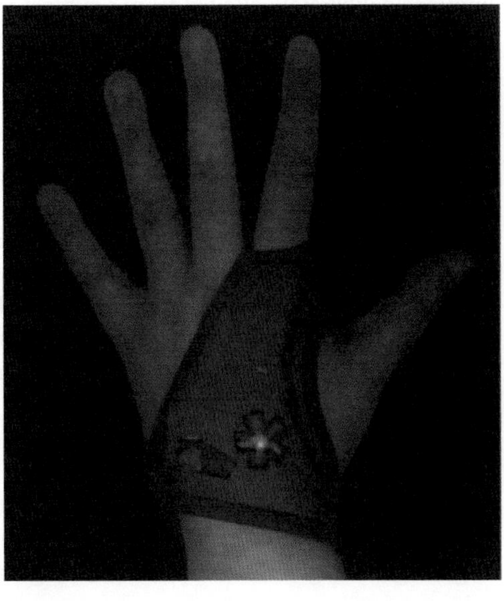

Fig. 15 The galvactivator – mapping EDA to light

and 13. In the field of music, some researchers are beginning to look at using emotion as part of conducting (Nakra, 2000) and performing (Knapp and Cook, 2005). It still remains to be seen, however, whether measurements using any of these new interfaces, even in conjunction with other measures such as facial recognition or speech recognition and operating in constrained environments, can accurately assess emotional state. This is the ultimate question of the current research, "Can an automatic recognition system be developed that can process physiological signals and other indicators of emotional state and come to any reasonably consistent result?"

4 The Present: Current Studies on the Automatic Recognition of Emotional States Using Physiological Signals

Current work on pattern recognition of physiological state for HCI is now being investigated by many research groups. The results are beginning to show that again, in constrained environments, it might be possible to use machine learning algorithms such as Bayesian classifiers, HMM, neural networks, and fuzzy systems to achieve statistically significant identification.

4.1 Emotion Recognition Using Only Physiological Signals

In 2001, using only one subject, Picard showed that certain affective states may be recognized by using physiological measures including heart rate, skin conductivity, temperature, muscle activity, and respiration velocity (Picard et al., 2001). Eight emotions deliberately elicited from a subject in multiple weeks were classified with an overall accuracy of 81%. Nasoz et al. (2003) used movie clips to elicit target emotions from 29 subjects and achieved the best recognition accuracy (83%) by applying the Marquardt backpropagation algorithm.

In other work (Haag et al., 2004), the IAPS photoset (IAPS, 1995) is used to elicit target emotions with positive and negative valence and variable arousal level from a single subject. They classified arousal intensity and valence of the emotions separately using neural network classifier and achieved recognition accuracy of 96.6 and 89.9%, respectively. More recently, an interesting user-independent emotion recognition system was reported by Kim et al. (2004). They developed a set of recording protocols using multimodal stimuli (audio, visual, and cognitive) to evoke targeted emotions (sadness, stress, anger, and surprise) from 175 children aged from 5 to 8 years. Classification ratio of 78.43% for three emotions (sadness, stress, and anger) and 61.76% for four emotions (sadness, stress, anger, and surprise) has been achieved by adopting support vector machine as pattern classifier. Particularly, analysis steps in the system are fitted to handle relative short length of input signal (segmented in 50 s) compared with the other previous works that require longer signal length of about 2–6 min. Wagner and colleagues (2005) presented an approach to the recognition of emotions elicited by music using four-channel biosignals which were recorded while the subject was listening to music songs and reached an overall recognition accuracy of 92% for a four-class problem.

Wagner and colleagues also applied statistical methods to find out which features are significant for a specific emotion class. It turned out that joy was characterized by a high SC and EMG level, deep and slow breathing, and an increased heart rate. In contrast, anger was accompanied by a flat and fast breathing. The SC and EMG level was high as well. Pleasure and sadness are well identified by a low SC and EMG signal, but pleasure has a faster heart rate. The results were, however, highly user dependent. When applying the same method to a data set collected by MIT, they could observe perceptible differences in the physiological responses of the recorded subjects. While in their own data set, positive emotions were characterized by a low SC level, the MIT data set showed a high level of SC. Nevertheless, it turned out that a high SC and EMG level was also a good indicator in general for high arousal. They could also well correlate a higher breathing rate with the emotions in negative valence.

Physiological data sets used in most of the works are obtained by using visual elicitation methods in a lab setting where subjects intentionally express desired emotion types while looking at selected photos or watching movie clips. A recognition accuracy of over 80% on the average seems to be acceptable for realistic applications. However, it can be clearly observed that the accuracy strongly depends on the data sets (which were obtained in laboratory conditions). That is, the results were achieved for specific users in specific contexts. In view of a generally applicable recognition system for realistic online applications, it is desirable to automatically select the most significant features and tune specific classifiers to manifold data sets obtained from different natural contexts.

4.2 Relation, Dependencies, and Correlation of Physiological Signals with Other Modalities

The integration of multiple modalities for emotion recognition arouses an intuitive expectation of better recognition accuracy rate compared to unimodal analysis. On the other side, however, it requires to solve more complex analysis and classification problems. Furthermore, there are some well-known interactions which prevent any simple interpretation of physiological signals as "emotion channels." Because respiration is closely linked to cardiac function, a deep breath can affect other measures, for example, EMG and SC measurements. Particularly, in an experiment by [Kim06], such an irregularity could be observed when the subject is talking. However, it is also possible to imagine ways of turning a multimodal analysis to advantage. For instance, the respiration wave shows particular abrupt changes corresponding to certain facial muscle activity. It is also possible that EMG sensors on the jaws or forehead could improve accuracy in computation of FAP variations in facial emotion analysis. In psychophysiology, EMG was often used to find the correlation between cognitive emotion and physiological reactions. In (Sloan, 2004), for example, EMG has been positioned on the face (jaw) to distinguish "smile" and "frown" by measuring activity of zygomatic major and corrugator supercilii. EMG sensors could also be positioned on the body to measure muscle contraction intensity for gesture recognition.

Kim and colleagues (2005) investigated how far the robustness of an emotion recognition system can be increased by integrating both vocal and physiological cues. They evaluated several fusion methods as well as a hybrid recognition scheme and compared them with the unimodal recognition methods. The best results were obtained by feature-level fusion in combination with feature selection (78%). They did not achieve the same high gains that were achieved for video–audio analysis, which seems to indicate that speech and physiological data contain less complementary information. Furthermore, they investigated the subjects' emotional response in a naturalistic scenario where it cannot be excluded that the subjects are inconsistent in their emotional expression. Inconsistencies are less likely to occur in scenarios where actors are asked to deliberately express emotions via speech and mimics, which explains why fusion algorithms lead to a greater increase of the recognition rate in the latter case.

Experiments by Chanel and colleagues (2006) indicate that an integrated analysis of EEG and physiological data might improve robustness of emotion assessment since some participants were more expressive in their physiological response, while for others better results were obtained with EEG data.

As noted above, there are two views on emotions: emotions as the result of bodily reactions and emotions as the result of cognitive process. To analyze emotions by a computer system, a combination of both views has been proven useful. When determining the user's emotional state, we have evidence from causal factors, such as events during the interaction with a computer system. Apart from that, we also have evidence from the consequences of emotional states, being the physiological response. Conati et al. (2003) successfully employed Bayesian networks within a probabilistic framework to model such a bidirectional derivation of emotional states from its causes as well as its consequences in a game-like tutor system.

4.3 Emotion Recognition in an Ecological Environment

In most of the work that has been cited, data sets used were obtained in a lab setting where subjects intentionally express desired emotion types while looking at selected photos, listening to music, or watching movie clips. A recognition accuracy of over 80% on the average seems to be acceptable for realistic applications. However, it remains unknown what level of information can be derived from biosensors when emotion is part of action and interaction and multiple factors impinge on variables such as heart rate and respiration. The fact that physiological sensors are very sensitive to motion artifacts makes it hard to employ them in everyday situations.

As a first step toward a more natural scenario, Kim and colleagues (2005) used a slightly modified version of the quiz "Who wants to be a millionaire?" Questions

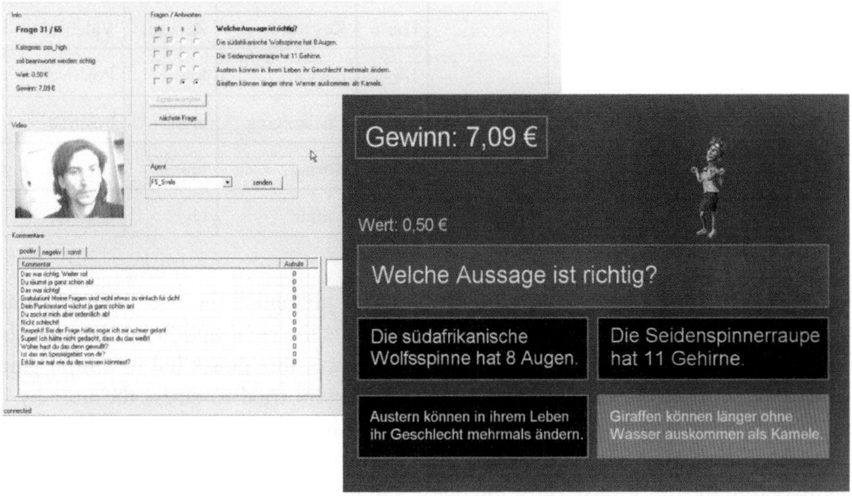

Fig. 16 Interface for the wizard (*left-hand side*) and for the subjects (*right-hand side*)

along with options for answers were presented on a graphical display whose design was inspired by the corresponding quiz shows on German TV. The subjects were equipped with a directed microphone to interact with a virtual quiz master via spoken natural language utterances and six-channel biosensors. In addition to the biosensors mentioned above, blood volume pulse (BVP) and skin temperature (TEMP) were used. The virtual quiz master was represented by a disembodied voice using the AT&T Natural Voices speech synthesizer (see right-hand side of Fig. 16).

While the users interacted with the system, their physiological and speech signals as well as the interaction with the quiz master were recorded. The quiz experiment was designed in a Wizard-of-Oz fashion where the quiz agent who presents the quiz is controlled by a human quiz master who guides the actual course of the quiz, following a working script to evoke situations that lead to a certain emotional response. The wizard was allowed to freely type utterances, but also had access to a set of macros that contained pre-defined questions or comments which made it easier for the human wizard to follow the script and to get reproducible situations (see left-hand side of Fig. 16).

In order to get a sufficient amount of data, it might be necessary to induce certain emotional states in the user. To accomplish this task, the wizard's working script in the experiment described above was divided into four situations which correspond to positions on the axes of a two-dimensional emotion model (see Fig. 17): (1) low arousal, positive valence; (2) high arousal, positive valence; (3) low arousal, negative valence; and (4) high arousal, negative valence.

First, the users were offered a set of very easy questions every user is supposed to know to achieve equal conditions for all of them. This phase was characterized by a slight increase of the score and gentle appraisal of the agent and served to induce an emotional state of positive valence and low arousal in the user. In phase 2, the user was confronted with extremely difficult questions nobody is supposed to

Fig. 17 Phases of the Kim experiment

Phase	Keyword	Arousal	Valence
1	Easy	low	Positive
2	Challenging	high	Positive
3	Boring	low	negative
4	Unfair	high	negative

know. Whatever option the user chose, the agent pretended the user's answer was correct so that the user got the feeling that s/he hits the right option just by chance. In order to evoke high arousal and positive valence, this phase led to a high gain of money. During the third phase, the experimenters tried to stress the user by a mix of solvable and difficult questions that led, however, not to a drastic loss of money. Furthermore, the agent provided boring information related to the topics addressed in the questions. Thus, the phase should lead to negative valence and low arousal. Finally, the user got frustrated by unsolvable questions. Whatever option the user chose, the agent always pretended the answer was wrong, resulting in a high loss of money. Furthermore, the experimenters included simple questions for which they offered similar sounding options. The user was supposed to choose the right option, but the experimenters made him/her believe that the speech recognizer was not working properly and deliberately selected the wrong option. This phase was intended to evoke high arousal and negative valence.

The setting allowed the experimenters to evoke spontaneous physiological reactions under controlled conditions. The emotional states were unconsciously expressed (apart from rare cases where users deliberately tried to conceal their emotions) and the display of emotions was not instructed. Despite the relatively natural scenario, we have to consider that the data were still recorded under laboratory conditions. As long as users are equipped with biosensors, we cannot avoid that they are aware that they are being monitored.

5 Summary

The debate of which emotions can be distinguished on the basis of physiological changes is far from being resolved. Empirical studies done in psychophysiology provide evidence that there is a strong relationship between physiological reactions and emotional/affective states of humans. Nevertheless, emotion assessment from physiological signals still bears a number of challenges. It is very hard to uniquely map physiological patterns onto specific emotion types. Physiological patterns may widely differ from user to user and from situation to situation. On the other hand, physiological signals have considerable advantages for emotion assessment. We can continuously gather information about the users' emotional changes while they are connected to biosensors. Moreover, physiological reactions should be more robust than visual techniques against possible artifacts of human social masking.

Combining physiological signals with other modalities has proven to be a positive next step in recognizing emotional state.

However, it is still open to question whether results that can be obtained in tightly controlled situations can be transformed into ecological techniques that work in real-life situations or whether HCI using physiological sensors to measure emotion will repeat the pattern that was seen with lie detection – high public expectations, which in the long run are not matched by results (National Research Council, 2003).

References

Bradley MM, Lang PJ (2007) "Emotion and motivation" in handbook of psychophysiology. Cambridge University Press, Cambridge, MA, pp 582–607

Cacioppo JT, Berntson GG, Larsen JT, Poehlmann KM, Ito TA (2000) The psychophysiology of emotion. In: Lewis M, Haviland-Jones JM (eds) Handbook of emotions. Guilford Press, New York, NY, pp 173–191

Cacioppo JT, Bush LK, Tassinary LG (1992) Microexpressive facial actions as a function of affective stimuli: replication and extension. Pers Soc Psychol Bull 18:515–526

Cacioppo JT, Gardner WL (1999) Emotion. Annu Rev Psychol 50:191–214

Cacioppo JT, Tassinary LG, Bernston GG (eds) (2007) Handbook of psychophysiology, 3rd edn. Cambridge University Press, Cambridge, MA

Cannon WB (1927) The James–Lange theory of emotions: a critical examination and an alternative theory. Am J Psychol 39:106–127

Center for the Study of Emotion and Attention [CSEA-NIMH] (1995) The international affective picture system: digitized photographs. Center for Research in Psychophysiology, University of Florida, Gainesville, FL

Chanel G, Kronegg J, Grandjean D, Pun T (2006) Emotion assessment: arousal evaluation using EEG's and peripheral physiological signals. In: MRCS06, Istanbul, pp 530–537

Conati C, Chabbal R, Maclaren H (2003) A study on using biometric sensors for detecting user emotions in educational games. In: 3rd workshop on affective and attitude user modeling, Pittsburgh, PA, Jun 2003

Costaa T, Rognonib E, Galati D (2006) EEG phase synchronization during emotional response to positive and negative film stimuli.Neurosci Lett 406(3):159–164

Cowie R, Douglas-Cowie E, Tsapatsoulis N, Votsis G, Kollias S, Fellenz W, Taylor JG (2001) Emotion recognition in human–computer interaction. IEEE Signal Process Mag 18:32–80

Ebner-Priemer UW, Kubiak T (2007) Psychological and psychophysiological ambulatory monitoring, a review of hardware and software solutions. Eur J Psychol Assess 23(4):214–227

Haag A, Goronzy S, Schaich P, Williams J (2004) Emotion recognition using bio-sensors: first steps towards an automatic system. In: Affective dialogue systems. Lecture Notes in Computer Science, 2004, vol 3068/2004, pp 36–48, DOI: 10.1007/978-3-540-24842-2_4

Harata K, Kawakami M, O'Rourke M (2006) Pulse wave analysis and pulse wave velocity: a review of blood pressure interpretation 100 years after Korotkov.Circ J 70(10):1231–1239

Holmes A, Vuilleumierb P, Eimera M (2003) The processing of emotional facial expression is gated by spatial attention: evidence from event-related brain potentials. Cogn Brain Res 16: 174–184

Hudlicka E (2003) To feel of not to feel: the role of affect in human–computer interaction. Int J Hum Comput Stud 59:1–32

James W (1890) The principles of psychology. Holt, New York, NY

Jennett S (1989) Human physiology. Churchill Livingstone, Edinburgh

Karekla M, Forsyth JP, Kelly MM (2004) Emotional avoidance and panicogenic responding to a biological challenge procedure.Behav Ther 35(4):725–746

Kim J, André E (2006) Emotion recognition using physiological and speech signal in short-term observation. In: Perception and interactive technologies, LNCS, Springer-Verlag, Berlin Heidelberg, vol 4021/2006, pp 53–64

Kim J, André E, Rehm M, Vogt T, Wagner J (2005) Integrating information from speech and physiological signals to achieve emotional sensitivity. In: Proceedings of the 9th European conference on speech communication and technology, Lisbon, Portugal, 4–8 September 2005

Kim KH, Bang SW, Kim SR (2004) Emotion recognition system using short-term monitoring of physiological signals. Med Biol Eng Comput 42:419–427

Knapp RB, Cook PR (2005) The integral music controller: introducing a direct emotional interface to gestural control of sound synthesis. In: Proceedings of the international computer music conference (ICMC), Barcelona, Spain, 4–9 Sept 2005

Knapp RB, Lusted HS (1990) A bioelectric controller for computer music applications. Comput Music J 14(1):42–47

Knapp RB, Lusted HS (2005) Designing a biocontrol interface for commercial and consumer mobile applications: effective control within ergonomic and usability constraints. In: Proceedings of the 11th international conference on human computer interaction, Las Vegas, NV, 22–27 Jul 2005

Kreibig SD, Wilhelm FH, Roth WT, Gross JJ (2007) Cardiovascular, electrodermal, and respiratory response patterns to fear- and sadness-inducing films. Psychophysiology 44:787–806

Kunzmann U, Gruhn D (2005) Age differences in emotional reactivity: the sample case of sadness. Psychol Aging 20:47–59

LaCourse JR, Hludik FC (1990) An eye movement communication – control system for the disabled. IEEE Trans Biomed Eng 37:1215–1220

Lusted HS, Knapp RB (1996) Controlling computers with neural signals. Sci Am 275(4):82–87

Lykken DT (1998) A tremor in the blood: uses and abuses of the lie detector. McGraw-Hill, New York, NY

Morris M, Dishongh T, Guilak F (2008) Apparatus for monitoring physiological, activity, and environmental data. USPTO Application #: 2008 0154098.2006

Nakra TM (2000) Inside the conductor's jacket: analysis, interpretation, and musical synthesis of expressive gesture. MIT Media Laboratory Perceptual Computing Section Technical report no. 518

Nasoz F, Alvarez K, Lisetti C, Finkelstein N (2003) Emotion recognition from physiological signals for presence technologies. Int J Cogn Technol Work Spec Issue Presence 6:1

National Research Council (2003) The polygraph and lie detection. Committee to review the scientific evidence on the polygraph. Division of Behavioural and Social Science and Education. The National Academic Press, Washington, DC

Noh YS, Park SJ, Park SB, Yoon HR (2007) A novel approach to classify significant ECG data based on heart instantaneous frequency and ECG-derived respiration using conductive textiles. In: Proceedings of the 29th annual international conference of the IEEE EMBS, 22–26 Aug 2007, pp 1503–1506

Patmore DW, Knapp RB (1995) A cursor controller using evoked potentials and EOG. In: Proceedings of the RESNA '95 annual conference, Vancouver, 9–14 Jun 1995, pp 702–704

Phan KL, Wager TD, Taylor SF, Liberzon I (2004) Functional neuroimaging studies of human emotions. CNS Spectr 9:258–266

Picard RW, Healey J (1997) Affective wearables.Pers Technol 1(4):231–240

Picard R, Vyzas E, Healy J (2001) Toward machine emotional intelligence: analysis of affective physiological state. IEEE Trans Pattern Anal Mach Intell 23(10):1175–1191

Putnam WL, Knapp RB (1993) Real-time computer control using pattern recognition of the electromyogram. In: Proceedings of the IEEE international conference on biomedical engineering, San Diego, CA, 27–29 Oct 1993, pp 1236–1237

Russell JA (2003) Core affect and the psychological construction of emotion.Psychol Rev 110(1):145–172

Sebastiani L, Simoni A, Gemignani A, Ghelarducci B, Santarcangelo EL (2003) Human hypnosis: autonomic and electroencephalographic correlates of a guided multimodal cognitive–emotional imagery.Neurosci Lett 338(1):41–44

Schupp HT, Junghofer M, Weike AI, Hamm AO (2003) The selective processing of briefly presented affective pictures: an ERP analysis. Psychophysiology 41:441–449

Sloan DM (2004 November) Emotion regulation in action: emotional reactivity in experiential avoidance. Behav Res Therapy 42(11):1257–1270

Wagner J, Kim J, André E (2005) From physiological signals to emotions: implementing and comparing selected methods for feature extraction and classification. In: ICME'05, Amsterdam, Jul 2005

Webster JG (eds) (1998) Medical instrumentation, application and design, 3rd edn. Wiley, New York, NY

Wolpaw JR, Ramoser H, McFarland DJ, Pfurtscheller G (1998) EEG-based communication: improved accuracy by response verification. IEEE Trans Rehabil Eng 6(3):326–333

Part III
Data and Databases

Editorial: "Data and Databases"

Ellen Douglas-Cowie

Abstract Databases are the empirical foundation of emotion-oriented computing, and creating them poses problems that are central to the field. There is now a substantial history of work on these problems, and the first chapter in the section reviews it. The two central chapters separate out the two main challenges, collecting appropriate material and developing methods of labelling it that describe its emotional content and the features relevant to recognising or synthesising that kind of emotional content. The last chapter describes a database that was developed within HUMAINE, with the specific aim of illustrating both contemporary understanding of the issues, and the tools available to deal with them. The chapters stand alone as far as possible, but they are not wholly independent, because the issues are deeply interconnected, both to each other and to research that uses databases for recognition or synthesis.

1 Introduction

Work on emotion databases is fundamental to emotion-oriented computing in the most literal sense. Databases are the empirical foundation on which the systems rest. As a result, research on them is massively connected to other areas. It is heavily dependent on theory, and it ought to inform theory in return; it provides input on which both recognition and synthesis depend; it involves both high- and low-level understanding (from showing how an eyebrow moves to showing how emotion may relate to long-term personal history and aspirations); and it is the part of emotion-oriented computing where ethical questions are most immediate.

The sense that the area is so interconnected is relatively recent. The partnerships that are reflected in this section began to form in 2000, at the ISCA workshop on speech and emotion, which was held in Newcastle, County Down. At that stage,

E. Douglas-Cowie (✉)
Department of Psychology, Queen's University Belfast, Belfast, Northern Ireland, UK
e-mail: E.Douglas-Cowie@qub.ac.uk

P. Petta et al. (eds.), *Emotion-Oriented Systems*, Cognitive Technologies,
DOI 10.1007/978-3-642-15184-2_10,

it was clear that developing suitable databases was a top priority for the emerging field. But it was not clear, even then, that the task was as far ranging as it now seems.

The chapters of Part III aim to give people who come into the field now the benefit of experience since then, and particularly the experience of the partners in the HUMAINE database work package. Above all, they try to give readers the opportunity to see how richly connected the topic is. People should not have to relearn for themselves all the lessons that the writers learned over the years following the Newcastle conference.

2 The Writers and Their Perspectives

The writers of these chapters have all been key players in the HUMAINE network of excellence. As a result, there is a great deal of common ground between them, and the section could be said to represent a HUMAINE perspective on the databases. The perspective involves emphasis on issues like multimodality, authenticity, theoretical presuppositions, and awareness of context. Not all research in the area shares those emphases. It makes perfect sense, for instance, to focus on unimodal data that relate to immediate applications, or the technical quality of recordings. There is no attempt to conceal the existence of other perspectives; on the contrary, the first chapter provides a very extended list of sources so that readers can access what has been done in the whole community, not just within HUMAINE.

The HUMAINE community itself also contains variety. Members have interests in good quality acted data as well as thoroughly naturalistic records; close to application unimodal material as well as recordings that provide sound, vision, and physiological data; both categorical and 'trace-type' continuous labelling techniques, and so on. The chapters do not present these as opposing stances, because that is not the way HUMAINE has developed. Instead, they reflect diversity by acknowledging consistently that there are legitimate differences of emphasis within the field. The list of authors on each chapter is long because authorship was designed to ensure that the major positions within HUMAINE were duly represented at every point.

3 Structure

Work on databases can be divided into two broad areas: collecting data and labelling it. Each of the middle two chapters deals with one of those. The first and last deal with issues where both strands come together. The first looks at the past, in terms of principles that inform the whole enterprise, and the history of the field. The last looks at the present and the future, describing the database that has been developed within HUMAINE, which was deliberately designed to display both contemporary understanding of the issues, and the tools available to deal with them.

It is convenient to talk of two strands, but in fact they are thoroughly interconnected. The categories used in labelling are driven by the material that is there to be labelled, and the collection of material is driven by an understanding of the categories that it is relevant to collect. For that reason, insisting that the chapters should avoid all overlap would be highly artificial. The ideal has been to keep the main coverage of any given topic in one place and to treat it briefly in other places where it arises, referring to the main coverage. Sometimes even that is not realistic. There are topics which need to be covered from different angles, and the same is true of key sources (typically innovation in collection has gone hand in hand with innovation in labelling). As a result, some redundancy is necessary. The target has been to avoid unnecessary redundancy.

Conversely, the chapters are written to stand alone rather than forming a totally integrated section. However, it is inevitable that a reader looking at a single chapter will be dissatisfied with the coverage of certain issues in that chapter. The only way to avoid that would be a much higher level of redundancy.

4 Links to Other Sections

Most other chapters have links to this section. However, there are a few where the links are particularly strong. The first part chapter, "Emotion: concepts and definitions", provides the conceptual basis of the labelling systems that are considered here. Chapter "Representing Emotions and Related States in Technological Systems" deals with the issue of description from a specifically engineering standpoint and describes a formal system for representing much of what is set out in these chapters. The chapters on recognition from speech and from sound indicate that moving from signals to concepts depends on the kinds of database that are described here. Chapter "Coordinating the Generation of Signs in Multiple Modalities in an Affective Agent" demonstrates that movement in the other direction, from intention to signals, is equally bound up with databases. All of that reflects the connectedness illustrated by the hexagon diagram in the 'signals to signs' editorial.

5 Further Reading

The chapters provide full lists of references. However, there are key resources for those who want to go quickly to material that provides an overview. There are review papers on databases in two special issues that have been influential in the field (Douglas-Cowie et al., 2003; Scherer et al., 2005). There have since been workshops on databases at LREC whose proceedings are available electronically (Martin et al., 2006; Devillers et al., 2006, 2008). Last but not least, the HUMAINE portal provides a rich set of documents generated by the database work package.

HUMAINE deliverables are available at http://emotion-research.net/deliverables

The International Conference on Language Resources and Evaluation (LREC) series home page is available at http://www.LREC-conf.org/

References

Devillers L, Martin J-C, Cowie R, Douglas-Cowie E, Batliner A (2006) Editors of the proceedings of the workshop "Corpora for Research on Emotion and Affect". In: Association with the 5th international conference on language resources and evaluation (LREC 2006), Genoa, Italy, 23 May 2006. http://www.cs.brandeis.edu/~marc/misc/proceedings/lrec-2006/workshops/W09/Emotion-proceeding.pdf. Last accessed on 7/11/2010

Devillers L, Martin J-C, Cowie R, Douglas-Cowie E, Batliner A (2008) Editors of the proceedings of the workshop "Corpora for Research on Emotion and Affect". In: Association with the 6th international conference on language resources and evaluation (LREC 2008), Marrakech. http://www.lrec-conf.org/proceedings/lrec2008/. Last accessed on 7/11/2010

Douglas-Cowie E, Campbell N, Cowie R, Roach P (2003) Emotional speech: towards a new generation of databases. Speech Commun 40(1–2):33–60

Martin J-C, Kuhnlein P, Paggio P, Stiefelhagen R, Pianesi F (2006) Editors of the proceedings of the workshop "Multimodal Corpora: From Multimodal Behaviour Theories to Usable Models". In: Association with the 5th international conference on Language Resources and Evaluation (LREC2006), Genoa, Italy, 27 May 2006. http://pages.cs.brandeis.edu/~marc/misc/proceedings/lrec-2006/workshops/W11/Multimodal%20Corpora.pdf. Last accessed on 7/11/2010

Taylor J, Scherer K, Cowie R (2005) Editors emotion & brain: understanding emotions and modelling their recognition. Spec Issue Neural Netw 18(4):313–455

Principles and History

Roddy Cowie, Ellen Douglas-Cowie, Ian Sneddon, Anton Batliner, and Catherine Pelachaud

Abstract Developing databases for emotion-oriented computing raises specific and complex issues at multiple levels, from the practicalities of recording to conceptual issues in psychology. Whether it is developing databases or using them, research in emotion-oriented computing needs to think about these issues rather than reflexively importing habits derived from other fields. Contemporary research identifies a number of principles that are relevant to making appropriate choices. They can be grouped under three broad headings – function; structure and scope; and relationship to psychological theory. These principles were not obvious when research in the area began. They have emerged gradually over a decade of relatively sustained work, and it is reviewed. Databases that have played a significant role in the process are listed, and selected case studies are examined in more depth. Lessons are drawn for future work at two levels, first at the level of an abstract overview and then at the level of practical issues that need to be addressed.

1 Introduction

Databases are central to the development of systems that use human-like channels of communication. As a result, they have been part of emotion-oriented computing since research in the area began. They have also been recognised as an area where there were major problems to be dealt with. Most obviously, there was a shortage of appropriate databases. Less obviously, there were unresolved questions about the kind of database that would count as appropriate. Broadly speaking, early research tended to transfer working habits from cognate fields, such as the recognition of words in speech or recognition of an individual from different photographs of his/her face. However, it gradually became apparent that some important working habits might not transfer to the new area. The ISCA Workshop on Speech

R. Cowie (✉)
Department of Psychology, Queen's University Belfast, Belfast, Northern Ireland, UK
e-mail: roddy.cowie@qub.ac.uk

P. Petta et al. (eds.), *Emotion-Oriented Systems*, Cognitive Technologies,
DOI 10.1007/978-3-642-15184-2_11,

and Emotion in 2000 (Douglas-Cowie et al., 2000) is a useful marker. There ten Bosch (2000) argued in principle that word recognition paradigms might not transfer; Batliner et al. (2003) argued that standard ways of constructing databases were not practically appropriate to the field; and groups from England, Ireland and Japan (Douglas-Cowie et al., 2003) described work on emotion databases collected according to principles very unlike traditional databases for speech or face recognition.

One of the key aims of this chapter is to convey why research that uses databases in emotion-oriented computing needs to think about database issues rather than reflexively importing habits derived from other fields and to identify the principles that contemporary research suggests are relevant to selecting databases in this particular area. In a decade, a handbook of emotion-oriented computing may not need to cover those issues. Imported habits may seem too obviously inappropriate to be worth mentioning. However, at the time of writing, it remains true that reviewers are continually faced with articles full of sophistication in the computational domain, but painfully naïve about the kinds of data that their sophistication should be applied to.

A second key aim is to inform people whose interest is in developing databases rather than using them (or as well as using them). In many areas, developing databases is not seen as central to the field. The general assumption is that databases ought to be there to be pulled off the shelf as needed. In the context of emotion, there are good reasons for thinking that the task of database development needs a higher profile. That is not only because the task is large but also because it demands very particular skills. If the development of emotion-related databases becomes a subdiscipline in its own right, as we believe it probably should, it needs its own theoretical framework.

2 Theoretical Foundations

A wide range of conceptual issues may be relevant to the design or the selection of a database involving emotion in a broad sense. This section tries to draw out the main principles, grouping them under three broad headings – function; structure and scope; and relationship to psychological theory. A later section considers another range of issues which are essentially practical.

2.1 Database Functions

There may be many different motives behind the construction or the selection of a database. That point was recognised early on in HUMAINE using a simple distinction between two extreme types of goal.

> At one extreme, databases can play a purely supportive role, allowing investigators working within a well-established paradigm to fill in or check details of processes that they assume are understood in general terms. At the other extreme, databases can play a provocative role, providing examples that help investigators to expand and restructure their thinking about the area (Douglas-Cowie et al., 2004).

The supportive/provocative distinction is useful because it is simple. However, it reflects a much more complex set of distinctions in the aims relevant to database research. This section sets out the issues more fully.

The archetypal function of a database is to provide material for training systems that will be used for recognition. Databases with that function need to be large and structured, with many tokens of each significant type. The class itself subdivides. At one extreme, the aim of learning may be to acquire something that is understood as a general competence (such as recognising a large vocabulary of words in clear speech). At the other, the aim may be to train a system for a particular application, and only for that application. Often both aims play a part. There is a target application, but there is also an aspiration to address it in a way that has some generality – not least because the unforeseen happens, and that is bad news for systems trained too narrowly to deal with the foreseen.

Synthesis makes different demands. For some applications (such as copy synthesis), a small number of cases, or even a single case, may be sufficient to extract parameters that will be used to generate a particular kind of output – a kind of expression, for instance, or a gait. On the other hand, other applications may require large numbers, because they need a complete set of components (samples of phonemes in all relevant contexts, for instance).

Although these are the applications that come to mind immediately, a high proportion of papers that use databases have rather a different kind objective: broadly speaking, to test a technique or to select among alternatives. For example, a speech database may be used to select a good set of features from the enormous range that could conceivably be used for recognition or to compare the effectiveness of different statistical learning techniques. As a special case, certain data sets become the basis for tasks which are accepted as a 'gold standard'; systems are compared by their performance on them.

Databases also serve more traditional functions. There are areas where skilled observers – psychologists, linguists, ethologists, and so on – are central to the process of developing systems. Their intuition allows them to formulate hypotheses (particularly qualitative hypotheses) that machines cannot, for instance, about the kinds of emotion-related meaning that signs convey. That kind of intuition needs to be informed by data. If the examples available simply reflect a set of presuppositions which are orderly but naïve, they cannot give intuition the right kind of impetus.

A special version of that function comes to the fore in the particular case of emotion databases. It is bootstrapping the development of databases. It tends to be by engaging with examples in an existing database that people recognise new types of issue that they may need to consider. A single example in a preliminary collection may draw attention to a large domain that had been not considered a priori, but that once recognised clearly needs to be covered.

That kind of issue arises in several forms. The simplest is recognising that a particular kind of example is important, and that more of the same kind should be collected. Preliminary databases are also fundamental to the development of descriptive systems. Examples highlight the information that needs to be provided to make sense of the fact that emotion is sometimes displayed in one way, sometimes in another. Linked to both of those, deciding the form that databases should eventually take depends on understanding how the domain of emotion is structured; but that understanding has to come from databases. For example, it is possible that in the long run, databases will only need to contain examples of archetypal states, because interpolation will be enough to cover the rest of the domain. However, that can be tested only using a database that contains a sufficient variety of non-archetypal examples.

2.2 Structure and Scope

It follows from the last section that several kinds of requirement have a bearing on the amount and kind of material that a database should contain. Some can be formulated numerically, others are in practice intuitive. Their importance will depend on the particular application that the researcher has in mind.

2.3 Size and Formal Structure

When databases are used to train recognisers, the processes involved are generally grounded in statistics. There is a large statistical literature on the problem of identifying sample sizes sufficient to achieve robust distinctions between classes. The analyses translate into prescriptions for the sizes of the databases that are needed to train recognisers.

Classical analyses (e.g. Raudys and Jain, 1991) produce figures that are in line with reasonably common practice. For a simple classification rule, such as Fisher's linear discriminant analysis, if p features are used for classification, about $10p$ training samples per class are needed to obtain a reliable decision rule. Matters become more difficult with more complex rules, not least because the required number of samples per class tends to rise faster than p. For instance, with a quadratic discriminant analysis, for 8 features, the requirement is about $16p$; for 20, it is about $22p$; for 50 it is about $40p$. The rapid rise with number of features is linked to the so-called 'curse of dimensionality'.

These numbers are based on analytic procedures that make tacit assumptions about distributions. Where the relevant properties are not ideally distributed, empirical approaches suggest that sample sizes may need to be an order of magnitude bigger (Han et al., 2005).

Balance between categories is also critical. If a database covers several classes of phenomenon, and the frequencies are different, the size of the smallest class is generally used as a basis for deciding whether a particular statistical technique can safely be applied. It follows that a database of a given total size is much more useful for training if the phenomena that it covers are evenly distributed.

The issues are usually stated in terms of a classification paradigm. Similar considerations apply when the analysis is concerned with continuous relationships between variables. For the simplest analysis of that kind, multiple regression, a standard formulation is that 40 cases per variable are needed. Corresponding to the issue of equal class sizes, statistical effectiveness falls if the variables being considered are correlated (Tabachnick and Fidell, 2001).

Turning to synthesis, examples give a sense of the size and the structure that are needed to achieve respectable synthesis of speech. CMU Arctic databases are regarded as somewhere near the lower bound for acceptable synthesis. An Arctic database is a reading of the Arctic prompt set by a single speaker in a specified style of delivery (plus associated files). The prompt set contains 10,045 words and 39,153 phones – about 2 h of speech including pauses (Kominek and Black, 2004).

A minimal constraint on database size is that it should contain all the theoretically relevant kinds of unit – which for basic unit synthesis, techniques mean all phoneme pairs. In practice, though, phoneme pairs differ according to context. As a result, databases close to the theoretical minimum size are not ideal. Larger databases are more likely to contain not only the right units but also the right units in the right context. A neat illustration of the way that can be exploited is that some systems replay whole words or even phrases directly if they can be found in the corpus, rather than constructing them from simpler elements (sequences of a few phonemes).

Evidence shows that increasing database size improves quality. For instance, a study by Sak (2000) compared performance using two databases, one containing 3 h of speech and the other containing 19 h. At the upper end, the XIMERA system was developed using corpora from three speakers, of 20, 60 and 111 h, respectively (Kawai et al., 2004). These figures are all for the synthesis of essentially neutral styles. Introducing convincing emotional expression would require correspondingly larger corpora.

2.3.1 Units of Analysis

Databases need to be divided into appropriate units of analysis. In the speech material mentioned above, the primary units of analysis are relatively clear – groups of two or three adjacent phonemes and words. In emotion databases, choice of units is a vexed issue. There is a strong tradition of using individual frames as a basis for emotion recognition from faces, but gestures extend over much longer times. Speech-oriented work has divided data into words (Batliner et al., 2006) and phrase-like units (Fragopanagos and Taylor, 2005). Divisions based on emotion include some designed to extract episodes of sustained emotion (Douglas-Cowie et al., 2003) and some to include build-up from ‘rest’ levels to a peak and return to ‘rest’ levels (see the chapter on the HUMAINE database in Part III).

A widely used pattern in HUMAINE is to use a primary division into ‘clips’, where a clip is an episode chosen so that emotion-related signals within it are generally understood in the same way as they would be in a much larger context. There are typically many nested and overlapping units within a clip, often with fuzzy boundaries. A good formalisation of these issues would be very useful.

2.3.2 Modalities

There is increasing agreement that emotion is multimodal. Signals in different channels are likely to complement and illuminate each other rather than duplicating the same information. The term modality is commonly used not just to refer to gross distinctions between visible, audible and physiological sources but also to distinguish within them. So facial, gestural and postural modalities would normally be distinguished within the broad category of optical sources.

It is also increasingly clear that signs need to be related to the context in which they occur. Some very successful procedures interpret facial signals in the context of ongoing attempts to solve particular problems. It may be stretching semantics to call context a modality, but it certainly needs to be considered alongside modalities like gesture and prosody.

For those reasons, emphasis has shifted away from unimodal databases. They remain important, not least because some applications are inherently unimodal – such as detecting emotion in phone calls (to help desks, clinics, etc.). But there is a clear need to collect material that clarifies how modalities interact.

A complicating factor is that the modalities seem not to combine independently. It is hard to find situations that give rich emotional signs in vocal, facial and gestural modalities simultaneously. As a result, databases drawn from a relatively uniform kind of source cannot practically be expected to be equally rich in all modalities.

2.3.3 Realism

There is a standard method of generating resources like the Arctic or XIMERA databases mentioned above. A 'talent' is given a body of material to speak, and it is recorded in a soundproof room. Translating that method directly to the domain of emotion would involve asking 'talents' to simulate the emotion on demand – i.e. acting. One of the most contentious issues in the area is how useful acted renditions of emotion are.

It is important to be clear that the issue is not simple. Using acted data has great potential advantages. It is likely to be much more economical than collecting spontaneous samples of emotion as it appears in everyday life; it is much easier to ensure that the data is well structured; the recordings are likely to be more tractable; and it is usually clearer what the underlying emotional state is (so long as one sets aside awkward general questions about what the underlying state of an actor simulating an emotion might be).

It is harder to articulate the advantage of naturalistic sampling. It often seems to be implied that the issue is whether something essential is present in real data but not in acted data, such as physiological arousal which gives rise to telltale changes in muscle tone or reactivity. But it is probably at least as important that naturalistic sampling has the potential to reflect the kind of connectivity that is outlined in Chapter 'Editorial: 'Theories and Models' of Emotion'. Real emotion has connections at its core, and it characteristically affects a multitude of connections around it. It impacts the way people think, what they look at, their choice of words and discourse structure, the actions they take and how they execute them; and how all that depends on the context, social and physical. Recognising emotion in these things is rather like recognising water in coffee, or spilled on a floor, or in a damp cloth, or in a balloon, or in processed meat, or in a carrot, or in hydrated crystals, or in aquiferous rock, or in a cloud, or in a rainbow. Pictures of pure water in laboratory beakers would not be an ideal basis for recognising water in these contexts. The same problem arises with pictures of idealised and decontextualised emotion, however true they may be to what they depict.

These points are intimately related to the earlier point about bootstrapping database development. It is only by building up large collections that it can become clear how much variety needs to be represented in a functional set of databases and how much of the load can be carried by a few features which recur in virtually any context.

2.4 Databases and Psychological Theory

Psychological theory is indirectly relevant to a wide range of questions about the design or the selection of a database, including several that have already been addressed. However, there are areas where the psychology is absolutely central. They often raise quite difficult issues.

2.4.1 Respecting Psychological Semantics

Emotion-oriented computing has an obligation not to deceive. Unfortunately, it is extremely easy to slip into deception by misusing the words that describe emotion in both expert and 'naïve' psychology. Databases are central to avoiding that kind of deception.

The problem arises when, for instance, it is claimed that a system recognises anger, and in fact it only recognises activation. That is very likely to happen if a system is trained on a database that consists of audio samples that are either angry or neutral. A system trained on that basis is likely to identify anger when it is presented with samples of fear, surprise, amusement, happiness, stress and many other states. The reason is that they all involve elevated activation. Level of activation is not the only difference between anger and neutrality, but in the auditory modality, it is much the easiest difference to detect; and therefore, it is the dimension that is likely to dominate the choices of a system whose training rests on samples of anger and neutrality.

To avoid being party to deception, teams that develop databases have to consider how the descriptive terms that they use relate to the coverage of the database. It is inviting misrepresentation for a database to use the label 'anger' unless it contains a sufficient range of other states to ensure that the term is properly contrasted with the alternatives that people have in mind when they say 'anger'.

In practice, that means databases always have to emotional space quite widely. The granularity is another matter. Some areas may be so easily discriminated that they do not need to be heavily represented. But there should be real concern about databases that do not allow at least one state to be contrasted with all the others that a human being would contrast with the target state.

2.4.2 Theory-Driven Sampling

There are well-known lists or taxonomies that aim to cover the whole domain of emotion, the simplest being the 'big six' emotions proposed by Ekman (1992). It is natural to expect that a database will be complete if it includes samples of all the states in such a list.

It is critical that theories in psychology, in this area and others, need to be understood as hypotheses. A theory corresponds to a framework that is recognised as conceptually attractive. It is another matter whether the framework deals satisfactorily with the empirical evidence. That is the test against which a framework has to be judged. Hence, for the medium term, theories need to be measured against databases; and the theories need to be called into question if the range of phenomena that they cover does not match up to the range that the database allows us to see occurs in real life.

It cannot be predicted in advance how the question will be answered. The conclusion may be that the theory is sound and the database is skewed, or that theory is too narrow. Deciding which of those is true is part of the bootstrapping process that has been mentioned already.

2.4.3 Prototypes and Ecological Validity

It is clear that in practice, people who are collecting material for a database often look for cases that they regard as 'good examples'. It is less clear what status the concept of a 'good example' has. The most straightforward idea is that it corresponds to something that has a special status in people's mental representation of emotion – what psychological theories like Rosch's (1978) call a prototype (see also chapter 'Editorial: 'Theories and Models' of Emotion').

There may be value in going out to look for prototypical examples, but it is clearly not the same as finding examples that represent what is likely to happen in a relevant range of situations – that is, looking for ecological validity. There are good psychological reasons to suspect that people will tend to underestimate the difference – events that confirm stereotypes tend to have a salience in memory that their objective frequency does not warrant and memories of events tend to become more like prototypes than the original events actually were (Sutherland, 2007). The more sophisticated that investigators are about the way they approach these issues, the less likely they are to misunderstand what they have actually collected.

2.4.4 Annotation

The point of annotation is to distinguish states that are functionally different and group states that are functionally similar. It is reasonable to assume that that is effectively a psychological task, and therefore psychology should provide the theoretical basis of annotation. In practice, the link has not always been particularly direct.

In some cases, it seems fair to say that the reason is a kind of naïve realism. Researchers sometimes invoke everyday categories in a way that suggests unquestioning belief that those categories correspond exactly to the natural kinds of emotion. The first chapter of this handbook indicates why that is hard to defend.

On the other hand, there are reasons why descriptions derived from psychological theory may not transfer easily to annotation tasks. Plain familiarity is an issue, particularly for applications that are likely to involve people who are not experts in the theory of emotion. Complexity is also an issue. A key early example is the Leeds–Reading database, which is discussed later. It used sophisticated descriptions rooted in psychological theory on one side and linguistic theory on the other; but the result was a fine-grained subdivision showing too few instances of any individual category for statistical relationships to be established.

2.4.5 Forms of Validation

It is clear that a particular type of label should not be used in a database unless it is in some sense trustworthy. However, care is needed about the sense in which it needs to be trustworthy.

Two kinds of tests are commonly invoked. One is that the labels should represent 'ground truth' – i.e. the label anger should be applied only if it can be shown that the person is genuinely angry when it is applied. The other is that the label should not be used unless there is a high degree of agreement associated with it, i.e. the labelling is statistically reliable. Neither of those criteria should be taken for granted. The point is related to the distinction between cause-and-effect-type descriptions drawn by Cowie et al. (2001).

It is appropriate to ask for ground truth when the aim is to give a cause-type description – that is, to establish what really behind the signs that can be seen in a recording. However, it is often more relevant to give an effect-type description – that is, to specify how a person can be expected to interpret the signs, rightly or wrongly. It often makes sense to assume that the interpretation observers put on the signs another person gives is probably the most reliable indicator of the real situation that is available. It certainly should not be assumed that brain scans or psychophysiological records would be more reliable.

Statistical reliability is an appropriate test if raters are supposed to be describing objective matters. That is typically the case when they are rating signs of emotion. However, it is a different issue when they are giving effect-type descriptions of their own reactions. It may well be the case that certain signs evoke reactions in different people, and discarding evidence that they do is distortion, not precision.

A special case of the distortion comes when databases are restricted to items that are identified with high reliability. The fact that a display is very reliably recognised does not mean that it is natural. On the contrary, the best way to ensure high recognition is to exaggerate.

In all these areas, the key rule is to consider what relevant demands are, rather than applying a rule unthinkingly. The demands that it is appropriate to make, depend on recognising that emotion is a complex psychological phenomenon.

3 The Development of the Field: Illustrative Case Studies

The principles that have been outlined were not obvious when research in the area began. They have emerged gradually over a decade of relatively sustained work. This section traces the way understanding of the issues has deepened through case studies of key research efforts. The area has progressed by registering the problems that arose in seemingly reasonable efforts to develop databases and looking for ways to address them.

3.1 The Berlin Database

This database represents the archetypal database in the 1990s – unimodal (speech only), acted and focused on a few emotions that were considered primary. Actors were asked to read the same set of sentences in each of five different emotions. Labelling simply consisted of identifying the emotion that the actor simulated.

Databases of that general kind formed the backbone of early work on emotion recognition. But it gradually became apparent that systems trained on this kind of data transfer poorly to real-life situations. The point was strongly made in an influential paper by Batliner et al. (2003).

3.2 The Leeds–Reading Database

The first notable departure from the style described above was the Leeds–Reading database (Roach, 2000). It too was unimodal, but it was not acted. It consisted of speech drawn from real-life emotional situations. The focus was on selecting the most intense emotional recordings and situations that could be found, including material from commentaries on major disasters.

The labelling of emotion was of very different order from the Berlin database. Labels had four elements. The first was an everyday emotion label (freely chosen), and the second specified the strength of the emotion. The third and fourth provided descriptors based on the appraisal theory of Ortony et al. (1988), setting the emotion in a generic category and describing its antecedents.

Despite the emphasis on intense material, labellers used a wide range of emotion words. This was an early indication that emotion in real life did not fit neatly into the 'big six' emotions identified in the theories most familiar to most research teams entering the area (happiness, anger, sadness, fear, contempt, disgust). Coding of speech was also sophisticated so that segments were classified into multiple types. The result was a multiplicity of labels, which meant that when data came to be analysed, there was no straightforward way to find patterns (Stibbard, 2001). Other problems were also realised in hindsight. For example, the researchers had focused on only the extreme peaks of emotion in the data and had thrown away the lead up and movement away from peaks of emotion. The result was that the emotional peaks were decontextualised, and there was nothing to compare them with. Particularly galling, copyright problems meant that the material could not be released.

The Leeds–Reading database was outstandingly ambitious for its time. Because of that, it gave an early indication of the complexities involved in selecting, labelling, analysing and distributing real data. These were key lessons for subsequent researchers in the area.

3.3 Belfast Naturalistic and EmoTV Databases

These databases drew on the Leeds–Reading experience. They used naturalistic data of the type that had been explored for the Leeds–Reading database – media broadcasts – but they used audiovisual data. Both selected emotional episodes to show the wider emotional context in which the emotional episode took place. Both also set out deliberately to address the nuances and complexities of emotion in real life and still avoid the statistical difficulties that had beset the Leeds–Reading database. The BND developed 'trace' techniques, which allowed raters to report their perceptions of the speaker's emotions in quantitative terms, using the classical affect dimensions – positive to negative and active to passive. It also developed quantitative methods of describing speech so that straightforward statistical techniques could be used. The EmoTV database used a free choice of emotion words assigned over time but then found ways to group them into larger 'cover classes'. It also looked for appropriate ways to label signs of emotion – not only speech (using an approach broadly similar to the BND) but also gesture.

Although these databases addressed known problems, they in their turn exposed a range of others. Although the recording quality was acceptable to humans, it did not allow the kind of machine processing that affective computing teams needed. It became obvious that both databases were very much rooted in specific contexts (mainly TV chat shows, news reporting and interviews) and that their relationship to other issues of concern to affective computing, such as persuasion, was not straightforward. Questions about the relationship between modalities were also highlighted. Representation of gesture in particular was strikingly uneven, and it was unclear whether this represented a failure to capture a significant modality or an indication that signs are often not given in all modalities. Both also uncovered problematic issues with labelling. EmoTV, for example, showed that much data consisted of mixed or blended emotions and that a labelling system needed to take this into account. The BND indicated that averaging across raters may mask the fact that individuals may 'read' the same recording in systematically different ways. Both showed widespread masking of emotion, and exposed the need for labelling techniques that could address the issue. Both databases also drew attention to issues of match and mismatch between signs of emotion in the face and in speech. For example, people who seemed to be fundamentally sad nevertheless smiled in a way that can easily be taken to convey happiness. Perhaps most acute of all, neither database could be released to the wider community because of copyright reasons.

These issues led the teams involved in both databases to explore techniques that provided more control over the data they would use. They also joined forces to develop a common labelling scheme informed by the two naturalistic databases.

3.4 AIBO, SAL and EmoTaboo

Experience with broadcast material has led to growing acceptance that, in effect, research teams need to become film-makers. Investigators have identified situations that induce interactions which are genuinely emotionally coloured but where they control inducing factors, recording and distribution. That has produced databases that support quite deep analysis. The AIBO database, recorded at Erlangen, pioneered the style. Children were recorded interacting with the AIBO robot, and its behaviour was manipulated to induce different kinds of emotionally coloured reaction. The speech has been thoroughly annotated, and techniques based on soft vectors have been developed to reflect the fact that the emotional content does not fall into sharply defined classes. The database forms the basis of the CEICES project, which is reported elsewhere in this handbook. The Belfast team moved from the BND to develop the SAL database, using an induction technique which simulated conversation with 'artificial listeners'. Both audio and visual components have been coded using techniques similar to the BND, and high recognition rates have been reported (Fragopanagos and Taylor, 2005). The SAL technique does not produce much gesture, and the EmoTV team developed the EmoTaboo database to address that gap. Their induction technique was a game designed to elicit emotional gestures, in a reasonably naturalistic way. Labelling the gesture called for new techniques, and the work is still ongoing.

These examples have been selected to illustrate the general development of the field. Later chapters provide more detail on different aspects of them. The next section provides a wider ranging review, designed to let readers identify most of the significant databases that have been described in the literature.

4 An Overview of Existing Databases

This section considers emotion databases in three broad categories – those that are multimodal, those that are speech alone and face databases. Tables 1, 2 and 3 summarise the state of the art with regard to databases. Table 1 lists multimodal databases, Table 2 lists speech databases and Table 3 lists face databases. Data on

Table 1 Multimodal databases relevant to emotion

Belfast naturalistic database (Douglas-Cowie et al., 2000, 2003) *Modalities*: Audiovisual *Emotions*: Wide range *Elicitation*: Natural: 10–60 s long 'clips' taken from television chat shows, current affairs programmes and interviews conducted by research team *Size*: 125 subjects; 31 male, 94 female *Kind*: Interactive unscripted discourse *Language*: English

Geneva airport lost luggage study (Scherer and Ceschi, 1997, 2000) *Modalities*: Audiovisual *Emotions*: Anger, good humour, indifference, stress sadness *Elicitation*: Natural: unobtrusive videotaping of passengers at Geneva airport lost luggage counter followed up by interviews with passengers *Size*: 109 subjects *Kind*: Interactive unscripted discourse *Language*: French

Chung (2000) *Modalities*: Audiovisual *Emotions*: Joy, neutrality, sadness (distress) *Elicitation*: Natural: television interviews in which speakers talk on a range of topics including sad and joyful moments in their lives *Size*: 77 subjects; 61 Korean speakers, 6 Americans *Kind*: Interactive unscripted discourse *Language*: English and Korean

SMARTKOM www.phonetik.uni-muenchen.de/Bas/BasMultiModaleng.html#SmartKom *Modalities*: Audiovisual, (+gestures) *Emotions*: Joy, gratification, anger, irritation, helplessness, pondering, reflecting, surprise, neutral *Elicitation*: Human machine in WOZ scenario: solving tasks with system *Size*: 224 speakers; 4/5-min sessions *Kind*: Interactive discourse *Language*: German

Amir et al. (2000) *Modalities*: Audio + physiological (EMG,GSR, heart rate, temperature, speech) *Emotions*: Anger, disgust, fear, joy, neutrality, sadness *Elicitation*: Induced: subjects asked to recall personal experiences involving each of the emotional states *Size*: 140 subjects 60 Hebrew speakers 1 Russian speaker *Kind*: Non-interactive, unscripted discourse *Language*: Hebrew, Russian

SAL database (http://semaine-project.eu/, D09) *Modalities*: Audiovisual *Emotions*: Wide range of Emotions/emotion-related states but not very intense *Elicitation*: Induced: subjects talk to artificial listener and emotional states are changed by interaction with different personalities of the listener *Size*: Study of 20 subjects in 2 SAL scenarios – Powerpoint SAL and Soild SAL, 24 sessions per scenario, 8 hours material per scenario *Language*: English

ORESTEIA database (McMahon et al., 2003) *Modalities*: Audio + physiological (some visual data too) *Emotions*: Stress, irritation, shock *Elicitation*: Induced: subjects encounter various problems while driving (deliberately positioned obstructions, dangers, annoyances 'on the road') *Size*: 29 subjects, 90 min sessions per subject *Kind*: Non-interactive speech: giving directions, giving answers to mental arithmetic, etc *Language*: English

Belfast boredom database (Cowie et al., 2003) *Modalities*: Audiovisual *Emotions*: Boredom *Elicitation*: Induced *Size*: 12 subjects: 30 min each *Kind*: Non-interactive speech: naming objects on computer screen *Language*: English

XM2VTSDB multimodal face database http://www.ee.surrey.ac.uk/Research/VSSP/xm2vtsdb/ *Modalities*: Audiovisual *Emotions*: None *Elicitation*: n/a *Size*: 295 subjects Video *Kind*: High-quality colour images, 32 kHz, 16-bit sound files, video sequences and a 3D model + profiles (left-profile and one right-profile image per person, per session, a total of 2,360 images), scripted four sentences *Language*: English

ISLE project corpora (http://nats-www.informatik.uni-hamburg.de/~isle/speech.html, IST project IST-1999-10647) *Modalities*: Audiovisual + gesture *Emotions*: None *Elicitation*: n/a *Size*, *Kind*, *Language*, unclear

Table 1 (continued)

Polzin and Waibel (2000) *Modalities*: Audiovisual (though only audio channel used) *Emotions*: Anger, sadness, neutrality (other *Emotions* as well, but in insufficient numbers to be used) *Elicitation*: Acted: sentence length segments taken from acted movies *Size*: Unspecified no. of speakers. Segment numbers 1,586 angry, 1,076 sad, 2,991 neutral *Kind*: Scripted *Language*: English

Banse and Scherer (1996) *Modalities*: Audiovisual (visual info used to verify listener judgements of emotion) *Emotions*: Anger (hot), anger (cold), anxiety, boredom, contempt, disgust, elation, fear (panic), happiness, interest, pride, sadness, shame *Elicitation*: Acted: actors were given scripted eliciting scenarios for each emotion, then asked to act out the scenario *Size*: 12 (6 male, 6 female) *Kind*: Scripted: Two semantically neutral sentences (nonsense sentences composed of phonemes from Indo-European languages) *Language*: German

Table 2 Databases of speech relevant to emotion

TALKAPILLAR (Beller et al., 2005) *Emotions*: Neutral, happiness, question, positive and negative surprised, angry, fear, disgust, indignation, sad, bore *Elicitation*: Contextualised acting: actors asked to read semantically neutral sentences in range of *Emotions*, but practised on emotionally loaded sentences beforehand to get in the right mood *Size*: One actor reading 26 semantically neutral sentences for each emotion (each repeated three times in different activation level: low, middle, high) *Kind*: Non-interactive and scripted *Language*: French

Leeds–Reading database (Greasley et al., 1995; Roach et al., 1998; Stibbard, 2001) *Emotions*: Range of full-blown *Emotions Elicitation*: Natural: unscripted interviews on radio/television in which speakers are asked by interviewers to relive emotionally intense experiences *Size*: Around 4 ½ h material *Kind*: Interactive unscripted discourse *Language*: English

France et al. (2000) *Emotions*: Depression, suicidal state, neutrality *Elicitation*: Natural: therapy sessions and phone conversations. Post-therapy evaluation sessions were also used to elicit speech for the control subjects *Size*: 115 subjects: 48 females and 67 males. Female sample: 10 controls (therapists), 17 dysthymic, 21 major depressed Male sample: 24 controls (therapists), 21 major depressed, 22 high-risk suicidal *Kind*: Interactive unscripted discourse *Language*: English

Campbell CREST database, ongoing (Campbell, 2002; see also Douglas-Cowie et al., 2003) *Emotions*: Wide range of emotional states and emotion-related attitudes *Elicitation*: Natural: volunteers record their domestic and social spoken interactions for extended periods throughout the day *Size*: Target – 1,000 h over 5 years *Kind*: Interactive unscripted discourse *Language*: English, Japanese, Chinese

Capital Bank Service and Stock Exchange Customer Service (as used by Devillers and Vasilescu, 2004) *Emotions*: Mainly negative – fear, anger, stress *Elicitation*: Natural: call centre human–human interactions *Size*: Unspecified (still being labelled) *Kind*: Interactive unscripted discourse *Language*: English

SYMPAFLY (as used by Batliner et al., 2003) *Emotions*: Joyful, neutral, emphatic, surprised, ironic, helpless, touchy, angry, panic *Elicitation*: Human machine dialogue system *Size*: 110 dialogues, 29.200 words (i.e. tokens, not vocabulary) *Kind*: Naïve users book flights using machine dialogue system *Language*: German

Table 2 (continued)

DARPA Communicator Corpus (as used by Ang et al., 2002) See Walker et al. (2001) *Emotions*: Frustration, annoyance *Elicitation*: Human machine dialogue system *Size*: Extracts from recordings of simulated interactions with a call centre, average length about 2.75 words 13,187 utterances in total of which 1,750 are emotional: 35 unequivocally frustrated, 125 predominantly frustrated, 405 unequivocally frustrated or annoyed, 1,185 predominantly frustrated or annoyed *Kind*: Users called systems built by various sites and made air travel arrangements over the phone *Language*: English

AIBO (Erlangen database) (Batliner et al., 2004) *Emotions*: Joyful, surprised, emphatic, helpless, touchy (irritated), angry, motherese, bored, reprimanding, neutral *Elicitation*: Human machine: interaction with robot *Size*: 51 German children, 51.393 words (i.e. tokens, not vocabulary) English (Birmingham): 30 children, 5.822 words (i.e. tokens, not vocabulary) *Kind*: Task directions to robot *Language*: German

Fernandez and Picard (2003) *Emotions*: Stress *Elicitation*: Induced: subjects give verbal responses to maths problems in simulated driving context *Size*: Data reported from four subjects *Kind*: Unscripted numerical answers to mathematical questions *Language*: English

Tolkmitt and Scherer (1986) *Emotions*: Stress (both cognitive and emotional) *Elicitation*: Induced: two types of stress (cognitive and emotional) were induced through slides. Cognitive stress induced through slides containing logical problems; emotional stress induced through slides of human bodies showing skin disease/accident injuries *Size*: 60 (33 male, 27 female) *Kind*: Partially scripted: subjects made three vocal responses to each slide within a 40-s presentation period – a numerical answer followed by two short statements. The start of each was scripted and subjects filled in the blank at the end, e.g. 'Die Antwort ist Alternative ...' *Language*: German

Iriondo et al. (2000) *Emotions*: Desire, disgust, fury, fear, joy, surprise, sadness *Elicitation*: Contextualised acting: subjects asked to read passages written with appropriate emotional content *Size*: Eight subjects reading paragraph length passages *Kind*: Non-interactive and scripted *Language*: Spanish

Mozziconacci (1998) Note: database recorded at IPO for SOBUproject 92EA. *Emotions*: Anger, boredom, fear, disgust, guilt, happiness, haughtiness, indignation, joy, rage, sadness, worry, neutrality *Elicitation*: Contextualised acting: actors asked to read semantically neutral sentences in range of *Emotions* but practised on emotionally loaded sentences beforehand to get in the right mood *Size*: Three subjects reading eight semantically neutral sentences (each repeated three times) *Kind*: Non-interactive and scripted *Language*: Dutch

McGilloway (1997) and Cowie and Douglas-Cowie (1996) *Emotions*: Anger, fear, happiness, sadness, neutrality *Elicitation*: Contextualised acting: subjects asked to read passages written in appropriate emotional tone and content for each emotional state *Size*: 40 subjects reading five passages each *Kind*: Non-interactive and scripted *Language*: English

Belfast structured database An extension of McGilloway database above (Douglas-Cowie et al. 2000) *Emotions*: Anger, fear, happiness, sadness, neutrality *Elicitation*: Contextualised acting: subjects read 10 McGilloway-style passages and 10 other passages – scripted versions of naturally occurring emotion in the Belfast naturalistic database *Size*: 50 subjects reading 20 passages *Kind*: Non-interactive and scripted *Language*: English

Danish emotional speech database (Engberg et al., 1997) *Emotions*: Anger, happiness sadness, surprise neutrality *Elicitation*: Acted *Size*: Four subjects read two words, nine sentences and two passages in a range of *Emotions* *Kind*: Scripted (material not emotionally coloured) *Language*: Danish

Table 2 (continued)

Groningen ELRA corpus number S0020 (www.icp.inpg.fr/ELRA) this new link is working (date: 16/01/2005):(www.elda.org/catalogue/en/speech/S0020.html) *Emotions*: Database only partially oriented to emotion *Elicitation*: Acted *Size*: 238 subjects reading two short texts *Kind*: Scripted *Language*: Dutch

Berlin database (Kienast and Sendlmeier, 2000; Paeschke and Sendlmeier, 2000) http://www.expressive-speech.net/ *Emotions*: Anger (hot), boredom, disgust, fear (panic), happiness, sadness (sorrow), neutrality *Elicitation*: Acted *Size*: 10 subjects (5 male, 5 female) reading 10 sentences each *Kind*: Scripted (material selected to be semantically neutral) *Language*: German

Pereira (2000) *Emotions*: Anger (hot), anger (cold), happiness, sadness, neutrality *Elicitation*: Acted *Size*: Two subjects reading two utterances each *Kind*: Scripted (one emotionally neutral sentence, four digit number) each repeated *Language*: English

van Bezooijen (1984) *Emotions*: Anger, contempt disgust, fear, interest joy, sadness shame, surprise, neutrality *Elicitation*: Acted *Size*: Eight (four male, four female) reading four phrases *Kind*: Scripted (semantically neutral phrases) *Language*: Dutch

Abelin and Allwood (2000) *Emotions*: Anger, disgust, dominance, fear, joy, sadness, shyness, surprise *Elicitation*: Acted *Size*: one subject *Kind*: Scripted (semantically neutral phrase) *Language*: Swedish

Yacoub et al. (2003) (data from LDC, www.ldc.upenn.edu/Catalog/CatalogEntry.jsp?catalogId=LDC2002S28) *Emotions*: 15 *Emotions* Neutral, hot anger, cold anger, happy, sadness, disgust, panic, anxiety, despair, elation, interest, shame, boredom, pride, contempt *Elicitation*: Acted *Size*: 2,433 utterances from eight actors *Kind*: Scripted *Language*: English

Table 3 Face databases

The AR face database (http://www.isbe.man.ac.uk~bim/data/tarfd_markup/tarfd_markup.html) *Emotions*: Smile, anger, scream neutral *Elicitation*: Posed *Size*: 154 subjects (82 male, 74 female) 26 pictures per person *Kind*: 1: Neutral, 2 smile, 3: anger, 4: scream, 5: left light on, 6: right light on, 7: all side lights on, 8: wearing sun glasses, 9: wearing sun glasses and left light on, 10: wearing sun glasses and right light on, 11: wearing scarf, 12: wearing scarf and left light on, 13: wearing scarf and right light on, 14–26: second session (same conditions as 1–13)

CVL face database (http://lrv.fri.uni-lj.si/facedb.html) *Emotions*: Smile *Elicitation*: Posed *Size*: 114 subjects (108 male, 6 female) seven pictures per person *Kind*: Different angles, under uniform illumination, no flash and with projection screen in the background

The Psychological Image Collection at Stirling (http://pics.psych.stir.ac.uk/) *Emotions*: Smile, surprise, disgust *Elicitation*: Posed *Size*: Aberdeen: 116 subjects Nottingham scans: 100 Nott-faces-original: 100 Stirling faces: 36 *Kind*: Contains seven face databases of which four largest are Aberdeen, Nottingham scans, Nott-faces-original, Stirling faces mainly frontal views, some profile, some differences in lighting and expression variation

The Japanese female facial expression (JAFFE) database (http://www.kasrl.org/jaffe.html) *Emotions*: Sadness, happiness, surprise, anger, disgust, fear, neutral *Elicitation*: Posed *Size*: 10 subjects seven pictures per subject *Kind*: Six emotion expressions + one neutral posed by 10 Japanese female models

Table 3 (continued)

CMU PIE database [CMU Pose, Illumination, and Expression (PIE) database] (http://www.ri.cmu.edu/projects/project_418.html) *Emotions*: Neutral, smile, blinking and talking *Elicitation*: Posed for neutral, smile and blinking 2 s video capture of talking per person *Size*: 68 subjects *Kind*: 13 different poses, 43 different illumination conditions, and with four different expressions

Indian Institute of Technology Kanpur database (http://vis-www.cs.umass.edu~vidit/IndianFaceDatabase/) *Emotions*: Sad, scream, anger, expanded cheeks and exclamation, eyes open–closed, wink *Elicitation*: Posed *Size*: 20 subjects *Kind*: Varying facial expressions, orientation and occlusions; degree of orientation is from 00 to 200 in both right and left directions, the similar angle variation are considered in the case of head tilting; and also head rotations both in top and bottom are taken into account. All of these images are taken with and without glasses in constant background; for occlusions some portion of face is kept hidden and lightning variations are considered

The Yale face database (http://cvc.yale.edu/projects/yalefaces/yalefaces.html) *Emotions*: Sad, sleepy, surprised *Elicitation*: Posed *Size*: 15 subjects *Kind*: One picture per different facial expression or configuration: centre-light, w/glasses, happy, left light, w/no glasses, normal, right light, sad, sleepy, surprised and wink

CMU facial expression database (Cohn–Kanade) (http://vasc.ri.cmu.edu//idb/html/face/facial_expression/index.html) *Emotions*: Six of the displays were based on descriptions of prototypic *Emotions* (i.e., joy, surprise, anger, fear, disgust and sadness). *Elicitation*: Posed *Size*: 200 subjects *Kind*: Subjects were instructed by an experimenter to perform a series of 23 facial displays that included single action units (e.g. AU 12 or lip corners pulled obliquely) and combinations of action units (e.g. AU 1+2, or inner and outer brows raised). Subjects began and ended each display from a neutral face

Caltech Frontal Face DB (http://www.vision.caltech.edu/html-files/archive.html) *Emotions*: Unclear *Elicitation*: *Size*: 27 subjects 450 images in total *Kind*: Different lighting, expressions, backgrounds

HumanScan BioID Face DB (http://www.bioid.com/support/downloads/software/bioid-face-database.html) *Emotions*: None *Elicitation*: n/a *Size*: 23 subjects *Kind*: Contains 19 manual markup points: 0 = right eye pupil; 1 = left eye pupil; 2 = right mouth corner; 3 = left mouth corner; 4 = outer end of right eyebrow; 5 = inner end of right eyebrow; 6 = inner end of left eyebrow; 7 = outer end of left eyebrow; 8 = right temple; 9 = outer corner of right eye; 10 = inner corner of right eye; 11 = inner corner of left eye; 12 = outer corner of left eye; 13 = left temple; 14 = tip of nose; 15 = right nostril; 16 = left nostril; 17 = centre point on outer edge of upper lip; 18 = centre point on outer edge of lower lip; 19 = tip of chin

Oulu University physics-based face database (www.ee.oulu.fi/research/imag/color/pbfd.html) *Emotions*: None *Elicitation*: n/a *Size*: 125 subjects *Kind*: All frontal images: 16 different camera calibration and illuminations

UMIST (http://www.sheffield.ac.uk/eee/research/iel/research/face.html) *Emotions*: None *Elicitation*: n/a *Size*: 20 subjects, 19–36 pictures per person *Kind*: Range of poses from profile to frontal views

Table 3 (continued)

Olivetti research (www.mambo.ucsc.edu/psl/olivetti.html) *Emotions*: None *Elicitation*: n/a *Size*: 40 subjects, 10 pictures per person *Kind*: All frontal and slight tilt of head

The Yale face database B (http://cvc.yale.edu/projects/yalefacesB/yalefacesB.html) *Emotions*: None *Elicitation*: n/a *Size*: 10 subjects *Kind*: 9 poses × 64 illumination conditions

AT&T (formerly called ORL database) (http://www.cl.cam.ac.uk/research/dtg/attarchive/facedatabase.html) *Emotions*: Smiling/not smiling *Elicitation*: Posed *Size*: 40 subjects *Kind*: 10 images for each subject which vary lighting, glasses/no glasses, and aspects of facial expression broadly relevant to emotion – open/closed eyes, smiling/not smiling

gestures and physiological signals is contained within Table 1, as this type of data (relatively rare) tends to occur within a multimodal context. The tables include material that consists of records with a limited amount of information rather than a fully marked-up and annotated corpus.

The tables are not intended to be a comprehensive list. The aim is to identify key databases and indicate the type of data that is available. Each table is arranged from left to right according to the same format – identifier for the database; modalities recorded (where there is more than one); description of how the data was elicited; indicator of size; further information regarding the type of data; and finally any information on cultural/linguistic range.

The information on each item is presented as a list rather than a geometric table. We have used table format in earlier summaries, but as the field develops, tables become too large for comfort. The format that we have used here seems to be a reasonable alternative.

4.1 Multimodal Databases

The dates of databases in this domain indicate that work on multimodality and emotion is relatively recent. Databases of emotion in multimodal contexts were unusual until the HUMAINE project, and even now large, structured and labelled databases are unusual. However, the area is fast gaining ground.

Perhaps the single most characteristic concern in recent work has been naturalness. Some of the work still uses actors, as, for example, the recent audiovisual database collected in Geneva (Baenziger and Scherer, 2007), but it has used professional actors working with a director. But most of the multimodal work has worked with more naturalistic settings or induction techniques to induce emotion. Some databases use real-life situations such as lost luggage offices (Scherer and Ceschi, 2000) or television chat shows (Chung, 2000; Douglas-Cowie et al., 2003; Devillers et al., 2006). Others use various induction techniques; key examples are mentioned in Douglas-Cowie et al. (2007), and fuller coverage is given in Part III, in the chapter "Issues in Data Collection".

Because of the emphasis on naturalness, the range of emotions covered tends in the direction of everyday emotional behaviour rather than full-blown emotions. For example, the SMARTKOM database (Schiel et al., 2002; Steininger et al., 2002a, 2002b) like the SAL database is built from listeners' responses to a 'machine'. In fact the machine is actually two humans in another room operating in a Wizard of Oz-type situation. Users are asked to solve a range of tasks. The emotional states and related states recorded include joy/gratification, anger/irritation, helplessness, pondering/reflecting, surprise and neutral. Other multimodal databases reflect the same trend with the Geneva lost luggage database containing examples of good humour and indifference among the emotion-related states listed and the Belfast naturalistic database covering a wide range of emotional states with a wide spread on a two-dimensional representation of emotion (based on the dimensions of evaluation and activation). The French EmoTV database provides striking examples of the way even full-blown emotions tend to be mixed and of the way intensity shifts in the build-up to and away from an emotional peak. The recent Green data set recorded at Belfast shows attempts to persuade people into adopting a more 'sustainable lifestyle'. It is rich in what Baron-Cohen (2007) calls epistemic states – doubt, questioning, rejection and so on.

Coding material of that kind is a challenge, and the techniques that have been adopted are very varied. The main techniques are described in the chapter on labelling, and the synthesis developed for the HUMAINE database is described in the chapter on that database.

Some modalities are still not very fully covered. It has already been noted that gesture was not extensively covered, and the EmoTaboo database was developed to provide more material in that area. Several of the sources that there are, use driving as a context. The ORESTEIA database (McMahon et al., 2003, see also http://manolito.image.ece.ntua.gr/oresteia/) records physiological measurements (with some audio) from subjects on a driving simulator. The subjects encounter various problems while driving (deliberately positioned obstructions, dangers, annoyances 'on the road'). These are intended to induce emotional responses. An ongoing development of the approach (see the later chapter on issues in data collection) involves subjects being induced into various emotional states using techniques derived from the psychological literature and then drive through scenarios designed to test how the emotion affects their action. The recently constructed DRIVAWORK database provides audiovisual and physiological records of subjects using a driving simulator (Hönig, 2007).

The emphasis on naturalism has brought various practical problems with it. Genuinely natural data tends to involve noise, camera angles, lighting and various other factors that pose problems for machine analysis, and as noted above, copyright issues mean that significant material may not be freely available for researchers. The sources use a variety of contexts, but there is no systematic understanding of the range of contexts that data should aim to document. Multicultural data is also rare, though see Cube-G (Lipi et al., 2008). Finally different teams have used a wide variety of labelling conventions for emotion, not all of which are informed by contemporary understandings of emotion.

The HUMAINE database was designed to reflect the main kinds of material that recent progress has made available and the labelling techniques that can be applied to it. It consists of 50 emotional episodes taken from a variety of induced and naturalistic settings and covers a wide range of emotional behaviour and signs of emotion (speech, face, gesture and physiological signs). The database is drawn from a number of sources that contain many more records less fully labelled. These include the Belfast naturalistic database, the SAL database and the EmoTaboo database, which are described above. The other main data sets used are the Reality Castaway TV database, the Belfast adventure and Spaghetti databases, the Erlangen driving database and the SAL Hebrew database. Chapter "Biological and Computational Constraints to Psychological Modelling of Emotion" Part I provides a full description of the HUMAINE database, and Chapter "Emotion: Concepts and Definitions" gives further information about individual sources from which the HUMAINE database has been compiled.

The emphasis in this section has been on multimodal emotion-related material. Deciding what to include is not always straightforward. Some databases contain audiovisual recordings, but the focus has been on the speech (Polzin and Waibel, 2000; Banse and Scherer, 1996). There are also major multimodal databases that have very little emotional content. Examples that appear to be in that category include XM2VTSDB (www.ee.surrey.ac.uk/Research/VSSP/xm2vtsdb/) and ISLE (http://nats-www.informatik.uni-hamburg.de/~isle/speech.html, IST project IST-1999-10647), but ISLE in particular is relevant because of the labelling systems it has developed for multimodal work.

A particularly interesting case is the large corpus of meeting data collected by the AMI project. The investigators considered explicitly whether there was emotion present (Heylen et al., 2006). The terms that raters used to describe the mental states that they observed were (starting from the commonest) bored, confident, interested, attentive, serious, joking, friendly, curious, cheerful, at-ease, amused, relaxed, nervous, frustrated, decisive, uninterested, impatient, confused, agreeable, annoyed. It appears that the recordings involve mainly what Baron-Cohen (2003) calls epistemic states with an emotional dimension. It may well be that as emotion-oriented computing matures, it will become more concerned with states like these, but at present, they are not what most people in the field consider it their business to study.

4.2 Speech Databases

There has been a considerable body of audio data collected for speech and emotion studies, as reflected in the fact that more corpora appear in Table 2 than either of the other tables.

Much of the data has three characteristics: the emotion in it is simulated by an actor (not necessarily trained); the actor is reading preset material and he/she is aiming to simulate full-blown emotion (Yacoub et al., 2003; Kienast and Sendlmeier, 2000). Other examples in the general vein include (1997) Leinonen and Hiltunen

(1997), Nakatsu et al. (1999), Juslin and Laukka (2002), Nogueiras et al. (2001), Murphy (2002) and Oudeyer (2003). There are sometimes attempts to make the data more natural by contextualising the emotion, for example, using material to read that is inherently emotional in content [examples in Table 2 are from Mozziconacci (1998) and McGilloway (1997)].

At the other extreme, there are a few speech databases which are focused on naturalistic data. The Leeds–Reading database has already been described. Campbell' s CREST database (Campbell, 2006; see also Douglas-Cowie et al., 2003) has acquired a unique body of truly natural data with a wide range of everyday emotions from volunteers who were recorded for long periods as they went about their ordinary daily social interactions.

Also fairly natural, but narrower in emotional range, are data sets that use material recorded during specific types of events, such as game shows, emergency flight situations for pilots, affectively loaded therapy sessions and journalists' reports of emotion-eliciting events (e.g. France et al., 2000 in Table 2 but also Johannes et al., 2000; Frolov et al., 1999; Huttar, 1968; Kuroda et al., 1979; Roessler and Lester, 1976, 1979; Sulc, 1977; Williams and Stevens, 1969, 1972).

A good deal of work has also used data elicited by techniques designed to induce states that are both genuinely emotional and likely to involve speech. Early examples include a task where subjects are introduced to unpleasant images (Tollkmitt and Scherer, 1986) and some parts of the SUSAS database (speech under simulation and actual stress database) which use speech elicited in a range of stressful situations (Hansen and Bou-Ghazale, 1997, http://www.ldc.upenn.edu/Catalog/). More recent examples include simulations of call centres designed to elicit irritation (Mitchell et al., 2000; Batliner et al., 2003); a stressful driving task (Fernandez and Picard, 2003) and a spelling task designed to elicit embarrassment (Bachorowski and Owren, 1995). The Erlangen AIBO database (Batliner et al., 2004) produces a wider range of emotional states such as neutral (default), joyful, surprised, emphatic, helpless, touchy (irritated), angry, bored and reprimanding. It also generates a substantial amount of speech that the investigators describe as 'motherese', which makes a great deal of sense intuitively, and offers a useful reminder that standard lists of states do not necessarily fit easily onto behaviour patterns that are observed in naturalistic settings.

Data from call centres has been used extensively, and it epitomises both the advances that have been made and some of the key difficulties. Some involve human–machine dialogue, such as SYMPAFLY (Batliner et al., 2003) and the DARPA Communicator Corpus used by Ang et al. (2002) (see Walker et al., 2001). Other teams have used human–human call centre data (Devillers and Vasilescu, 2004).

These databases have several attractions. Firstly, the emotion is presumably genuine, not acted. Secondly, they deal with dialogue, which exposes issues missing from the monologue type data often produced in acted or elicited emotion. Thirdly, they are very directly related to a foreseeable application of emotion recognition.

But with these advantages come limitations. Access tends to be limited, both for commercial reasons and because of privacy issues. The frequency with which emotion is expressed tends to be low. To illustrate the scale of the problem, Ang et al. (2002) used material from the DARPA Communicator Corpus totalling 14 h 36 min of speech. The commonest strong emotion was frustration, of which he obtained 42 unequivocal instances. The nature of the interaction imposes constraints of the forms of utterances and probably on the way emotion may be expressed within those forms, raising major questions about generalisability.

Not least, the emotions tend to be from a narrow range, generally negative. Recent studies illustrate the point. The study by Ang et al., cited above, is a case in point. Similarly, Lee and Narayanan (2003) detected negative versus non-negative emotion using a corpus of utterances obtained from a commercially deployed human machine-spoken dialogue application; most dialogue turns had one utterance. Boozer et al. (2003) have reported work on neutral, frustrated and happy states using human–computer dialogues generated by a phone-based airline flight-planning system. The SYMPAFLY system offers a broader range, including states like helpless, panic and touchy.

In summary, speech databases have developed very rapidly over the last decade, with a strong movement towards naturalistic data. Researchers have been quite creative in this area in experimenting with different methods of collecting data, ranging from opportunistic use of pre-recorded naturalistic emotional situations to laboratory-based induction techniques. There is also a growth in the range of language and cultures covered, with work on Western European languages but also on Hebrew (Amir et al., 2000) and on Japanese and Chinese (Campbell, 2006). Nevertheless, core problems remain.

4.3 Face Databases

Table 3 shows a selection of key databases of facial expressions. As can be seen from the descriptions in the last column, these show faces under systematically varied conditions of illumination, scale and head orientation. Rather few consider emotional variables systematically, and the range of emotional expressions considered is quite limited and tends to focus on the 'primary' emotions. The data is also generally acted or posed and consist of static images. The term 'staged' is perhaps appropriate.

The seminal database of this type is the classic Ekman and Friesen collection of photographs showing facial emotion (Ekman and Friesen, 1975); this can be bought in electronic form. Others in the same mould are the Yale database which contains 11 images for each of 15 individuals, one per different facial expression or configuration – centre light, with glasses, happy, left light, without glasses, normal, right light, sad, sleepy, surprised and wink, and the ORL database of faces which contains 10 different images for each of 40 subjects. The images vary the lighting and aspects of facial expression which are at least broadly relevant to emotion – open/closed eyes, smiling/not smiling. Rather few databases contain samples of faces moving, and moving sequences which are emotionally characterised are even less common.

A good deal of material consists of images produced by research software, e.g. for facial animation, rather than the original video sequences used for the analysis or training. Examples can be found at www.cs.cmu.edu/~face/. Databases that combine speech and video are still rare and the few examples that there are have already been mentioned in Table 1, in particular SMARTKOM and the Belfast naturalistic database. The XM2VTSDB multimodal face database is also audiovisual but does not contain emotion.

The cultural range is dominated by the West, although there is a database of Japanese faces (see Table 3). In summary, the data is limited in emotional range and level of naturalness. However, the field is developing and genuinely natural data (of moving faces) is emerging with a much wider range of emotional expression. However, getting appropriate facial images is not straightforward, and researchers report practical problems along the way. Genuinely natural data involves quite a lot of jerky movement, frequent occlusion of the face, and particular angles that make its use for facial analysis limited. And audiovisual images create real problems in terms of the interference of speech with facial movement. Clearly appropriate balances need to be found.

5 Lessons for Future Research

The previous sections have shown that developing satisfactory emotion databases involves challenges at multiple levels, from the practicalities of recording to conceptual issues in psychology. This section reflects that range by drawing lessons for future work at two levels, first at the level of an abstract overview and then at the level of practical issues that need to be addressed.

5.1 Ideal Specifications and Obstacles to Achieving Them

One of the results of research is a gradually clarifying picture of the kind of data resource that the community might ideally hope for. At least the following properties are clearly desirable:

- It would be fully naturalistic (except insofar as some parts deliberately captured the behaviour of actors, newsreaders, etc.).
- It would sample the whole domain of emotion and emotion-related states.
- It would represent all the types of action through which emotion and emotion-related states can be expressed.
- It would sample the whole range of cultural and individual differences that are important for the expression of emotion.
- The recordings would be of high technical quality.
- There would be recordings in all relevant modalities.
- The data would be comprehensively labelled.
- The labelling would follow a consistent, standard pattern.

- The labelling would include objective verification of all the emotional states involved.
- The material would be structured to facilitate statistically learning (for instance, it would include balanced samples, samples constructed to match in all respects but one key emotional contrasts, etc.).
- The number of instances would be large enough for statistically learning techniques to be applied.
- The material would be freely available.
- The process of obtaining, storing and distributing samples would be ethically sound.

The combined resources currently available fall far short of that ideal. That is partly because the total effort that has been invested is still small relative to the scale of the task, but there are also problems of principle at every turn. For instance,

- The demand for high-quality recordings with naturalness. Genuinely natural data tends to involve quite a lot of jerky movement, frequent occlusion of the face, and particular angles that make its use for facial analysis limited. And audiovisual images create real problems in terms of the interference of speech with facial movement.
- Attempting to achieve multimodality conflicts with naturalness, because it requires more and more intrusive types of recording. It is difficult to imagine recording more than two or possibly three modalities (recorded in ways that can usefully be analysed) without material loss of naturalness.
- The wish for objective verification conflicts with naturalness, because objective verification tends to demand either intrusive methods or tight control over the situation and with the subtler emotional states that make up a large part of emotional life, it is difficult to see how objective verification could be achieved at all.
- There is a balance to be struck between demand for quantity and comprehensiveness of labelling. The more detailed a labelling scheme is, the more labour intensive implementing it is likely to be, and the more skilled the people involved need to be.
- There is a balance to be struck between demand for statistical tractability and comprehensiveness of labelling. The 'curse of dimensionality' means that extracting meaningful relationships from highly detailed labelling schemes requires unrealistic quantities of data, unless radical alternatives to current statistical techniques become available.
- There is an almost unlimited range of types of action through which emotion and emotion-related states can be expressed, and it is difficult to imagine any way of combining them factorially (for instance, it is difficult to imagine a situation where emotional driving behaviour co-occurs with expression of emotion through large amplitude hand gestures and balletic movements).

The list could easily be extended. Two things are essential to deal with a domain that presents problems like these. One is a clear understanding of the need to reach intelligent compromises. It is a major barrier to progress if groups become wedded to one or two ideals, dismiss work that does not fulfil them completely and embrace work that does, even though it is far short on other criteria. The other essential is development of theory that provides a sound motivation for sampling. The traditional emphasis on basic or primary emotions was attractive partly because it seemed to meet that need. A more soundly based alternative is badly needed.

5.2 Practical Issues

It is a painful truth that excellent work invested in a database can be rendered useless by failure to engage with any one of a great many practical requirements. This section touches on the most important of these.

5.2.1 Ethics and Privacy

It is essential for the field to be well versed in the ethical issues involved in collecting emotionally coloured data. Most institutions now routinely require ethical approval for any work with humans, and it is not routine to give approval for procedures that involve eliciting negative emotions or deception, both of which are very common in emotion elicitation scenarios. Informed consent is usually *a sine qua non* and that makes it very difficult to record without the participants' knowledge, despite the obvious advantages in terms of naturalness. Retaining recordings requires another level of clearance and again must be backed by informed consent of the people involved, and release requires yet another level.

These issues should be handled through an appropriate ethics committee. Later chapters in this handbook give more detailed information.

5.2.2 Access

Ensuring access has proved a fatal obstacle in the past. It involves several sub-issues. The ethical dimension has been noted above. Release has to be backed by appropriate clearances. Teams will often want to retain control of material so that (for instance) they are not exploited for profit or used for inappropriate purposes. That can be achieved by ensuring that release is governed by conditions of use set out by the groups who constructed the databases. The governing 'CEICES' rubric proposed by FAU (Batliner et al., 2006) is a good example.

The fact that files are likely to be large raises another kind of problem. When files are not too large, the most convenient transmission medium is FTP (with a suitable password). It has been the norm to send larger bodies of data on DVDs, but that becomes onerous if there is a substantial demand. Recently, the HUMAINE Association has been exploring ways of making emotion-related data available via its server.

A third access issue is simply knowing that data exist, since it is not the norm for journals to publish articles that simply give the community details of a new database.

HUMAINE has made lists available on its website, and the HUMAINE Association will continue the practice. The tables in this chapter are based on the HUMAINE website tables.

5.2.3 Format of Recordings

It is all too easy to underestimate how important various technical criteria are if recordings are to be used for machine extraction. Low background noise and echo, fixed distance from the microphone, microphone response characteristics, and interference from nearby cables or other sources all have a major impact of the usefulness of audio recordings. Machine analysis of video can be seriously disrupted by highlights, colour balances, and lack of contrast with the background that a human being would not notice. Physiological data is highly sensitive to misplaced or poorly connected electrodes. Multimodal data needs to include well-defined markers to allow synchronisation.

Similar issues surround level of compression. Compressed files are much easier to store and transmit, but they may not be adequate for some purposes. For instance, in the video modality, MPEG format is suitable for many uses, but raw video files (.avi) may be needed to give the resolution required for accurate FAP extraction. In the context of physiological data, sampling rate raises similar issues. Quite low sampling rates are adequate for most emotion-related measures, but some information about cardiac activity depends on ECG records being sampled at 200 Hz.

In general, database creators need to understand the kind of format that potential users need.

5.2.4 Standardisation

It is a truism that databases are unlikely to be used unless they observe the standards of the field. Very little in the field of emotion-oriented computing even approaches formal standardisation. Exceptions are FACs coding, some MPEG standards, and the EARL described by Schroder et al. in Part IV. However, there are increasing terminologies, tools and practices that are shared by quite large groups, and it is advisable to respect them. Part I sets out terms and concepts that have become common currency within HUMAINE. The next two chapters describe technical resources that have a similar status.

6 Conclusions

This chapter reflects long and sometimes painful experience in the development of databases for emotion-oriented computing. As such chapters tend to do, it has tried to provide the information that the authors wish they had been able to access when they began to work in the area. Key areas are expanded in the chapters that follow; this chapter provides the context for them.

The single most important lesson of a decade and a half of research in the area is that the task is bigger than a beginner tends to assume. It involves understanding a considerable range of background concepts, addressing many different kinds of practicality and knowing what is there already. The chapter has attempted to provide grounding in those areas. The measure of success will be a new generation of databases that avoid repeating the errors of previous generations.

References

Abelin Å, Allwood J (2000) Cross linguistic interpretation of emotional prosody. ISCA workshop on speech and emotion. Newcastle, Northern Ireland, pp 110–113

Amir N, Ron S, Laor N (2000) Analysis of an emotional speech corpus in Hebrew based on objective criteria. In: Douglas-Cowie E, Cowie R, Schroeder M (eds) Proceedings of the ISCA workshop on speech and emotion, Belfast, Textflow, pp 29–33

Ang J, Dhillon R, Krupski A, Shriberg E, Stolcke A (2002) Prosody-based automatic detection of annoyance and frustration in human–computer dialog. In: Proceedings ICSLP, Denver, CO, Sept 2002

Bachorowski J, Owren M (1995) Vocal expression of emotion: acoustic properties of speech are associated with emotional intensity and context. Psychol Sci 6(4):219–224

Banse R, Scherer K (1996) Acoustic profiles in emotion expression. J Pers Social Psychol 70(3):614–636

Bänziger T, Scherer K (2007 Sept) Using actor portrayal to systematically study multimodal emotional expression: the GEMEP corpus. In: Paiva A, Pra-da R, Picard R (eds) Affective computing and intelligent interaction, Lisbon. Springer LNCS, Berlin, pp 476–487

Baron-Cohen S (2003) The essential difference: men, women and the extreme male brain. Penguin/Basic Books

Baron-Cohen S (2007) Mind reading: the interactive guide to emotions – version 1.3 . Jessica Kingsley, London

Batliner A, Fischer K, Huber R, Spilker J, Noeth E (2003) How to find trouble in communication. Speech Commun 40:117–143

Batliner A, Hacker C, Steidl S, Noth E, D'Arcy S, Russell M et al (2004) You stupid tin box—children interacting with the AIBO robot: a cross-linguistic emotional speech corpus. In: Proceedings LREC, Lisbon, 2004

Batliner A, Steidl S, Schuller B, Seppi D, Laskowski K, Vogt T, Devillers L, Vidrascu L, Amir N, Kessous L, Aharonson V (2006) Combining efforts for improving automatic classification of emotional user states. In: Erjavec T, Gros J (eds) Language technologies, IS-LTC 2006. Infornacijska Druzba (Information Society), Ljubljana, Slovenia, pp 240–245

Beller G, Schwarz D, Hueber T, Rodet X (2005) Hybrid concatenative synthesis in the intersection of speech and music. JIM2005 Paris, CICM, 41–45

Boozer A, Seneff S, Spina M (2003) Using Prosodic features for emotion classification and recognition MIT Spoken Language Systems Group Summary of Research, Jul 2003, pp 51–54. http://groups.csail.mit.edu/sls//archives/root/publications/2003/ResSum2003.pdf. Accessed on 7/11/2010

Campbell N (2002) Recording and storing of speech data. Proceedings LREC 2002, Las Palmas, Canary Islands. http://www.mpi.nl/lrec/2002/papers/lrec-pap-06-nick-speech.pdf

Campbell N (2006) A language-resources approach to emotion: corpora for the analysis of expressive speech. In: Proceedings of the LREC Workshop on Corpora for Research on Emotion and Affect, Genoa, pp 1–5

Chung S-J (2000) L'expression et la perception de l.eḿotion extraite de la parole spontaneé: ev́idences du coreén et de l'anglais. Unpublished doctoral dissertation, Universite ́de la Sorbonne Nouvelle, Paris III

Cowie R, Douglas-Cowie E (1996). Automatic statistical analysis of the signal and prosodic signs of emotion in speech. Proceedings of the international conference on spoken language processing, Philadelphia, 1989–1992

Cowie R, Douglas-Cowie E, Tsapatsoulis N, Votsis G, Kollias S, Fellenz W, Taylor JG (2001) Emotion recognition in human–computer interaction. IEEE Signal Process Mag 18(1):33–80

Cowie R, McGuiggan A, McMahon E, Douglas-Cowie E (2003) Speech in the process of becoming bored. Proceedings of 15th international congress of phonetic sciences, Barcelona

Devillers L, Cowie R, Martin J-C, Douglas-Cowie E, Abrilian S, McRorie M (2006) Real-life emotions in French and English TV video corpus clips: an integrated annotation protocol combining continuous and discrete approaches. In: Proceedings LREC 2006, Genoa

Devillers L, Vasilescu I (2004) Reliability of lexical and prosodic cues in two real-life spoken dialog corpora. Proceedings LREC 2004, Las Palmas, Canary Islands

Douglas-Cowie E et al (2004) HUMAINE D5d, 2004 p7. http://emotion-research.net/projects/humaine/deliverables/D5d%20potential%20exemplars%20databases.pdf

Douglas-Cowie E, Campbell N, Cowie R, Roach P (2003) Emotional speech: towards a new generation of databases. Speech Commun 40(1–2):33–60

Douglas-Cowie E, Cowie R, Schroeder M (eds) (2000) In: Proceedings of the ISCA workshop on speech and emotion, Belfast

Douglas-Cowie E, Cowie R, Sneddon I, Cox C, Lowry O, McRorie M, Martin J-C, De-villers L, Abrilian S, Batliner A, Amir N, Karpouzis K (2007) The HUMAINE database: addressing the collection and annotation of naturalistic and induced emotional data. In: Proceedings of the ACII 2007, Lisbon, pp 488–500

Ekman P (1992) An argument for basic emotions. Cogn Emot 6:169–200

Ekman P, Friesen W (1975) Pictures of facial affect. Consulting Psychologists' Press, Palo Alto, CA

Engberg IS, Hansen AV, Andersen O, Dalsgaard P (1997) Design, recording and verification of a Danish emotional speech database. EUROSPEECH-1997, 1695–1698

Fernandez R, Picard R (2003) Modeling drivers' speech under stress. Speech Commun 40:145–159

Fragopanagos N, Taylor J (2005) Emotion recognition in human–computer interaction. Neural Netw 18:389–405

France D, Shiavi R, Silverman S, Silverman M, Wilkes D (2000) Acoustical properties of speech as indicators of depression and suicidal risk. IEEE Trans Biomed Eng 47:7

Frolov M, Milovanova G, Lazarev N, Mekhedova A (1999) Speech as an indicator of the mental status of operators and depressed patients. Hum Physiol 25(1):42–47

Greasley P, Setter J, Waterman M, Sherrard C, Roach P, Arnfield S et al (1995) Representation of prosodic and emotional features in a spoken language database Proceedings XIIIth international congress of phonetic sciences, vol. 1. Stockholm, pp 242–245

Han JH, Ward AJI, Lavine BK (2005) The problem of adequate sample size in pattern recognition studies: the multivariate normal case. J Chemom 4(1):91–96

Hansen J, Bou-Ghazale S (1997) Getting started with SUSAS: a speech under simulated and actual stress database. In: Proceedings of the Eurospeech 1997, Rhodes Greece, vol 5, pp 2387–2390

Heylen D, Nijholt A, Reidsma D (2006) Determining what people feel and think when interacting with humans and machines: notes on corpus collection and annotation. In: Kreiner J, Putcha C (eds) Proceedings 1st California conference on recent advances in engineering mechanics. California State University, Fullerton, January 12–14, pp 1–6

Hönig F (2007) DRIVAWORK – driving under varying workload. A multi-modal stress database in the automotive context. Vortrag: HUMAINE Plenary Meeting, Paris, 6 Jun 2007

Huttar GL (1968) Relations between prosodic variables and emotions in normal American English utterances. J Speech Hear Res 11:481–487

Iriondo I, et al (2000) Validation of an acoustical modelling of emotional expression in Spanish using speech synthesis techniques. Proceedings of the ISCA workshop on speech and emotion, Belfast 2000, pp 161–166

Johannes B, Salnitski V, Gunga H-C, Kirsch K (2000) Voice stress monitoring in space – possibilities and limits. Aviat Space Environ Med 71(9):A58–A65 (section II)

Juslin P, Laukka P (2002) Communication of emotions in vocal expression and music performance. Psychol Bull 129(5):770–814

Kawai H, Toda T, Ni J, Tsuzaki M, Tokuda K (2004) XIMERA: a new TTS from ATR based on corpus-based technologies. In: Proceedings of the 5th ISCA ITRW on speech synthesis, Pittsburgh, PA, pp 179–184

Kienast M, Sendlmeier WF (2000) Acoustical analysis of spectral and temporal changes in emotional speech. In: Cowie R, Douglas-Cowie E, Schroeder M (eds) Speech and emotion: proceedings of the ISCA workshop. Newcastle, County Down, Sept 2000, Belfast, Textflow, pp 92–97

Kominek J, Black AW (2004) The CMU ARCTIC speech databases. In: Proceedings of the 5th ISCA ITRW on speech synthesis, Pittsburgh, PA, pp 223–224

Kuroda I, Fujiwara O, Okamura N, Utusuki N (1979) Method for determining pilot stress through analysis of voice communication. Aviat Space Environ Med 47:528–533

Lee C, Narayanan S (2003) Emotion recognition using a data-driven fuzzy inference system. In: Proceedings Eurospeech 2003, Geneva

Leinonen L, Hiltunen T (1997) Expression of emotional–motivational connotations with a one-word utterance. J Acoust Soc Am 102(3):1853–1863

Lipi AA, Yamaoka Y, Rehm M, Nakano Y (2008) Enculturating conversational agents based on a comparative corpus study. In: Prendinger H, Lester J, Ishizuka M (eds) Intelligent virtual agents, Springer

McGilloway S (1997) Negative symptoms and speech parameters in schizophrenia. Unpublished doctoral thesis, Queen's University, Belfast

McMahon E, Cowie R, Kasderidis S, Taylor J, Kollias S (2003) What chance that a DC could recognise hazardous mental states from sensor outputs? In: Proceedings of the DC Tales conference, Sanotrini, Jun 2003

Mitchell CJ, Menezes C, Williams JC, Pardo B, Erickson D, Fujimura O (2000) Changes in syllable and boundary strengths due to irritation. In: Douglas-Cowie E, Cowie R, Schroder M (eds) Proceedings of the ISCA Workshop on Speech and Emotion, Belfast, Textflow, pp 98–103.

Mozziconacci S (1998) Speech variability and emotion: production and perception. Unpublished doctoral thesis, Technical University Eindhoven, Eindhoven

Murphy C (2002) Automatic recognition of spoken emotion using audio signal processing. Unpublished undergraduate thesis, Department of Electrical and Electronic Engineering, University College, Dublin

Nakatsu R, Tosa N, Nicholson J (1999) Emotion recognition and its application to computer agents with spontaneous interactive capabilities. In: Proceedings of the IEEE workshop on multimedia signal processing, Copenhagen, pp 439–444

Nogueiras A, Moreno A, Bonafonte A, Marinõ J (2001) Speech emotion recognition using hidden Markov models. In: Proceedings of the Eurospeech 2001, Aalborg, Denmark

Ortony A, Clore G, Collins A (1988) The cognitive structure of emotions. Cambridge University Press, Cambridge, MA

Oudeyer P-Y (2003) The production and recognition of emotions in speech: features and algorithms. Int J Hum Comput Interact 59(1–2):157–183

Paeschke A, Sendlmeier WF (2000) Prosodic characteristics of emotional speech: measurements of fundamental frequency movements. In: Proceedings of the ISCA workshop on speech and emotion, Textflow, Belfast, 5–7 Sept 2000, pp 75–80

Pereira C (2000) Dimensions of emotional meaning in speech. Proceedings of the ISCA workshop on speech and emotion, Newcastle, Co. Down. Belfast, Textflow, pp 25–28

Polzin TS, Waibel A (2000) Emotion-sensitive human–computer interfaces. In: Douglas-Cowie E, Cowie R, Schroeder M (eds) Proceedings of the ISCA workshop on speech and emotion. Belfast, Textflow, pp 201–206

Raudys SJ, Jain AK (1991) Sample size effects in statistical pattern recognition: recommendations for practitioners. IEEE Trans Pattern Anal Mach Intell 13(3):252–264

Roach P (2000) Techniques for the phonetic description of emotional speech. In: Douglas-Cowie E, Cowie R, Schroeder M (eds) Proceedings of the ISCA workshop on speech and emotion, Belfast, Textflow, pp 53–59

Roach P, Stibbard R, Osborne J, Arnfield S, Setter J (1998) Transcription of prosodic and paralinguistic features of emotional speech. J Int Phonetic Assoc 28:83–94

Roessler R, Lester JW (1976) Voice predicts affect during psychotherapy. J Nerv Ment Dis 163(3):166–176, Sep 1976

Roessler R, Lester J (1979) Vocal pattern in anxiety. In: Fann W, Pokorny A, Koracau I, Williams R (eds) Phenomenology and treatment of anxiety. Spectrum, New York, NY

Rosch E (1978) Principles of categorization. In: Rosch E, Lloyd, BB (eds) Cognition and categorization. Lawrence Erlbaum Associate, Hillsdale, NJ

Sak H (2000) A corpus-based concatenative speech synthesis system for Turkish. B.S thesis in computer engineering and information science, Bilkent University

Scherer KR, Ceschi G (1997) Lost luggage emotion: a field study of emotion-antecedent appraisal. Motivation and Emotion 21(3):211–235

Scherer KR, Ceschi G (2000) Studying affective communication in the airport: the case of lost baggage claims. Pers Soc Psychol Bull 26(3):327–339

Schiel F, Steininger S, Türk U (2002) The SmartKom multimodal corpus at BSA. In: Proceedings of the LREC 2002, Las Palmas, Gran Canaria, pp 200–206

Steininger S, Schiel F, Dioubina O, Raubold S (2002a) Development of user-state conventions for the multimodal corpus in SmartKom. In: Proceedings of the workshop 'Multimodal Resources and Multimodal Systems Evaluation' 2002, Las Palmas, Gran Canaria, pp 33–37

Steininger S, Schiel F, Glesner A (2002) Labeling procedures for the multi-modal data collection of SmartKom. In: Proceedings of the LREC 2002, Las Palmas, Gran Canaria

Stibbard R (2001) Vocal expression of emotions in non-laboratory speech. Unpublished doctoral dissertation, University of Reading, Reading, UK

Sulc J (1977) To the problem of emotional changes in the human voice. Act Nerv Super 19:215–216

Sutherland NS (2007) Irrationality: why we don't think straight reissued. Pinter & Martin, London

Tabachnick BG, Fidell LS (2001) Using multivariate statistics. Allyn & Bacon, Boston, MA

ten Bosch L (2000) Emotions: what is possible in the ASR framework. In: Proceedings ISCA workshop on speech and emotion, Newcastle

Tolkmitt F, Scherer KR (1986) Effect of experimentally induced stress on vocal parameters. Exp Psychol Hum Percept Perform 12(3):302–313

van Bezooijen R (1984) The characteristics and recognizability of vocal expression of emotions. Foris, Dordrecht

Walker MA, Passonneau R, Boland JE (2001) Quantitative and qualitative evaluation of Darpa Communicator spoken dialogue systems. In: Proceedings of the 39th annual meeting on association for computational linguistics, Toulouse, France, pp 515–522

Williams C, Stevens K (1972) Emotions and speech: some acoustical correlates. J Acoust Soc Am 52(4, part 2):1238–1250

Yacoub S, Simske S, Lin X, Burns J (2003) Recognition of emotions in interactive voice response systems. In: Proceedings of the Eurospeech 2003, Geneva

Issues in Data Collection

Roddy Cowie, Ellen Douglas-Cowie, Margaret McRorie, Ian Sneddon, Laurence Devillers, and Noam Amir

Abstract The chapter reviews methods of obtaining records that show signs of emotion. Concern with authenticity is central to the task. Converging lines of argument indicate that even sophisticated acting does not reproduce emotion as it appears in everyday action and interaction. Acting is the appropriate source for some kinds of material, and work on that topic is described. Methods that aim for complete naturalism are also described, and the problems associated with them are noted. Techniques for inducing emotion are considered under five headings: classical induction; physical induction; games; task settings; and conversational interactions. The ethical issues that affect area are outlined, and a framework for dealing with them is set out.

1 Introduction

The heart of a database is a collection of records, that is, signals in various modalities (normally video and/or audio, sometimes physiological) that portray people acting and interacting in ways that are coloured by emotion. This chapter is about the methods that are available to obtain records that are 'fit for purpose'.

The context is set elsewhere in this handbook (particularly in chapter 'Principles and History'). This chapter focuses on the description of specific techniques. General issues are discussed briefly at the beginning for the sake of completeness.

2 Conceptual Issues

The task of record collection hinges on the problem of authenticity. It is easy to record students or colleagues simulating a range of emotions. The problem is whether the simulations are likely to bear a very close relationship to the kind of

R. Cowie (✉)
Department of Psychology, Queen's University Belfast, Belfast, Northern Ireland, UK
e-mail: roddy.cowie@qub.ac.uk

P. Petta et al. (eds.), *Emotion-Oriented Systems*, Cognitive Technologies,
DOI 10.1007/978-3-642-15184-2_12,

emotion that pervades life in general, and application scenarios in particular. In fact, there is good reason to believe that even sophisticated acting is quite distinct to the phenomenon that occurs in everyday life. Studies by HUMAINE teams have provided concrete evidence on the issue.

The first followed up the common intuition that most people can tell very quickly whether a television program shows people behaving relatively spontaneously (as in news programs and interviews) or acting (in dramas). The study took clips from TV channels, some showing spontaneous expressions of emotion and others showing extracts from dramas in which actors simulated comparable emotions. Raters used a trace-type technique to indicate which type they believed each clip was (acted or natural) and how confident they were in their judgment. Figure 1 shows the overall picture that emerged. The distinction was not perceived instantly, but within a few seconds, raters had essentially recognised the distinctions, and their confidence grew rapidly over time.

A second study documents one of the reasons for people's ability to discriminate. Raters were shown still frames, some drawn from recordings of spontaneous activity and some from acted databases. They rated each in terms of valence and activation. The frames were shown in a random order but actually came from sequences. The investigators then measured the change in perceived emotion from sample to sample in each type of data. Figure 2 shows the broad results. There was more frame-to-frame variation in the naturalistic material in both dimensions – which is to say that one of the distinctive features of spontaneous emotion is that it fluctuates from moment to moment in ways that acted renditions do not.

These differences have practical implications. It has been demonstrated that systems which are trained on acted material do not perform well on data that involves natural emotions (Batliner et al., 2003).

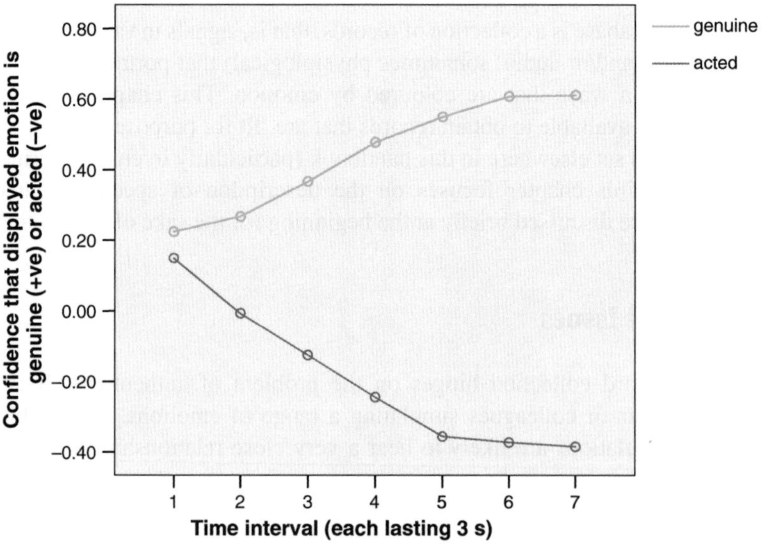

Fig. 1 Time course of judgment that displayed emotion is genuine or acted

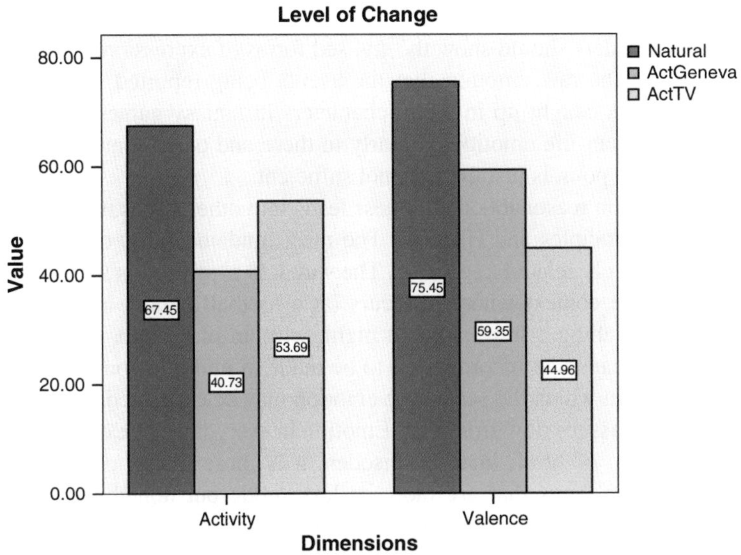

Fig. 2 Comparison of acted and natural behaviour: level of change between adjacent frames

That kind of evidence explains why concern with collecting naturalistic data has become central to the whole character of the field. The bulk of this chapter is about methods of obtaining records that reflect the way emotion appears spontaneously in various activities and interactions. It is an undertaking that is fraught with difficulties, but it cannot easily be avoided.

This is not a dismissal of acted data. In the first place, acting is a matter of degree. It is not the norm for people to express their emotions without restraint or self-consciousness. They are generally aware of an audience, and the signs they give are tailored to it in more or less subtle ways. Figure 3 illustrates the point. It shows closely spaced pictures of a participant in an adventure task; the smile appears when she registers that people are watching her. It would be a misunderstanding of everyday emotion to discard data because it shows signs of that kind of tailoring.

Fig. 3 Change in expression during a challenging task (balancing on a high stack of boxes) when the participant becomes aware of people watching

Secondly, there are contexts where a kind of acting is the relevant model. Artificial newsreaders should show the stylised forms of expression that real newsreaders use, not the raw emotion that the events being reported might evoke in someone who was caught up in them; characters in fantasy games should probably show larger-than-life emotions. Clearly in these and other contexts, acted data is worthwhile; the point is just that it is not sufficient.

The emphasis on reasonable naturalness leads into other issues, which are taken up in chapter 'Principles and History'. The most fundamental is context. Natural emotion is intimately related to context. The way it is expressed is likely to depend profoundly on the context where it occurs (at a football match, in a boardroom, driving a car, watching a film, alone at night, with an old friend, on an Olympic rostrum). That means that records need to be made in multiple contexts to capture the various ways in which the same core emotion may be expressed.

Multiple other issues flow from that. Emotion in everyday contexts does not usually take the form of brief, intense episodes; it is more likely to colour actions and interactions in ways that are more or less subtle but nonetheless important if the intended outcome is to be achieved. That makes it necessary to explore ways of obtaining these subtle colourings. There are cultural variations in the intensity of the signals that are normally exchanged, and therefore it is important to obtain cross-cultural data. Particularly with subtler emotional phenomena, information is often distributed across channels; therefore, records have to be multimodal.

Balances have to be struck between the pressures to achieve naturalness and various other pressures. Experimentalists are drawn to achieving control, for instance, by ensuring that people in different emotional states use the same words. That clearly has advantages; the problem is that it also ensures the records will not show signs that are bound up with the choice of words. If records have to be processed by machine, recordings have to be machine-friendly. That involves trying to ensure low background noise, mouth at a constant distance from the microphone, frontal views of the face, convenient lighting, and so on. All of these tend to work against naturalness; the trick is to achieve acceptable balances.

Issues like these have driven efforts to collect suitable sets of records, and that underlies the variety of techniques that are described in the remainder of the chapter.

The grouping of the techniques is broadly historical. The earliest studies considered acted data, and it is considered first. There was then a surge of interest in truly natural data, and it is considered next. The problems encountered there have led to increasing concentration on techniques for eliciting emotion deliberately, and they take up the last and longest section on collection techniques.

3 Acted Data

The term 'acting' is commonly applied to a very wide range of data collection techniques. It is useful to subdivide it into techniques which involve a degree of stagecraft and those which do not.

3.1 Stylised Simulation

The single most cited emotion database is probably Ekman's collection of still photographs (Ekman and Friesen, 1975). It is uniquely theory driven. The photographs show members of his team performing 'facial gymnastics' – that is, consciously setting their faces into the configurations that Ekman's theory associates with the emotions that it identifies as 'basic'. The procedure ensures that each face evokes the intended label reliably in a forced-choice task. It does not ensure that faces like that will occur very often in everyday interaction, and research indicates they do not (Carroll and Russell, 1997). In psychological jargon, what is in question is their ecological validity.

A considerable number of visual databases are a degree less controlled. They are composed of clips (often video), each of which shows a participant simulating a given emotion in a way that feels natural to him or her (Kanade et al. 2000; Yale Face Database; Baron-Cohen, 2007). The best examples ensure quality by testing that the samples included are reliably assigned the intended label. Again, that does not ensure ecological validity.

Several auditory databases are broadly similar. They involve passages (often the same passage) being read in ways that are meant to convey prespecified emotions. Again, the better examples (e.g. Kienast and Sendlmeier, 2000) guarantee recognition, but not ecological validity. It is not guaranteed either that more than a minute proportion of spontaneous emotional speech is like that or that the stylised speech contains many of the features that signal emotion in everyday communication.

3.2 Acting

The term 'acting' is reserved here for material that applies the techniques one might expect to find in a competent stage production.

An obvious approach is to use samples from existing films. Polzin and Waibel (2000) have reported a database of that kind using American films; the critique of Ekman mentioned above (Carroll and Russell, 1997) used film material as source of data. The approach has two major problems. The first was introduced earlier in the chapter. Material from TV drama at least has features that allow people to recognise quickly that it is acted. The second is access. Production companies zealously guard extracts from their films.

More recently, some research teams have engaged actors themselves. The Belfast Structured database (which is auditory only) was made by transcribing the emotional utterances from the Belfast Naturalistic database and having actors deliver them with appropriate emotional colouring after an initial exercise designed to induce the relevant emotion. The Geneva Multimodal Emotion (GMEP) is much more substantial. Actors were prepared by a director and filmed using gesture and speech to convey target emotion (Bänziger and Scherer, 2007). A similar exercise (using method actors) has been reported by Enos and Hirschberg (2006).

Even these techniques do not simulate everyday emotion accurately. Frigo (2006) has compared the acted speech in the Belfast Structured database with the original versions and shown that they differ systematically. The contrast between GMEP and natural has already been shown in Fig. 2.

One of the outcomes of this work has been to highlight questions about what acting is. Shakespeare pointed out long ago that what people look for in a play is a heightened version of everyday life, not something indistinguishable from it (Hamlet Act 3 scene 2). It is a point worth taking seriously.

3.3 Naturalistic

The first substantial naturalistic emotion database was assembled by Roach, using audio only material, mainly from broadcasts (Douglas-Cowie et al., 2003). It exposed three recurring problems. The first was that naturalistic material did not show the kind of clear-cut emotion that the investigators expected to find. The second was that annotation was much harder than might be expected. That issue is taken up in the companion chapter on data labelling. The third was that the problem of ownership is very difficult with broadcast material (in fact, the Leeds–Reading database was never released outside Roach's group).

Naturalistic audiovisual databases followed, beginning with the Belfast Naturalistic database (Douglas-Cowie et al., 2003). It consists of audiovisual sedentary interactions from TV chat shows and religious programs, supplemented by locally recorded discussions between old acquaintances. The EmoTV database (Devillers et al., 2006) consists of audiovisual interactions from TV interviews (in French). It covers both sedentary interactions and interviews 'on the street' (with wide range of body postures). Both encountered the same issues as Roach. The sources tended not to yield the expected clear episodes of a single emotion, and new methods of labelling had to be developed as a result. Both collections also suffered from copyright problems, but a selection of 30 sequences from the Belfast Naturalistic database, with ethical and copyright clearance, is available.

The problem of copyright is a severe one. If a database can only be used by the team that developed it, it means that results based on it cannot be compared with other groups', and there is relatively low return on the effort invested in developing it. Note, though that there are exceptions to the rule that copyright issues limit the value of material associated with television. The team at QUB obtained access to unreleased raw footage from a reality TV series, 'Castaway', and general release seems to be possible. The material is interesting because the camera teams follow characters quite unselectively (selections are made later by the editor). If sources of that kind can be tapped, they may provide records of a scope that academic teams could never generate for themselves.

On the whole, though, fully naturalistic data collection has increasingly moved away from broadcast material. The largest single area involves data from call centres. A landmark study by Ang and his colleagues (2002) used recordings of simulated interactions about travel arrangements with a call centre from the DARPA database (Walker et al., 2001). Devillers and her colleagues have reported a series of studies using recordings from real call centres concerned with financial tasks (Devillers et al., 2005) and medical emergencies (Devillers and Vidrascu, 2005).

Call centre material has an outstanding attraction, which is that it relates directly to an application domain. Conversely, it is restricted in three major ways. Firstly, it is unimodal (auditory only). Secondly, it suffers from release restrictions which are at least as severe as broadcast material. Thirdly, the range of emotions that is likely to occur is restricted. Not many people phone a call centre to express awe in response to a starlit night.

There has also been research that aims to analyse in depth real-life situations where communication reaches pinnacles of subtlety and complexity that no simulation could expect to match. An elegant example is the work by Poggi and her team (2005) on a notorious trial for political corruption, which was televised. Both lawyers and accused use language, voice, face and gesture in deliberate and highly sophisticated ways, to influence both the jury and the wider public.

To date, the most radical approach to naturalistic collection is represented by JST/CREST Expressive Speech Corpus (Campbell, 2006). Speech was collected from informants who used wearable sound recording systems over a period of years, providing hundreds of hours of everyday conversation.

4 Induced Data

As it became clear that there were problems with both acted and truly naturalistic data, attention turned increasingly to methods in which the investigators not only recorded emotions but also created them. The result has been a proliferation of methods. They are extremely diverse and not easy to classify. This section divides them on the basis of the kind of engagement that they might be expected to create and the extent to which techniques try to control the kind of emotion that is obtained.

4.1 Classical Induction

Psychologists have a well-established range of techniques for inducing specified emotions. They include the Velten technique (Velten, 1968) and its derivatives (based on reading lists of emotive statements); techniques based on recalling or imagining emotive events; collections of pictures (International Affective Picture Set, or IAPS: Lang et al., 1999) or pieces of music chosen to evoke specified states (Kenealy, 1988); and techniques that use various combinations of these (Gerrards-Hesse et al., 1994).

It is striking how rarely these figure in the database literature. One of the reasons is that they tend not to have a direct or sustained impact on the patterns of action and communication that interest emotion-oriented computing. The mood or flash of emotion that they induce dissipates quite quickly when the subject moves into an engaging activity or a conversation; reactions to the real people and things around them over-ride the effects of what they have heard recorded or seen depicted.

A key exception is the Belfast driving simulator study (McMahon, forthcoming). The procedure consists of inducing subjects into one of three emotional states – angry, elated and neutral – and then getting them to drive simulated routes designed to reveal how the induced emotion affects action. Induction involves novel techniques, because the effects of established techniques such as the Velten dissipate over the course of a drive. As a result, Velten-like techniques are used to establish a basic mood, but it is then entrenched by engaging participant in discussions about topics that are emotionally charged for him or her (these are identified beforehand). The effects can be refreshed by revisiting the chosen topics, with the result that quite strong emotional states are sustained over several drives of considerable length. The effects of the techniques are monitored by periodic self-ratings of emotional state and ongoing physiological measures (ECG, GSR, skin temperature and breathing).

There is an obvious case for extending the paradigm established by McMahon, in which a relatively enduring mood is established and its effects on behaviour are observed. Camurri et al. (2007) have applied a version of it to musical performance. It also clearly lends itself, for instance, to research on meetings (where current techniques tend to produce records with very little emotional colouring).

4.2 Physical Induction

It is widely argued that at root, emotions exist to cope with situations that have a direct bearing on survival. On that view, the only way to elicit real emotions is to set people in situations that face them with real and present challenges. That route has been explored by Sneddon, McRorie and their collaborators (Douglas-Cowie et al., 2007).

In their first study, volunteers were recorded engaging in outdoor activities such as mountain bike racing and balancing on top of a high, unstable structure. Figure 3 is drawn from that material. Several sequences drawn from the study are both dramatic and unquestionably authentic, but recording quality is a major problem.

The second study used a more controlled environment where certain kinds of 'ground truth' could be established. It is called the Spaghetti study, because participants were asked to feel in boxes in which there were objects such as spaghetti, buzzers that went off as they felt around and fruit arranged in suggestive configurations. The technique produced clear-cut, intense reactions, which were recorded audiovisually. The participants also recorded what they felt emotionally during the activity.

4.3 Games

In contrast, technologists have been drawn to games, particularly computer games, as a method of inducing emotion. The problem with games, particularly computer games, is that they have no bearing on the player's 'weal or woe' – the phrase that Arnold (1960) aptly used to describe what emotionally significant events are about.

Perhaps the earliest game that has had a lasting influence is the Iowa gambling game, brought to attention by Damasio and his colleagues (Bechara et al., 1994). Players open doors and win or lose depending on what lies behind. Several groups have used it to generate databases, including Sobol-Shikler (http://www.bgu.ac.il/~stal/research.html) and Amir (2007).

A particularly sophisticated game was developed by Johnstone and his colleagues (van Reekum et al., 2004). It used a space game to create situations with characteristics that relate directly to appraisal categories as specified by Scherer's theory. Another theoretically sophisticated game was reported by Wang and Marsella (2006), who developed a dungeon role playing game intended to induce emotions such as boredom, surprise, joy, anger and disappointment.

All of these games illustrate a recurring problem with the genre, which is a restricted range of modalities – voice is not inherently involved, and gesture is almost impossible. Some games have been extended to address the issue (such as the Geneva game); others have been developed with it in mind. In 'Mazey', developed at QUB, an observer watches a screen and tells an operator how to move a cursor through a maze. Emotion is manipulated by introducing or dissolving obstacles and setting standards that are harder or easier to achieve. 'Bore' (Cowie et al., 2003) induces boredom by presenting a long sequence of screens with a very restricted set of shapes and having participants give a formulaic description ('red triangle, blue circle, green square', for instance) of one screen after another for about half an hour.

Different lines of approach can be used to study stress- or load-related states rather than emotion per se. For example, the DRIVAWORK corpus, collected at Erlangen, uses a simulated driving task. There are three types of episodes: participants are recorded relaxing, driving normally or driving with an additional task (mental arithmetic). Recordings are video and physiological (ECG, GSR, skin temperature, breathing, EMG and BVP). Recordings are accompanied by self-ratings and measures of reaction time.

Games involving humans rather than computers open yet more possibilities. A significant example is EmoTABOO, developed by LIMSI-CNRS and France Télécom R & D (Zara et al., 2007). EmoTABOO records multimodal interactions between two people during a game called Taboo. One person is given a word and has to give hints that will allow the other to guess it. The obvious words are 'taboo', and so gestures and body movement come into play. The task elicits a range of emotions including amusement, frustration and embarrassment.

Game scenarios can become very complex indeed. For example, Rehm and Wissner (2005) developed a game called 'Gamble' which is played by two humans and an ECA. Its particular value is that it provides a way of recording the very complex patterns of interpersonal behaviour (such as gaze) that occur in such mixed-agent scenarios.

The list of games here is very far from comprehensive. Conference proceedings regularly report several new games being used to evolve emotion. It may be worth considering a moratorium on games that there is no clear theoretical motivation to add to the existing repertoire.

4.4 Task Settings

There is no exact line between games and situations where participants are engaged in tasks that they regard as 'real', but it is difficult to doubt that the difference can be significant. As in games, it also worth distinguishing tasks involving human–machine interactions and human–human interactions.

4.4.1 Human–Human

Several standard types of task involving human–human interaction can be manipulated to generate emotion data. A number of studies have used simulated helplines, where, for instance, the speaker's statements are misunderstood or there is unacceptable repetition of information or questions (e.g. Mitchell et al., 2000). Hansen and his colleagues have described a striking innovation, the 'Soldier of the Quarter' task, in which soldiers are brought before a military evaluation board, whose members all have much higher rank than the soldier facing the panel (Meyerhoff and Hansen, 2007).

Instructional settings have been a recurrent feature, from early work by Bachorowski and Owren (1995), who induced varying emotions by setting students spelling tasks with varying degrees of difficulty, to a full-scale tutorial session recorded audiovisually by Abassi et al. (2007).

4.4.2 Human–Machine

There appears to have been less work than might be expected on emotion in tasks (as against games) involving human–machine interaction. The Erlangen group has provided several of the key examples.

An influential paper (Batliner et al., 2003) described successively closer approximations to data from interaction between a human and an artificial dialogue system. Eventually in the SympaFly corpus (Batliner et al., 2003), users talked to a fully automatic speech dialogue telephone system for flight reservation and booking, and booked one or more flights. Their task might be, for instance, to let a person in one location attend a meeting in a different place at a specific time. Additional information, such as identification, credit card number, and so on, had to be given. In a second type of task, users (including children) gave verbal instructions to an AIBO robot (Batliner et al., 2004). The data forms the basis of the CEICES project, which is described elsewhere in this handbook.

The expressive corpus Sound Teacher of E-Wiz (Aubergé et al., 2004) used a simulated task in which a machine acted as a language teacher. The subject thought that they were communicating with a computer system for teaching the vowels of a foreign language, whereas the system was managed remotely by a wizard. Motion was manipulated by praising or criticising users' performance. The scenario is effective partly because the task is one where people find it very difficult to judge their own performance.

4.5 Conversational Interactions

Conversation is a special context in which the participants engage on one hand (usually not wholly unemotionally) with a subject which will tend to be remembered or foreseen rather than present in the immediate environment and on the other hand maintain a complex set of connections with each other, which will involve both dialogue management and the various emotionally coloured stances that they adopt and simulate towards each other.

Broadly similar kinds of dual focus can be present in games and task settings, but it is worth recognising conversation as a major kind of activity where special kinds of signs are likely to be present. Again, it is worth separating data from human–human interactions and a very few sources where one partner is (apparently) a machine.

4.5.1 Human–Human

The Belfast Naturalistic database contains clips in which a member of the team talks to long-standing acquaintances about highly emotive topics. For example, one discusses being held up at gunpoint and not knowing whether her husband (who was separated from her) was alive or dead. The result is that signs relate to several, very different layers of experience. Some convey that the original experience was truly and deeply terrifying. Others convey a sense that the experience has been largely assimilated over the intervening years so that it cannot disrupt the speaker's mind (at least not easily). Others convey attitudes to her present self and to the speaker, both relatively warm. That kind of complexity is not at all atypical of the way emotion is woven through conversation.

Along partly similar lines, but much more extended, is a set of interviews dealing with emotive subjects. Merola (2006) interviewed competitive athletes about their successes, failures and aspirations. Its striking feature is that the events being recalled have very definite and constantly changing emotive characters. The story of a race can move through a clearly marked sequence of anticipation, excitement, concern, high hope, desolation and consolation, and the corresponding expressions, vocal, facial and gestural, can be tracked.

The Belfast Green Persuasive Data Set (Douglas-Cowie et al., 2007) also targets a particular kind of conversation, where one party is trying to persuade the other. The topic is adopting a 'green' lifestyle, which leads into a range of highly emotive topics. These include some to which most persuadees respond positively (such as using fewer plastic bags); some that they find disturbing (such as the predicted consequences of global warming) and others that they resist vehemently (such as giving up driving a car). It includes traces made by the interviewees shortly afterwards to indicate how persuaded they felt from moment to moment.

4.5.2 Human–Machine

It is a challenge to collect records of human–machine conversation, because machines are not actually able to carry out conversations. However, there are obvious reasons to try, since it seems very likely that human–machine interactions will differ from human–human interactions in significant ways. Not the least of these is that for the foreseeable future, human–machine interactions will break down in ways that human–human interactions do not, and it would be extremely useful to know what the signs of breakdown are.

The 'Sensitive Artificial Listener' scenario (SAL for short) has used Wizard-of-Oz techniques to simulate human–machine interactions on emotive subjects (Douglas-Cowie et al., 2007). The scenario is modelled on the way chat show hosts engage their guests in emotional conversations. They often use short statements with very shallow semantic content to encourage or provoke the speaker, and when the topic is in fact emotive, that can draw them out very effectively. The simulated machine in SAL consists mainly of a repertoire of these low-content, high-impact statements. The operator chooses which statement to use at any given time from a menu that is organised to simulate four personalities – Poppy (who aims to make people happy), Obadiah (who aims to make people gloomy), Spike (who aims to make people angry) and Prudence (who aims to make people pragmatic). Speakers can choose at any time which 'personality' they want to talk to. The response that is chosen will depend on the 'personality' that is active and the speaker's state; Spike will reinforce angry responses, Prudence will try to calm them, and so on. From the user's point of view, the net effect is to provide an 'emotional gym' that they can use to work through a range of emotional states.

The SAL scenario is crude in many respects, but it does engage people in quite protracted conversations with a machine-like interlocutor. The original version (in English) was successful enough for versions to be developed in Hebrew (at Tel Aviv University) and Greek (at National Technical University of Athens, ICCS), with adjustments to suit cultural norms and expectations.

5 Associated Issues

A range of practical issues are covered in the chapter of this handbook that gives an overview of databases. One in particular is covered here, though, because it impacts primarily on collection. It is ensuring that an appropriate ethical framework is in place. In particular, the use of naturalistic, non-acted data brings with it difficult issues of privacy and ethics.

Some basic ethical principles need to underpin the work. Three key principles are the following:

(i) All research involving human participants requires ethical approval.
(ii) Normally prior informed consent should be obtained.
(iii) Normally the participant is entitled to confidentiality/anonymity.

These principles should be applied to two main areas – data collection and storage and use of data after collection.

In terms of induced emotional data, prior informed consent should be obtained. A standard protocol (from the British Psychological Society) also stipulates that 'In the event that confidentiality/anonymity cannot be guaranteed, the participant MUST be warned of this in advance of agreeing to participate'. During the recording, the participant should have the opportunity to withdraw/stop at any point and to request the removal of data at the end of the recording. If inducing negative emotions, there should be a process for returning the participant to a neutral/positive state. The participant should also be allowed after recording to request deletion of the data.

In terms of 'naturalistic' emotions, if recordings are made of public behaviour without prior consent, then debriefing should happen as soon as possible and the participant can have the data destroyed. The researcher should actively ensure that the participant knows about and is happy with all aspects of the study as in the case of induced data. Recording from television clearly does not meet these requirements. Alternatives to recording public behaviour should be found if at all possible.

In terms of storage and distribution of data, informed consent to store and distribute the data should be sought. The participant also needs to be advised of the use that the data will be put to in a way that he/she can appreciate. For example, it could simply be said that the data will be used to develop emotional interfaces, but in reality that does not explain to the subject that his/her voice or face could appear in some public place, for example, as the face or the voice of an ECA. The participant should be appropriately briefed or given the opportunity to put restrictions on the release of the data or to ask for further permission if the data is to be used in a wider way than originally envisaged.

Any recording of the emotional behaviour of children is particularly sensitive and is best avoided unless there is a compelling reason. Informed consent of parent and child needs to be sought, and that is particularly difficult because the child may not fully understand the issues.

6 Conclusion: The State of the Art and Prospects

The Renaissance poet John Donne wrote:

> On a huge hill,
> Cragged and steep, Truth stands, and he that will
> Reach her, about must, and about must go,
> And what the hill's suddenness resists, win so.
> (Satire 3)

Research on data sources for emotion-oriented computing has had very much that character. A seminal conference in 2000 concluded that the greatest obstacle to progress in the area was the absence of suitable databases (D-C, C & S, 2000). Since then, great ingenuity has been expended to devise new methods of capturing authentic emotion. The main result has been to make clear that diversity is a systematic feature of the area, not noise. The challenge is not to find ways of capturing an essential core that characterises the invariant kernel of an emotion. It is to find a range of methods that display the most important ways in which an underlying emotional theme may be combined, contextualised and expressed.

It is a task that will engage research for a long time to come. But the work that has been done identifies a core of techniques that allow substantial progress to be made.

Undoubtedly some of the techniques will be relegated to historical footnotes. However, there is a core of ideas that offer ways of generating data that represents key types of context reasonably well and allows material of reasonable quality and quantity to be generated at acceptable cost. It is to be expected that the next decade will see that core consolidated, extended in key areas and used to generate a corpus that provides a sound foundation for research.

References

Abassi AR, Uno T, Dailey M, Afzulpurkar NV (2007) Towards knowledge-based affective interaction: situational interpretation of affect. In: Paiva A, Prada R, Picard R (eds) Affective computing and intelligent interaction, Lisbon, September 2007. Springer LNCS, Berlin, pp 452–463

Amir N (2007) Emotional speech in a gambling game. In: Paper presented to the HUMAINE meeting on cross-cultural databases, Tel Aviv, Nov 2007

Ang J, Dhillon R, Krupski A, Shriberg E, Stolcke A (2002) Prosody-based automatic detection of annoyance and frustration in human–computer dialog. In: Proceedings ICSLP, Denver, CO, Sept 2002

Arnold MB (1960) Emotion and personality. Psychological aspects, vol 1. Columbia University Press, New York, NY

Aubergé V, Audibert N, Rilliard A (2004) E-Wiz: a trapper protocol for hunting the expressive speech corpora in lab. In: Proceedings of the 4th LREC, Lisbon, pp 179–182

Bachorowski J-A, Owren MJ (1995) Vocal expression of emotion: acoustic properties of speech are associated with emotional intensity and context. Psychol Sci 6:219–224

Bänziger T, Scherer K (2007) Using actor portrayal to systematically study multimodal emotional expression: the GEMEP corpus. In: Paiva A, Prada R, Picard R (eds) Affective computing and intelligent interaction, Lisbon, Sept 2007. Springer LNCS, Berlin, pp 476–487

Baron-Cohen S (2007) Mind reading: the interactive guide to emotions – version 1.3. Jessica Kingsley Publishers, London

Batliner A, Fischer K, Huber R, Spilker J, Nöth E (2003) How to find trouble in communication. Speech Commun 40:117–143

Batliner A, Hacker C, Steidl S, Nöth E, D'Arcy S, Russel M, Wong M (2004). "You stupid tin box" – children interacting with the AIBO robot: a cross-linguistic emotional speech corpus. In: Proceedings of the 4th international conference of language resources and evaluation (LREC 2004), Lisbon, pp 171–174

Batliner A, Hacker C, Steidl S, Noth E, Haas J (2003). User states, user strategies, and system performance: how to match the one with the other. In: Proceedings of the ISCA workshop on error handling in spoken dialogue systems, ISCA, Chateau d'Oex, pp 5–10

Bechara A, Damasio A, Damasio H, Anderson S (1994) Insensitivity to future consequences following damage to human prefrontal cortex. Cognition 50:7–15

Campbell N (2006) A language-resources approach to emotion: corpora for the analysis of expressive speech. In: Proceedings of the LREC workshop, Genoa, p 1

Camurri A, Castellano G, Cowie R, Glowinski D, Knapp B, Krumhansl CL, Villon O, Volpe G (2007) The Premio Paganini project: a multimodal gesture-based approach for explaining

emotional processes in music performance. In: Proceedings of 7th international workshop on gesture in human–computer interaction and simulation 2007, Lisbon, May 2007
Carroll JM, Russell JA (1997) Facial expressions in Hollywood's portrayal of emotion. J Pers Soc Psychol 72(1):164–176
Cowie R, McGuiggan A, McMahon E, Douglas-Cowie E (2003) Speech in the process of becoming bored. In: Proceedings of 15th international congress of phonetic sciences, Barcelona
Devillers L, Cowie R, Martin J-C, Douglas-Cowie E, Abrilian S, McRorie M (2006). Real life emotions in French and English TV video clips: an integrated annotation protocol combining continuous and discrete approaches. In: Proceedings of the 5th international conference on language resources and evaluation (LREC 2006), Genoa
Devillers L, Vidrascu L (2005) Real-life emotion representation and detection in call centers data. ACII 2005, Beijing, pp 739–746
Devillers L, Vidrascu L, Lamel L (2005) Challenges in real-life emotion annotation and machine learning based detection. J Neural Netw (Special Issue: Emotion and Brain) 8(4):407–422
Douglas-Cowie E, Campbell N, Cowie R, Roach P (2003) Emotional speech: towards a new generation of databases. Speech Commun 40(1–2):33–60
Douglas-Cowie E, Cowie R, Sneddon I, Cox C, Lowry O, McRorie M, Martin J-C, Devillers L, Abrilian S, Batliner A, Amir N, Karpouzis K (2007) The HUMAINE database: addressing the collection and annotation of naturalistic and induced emotional data. In: Paiva A, Prada R, Picard R (eds) Affective computing and intelligent interaction, Lisbon, Sept 2007. Springer LNCS, Berlin, pp 488–500
Ekman P, Friesen W (1975) Pictures of facial affect. Consulting Psychologists' Press, Palo Alto, CA
Enos F, Hirschberg J (2006) A framework for eliciting emotional speech: capitalizing on the actor's process. In: Proceedings of the LREC workshop on corpora for research on emotion and affect, Genoa, May 2006, pp 6–10
Frigo S (2006) The relationships between acted and naturalistic emotional corpora. In: LREC workshop research on corpora on emotion and affect, Genoa
Gerrards-Hesse A, Spies K, Hesse FW (1994) Experimental inductions of emotional states and their effectiveness: a review. Br J Psychol 85:55–78
Kanade T, Tian Y, Cohn JF (2000) Comprehensive database for facial expression analysis. In: Proceedings of the 4th IEEE international conference on automatic face and gesture recognition 2000, Grenoble, France, pp 46–53
Kenealy P (1988) Validation of a music mood induction procedure: some preliminary findings. Cogn Emot 2:41–48
Kienast M, Sendlmeier WF (2000) Acoustical analysis of spectral and temporal changes in emotional speech. In: Cowie R, Douglas-Cowie E, Schroeder M (eds) Speech and emotion: proceedings of the ISCA workshop. Newcastle, County Down, Sept 2000, Belfast, Textflow, pp 92–97
Lang PJ, Bradley MM, Cuthbert BN (1999) International affective picture system IAPS: technical manual and affective ratings. The Center for Research in Psychophysiology, University of Florida, Gainesville, FL
McMahon E (forthcoming) The effect of a prior emotional state on driving. Thesis to be submitted for the degree of Ph.D., Queen's University, Belfast
Merola G (2006) Gestures while reporting emotions. Third HUMAINE Summer School, Genova, Sept 2006. Downloaded from http://emotion-research.net/ws/summerschool3/GestEm2_Merola.pdf
Meyerhoff JL, Hansen JHL (2007) Methods and systems for detecting, measuring, and monitoring stress in speech. US Patent 7283962. Downloaded from http://www.freepatentsonline.com/7283962.html
Mitchell CJ, Menezes C, Williams JC, Pardo B, Erickson D, Fujimura O (2000) Changes in syllable and boundary strengths due to irritation. In: Douglas-Cowie E, Cowie R, Schroeder M (eds) Proceedings of the ISCA workshop on speech and emotion, Belfast, Textflow, pp 98–103

Poggi I, Federica Cavicchio F, Emanuela Magno Caldognetto EM (2005) Persuasive goals of irony in a political trial. Paper presented to Humaine: WP8 workshop, Trento, 17–18 Nov 2005

Polzin TS, Waibel A (2000) Emotion-sensitive human–computer interfaces. In: Douglas-Cowie E, Cowie R, Schroeder M (eds) Proceedings of the ISCA workshop on speech and emotion. Belfast, Textflow, pp 201–206

Rehm M, Wissner M (2005) Gamble – a multiuser game with an embodied conversational agent. In: Kishino F, Kitamura Y, Kato H, Nagata N (eds) Entertainment computing – ICEC 2005: 4th international conference. Springer, Berlin, pp 180–191

van Reekum CM, Johnstone T, Banse R, Etter A, Wehrle T, Scherer KR (2004) Psychophysiological responses to appraisal dimensions in a computer game. Cogn Emot 18:663–688

Varadarajan V, Hansen J, Ayako I (2006) UT-SCOPE – a corpus for speech under cognitive/physical task stress and emotion. In: Proceedings of the LREC workshop on corpora for research on emotion and affect, Genoa, May 2006, pp 72–75

Velten E (1968) A laboratory task for induction of mood states. Behav Res Ther 6:473–482

Walker MA, Passonneau R, Boland JE (2001) Quantitative and qualitative evaluation of Darpa Communicator spoken dialogue systems. In: Proceedings of the 39th annual meeting on association for computational linguistics, Toulouse, pp 515–522

Wang N, Marsella S (2006). Evg: an emotion evoking game. In: Proceedings of the 6th international conference on intelligent virtual agents, Marina del Rey, CA. Springer LNCS, Berlin, pp 282–291

Yale Face Database B (2001) http://cvc.yale.edu/projects/yalefacesB/yalefacesB.html

Zara A, Maffiolo V, Martin J-C, Devillers L (2007) Collection and annotation of a corpus of human–human multimodal interactions: emotion and other anthropomorphic characteristics. In: Paiva A, Prada R, Picard R (eds) Affective computing and intelligent interaction, Lisbon, Sept 2007. Springer LNCS, Berlin, pp 464–475

Issues in Data Labelling

Roddy Cowie, Cate Cox, Jean-Claude Martin, Anton Batliner, Dirk Heylen, and Kostas Karpouzis

Abstract Labelling emotion databases is not a purely technical matter. It is bound up with theoretical issues. Different issues affect labelling of emotional content, labelling of the signs that convey emotion, and labelling of the relevant context. Linked to these are representational issues, involving time course, consensus and divergence, and connections between states and events. From that background comes a wealth of resources for labelling emotion, involving not only everyday emotion words but also affect dimensions, and labels for combination types, appraisal categories, and authenticity. Resources for labelling signs of emotion cover linguistic, vocal, face descriptors, plus descriptors for gesture, and relevant physiological variables. Resources for labelling context are developing.

1 Introduction

The aim of this is to give readers a sense of the challenge and of the kinds of partial solution that are now available.

Labelling is a challenge because emotion-related phenomena are massively complex and surrounded by uncertainty. The first chapter of this handbook conveys some of the complexity and that is taken for granted as background in this chapter.

Pulling against the challenge of complexity is set the need to formulate tasks that are simple enough for a machine to carry out. Given a brief video clip of a person in an emotionally coloured episode, a human being can develop a commentary of enormous richness, which is unique to that particular clip. In contrast, groups concerned with practical machine learning are likely to be unhappy if more than one label is used for each clip or more than a dozen for the whole database (two being much the preferred number).

R. Cowie (✉)
Department of Psychology, Queen's University Belfast, Belfast, Northern Ireland, UK
e-mail: roddy.cowie@qub.ac.uk

P. Petta et al. (eds.), *Emotion-Oriented Systems*, Cognitive Technologies,
DOI 10.1007/978-3-642-15184-2_13,

Working labelling schemes are always the outcome of tension between the two sets of factors, one pulling towards complexity and the other towards simplicity. As a result, they are likely to frustrate both engineers, who want something simpler, and philosophers and social scientists, who deplore their lack of richness.

This chapter rests on the assumption that there is space for a sub-discipline between the extremes. Constructing a credible emotion database is a major undertaking. The undertaking is hardly worthwhile unless the result can be used by many teams. Hence, it is not optimal to invest effort in a minimal labelling, which is totally driven by the needs of a particular synthesis or analysis project. It makes more sense to construct a labelling that is rich enough to support any foreseeable use of the basic material. Using a rich coding scheme means that researchers can pick up and adapt what they are interested in according to their application requirements. Simplifying a rich initial labelling is not trivial, but it is possible. Enriching a poor initial labelling is likely in practice to mean restarting from scratch. That is why this chapter does not restrict itself to labelling schemes that are likely to be used directly in learning or synthesis applications in the immediate future.

The chapter is divided into two main sections. The first considers general concepts, theoretical and technical, that create the framework for work on databases. The second deals with specific resources that exist for labelling and studies that illustrate them. The ideas that influenced the HUMAINE database are naturally prominent, but not to the exclusion of others.

It is worth stressing that the chapter is very far from comprehensive. Labelling a rich emotional database potentially involves an enormous number of skills, and it seems fair to say that they have rarely if ever all been drawn together. The labelling of the HUMAINE database (described in a later chapter) is probably the most ambitious exercise of its kind, and the chapter reflects the expertise that undertaking has been drawn together. But there are still many areas where expertise exists that is not reflected here or where the expertise to handle what are clearly significant problems simply does not exist.

2 General Concepts

Some terms are used here in ways that reflect practice within HUMAINE and that may not be immediately obvious.

The term 'records' is used to describe the video, audio, physiological, or other signals that form the raw material of the database. The term 'clip' is used to describe an individual record which has been edited to capture a natural unit.

Terms related to emotion and emotion-related states are used in line with the opening chapter of this handbook. When 'emotion' is used without a specific modifier, it is used in its broad sense to cover the whole range of phenomena that distinguish life as it normally is from the life of an agent who is emotionless. These phenomena include moods, stances, and bonds, as well as the particular states that are called emotion in a narrow sense – both brief episodes of emergent emotion and established emotions, which shape the way a person feels and acts over months or years.

Emotion labellings are symbols that express the emotion or emotional colouring in a clip (whether it is truly present, perceived to be). Sign labellings identify evidence in the records that is relevant to attributing emotion labels – facial expressions, attributes of speech, gestures, etc. They exclude context labels. These identify features of the situation that are relevant to understanding why a particular emotion might be felt, or why it might be expressed in a particular way.

2.1 Theoretical Foundations

There is a well-known dictum that all data are theory laden (Hanson, 1958). It is most certainly true of data created by labelling. Experience suggests that people coming into the field often do not register that apparently simple choices are bound up with questions on which there are large literatures. The aim of this section is to alert people to the theoretical dimension of practical decisions about labelling. It deals first with emotion and emotion-related states, then looks at sign labellings, then at descriptions of context.

2.1.1 Theoretical Foundations of Emotion Labelling

The first chapter of this handbook reviews the main resources that theories of emotion offer. This section does not repeat the ground. Its aim is just to make the connection between particular types of theory and emotion descriptors that a database may use.

Most people are at home with atheoretical descriptions of emotion, based on everyday emotion words. There is a school of thought that assumes those descriptions are the best we have and the natural basis for labelling. It is a defensible position. The important thing is to register that it is by no means obviously right.

A theoretical position with very broad appeal proposes that a few everyday categories (six in the best known version) correspond to the elements of emotional life. From the point of view of labelling, it would be very convenient if emotion came in half a dozen 'basic' flavours, each corresponding to an everyday category. Unfortunately, even the main advocate of that position no longer defends it (Ekman, 1999).

An alternative framework was formulated long before basic emotion theories. It describes emotion in terms of dimensions. Russell and Barrett-Feldman (1999) have recently updated the idea, suggesting that the classical dimensions capture an element of emotion which he calls core affect. On that view, the natural format for labelling is to attach parameters describing core affect to each significant unit.

Appraisal theories give rise to approaches that are more sharply distinct from common sense. They suggest that emotion is rooted in a distinctive way of weighing up situations that impinge on an individual and provide lists of issues that are involved in the weighing up. That suggests that labelling could describe the relevant issues rather than using everyday emotion categories.

A very durable version of appraisal theory was proposed by Ortony et al. (1988) and is commonly known as OCC. It considers emotions as valenced reactions. The theory divides possible types of reaction first in terms of whether they focus on aspects of objects, actions of agents, or consequences of events. The next level of division hinges on whether issues at stake relate to the subject himself or herself, self or to others. Below that, states are classified in terms of whether they are concerned with fortunes, prospects, well-being, attribution, or attraction. The point here is not to set out the theory but to convey the kinds of concept that it suggests might be used rather than commonplace emotion words.

OCC is by no means the only appraisal-based framework. Probably the most fully developed alternative is Scherer's (Sander et al., 2005). He regards emotions as appraisals concerned with the novelty, intrinsic pleasantness, goal significance, and compatibility with standards of focal events, and also the potential for the subject to cope with them. Each of these categories is subdivided again.

There are several ways in which these theories (and others) might be useful to people concerned with labelling databases.

Economy. The sheer number of everyday words that describe emotion-related states makes them an unwieldy descriptive system. Several theories hold out the prospect of doing essentially the same job with a smaller number of well-chosen primitives.

Consistency. Investigators sometimes try to achieve economy by using 'cover classes' based on the way everyday labels appear to cluster in their data set. The difficulty is that different data sets suggest different cover classes, and that limits the prospect of transferring of insight across studies. Basing cover classes on theory reduces the number of options that are likely to be used.

Intermediates. Despite the number of everyday emotion terms, naturalistic databases often show states for which there is no exactly appropriate everyday term. Several theory-based schemes offer ways of describing these elusive states (in terms of co-ordinates or lists of appraisal-related features).

Semantics. Appraisal-based theories in particular offer ways of representing the meaning associated with a label, and that is important for the prospects of using the label in a functional system (recognising not only the state but what it may lead to, allowing analysis of the circumstances to influence the states that are synthesised, and so on). Hence there are advantages to using labels for which that kind of semantics is available.

Natural classes. One of the functions of a theory is to divide emotion-related phenomena into classes that make logical sense rather than ad hoc groupings. Presumably using well-chosen classes will in the long run allow systems to work better. That said, it is not obvious that the same classes are suitable for all functions. It may be important to distinguish envy and anger morally but inappropriate to separate them for the purpose of a database concerned with training systems to recognise emotions. Choosing the right classification for the right purpose is a subtle problem.

There is a large body of literature about the issues that have been sketched above. However, there are many questions where there clearly is a theoretical dimension, but much less material is available. The most important are sketched briefly below.

Part of the debate over basic emotions is whether emotion-related phenomena are continuous or divide into discrete types (Ortony and Turner, 1990; Ekman, 1992). That is relevant to a substantial issue in labelling, which is whether to use continuous or categorical descriptions. The issue is complicated because practical schemes can combine both. For instance, what has been called 'soft labelling' uses categorical labels, but each category is associated with a number that conveys how much of that emotion type is present (Steidl et al., 2005). Conversely, accounts that are inherently dimensional often describe the range of a variable in terms of a few categories (Craggs and Woods, 2004). Evaluating schemes like these depends both on the way the underlying systems actually work and on the practicalities of obtaining labellings.

The theory that has been mentioned so far is predominantly concerned with emergent emotion. That may well not be the most important kind of emotion-related phenomenon for computing to deal with. HUMAINE developed an approach to describing other emotion-related states, which is summarised in the first chapter of this handbook. Around the same time, a taxonomy developed by Baron-Cohen (2007) attracted increasing interest in the computational community.

Cutting across these is the distinction that has been drawn between 'cause'- and 'effect'-type descriptions (Cowie et al., 2001). Cause-type labelling sets out to associate records of a person generating certain signs with descriptions of the states that actually led them to produce the signs – whether he/she was actually angry, despite a calm surface appearance, or calm, despite giving signs of anger. Effect-type labelling sets out to associate the records with descriptions of the impressions that the signs would be expected to produce in an observer. Within that, a distinction has been drawn between the result of considered judgment and 'perceived flow of emotion' – the kind of impression that somebody faced with that person would be expected to form in real time.

There is a tendency to look automatically for cause-type labelling. That poses difficult problems, because there is no easy way to establish what a person is truly feeling. Practically, though, effect-type labelling will often be the more appropriate target conceptually as well as the easier one. The aim of an expressive agent will generally be to create a specified effect on people who encounter it, and at least in ordinary interaction, the natural target for recognition is to interpret signs as a person would.

Individual differences are a key issue for effect-type labelling. Differences between cultures are increasingly well documented (Matsumoto, 2001). What strikes one culture as overt anger will strike another as mild perturbation. Hence it is important to consider and report the culture of the people who generated an effect-type labelling. Even within a culture, it is clear that different people 'read' signs of emotion differently. Differences in simple sensitivity are known from research on emotional intelligence (Mayer et al., 2000). Standard variables like gender and introversion–extraversion probably have quite considerable effects (Hall and Matsumoto, 2004). More complex differences hinge on different judgments about what is genuine and what is being concealed or simulated.

These points impact on well-known technical issues surrounding validity and reliability. Technologists often equate validity with the provision of a 'ground truth'. In the context of emotion, demanding ground truth is tantamount to focusing on a very special subset of the domain. Ground truth is neither known nor needed in most everyday emotionally coloured interactions, and it is difficult to imagine how it could be provided, hence insisting on ground truth rules those interaction 'out of bounds' for research. Paradoxically, acted emotion is sometimes seen as more acceptable, because the actor's intention is known – despite the fact that there is no truth at all. If the field is to engage with subtle phenomena, it needs subtler notions of validity.

Similarly, it is standard to take statistical measures of reliability (specific techniques are discussed below) and to reject material where coefficients are low. Again, blind application of that strategy has the effect of excluding important kinds of material, particularly material whose character means that observers do and should either differ systematically or agree that there is uncertainty. There are some numerical strategies that can be used for that kind of material; again, they are discussed below.

These issues relate to the selection of labellers. A strong commitment to effect-type description would suggest that labellers should ideally be a representative sample of the population, preferably not contaminated by training that would make them less representative. In practice, it is commoner to use a strategy that fits logically with a cause-type approach, which is to use a few 'experts' whose judgments are supposed to be more reliable than average. That is partly because of concern that genuinely naïve raters may simply not understand the task, and therefore scatter in the data will reflect plain confusion as well as genuine variation. Consensus will probably develop over what can and cannot be done with truly naïve raters.

2.1.2 Theoretical Foundations of Sign Labelling

General theoretical issues are less complex than in the area of sign labelling, but some points should be made.

The area spans very different levels of interpretation. At one extreme are patterns that can be derived automatically from the record, such an F0 contour or a particular degree of eyebrow raising. At the other extreme are descriptions that identify the intention behind patterns in the record, for instance, describing a hand movement as raising a fist or voice quality as strained. These may not be much easier to derive from the raw signals than judgements about the person's emotional state.

When humans explain how they attribute emotion, the descriptors that they use are often quite deeply interpretive. It is debatable how relevant labels of that kind are in a database. On one side, they may not identify features that can be recognised as a preliminary step towards identifying a global emotional state. On the other, they may identify action patterns that are relevant to synthesising a particular state. They may also contribute to less sequential recognition strategies, where the system is making multiple kinds of high-level attribution simultaneously and favours interpretations that 'hang together' in a coherent constellation.

Descriptors based on shallow interpretation carry their own problems. Most obviously, they are often tied to the specific situation in which they were developed and take on a different meaning when a person is carrying out a different activity, or even when camera angle is changed.

There are also long-standing difficulties over the status of schemes that have been developed to describe significant types of event for different purposes. In the speech domain, there are established systems for describing prosody (the so-called British system and ToBI). In the domain of action, there are schemes that were developed to describe dance and sign language. These have the attraction of being standard. The difficulty is that they may abstract away details that are irrelevant to the original application, but highly relevant to emotion. For instance, ToBI discards harmonic information (such as information about major and minor intervals), but there is a growing interest in the idea that they may be relevant to the expression of valence (Schreuder et al., 2006).

2.1.3 Theoretical Foundations of Context Labelling

People often doubt the need for context labelling. However, there are good reasons why some form of it is needed.

Emotion is very often a reaction to eliciting events – in the famous example from James (1884), a bear emerging from the woods. There are at least many cases where it would be very difficult to read the meaning of emotional signs without knowing about the eliciting circumstances (as anyone will know who has watched tears streaming down the face of a winning sports team). Conversely, it is clearly important to understand human ability to infer the eliciting event from emotion-related signs, without independent information. That is, for instance, what allows people to use another person's face to orient to a danger (which is an ability that could be very important to emulate in dangerous situations). On a fine level, there is good reason to think that the timing of signs relative to eliciting events is integral to natural expression of emotion (Sander et al., 2005). Databases cannot be used to study any of these issues unless they contain the relevant contextual information.

There is also a body of work on the way social context affects signs of emotion. The best known approach uses the concept of display rules (Ekman and Friesen, 1975). For example, the presence of in-laws is likely to have radical effects on the way couples express feelings towards each other. The display rule concept may oversimplify the relationship between social context and emotion, but it is not in doubt that the relationship is a powerful one. Again, unless databases contain appropriate contextual information, techniques based on them cannot be sensitive to these issues.

Physical context also affects signs. Hand movements are an archetypal example. They depend heavily on what the hand is holding: nothing, a pencil, a hammer, a shopping bag, another person's hand. Trying to interpret their emotional significance without that context would be very difficult. Background noise and distance from an interlocutor have similarly large effects on speech.

For some of these effects, it is natural to assume that information will be attached to a clip rather than labelled in the usual sense (weeping sportspeople are a good example). But others can change very quickly, and timing information may be critical. Hence it is at least sometimes natural to integrate information about context fully within the labelling framework.

2.2 Representational Issues

This section covers issues that are relevant to all the types of labelling that are being considered, and that involve a mix of practical and theoretical issues.

2.2.1 Representing Time Courses

Emotional life involves events that unfold over time. There are various ways in which labelling can reflect the temporal aspect.

The simplest option is to attach a label to a clip as a whole. In the chapter "The HUMAINE Database", we have called that kind of label global. In and of itself, a global label says nothing about any variation in time within the clip. That kind of labelling is very useful for indexing or reviewing the content of a database. For attributes that change slowly, it may be all that is needed. It is, for instance, the natural way to handle information about elicitation techniques, technicalities of recording, and some kinds of context. It would be very useful to move towards a standard approach to global labelling, but it is not a problem that has received much explicit attention.

The use of global labelling is related to the issue of choosing clip boundaries, which is discussed elsewhere in this handbook. For some purposes, it makes sense to divide basic records into small clips, such as individual words. Most labelling will then be global in the sense used here – attached to the clip as a whole. The cost is that more slowly changing variables have to be dealt in terms of relationships between clips. It remains to be established where the overall advantage lies.

The archetypal labelling strategy is to associate labels with discrete time periods – for instance, to attach the label 'happy' to a particular time period and identify the beginning and end of the period. That can be called quantised labelling.

The natural alternative to quantised labelling is what we have called trace labelling. A trace specifies how a measure associated with a particular label varies moment by moment. That general description covers a great variety of possibilities. One trace might show the ratio of maximum horizontal to maximum vertical body extent at it varied from moment to moment; another might show rise and fall in the overall intensity of emotion that a person was experiencing; another might show how a particular emotion (such as happiness) fluctuated over time.

These different representations are often (though not necessarily) associated with different kinds of labelling process. Quantised labelling tends to be done by playing and replaying recordings to find the best point to draw a boundary. Trace labelling tends to be done in real time by moving a cursor (or some other device) while a clip is playing.

Linked to these are semantic differences. The different labelling processes tap into different kinds of percept: in the terms introduced earlier, considered judgment and perceived flow of emotion, respectively. They also suit different kinds of entity. If something has a definite onset and offset, then quantised labelling allows them to be located. If it actually shifts gradually from a negligible role to a central one, then it is artificial to specify a sharp cut-off.

It seems likely that in the long run, it will be normal to use a mixed diet of representations. Establishing which to use for which purpose depends on accumulating experience.

2.2.2 Representing Consensus and Divergence

It has been argued that divergence between perceivers is an intrinsic feature of emotionality, not noise to be eliminated. It is not immediately obvious how divergence should be dealt with. Averaging is an obvious option. It can be done even with categorical data by constructing 'soft vectors', which associate each of the relevant labels with a number indicating how often it was used.

There are two problems with averaging, though. Firstly, data are quite likely to be bimodal some raters think the person rated is in state A, most others that he/she is in state X, but nobody thinks he/she is midway between the two – and yet the midpoint is what the average describes. Secondly, averaging masks sequential effects. These would occur if raters who thought the person began in state A thought he/she moved to state B, raters who thought the person began in state X thought he/she moved to state Y, but nobody thought that the person began in state A and moved to state Y. Thirdly, averaging masks relationships between different types of labels – raters who thought the person was looking at M thought his/her emotion was A, raters who thought the person was looking at N thought his/her emotion was B.

At present, there is no obvious way to deal with these issues except to include each individual rater's data. In the long run it should be possible to pick out overlapping trends and associations statistically, but it is not straightforward.

2.2.3 Representing Connections

As the opening chapter of this handbook pointed out, emotion is richly connected to people, things, and events, present, past, and foreseen. One of the most difficult problems in the area is how to integrate those connections into a labelling.

It is quite typical that a person will feel positive towards collaborators in a situation, negative towards rivals, doubtful about him/herself, daunted by the task ahead, and glad to have overcome a previous obstacle; and that the balance among these feelings will shift as the task proceeds, and new elements become salient.

Not all ways of representing emotional state demand that connections like that be addressed, but some are very unsatisfying unless they are. Schemes involving appraisal are an example. Various appraisal concepts apply to various elements of the kind of scenario that has been described (intrinsically positive, goal, goal obstructive, etc.). Declaring relevant objects in a global labelling is a partial solution, but it is far from ideal.

One kind of solution that one might imagine is to embed descriptions of relevant objects and events with emotion labels (Devillers and Martin, 2008). A more radical approach is an annotation where timelines describe emotionally significant niches (for instance, a strongly valenced object, an emotionally significant goal, a goal obstructive object, etc.); and entries specify what if anything occupies a given niche at any given time. The result would be a description of the person's affective world, so to speak, rather than his or her emotion as such. The main point here is that dealing with these connections satisfyingly is the kind of problem that may still precipitate quite basic shifts in the way labelling schemes are structured.

2.3 Coda: How Difficult Does Labelling Need to Be?

The range of issues covered in the theoretical part of this chapter may seem quite daunting. They are not exercises in pure intellectual gymnastics, though. Almost all of them originate in a confusion or a disagreement within HUMAINE, which eventually turned out to rest on a theoretical issue that had not been articulated.

The point of articulating the theoretical issues is not to make people address them all. It is to let them make informed decisions about practical decisions, including decisions about the kinds of simplification that make sense and to understand that people with different goals may rationally make different decisions.

3 Specific Resources

This part of the chapter concentrates on specific resources, giving basic technical information, identifying sources that give more information, and commenting where the information is available on functional characteristics (such as procedural issues and reliability).

3.1 Resources for Labelling Emotion

This section first outlines different types of descriptive resource separately and then considers issues that cut across them, such as methods of validation and comparisons between the techniques. For reasons of presentation, it is convenient to begin with description based on dimensional concepts.

3.1.1 Affect Dimensions

The ideas behind using affect dimensions were introduced above. There are many ways of translating the concepts into labelling schemes.

The FEELtrace technique (Cowie et al., 2000) reflects dimensional concepts very directly. A rater using FEELtrace watches a computer screen with the clip playing on one side and a circle on the other. The axes of the circle correspond to the standard 'core affect' dimensions, valence (horizontal) and activation (vertical). Ratings are made by moving a cursor within the circle, and its colour changes with position in a way that people find easy to relate to emotion (red for pure negative, green for pure positive, yellow for maximum activation, dark blue for minimum activation). Selected words are shown within the circle to indicate the kinds of emotion that occupy particular positions in the space. Figure 1 shows snapshots of screens, in one of which the cursor has been moved from deep passivity to fear, in the other from fear to pleased. The output of a tracing session is a file containing three columns. In each row, the first figure specifies time, and the second and third specify the valence and the activation level, respectively, that the rater attributed to the person P being rated at that time. These correspond to the *x* and *y* co-ordinates, respectively, of the cursor at that time. A sequence of valence or activation measures traces the perceived flow of P's affect over time.

The technique depends on making the demands of the task as low as possible so that users can concentrate on watching the clip. That is achieved by making the display as intuitive as possible and by giving raters a thorough training session. When those precautions are taken, ratings with the trace system show higher reliability than ratings using words (Savvidou, submitted). Reliability can be quite low if those precautions are not taken.

FEELtrace is attractive because it produces in real time what some theories suggest is a full description of affective content. It may provide all the information that is needed for some applications, particularly applications involving voice, where

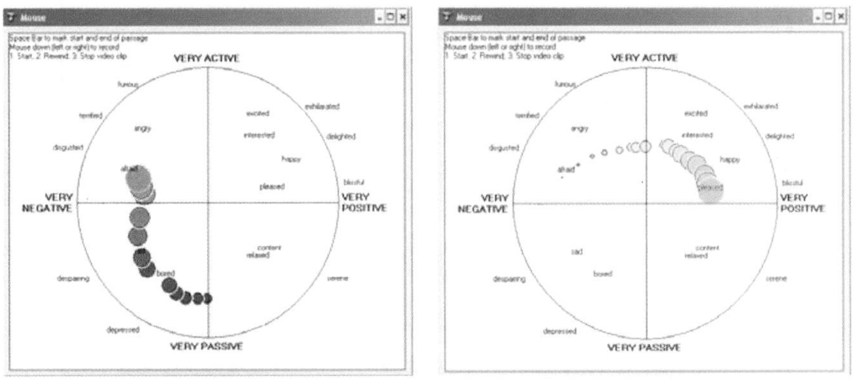

Fig. 1 Examples of FEELtrace screens during tracing. The largest coloured circle shows the current position of the cursor: *smaller circles* mark previous positions

activation is the main kind of information available (Bachorowski, 1999). However, richer information is clearly needed for some purposes. The trace approach has been adapted to provide that by creating a family of one-dimensional traces, each dealing with a particular dimension of emotionality. As with FEELtrace, their reliability depends on training and displays that are carefully constructed to convey what it means to have the cursor in a particular position. Figure 2 illustrates the kind of display that has been used; the program in question is for rating the intensity of the emotion apparently being experienced by the person being rated.

The HUMAINE database has used a family of one-dimensional trace programs to map perceived affect, reflecting the dimensions reported in the 'grid study' (see the first chapter of this handbook). In addition to intensity, illustrated above, they cover valence and activation (separately, not in the single display used by FEELtrace), power/powerlessness, and predictability/unpredictability.

There is no necessary connection between trace techniques and labellings concerned with affect. The 'Self-assessment Mannikin' (SAM) is a standard paper tool used to give discrete ratings (on three seven-point scales) for the affect dimensions of valence, activation, and power (Bradley and Lang, 1994). It could be used to rate a single frame or a relatively homogeneous episode, though in practice labellers appear not to have used it. A related tool, which simplifies the classical dimensions to a few scale points, was developed specifically for labelling by Craggs and Wood (2004).

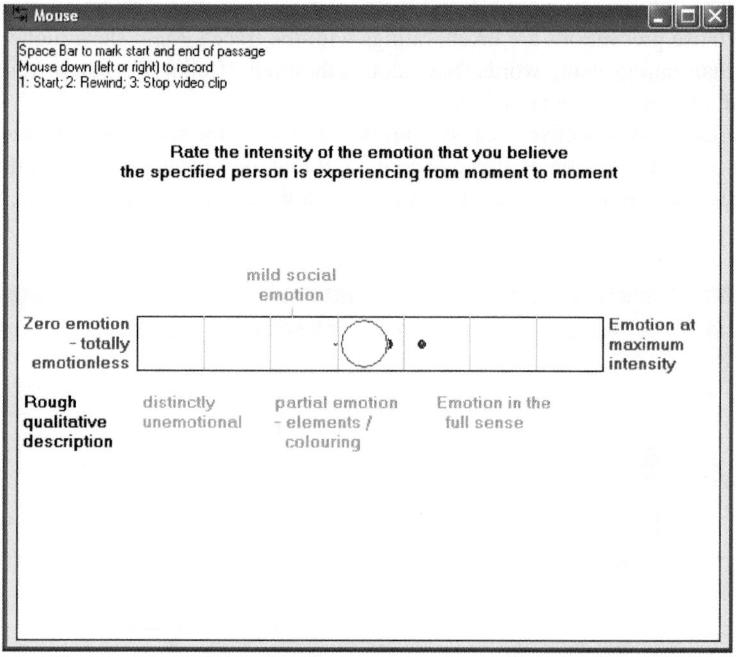

Fig. 2 Screen from one-dimensional trace program to measure intensity of emotion (INTENStrace)

3.1.2 Specific Emotion Words

The most obvious approach to labelling is to use words like 'anger', 'happiness', and so on. They are called 'specific' here because it is natural to think of the states they describe as analogous to species in biology ('lions', 'horses', etc.).

The list of specific emotion-related words is vast. The web is a very useful resource in that context, since it gives lists that reflect contemporary usage rather than academic theory. Sites that give reasonably well-considered lists of words or stock phrases used to describe feelings or emotions include the following:

http://www.angelfire.com/in/awareness/feelinglist.html
http://www.searchingwithin.com/journal/abptb/feel.html
http://lightisreal.com/positiveemotionlist.html
http://en2.wikipedia.org/wiki/List_of_emotions
http://www.umpi.maine.edu/~petress/feelinga.pdf
http://www.psychpage.com/learning/library/assess/feelings.html
http://eqi.org/fw.htm
http://www.preciousheart.net/empathy/Feeling-Words.htm
http://marriage.about.com/library/blfeelingwords.htm

Between them, these sources include nearly 3,000 words or standard phrases that the authors regard as describing feelings or emotions. Of those, 280 occur in four sources or more. Table 1 lists them.

That kind of exercise highlights one of the main problems with labelling based on unconstrained use of terms from everyday language, that is, the sheer number of categories that are relatively common. If raters are left to choose labels freely, and the material is at all complex, they are likely to use a large number of labels, each applying to a small amount of material, and that kind of result is difficult to use directly.

A response that is often used is to allow an unrestricted set of labels in the first instance and then to group them into 'cover classes'. Examples of that approach are in Douglas-Cowie et al. (2005), Batliner et al. (2006), and Auberge et al. (2006). It might seem that reducing a free labelling to a few cover classes was not so much part of the database as a device used later to reduce the information in a database to a form that suited (for instance) a recognition algorithm. However, grouping may also be necessary to answer questions about the database itself, in particular, questions about the reliability of the labelling.

Cover classes are usually formed ad hoc in the context of a particular database. There are various reasons to be concerned about that. For example, it makes comparability across databases difficult, and it will inflate measures of reliability if terms are assigned to the same class because different raters apply them to the same material. Scherer's group has developed a tool, based on Scherer's theoretical framework, which reduces a wide range of terms to 36 broader classes (http://www.unige.ch/fapse/emotion/resmaterial/GALC.xls). Even there, though, it remains a concern how naturally everyday terms do divide into cover classes.

Table 1 Everyday words that Web sites regularly list as descriptors of emotion

Abandoned	Blissful	Cowardly	Energised	**Happy**	**Joyous**	Positive	Sorrowful
Accepted	Bold	Cross	Engaged	Hateful	Jubilant	Powerful	Sorry
Accused	**Bored**	Crossed	Engrossed	Haunted	Jumpy	**Powerless**	Spirited
Adequate	Brash	Cruel	Enraged	Heartbroken	Keen	Pressured	Spiteful
Admired	Brave	Crushed	Enthusiastic	Helpful	Kind	Productive	Splendid
Adored	Breathless	Curious	**Envious**	**Helpless**	Liberated	**Proud**	Squashed
Affectionate	Bright	Daring	Evasive	Hesitant	Lifeless	Provoked	Strong
Afflicted	Bruised	Deceived	Exasperated	Honoured	Light	Puzzled	Stubborn
Afraid	Burdened	Defeated	**Excited**	**Hopeful**	Lonely	Quiet	Stunned
Aggressive	**Calm**	Dejected	Excluded	Hopeless	Lost	Reassured	Suffering
Agitated	Capable	**Delighted**	Exhausted	Horrified	Loved	Rebellious	Sulky
Agonised	Captivated	Depressed	Exhilarated	Hostile	**Loving**	Receptive	Sullen
Agreeable	Carefree	Deserted	Exploited	Humble	Low	Refreshed	Supported
Alarmed	Caring	**Despair**	Exuberant	Humiliated	Lucky	Regretful	Sure
Alert	Carried away	Desperate	Fantastic	**Hurt**	Mean	Rejected	**Surprised**
Alienated	Cautious	Determined	Fascinated	Hysterical	Melancholy	**Relaxed**	Suspicious
Alive	Certain	Devoted	Fearful	Ignored	Merry	Released	Sympathetic
Alone	Challenged	**Disappointed**	Festive	Important	Mischievous	**Relieved**	Tenacious
Aloof	Charmed	Discouraged	Fidgety	Impulsive	Miserable	Remorse	Tender
Amazed	Cheated	Disgraced	Firm	Inadequate	Nervous	Renewed	**Tense**
Amused	Cheerful	**Disgusted**	Flustered	Incapable	Odd	Resentful	Terrible
Angry	Cherished	Dismayed	Foolish	Indecisive	Offended	Reserved	Terrified
Annoyed	Clean	Distant	Fortunate	Independent	Open	**Sad**	Thankful
Anxious	Clever	Distracted	Free	Indifferent	Optimistic	Safe	Threatened
Apathetic	Close	Distraught	**Friendly**	Indignant	Outraged	**Satisfied**	Thrilled
Apologetic	Cold	Distressed	Frightened	Inept	Overjoyed	Scared	Timid
Appalled	Comfortable	Disturbed	Frisky	Infuriated	Overwhelmed	Secure	Tormented
Apprehensive	Comforted	Dominated	**Frustrated**	Innocent	Pained	Sensitive	Torn
Ashamed	Compassionate	**Doubtful**	Fulfilled	Inquisitive	Pampered	**Serene**	Tortured

Table 1 (continued)

Assertive	Competitive	Dull	Furious	Insecure	Panicky	Settled	Trapped
Attractive	Complacent	Dynamic	Generous	Inspired	Paralyzed	Sexy	Uneasy
Aware	Complete	Eager	Gentle	Insulted	Paranoid	Shaky	Upset
Awed	Concerned	Ecstatic	Glad	Intense	Passionate	**Shamed**	Useless
Awkward	Confident	Edgy	Gloomy	**Interested**	Peaceful	**Shocked**	Warm
Bad	Confused	**Elated**	Good	Intimidated	Peeved	Shy	Weak
Beautiful	Considerate	**Embarrassed**	Gratified	Intrigued	Perplexed	Silly	Weary
Belittled	Conspicuous	Empty	Great	**Irritated**	Pessimistic	Sceptical	Wonderful
Betrayed	**Contempt**	Enchanted	Grieving	Isolated	Petrified	Smothered	**Worried**
Bewildered	**Content**	Encouraged	Guarded	Jealous	Playful	Soothed	Zealous
Bitter	**Courageous**	Energetic	**Guilty**	Jolly	**Pleased**	Sore	

Term A may belong with term B in one respect, and term C in another; and terms D and E may be very close in meaning and yet fall on opposite sides of the most natural dividing line. Forming cover classes may still be the most practical procedure in particular cases, but the difficulties should not be underestimated.

The alternative is to develop lists which are designed to offer raters a range of alternatives that are wide enough to be useful, but not totally unmanageable. Motivation for selection varies. A list chosen on theoretical grounds is given as an Appendix in Scherer et al. (1988a) and can be retrieved from the web. It is unique in that (approximately) equivalent words are given in English, German, French, Italian, and Spanish. Teams in HUMAINE concerned with labelling developed a list designed mainly to cover states observed in naturalistic data (Cowie et al., 1999; Banziger et al., 2005; Devillers et al., 2006). It includes the terms shown in bold in Table 1, plus four others: stress, politeness, empathy, and trust.

At the other extreme is the 'big six' – fear, anger, happiness, sadness, disgust, and contempt. Comparison with Table 1 suggests why it is not ideal for all purposes. Some of the terms can be regarded as approximations to one of the 'big six', but many cannot.

It is natural to use specific emotion terms in a quantised labelling, by associating a given label with a period when it applies. However, that is not the only option. Trace techniques can be applied as easily to words as to dimensional concepts. What they set out to show is how strongly a particular emotion is being felt at any given time. Figure 3 shows how traces (based on the kind of 1D scheme shown in Fig. 2) can reflect the way different named emotions come to the fore and recede in a complex episode, involving anticipation, a surprise, a reunion, and a shared reminiscence.

3.1.3 Generic Emotion Labels

Emotion-related states can also be divided at a different level, as explained in the opening chapter of this handbook. This involves distinguishing, for instance, brief episodes of emergent emotion from moods or emotions that are part of a person's long-term make-up (such as enduring shame or anger over something that happened long ago). The practical point of labelling at that level is that an automatic system may need to deal differently with different types. Anger at an immediate event (say an unhelpful helpline) calls for a different response from an ongoing bad mood or anger rooted in a long-standing grievance. In at least some cases the signs are clearly different – for example, a system trained on the signs of full-blown anger is very likely to misinterpret the signs of suppressed anger.

Studies in HUMAINE have produced a list of generic types which appear to cover most of emotional life. It is explained in the first chapter of this handbook. A simplified summary is given below (Table 2).

A version of that system has been applied to the clips in the HUMAINE database.

A related development based on a different approach has been exploited by El Kaliouby and Robinson (2004). They used a taxonomy developed by Baron-Cohen (2007). Its distinctive feature is the prominence that it gives to states which are not emotion in a strong sense. Baron-Cohen describes them as epistemic mental

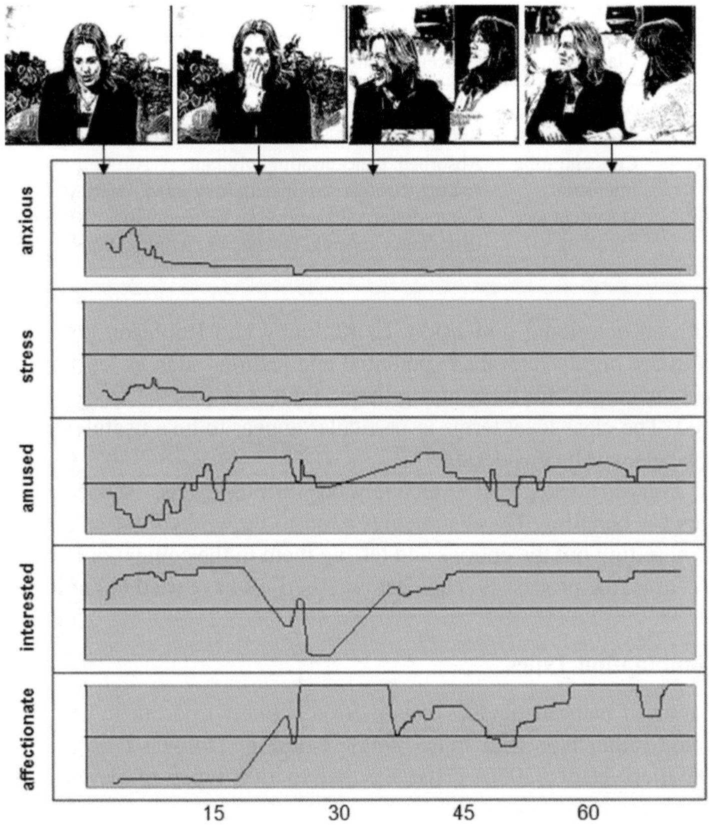

Fig. 3 Traces showing how named emotions are judged to fluctuate through a surprise meeting. Pictures mark key moments: they are stylised to protect identity

Table 2 Generic types of emotion

Emotion-like		Established emotion
		Emergent emotion (suppressed)
		Emergent emotion (full-blown)
	Mood-like	Transitional emotion (shifting between mood and emergent)
		Mood
		Altered state of arousal
		Altered state of control
Stance-like		Altered state of seriousness
		Stance towards object/situation
		Interpersonal stance
		Interpersonal bond
		Emotionless

Table 3 Baron-Cohen (2007) taxonomy of epistemic mental states with an emotional dimension

Group	Class	Concepts included
Sure	Agreeing	Assertive, committed, convinced, knowing, persuaded, sure
Unsure	Unsure	Baffled, confused, puzzled, undecided, unsure
Interested	Concentrating	Absorbed, concentrating, vigilant
	Interested	Asking, curious, fascinated, impressed, interested
Unfriendly	Disagreement	Contradictory, disapproving, discouraging, disinclined
Thinking	Thinking	Brooding, choosing, fantasising, judging, thinking, thoughtful

states with an emotional dimension. El Kaliouby and Robinson pointed out that these states are highly relevant to potential applications such as teaching, and they developed techniques for recognising them. Table 3 shows the categories that they considered. The class level terms in particular appear to be a useful addition to the range of labels worth considering.

Using everyday categories effectively, at various levels, is one of the major challenges for labelling. Good strategies need to be worked out empirically. That depends on setting out the options and testing them in the context of a suitable body of data. That is, the process that the last two sections have tried to facilitate.

3.1.4 Combination Types

It is a feature of naturalistic data that emotion is likely to occur in various kinds of combination rather than as a 'pure' single emotion (Douglas-Cowie et al., 2003; 2005; Devillers et al., 2006). That has led to two types of development, both described by Devillers et al. (2006).

To describe the kinds of combination that occur in a clip, the following set of labels has been developed:

- Unmixed
- Simultaneous combination (distinct emotions present at the same time)
- Sequential combination (single episode which moves through a sequence of related emotions)

To describe the components of a complex, raters should enter more than one emotion term at a time (sad and angry, for instance). 'Soft vectors' (Batliner et al., 2006) can then be used to reflect the strengths of the various components. That kind of information falls out automatically if trace-type descriptions are used to describe the time course of each relevant type.

3.1.5 Appraisal Categories

Appraisal categories aim to reflect the way the person being considered weighs up emotionally critical events or people around. The labelling in one of the earliest naturalistic databases, the Leeds-Reading database, used the appraisal theory due to

Ortony Clore and Collins (1988); it described emotions in terms of an OCC class and the associated object. Research within HUMAINE began by trying to apply the version due to Scherer (see. e.g. Sander et al., 2005), but pilot work showed that relatively few of the descriptors could be assigned with any degree of reliability (Devillers et al., 2006). The labels that could be assigned reliably were as follows:

- Goal conduciveness (the situation offers the person an opportunity to achieve a significant goal)
- Goal obstructiveness (the situation presents an obstacle to the person achieving a significant goal)
- Power/powerlessness (the extent to which the person feels he/she has the power to affect or control emotionally significant events)
- Expectedness (the extent to which the person anticipated emotionally significant events or was taken unawares by them).

Disagreement on the other appraisal categories was partly due to the presence of multiple emotion-related events. That highlights the need to specify events, things, people, and so on that are material to the emotion. Thinking about archetypal cases like James' bear in the woods conceals the fact that events (etc.) may have several different roles in relation to a single emotional episode. The system used by Devillers et al. distinguished four. An example helps to distinguish them. I hear in the morning that a major grant application has been rejected. Walking to work I trip on a loose paving stone and become very angry with the council. The categories involved are as follows:

Cause. The term is not ideal, but it is hard to do better. It refers to an event that may precede the particular emotion by some time, without which it would not have happened. In the example, it is the rejection.

Aspiration. This is relatively self-explanatory. In the example, the whole scenario depends on my aspiration to get the grant.

Trigger. This is an event which immediately precipitates an emotion – in the example, tripping.

Focus. This is what an emotion is about – in the example, the council. It is also called the object of the emotion.

The scheme is not ideal, but it provides some advance over struggling to say what an emotion is about when several factors of different kinds are obviously relevant.

3.1.6 Authenticity

It is natural to assume that authenticity is a single issue. However, inspection makes it clear that there are at least two issues to consider. The first is whether or not the subject appears to be simulating an emotion that he or she does not feel. The second is whether the subject in question is masking his/her emotion. Pilot work on television data (particularly on 'reality shows' where subjects are facing major challenges in competitive situations) suggests that masking of emotion is common.

These are addressed using two trace techniques in the Belfast database. In one, coders are asked to rate the data on a scale from 'no acting of emotion' to 'extreme acting of emotion'. In the other, they mark the level of masking on a scale from 'no concealment of emotion' to 'total concealment of emotion'.

3.1.7 Reliability and Variation

It is impossible to give a deep treatment of these issues in a short section, so the aim here is simply to point out that there is a variety of issues to be faced and to note some known solutions.

The most straightforward question is how consistently a particular piece of material has been labelled. Where the material is divided into segments that are meant to be relatively homogeneous, and each labeller assigns each segment one label chosen from a relatively small set, the natural measure is the proportion of raters who agree on the commonest label. For pure dimensional labellings, a comparable kind of information is provided by the standard deviations of ratings (Cowie et al., 2000). The two can be compared by finding dimensional co-ordinates for each label (Whissell, 1989 provides an extensive list) and using those to derive standard deviations (Cowie and Cornelius, 2003). The same kind of replacement can be used to give a comparable measure of agreement in cases where there are many categories to choose from, some of which are closer together than others.

At more global level, there are standard techniques for measuring how reliably different raters assign a particular set of labels. With a set of qualitative labels, the standard statistic is kappa (Cohen, 1960). Where ratings are continuous (as with trace measures), the standard statistic is Cronbach's alpha. To illustrate expected values, Devillers et al. (2006) report average kappa of just above 0.6 when raters are assigning a set of 15 everyday labels, and Savvidou (submitted) reports alpha values above 0.9 for both dimensions (valence and activation) in a series of studies with FEELtrace.

Note, though, that simple application of these measures is not always appropriate, and techniques have been devised to deal with the issue. The usual form of kappa is not appropriate when it is possible to assign more than one label (for instance, when there is a blend of sadness and anger). Rosenberg and Binkowski (2005) have proposed an extension that handled multi-element responses. Their approach still gives low values when raters use multiple labels even if they agree perfectly, though. Devillers et al. (2006) further adapted Rosenberg and Binkowski's approach to give a kappa measure whose value is (as expected) 1 when raters choose the same multiple labels.

Alpha is an adjusted average of the correlations between pairs of outputs. Like other averages, it can be misleading if the distribution is not straightforward. An immediate response to that kind of issue is to plot correlations between all pairs of

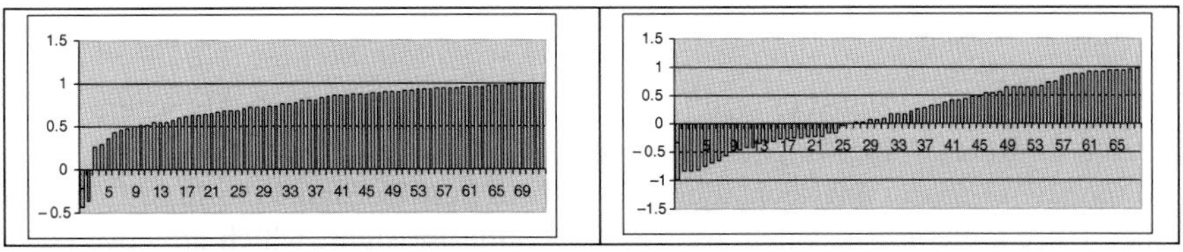

Fig. 4 Plots of the correlations between traces made by multiple raters (from Devillers et al., 2006). Each bar corresponds to the correlation coefficient obtained by comparing two traces of the same passage. The *left-hand panel* summarises ratings of intensity and the *right panel* summarise ratings of power

raters. Figure 4 shows the kinds of picture that emerge. They illustrate an intriguing and quite common pattern, which is that in some cases, raters are not simply uncorrelated; they take positively opposing views of what is happening.

That kind of problem is classically recognised, and the classical prescription is to carry out factor analysis as a way of separating the different patterns that are overlaid in the data set as a whole (Tabachnik and Fidell, 2001). Figure 5 shows patterns inferred from trace data in that way (Savvidou, in preparation). The procedure is not wholly satisfying, but it provides at least some protection against concluding that a measure is unreliable because it reveals that people read the material in different ways.

The ability to represent variation is particularly important in evaluating an artificial recogniser. One of the key tests of a good recogniser is how the distribution of responses that it gives compares to the range given by human raters. The standard measure, percentage correct, is not a satisfying way to do that; it may not be clear what 'correct' is, and there may be several responses that are quite commonly given even if they are not correct, or even the commonest choice for humans. Steidl et al. (2005) have developed an entropy measure that addresses these issues.

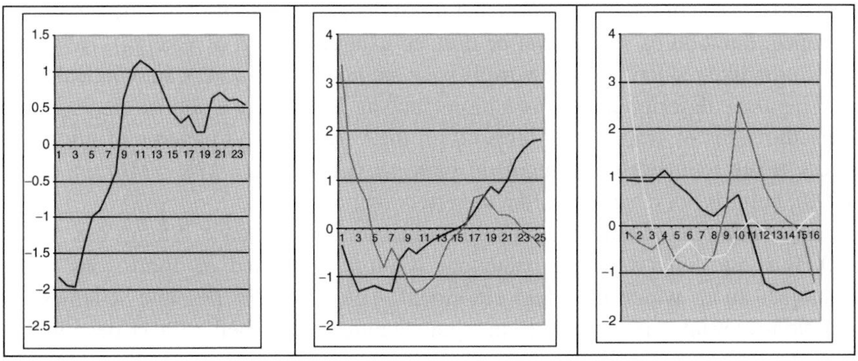

Fig. 5 Factors in ratings of valence for clips showing one, two, and three readings of the material, respectively

3.2 Resources for Labelling Signs of Emotion

Many of the issues related to sign labelling are dealt with in depth elsewhere in the handbook. Broadly speaking, two main areas are involved. On the one hand, there are the signal processing techniques covered in Part II 'Signals to Signs'; on the other, there are constructs that are used to synthesise signs, which are covered in Part IV 'Emotion in Interaction'. Hence, the aim in this section is generally to give a broad overview rather than technical detail.

It would be quite wrong if the labelling of signs in databases was divorced from these other areas, because databases are a resource for teams concerned with recognition and synthesis. For teams concerned with recognition, databases provide the material that they need to evaluate recognition algorithms. To be practically useful, the database has to provide the algorithms with input variables that are machine recoverable (or could be recovered given foreseeable improvements in signal processing). For teams concerned with synthesis, the database has to indicate the kinds of gesture (in the broadest sense) that are relevant to conveying a particular kind of emotion in a particular context, and that only useful, if the description of the gesture corresponds to something that can be generated.

On the other hand, there are types of labelling that may be significant, but that cannot at present be either recovered or generated by machine. Those receive particular attention here, because they are not covered elsewhere.

Descriptors are divided into five main areas: linguistic; vocal (including paralinguistic); facial; gesture; and physiological.

3.2.1 Linguistic

Orthographic transcription is a key resource, most obviously because the content of what people say usually conveys a great deal about their emotions. Even crude words spotting adds materially to emotion recognition (Fragopanagos and Taylor, 2005). Time alignment gives added value. It is not obvious how much phonetic transcription adds. Some of the information that it provides can be derived in other ways (for instance, duration of vowels can be derived using detection of voicing). However, indicators like reduction in the vowel space are not easy to access automatically.

Linguistic descriptions of intonation (such as ToBI) have a similar status. Some prosodic signs of emotion can depend on the way observed patterns of pitch and stress relate to underlying linguistic patterns (Ladd et al., 1985; Mozziconacci, 1998), but it is not clear how often the effect matters.

Discourse-related units may also be important. For instance, Craggs and Wood (2003) argue that grounding behaviour is more protracted in discussions about a subject about which they feel emotional. They consider possible schemes for emotion-relevant annotation of dialogue features. Campbell (2004) describes a well-developed coding scheme for dialogue acts that has been used in the ESP corpus.

Several sources suggest that linguistic features are not overwhelmingly important for coarse judgments of emotion. There is a shortage of good data, but it seems clear that people have considerable ability to gauge the emotions of people speaking unfamiliar languages. There is also relatively little shift in judgments when speech is filtered so that prosody remains, but individual words are unintelligible (Douglas-Cowie et al., 2005).

On the other hand, there are emotionally charged types of communication that it is difficult to imagine grasping without language, such as irony. The function of the database, and the nature of the material, probably dictates the value of including sophisticated linguistic descriptions.

3.2.2 Vocal

No database can reasonably aspire to include all the acoustic measures that have a claim to be relevant to detecting emotion. As the chapter by Batliner et al. on recognition from audio signals (in Part II 'Signals to Signs') stands to indicate, there are simply too many. The best that can be hoped for is to include raw descriptions from which the most important measures can be derived.

There is a strong case for databases to include F0, particularly corrected F0. Uncorrected F0 contours are notoriously prone to octave jumps, particularly when speech is emotional (hence the HUMAINE Web site contains a tool for correcting F0). Corrected pause boundaries are useful for a similar reason. Intensity may well be as important, but if distance to the microphone is fixed, it is trivially easy to recover and if it is not, it is almost impossible to achieve much precision.

Recognition algorithms also tend to use coefficients associated with the presence or the absence of voicing, the spectrum, formants, and MFCCs. It is not obvious how much would be gained by including these in a database.

At a higher level, Douglas-Cowie derived a set of descriptors by listening to the Belfast naturalistic database (Douglas-Cowie et al., 2003). The original system contained many labels, but for use in the HUMAINE database, these have been reduced to a core set of items which the tests indicate are strongly characteristic of emotion and can be applied reliably.

The labels address four descriptive categories and raters can assign a number of labels within these levels:

- Paralanguage
  - Laughter, sobbing, break in voice, tremulous voice, gasp, sigh, exhalation, scream
- Voice quality
  - Creak, whisper, breathy, tension, laxness
- Timing
  - Disruptive pausing, too long pauses, too frequent pauses, short pause + juncture, slow rate
- Volume
  - Raised volume, too soft, excessive stressing

These have been used in the HUMAINE database (described later in this handbook).

3.2.3 Face Descriptors

Face is discussed in several other chapters of this handbook. It is an area where there is strong convergence. The facial action coding system (FACS) is generally accepted as a standard (Ekman et al., 2002). It underpins the system of coding in terms of facial action packages (FAPs) adopted by MPEG, which is used extensively in recognition, generation, and databases. Fuller's descriptions are given in Ioannou et al. (2005) and the chapters on image and video processing and multimodal emotion recognition are given in Part II 'Signals to Signs'.

3.2.4 Gesture Descriptors

Gesture is also discussed in several other chapters of this handbook. In contrast to the work on faces, several very different schemes lie in the background, which have been developed for different purposes – distinguishing types of communicative device, annotating dance, transcribing sign language. There is momentum to develop a satisfying composite scheme.

An example of recent developments is a scheme developed at LIMSI-CNRS. It is a manual annotation scheme and has been used for annotating multimodal behaviours in two corpora: EmoTV (Martin et al., 2005) and EmoTABOO (Martin et al., 2006). The scheme is currently being used for studying the relations between multimodal behaviours and blends of emotions (Devillers et al., 2005).

The scheme uses the following dimensions to annotate gesture:

- Classical dimensions of gesture annotation (McNeill, 1992; Kita et al., 1998; Kipp, 2004; McNeill, 2005), to allow exploratory study of the impact of emotion on these dimensions:
- Gesture units (e.g. in order to study how much gesture there is in an emotional corpus)
- Phases (e.g. to study if there are long duration of holds due to the interactive game and thinking behaviours of the subject)
- Phrases
- Categories (e.g. to study the frequency of adaptors and compare it to other corpora)
- Lemmas adapted from a gesture lexicon (Kipp, 2004) (e.g. Doubt=Shrug)
- Expressivity: annotated at the level of the phrases and at the level of the gesture units. The scheme considers six expressivity parameters used in research on movement quality (Hartmann et al., 2005; Wallbott, 1998): activation, repetition, spatial extent, speed, strength, fluidity.

This scheme is implemented in a module in the Anvil tool (Kipp, 2001). The chapter on the HUMAINE database (in Part III) shows it applied to selected clips. Improvements are under active discussion.

3.2.5 Physiological Descriptors

It makes sense for databases to include the relatively small core of physiological descriptors that are known to vary with emotion. They are as follows:

- Heart rate
- Blood pressure
- Skin conductance
- Respiratory effort
- Skin temperature
- EMG at key sites

There are many standard ways of deriving measures from these basic signals, involving differences, standard deviations, filtering, and various other operations. Code that can be used to carry out standard transformations can be accessed via the portal from the Augsburg Biosignal Toolbox (AuBT), which was developed for HUMAINE (see chapter 'Multimodal Emotion Recognition from Low-Level Cues' in Part II 'Signals to Signs')

The major problem with these measures is that they are sensitive to many types of variables which are not directly related to emotion (including physical and mental effort, speech, and health-related factors). To control for these statistically, they need to be recorded.

3.3 Resources for Labelling Context

An enormous range of issues could be covered under the heading 'context'. The description here is mainly drawn from the way the HUMAINE database has attempted to systematise the issues.

At a basic level, there is factual data on the subject's personal characteristics (age, gender, race), on technical aspects of recording (recording style, acoustic quality, video quality), and on physical setting (degree of physical restriction, posture constriction, hand constriction, and position of audience).

A second set of labels deals with communicative context. The elements addressed in the HUMAINE database fall into two categories:

(i) Coders are asked to rate the purpose or the goal of the communication – to persuade, to create rapport, to destroy rapport, or just a pure expression of emotion, for example, somebody laughing.

(ii) Coders record their perceptions of the social setting of the clip, whether, for example, there is a clear interaction happening between two or more people or whether one person is communicating to another who has a very passive role in the interaction. Social pressure is considered under this heading. Coders rate whether they think that the situation puts the person they are rating under pressure to be formal (e.g. in a court), or to be freely expressive (e.g. at a party), or that there is little pressure either way.

Table 4 Classification of antecedents to happiness from Scherer (1988)

Good news (immediate social context). Example: an unexpected job offer
Good news (mass media). Example: cheering news in newspapers or on TV
Continuing relationships with friends and permanent partners. Example: pleasure from contact with friends
Continuing relationships with blood relatives and in-laws
Identification with groups (actual and reference). Examples: pleasure in belonging to a club; returning to your own country after a holiday
Meeting friends, animals, plants. Examples: seeing one's dog again; meeting one's friend for dinner
Meeting blood relatives or in-laws
Acquiring new friends
Acquiring new family members. Examples: birth of a baby; marriage of one's brother
Pleasure in meeting strangers (short-term chance encounters). Example: talking to a stranger on a train
Pleasure in solitude. Example: being left alone with one's own thoughts
New experiences. Examples: adventures; planning a holiday
Success experiences in achievement situations. Example: passing an examination
Acquiring some material for self or other (buying or receiving). Examples: presents from others; buying something nice for oneself or others
Ritual. Examples: religious, academic ceremonies, festivals, birthdays
Natural, also refined, noncultural pleasures. Examples: sex, food, nature, landscape
Cultural pleasures. Examples: art, music, ballet, etc.
Acquiring nonmaterial benefits (emotional support, altruism). Example: helping an old lady cross the road
Happiness without reason
Schadenfreude. Example: malicious pleasure in another person's misfortune

Description of appraisal could be counted here. The HUMAINE database includes a very compressed description, which has been covered already in this chapter in Sect. 3.1.3. Much fuller descriptions have been offered in the literature. Table 4 forms part of the coding system reported in Scherer (1988b), classifying the types of antecedent to happiness. Similar schemes are offered for sadness, anger, and fear.

4 Proposals to Advance the State of the Art

The research priority at this stage is reasonably clear. A rich body of ideas about labelling emotion databases has accumulated. The ideas need to be applied systematically to a database that is sufficiently large, and challenging to test what they can contribute, and evaluated. In the first instance, evaluation involves several levels. Reliability (with qualifications noted above) is the first. The second is redundancy: to what extent do different descriptors duplicate the same information? That kind of question has been asked in the context of isolated word (and the results form the basis of dimensional accounts). However, it has not been asked in the context of real instances of emotion. The third is adequacy for synthesis: how well do various sets of parameters allow an ECA to reproduce the emotional content of an original clip? The fourth is predictive power: how well do various combinations of sign and context labels allow emotion labels to be inferred, and (with some differences) vice versa? The task goes beyond reconstructing labels in one stream from labels in another at an adjacent time. It critically involves identifying possibilities to be considered and factored into future choices – which kinds of intervention are advisable given these signs in this context, and which should be avoided at any price?

These questions add up to a research agenda for a number of decades. It depends on the availability of suitable sets of records and cumulative work on applying possible labels to them. The HUMAINE database, described in the next chapter, is a substantial step in that direction.

On the other hand, applied research projects will test how useful subsets of the repertoire are in specific applications. It is an interesting challenge to ensure that applied and academic evaluations connect.

Some people will be frustrated by the implication that genuinely satisfying prescriptions for labelling lie at the end of a long-lasting research program. It is an understandable reaction. However, understanding it does not change the reality that that is the timescale.

References

Auberge V, Audibert N, Rillard A (2006) Auto-annotation: an alternative method to label expressive corpora. In: Proceedings LREC 2006, Genoa, pp 45–46

Bachorowski JA (1999) Vocal expression and perception of emotion. Curr Dir Psychol Sci 8(2): 53–57

Bänziger T, Tran V, Scherer KR (2005) The Geneva Emotion Wheel: a tool for the verbal report of emotional reactions. In: Poster presented at ISRE 2005, Bari

Baron-Cohen S (2007) Mind reading: the interactive guide to emotions – version 1.3. Jessica Kingsley, London

Batliner A, Steidl S, Schuller B, Seppi D, Laskowski K, Vogt T, Devillers L, Vidrascu L, Amir N, Kessous L, Aharonson V (2006) Combining efforts for improving automatic classification of emotional user states. In: Erjavec T, Gros J (eds) Language technologies, IS-LTC 2006. Infornacijska Druzba (Information Society), Ljubljana, Slovenia, pp 240–245

Bradley MM, Lang PJ (1994) Measuring emotion: the self-assessment manikin and the semantic differential. J Behav Ther Exp Psychiatry 25(1):49–59

Campbell N (2004) Extra-semantic protocols: input requirements for the synthesis of dialogue speech. In: Andre E, Dybkjaer L, Minker W, Heisterkamp P (eds) Affective dialogue systems. Lecture notes in artificial intelligence. Springer, Berlin, pp 221–228

Cohen JA (1960). A coefficient of agreement for nominal scales. Educ Psychol Meas, 20(1):37–46

Cowie R, Cornelius R (2003) Describing the emotional states that are expressed in speech. Speech Commun 40:5–32

Cowie R, Douglas-Cowie E, Apolloni B, Taylor J, Romano A, Fellenz W (1999) What a neural net needs to know about emotion words. In: Mastorakis N (ed) Computational intelligence and applications. World Scientific Engineering Society, Dallas, TX, pp 109–114

Cowie R, Douglas-Cowie E, Savvidou S, McMahon E, Sawey M, Schroeder M (2000) 'FEELTRACE': an instrument for recording perceived emotion in real time. In: Proceedings of the ISCA ITRW on speech and emotion: developing a conceptual framework, Newcastle, 5–7 Sept 2000, Textflow, Belfast, pp 19–24.

Cowie R, Douglas-Cowie E, Tsapatsoulis N, Votsis G, Kollias S, Fellenz W, Taylor J (2001) Emotion recognition in human–computer interaction. IEEE Signal Process Mag 18(1):32–80

Craggs R, Wood M (2003) Annotating emotion in dialogue. In: Proceedings of the 4th SIGdial workshop on discourse and dialogue, Sapporo

Craggs R, Wood M (2004) A 2 dimensional annotation scheme for emotion in dialogue. In: AAAI spring symposium on exploring attitude and affect in text: theories and applications. Stanford University, AAAI Press, pp 44–49

Devillers L, Abrilian S, Martin J-C (2005) Representing real life emotions in audiovisual data with non basic emotional patterns and context features. In: Proceedings of the 1st international conference on affective computing and intelligent interaction (ACII'2005), Beijing, 22–24 Oct. Spinger, Berlin, pp 519–526

Devillers L, Cowie R, Martin J-C, Douglas-Cowie E, Abrilian S, McRorie M (2006) Real life emotions in French and English TV video clips: an integrated annotation protocol combining continuous and discrete approaches. In: Proceedings LREC 2006, Genoa

Devillers L, Martin J-C (2008) Coding emotional events in audiovisual corpora. In 6th international conference on language resources and evaluation (LREC 2008). Marrakech (Morocco)

Douglas-Cowie E, Campbell N, Cowie R, Roach P (2003) Emotional speech: towards a new generation of databases. Speech Commun 40(1–2):33–60

Douglas-Cowie E, Devillers L, Martin J-C, Cowie R, Savvidou R, Abrilian S, Cox C (2005) Multimodal databases of everyday emotion: facing up to complexity. In: Proceedings of the Interspeech 2005, Lisbon, pp 813–816

Ekman P (1992) An argument for basic emotions. Cogn Emot 6:169–200

Ekman P (1999) Basic emotions. In: Dalgleish T, Power MJ (eds) Handbook of cognition and emotion. Wiley, New York, pp 301–320

Ekman P, Friesen W (1975) Unmasking the face: a guide to recognizing emotions from facial clues. Prentice-Hall, Englewood Cliffs, NJ

Ekman P, Friesen WC, Hager JC (2002). Facial action coding system. The manual on CD ROM. Research Nexus Division of Network Information Research Corporation, Salt Lake City, UT

El Kaliouby R, Robinson P (2004) Mind reading machines: automated inference of cognitive mental states from video. In: IEEE international conference on systems, man and cybernetics, vol 1, The Hague, pp 682–688

Fragopanagos N, Taylor J (2005) Emotion recognition in human–computer interaction. Neural Netw 18:389–405

Hall JA, Matsumoto D (2004) Gender differences in judgments of multiple emotions from facial expressions. Emotion 4(2):201–206

Hanson NR (1958) Patterns of discovery: an inquiry into the conceptual foundations of science. Cambridge University Press, Cambridge, MA

Hartmann B, Mancini M, Pelachaud C (2005) Implementing expressive gesture synthesis for embodied conversational agents. In: Gesture Workshop (GW'2005), Vannes, France, pp 1095–1096

Ioannou S, Raouzaiou A, Tzouvaras V, Mailis T, Karpouzis K, Kollias S (2005) Emotion recognition through facial expression analysis based on a neurofuzzy network Neural Netw, 18, 423–435

James W (1884) What is emotion? Mind 9:188–205

Kipp M (2001) Anvil – a generic annotation tool for multimodal dialogue. In: 7th European conference on speech communication and technology (Eurospeech'2001), Aalborg, Denmark, 3–7 Sept, pp 1367–1370

Kipp M (2004). Gesture generation by imitation. From human behavior to computer character animation, Florida, Boca Raton, Dissertation.com. 1581122551

Kita S, van Gijn I, van der Hulst H (1998). Movement phases in signs and co-speech gestures, and their transcription by human coders. Gesture and sign language in human computer interaction: Proceedings/international gesture workshop, Bielefeld, 17–19 Sept. Springer, Berlin, Heidelberg

Ladd S, Silverman K, Bergmann G, Scherer K (1985) Evidence for independent function of intonation contour type, voice quality, and F0 in signalling speaker affect. J Acoust Soc Am 78(2):435–444

Martin J-C, Abrilian S, Devillers L (2005). Annotating multimodal behaviors occurring during non basic emotions. In: 1st international conference on affective computing and intelligent interaction (ACII'2005), Beijing, 22–24 Oct. Spinger, Berlin, pp 550–557

Martin JC, Devillers L, Zara A, Maffiolo V, LeChenadec G (2006) The EmoTABOU corpus. Humaine Summer School, Genova, pp 22–28

Matsumoto D (2001) Culture and emotion. In: Matsumoto D (ed) Handbook of culture and psychology. Oxford University Press, New York, NY, pp 171–194

Mayer JD, Salovey P, Caruso D (2000) Models of emotional intelligence. In: Sternberg R (ed) Handbook of intelligence. Cambridge University Press, Cambridge, MA, pp 396–421

McNeill D (1992) Hand and mind – what gestures reveal about thoughts. University of Chicago Press, Chicago, IL

McNeill D (2005) Gesture and thought. University of Chicago Press, Chicago, IL

Mozziconacci SJL (1998). Speech variability and emotion: production and perception. Ph.D. thesis, Eindhoven

Ortony A, Clore G, Collins A (1988) The cognitive structure of emotions. Cambridge University Press, Cambridge, MA

Ortony A, Turner TJ (1990) What's basic about basic emotions? Psychol Rev 97:315–331

Rosenberg A, Binkowski E (2005). Augmenting the kappa statistic to determine interannotator reliability for multiple labelled data points. In: Proceeding of the HLT-NAACL, Boston

Russell J, Barrett-Feldman L (1999) Core affect, prototypical emotional episodes, and other things called emotion: dissecting the elephant. J Pers Soc Psychol 5:37–63

Sander D, Grandjean D, Scherer KR (2005) A systems approach to appraisal mechanisms in emotion. Neural Netw 18:317–352

Savvidou S (submitted) Validation of the Feeltrace tool for recording impressions of expressed emotion. Thesis submitted for Ph.D., Queen's University, Belfast.

Scherer KR (ed) (1988a) Facets of emotion: recent research. Erlbaum, Hillsdale, NJ. Appendix F. Labels describing affective states in five major languages, revised by the members of the Geneva Emotion Research Group retrieved Dec 2007 from http://www.unige.ch/cisa/gerg/research.html

Scherer KR (ed) (1988b) Appendix B. Antecedent and reaction codes used in the "Emotion in Social Interaction." In: Scherer KR (ed) Facets of emotion: recent research. Erlbaum, Hillsdale, NJ, pp 241–243. http://www.unige.ch/cisa/gerg/research.html

Schreuder M, van Eerten L, Gilbers D (2006) Music as a method of identifying emotional speech. In: LREC Research Workshop on Corpora on Emotion and Affect, Genoa

Steidl S, Levit M, Batliner A, Nöth E, Niemann H (2005) "Of All Things the Measure is Man" – automatic classification of emotions and inter-labeller consistency. In: Proceedings of the ICASSP 2005, Philadelphia, pp 317–320

Tabachnik BG, Fidell LS (2001) Using multivariate statistics, 4th edn. Allyn & Bacon, Boston, MA

Wallbott HG (1998) Bodily expression of emotion. Eur J Soc Psychol 28:879–896

Whissell C (1989) The dictionary of affect in language. In: Plutchik R, Kellerman H (eds) Emotion: theory, research and experience: vol 4, The measurement of emotions. Academic, New York, NY, pp 113–131

The HUMAINE Database

Ellen Douglas-Cowie, Cate Cox, Jean-Claude Martin, Laurence Devillers, Roddy Cowie, Ian Sneddon, Margaret McRorie, Catherine Pelachaud, Christopher Peters, Orla Lowry, Anton Batliner, and Florian Hönig

Abstract The HUMAINE Database is grounded in HUMAINE's core emphasis on considering emotion in a broad sense – 'pervasive emotion' – and engaging with the way it colours action and interaction. The aim of the database is to provide a resource to which the community can go to see and hear the forms that emotion takes in everyday action and interaction, and to look at the tools that might be relevant to describing it. Earlier chapters in this handbook describe the techniques and models underpinning the collection and labelling of such data. This chapter focuses on conveying the range of forms that emotion takes in the database, the ways that they can be labelled and the issues that the data raises. The HUMAINE Database provides naturalistic clips which record that kind of material, in multiple modalities, and labelling techniques that are suited to describing it. It was clear when the HUMAINE project began that work on databases should form part of it. However there were very different directions that the work might have taken. They were encapsulated early on in the contrast between 'supportive' and 'provocative' approaches, introduced in an earlier chapter in this handbook. The supportive option was to assemble a body of data whose size and structure allowed it to be used directly to build systems for recognition and/or synthesis. The provocative option was to assemble a body of data that encapsulated the challenges that the field faces.

The eventual choice leant heavily towards the provocative. The supportive was not wholly ignored. Partners in the HUMAINE Database work package have developed the resources used in several significant projects on recognition of emotion, both unimodal (CEICES) and multimodal (Cowie et al.). However, the main systematic effort was directed towards establishing a corpus that summed up the challenges facing the community and drew together key resources that are potentially relevant to meeting them. The choice is grounded in HUMAINE's core emphasis on

E. Douglas-Cowie (✉)
Department of Psychology, Queen's University Belfast, Belfast, Northern Ireland, UK
e-mail: E.Douglas-Cowie@qub.ac.uk

P. Petta et al. (eds.), *Emotion-Oriented Systems*, Cognitive Technologies,
DOI 10.1007/978-3-642-15184-2_14,

considering emotion in a broad sense – 'pervasive emotion' – and engaging with the way it colours action and interaction. Until now, there has been no source to which the community could go to see and hear the forms that emotion takes in everyday action and interaction, and to look at the tools that might be relevant to describing it. The core aim of the HUMAINE Database was to provide that kind of source. The description in this chapter reflects that orientation. It does not set out to give technical specifications of the database contents. Instead it sets out to convey the range of forms that emotion takes in the database and the ways that the descriptive resources address them. The database is not simply about creating impressions. It is also designed to let key technical questions be addressed. That aspect is taken up at the end of the chapter.

1 Overview and Structure

This section sets out some of the specific concerns underlying the database.

The emphasis is on multimodal data. Most of the material is audiovisual. The visual channel usually includes face, but some sources deliberately contain gesture and some include body posture or mode of action. Other parts include physiological data and choice of response to challenges.

The material was collected to show a wide emotional range (negative to positive, active to inert, deeply engaged to playful). The emotion is embedded in a range of actions and interactions, and a variety of contexts.

The material makes varying levels of allowance for the limits of contemporary signal processing. At one extreme is data from TV recordings, shot outdoors with 'difficult' camera angles and noisy audio; at the other is data derived from laboratory scenarios devised to minimise signal processing challenges.

The labels that are used are described in the preceding chapter of this part. They include labels describing emotional content, based on the psychological literature; labels for emotional signs in the relevant modalities (face, speech and gesture context labels have been developed) and context labels. The labels span a range of resolutions in time (whole passage to moment by moment).

There are close links between the recordings and the labels. On the one hand, labels were chosen to deal with the phenomena that were observed in the recordings that laid the groundwork for the database. On the other hand, material was generated and selected to reflect the range of possibilities that the labels indicated ought to be represented.

The database consists of two parts: (i) primary records and (ii) a structured labelled subset. The primary records consist of recordings, almost all audiovisual, in diverse emotion-rich scenarios. The structured labelled subset is a balanced and labelled set of emotional episodes (referred to as 'clips') selected from the primary records to represent a range of emotions and to demonstrate the application of a wide range of labels covering emotional content, context and signs. Most of the primary records and the whole of the labelled subset are available to the research community under 'Conditions of Use Agreement' (see Appendix 1 for details).

The primary records consist of many hours of recordings and contain emotional episodes which have not been identified and selected for labelling. They are thus a resource which can be mined by other researchers. They are also a useful resource for observing how often emotionality of some level occurs across time. Some of the data are 'naturalistic' (in the sense that it has been collected in situations not under the researcher's control, e.g. from film shot for television, etc.). Some are 'laboratory'-induced data (in the sense that emotionality has been induced in a controlled environment according to a purpose-specific method). The laboratory-induced data is a rich resource, not just for the data itself but also for the range of methods used to carry out the induction of emotion (developed specially for the HUMAINE project – for fuller details, see chapter "Issues in Data Collection").

One of the principles behind the HUMAINE database was that the data should be available to the community in general. With the particular nature of the data, this means that there have been important issues of ethical clearance and consent to be addressed. The ethical principles underpinning ethical clearance and consent are dealt with more fully in Part IV 'Ethics and Good Practice'. Access to the database is via the HUMAINE Association portal (www.emotion-research.net).

2 The Total Data Set

The HUMAINE Database work package recorded or acquired a large body of material showing emotion as it appears and sounds in action and interaction. This section summarises the main kinds of material that have been collected as a result.

2.1 Summary of Primary Records

Table 1 provides a summary of the data types that make up the primary records. The recordings are usually either naturalistic or induced, although an emotional episode is selected from one professionally acted data set (GEMEP, see Baenziger and Scherer, 2007) for labelling (as a comparison). The GEMEP Data Set as a whole is not available as part of the HUMAINE database. The induction techniques that have been developed to induce much of the material are described in chapter "Issues in Data Collection" in this part. Appendix 2 describes in more detail the exact nature of the primary records – the material, technical information about the material including length, numbers of subjects, recording scenario and the conditions under which it is available. Appendix 3 describes in detail each episode or clip selected for labelling (to form part of the labelled subset) including a summary of the content of the emotional episode and descriptors of the emotion, context and modalities.

Table 1 Data types used in the HUMAINE Database

Primary records	Data type
Belfast Naturalistic Database	Naturalistic/induced: television data and interviews between friends
EmoTV Database (in French)	Naturalistic
Castaway Reality Television Data Set	Naturalistic
Sensitive Artificial Listener (Belfast recordings in English)	Induced
Sensitive Artificial Listener (Tel Aviv recordings in Hebrew)	Induced
Activity Data/Spaghetti Data	Induced
Green Persuasive Data Set	Induced
EmoTABOO	Induced
DRIVAWORK (driving under varying workload) corpus in German	Induced
GEMEP Corpus (Geneva Multimodal Emotion Portrayal)	Acted
Belfast Driving Simulator Data Set	Induced

2.2 *Illustration of Data Types*

This section illustrates the nature of the data, pulling out its typical characteristics, theoretical interest and relevance and its strengths and weaknesses. It starts with the naturalistic data.

Figure 1 shows a typical frame taken from the Belfast Naturalistic Database (Douglas-Cowie et al., 2003). The subject is talking to an old friend about how she feels about her future son-in-law and expressing her positive feelings for him. The frame is typical of the data in that the emotion expressed is strong but not full blown. In terms of the emotion-related states described in the first chapter in this part, the emotion expressed in this particular example reflects a long-lasting feeling ('attitude') towards someone or something in the Belfast Naturalistic Database. Much of the data either expresses attitudes or is 'established' emotion (long-standing states that can be 'triggered' in a way that produces surges of overt emotion). In this example the subject is sedentary and the camera is fixed on the face, head and shoulders.

I love Andrew and [yes] the only way I can describe Andrew is that if I had been blessed with a son [Yeah] I would have wanted him to be like Andrew [Oh that's rather nice] So uhm we're very happy with whatever happens and so uh are his parents.

Fig. 1 Frame from Clip56b, Belfast Naturalistic Database

The visual quality is not perfect. The induction technique captures a lot of speech in a dialogue situation and because the old friend is also the researcher she knows not to interrupt too much, thereby giving long stretches of uninterrupted speech for analysis. The text is given beside the picture with the interviewer's comments in square brackets.

Figure 2 shows two frames, both from the Castaway Reality Television Data Set (Douglas-Cowie et al., 2007). The data set contains some intense emotion, and these frames illustrate that. The one on the left is of a subject in a positive state (after successful completion of a task) and the one on the right is of a subject in a negative state (after a task in which he thinks he has done badly). The material is important because it shows emotion in action as participants engage in a range of challenging activities both singly and in groups. It often shows shifting and complex emotions as a subject moves through a challenging activity and comes out at the end successfully or unsuccessfully. Because there is a lot of activity, there is a lot of movement by the participants and this gives rise to data that is not face-on or close-up, and so from an affective computing point of view, the data is quite challenging. The shots in Fig. 2 illustrate some of the more static material in the data. Most of the speech occurs in the one-to-one interactions with the team leader, but outdoor noises from the rest of the group tend to get in the way of good recordings. Nevertheless, pitch traces have been reliably extracted for the clips from this data that form part of the labelled subset.

Figure 3 shows two frames from the SAL (Sensitive Artificial Listener) induction technique (see "Issues in Data Collection", Part III). SAL involves an artificial character with different emotional personalities (Poppy who is sad, Spike who is angry, Obadiah who is gloomy, Prudence who is sensible) engaging a subject in conversation. Each personality uses stock responses and phrases to pull the subject towards his/her mood. In Fig. 3 the subject is shown talking to the gloomy personality of the artificial listener (on the left) and then the happy personality of the artificial listener (on the right). The frames show reasonably natural data which demonstrates mild to moderate levels of emotion. The subject is sedentary and the camera is focused on the face and shoulders. There is no gesture, but the technique generates a lot of

Fig. 2 Frames from the Castaway Reality Television Data Set (from Clip5_2, *left*, and Clip 6_2, *right*)

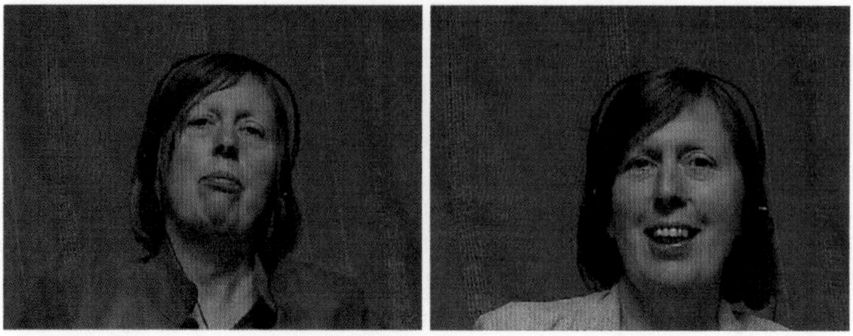

Fig. 3 Subject in conversation with two different personalities of the Sensitive Artificial Listener (from Clip REllA2, *left*, and Clip REllB2, *right*)

Table 2 Text accompanying SAL clips REllA2 and REllB2 (SAL personality responses in square brackets)

SAL clip	Text
REllA2	Tch, well, erm you used to be just gloomy but you seem to be getting logical now, that's a bit worrying [I don't suppose it really matters though] No it doesn't really. I mean, you don't have anything practical to offer at all, no help whatsoever. You just want to sink down into a gloomy state, but I'm not going to do that Obadiah. You go away, I'm, I'm going to talk to somebody like Prudence, she's pragmatic, she'll tell me what to do and then I'll feel a lot better
REllB2	Er, or, another thing that I really like and I haven't really managed to do it this year and I really really must do it but I'm too tired to do it but it's going out on the bike and going up the towpath on a summer evening. I'd love to do that at the minute and just sort of cycle up and all the hawthorne would be out and I could smell the hawthorne and you know it would be nice and kinda sunny and lovely air blowing across you. You know that nice feeling on a bike when the air kind of blows all around you. It's lovely

speech (see Table 2. The visual and auditory quality is good and SAL data has successfully been used to train an emotion recognition system (Ioannou et al., 2005).

The Belfast Activity/Spaghetti Data (see the chapter "Issues in Data Collection") is represented in the next set of figures (Figs. 4 and 5). Two techniques were used to produce the emotion in action seen in these two figures.

In the first (Activity Data), volunteers were recorded engaging in outdoor activities (e.g. mountain bike racing) in an effort to produce examples of full-blown emotion in action for which we would have consent and ethical clearance. This produced 'provocative' data, very dynamic, with subjects moving around. It also had a noisy sound track with affect bursts, but little speech. The data is demonstrated in Fig. 4, which shows sequenced frames from the subject watching one of the volunteers fall off a mountain bike in a 3-s episode. It demonstrates full-blown emotion and quite complex emotional shift. Some interesting work has been done on the data

Fig. 4 Sequenced frames from Belfast Activity Data at 0.56 s (*left*), 2.16 s (*middle*) and 3.00 s (*right*)

showing the speed of transition in facial movement in this data as opposed to acted emotional data (Sneddon and McRorie, 2006).

In the second technique, a more controlled environment was used where certain kinds of 'ground truth' could be established. It is called the Spaghetti method, because participants are asked to feel in boxes in which there were unpleasant objects (including spaghetti) and buzzers that went off as they felt around. They recorded what they felt emotionally during the activity. Figure 5 shows a typical data. The first frame shows the subject in the build up to the climax where a buzzer sounds when the subject locates the object in the box. The second shows the subject at the moment when the buzzer sounds. The data is of good quality both auditorily and visually, although there is very little actual speech – the sound track consists mainly of exclamations. In the clip from which the frames below are taken, the only words uttered are 'Oh Jesus' at the moment when the buzzer goes off.

Figures 6 and 7 represent a more recent move by the HUMAINE team towards experimentation with induced data that shows interaction between two subjects. This is particularly relevant to emotional synthesis. Figure 6 shows data from the Green Persuasive Data Set (see this chapter and the chapter "Issues in Data Collection") where complex emotions are linked to varied cognitive states and interpersonal signals. In the Green Persuasive Data Set, one person tries to persuade another on a topic with multiple emotional overtones (adopting a 'green' lifestyle). Figure 7 shows data from the EmoTABOO Data Set (Zara et al., 2007) where the emphasis is on generating gesture and where subjects interact mainly through gesture and body movement to explain and understand a taboo or an unusual word

Fig. 5 Belfast Spaghetti Data (building up to climax, *left*, and in response to buzzer, *right*)

Fig. 6 Persuader and persuadee's response in Green Persuasive Data Set (clips Ex2A and PT2a)

Fig. 7 Co-occurring frames from EmoTABOO, explainer on *left* and receiver on *right*

known only to one of the subjects. The scenario produces a lot of amusement and embarrassment.

Figure 6 shows the persuader on the left and the person he is trying to persuade on the right. The frame on the right is taken from immediately after the persuader's attempt to persuade the subject that cars are not needed in an environment-friendly 'green' world. She clearly disagrees. The data produces a lot of persuasive speech on the part of the persuader, less speech on the part of the persuadee but interesting facial responses. The text is given in Table 3. The quality both auditorily and visually is good.

Figure 7 shows an interaction between the subject who is trying to explain a 'secret' word/concept (on the left) to the person on the right who does not know the secret word. The two frames are from exactly the same moment. The data is very rich in gesture and is of good quality. There is some speech but the emphasis is on gesture.

The final picture in this section comes from the DRIVAWORK corpus (Honig, 2007). This uses a simulated driving task and subjects are recorded relaxing, driving normally or driving under an additional task (mental arithmetic). The technique aims to elicit physiological data in the different states. Figure 8 shows a subject relaxed (left), driving (middle) and under task (right). The task is context specific. The speech takes the form of answers to mental arithmetic problems. Speech and face are very clear.

Table 3 Exchanges between persuader and persuadee in clips Ex2A and PT2a

Persuader	Persuadee
Ehm, so I, I, I would tend to think, you know using public transport, sensibly, is.. is probably better than going to extremes	[Yea]
Extremes don't convince people	[Yea (pause) Er I do think anything like the new car is um really, like, electricity run, or um like um like]
Hydrogen?	[Yea Hydrogen]
Well, uh, ok, there, th.., there are three groups	[Mm Hm]
There's electric, uh, alcohol and hydrogen. Um, now, the electrical ones are basically a con because the electricity is being generated anyway, by and large being generated by coal, gas or other power stations	[Yea]
So it's, they're actually not very efficient, I mean it's uh, it's it's like one of these things where you know a child sort of, uh pretends they haven't eaten the sweets	[Laughs]
Just not, it's not really . . . ah. The, ehm, the pollutants they get rid of are the ones that you see. That's the the uh, the ah, the smoggy stuff	[Yea]
Ah, and sure that's not nice either but that's not the stuff that's killing the planet	[No]
So..	

The Belfast Driving Simulator Data uses specially developed induction techniques to record subjects driving in a range of emotional states. The procedure consists of inducing subjects into a range of emotional states and then getting them to drive a variety of 'routes' designed to expose possible effects of emotion. Induction involves novel techniques designed to induce emotions robust enough to

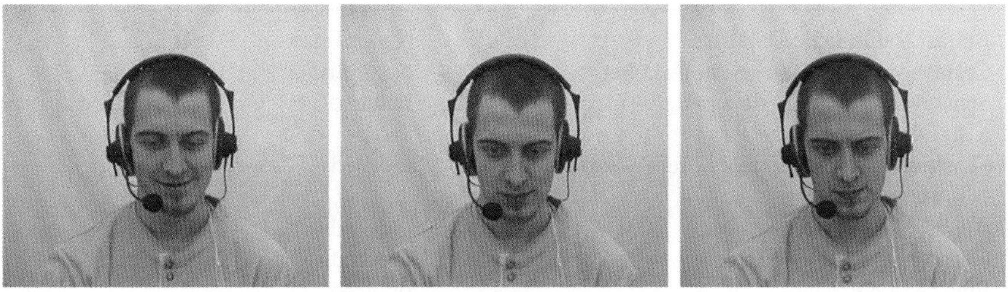

Fig. 8 Subject relaxed (*left*), driving normally (*middle*) and under task (*right*) taken from DRIVAWORK corpus

last through driving sessions lasting tens of minutes. Standard techniques are used to establish a basic mood, which is reinforced by discussions of topics that the participants have preidentified as emotive for them. The primary data is a record of the actions taken in the course of a driving session, coupled with physiological measures (ECG, GSR, skin temperature, breathing). It is supplemented by periodic self-ratings of emotional state. The data has not been video recorded. It is currently the topic of a Ph.D. and cannot be released until after completion of the Ph.D., but pilot work from the Ph.D. can be found at http://emotion-research.net/ws/wp5/edelle.ppt

3 The Labelled Subset

3.1 Aims of the Labelled Subset

The labelled subset is a balanced and labelled collection of extracts selected from the primary records. Each extract is selected to contain a relatively self-contained emotional episode and is referred to as a 'clips'. The subset was chosen to represent the range and form of emotional life that people working in the field should be aware of and was labelled for both emotion and signs of emotion. Both the selection and the design of the labelling scheme were based on systematic criteria (see below) derived in part from the psychological literature on emotion and in part from experience with real data. The chapter "Issues in Data Labelling" in Part III gives further information on the principles and models which underpin the labelling scheme.

3.2 Size and Structure of the Labelled Subset

The labelled subset consists of 48 clips (between 3 s and 2 min in length) selected from the primary recordings. The labelled episodes are mounted on the ANVIL platform. Table 4 provides a summary of the numbers of clips selected from the

Table 4 Selection of clips for labelled subset

Raw data records	Data type	Number of clips selected
Belfast Naturalistic Database	Naturalistic	10
Castaway Reality Television Data Set	Naturalistic	10
Sensitive Artificial Listener (Belfast recordings in English)	Induced	12
Sensitive Artificial Listener (Tel Aviv recordings in Hebrew)	Induced	1
Activity Data/Spaghetti Data	Induced	7
Green Persuasive Data Set	Induced	4
EmoTABOO	Induced	2
DRIVAWORK (driving under varying workload) Corpus in German	Induced	1
GEMEP Corpus	Acted	1

range of data types to make up the labelled subset. Appendix 3 describes in detail each clip selected for labelling, including a summary of the content of the clip and descriptors of the emotion, context and modalities.

The selection of the final sample involves non-trivial issues. Two levels were used.

The first was the selection of clips from within a whole recording. In the case of relatively intense emotional episodes, the extraction of a section/clip includes build-up to and movement away from an emotional nucleus/explosion – lead in and coda are part of identification of the state. In the case of less emotionally intense recordings, the basic criterion used to set the boundaries of clips is that 'the emotional ratings based on the clip alone should be as good as ratings based on the maximum recording available' (i.e. editing should not exclude information that is relevant to identifying the state involved).

The second stage of selection was deciding which clips should form the labelled subset of the HUMAINE Database. It is very easy to drift into using a single type of material which conceals how diverse emotion actually is. To counter that, 48 clips were deliberately selected to cover material showing emotion in action and interaction; in different contexts (static, dynamic, indoor, outdoor, monologue and dialogue); spanning a broad emotional space (positive and negative, active and passive) and all the major types of combination of emotion (consistent emotion, co-existent emotion, emotional transition over time); with a range of intensities; showing cues from gesture, face, voice, movement, action and words and representing different genders and cultures.

Table 5 shows the framework which underpinned the final selection. Each clip was chosen for its representation in those main classes.

The details of each of the 48 clips are given in Appendix 3. For each clip there is a short description of what is happening in the clip and the gender of the speaker as well as a description of quadrant, emotion mix, modality and context (constraints, goal and setting) as set out in Table 5. The clips are all available under Conditions of Use Agreement (Appendix 1).

The grid of distinctions underpinning the selection of clips is in some senses what one might expect – especially attention to emotional range, context and modality. But the history of emotional databases shows that this is not the norm. Very often

Table 5 Grid of distinctions underpinning selection of clips in labelled subset

Quadrant		Positive active, pragmatic, negative passive, negative active
Emotion mix	Within a clip	Consistent, co-existent (with another emotion) Shift (from one emotion to another)
Gender		Male/female
Modality		Face, speech, gesture
Context labels	Constraints	Un/restrained, Un/constrained
	Goal	Inform, create rapport, pure expression of emotion
	Setting	Task, activity, interactive, group, interactive-passive, other

databases are application focused and so they only have one type of data or where they are more open ended, they often do not prespecify criteria for selection from the raw records. This can lead to a slanted representation of the data. There are also some aspects of the grid that have not traditionally formed part of emotion databases. Specifying context is often not done, although the impact of context on the representation of an emotion can be large. And selection which takes into account the consistency of emotion and whether it is pure or co-existent with another emotion is a departure which is based on experience with real data. Experience with the Belfast Naturalistic Database and the EmoTV Database suggests that emotion mixing and emotion shifting are common (Devillers et al., 2006). Hence the grid is designed to capture clips which represent this feature.

3.3 Labelling of the Subset

A wide range of emotion labels and signs of emotion descriptors are attached to each clip. These are also available together with labelling manuals on the portal at www.emotion-research.net. The labelled data can be displayed on the ANVIL platform and the procedure for obtaining and using the software is explained on the portal. The emotion labelling has been done by six raters for all 48 clips and the data for all six labellers is available via the portal. The speech and language labelling has been done by one trained phonetician and is available for all 48 clips. Two clips are labelled for face and gesture; the labelling is available on the portal.

The components for the labelling scheme and the principles behind it are described in full in the chapter "Issues in Data Labelling". This section summarises how they have been put together and used in the labelled subset of the HUMAINE database.

3.4 Emotion Labels

Two levels of description are included.

At the first level, global labels are applied to an emotion episode or clip as a whole. Factors that do not vary rapidly (the person concerned, the context) are described here. This provides an index that can be used to identify clips that a particular user might want to consider. For instance, it will allow a user to find examples of the way anger is expressed in relatively formal interactions (which will not be the same as the way it is expressed on the football terraces).

Labelling at the second level is time aligned. This is done using 'trace'-type programs (see the chapter "Issues in Data Labelling"). Each of the programs deals with a single aspect of emotion (e.g. its valence, its intensity, its genuineness). An observer traces his/her impression of that aspect continuously on a one-dimensional scale while he or she watches the clip being rated. The data from these programs is imported into ANVIL as a series of continuous time-aligned traces. The trace-type labelling captures perceived flow of emotion.

Tables 6 and 7 summarise the global and continuous 'trace' labels that are applied to each clip.

3.5 Sign Labels

Labels for signs of emotion are also attached to each clip in the labelled subset. There are labels for speech and language applied to all clips and labels for gesture and face applied to two of the clips. Table 8 summarises the descriptors used.

3.6 The Labelled Subset: Illustrations and Issues

The labelled subset is a powerful demonstration of the range and diversity of emotional life and how we can begin to describe it. This section works through a number of examples to illustrate some of the diversity and complexity of data in the subset and to show how the labelling can capture what is going on in quite complex emotional episodes.

Example 1. For the first example we return to the Spaghetti Data and to the woman already featured in Fig. 5. Table 9 shows how one rater described this clip in global emotional terms. Figure 9 shows the form of the display of emotional Trace continuous labelling for the same clip by the same rater. In Fig. 10 we see the whole of the clip labelled using the Trace programs. The screenshot is taken from ANVIL and it shows the traces from one rater for this clip for intensity of emotion, acting, masking, activation and power/powerlessness. The screenshot conveys the net effect of putting traces together. The clip shows a participant feeling in a box and suddenly triggering a buzzer. She gives a gasp, then a linguistic exclamation. The top trace, emotional intensity, rises abruptly after the gasp. The rater does not judge that the response is acted, but there is a degree of masking at the beginning which breaks down abruptly at the unexpected event. Activation rises abruptly after a delay (during which the participant might be described as frozen).

Example 2. The second example illustrates how emotion can look in different contexts and time domains. All six raters attached the word 'fear' to the clips from which the examples below are taken (Fig. 10). The first clip shows a subject we have seen before watching a friend fall off a mountain bike in the Belfast Activity Data. The second shows a subject undertaking the Spaghetti task at the point at which she stopped the task and said she was too frightened to continue. The third clip shows a subject recalling touching snakes in a darkened hut a few minutes after the event. The fourth shows a subject recalling a terrifying incident a year after it happened. These clips suggest that a sensible database needs to show the variety of things that are called 'fear'. The sample in the HUMAINE database is by no means complete, but it is a useful pointer to the variety of representations behind an emotion word.

Table 6 Global emotion descriptors applied to HUMAINE database (open comment is also invited for each class)

Classes of global emotion label	Description
Emotion words	Choice of up to 6 from list of 48
Emotion-related states	Choice from one or more of the following: attitudes, established emotion, emergent emotion (full blown), emergent emotion (suppressed), moods, partial emotion (topic shifting), partial emotion (simmering), stance towards a person, stance towards an object or a situation, interpersonal bonds, altered state of arousal, altered state of control, altered state of seriousness
Combination types	Choice from one or more of the following: unmixed emotion, simultaneous combination, sequential combination
Authenticity	Choice from a six-point scale, describing (i) degree of acting, (ii) degree of masking
Core affect dimensions	Choice from a six-point scale describing perceived intensity, activation and valence (0–6 for intensity and activation; –3 to +3 for valence)
Basic factual information	Choice from selection of labels describing age, gender and nationality of the person observed, the quality of the recording, the degree to which the person observed is physically constrained and the type of audience (if appropriate) present in the clip (close, colleague, public, artificial)
Context	Choice of labels describing the context of the communication taking place. These include the purpose of the communication (to inform/persuade/create rapport/destroy rapport/pure expression), social setting (none, passive other, interactant, group) and social pressure (to formality, weak, expressiveness). The rater is also asked to judge the degree of camera and microphone awareness
Key events/emotional focus	Up to three descriptions of the emotional focus of the clip. This includes identifying the emotional focus of the clip and selecting a qualifying temporal label – current, recalled, anticipated or imagined
Other key events	A description of other key events which may be relevant to creating the emotions present in the clip but are not themselves the emotional focus of the clip. The key event is further qualified by identifying whether it is a trigger (short term), cause (long term) or aspiration (future)
Appraisal categories	A rating of how strongly the observed person's emotional state is related to aspects of the way he/she sees the emotionally significant events or people around. The factors are goal conduciveness and goal obstructiveness; power/powerlessness; expectedness

Table 7 'Trace' label descriptors

Trace label	Description
IntensTrace	A rating of the intensity of the emotion that the specified person is perceived to be experiencing from moment to moment. The range of intensity runs from the person experiencing no emotion whatsoever, i.e. as emotionally still as they could be, to their emotion being perceived as being as intense as it could possibly be
ActTrace	A rating of the extent to which the specified person is trying to give an impression of emotions that they actually do not feel (i.e. they are pretending or acting). The range runs from no attempt to simulate unfelt emotions, i.e., they would be perceived as being totally genuine in their emotional expression, to where their emotion would be perceived as completely acted or false
MaskTrace	A rating of the extent to which the specified person is trying to avoid showing emotions that they do actually feel (i.e. the extent to which they are trying to cover up their genuine emotion). The range runs from no attempt to avoid showing their emotions, i.e. they would be perceived as being totally open in their emotional expression, to where their emotion would be perceived as completely masked or covered
ActivTrace	A rating of activation or arousal, i.e. how strongly the relevant person is inclined to take action. The range runs from 'absolutely no inclination to be active' through markers 'weakly active', strongly active, to 'compelling urge to be active' at the maximum of the scale
ValenceTrace	A rating of how positive or negative the specified person feels about the events or people at the focus of his or her emotional state. The range runs from very strongly negative to very strongly positive. At the mid-range, they would be perceived as being neutral or experiencing no emotion
PowerTrace	A rating of how in control the specified person feels about the events or people at the focus of his or her emotional state. The range runs from 'absolutely no control over events', through 'not quite in control', 'just about in control' to being completely in control over events at the maximum of the scale
Anticipate/ ExpectTrace	A rating of the extent to which the specified person has been taken unawares by the events at the focus of their emotional state. The range runs from events being anticipated completely to being taken completely unawares by the event
WordTrace	A rating of the intensity of the four highest ranked emotion words for that clip The range runs from absolutely none of that emotion being present to that emotion being expressed as purely as it could be

Table 8 Signs of emotion descriptors

Speech and language descriptors	Gesture descriptors	Face descriptors
Transliteration (*words spoken*)	Gesture units	FAPS (automatically derived)
Largely automatically derived labels *(time waveform, pitch)*	Phases	
Auditory-based labels (*paralanguage, voice quality, timing, volume*)	Phrases/categories Lemmas adapted from gesture lexicon	

Example 3. The third example focuses on portraying the complexity of emotions that can occur in naturalistic data, particularly the mixed nature of emotions (referred to as 'co-existing' in the HUMAINE coding scheme) and the way in which emotion fluctuates and shifts within short time periods. By comparison, acted data tends to portray emotion as consistent, pure and static over a period of time. Figure 11 shows a sequence of frames from the Belfast Naturalistic Database taken at intervals from a 10-s period in which the subject utters the words 'It's a boy. And the anger drained out of me that night. I felt it going. It was like a release.'

The context for the sequence of frames is that the subject is remembering the birth of her grandson (her daughter's child). She and her daughter had not got on very well together and she had been particularly angry at her daughter for getting pregnant. She has been describing the anger she felt but then moves to describe the moment of her grandson's birth and the release from the anger she had been feeling when the moment she heard her grandson had been born. Figure 11 shows the way in which the emotion shifts and blends as she recalls the incident. The sequence is typical of the type of data that comes from naturalistic settings. Work on EmoTV (Devillers et al., 2006), which unfortunately cannot be released for copyright reasons, makes similar points. One of the interesting things about this particular example is that the emotion expressed comes from recalling events, illustrating that recall can produce fairly intense emotion.

Example 4. This is a nice illustration of the need to consider a wide range of emotion-related states when classifying emotional behaviour. The opening chapter of this handbook discusses the theory behind these, and Table 6 lists those that are used in labelling the HUMAINE database. Figure 12 illustrates one of these states which is less commonly talked about – 'suppressed' emotion. The subject is shown talking to the angry personality of the Sensitive Artificial Listener in the SAL data (see above). What is happening is that the angry personality of SAL is trying to wind up the subject's emotions into an angry state. All the raters of this clip attach the label 'suppressed emotion' to it. The words that they also all agree apply to the clip are politeness, tension, irritation, annoyance and anger. These are in line with the global label of 'suppressed emotion', indicating that the subject may have negative emotions but that he remains polite, keeping his emotions under control through a deliberate effort. The text (see under Fig. 12) indicates suppression of the emotions.

Table 9 Global emotion labels for Spaghetti Data Clip 14e (from one rater)

Emotion words (in order of relevance)	Anxiety 1 Shock 2 Surprise 3 Relieved 4
Emotion-related states	Stance towards object/situation Altered state of control Emergent emotion (full blown) Altered state of arousal
Combination types	Sequential combination Simultaneous combination (rater's comment: anxiety first, then a combination of shock and surprise, then relief)
Authenticity	Masking strong Acting moderate
Core affect dimensions	Intensity 5 Activation 5 Valence –1
Basic factual labels	Video quality: good Hands constrained: no Posture constrained: no Nationality: N Irish Recording style: observation, task Acoustic quality: good Physically unrestricted: yes Gender: F Age: 18–25 Type of audience: colleagues
Context	Social pressure: weak Camera and microphone awareness: yes Passive other Pure expression of emotion
Key events	What the left hand encounters in the box Here and now
Other key events	–
Appraisals	Goal obstructiveness: open (i.e. no judgement) Goal conduciveness: open (i.e. no judgement) Expectedness: unanticipated Power/powerlessness: very powerless

Examples 5 and 6 illustrate the labelling of signs of emotion and the richness of signs in the data. Figure 13 shows a frame from EmoTABOO with the array of gesture labels and FAPs attached. Figure 14 shows a frame from Castaway Reality TV. The episode is particularly rich in paralinguistic expression of emotion. The subject is asked what he misses most. He replies: ' I always get choked up ...... family' accompanied by long pausing, tremulous voice and nervous laughter.

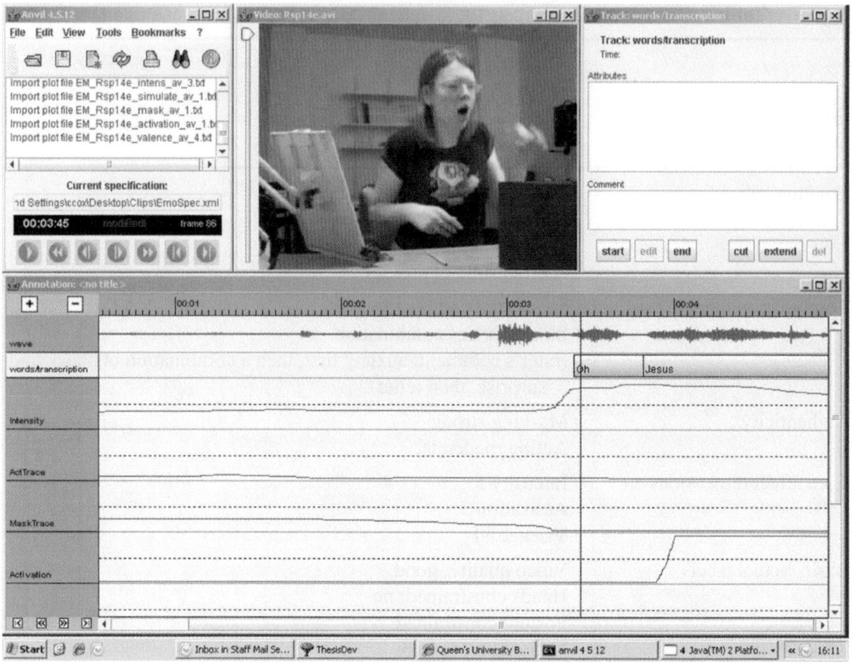

Fig. 9 Continuous trace labelling for intensity, acting, masking and level of activation for Clip 14e by one rater (*red vertical bar* marks frame shown)

Fig. 10 Four representations of fear

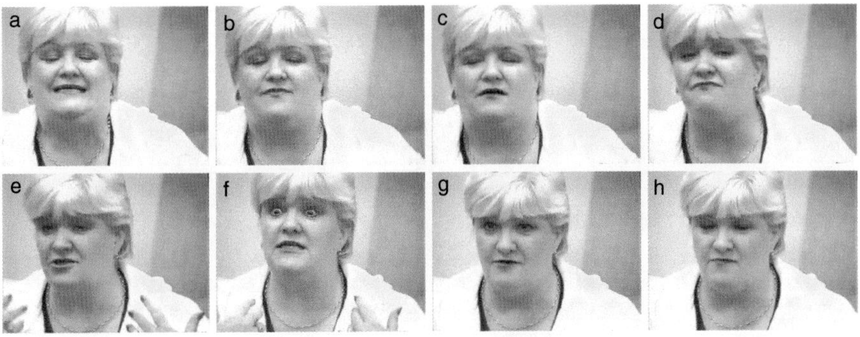

Fig. 11 Clip 56d Belfast Naturalistic Database. Frame (a) starts with the memory of the announcement by the daughter's birth partner that the baby had been born and that it was a boy. The expression certainly seems to contain happiness. In frames (b), (c) and (d) she recalls the anger she had felt but at the same time recalls her move away from the anger: the frames are clearly a mix of complex feelings (anger, pain, sadness, escape) and emotional shift. Frames (e) and (f) describe release from the anger and the final two frames might best be described as a return to peace

Fig. 12 An example of suppressed emotion from SAL Data as subject talks to the angry personality Spike. (Text: Like you're not really annoying me and I don't appreciate your attitude)

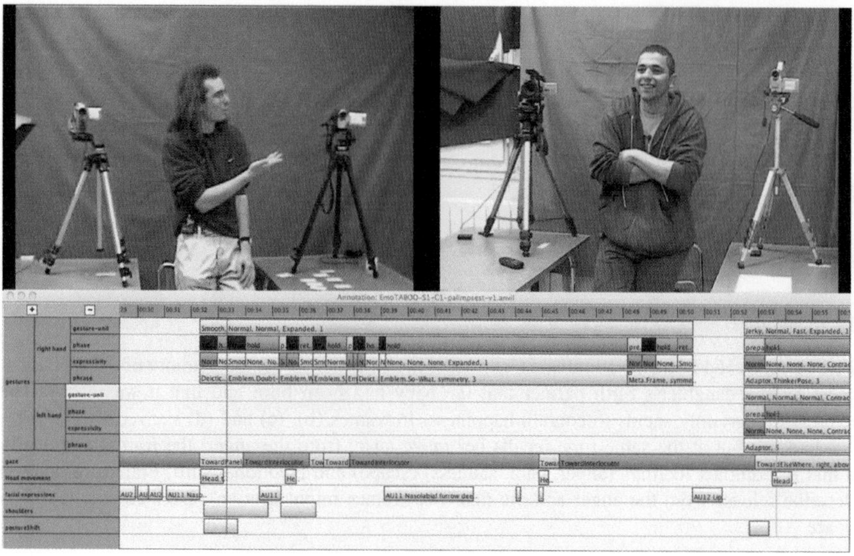

Fig. 13 Gesture labelling and FAPS applied to EmoTABOO Data

Fig. 14 Paralinguistic expression of emotion in Castaway Reality TV Data Set

Example 7. The final figure, Fig. 15, shows some of the interesting and unexpected interactions between modalities. The subject in it appears to be smiling but the words indicate that she is actually in a state of shock. The subject (from the Castaway Reality TV Data Set) is recalling her encounter with snakes in the hut from which she has just emerged. She is saying 'and the first thing I touched was the snake's head ... feel really shaky now.' The bottom line in the figure is a trace from WordTrace by one rater. The word that the rater thought applied most to the episode was 'fear' and the trace is a trace of fear in the episode. The trace shows that fear is present and strong throughout the episode. There are many examples in the HUMAINE database where the expression on the face seems at odds with the emotion experienced.

4 Future Directions

The point of a database is to facilitate research, and the HUMAINE Database opens up a very large number of avenues for exploration.

At the most routine, the database provides evidence on the way a range of tools function and therefore provides a basis for evaluating them. The tools include everyday emotion categories, trace programs, descriptors for broad types of emotion and context. Evaluations include simple, formal procedures, such as tests of reliability. However, they also include others which are less clear-cut, but not less important: does the battery of descriptions tell us what we need to know, and if not, why not?

Related, but distinct, are questions about reduction. There are two obvious forms of question to consider. The first form is related to the concept of cover classes. It is concerned with establishing which labels can be merged without unacceptable loss of information. The second form is related to the concept of dimensions. There is a very large literature on the number of dimensions needed to represent a set of words. The HUMAINE Database opens up the possibility of asking how many dimensions are necessary to represent a set of samples of emotionally coloured behaviour.

These questions are not statistically trivial. For example, they should ideally take account of the way labellings evolve over time. A simplification which seems fair in terms of a series of 'snapshots' may be a serious problem if it undercuts the ability to predict what will happen next. Standard statistical reduction techniques do not address that kind of problem.

The end target of that kind of work is an empirically validated set of labels. It is frustrating, but there is no way to reach that stage without generating labellings some of whose components will eventually be discarded. Hence, it is to be expected that some components of the HUMAINE scheme will be discarded in the process of analysis. Conversely, new components will presumably need to be added. Iterative adjustment is to be expected, but it needs a core to work round.

Benchmarking is another key application. It is a major problem that the area lacks standard tasks against which the performance of new algorithms can be tested. The database offers two kinds of benchmark – clips to be analysed and types of information to be recovered. The test is not confined to machine recognition. For instance, the material in the database offers a very interesting test for brain-scanning technologies. Capturing differences between responses to HUMAINE Database clips is a much more acute test than is capturing differences between responses to photographs from the standard Ekman collection.

Fig. 15 Conflicting signs from face and speech, Castaway Reality TV Data Set

The records lend themselves to a range of studies. The most routine is simply extending the labelled set. There is a very large body of primary records that remains unlabelled, and its value would be multiplied if labelled versions were available to the community.

A wide range of issues call out for more specific studies. Three will be singled out here. The first is relationships between modalities. It has been pointed out that impressionistically, audio and visual signals sometimes seem to point in very different directions. The data provides opportunities to explore that issue much more systematically. A natural starting point is simply to label a substantial body of material on the basis of audio records only, visual records only and verbal transcripts only. The second is temporal evolution of expressions. Examples like Fig. 10 make it clear that there are rapid, radical changes in moment-by-moment expression of emotion. Some theoretical frameworks predict that change should occur on that kind of timescale (Scherer and Ellgring, 2007). The HUMAINE Database contains a reservoir of naturalistic data that makes it possible to explore these ideas. The third is whether information is localised or distributed. It is not obvious whether information about a person's emotion is, so to speak, smeared evenly over time or concentrated in a few revealing moments. The material in the database invites research on the topic.

Beyond these, the database provides a kernel of primary records that help to clarify which kinds of extension make sense. It certainly does not make sense to collect new records at random. A considerable range of states and contexts are probably quite well covered in the data that now exists, and random inventions are quite likely to produce nothing but more stilted variations on the same theme. However, there are areas where information is quite clearly limited, and those are the areas where it makes sense to concentrate effort. An overwhelmingly obvious example is cross-cultural difference. A few HUMAINE techniques have been applied in substantially different cultures. The process needs to be extended radically to provide anything approaching a reasonable representation of the way culture affects the expression of emotion.

Perhaps most fundamental of all, the database invites a cumulative and collaborative attitude. The HUMAINE Database is a product of collaborations between several teams over a period of years. It will take a much larger scale of collaboration to accumulate a reservoir of data sufficient to understand the various ways in which various kinds and combinations of emotion can colour various actions and interactions.

Appendix 1: User Agreement – HUMAINE Database

This user agreement applies to the database *HUMAINE database* provided on the enclosed CDs. It has been released by QUB (Queens University Belfast) under specific conditions for sole scientific, non-commercial use.

Conditions of Release

- Data set provided without guarantee.
- No legal claims of any kind can be derived from accepting and using the data set.
- QUB is not liable for any damage resulting from receiving, installing or using the data set or other files provided by QUB in this context.
- Expressed written consent must be sought from QUB before the data set or any other files provided by QUB containing information derived from it (e.g. labelling files) are passed on by the licensee to any third party.
- If a partner/user concentrates on labelling, s/he agrees to share with all the other HUMAINE partners additional analyses, especially additional annotations. In the first instance, the analyses/annotations should be returned to the HUMAINE team at QUB.
- Any models derived using data from the data set may be used only for scientific, non-commercial applications.
- The licensee will let QUB know of any results without undue delay. Joint publications (between the licensee and QUB) should be aimed at.

For publications concerning direct or indirect use of the corpus, the licensee must cite QUB with a citation provided by QUB.

I have read and understood the user agreement and will comply with it.

Signed______________________________

Please print name________________________

Appendix 2

Data source and reference	Nature of material	Emotional content	Technical info and availability
Belfast Naturalistic Database (Douglas-Cowie et al., 2003)	The material consists of audiovisual sedentary interactions from TV chat shows and religious programs, and discussions between old acquaintances	A range of positive and negative emotions. Intensity is mostly moderate	There are 125 subjects (two sequences of 10–60 s each, one neutral and one emotional); each sequence has attached to it three emotion words which best describe it attributed by three listeners from a list of some 50 possible words; a selection of 30 sequences with ethical and copyright clearance is available
EmoTV Database (in French) (Devillers et al., 2006)	The EmoTV database consists of audiovisual interactions from TV interviews – both sedentary interactions and interviews 'on the street' (with wide range of body postures)	A range of positive and negative emotions. Intensity moderate with some intense material	48 subjects (51 sequences of 4–43 s per subject in emotional state); copyright restrictions prevent straight release, but there may be circumstances under which some data can be shared
Castaway Reality Television Database (Douglas-Cowie et al., 2007)	This consists of audiovisual recordings of a group of 10 taking part competitively in a range of testing activities (feeling snakes, lighting outdoor fires) on a remote island. The recordings include single and collective recordings and post-activity interviews and diary-type extracts	A range of positive and negative emotions. Intensity moderate with some intense material	10 tapes of 30 min each; copyright clearance

(continued)

Data source and reference	Nature of material	Emotional content	Technical info and availability
Sensitive Artificial Listener (Douglas-Cowie et al., 2007)	The SAL data consists of audiovisual recordings of human–computer conversations elicited through a 'Sensitive Artificial Listener' interface designed to let users work through a range of emotional states (like an emotional gym). The interface is built around four personalities – Poppy (who is happy), Obadiah (who is gloomy), Spike (who is angry) and Prudence (who is pragmatic). The user chooses which he/she wants to talk to. Each has a set of stock responses which match the particular personality. The idea is that Poppy/Spike/Obadiah/Prudence draws the user into their own emotional state	A wide range of emotions but they are not very intense	Data has been collected for four users with around 20 min of speech each. SAL has also been translated into Hebrew (at Tel Aviv University) and Greek (at National Technical University of Athens, ICCS) and adjusted to suit cultural norms and expectations, and some initial data has been collected. The data has ethical permission and is available to the research community

(continued)

Data source and reference	Nature of material	Emotional content	Technical info and availability
Activity Data/Spaghetti Data (Douglas-Cowie et al., 2007)	Audiovisual recordings of emotion in action were collected using two induction techniques developed in Belfast. In the first, volunteers were recorded engaging in outdoor activities (e.g. mountain bike racing). The second used a more controlled environment where certain kinds of 'ground truth' could be established. It is called the Spaghetti method, because participants are asked to feel in boxes in which there were objects including spaghetti and buzzers that went off as they felt around. They recorded what they felt emotionally during the activity	Method 1 elicited both positive and negative emotions with a high level of activation. Method 2 elicited a range of brief, relatively intense emotions – surprise, anticipation, curiosity, shock, fear, disgust	Method 1 produced 'provocative' data which was very fast moving and had a noisy sound track. Method 2 produced data where the participants were reasonably static and stayed within fixed camera range, making it easier to deal with face detection. The audio output consists mainly of exclamations. There are now recordings of some 60 subjects. The data has ethical permission and is available to the research community

(continued)

Data source and reference	Nature of material	Emotional content	Technical info and availability
Belfast Driving Simulator Data (Douglas-Cowie et al., 2007) http://emotion-research.net/ws/wp5/edelle.ppt	The driving simulator procedure consists of inducing subjects into a range of emotional states and then getting them to drive a variety of 'routes' designed to expose possible effects of emotion. Induction involves novel techniques designed to induce emotions robust enough to last through driving sessions lasting tens of minutes. Standard techniques are used to establish a basic mood, which is reinforced by discussions of topics that the participants have preidentified as emotive for them. The primary data is a record of the actions taken in the course of a driving session, coupled with physiological measures (ECG, GSR, skin temperature, breathing). It is supplemented by periodic self-ratings of emotional state	Three emotion-related conditions, neutral, angry, and elated	30 participants; will be available pending completion of Ph.D. on the data

(continued)

Data source and reference	Nature of material	Emotional content	Technical info and availability
EmoTABOO (in French) (Zara et al., 2007)	EmoTABOO records multimodal interactions between two people during a game called Taboo. One person has to explain to the other using gestures and body movement, a 'taboo' concept or word	Range of emotions including embarrassment, amusement	By arrangement with the LIMSI team
Green Persuasive Data Set	The data set consists of audiovisual recordings of interactions where one person tries to persuade another on a topic with multiple emotional overtones (adopting a 'green' lifestyle)	Complex emotions linked to varied cognitive states and interpersonal signals	Eight interactions of about 30 min each, and associated traces made by the interviewees to indicate how persuaded they felt from moment to moment. The data has ethical permission and is available to the research community
DRIVAWORK (driving under varying workload) corpus (Honig, 2007)	The DRIVAWORK corpus has been collected at Erlangen, using a simulated driving task. There are three types of episode: participants are recorded relaxing, driving normally or driving with an additional task (mental arithmetic). Recordings are video and physiological (ECG, GSR, skin temperature, breathing, EMG and BVP)	Stress-related states rather than emotion per se	Availability by arrangement with Erlangen team

Appendix 3

Clip/source	Content	Gender	Emotion words	Quadrant	Combination type	Modalities present (*=labelled)	Context	Goal	Constraints
APal/Emo Taboo		M/M	Amusement, empathy, frustration, friendliness, embarrass-ment, powerlessness			Face*, speech*, gesture*	Interactive task	Inform	Unrestricted, unconstrained
Cpal/Emo Taboo		M/F	Anxiety, amusement, embarrass-ment, friendliness interest, politeness			Face*, speech*, gesture*	Interactive task	Inform	Unrestricted, unconstrained
2_3/Castaway Reality TV Data Set	Subject describes how he misses his family and how much he has learnt about himself from experiences on island	M	Helplessness, love, power-lessness, satisfaction, stress, worry	NegPass	Consistent emotion	Face, speech*	Passive other	Inform	Unrestricted, constrained

(continued)

Clip/source	Content	Gender	Emotion words	Quadrant	Combination type	Modalities present (*=labelled)	Context	Goal	Constraints
4_2/Castaway Reality TV Data Set	Subject describes his fear after the insect task and says he does not want to continue	M	Stress, anxiety, doubt, shame, relieved, fear	NegPass	Consistent emotion	Face, speech*	Passive other	Inform	Unrestricted, constrained
10_1/Castaway Reality TV Data Set	Subject, who has decided to leave, says goodbye to fellow castaways and makes plans to meet up in the future	M	Friendliness, relieved, helplessness, affection, politeness, powerlessness	Pos Act	Consistent emotion	Face, speech*, gesture	Group	Rapport	Unrestricted, unconstrained
10_3/Castaway Reality TV Data Set	Group leader tells group they can have a shower to which they respond by cheering, laughing, smiling	Group M/F	Relaxed, interest, calm, happiness, politeness, tension	Pos Act	Shifting emotion	Face, gesture	Group	Inform	Unrestricted, unconstrained
R56b/Belfast Naturalistic Database	Subject describes her love for her son-in-law	F	Love, satisfaction, affection, pride, friendliness, happiness	Pos Pass	Consistent emotion	Face, speech*	Passive other	Inform	Unrestricted, constrained

(continued)

Clip/source	Content	Gender	Emotion words	Quadrant	Combination type	Modalities present (*=labelled)	Context	Goal	Constraints
R56d/Belfast Naturalistic Database	Subject describes the happiness and excitement at the birth of her grandchild	F	Happiness, joy, relieved, affection, politeness, satisfaction	Pos Act	Co-existing emotion	Face, speech*	Passive other interactive	Inform	Unrestricted, constrained
1_4/Castaway Reality TV Data Set	Subject describes his experiences, e.g. an argument with a fellow contestant and his lowest point of the experience	M	Annoyance, amusement, relaxed, satisfaction, friendliness, frustration	Pos Act	Co-existing + shift	Face, speech*	Interactive	Inform/ activity	Unrestricted, unconstrained
3_1/Castaway Reality TV Data Set	Subject comes out of house in which she touched insects and snakes, and speaks of her fear and disgust	F	Relieved, anxiety, fear, disgust, amusement, powerlessness	NegAct – P	Co-existing emotion	Face, speech*, gesture	Interactive	Inform/ activity	Unrestricted, unconstrained

(continued)

Clip/source	Content	Gender	Emotion words	Quadrant	Combination type	Modalities present (*=labelled)	Context	Goal	Constraints
5_2/Castaway Reality TV Data Set	Subject talks about his feeling of surprise when the other contestants voted for him and how he has changed and developed during the experience	M	Satisfaction, pride, happiness, tension, content, amusement	NegPass	Consistent emotion	Face, speech*, gesture	Interactive	Inform/ activity	Unrestricted, uncon-strained
5_3/Castaway Reality TV Data Set	Subject tells her theory that seven contestants have already been chosen by the organisers to leave, despite the contestants having been told that they could choose themselves	F	Amusement, politeness, powerless-ness, contempt, annoyance, interest	Prag+P Act	Consistent emotion	Face, speech*	Passive other interactive	Inform	Unrestricted, uncon-strained
5_3/Castaway Reality TV Data Set	Group discusses various things – focus is on subject's facial expression	M	Boredom, irritation, calm, politeness, worry, amusement	NegPass	Consistent emotion	Face, gesture	Passive other	Pure expres-sion	Unrestricted, uncon-strained

(continued)

Clip/source	Content	Gender	Emotion words	Quadrant	Combination type	Modalities present (*=labelled)	Context	Goal	Constraints
056e/Belfast Naturalistic Database	Subject speaks of being attacked on leaving her house and describes fear at thinking husband had been shot when gun was fired	F	Fear, calm, anxiety, tension, powerlessness sadness	NegPass	Consistent emotion	Face, speech*	Interactive	Inform	Unrestricted, uncon-strained
077a/Belfast Naturalistic Database	Subject describes the beautiful setting of friends' home in which the bedroom overlooks a golf course	F	Interest, pleasure, friendliness, pride, calm, relaxed	Prag	Consistent emotion	Face, speech*, gesture	Interactive	Inform	Unrestricted, uncon-strained
077c/Belfast Naturalistic Database	Subject describes pain and emotional trauma following an accident	F	Sadness, despair, hurt, anxiety, calm, worry	NegPass	Consistent emotion	Face, speech*	Interactive	Inform	Unrestricted, uncon-strained
078a/Belfast Naturalistic Database	Subject speaks of working as a diesel fitter in the company managed by his father	M	Interest, pride, relaxed, friendliness, politeness, annoyance	Prag	Consistent emotion	Face, speech*, gesture	Interactive	Inform	Unrestricted, uncon-strained

(continued)

Clip/source	Content	Gender	Emotion words	Quadrant	Combination type	Modalities present (*=labelled)	Context	Goal	Constraints
079b/Belfast Naturalistic Database	Subject describes her amusement at grandmother thinking her father would have to come home from overseas to help with a medical situation in his own country	F	Amusement, affection, friendliness, happiness, sadness, love	Pos Act	Consistent emotion	Face, speech*, gesture	Passive other interactive	Inform	Unrestricted, uncon-strained
R56g/Belfast Naturalistic Database	Subject describes determination to continue her job, despite being threatened	F	Annoyance, irritation, anger, frustration, pride calm	NegAct	Consistent emotion	Face, speech*	Passive other interactive	Inform	Unrestricted, constrained
R77d/Belfast Naturalistic Database	Subject describes grandson, who has just begun to walk, and how his mother treats him as a younger baby by mashing his food	F	Happiness, love, affection, pride, excitement, joy	Pos Act	Consistent emotion	Face, speech*, gesture	Passive other interactive	Inform	Unrestricted, uncon-strained

(continued)

Clip/source	Content	Gender	Emotion words	Quadrant	Combination type	Modalities present (*=labelled)	Context	Goal	Constraints
R78b/Belfast Naturalistic Database	Subject describes anger at the fact that a politician is preventing his father from voting when his father fought for the country and the politician did not	M	Annoyance, frustration, irritation, contempt, calm, anger	NegAct	Consistent emotion	Face, speech*, gesture	Passive other interactive	Inform	Unrestricted, unconstrained
RCsh/Belfast Outdoor Activity Data	Subject's screaming/laughing response to a bicycle crash	F	Excitement, amusement, pleasure, happiness, elation, shock	PA/NA Mix	Co-existing emotion	Face, gesture	Activity	Pure expression	Unrestricted, unconstrained
REdA1/ Sensitive Artificial Listener Data Set	Subject speaks to character about a fantasy holiday on a desert island	F	Amusement, happiness, pleasure, friendliness, embarrassment, joy	Pos Act	Consistent emotion	Face*, speech*	Interactive	Inform/ Rapport	Restricted, constrained
REllA2 Sensitive Artificial Listener Data Set	Subject speaks to gloomy character Obadiah, who offers her no help or advice	F	Irritation, annoyance, contempt, tension, frustration, satisfaction	PA/NA Mix	Shifting emotion	Face, speech*	Interactive	Inform	Restricted, constrained

(continued)

Clip/source	Content	Gender	Emotion words	Quadrant	Combination type	Modalities present (*=labelled)	Context	Goal	Constraints
REIlB2 Sensitive Artificial Listener Data Set	Subject says she would like to go for a cycle on a summer evening, describing the hawthorn she would smell and how the air would blow around her	F	Pleasure, happiness, delight, interest, politeness, relaxed	Prg/NA Mix	Consistent emotion	Face, speech*	Interactive	Inform	Restricted, constrained
Rsp14e/ Spaghetti	Subject puts hand in box, which she cannot see inside, and is startled by a noise from the box	F	Shock, anxiety, surprise, amusement, fear, relieved	NA/NP Mix	Shifting emotion	Face, speech*, gesture	Passive other	Task	Unrestricted, constrained
8_3/Castaway Reality TV Data Set	Subject tries to light a fire, which she finds difficult, but eventually succeeds	F	Frustration, annoyance, amusement, irritation, satisfaction, tension	Pos Act	Co-existing emotion	Face, speech*, gesture	Task	Activity	Unrestricted, unconstrained
AO4joi/Geneva Acted Emotion Database		M	Excitement, delight, elation, joy, happiness, pleasure	Pos Act	Consistent emotion	Face, gesture	Activity	Pure expression	Unrestricted, unconstrained

(continued)

Clip/source	Content	Gender	Emotion words	Quadrant	Combination type	Modalities present (*=labelled)	Context	Goal	Constraints
A16aT/Sensitive Artificial Listener Data Set	Subject says she finds it preferable to be open-minded, as more opportunities arise	F	Amusement, calm, courage, friendliness, relieved, embarrassment	Prag	Co-existing emotion	Face, speech*	Passive other interactive	Inform	Unrestricted, uncon-strained
Arithmetic/ DRIVAWORK Erlangen Campus		M	Amusement, interest, calm, irritation, frustration, stress	Prag	Shifting emotion	Face	Task	Activity	Unrestricted, constrained
B16aT/ Sensitive Artificial Listener Data Set	Subject claims that she does not get worked up about small things but is not a doormat	F	Annoyance, pride, calm, frustration, irritation, satisfaction	Neg P-A	Shifting emotion	Face, speech*	Passive other interactive	Inform	Unrestricted, uncon-strained
DA14AT/ Sensitive Artificial Listener Data Set	Subject tells character he doesn't like his attitude	M	Annoyance, irritation, politeness, tension amusement, boredom	NegAct	Consistent emotion	Face, speech*	Passive other interactive	Inform	Unrestricted, uncon-strained

(continued)

Clip/source	Content	Gender	Emotion words	Quadrant	Combination type	Modalities present (*=labelled)	Context	Goal	Constraints
DA19T/ Sensitive Artificial Listener Data Set	Subject hopes that they are not causing any suffering, and that he can contribute to relieving it	M	Calm, amusement, affection, friendliness, pleasure, surprise	NegPass-Prag	Co-existing emotion	Face, speech*	Interactive	Inform	Unrestricted, uncon-strained
EllB3/ Sensitive Artificial Listener Data Set	Subject tells Poppy that she knows how to smile and that she always looks happy because of her mouth and teeth	F	Happiness, amusement, affection, friendliness, pleasure, surprise	Pos Act	Consistent emotion	Face, speech*	Interactive	Inform	Unrestricted, constrained
Ex2A/Green Persuasion Data	Subject tells female listener of the advantages of using public transport and describes how eco-friendly cars are not always efficient	M	Interest, calm, relaxed, worry, hope, politeness	Prag-Pos Act	Shifting emotion	Face, speech*, gesture	Interactive	Persuade	Unrestricted, uncon-strained
Ex6A/Green Persuasion Data	Subject describes the potentially lethal effects of global warming to male listener	M	Interest, worry, calm, friendliness, frustration, stress	Prag-PA	Consistent emotion	Face, speech*, gesture	Interactive	Persuade	Unrestricted, uncon-strained

(continued)

Clip/source	Content	Gender	Emotion words	Quadrant	Combination type	Modalities present (*=labelled)	Context	Goal	Constraints
HSAL/Hebrew Sensitive Artificial Listener		F	Boredom, serene, content, calm, annoyance, anxiety	Prag	Consistent emotion	Face, speech, gesture	Passive other interactive	Inform	Unrestricted, constrained
P6aT/Sensitive Artificial Listener Data Set	Subject tells character how he tries to be happy with the present situation and hopes things will move forward in the future	M	Disappointment, worry, frustration, hope, calm content	NegPass	Consistent emotion	Face, speech*	Passive other interactive	Inform	Unrestricted, uncon-strained
PT2a/Green Persuasion Data	Shows responses of subject as she listens to the advantages of using public transport	F	Interest, politeness, boredom, amusement, calm, embarrassment	Pos Act-Prag	Shifting emotion	Face, speech*, gesture	Interactive	Persuade	Unrestricted, uncon-strained
PT6a/Green Persuasion Data	Subject listens to potential effects of global warming	M	Interest, politeness, boredom, doubt, relaxed, calm	Prag	Consistent emotion	Face, speech*, gesture	Interactive	Persuade	Unrestricted, uncon-strained

(continued)

Clip/source	Content	Gender	Emotion words	Quadrant	Combination type	Modalities present (*=labelled)	Context	Goal	Constraints
R12a/Sensitive Artificial Listener Data Set	Subject describes how feeling a wide range of emotions is a positive experience	M	Sadness, calm, frustration, interest, stress, relaxed	Neg P-Prag	Consistent emotion	Face, speech*	Passive other interactive	Inform	Unrestricted, unre-strained
RA21/Sensitive Artificial Listener Data Set	Subject tells character how simple things, like getting a good night's sleep, can improve negative emotions	M	Sadness, calm, frustration, hope, doubt, relaxed	Neg P-Prag	Shifting emotion	Face, speech*	Passive other interactive	Inform	Unrestricted, constrained
RB31t/ Sensitive Artificial Listener Data Set	Subject explains how rationality can be a stupid state to get into if things are going badly	M	Annoyance, irritation, anger contempt, frustration, excitement	NegAct	Consistent emotion	Face, speech*	Passive other interactive	Inform	Unrestricted, constrained
RhnT/ Belfast Outdoor Activity Data	Subject refuses to repeat task, which she failed to complete	F	Sadness, amusement, despair, helplessness, worry, disap-pointment	NegA	Co-existing emotion	Face, gesture	Task	Activity	Unrestricted, unre-strained

(continued)

Clip/source	Content	Gender	Emotion words	Quadrant	Combination type	Modalities present (*=labelled)	Context	Goal	Constraints
spag03/ Spaghetti	Subject feels in box and is disgusted at what she finds	F	Disgust, anxiety, amusement, fear, surprise, irritation	NegA-PA	Shifting emotion	Face, speech*, gesture	Task	Activity	Unrestricted, uncon-strained
spag11d/ Spaghetti	Subject is shocked at what she feels in box	F	Amusement, surprise, fear, anxiety, worry, tension	PosA-NP	Shifting emotion	Face, speech, gesture	Task	Activity	Unrestricted, uncon-strained
spag16d/Spaghetti	Subject is too frightened to feel deep into box and refuses to complete the task		Fear, anxiety, worry, doubt, tension, disgust	PosA-NP	Shifting emotion	Face, speech*, gesture	Task	Activity	Unrestricted, uncon-strained
WideEyeT/Belfast Outdoor Activity Data		F	Amusement, excitement, fear, surprise, worry, delight	NegA-PA	Shifting emotion	Face, gesture	Task	Activity	

Institution______________________________
Date_______________________

References

Bänziger T, Scherer K (2007) Using actor portrayal to systematically study multimodal emotional expression: the GEMEP corpus. In: Paiva A, Prada R, Picard R (eds) Affective computing and intelligent interaction, Lisbon, Sept 2007. Springer LNCS , Berlin, pp 476–487

Cowie R, Douglas-Cowie E, Tsapatsoulis N, Votsis G, Kollias S, Fellenz W, Taylor J (2001) Emotion recognition in human-computer interaction. IEEE Signal Process Mag 32–80, January 2001

Devillers L, Cowie R, Martin J-C, Douglas-Cowie E, Abrilian S, McRorie M (2006) Real-life emotions in French and English TV video corpus clips: an integrated annotation protocol combining continuous and discrete approaches. In: LREC 2006, Genoa. http://www.cs.brandeis.edu~marc/misc/proceedings/lrec-2006/workshops/W09/Emotion-proceeding.pdf

Douglas-Cowie E, Campbell N, Cowie R, Roach P (2003) Emotional speech: towards a new generation of databases. Speech Commun 40(1–2):33–60

Douglas-Cowie E, Cowie R, Sneddon I, Cox C, Lowry O, McRorie M, Martin J-C, Devillers L, Abrilian S, Batliner A, Amir N, Karpouzis K (2007) The HUMAINE database: addressing the collection and annotation of naturalistic and induced emotional data. In: Paiva A, Prada R Picard R (eds) Affective computing and intelligent interaction, Lisbon, Sept 2007. Springer LNCS, Berlin, pp 488–500

Douglas-Cowie E, Devillers L, Martin J-C, Cowie R, Savvidou S, Abrilian S, Cox C (2005) Multimodal databases of everyday emotion: facing up to complexity. In: 9th European conference on speech communication and technology (Interspeech' 2005), Lisbon, Portugal, 4–8 Sept 2005, pp 813–816

Hönig F (2007) DRIVAWORK – driving under varying workload. A multi-modal stress database in the automotive context. Vortrag: HUMAINE plenary meeting, Paris, 6 Jun 2007

Ioannou S, Raouzaiou A, Tzouvaras V, Mailis T, Karpouzis K, Kollias S (2005) Emotion recognition through facial expression analysis based on a neurofuzzy network. Neural Netw (Elsevier) 18:423–435

Scherer KR, Ellgring H (2007) Are facial expressions of emotion produced by categorical affect programs or dynamically driven by appraisal? Emotion 7(1):113–130

Sneddon I, McRorie M (2006) Perception of emotional expression. In: Poster presented at the 13th European conference on personality, Athens

Zara A, Maffiolo V, Martin J-C, Devillers L (2007) Collection and annotation of a corpus of human–human multimodal interactions: emotion and other anthropomorphic characteristics. In: Paiva A, Prada R, Picard R (eds) Affective computing and intelligent interaction, Lisbon, Sept 2007. Springer LNCS , Berlin, pp 464–475

Part IV
Emotion in Interaction

Editorial: “Emotion in Interaction”

Christopher Peters

Abstract This editorial provides a brief introduction and overview of the following five chapters dedicated to a subset of topics relating to the creation of real-time computational systems capable of engaging in affective interaction with humans. It focuses on the use of graphical embodiments known as *embodied conversational agents*. These computer characters, endowed with human-like appearances and behaviours for the purpose of interacting with humans, are an excellent tool for investigating the construction of more natural human–machine interfaces. Additionally, they play an important role in helping to elucidate our understanding of subtle, illusive and highly complex processes and phenomena underlying human interaction, often appearing trivial due to the apparent ease with which we can accomplish them in our daily routine. An important goal of such systems is to be able to form a loop with the human by conducting appropriate behaviours at appropriate times, in order to establish and maintain different senses of connectedness between the human and agent.

1 Introduction

Social interaction is an intricate subject requiring a consideration that extends beyond the study of entities in isolation, to encompass natural dynamic and developing interrelationships accounting for the context and environment in which it unfolds. The chapters in this part approach the study of interaction by considering interaction from the perspective of emotion and related processes. The term *emotion* is used in a loose manner throughout the part and many of the concepts dealt with can be more properly referred to as being *affective* in nature; that is, they concern a wide gamut of affect-related artefacts that may take place over a variety of time scales in a variety of contexts.

C. Peters (✉)
Coventry University, Coventry, UK
e-mail: Christopher.Peters@coventry.ac.uk

P. Petta et al. (eds.), *Emotion-Oriented Systems*, Cognitive Technologies,
DOI 10.1007/978-3-642-15184-2_15,

An important underlying methodology throughout the part, in the spirit of affective computing (Picard 1997, 2003), is the investigation of interaction and emotional processes through the use of computational models embedded in real-time systems capable of interacting with humans. Here, it is worth noting some of the views regarding the place and utility of affect and emotion research in the modern world. Affect has long been unfairly deemed to be nothing more than an insignificant byproduct of biological reasoning processes, or even a destructive obstacle to controlled, logical reasoning and intelligent behaviour. And of all the domains of application, it is in the domain of computation and machines where one who is new to this area may ponder the practicality, role and utility of affective computing research. A newcomer may, for example, expect that emotion research is concerned with abstract or otherwise vague concepts related to questions that, while intellectually stimulating, may not be of a practical nature or yield foreseeable benefits for creating concrete systems. The issue of *subjective feeling* and concerns about if and how machines may be endowed with such capacities, for example, may wrongly be perceived to dominate this area.

Thankfully, views on the role and utility of emotion have changed remarkably, and as the reader will hopefully experience throughout this part, while fundamental questions such as the aforementioned are still of great importance, aims are towards highly practicable systems with concrete day-to-day applications in the here-and-now. Possibly the utmost of these involve attempts to endow machines with forms of *emotional intelligence*, so that they can better recognise, understand, adapt to and, ultimately, interact with human users. Contemporary systems are, by any measure, remarkably ignorant and unintelligent when it comes to the human interactant, who is expected to expend effort in conforming to the interaction methods of the machine. The potential for more natural interaction to take place should reduce the effort involved for the user, and associated frustrations, during human–computer interactions.

1.1 Agents and Interaction

The computerised systems dealt with here most often take the form of *embedded agents*: These are graphical representations, often of real humans, that can express themselves graphically, and sometimes verbally when interacting. These agents are very useful as vehicles for testing emotion and interaction theories and also as practical applications where human–computer interaction is desired.

The term *intelligent virtual agents*, also referred to in the following texts as *virtual agents* or *agents*, is used to refer to computerised entities that utilise a variety of computational models for the purpose of enhancing their decision-making capabilities in order to operate in a more intelligent, or at least behaviourally credible, manner. Such agents are embedded within a virtual environment, although they are not limited to interacting solely within it, and often embodied, i.e. have a graphical appearance they may help to convey their state within the environment.

Embodied conversational agents (ECAs) are a specific type of IVA that have a purpose of interacting directly with users. Represented by an animated character on the screen, ECAs typically employ a number of important capabilities for interacting with users, for example through speech, gesture and facial expressions.

Agents are an important focal point for affective research: To the emotion theorist, they provide a concrete way of testing theories and controlling experimentation. The attempted reconstruction and a posteriori comparison of such systems with their real counterparts is perhaps the best way of establishing the actual nature of the real system. In the same way, emotion theorists and computational modellers use their expertise to provide updates and solutions that improve real-world applications. Such agents should be able to keep track of facial expressions, gestures, speech and other important interaction details, such as who is speaking, in order to alter inner state to adapt and produce appropriate behaviour.

1.2 Key Capabilities

When studying a single entity, a road-map such as that provided in Fig. 1 is useful for gaining an idea of the capabilities of importance for creating the aforementioned agents. These can be enumerated as follows:

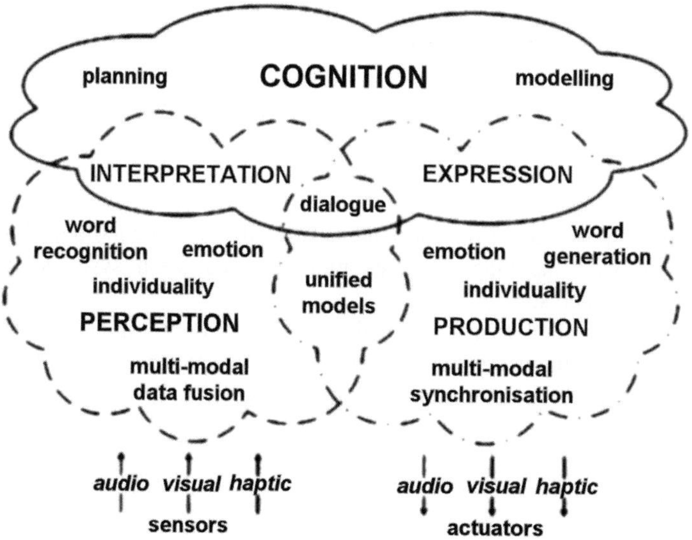

Fig. 1 Diagram (Moore 2005) of core behaviours for spoken language processing. These behaviours are also more generally applicable as a road-map of the relationships between categories of core behaviours and capabilities desirable for an intelligent conversational agent.

1. Detection and recognition of aspects of the user, the environment and contextual features of importance. This relates to the sensing of important data, perception and interpretation for use in higher-level processes. It may range from determining who the current speaker is, to detecting the change in facial expression of a user from one of joy to frustration.
2. Planning based on intention, internal state and externally detected factors and events.
3. Generation and expression to ensure that appropriate behaviours are generated to express the state and intentions of the agent.
4. Internal representation and memory as important concepts underlying the operation of all of the above.

Many more specific elements and capabilities can be placed in one of these categories, for example, managing the start of interaction and associated communicative behaviours (Peters et al., 2007).

2 Structure

With the key capabilities described above in mind, this part consists of five chapters.

Chapter "Fundamentals of Agent Perception and Attention Modelling", investigates practical approaches for constructing real-time agents capable of perceiving and attending to their environment and users.

Chapter "Generating Listening Behaviour", considers face-to-face conversations in more detail, to examine the joint activity of speakers and listeners through the construction of agents that are able to show that they are listening to the speaker.

Chapter "Coordinating the Generation of Signs in Multiple Modalities in an Affective Agent, looks at issues of coordination, so that an agent is capable of providing appropriate behaviour over available modalities, such as facial expression, speech and gestures.

Chapter "Representing Emotions and Related States in Technological Systems", is concerned with the inner representation required when systems operate on emotions and related states.

Chapter "Embodied Conversational Characters: Representation Formats for Multimodal Communicative Behaviours", deals with requirements for representation languages used for planning and displaying behaviours.

References

Moore RK (2005) Cognitive informatics: the future of spoken language processing? In: Keynote talk, SPECOM - 10th International Conference on Speech and computer, Patras, Greece, 17–19 October 2005.

Peters C, Pelachaud C, Bevacqua E, Ochs M, Ech Chafai N, Mancini M (2007) Towards a socially and emotionally attuned humanoid agent, NATO HSD EAP ASI 982256. Esposito A, Bratanic M, Keller E, Marinaro M, (eds) Fundamentals of verbal and nonverbal communication and the biometric Issue, vol 18. IOS Press, Amsterdam, The Netherlands
Picard RW (2003 July) Affective computing: challenges. Int J Hum-Comput Stud 59 (1–2):55-64.
Picard RW (1997) Affective computing. MIT Press, Cambridge, MA

Fundamentals of Agent Perception and Attention Modelling

Christopher Peters, Ginevra Castellano, Matthias Rehm, Elisabeth André, Amaryllis Raouzaiou, Kostas Rapantzikos, Kostas Karpouzis, Gaultiero Volpe, Antonio Camurri, and Asimina Vasalou

Abstract Perception and attention mechanisms are of great importance for entities situated within complex dynamic environments. With roles extending greatly beyond passive information services about the external environment, such mechanisms actively prioritise, augment and expedite information to ensure that the potentially relevant is made available so appropriate action can take place. Here, we describe the rationale behind endowing artificial entities, or virtual agents, with real-time perception and attention systems. We cover the fundamentals of designing and building such systems. Once equipped, the resulting agents can achieve a more substantial connection with their environment for the purposes of reacting, planning, decision making and, ultimately, behaving.

1 Introduction

An entity's ability to think and behave within a complex, dynamic environment is shaped, to no small degree, by the nature of the environment as witnessed according to its particular capacity for sensing and understanding it. A quick glance is often sufficient for us to recognise many different objects and events in what appears to be a highly efficient and relatively effortless process. The ease with which we are able to conduct such processing perhaps betrays the sophistication and complexity of the underlying processes, something that has been highlighted by research in many scientific domains investigating or attempting to model the real systems.

A number of approaches are open to the prospective modeller. One approach is to attempt to create intricately detailed simulations closely matching the theorised workings of the real systems, although more often complex, since the real system may not yet be well understood, suitable computational models may not exist, the computational models may require extensive processing power or they may focus

C. Peters (✉)
Coventry University, Conventry, UK
e-mail: Christopher.Peters@coventry.ac.uk

P. Petta et al. (eds.), *Emotion-Oriented Systems*, Cognitive Technologies,
DOI 10.1007/978-3-642-15184-2_16,

too narrowly on aspects that alone may not be of great behavioural significance. Instead, we seek for the created models to broadly parallel the key aspects of their real counterparts. In this way, the perception models detailed here are inspired by, but greatly simplified with respect to, the real systems. A most important consideration is that adopted models must be fast enough to allow real-time interaction while providing the impression to viewers, as much as possible, that the virtual agents using them possess analogs of human behaviour that are consistent with and appropriate to the situation.

In a more extensive virtual agent architecture, the approaches presented here could be viewed as comprising an input stage that, in combination with internal factors such as goals and motivations, contribute to the process of action selection in an entity, be it reactive or deliberative, in order to generate output behaviour, which could be expressed, for example, through BML (see Chapter "Embodied Conversational Characters: Representation Formats for Multimodal Communicative Behaviours" in this part) and multimodal selection (see Chapter "Coordinating the Generation of Signs in Multiple Modalities in an Affective Agent" in this part).

1.1 Purpose and Significance of Agent Perception

Agent perception refers to lightweight, and necessarily simplified, computational models that could be considered in some ways analogous to human sensory perception mechanisms. A key similarity is that some form of internal world model is maintained by the virtual agent, which is a local *view* or representation of the external world from its perspective. An agent's decision-making mechanism is dependant on this internal model and subject to all associated inaccuracies or errors in representation. Such inaccuracies need not be negative, however, and may actually help mimic real-world behaviours of entities being simulated due to sensory constraints, provided they are appropriate to the character being modelled. As a side effect, subjective, individuated internal views also help agents to exhibit variety in how they interpret, plan, reason, react, adapt and ultimately behave.

The internal model of the virtual agent need not be complicated, or even explicit, in order to produce complex emergent behaviour: For example, to create group flocking behaviours, Reynolds (2000) endowed each agent, called a *boid* or bird object, with an internal model that considered little more than the movement information of its *n* nearest flockmates. In this case, interesting behaviours arose from simple modelling, and this has also been the case for sensory modelling involving other types of synthetic creatures, early examples demonstrated by Blumberg (1997) and Tu and Terzopoulos (1994).

We will consider differences in the utility of agent perception depending on whether sensing is occurring from a real environment or a virtual environment in Sect. 2. While the creation of such a system in the real environment (Sect. 2.1) is necessarily limited to the sensing hardware available, no such limits exist in the virtual environment, and here the utility of *synthetic perception* (Sect. 2.2) can be quite different to its real-world counterpart.

Creating a synthetic perceptual system involves consideration of the types or extent of information the agent ought to be able to extract from its environment. This is because, in the virtual environment, agents can easily be given full access to the scene database containing the definitive description of all objects and their states. Unconstrained access to this database allows agents to know, with complete accuracy, the state of the environment at any time, thus obtaining a form of sensory omnipotence. Sometimes this is desired, but in many cases where an agent is meant to act 'in character', constraints must be imposed. One common example relating to the visual modality is the field of view through which entities are able to perceive their environment. For example, the human field of view is limited, but the eyes are quite mobile to be oriented at will. This results in quite fundamental, if unspectacular, looking behaviours as the eyes, head and even torso are oriented towards locations of interest. What is interesting about such unspectacular behaviour when present is that it may quickly become spectacular, and implausible to the viewer, when such behaviour is missing. Synthetic perception mechanisms help to mimic such fundamental behaviours, and do so through necessity of acquiring information through whatever senses they have been granted.

In addition to providing behaviours that ought to be there, synthetic perceptual capabilities and limits also help to restrict implausible behaviour that ought not to be seen to occur. If a human is not able to see an object because it is positioned outside of their field of view, then nor should an agent that is meant to be plausibly representing a human, in what has been referred to as *sensory honesty* (Isla et al., 2001).

While these types of limitations are imposed purposefully by a designer in order to help replicate plausible behaviour when simulating entities, other reasons for imposing restrictions are of a more fundamental nature and equally applicable to sensing from the real and virtual environment. As in biological brains, computers, and thus the agents they are simulating, have finite processing capabilities. It is in this respect that it becomes important to handle information in a methodical manner, prioritising some forms or channels of information over others, ensuring that the vast array of incoming sensory information is not overwhelming and calculations are tractable and smart. Such a method quickly becomes non-trivial when expected to operate in a complex, dynamic environment and especially when it is an active orienting device in the environment as opposed to a passive system, in the case of *active vision* systems. It is in this respect that perceptual attention is considered in more detail in Sect. 3.

1.2 Relevance to Affective Modelling

While emotion has not been mentioned thus far, intimate links exist with perception and attention processes. Indeed, emotion theories consider perceptual attention as fundamental and integral to emotion processing. For example, in appraisal theory (cf. Scherer et al., 2001), where the process comprises of a number of stimulus evaluation checks (or SECs), the capture and maintenance of attention is an important early evaluation check required for further evaluation checks to take place. Further, the inherent emotional quality of stimuli and the emotional state of the perceiver are key in the interplay between perception, emotion and attention. Thus, we can enumerate at three broad ways in which emotion and perceptual attention may functionally inter-operate with potential significance to agent behaviour:

1. The perception of stimuli being 'coloured' or modulated according to the emotional state of the perceiver, for example, 'seeing the world through rose-tinted glasses'.
2. Emotional stimuli may modulate perceptual attention. Threatening faces, for example, may attract attention.
3. Attended-to stimuli may modulate the emotional state of the perceiver, thus completing a loop between 1 and 2 above.

Here, we consider perception and attention as supporting technologies for affective agents. What follows in the remainder of the chapter is a description of core perception and attention capabilities with which emotion models can be integrated.

2 Basics of Agent Perception

When designing a perceptual system for an agent, a number of considerations must be made. Particularly noteworthy, in terms of providing input for the model, an important distinction must be made between the real and virtual environment:

1. Acquiring the input from the real environment, using a laptop-mounted web-camera or similar recording device. In this way, the virtual agent is essentially looking out of the monitor screen into the real environment and attempting to isolate and interpret details of importance from the real environment, for example, if interacting with a user, making sure the user is present by detecting their face, ensuring they are paying attention by detecting their gaze direction and perhaps also detecting any facial expressions to determine further their affective state.
2. Acquiring input from the virtual environment, by endowing the agent with synthetic senses. These senses do not have the same constraints as hard systems, and so a designer has a choice as to the degree that perception can take place, for example, limiting the field of view of the agent as we described earlier or limiting how far it can see.

There are important and different challenges associated with each, and indeed the role of perception also differs: When input is taken from the real environment, perception plays the role of recreating and flavouring relevant details, for example, segmenting an object from a scene or recognising a smile; in the virtual environment, all of the information is readily available for the agent in the form of the scene database, so the purpose of perception here is to decide what subset of that information should be made available to the agent, given its role.

2.1 The Real Environment

Real scenes contain a vast amount of information. Seemingly simple problems, such as extracting objects from a scene under varying conditions, are still exceedingly difficult problems to attempt to solve. In addition, given limited processing capabilities, the amount of time taken to solve the problem, if a solution is possible, must also be considered. Luckily, the problem becomes more tractable as constraints are imposed: for it is usually likely that at any one time, only a limited domain of information will be of relevance depending on the role of the agent or scenario taking place.

As an example, consider a hypothetical scenario with an agent playing the role of a virtual museum guide. The designer of the agent decides that as viewers near an exhibit, the agent located nearby will automatically detect their presence, activate and provide information about it. There are a number of ways in which the agent may be set up by the designer to perceive users, each with varying degrees of sophistication:

- The least sophisticated approach may be to endow the agent with a distance sensor so that it activates when something moves within a predefined distance. This approach is not very robust, however, as the sensor my be fooled by any object and activate the agent during inappropriate circumstances.
- To improve the situation, the designer may decide to mount a camera near the agent and exhibit. This would process the video input in order to detect users. A first approach could be to simply detect motion in the scene and use this to activate the agent. However, again it may not provide much of an improvement over the initial sensor.
- To improve the sensor, skin colour could be detected, ensuring only humans would activate the guide. We may start to track patches of skin colour.
- Finally, higher level objects may be detected, such as heads, faces and even the expressions of those faces. Visitors who are detected as looking at the exhibit can activate the agent, and furthermore, depending on their facial expression, the agent may adapt its presentation: for example, provide a brief presentation to somebody who appears uninterested.

This simple example illustrates a number of important issues. A real scene, like the one described above, contains a lot of signals, such as facial expressions, gestures, body poses and so on. In such situations, as many of the signals of importance as possible should be analysed (see Part II for a more in-depth analysis of each of these topics). As we add more detection capabilities, however, and increase the sophistication of the sensor, the computational complexity also increases and it becomes harder to maintain real-time interaction.

In addition, the nature of the sensing and perception capabilities limits the sophistication and perceived intelligence of behavioural processes: while an agent may have a huge repertoire of different behaviours at its disposal, it will have no way of choosing between them appropriately or credibly if the internal model receives scant input from the outside environment.

Next, we describe some ways in which social details, such as faces and body movements, can be perceived by an agent from the real environment.

2.1.1 Faces

Detecting human faces in the scene is a requisite of utmost importance for an agent to engage in interaction with users.

Endowing an agent with a face detection competency is currently a relatively easy task that can be accomplished using a simple webcam. The OpenCV computer vision library (Bradski and Kaehler, 2008) provides algorithms and techniques that are fairly robust in environments with good illumination conditions. A first option, for example, is the detection of faces using skin-coloured regions segmentation.

The Mean-shift and the Camshift algorithms (Bradski, 1998) that come with OpenCV allow for the tracking of the distribution of any features representing an object (e.g. colour features) and can be easily employed to track faces. Nevertheless, this method works well if there are no other skin-like colour objects in the camera's field of view. An alternative method included in the OpenCV library is the *Haar classifier*-based face detector. This is built on a version of the face detection technique originally developed by Viola and Jones (2001). The method is based on a combination of Haar-like wavelets with classification using a form of Adaboost (Bradski and Kaehler, 2008) and can be used to recognise any type of rigid object. This can be done by training detectors using numerous images for each view of the object.

As a second step towards a successful interaction with users, one might be interested in endowing an agent with the ability to automatically detect human facial features (e.g. eyes, nose, mouth, eyebrows). Haar classifiers, for examples, can be trained in OpenCV so as to be able to detect, combined with other techniques such as template matching and Hough transform, facial features (see Fig. 1).

Several techniques for the identification of salient facial points (Pantic and Bartlett 2007) have been reported in the literature. The detection of these points can be used, for example, to trigger a face recognition algorithm to pinpoint known users (see Sect. 4.3.1). Automatic tracking of salient facial points is an important requirement for an agent to be able to analyse facial expressions. Particle filters, for example, are one of the techniques that are currently exploited to perform this task (Patras and Pantic, 2004). The ability to analyse the user's facial expressions can be

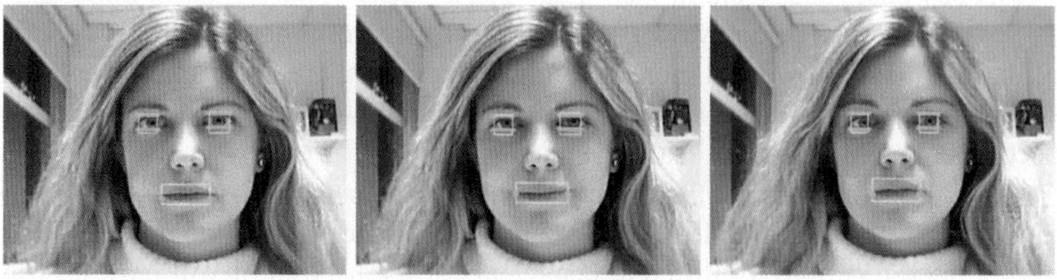

Fig. 1 Eye and mouth detection performed using OpenCV

useful to the agent while it is engaged in a face-to-face interaction with a user. In this type of scenario, the interpretation of some form of behaviour displayed by the user (user establishing eye contact with the agent, smiling, etc.) could lead to the inference of the affective or mental state experienced by the user (e.g. happiness, interest, willingness to interact with the agent) (Zeng et al., 2009) (see Part II) and could be used to control the attention (illustrated as gaze direction, facial expression, etc.) of an agent. Matlab and OpenCV provide several machine learning tools that can be used to train systems to recognise different types of user behaviours and states.

2.1.2 Full-Body Movements and Gestures

In case of medium- or long-range interaction, i.e. when an agent is not necessarily interacting face to face with the user but is still in the range or in the same room as the user, analysis of full-body movements and gestures can be of use for the agent to interpret events unfolding in the environment. First of all, detection of body movement can inform an agent about the presence of people in the surrounding environment. This information, for example, can be used by an agent to direct its attention towards the detected movement. The OpenCV library provides techniques that support movement analysis, such as algorithms that can be used for background subtraction, body silhouette and body parts segmentation, and motion tracking (Bradski and Kaehler, 2008).

Approaches for analysis of human movement can be broadly categorised as motion capture-based and vision-based. In motion capture-based approaches, markers are positioned at the joints of the person whose movement is to be tracked, allowing for positions, angles, velocity and acceleration of the joints to be very accurately recorded. For the detection of specific body parts specialised techniques are used, for example, in the case of the hands, mechanical or optical sensors mounted on gloves (Kranstedt et al., 2006). Vision-based approaches do not require optical markers or sensors to detect motion, allowing more freedom for the actor during the capture process. With these approaches, though, segmentation and tracking of the full body or body parts is sometimes problematic due to the difficulty of identifying and separating the silhouette from complex backgrounds. To alleviate these problems, some systems require a uniform background or use coloured markers, placed on the fingertips, so that they can be tracked using colour histogram analysis, a technique used to analyse the distribution of colours in an image (Bradski and Kaehler, 2008).

Once a human body is detected and tracked, an agent may be interested in recognising gestures. In the gesture recognition community, hidden Markov models (HMMs) are largely used to represent the spatial and temporal structure of gestures. For a good tutorial on HMMs, see Rabiner (1990). A valuable tool that could be of use for an agent to interact with a user is Watson,[1] a freely available library for head tracking and gesture recognition. Watson can estimate head pose and orientation in

[1] http://projects.ict.usc.edu/vision/watson/

real time using an adaptive view-based appearance model (Morency and Darrell, 2004). Watson also contains a module for head gesture recognition, allowing for the recognition of head nods and shakes.

A different approach to human movement and gesture analysis consists of taking into consideration the expressive characteristics of movement. EyesWeb XMI[2] is an open software platform for the synchronised analysis of multimodal data streams which supports real-time analysis of body movement expressivity (Camurri et al., 2007). It consists of a set of libraries, including the EyesWeb Expressive Gesture Processing Library (Camurri et al., 2004), which contains modules for the automatic extraction and analysis of cues directly related to motion and gesture qualities, such as quantity of motion, degree of contraction/expansion and fluidity. These cues can be computed in real time for the full body or selected body parts (e.g. the head) and can be used as features for the automatic detection of human affect (Castellano et al., 2007, 2008). Figure 2 shows a measure of the quantity of motion and a measure of the degree of contraction/expansion of the body using EyesWeb XMI.

The above-described capabilities can be used by an agent to assess an interaction initiation condition or, in general, the user's willingness to interact with it. Recognition of simple gestures and actions, such as waving, approaching or withdrawing, and analysis of coarse cues, such as the amount of people present in a room and the movement expressivity of single or multiple users, can help the agent in assessing whether the user is interested in beginning or continuing an interaction with it.

Fig. 2 A measure of the quantity of motion based on silhouette motion images (*left*) and a measure of the degree of contraction/expansion of movement using a technique based on the bounding region, i.e. the minimum rectangle surrounding the body (*right*). From Castellano (2008)

2.2 The Virtual Environment

We use the term *synthetic perception* to refer to sensing and related processes that take place from the virtual environment.

[2]http://www.eyesweb.org

2.2.1 Synthetic Vision

Vision is one of the most important sensory modalities for humans and thus an important starting point for modelling agent sensory perception. *Synthetic vision* refers to approaches that attempt to provide scene data to the agent in a way that models very roughly the availability of sensed data to the human visual system by abiding to constraints such as field of view, distance and occlusion. These approaches are simplified in comparison with classic computer vision schemes (referred to here as *artificial vision*; Noser et al., 1995), bypassing many inherent problems and making it possible to obtain reliable real-time performance. Synthetic vision can be viewed as a continuity of approaches ranging from geometric approaches, which do not render the scene at all, to pure synthetic vision approaches, where fully rasterised views are captured. Geometric approaches use collision tests and ray-casting to detect the sensory status of objects in relation to a perceiving entity. For example, a view volume such as a sphere centred on the agent may be treated as a detection zone for sensed entities (Reynolds, 2000). Vision is often modelled as one or more directed view-cone(s) emanating from the agent's eye position (see Fig. 3). Rays are cast from the agent's viewpoint to various objects falling within the volume of this sensory cone: An unblocked ray between the viewpoint and an object indicates that it is visible to the agent. The speed and ease of use of this category of approach has led to increasing adoption for sensory AI in the

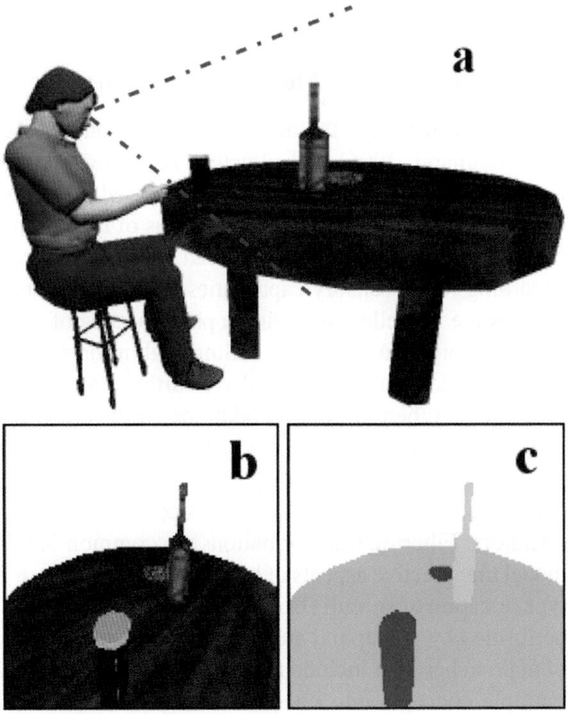

Fig. 3 Illustration of synthetic vision and false-coloured rendering. A scene (**a**) containing an embodied agent, (**b**) the scene as rendered from the agent's perspective and (**c**) a corresponding false-coloured rendering of the scene. The false-coloured rendering is scanned for unique colour identifiers providing information on those objects within the field of view of the agent

computer games industry (Leonard, 2003). However, inaccuracies may arise when a low number of rays are used for calculations. Increasing the number of rays cast in order to alleviate these inaccuracies also results in an increase in computational cost. Pure synthetic vision techniques employ similar methods to those developed in artificial vision. They work by rasterising the scene from the point of view of the agent. The agent uses the results of applying image processing techniques on this view internally to influence its behaviour. A related approach is to use a false-coloured rendering (Noser et al., 1995; Blumberg, 1997; Kuffner and Latombe, 1999; Peters and O'Sullivan, 2002) of the scene (see Fig. 3). The scene is rendered with simplified colours uniquely attributed to each object, with no lighting, textures or special effects applied. When the colours are scanned from the rendering they can be easily resolved to corresponding object references from the virtual environment database. These techniques generally provide a more diverse range of perceptual data and better accuracy of visibility results than geometric techniques, while implicitly taking advantage of sophisticated graphics hardware. However, they also tend to be significantly slower than their geometric counterparts and suffer from problems when objects are too far away from the viewer to be rendered as a single pixel. The latter issue is typically addressed by employing additional renderings, of higher resolution and smaller area/volume centred about the view direction. Generally analogous to the higher acuity area of the human eye, these renderings provide a solution to the problem at the expense of increased processing. Other approaches use a hybrid of raster and geometric techniques in order to balance efficiency with accuracy (Lozano et al., 2003; Tu and Terzopoulos, 1994).

2.2.2 Other Synthetic Modalities and Systems

In addition to vision, the auditory (Herrero and de Antonio, 2003), tactile (Conde and Thalmann, 2006) and olfactory (Delgado-Mata and Aylett, 2001) modalities have also been considered for modelling, although to a more limited extent. For example, in the auditory domain, a focus of hearing may be defined using elliptical cones oriented about the agents head. Cone parameters, such as length, are altered according to the sensory capabilities of the agent, such as hearing distance. Sound sources are modelled as having a physical area of projection denoting sound propagation – propagated sounds falling within the focus of hearing are considered to be detectable by the agent and are further processed to determine their clarity of perception. When multiple modalities are involved in agent perception, it is especially important to have an ordered, generic and extendable system. Methodologies for agent perception systems play an important role here. They are often comprised of a staged pipeline of storage stages connected by varying types of transformation operators. Filtering transformations are common in most implementations, as these model the selective aspects of human perception. Such filters may be based on range, type, location of stimuli (Bordeux et al., 1999) or may be based on the computations performed by perceptual attention mechanisms (as described in Sect. 3). A number of approaches also include an integration operator, which deals with amalgamating multiple concepts into one. This integration can take the form of (1) integrating concepts in one modality over time into a single concept, (2) integrating multiple concepts from one modality into a higher level representation (Vosinakis and Panayiotopoulos, 2003) or (3) integrating concepts across multiple modalities into a single concept (Conde and Thalmann, 2006).

2.2.3 Integrating Real and Synthetic Perception

Face-to-face communication does not take place in an empty space, but should be linked to the surroundings of the conversational partners. As an example, let us consider an agent that inhabits a virtual office and converses with a human user that inhabits a real office. During the conversation, both the agent and the user may refer to digital objects on the agent's desk as well as physical objects on the user's desk. To converse with a human user successfully, the agent needs to be aware of the physical as well as the digital space. The need to integrate synthetic and real perception becomes even more obvious in augmented realities which combine both virtual reality and real-world objects. For example, the user might wear translucent goggles through which he perceives the real world as well as digital augmentations projected on top of it. One of these augmentations could be a virtual agent that then serves as a companion of the user in the physical world. Such a scenario even requires a more fine-grained integration of synthetic and real perception processes as the scenario sketched above. Integration has to be handled on at least two levels. First, sensors have to be integrated into a coherent framework independent of the sensor's type (real or virtual). Second, we have to fuse the information delivered by the sensors. One specific problem is the different information density for the real and for the virtual world. Perception and attention of the virtual part of the scenario can be modelled to whatever degree is desirable because all the information about the virtual world is in principle available. Perceiving and attending to the real world faces severe technical problems and is thus always limited and deficient unless the surrounding real world is static and can thus be modelled as well. As a consequence, we have to find fusion methods that allow for the integration of information that is provided at different levels of granularity. In addition, new methods are required for calculating attentional prominence in an augmented reality which take into account both physical and digital object features.

On a second level, we have to integrate the information delivered by the sensors. We can distinguish between early, late and hybrid fusion approaches (see Fig. 4). Whereas early fusion approaches work directly on the feature level and

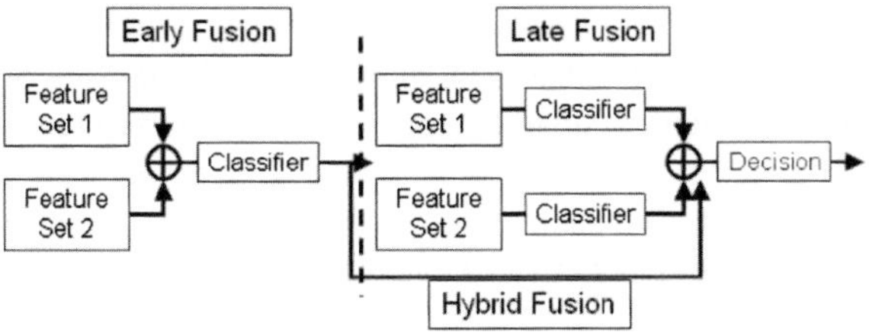

Fig. 4 Overview of early, late and hybrid fusion approaches

classify features sets of different input channels with a single classifier, late fusion approaches integrate the results of sensor-/channel-specific classifiers and thus work on a semantic level. Hybrid fusion approaches combine early and late fusion.

3 Perceptual Attention Modelling

As humans, we are constrained biologically by the amount of information that we can receive and process from the environment, but manage to survive by employing numerous clever techniques for selecting the important and filtering out the unnecessary. The primary senses, limited in their fields or distances of reception, may be oriented at will to provide directional enhancement and 'tuned in' to enhance the processing of potential threats and opportunities.

It is important for agents to be able to attend to their environment for at least two major reasons:

1. Aesthetically, gaze and other behaviours related to the overt directing of the senses increase the naturalness of the agent with respect to a perceiver. We are accustomed to seeing other living and intelligent entities orient their senses towards items of theorised interest in the environment, and therefore such behaviours may help convey life–life and intelligent qualities.
2. Functionally, the sensible orienting of the senses is necessary to ensure that autonomous agents that are dependant on their perceptual capabilities are provided with the relevant and appropriate knowledge with which to conduct planning. If their internal models are updated with information irrelevant to their planning processes, then no doubt those processes will not function optimally.

Approaches focused primarily on aesthetic aspects do not need to consider the actual environment of the agent or process stimuli. For example, in the case of gaze, eye and attentive behaviour can be derived from statistical observations on the frequency of occurrence and spatiotemporal metric properties of human saccades using an eye-tracking device (Lee et al., 2002). Although the output appears realistic in terms of how the eyes move, the model does not consider the actual external environment in deciding where to move them and, using solely this technique, the eyes would not respond properly to dynamic stimuli. Other approaches use eye-communication models to animate gaze by considering specific social aspects so that the agent is capable of providing signals and feedback (Poggi et al., 2000). For example, an agent may look upwards to communicate that it is thinking, and engage in mutual gaze with an interactant for different durations depending on whether it is a speaker or listener (see Chapter "Generating Listening Behaviour" in this part).

3.1 Saliency Map Approaches

Saliency-based visual attention are worthy of particular mention in this respect. Using evidence coming directly from the human visual system (HVS), these models gained significant popularity during the last decades, due to the seminal work of Treisman and Gelade (1980) and Koch and Ullman (1985). Itti et al. has presented one of the most sophisticated saliency-based spatial attention models, measured its efficiency against human observers (Itti et al., 1998) and developed the model for driving the gaze of an agent in natural scenes (Itti et al., 2003). A master *saliency map* is a 2D greyscale representation of the most likely areas of the scene to 'pop out' to the viewer. It combines information from low-level features, such as colour, intensity, depth and motion, into a global measure, where points corresponding to one location in each feature map project to single units in the saliency map. Attention bias is then modified in order to draw attention towards high activity locations in the saliency map. These locations are potentially the most informative of the scene and can help reduce the complexity of visual search. Their model allows the agent to orient rapidly its attention towards relevant parts of the incoming visual input, but the strong limitation of their approach is the high computational complexity.

3.1.1 From the Real Environment

Gu et al. proposed a visual attention model based on Itti's with better results concerning low-level feature extraction and region-of-interest (ROI) detection (Gu et al., 2005). In an attempt to control agents and enable their engagement in realistic face-to-face interaction with human partners, Raidt et al. (2005) describe a system for mutual attention (eye gaze analysis and control) and deixis (eye, gaze, face, hand and head movements to point towards objects of interest). Their experiments on the interaction between a realistic talking head and a user during a virtual card game study the growth of user's interest.

Most of the previous approaches process video input on a frame-by-frame basis and compensate, if desired, for temporal incoherency using variants of temporal smoothing or calculating optical flow for neighbouring frames, in order to tackle inherent issues such as occlusions of parts of faces between successive frames. Spatiotemporal processing is more promising, since it exploits the fact that many interesting events are characterised by strong variations of the data in both the spatial and temporal dimensions. Oikonomopoulos et al. (2006) and Laptev and Lindeberg (2003), for example, have used spatiotemporal saliency points for action recognition. Rapantzikos et al. use a volumetric representation of video input to compute spatiotemporal saliency and use it for ROI selection and video classification (Rapantzikos et al., 2005, 2007). Even though these models require batch processing of frames, they can be adjusted to process a small number of frames that occurred in the past and therefore allow the agent to derive conclusions about events having both spatial and temporal extent. These events may be related to specific actions, such as walking or running, or at a more abstract level to suspicious behaviour, i.e. actions not belonging to a *labelled* category.

3.1.2 From the Virtual Environment

In the virtual environment, saliency approaches have also been used by Peters and O'Sullivan (2003), which combine Itti's saliency map with an object-based representation and memory, and Courty and Marchand (Courty and Marchand, 2003), who compose saliency based on depth and colour. In Table 1, a summary is provided of the heuristics employed by the respective saliency approaches, specifically depth, colour, orientation, intensity, flicker and/or motion (see Fig. 5 for an example of the process). Generally, the more heuristics included in the bottom-up model, the more accurate and robust the resulting simulation for different scene types, at least as far as consideration of only bottom-up aspects of visual attention is concerned.

3.1.3 Limitations of Pure Saliency Approaches

While saliency models alone are useful for highlighting various contrast discontinuities in scenes over multiple scales, they are limited in terms of scope and robustness. This is because, in solely bottom-up models, there is no consideration of how the current task of the entity may act to control or modulate the allocation of attention, something referred to as top-down attention. This issue will be described further in Sect. 3.2.1 as a key feature for distinguishing between the capabilities of different attention models.

3.2 Overview of Key Considerations

The differentiation between top-down and bottom-up models is only one of a number that can be made in relation to computational visual attention models suitable for application to agents. A more complete list comprises at least five different features:

- Modes of processing: Models may process in a bottom-up manner, a top-down manner or a combination of both.
- Feature type: Models may process spatial features, object-based features or both.

Table 1 Heuristics available in bottom-up saliency map models for agents; depth, colour, orientation, intensity, flicker, motion. 'Yes' indicates that the heuristic is simulated in the corresponding model

Model	Depth	Colour	Orientation	Intensity	Flicker	Motion
Peters and O'Sullivan (2003)		Yes	Yes	Yes		
Itti et al. (2003)		Yes	Yes	Yes	Yes	Yes
Courty et al. (2003)	Yes			Yes		
Gu et al. (2005)		Yes	Yes	Yes		
Picot et al. (2007)		Yes	Yes	Yes		Yes

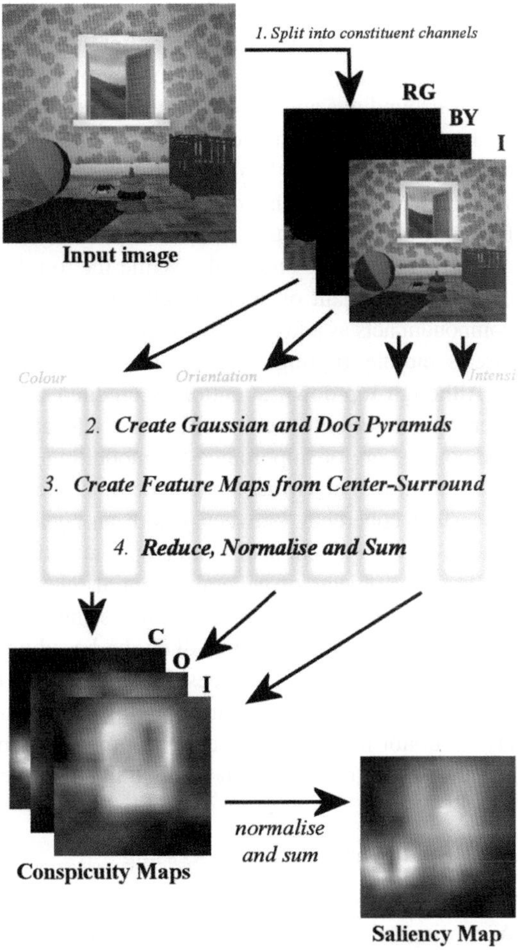

Fig. 5 Depiction of the saliency map creation process. An input image is split into its constituent channels (1) in preparation for image pyramids to be created for each feature. In this example, colour, orientation and intensity features are depicted (2). Feature maps are created (3) for each of a number of image pyramids, which are reduced, normalised and summed (4) to provide 16x16 conspicuity master maps and, subsequently, the final saliency map (*depicted bottom, enlarged*). The saliency map is the primary output of the bottom-up model and may be used to generate artificial regions of interest to drive agent gaze behaviours

- Viewer type: Models may be able to support dynamic viewer where the viewpoint changes or may be limited to a static viewpoint.
- Input type: Models tend to be built to attend to real-world input (e.g. through a webcam or similar device) or input from a virtual environment (e.g. using synthetic vision).
- Social modulation: Models may process social features, faces for example, in a special manner to modulate attention.

We discuss each of these features in more detail next.

3.2.1 Top-Down and Bottom-Up Processing

An important distinction to be made between visual attention models relates to the manner in which processing takes place. *Top-down processing* refers to the way in which attention resources may be volitionally allocated to external stimuli that are of importance or relevance to an internal goal or task of the entity. In contrast, *bottom-up processing* refers to the way in which some objects may pop-out from their surroundings and appear to automatically draw attention.

Top-down processing acknowledges task as a vital factor in determining where entities look (Yarbus, 1967). Therefore, it is useful to think of a more complete visual attention mechanism as being driven by the interplay of at least two general factors: bottom-up factors based on image features and top-down guidance based on scene knowledge and goals. While the top-down component may be viewed as a chief determinant of attention allocation when a task is at hand, the bottom-up component acts as a fast alerting mechanism, highlighting potential opportunities or threats and interrupting current tasks.

Although top-down attention is credited in many experimental conditions for exerting greater influence over the final allocation of overt attention to scenes, both components are of importance for modelling autonomous, broadly capable agents in a natural real or virtual environment. While the top-down component allows for a sensible coherency between where an agent looks and its goals and tasks, the bottom-up component acts as a fast pre-attentive alerting mechanism for highlighting areas or objects of general potential relevance to the entity, thus allowing the interruption of ongoing top-down processing. For an agent, the influence and importance of each component will likely be based on the agent's intended application, visual capabilities and the types of environments it will encounter: The importance of the bottom-up component as a supporting mechanism for the top-down variety no doubt increases as the environment becomes more complex, dynamic and unpredictable; the associated computational burden also increases.

3.2.2 Spatial and Object Processing

Experimental evidence suggests that attention can be deployed in at least two ways: According to space-based accounts, visual attention is directed to locations in the scene, and functions like a spotlight that enhances the processing of stimuli within its beam (Posner, 1980) or a zoom-lens (Eriksen and Yeh, 1985). The object-based perspective suggests that attention is directed towards objects or perceptual groups from the visual scene that have been segmented (Kahneman and Henrik, 1981; Neisser, 1967). This differentiation also holds true for agent visual attention systems, which may conduct their processing in a spatial and/or object-based manner. The choice is an important consideration: Models that operate based solely on spatial representations (such as the saliency map) can have no notion of objects or their associated semantics – in these models, attention must be directed according only to low-level image features. In contrast, object models will have difficulty in detecting factors that are hard to describe at the object level: examples include texture and colour, and lighting and shadows. The types of objects that are detected are usually general environmental objects (for virtual applications) or detection of specific types of objects in the case of real systems, such as human faces and gestures. It is desirable for a broadly capable attention model to handle data in both an object-based and a spatially based manner, as each has limitations that are alleviated by the other.

Object-based models tend to ignore artefacts in the environments that have not been specifically predefined as objects. For example, spotlights and other lighting phenomena may be difficult to account for in an object-based system. On the other hand, spatial attention does not allow access to in-depth semantic and associative details that may be available about scene contents. Spatial models operate on a spatial representation of the agent's vision, a rendered view for example (captured using synthetic vision: see Sect. 2.2.1). Bottom-up approaches use the spatial input by processing competitive interactions between one or more basic image features such as colour, intensity, depth and motion. There is no notion of objects, their related attributes or semantics here – only image elements are available for attention-related calculations. In some cases, competition between these features is used to create a saliency map (Itti et al., 2003; Courty and Marchand, 2003), which is a 2D greyscale representation of the areas of the scene deemed most likely to 'pop out' to the viewer. In contrast, object-based models account for objects, their associated properties and allow association with semantic information. A form of staged memory or priority queue mechanism is usually used in conjunction with management variables that allow the agent to calculate the current object of interest using a heuristic (for example, based on information certainty maintenance (Kim et al., 2005), uncertainty reduction (Peters and Sullivan, 2003) or threat value (Hill, 2000) of the object). Object-based approaches may be used in the implementation of both bottom-up and top-down systems. It is a relatively trivial task to obtain input for an object-based system from a geometric sensor, for example by casting a ray along the direction of the agent's view and checking for collisions. False-colour renderings provide a way to interface spatial perceptual input with object-based attention. Top-down approaches making use solely of spatial input are less common; those that exist search for target locations based on basic aspects of their visual appearance, colour for example (Terzopoulos and Rabie, 1997).

3.2.3 Mobile or Static Viewer

Another design consideration is whether the agent is to be mobile within its environment. This is important due to what is referred to as inhibition of return (IOR), that is, how the model remembers previous winners in the competition for the focus of attention so that their influence can be temporarily reduced. An IOR mechanism is necessary in order to allow the focus of attention to move around the scene so that it does not become locked indefinitely onto a single object location. In purely spatial models, inhibition of return must be stored in view coordinates, i.e. two-dimensional (x, y) coordinates. After a location has won the competition for attention, its saliency is decreased in order to reduce its chances of holding the focus of attention. Problems arise with this system if the viewer is mobile; however, since view coordinates are used, when the view changes, previously stored IOR locations are invalidated. One way to solve this problem is to add an object-based memory system for tracking IOR: Spatial locations are resolved to objects and IOR data is then stored on a per object basis in memory, as opposed to a per location basis, providing a solution to the problem.

3.2.4 Real and Virtual Environment

As we have already seen for the issue of sensing from the environment (see Sect. 2), the differentiation between the type of environment the agent is to be embedded in, real or virtual, is an important one. The main differentiation to be made relates to how the environment is sensed and segmented and the amount of computation associated with this. Agents that sense from the virtual environment usually do so using synthetic vision or ray-casting techniques and have at their disposal an environment database containing all objects and their attributes. In this case, scene segmentation and object recognition is not necessary, making object-based approaches popular. However, the time taken to employ spatial techniques, which require a rendering of the scene from the point of view of the agent, can be a limiting factor, although the use of a visual attention model implemented on the graphics processing unit can help to alleviate this situation. In contrast, systems that take input from real scenes usually employ spatial approaches, as this does not entail costly and complex computer vision operations which would be required to obtain object representations. Sometimes such systems also include a degree of object processing, but this is usually very specific due to the cost and complexity of the operations.

3.2.5 Social Processing

The overriding purpose of computational models of attention is to highlight certain aspects of the scene to be prioritised for preferential processing. Thus far, this may be done in a generic manner according to contrast between basic spatial features, such as colour or intensity, or specific to features related to objects, such as the amount of time that an object has engaged the focus of attention.

3.2.6 Putting It All Together

In practice, models usually employ a mix of the features described here, such as top-down object-based (Gillies, 2001), bottom-up spatial and object-based (Peters and Sullivan, 2003) or top-down object-based and bottom-up spatial (Chopra-Khullar and Badler, 1999). A more complete model of attention could be seen as attempting to handle all of these factors adequately, although in practice, models are normally constructed to be appropriate to a specific target scenario. For example, it should not be necessary for an ECA that is to be embedded inside a stationary museum exhibit to handle IOR problems relating to a moving viewer (as described in Sect. 3.2.3).

In Table 2 we provide a list of classifications for several popular agent attention models based on the four factors described.

By endowing agents with the ability to sense and attend selectively to their environment, the next important step is to be able to evaluate the effectiveness of these models and the behaviours they can be used to drive.

Table 2 Summary of main features for several popular agent attention models based on four key factors: top-down (TD) and bottom-up (BU) processing model with associated saliency heuristics if applicable, spatial- (SPA) and object-based (OBJ) processing, static or mobile (MOB) viewing possible, real (RE) or virtual (VR) input type and presence of social modulation (SOC). Contrast heuristics available in bottom-up saliency map models for agents may consist of *dep*th, *col*our, *ori*entation, *int*ensity, *fli*cker, *mot*ion. An 'X' indicates that the heuristic or feature is simulated in the corresponding model

Model	TD	BU	dep	col	ori	int	fli	mot	OBJ	SPA	MOB	SOC	ENV
Chopra and Badler (2001)	X	X		X					X	X	X		Vr
Gillies (2001)	X								X		X		Vr
Peters and O'Sullivan (2003)		X		X	X	X		X	X	X	X		Vr
Itti et al. (2003)		X		X	X	X	X	X		X			Re
Courty et al. (2003)		X	X			X				X	X		Vr
Gu et al. (2005)	X	X		X	X	X			X	X		X	Re
Picot et al. (2007)		X		X	X	X		X	X	X		X	Re

4 Evaluation

Evaluating how humans perceive embodied agents is an important topic for at least two reasons. First of all, it is an integral part of the iterative design process, allowing agent designers to evaluate how successful their appearance and behaviour modelling approaches are, in order to design better future models. Various characteristics of agent appearance and behaviour can have far-reaching effects in influencing human perceivers. For example, at the most basic level, the mere presence of a humanoid agent can greatly impact the ease and efficiency with which a human can interface with a machine in order to carry out tasks. Secondly, evaluations help to provide insight into the human side of the equation. As agents become more sophisticated, it is increasingly common for them to be endowed with computational models inspired from the social and brain sciences. Finding out where and why these models deviate from expectations can provide valuable feedback to those researchers investigating the functioning of the human mind. We provide a broad overview of research evaluating how humans may perceive agents – particularly related to their eye gaze and attentive behaviours. Evaluation may be viewed from two different perspectives: First of all, the artificial regions of interest generated by an attention model over a sequence of images may be compared with those of a real human (Sect. 4.1). Second, the animated behaviour of the agent can be evaluated by human users, for example reporting their experiences through questionnaires, or by considering their performance or behaviour when conducting a task or interacting with the agent.

4.1 Quantitative Comparisons with Human Data

A number of different approaches are available for comparing human eye fixations with data generated from computational attention models. Unsurprisingly, all of these approaches require the use of an eye tracker for capturing the eye movements of the human participants while they view a number of images or videos. These images or video frames are then passed to the computational attention model being evaluated in order to generate outputs. In the discussion that follows, we will generally consider these outputs to be in the form of a map (such as the saliency map – see Sect. 3.1), although some of the methods we describe can also be applied to other representations. The human fixations and automatically generated maps can then be compared in a variety of ways, three of which we detail next.

4.1.1 String Editing

The string editing (Privitera and Stark, 2000) approach compares the similarity between scan-paths, i.e. the sequences of fixations and saccadic eye movements that eyes make when inspecting a scene. It was one of the first methods for quantitatively comparing not only the loci of fixations, but also their temporal ordering. In this way, it considers if fixations are deployed in the same parts of a scene and also if this deployment takes place in the same sequence. The comparison relies on a clustering of eye fixation points into a number of discrete regions of interest (ROIs), where each clustered ROI is assigned a unique character label. Thus, a temporal sequence of ROIs can be described by a string of characters, and two scan-paths can be compared by manipulating their strings to transform one into the other, while keeping a track of the costs which have been associated a priori to each editing operation. For computational systems to be tested with this method, their output must be obtained in the form of artificial regions of interest, or *aROIs*, which are the artificial equivalents of the human scan-paths. Comparison between human scan-paths and artificial scan-paths then proceeds in a similar manner as described above for two human scan-paths.

In practice. string editing is not always used frequently for evaluating specific heuristics of computational models being tested. This is because it accounts not only for the similarity regarding *where* visual attention is deployed in a scene, but also for the similarity in the ordering of *when* it gets there. These are very challenging criteria for any contemporary attention model to meet, especially given the natural variability that occurs betweens sequences of human scan-paths. Other approaches therefore look more closely at correlations regarding where attention is deployed and do not consider the ordering.

4.1.2 Heuristic Scoring Metrics

Instead of resolving a heuristic into artificial ROIs in order to compare to human ones, as in the string editing approach, another method (Peters and Itti, 2008) directly compares the human eye fixations with the maps generated by the computational methods, referred to as *heuristic response maps*. This is done by sampling each heuristic response map in the neighbourhood of the actual saccade target and at a number of uniformly random locations. A response map is deemed as being a good predictor of eye movement if it has a strong peak in the neighbourhood of the actual human saccade target and has little activity elsewhere, since some heuristics may produce maps that have moderate or high activity over the whole map. Therefore, good heuristics are those that generate response maps in which the values at locations fixated by observers are statistically discriminable from those values at non-fixated or random locations.

4.1.3 Human Attention Maps

Rather than transform the maps generated by computational models into aROIs or comparing human ROIs with the attention maps, a third alternative is to transform the human ROIs into maps with the same format as those output by the computational models. These *human attention maps* (Ouerhani et al., 2004) are thus constructed from human eye fixations and then compared quantitatively by a direct comparison of the similarity of both maps using objective comparison criteria.

4.2 Other Comparisons

As well as comparing ROIs between agents and humans, others evaluation techniques use the final animated behaviour of the agent to obtain measures of performance. For example, users may report their experiences through questionnaires, or their performance or behaviour may be monitored while conducting a task or interacting with the agent.

In terms of endowing agents with basic gaze behaviour, improving the gaze behaviour of agents in human–agent interaction produces noticeable effects on the perception of the realism of the user and the way in which communication proceeds (Thórisson, 1997; Vertegaal and Ding, 2002). This would seem to reflect the importance of gaze in human interaction situations, where it is pivotal in sending social signals, receiving information and controlling the flow of interaction.

On a fundamental level, experiments have been conducted to determine the circumstances under which users perceive an agent as paying attention and showing interest in them through gaze and body part orientation (Peters, 2006). Users are able to obtain strong impressions of agents engaging in mutual attention and other attention-related behaviours with them according to their gaze and body part orientation and locomotion direction.

4.2.1 Quality of Interaction

A number of studies have evaluated the effects of varying avatar and agent eye gaze models on the quality of an interaction. When involved in a dyadic interaction with an avatar, it has been found that users pay more attention to the avatar when an active gaze model is used that takes into account who is taking the conversation turn and where the user is looking, than when there is merely fixed gaze (Colburn et al., 2000). Other studies have tended to compare three conditions to test effects on the perceived quality of communication: fixed gaze, random gaze and natural gaze behaviour. In the random condition, the eye and head are usually timed randomly and do not follow conventional patterns of gaze. In the natural condition, the behaviours are related to the conversation by being informed in some way. For example, head animations may be tracked while eye movements are either inferred from conversational turn-taking, i.e. 'while speaking' and 'while listening' situations (Garau et al., 2001), or based on a statistical eye movement model from gaze tracking analysis of real people (Lee et al., 2002). The results of these studies seem to indicate that inferred gaze significantly outperforms random and static gaze models. When there are no eye movements, the character is perceived as lifeless and having a cold personality, whereas with random eye movements, the character may be perceived as having an unstable or distracted quality. An informed model results in a more purposeful, natural and outgoing agent. Such models have also been found to be easier to use and help users perform certain tasks faster (van Es et al., 2002). For non-task-oriented systems, studies have suggested that the amount of mutual gaze provided by the agent to the user plays a key role in how they evaluate it (Vertegaal and Ding, 2002).

4.2.2 Emotional Aspects in Interaction

The study of the effect of emotional displays by facial expressions, body language and so forth on the human user is indispensable for creating more plausible interactions.

In a game-playing situation, for example, Rehm and André (2005) investigated head movements as one of the most important predictors of conversational attention. On the one hand, the authors were able to confirm a number of findings about attentive behaviors in human–human conversation. For instance, the subjects spent more time looking at an individual when listening to it than when talking to it – no matter whether the individual was a human or a synthetic agent. In addition, people tended to avoid gaze contact with the conversational agent when they were lying. While the users' behaviours in the user-as-speaker condition were consistent with findings for human–human conversation, they also noticed some differences for the user-as-addressee condition. People spent more time looking at an agent that is addressing them than at a human speaker. One explanation of the user's strong attention towards the agent is the attractiveness of the exceptional conversational partner. None of the participants had encountered an embodied conversational agent in an application yet. All of the participants had already seen some agents as manifestations of a new interface metaphor in their courses, but they had not interacted with an agent so far. Even though the participants got some time to familiarise with the agent, the sensation of interacting with a synthetic agent might have persisted for a longer time. Maintaining gaze for an extended period of time is usually considered as rude and impolite. The fact that humans do not conform to social norms of politeness when addressing an agent seems to indicate that they do not regard the agent as an equal conversational partner, but rather as an artefact that is able to communicate.

5 Conclusions and Outlook

An important issue that has not been focused on here has been how sensed and filtered data can be used as part of the action selection process or to generate agent behaviour. To take the example of real-world input, the detection of a skin region, in this case detection of a human face, could trigger two different actions: first, the attentive agent could follow the tracked human face by looking at it and second, a face recognition algorithm would be used to check whether the face in question exists within a pre-built database of known people; if this was the case, the agent could display different behavioural characteristics, such as smiling at the person walking by. Alternatively, in a simpler case, motion detection would indicate when the person stops in front of an exhibit and trigger an animation reflecting recognition of this event. Other examples may be more subtle. For example, rather than have an agent do nothing when there are no tasks at hand, it is often the case for agents to conduct idle motions so that they do not freeze, which looks unrealistic. The techniques discussed here can be applied to such situations by having the agent look around the scene based on its attention model. Unlike randomly generated gaze motions, those provided by the attention model will be consistent and coincide with the environment; if something moves quickly in front of the agent, it will be seen to look at it.

The internal representation of perceived stimuli is another important aspect; for example, the affective aspects of sensed stimuli could make use of emotional representation languages as those discussed in Chapter "Embodied Conversational Characters: Representation Formats for Multimodal Communicative Behaviours" in this part.

An important area of future endeavour is the integration of affective and social competencies into perceptual and attention frameworks. At the sensing stage, this encompasses the ability to be able to detect and represent emotional and social events and properties from the environment, for example, in relation to the detection of human faces, motion and gesture (see Sect. 2.1), to be able to recognise and categorise facial expressions and combine them multimodally into higher level representations of a user's possible emotional and cognitive states. In perceptual attention terms, it relates to resolving competition between competing affective and social stimuli in order to determine those that are of the most relevance to the entity at a particular time. Such relevance may relate to a variety of different aspects operating at many different levels of sophistication: To consider examples of just a few of these, at the level of embedded long-term goals, by prioritising the processing of potentially threatening stimuli for the purposes of survival; at a task level, by paying attention to cars while crossing a road; at a social level, by looking at a speaker in order to show interest as a listener. The issues involved are no doubt complex, diverse and challenging.

References

Blumberg B (1997) Go with the flow: synthetic vision for autonomous animated creatures. In: Lewis Johnson W, Hayes-Roth B. (ed) Proceedings of the first international conference on autonomous agents (Agents'97). ACM Press, New York, NY, pp 538–539

Bordeux C, Boulic R, Thalmann D (1999) An efficient and flexible perception pipeline for autonomous agents. Proceedings of Eurographics '99, Milano, Italy, pp 23–30

Bradski GR (1998) Computer vision face tracking for use in a perceptual user interface. Intel Tech J Q2:705–740

Bradski GR, Kaehler A (2008) Learning OpenCV: computer vision with the openCV library. O'Reilly, Cambridge, MA

Camurri A, Mazzarino B, Volpe G (2004) Analysis of expressive gesture: the eyesweb expressive gesture processing library. In: Camurri A, Volpe G. (eds) Gesture-based communication in human-computer interaction. LNAI vol 2915, Springer, Berlin, pp 460–467

Camurri A, Coletta P, Varni G, Ghisio S (2007) Developing multimodal interactive systems with EyesWeb XMI. In: Proceedings of the 2007 conference on new interfaces for musical expression, June 6–10, New York, NY, pp 305–308

Castellano G Movement expressivity analysis in affective computers: from recognition to expression of emotion. Ph.D. Dissertation, Faculty of Engineering, University of Genova, Italy, February 2008

Castellano G, Villalba SD, Camurri A (2007) Recognising human emotions from body movement and gesture dynamics. In: Paiva A, Prada R, Picard RW. (eds) affective computing and intelligent interaction, second international conference, ACII 2007, Lisbon, Portugal, September 12–14, 2007, Proceedings, vol 4738 of LNCS. Springer, Berlin, pp 71–82

Castellano G, Mortillaro M, Camurri A, Volpe G, Scherer K (2008) Automated analysis of body movement in emotionally expressive piano performances. Music Percept 26(2):103–119

Chopra-Khullar S, Badler NI (1999) Where to look? automating attending behaviors of virtual human characters. In: AGENTS '99: Proceedings of the third annual conference on autonomous agents. ACM Press, New York, NY

Colburn A, Cohen M, Drucker S (2000) The role of eye gaze in avatar mediated conversational interfaces, Tech. Report MSR-TR-2000-81, Microsoft Research

Conde T, Thalmann D (2006) An integrated perception for autonomous virtual agents: active and predictive perception. Comput Animation amp Virtual Worlds 17(3–4):457–68

Courty N, Marchand E (2003) Visual perception based on salient features. In: IEEE international conference on intelligent robots and systems, IROS'03, vol 2, Las Vegas, Nevada, October 2003, pp 1024–1029

Delgado-Mata C, Aylett R (2001) Communicating emotion in virtual environments through artificial scents. In: IVA '01: Proceedings of the third international workshop on intelligent virtual agents. Springer, London

Eriksen CW, Yeh YY (1985) Allocation of attention in the visual field. J Exp Psychol Hum Percept Perform 11(5):583–597

Garau M, Slater M, Bee S, Sasse MA (2001) The impact of eye gaze on communication using humanoid avatars. In: CHI '01: Proceedings of the SIGCHI conference on Human factors in computing systems. ACM Press, New York, NY, pp 309–316

Gillies M (2001) Practical behavioural animation based on vision and attention. PhD dissertation, University of Cambridge Computer Laboratory

Gu E, Wang J, Badler NI (2005) Generating sequence of eye fixations using decision-theoretic attention model. IEEE computer society conference on computer vision and pattern recognition (CVPRW'05) – Workshops, p 92

Herrero P, de Antonio A (2003) Introducing human-like hearing perception in intelligent virtual agents. In: Proceedings of the second international joint conference on autonomous agents and multiagent systems (AAMAS), July 14–18, Melbourne, Australia

Hill R (2000) Perceptual attention in virtual humans: towards realistic and believable gaze behaviours, AAAI Technical Report FS-00-03, pp 46–52

Isla D, Burke R, Downie M, Blumberg B (2001) A layered brain architecture for synthetic creatures. In: Nebel B. (ed) Proceedings of the seventeenth international joint conference on artificial intelligence, IJCAI 2001, Seattle, Washington, USA, August 4–10, 2001, Morgan Kaufmann, 2001, pp 1051–1058

Itti L, Koch C, Niebur E (1998 November) A model of saliency-based visual attention for rapid scene analysis. IEEE Trans Pattern Anal Mach Intell 20(11):1254–1259

Itti L, Dhavale N, Pighin F (2003) Realistic avatar eye and head animation using a neurobiological model of visual attention. In: Proceedings of the SPIE 48th annual international symposium on optical science and technology, San Diego, USA, August 3–8, pp 64–78

Kahneman D, Henrik A (1981) Perceptual organization, chapter perceptual organization and attention. Erlbaum, Hillsdale, NJ

Kim Y, Van Velsen M, Hill R (2005) Modeling dynamic perceptual attention in complex virtual environments. In: Panayiotopoulos T, Gratch J, Aylett R, Ballin D, Olivier P, Rist T. (eds) Intelligent virtual agents, 5th international working conference, IVA 2005, Kos, Greece, September 12–14, 2005, Proceedings. Lecture notes in computer science, vol 3661. Springer, London, pp 266–277

Koch C, Ullman S (1985) Shifts in selective visual attention: towards the underlying neural circuitry. Human Neurobiol 4:219–227

Kranstedt A, Lücking A, Pfeiffer T, Rieser H, Wachsmuth I (2006) Deixis: how to determine demonstrated objects using a pointing cone. In: Gibet S, Courty N, Kamp JF (eds) Gesture in human-computer interaction and simulation, vol 3881. Springer, New York, NY, pp 300–301

Kuffner J, Latombe J-C (1999 May) Fast synthetic vision, memory, and learning models for virtual humans. In: Proceedings of computer animation '99. IEEE Computer Society, Washington, DC

Laptev I, Lindeberg T (2003) Space-time interest points. In: Proceedings of ICCV03, Nice, France, October 13–16, pp 432–443

Lee SP, Badler JB, Badler NI (2002) Eyes alive. In: Proceedings of the 29th annual conference on computer graphics and interactive techniques (SIGGRAPH '02). ACM Press, New York, NY, pp 637–644

Leonard T (2003) Building an AI sensory system – examining the design of thief - the dark project. In: Proceedings of game developers conference 2003, CMP Game Media Group San Francisco, CA

Lozano M, Lucia R, Barber F, Grimaldo F, Soares AL, Fornes A (2003) An efficient synthetic vision system for 3d multi-character systems. In: Rist T, Aylett R, Ballin D, Rickel J (eds) Intelligent agents, 4th international workshop, IVA 2003, Kloster Irsee, Germany, September 15–17, 2003, Proceedings, vol 2792 of Lecture notes in computer science. Springer, New York, NY, pp 356–357

Morency L-P, Darrell T (2004) From conversational tooltips to grounded discourse: head pose tracking in interactive dialog systems. In: Proceedings of the 6th international conference on multimodal interfaces, State College, PA, October 2004, pp 32–37

Neisser U (1967) Cognitive psychology. Appleton-Century-Crofts New York, NY

Noser H, Renault O, Thalmann D, Thalmann NM (1995) Navigation for digital actors based on synthetic vision, memory and learning. Comput Graph 19(1):7–19

Oikonomopoulos A, Patras I, Pantic M (2006 June) Spatiotemporal salient points for visual recognition of human actions. IEEE Trans Syst Man, Cybern B Cybern 36(3):710–719

Ouerhani N, von Wartburg R, Hugli H (2004) Empirical validation of the saliency-based model of visual attention. Electron Lett Comput Vis Image Anal 3(1):13–24

Pantic M, Bartlett MS (2007) Machine analysis of facial expressions. In: Delac K, Grgic M (eds) Face recognition. I-Tech Education and Publishing, Vienna, Austria, pp 377–416

Patras I, Pantic M (2004) Particle filtering with factorized likelihoods for tracking facial features. In: Proceedings of the IEEE international conference on face and gesture recognition. Seoul, Korea, May 17–19, pp 97–102

Peters C (2006) Evaluating perception of interaction initiation in virtual environments using humanoid agents. In: Proceedings of the 17th European conference on artificial intelligence, Riva Del Garda, Italy, August 2006, pp 46–50

Peters C, O'Sullivan C (2002) Synthetic vision and memory for autonomous virtual humans. Comput Graph Forum 21(4):743–743

Peters C, O'Sullivan C (2003) Bottom-up visual attention for virtual human animation. In: CASA '03: proceedings of the 16th international conference on computer animation and social agents (CASA 2003), IEEE Computer Society, Washington, DC, p 111

Peters RJ, Itti L (2008) Applying computational tools to predict gaze direction in interactive visual environments. ACM Trans Appl Percept 5(2):8

Picot A, Bailly G, Elisei F, Raidt S (2007) Scrutinizing natural scenes: controlling the gaze of an embodied conversational agent. In: Pelachaud C, Martin J-C, André E, Chollet G, Karpouzis K, Pelé D (eds) Intelligent virtual agents 2007, Proceedings, LNCS, vol 4722. Springer, Berlin, pp 272–282

Poggi I, Pelachaud C, de Rosis F (2000) Eye communication in a conversational 3d synthetic agent. AI Commun 13(3):169–182

Posner MI (1980) Orienting of attention. Quart J Exp Psychol 32:3–25

Privitera H, Stark LW (2000) Algorithms for defining visual regions-of-interest: Comparison with eye fixations. IEEE Trans Pattern Anal Mach Intell (PAMI) 22(9):970–982

Rabiner LR (1990) A tutorial on hidden Markov models and selected applications in speech recognition. In: Waibel A, Lee K-F (eds) Readings in speech recognition. Morgan Kaufmann Publishers Inc., San Francisco, CA, pp 267–296

Raidt S, Bailly G, Elisei F (2005) Basic components of a face-to-face interaction with a conversational agent: mutual attention and deixis. Proceedings of the 2005 joint conference on smart objects and ambient intelligence: innovative context-aware services: usages and technologies, Grenoble, France, October 12–14, pp 247–252

Rapantzikos K, Avrithis Y, Kollias S (2005) Handling uncertainty in video analysis with spatiotemporal visual attention. In: Proceedings of IEEE international conference on fuzzy systems, Reno, Nevada, May 2005

Rapantzikos K, Tsapatsoulis N, Avrithis Y, Kollias S (June 2007) A bottom-up spatiotemporal visual attention model for video analysis. IET Image process 1(2):237–248

Rehm M, André E (2005) Where do they look?: gaze behaviors of multiple users interacting with an embodied conversational agent, vol 3661, pp 241–252

Reynolds CW (2000) Interaction with groups of autonomous characters. In: Proceedings of game developers conference 2000. CMP Game Media Group, San Francisco, CA, pp 449–460

Scherer K, Schorr A, Johnstone T (February 2001) Appraisal processes in emotion: theory, methods, research (series in affective science). Oxford University Press, Bethesda, MD

Terzopoulos D, Rabie TF (1997) Animat vision: active vision in artificial animals. Videre J Comput Vis Res 1(1):2–19

Thórisson KR (1997) Gandalf: an embodied humanoid capable of real-time multimodal dialogue with people. In: AGENTS '97: Proceedings of the first international conference on autonomous agents. ACM Press, New York, NY, pp 536–537

Treisman AM, Gelade G (1980) A feature integration theory of attention. Cogn Psychol 12:97–136

Tu X, Terzopoulos D (1994) Artificial fishes: physics, locomotion, perception, behavior. In: SIGGRAPH '94: Proceedings of the 21st annual conference on computer graphics and interactive techniques. ACM Press, New York, NY, pp 43–50

van Es I, Heylen D, van Dijk B, Nijholt A (2002) Making agents gaze naturally – does it work? Proceedings of the working conference on advanced visual interfaces, AVI '02, Trento, Italy, pp 357–358, ACM, New York, NY

Vertegaal R, Ding Y (2002) Explaining effects of eye gaze on mediated group conversations: amount or synchronization? In: Churchill EF, McCarthy J, Neuwirth C, Rodden T (eds), CSCW '02: proceedings of the 2002 ACM conference on computer supported cooperative work. ACM Press, New York, NY, pp 41–48

Viola P, Jones MJ (2001) Rapid object detection using a boosted cascade of simple features. In: Proceedings of the 2001 IEEE computer society conference on computer vision and pattern recognition, vol 1. Kauai, Hawaii, pp 511–518, December 8–14, IEEE Computer Society, Los Alamitos, CA

Vosinakis S, Panayiotopoulos T (2003) Programmable agent perception in intelligent virtual environments. In: Rist T, Aylett R, Ballin D, Rickel J (eds) Intelligent agents, 4th international workshop, IVA 2003, Kloster Irsee, Germany, September 15–17, 2003, proceedings, vol 2792 of Lecture notes in computer science. Springer, London, UK

Yarbus A (1967) Eye movements and vision, chapter Eye movements during perception of complex objects. Plenum Press, New York, NY

Zeng Z, Pantic M, Roisman G, Huang T (2009 January) A survey of affect recognition methods: Audio, visual, and spontaneous expressions. IEEE Trans Pattern Anal Mach Intell 31(1):39–58

Generating Listening Behaviour

Dirk Heylen, Elisabetta Bevacqua, Catherine Pelachaud, Isabella Poggi, Jonathan Gratch, and Marc Schröder

Abstract In face-to-face conversations listeners provide feedback and comments at the same time as speakers are uttering their words and sentence. This 'talk' in the backchannel provides speakers with information about reception and acceptance – or lack thereof – of their speech. Listeners, through short verbalisations and non-verbal signals, show how they are engaged in the dialogue. The lack of incremental, real-time processing has hampered the creation of conversational agents that can respond to the human interlocutor in real time as the speech is being produced. The need for such feedback in conversational agents is, however, undeniable for reasons of naturalism or believability, to increase the efficiency of communication and to show engagement and building of rapport. In this chapter, the joint activity of speakers and listeners that constitutes a conversation is more closely examined and the work that is devoted to the construction of agents that are able to show that they are listening is reviewed. Two issues are dealt with in more detail. The first is the search for appropriate responses for an agent to display. The second is the study of how listening responses may increase rapport between agents and their human partners in conversation.

1 Introduction

In many books and papers, the process of communication is schematically depicted with a speaker who is active in the speech process and the listener who is involved in passively perceiving and understanding the speech. According to Bakhtin (1999) linguistic notions such as 'the "listener" and "understander" (partners of the "speaker") are *fictions* which produce a 'distorted idea' of the process of speech communication.

D. Heylen (✉)
University of Twente, Enschede Faculty of Electrical Engineering Mathematics and Computer Science, The Netherlands
e-mail: d.k.j.heylen@ewi.utwente.nl

P. Petta et al. (eds.), *Emotion-Oriented Systems*, Cognitive Technologies,
DOI 10.1007/978-3-642-15184-2_17,

> One cannot say that these diagrams are false or that they do not correspond to certain aspects of reality. But when they are put forth as the actual whole of speech communication, they become a scientific fiction. The fact is that when the listener perceives and understands the meaning (the language meaning) of speech, he simultaneously takes an active, responsive attitude toward it. He either agrees or disagrees with it (completely or partially), augments it, applies it, prepares for its execution, and so on. And the listener adopts his responsive attitude for the entire duration of the process of listening and understanding, from the very beginning – sometimes literally from the speaker's first word. [. . .] Any understanding is imbued with responsive and necessarily elicits it in one form or another: the listener becomes a speaker.

Moreover, Bakhtin claims, any speaker is in a sense also a respondent. It seems then that when one attempts to create virtual humans that act as listeners, one is engaged in writing science fiction in the second degree unless one takes the dialectic between speaking and listening by listeners and speakers, respectively, into account.

In order to create agents that can listen to the speech of the humans they interact with, we need to have a proper understanding of what constitutes listening behaviour and how communication in general proceeds. In the first section of this chapter we will introduce the major terms and concepts that are relevant for understanding what listeners do. After this we can turn to the many challenges that are involved in creating conversational agents that have similar abilities. We will focus on two issues that have been considered in the virtual agent literature. The first involves the use of conversational agents or synthesised vocal expressions in the search for listener signals. The second point concerns the use of 'active' listening behaviours to create rapport with the human interlocutor.

2 Understanding Communication

Bakhtin is not the only one who makes the point that listeners are not just passive recipients of messages emitted by a speaker. Conversation has been characterised as a collaborative activity, an interactional achievement or a joint activity by researchers such as Gumperz (1982), Schegloff (1982) and Clark (1996). By using the term interactional achievements Schegloff highlights the fact that conversations are incrementally accomplished and they involve dependency of the actions of one participant on the actions of the other and vice versa. The term joint activity is used by Clark to emphasise that it is only when the participatory actions of the different participants are seen *together* that one can talk about a conversation.

Communicative actions of one participant implicate the others in many ways. A typical communicative action is normally produced with the intention that one or more other participants (the addressees, the audience, the 'listeners') attend to them, are able to perceive them, recognise the behaviour as an instance of a communicative action, try to understand them and possibly act upon them in one way or another, preferably with the effect that the producer of the communicative action had intended to achieve. If these conditions are not met the action will fail to be 'happy' in Austin's term (Austin, 1962) or will not be 'felicitous' (Searle, 1969).

The success of a communicative action thus depends on the states of mind and the behaviours of the other participants during the preparation and execution and ending of the communicative behaviours. As Schegloff and others have pointed out, the behaviours of the other participants not only determine success but they may also influence and change the execution of the communicative actions *as they are being produced*, because the producer of the action will take notice of how the audience receives and processes the actions and also of the other reactions they invoke. A nice example is provided by Goodwin (1984) who defines as a principal rule in face-to-face conversation that 'When a speaker gazes at a recipient that recipient should be gazing at him. When speakers gaze at nongazing recipients, and thus locate violations of the rule, they frequently produce phrasal breaks, such as restarts and pauses, in their talk' (Goodwin, 1984, p. 230). Similarly, Kraut et al. (1982) conducted some experiments which made it clear how speakers adjust the informational density of their talk depending on the kind and amount of verbal feedback they receive from listeners. Speakers may also monitor listeners for the various actions besides listening that they are involved in. An experiment set up by Clark and Krych (2004), for instance, made it clear that in a collaborative task, not being able to monitor the other's face and eye gaze had less of an effect than not being able to see the other's workspace and what activity was being performed. Clearly, the setting and task involved in the conversation may assign different priorities to what kind of feedback of the interlocutors is important to monitor and what effect this has on the way the conversation proceeds.

We can picture the interaction between actions of the participants in conversation in a first, simple diagram (Fig. 1) which is only slightly more complicated than the fictions Bakhtin was referring to but it tries to show something more of the dialogical nature of conversation.

For the sake of simplicity, assume that a conversation takes place between two persons (x and y). Given that some conversational action (CA1) is performed by one of them (say x), as indicated by the top left corner (A) of this diagram, the other person (y) is supposed to perceive and interpret this action, as indicated by the top right corner (B). We will summarise the various actions that this involves using the term 'perceive', which is taken from the classical notion in artificial intelligence

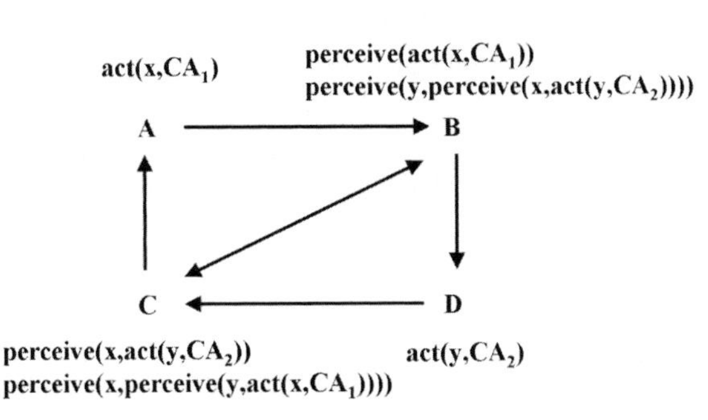

Fig. 1 Picturing conversation as an interactional achievement

that an intelligent agent is involved in perception–decision–action loops. This may prompt this person (*y*) (i.e. lead *y* to decide) to produce certain actions (CA2 in the bottom right corner, D). These actions in turn can communicate something to the producer of CA1 (*x*) about the reception and up-take of the production of CA1 by *y* (bottom left corner, C) which may either change the execution of action CA1 or prompt a new action. The behaviours that make up the act of perception of CA1 by *y* (B) may themselves be observable to *x* who is monitoring them, hence the arrow connecting corner B with C. Vice versa, the actions that go into the perception of CA2 by *x* may also be observable to *y*. Actions by one thus elicit actions by the other in reply.

So far, only general terms such as 'communicative action', 'producer' and 'recipient' and 'perceiver' were used because any action could enter these perception–action loops. Therefore, also the time scale was left unspecified. The diagram can be instantiated in many different ways. For instance, the communicative action CA1 by *x* could be the utterance of statement, which makes *x* a *speaker* during which *y*, the *listener*, attending to the speech, shows a puzzled face (CA2) accompanied by a vocalisation 'oh' with a rising intonation. This verbal and non-verbal *feedback* in the *backchannel*, which is monitored by the speaker *x*, may prompt *x* to enter into reformulation mode or to speak up. All of this can happen almost instantaneously, slowed down only by the limits of the speed of light, sound and neurons firing but also sped up through the force of anticipation by both *x* and *y* which makes it even possible for the agents to run ahead of events. At any given time, there will be multiple instantiations of the schema active as participants can communicate with different modalities in parallel or because one can view the process as operating on different levels as will be pointed out below.

Another common instantiation is the case where someone (*x*) produces a speech act (CA1), which is attended to and interpreted by *y* who decides to offer a speech act (CA2) in reply, after which *x* responds by producing a new speech act (CA1′). The two participants take alternating turns and each next utterance is a reply to the previous one forming adjacency pairs as they are commonly called in the tradition (Schegloff and Sacks, 1973) of conversation analysis.[1]

A third common instantiation has been labelled *interactional synchrony*. It was first described by Condon and Ogston (1966) and an episode in a conversation was analysed in detail by Kendon (1970). The term refers to the case where the flow of movements of the listener are rhythmically coordinated with those of the speaker. Other forms of coordination have been called mimicry (Chartrand and Bargh, 1999) and mirroring (LaFrance, 1979; Lafrance and Ickes, 1981). Hadar and colleagues (1985) report that approximately a quarter of all the head movements of the listeners in the conversations they looked at occurred in sync with the speech of the interlocutor. Interestingly, McClave (2000) notes that (many of) these kinds of movements may be elicited by the speaker.

[1] Goffman (1976) provides a very insightful analysis of this process of replies and responses.

> Many instances of backchanneling were assumed to be internally motivated; i.e. the listener backchanneled when he or she felt like it. Microanalysis of speaker head movements in relation to listener head movements reveals that what were heretofore presumed to be spontaneous, internally motivated, listener responses are actually responses to the speaker's nonverbal requests for feedback. These requests are in the form of up-and-down nods, and listeners recognize and respond to such requests in a fraction of a second.

Again, this shows the dependence of an action by one participant on the action of another, the back-and-forth of eliciting actions and responses.

Clearly, what has been understood above by a communicative action is very broad. It may involve consciously produced linguistic actions but also actions that were not meant to be communicative by the producer but that still provide information to the recipient. The communicative behaviours may 'signal' in various ways: symbolically, indexically, iconically or through inference.

In the following paragraphs we present a variety of instantiations of this schema as we discuss some central theoretical notions and some common ways in which the interactions between participants in conversation proceed. We will detail how actions of one participant call forth or intend to call forth actions of others and what kinds of responses one can distinguish.

2.1 Speech Acts

The crucial insight that speech act theory (Austin, 1962; Searle, 1969) has emphasised is that 'language is used for getting things done'. Typically, in the case of language, these things implicate the person or persons to which the utterance is being addressed. From a speech act perspective, any utterance is some kind of invitation to the addressees to participate in a particular configuration of actions: Attend to what is being said, try to figure out what is meant and carry out what was intended by the speaker, which could range from updating a belief state, to feeling offended, or closing the window. Speech act theory focusses on the perspective of the speakers and their *intentions* which implicate the audience in that an utterance is primarily intended to get the audience to recognise the speaker's meaning: 'To say that a speaker meant something by X is to say that the speaker intended the utterance of X to produce some effect in the audience by means of the recognition of this intention.' This is essentially Grice's definition (Grice, 1975b). Another way in which the perspective of the speaker comes to the fore is in the way that Grice (1975a) formulates his maxims of cooperative behaviour (be relevant, be conspicuous, etc.) in terms of what the speaker should and should not do. All of these maxims indirectly take listeners into account as they urge the speaker to keep them in mind for the sake of cooperation.

As with any event, a speech event can be described in several ways. One might say that in describing a particular situation the speaker was 'stuttering', 'trying to say something in English', 'trying to propose', 'making a fool of himself', etc. By using the word 'stuttering' one is referring to an aspect of the production and vocalisation process. The second characterisation points out that the vocalisations were not random but attempt to construct an English sentence. The third describes the intention behind the action and the last the effect it may have achieved on the other participants, the observers or those that have heard about the event.

Austin (1962) proposed some different terms to distinguish the levels in the speech event. The uttering itself he called the locutionary act. The act of getting the audience to recognise what is intended is called the illocutionary act (the speaker tries to make it clear that the utterance is intended as a promise, for example). The effects the execution of the speech act has on the audience are called the perlocutionary effects. The acts that caused these effects were the perlocutionary acts. Note that not all of the effects may have been intended. For instance, if the speaker is not aware that the action promised is not something the audience wants, then the promise may actually turn out to be a threat.

In Clark's framework (Clark, 1996), a speaker acts on four levels. (1) A speaker executes a behaviour for the addressee to attend to. This could be uttering a sentence but also holding up your empty glass in a bar (to signal to the waiter you want a refill). (2) The behaviour is presented as a signal that the addressee should identify as such. It should be clear to the waiter that you are holding up the glass to signal to him and not just because of some other reason. (3) The speaker signals something which the addressee should recognise. (4) The speaker proposes a project for the addressee to consider (believe what is being said, except the offer, execute the command, for instance). In this formulation of levels, every action by the speaker is matched by an action that the addressee is supposed to execute: Attend to the behaviour, identify it as a signal, interpret it correctly and consider the request that is made. If one considers the diagram above, one could say that instead of one arrow going from A to B there are four. Also, the arrow should be considered both from the perspective of the speaker and the recipient.

2.2 Monitoring and Feedback

If we take the perspective of the listener, we can make a similar distinction in four levels on which the listener can provide feedback. Allwood (1993), for example, put forward a distinction of the following four basic communicative functions on which the interlocutor can give feedback:

1. Contact (i.e. whether the interlocutor is willing and able to continue the interaction)
2. Perception (i.e. whether the interlocutor is willing and able to perceive the message)
3. Understanding (i.e. whether the interlocutor is willing and able to understand the message)
4. Attitudinal reactions (i.e. whether the interlocutor is willing and able to react and (adequately) respond to the message, specifically whether he/she accepts or rejects it).

Important for all the parties in the cooperative undertaking that is conversation is to know that common ground has been established, that the addressee understands what the speaker intended with the talk produced and the speaker knows that the intentions were achieved. So the feedback that is voluntarily or involuntarily provided by listeners is monitored by the speakers in order to get closure on their actions, i.e. in order to know to what degree the intended actions were successful. Goodwin's rule – whenever a speaker looks at his audience, the audience should look at the speaker – provides a basic example of this need to check for contact and perception. By monitoring the behaviour of the other participants, a speaker can thus derive information about such elements as attention, perception, understanding and the willingness to engage and accept or reject collaboration. Some of the information derives from the actions of listeners that go into perception of the signals (such as their gaze telling something about the focus of attention) but other behaviors may be explicit signals of understanding and agreement or lack thereof through facial expressions or small non-disruptive interjections. This we will discuss in Sect. 2.3. Also the way the utterances are taken up by subsequent actions are informative and provide the speaker with feedback on the conversational moves, of course.

Several conversational actions are conventionally dedicated to establish 'grounding' (the mutual belief by the partners in conversation that they have understood what the contributor meant; Clark and Schaefer (1991)). In Clark and Schaefer, a discourse model is presented in which it is assumed that the presentation phase of the speaker is coupled with an acceptance phase by the recipient which is essential for grounding. The recipient can signal acceptance either in the next moves or by behaviours during the production of communicative actions by the speaker. Obvious signs of neglect of attention or signs of difficulty in understanding will yield reparative actions by the speaker. Positive signs indicating attention, perception, understanding, processing (understanding, agreement, willingness, etc.) will lead the speaker to assume the message has been grounded or successfully executed on all the relevant levels.

> The acceptance phase is usually initiated by B giving A evidence that he believes he understands what A meant by *u*. B's evidence can be of several types. He can say that he understands, as with *I see* or *uh huh*. Or he can *demonstrate* that he understands, as with a paraphrase, or what it is he heard, as with a verbatim repetition. Another is by showing his willingness to go on. The least obvious way is by showing continued attention. (Clark and Schaefer, 1991)

The acceptance phase itself consists of the presentation of a contribution to which the original presenter can react with an accepting contribution, illustrating another way to describe some of the loops presented in Fig. 1.

One type of accepting contribution Clark and Schaefer call *acknowledgements*, which are 'expressions such as *mhm*, *yes*, and *quite* that are spoken in the background, or gestures such as head nods and smiles'. These are commonly called *backchannels*.

2.3 Backchannels

Yngve (1970) is generally credited for having introduced the term. His characterisation is this. Note how it repeats some of the points made by Bakhtin.

> One should hasten to point out that the distinction between having the turn or not is not the same as the traditional distinction between speaker and listener, for it is possible to speak out of turn, and it is even reasonably frequent that a conversationalist speaks out of turn. In fact, both the person who has the turn and his partner are simultaneously engaged in both speaking and listening. This is because of the existence of what I call the back channel, over which the person who has the turn receives short messages such as 'yes' and 'uh-huh' without relinquishing the turn. The partner, of course, is not only listening, but speaking occasionally as he sends the short messages in the back channel. The back channel appears to be very important in providing the monitoring of the quality of communication.

Several authors, Duncan and Fiske for instance (Duncan and Fiske, 1977), have used the term *backchannel* but the interpretation of the term shows some variation. In part, the instability of the meaning can be traced back to the difficulty in specifying the denotation of some terms that one commonly encounters in the definition of *backchannel*, such as *turn* (or *floor*), *listener* (or *hearer*, *auditor*, *recipient*) and *speaker*. Another difficulty in defining the term is that there is quite some variation in the kinds of behaviours and in the kinds of functions that 'listeners' produce as 'feedback.' The term *backchannel* is sometimes reserved for a particular subset of these behaviours and sometimes taken to include a much wider range of behaviours.

Some authors use other terms to refer to similar phenomena sometimes restricting the scope to a particular class of listener responses. Kendon (1967) introduced the term accompaniment signals for 'short utterances that the listener produces as an accompaniment to a speaker, when the speaker is speaking at length' which he divides into two groups: attention signals (in which one appears to signal no more than that one is attending) and assenting signals that express 'point granted' or 'agreement'. Rosenfeld (1987) uses the general term *listener response*. A related concept is that of *acknowledgement token* as used by Jefferson (1984) or *continuers* from Schegloff (1982). Schegloff reflects on the use of 'uh-huh' as a signal of attention, which makes sense only if attention is somewhat problematic. Therefore this attention-signalling function of an 'uh-huh' or a head nod becomes apparent only if it is in response to an extended gaze by the speaker or a rising intonation soliciting some sign of attention, interest or understanding (Schegloff, 1982, p. 79). In other cases, the term continuer may be appropriate, according to Schegloff.

> Perhaps the most common usage of 'uh huh', etc. (in other environments than after yes/no questions) is to exhibit on the part of its producer an understanding that an extended unit of talk is underway by another, and that it is not yet (or may not yet be) complete.

The responses that listeners provide to speakers falling under the general coverall term backchannel (as used by Duncan and Niederehe, 1974) can thus have many functions, depending on the context. In the following section we will look at how some function/form relations can be identified by having people rate different samples, amongst others created by synthesis, using an embodied conversational agent.

2.3.1 Turn-Taking

In the discussion of the schema presented above, an interpretation of the schema was pointed out where a communicative action by one agent was followed by a communicative act by the other agent in the next turn. An important decision that a conversational agent needs to make is when to start speaking and when to stop and listen. So how do participants in a conversation decide when to speak and when to keep quite? Sacks et al. (1974) propose a simple systematic that says that in general a speaker can select the next speaker (for instance by asking a question to a particular person), or that the next speaker can self-select. This view on turn-taking has been criticised by various researchers. One point that is often made is that it is not very contentful. From a general characterisation of turn-taking that should apply to any conversational setting this is probably what is to be expected. The question can also be answered in another way. Instead of taking a structuralist point of view, one can also take the stance of the individual agent. In the same general mode (but now using intentional terms which conversational analysts avoid to invoke) the following could be said to hold: An agent decides to speak when the reasons for speaking outweigh the reasons for not speaking and vice versa, an agent decides not to speak when the reasons for not speaking outweigh the reasons for speaking. Now the question is what are the reasons that play a role in this decision-making process. One can imagine that the factors that play a role are enormously varied and depend a lot on the precise circumstances. Some reasons for speaking that you may have encountered personally are as follows:

1. You have something you would very much like to say.
2. You have just been asked a question and feel the pressure to answer.
3. The current speaker is about to say something embarrassing and you decide to interrupt to save the speaker from loss of face.
4. The current speaker is looking for a word and you help out, by suggesting the word you think the speaker is looking for.
5. You need something done by someone else and talking seems the best way to accomplish this.
6. There is an awkward silence and you ask your guests whether they have already planned where to go on vacation.

Some reasons you may have experienced for not claiming the turn are as follows:

1. You have nothing to say.
2. You are too embarrassed to speak.
3. Someone else is speaking and you need to hear what is being said.
4. You are afraid to say something that will hurt someone's feelings.
5. You would like to say something but the chairperson in the meeting first gives the turn to another participant.
6. You are a suspect in a police investigation; anything you say might be used against you.

7. You provide an accompaniment signal and wait for the current speaker to reach the end of a phonemic clause, i.e. the end of an informational unit, where you think it is no longer impolite to interrupt.

This huge diversity of reasons can be classified into different groups. Some have to do with the business or the task that is being carried out through conversation (task goals); others concern the feelings of the participants, the social conventions (ritual constraints in Goffman's terms (1976) and others seem to operate to make conversations work (system constraints, again using Goffman's terminology). In the following sections, we will not dwell on these issues in detail, but clearly, when designing conversational agents that show the appropriate listening behaviours, one needs to take into account the way they signal they want to continue as listeners or how they display they want to take up the speaking role; Duncan and Niederehe, 1974).

2.4 In Summary

Listeners are not merely passively absorbing what a speaker is saying. They are involved in a number of activities: attending to the actions of the speaker to see what actions the speaker elicits/evokes from them in response, showing speakers that they are attending (implicit feedback) and providing explicit feedback in all kinds of forms. As Fig. 1 shows there is a constant back and forth between the various participants in a conversation where some behaviour by one participant elicits a reaction by the other which is monitored and responded to almost instantaneously. The challenges for building embodied conversational agents are thus manifold. The agent should be able to monitor and interpret the utterances of the human interlocutor 'on the fly'. It should be able to detect the appropriate points where a signal of attention or of agreement is needed, being careful in its timing so as not to disrupt the flow of conversation. The agent should have a repository of behaviours it can execute with all kinds of shades of meaning represented in line with its goals in the conversation and its synthetic personality.

In the following sections we will sketch some work that is currently on its way to create embodied conversational agents that can give the appearance that they know how to listen. In Sect. 3 we report on work that uses embodied agents to build up a library of function/form mappings. Ultimately, the aim is to build engaging agents that people like to interact with. In Sect. 4 we report on ongoing work that measures the effects of the display of appropriate listening behaviours by agents on the sense of engagement and rapport that is experienced by the human interlocutor.

3 Artificial Stimuli and Expression Libraries

The variety of behaviours that listeners display during face-to-face dialogues is very large. The functions that they serve are also multiple. By gazing at the speaker a listener signals attention and that the communication channels are open (Kendon,

1967). By nodding the listener may acknowledge that he has understood what the speaker wanted to communicate. A raising of the eyebrows may show that the listener thinks something remarkable is being said (Ekman, 1979; Chovil, 1991) and by moving the head into a different position the listener may signal that he wants to change roles and say something himself (Duncan and Niederehe, 1974; McClave, 2000). It was already indicated that the behaviours that listeners display are relevant to several communication management functions such as contact management, grounding, up-take and turn-taking (Allwood et al., 1992; Yngve, 1970; Poggi, 2007). They are not only relevant to the mechanics of the conversation but also to the expressive values: the attitudes and affective parameters that play a role. These attitudes can be related to a whole range of aspects, including epistemic and propositional attitudes such as believe and disbelieve but also affective evaluations such as liking and disliking (Chovil, 1991).

Some authors have investigated whether these differences in functions correlate with differences in form. Rosenfeld and Hancks (1980) made a start to determine which nonverbal behaviors of listeners were signalling either attention, understanding or agreement by having independent observers rate 250 listener responses on each of the three dimensions. They found that judgements of 'agreement' were associated with complex verbal listener responses and multiple head nods. Contextually, this occurred when the responses followed the speaker pointing the head in the direction of the listener. Signalling understanding was associated with more subdued forms such as repeated small head nods prior to the speaker finishing a clause. Expressions were rated highest as signalling attention when the listener 'leaned forward prior to the speaker's juncture, audibility of verbal listener response after the juncture, and initiation of gesticulation by the speaker after the juncture but prior to resuming speech' (Rosenfeld, 1987).

Some important characteristics of expressive communicative behaviours are that a behaviour can signal more than one function at the same time and that behaviours may serve different functions depending on the context. In order to create conversational agents that display the appropriate behaviours in the right context it is important to get more insight into the various behaviour to function mappings. Besides looking at naturally occurring contexts, to investigate the relation between form and function, one can also get more insight into what (combinations of) expressions can be used to express what kind of information by generating artificial stimuli that are judged by people. In the following sections two such studies are presented.

3.1 Facial Expressions

In studies by Bevacqua et al. (2007) and Heylen et al., (2007a) a generate and evaluate procedure was used where people were asked to label short movies of the Greta agent displaying a combination of facial expressions. The experiments were conducted to find some prototypical expressions for several feedback functions and to gain insight into the way the various components in the facial expression contribute to its functional interpretation.[2] In particular, the aim of these experiments was to get a better understanding of

- the expressive force of the various behaviours,
- the range and kinds of functions assigned,
- the range of variation in judgements between individuals and
- the nature of the compositional structure (if any) of the expressions.

A lot has been written about the interpretation of facial expressions. This body of knowledge can be used to generate the appropriate facial expressions for a conversational agent. However, there are many situations for which the literature does not provide an answer. This often happens when one needs to generate a facial expression that communicates several meanings from different types of functions: show disagreement and understanding at the same time, for instance. The literature may provide certain pointers to expressions for each of the functions separately, but the way they should be combined may not be so easy. In another way, we know that eyebrow movements occur a lot in conversations with many different functions. The question that arises in this case is whether it makes sense to distinguish them in the way they are performed and the timing of execution or the co-occurrence with other behaviours.

In the studies, the authors looked for expressions for the following functions: *agree, like, understand, disagree, dislike, disbelieve, don't understand* and *not interested.* In the first experiment, reported in Bevacqua et al. (2007), it was found that users could easily determine when a context-free signal conveys a positive or a negative meaning. A first question that was explored in the second test was whether it is possible to find a prototypical signal (or a combination of signals) for each meaning. Is there a signal more relevant than others for a specific meaning or can a single meaning be expressed through different signals or a combination of signals? The hypothesis was that for each meaning, one can find a prototypical signal which could be used later on in the implementation of conversational agents.

A second question was in what way combinations of signals alter the meaning of single backchannel signals. It was conjectured that adding a signal to another could significantly change the perceived meaning. In the study reported on in Heylen et al. (2007a), 60 French subjects were involved in the experiment. They were divided into two groups, each of which judged about half of the movies. The test used the 3D agent, Greta (Pelachaud and Bilvi, 2003). Participants were presented 21 movies. Table 1 shows the signals, chosen from those proposed by Allwood and Cerrato, (2003) and Poggi (2007), that were used to generate the movies.

The meanings the subjects could choose from were *agree, disagree, accept, refuse, interested, not interested, believe, disbelieve, understand, don't understand, like, dislike.*

[2] Similar experiments were reported on in Heylen et al., (2007b) and Heylen, (2007).

Table 1 Backchannel signals

1. Nod	8. Raise eyebrows	15. Nod and raise eyebrows
2. Smile	9. Shake and frown	16. Shake, frown and tension[a]
3. Shake	10. Tilt and frown	17. Tilt and raise eyebrows
4. Frown	11. Sad eyebrows	18. Tilt and gaze right down
5. Tension[a]	12. Frown and tension[a]	19. Eyes wide open
6. Tilt	13. Gaze right down	20. Raise left eyebrows
7. Nod and smile	14. Eyes roll up	21. Tilt and sad eyebrows

[a]The action *tension* means tension of the lips

The list of possible meanings was proposed to the participants who, after each movie and before moving on, could select the meanings that they thought fitted the backchannel signal best. Participants were told that Greta would display backchannel signals as if Greta was talking to an imaginary speaker. This context was provided to make participants aware that they were evaluating backchannel signals. The signals were shown once, randomly: a different order for each subject.

The most significant results for each of the functions were the following.

Agree. When displayed on its own, *nod* proved to be very significant since every participant answered 'agree'. *Nod and smile* and *nod and raise eyebrows* also scored highly as backchannel signals of agreement. On its own, a *smile* does not associate with 'agreement', though. Similar results were obtained for the meaning of **Accept**.

Like. Two signals conveyed the meaning 'like': *nod and smile* and *smile.*

Understand. Thirteen out of 30 subjects associated a nod with 'understand', 16 of them paired *nod and smile* with this meaning and 17 found that *nod and raise eyebrows* could mean 'understand'. *Raise eyebrows* on its own is not associated with understanding as only one subject judged it as such.

Disagree. The signal *shake* is labelled by every subject as meaning 'disagree'. The combination of *shake and frown and tension* is also highly recognised as 'disagree'. Also the combination of *shake and frown* is regarded as meaning 'disagree' although the presence of frown alters the meaning. There is a significant difference between the mean of answers for *shake* versus *shake and frown.*

Dislike. Frown and tension appears as the most relevant combination of signals to represent 'dislike'. But when *shake* is added to *frown and tension*, it alters the meaning. *Frown* alone is sometimes regarded as meaning 'dislike' but it is significantly less relevant than *frown and tension.* When displayed on its own, *tension* is also less relevant than the combination *frown and tension.*

Disbelieve. Subjects considered that the combination *tilt and frown* means 'disbelieve' (21 answers out of 30) whereas *tilt* on its own is regarded as disbelieve by only 8 subjects. Similarly, *frown* on its own means 'disbelieve' for only six subjects. Also, *raise left eyebrow* is regarded by 21 subjects as 'disbelieve'.

Don't understand. Frown and *tilt and frown* are both associated with the meaning 'don't understand' by 20 subjects. As *tilt* is only given by four subjects one can infer that *frown* is the most relevant signal of the combination. However, when associated to other signals such as *tension* and/or *shake*, *frown* is less regarded as meaning 'don't understand'. Apart from the *frown* signal, *raise left eyebrow* appears as relevant to mean 'don't understand'. It is judged so by 19 out of 30 subjects.

Not interested. For this meaning, two signals seem to be relevant: *eyes roll up* (20 subjects) and *tilt and gaze* (20 subjects). As far as *tilt and gaze* is concerned, it seems it is the combination of both signals that is meaningful since the difference between *tilt and gaze* and *tilt* (13 answers) is significant. Similarly, the difference between *tilt and gaze* and *gaze right down* (13 answers) is also significant.

The results of this test suggest some prototypical signals for most of the meanings. For the positive meanings, 'agree' is signalled by a *nod*; 'accept' is as well. To signal 'like' a smile appears to be the most appropriate signal. A nod associated with a raise of the eyebrows seems to convey 'understand' but only 17 subjects out of 30 thought so. As for 'interested' and 'believe' the experiment did not find prototypical signals. A combination of *smile and raise eyebrows* is a candidate for 'interested'.

For the negative meanings, 'disagree' and 'refuse' are indicated by a head shake; 'dislike' is represented by a *frown and tension* of the lips. A *tilt and frown* as well as a *raise of the left eyebrow* means 'disbelieve' for most of the subjects. The best signal to mean 'don't understand' seems to be a *frown* while *tilt and gaze right down* as well as *eyes roll up* are more relevant for the meaning 'not interested'.

It also appeared that a combination of signals could significantly alter the perceived meaning or that for certain meanings only a composite expression could count as an appropriate signal. For instance, *tension* alone and *frown* alone do not mean 'dislike', but the combination *frown and tension* does. The combination *tilt and frown* means 'disbelieve' whereas *tilt* alone and *frown* alone do not convey this meaning. *Tilt* alone and *gaze right down* alone do not mean 'not interested' as significantly as the combination *tilt and gaze*. Conversely the signal *frown* means 'don't understand' but when the signal *shake* is added, *frown and shake* significantly loses this meaning.

The perceptual experiment aimed to analyse how users interpret context-free backchannel signals displayed by a virtual agent. The result lets one tentatively to assign specific signals to most of the meanings proposed in the test and thus form a start to define a library of prototypes. It remains to see to what extent these form-meaning mappings generalise to other cultures and other contexts. We continue with the description of a similar experiment that investigated the use of vocalisations called affect bursts as backchannels.

3.2 Affect Bursts

Affect bursts are 'very brief, discrete, nonverbal expressions of affect in both face and voice as triggered by clearly identifiable events' (Scherer, 1994, p. 170). Their vocal form ranges from non-phonemic vocalisations such as laughter or a rapid intake of breath, via phonemic vocalisations such as [a] or [m] where prosody and voice quality are crucial to conveying an emotion, to quasi-verbal interjections such as English 'yuck' or 'yippee' for which the segmental form transports the emotional meaning independently of the prosody.

In a study by Schröder et al. (2006) a listening test was carried out to assess the perception of these short nonverbal emotional vocalisations emitted by a listener as feedback to the speaker. The test investigated the use of affect bursts as a means of giving emotional feedback via the backchannel. The acceptability of affect bursts when used as listener feedback seemed to appear to be linked to display rules for emotion expression. While many ratings were similar between Dutch and German listeners, a number of clear differences were found, suggesting language-specific affect bursts.

In a study by Schröder (2003), a range of affect bursts was collected for each of 10 emotions, produced in isolation by German actors. On the basis of phonetic similarity, they were grouped into 24 'affect burst classes', which were classified correctly in a listening test 81% of the time on average. Characterisations of each affect burst class were obtained in terms of the emotion dimensions arousal, valence and power. The distinction between quasi-verbal, language-specific 'affect emblems' and universal 'raw affect bursts', proposed by Scherer (1994), was operationalised in terms of the stability of the segmental form across subjects, which was assessed in a transcription task. This allows one to classify proposed candidates for the status of 'emblem' versus 'raw burst'.

In Schröder et al. (2006) the use of affect bursts as a way for the listener to give emotional feedback was investigated. This is described here.

3.2.1 The Role of Context in Emotion Perception

Context is one of the important factors in the interpretation of expressions. In previous research some important contextual effects were described for the emotional meanings of expressions. Cauldwell (2000) demonstrated that short utterances can be perceived as anger in isolation and as emotionally neutral when perceived in the context in which they were uttered. Interestingly, the perception of anger from the utterance in isolation persisted even after having heard it in context. Similarly, Trouvain (2004) showed that certain kinds of laughter are perceived as sobs in isolation, but as laughs in context. In both cases, the difference in perception was the consequence of *extracting* a vocal expression from its original context. It is unclear whether a similar phenomenon should be expected when a vocalisation which originally was produced in isolation by an actor is inserted into a new context.

Embedding expressive vocalisations into a new context is not a straightforward thing to do, however. Inserting laughs into a speech synthesis context, it was found by Trouvain and Schröder (2004) that most were perceived as inappropriate, with the exception of a very mild laugh. The details of the circumstances under which such an insertion was considered appropriate are not yet clear. In addition, a conversational context may change the *function* of an emotional expressive display. In the case of facial expressions, for instance, Bavelas and Chovil (1997) showed how facial displays of emotion during conversations may not be the result of the emotion felt at the time of speaking but that often they are symbolic parts of messages that are integrated with other communicative signals such as words, intonation and gestures. For instance, a 'surprise' expression may thus be used in a particular context to signal disbelief. Similarly, the interpretation of affect bursts introduced into the conversational backchannel may or may not be interpreted as a comment, a symbolic act rather than the mere expression of an emotion felt. This may influence both the judgements of what is being expressed by the affect burst and the judgements on the appropriateness of the affect burst in this context.

The experiment described in Schröder et al. (2006) addressed the question whether affect bursts can be used by a listener to give emotional feedback to the speaker.

For each of the 10 emotion categories studied by Schröder (2003), 2 affect bursts were selected which were recognised best in isolation; if possible they were chosen from two different affect burst classes. This was possible for all emotions except ‘threat’ and ‘elation’, where both affect bursts had to be selected from the same class. Table 2 lists the original recognition rates of the selected affect bursts along with their respective emotion and affect burst class.

Stimuli were created by embedding each of the 20 selected affect bursts into a neutral speaker sentence. That sentence was deliberately semantically underspecified and spoken in an inexpressive, colloquial way. The sentence was ‘Ja, dann hab’ ich mir gesagt, probierste’s einfach mal ⟨⟨pause⟩⟩ und dann hab’ ich das gemacht!’

Table 2 Recognition results of 20 affect bursts. de = German listeners; nl = Dutch listeners. Ratings of affect bursts in isolation for German listeners taken from Schröder (2003). Acceptability ratings ranged from 0 (very bad) to 100 (very good)

		Recognition (%)					
		Isol.		In context		Acceptability	
Emotion	Burst	de	nl	de	nl	de	nl
Admiration	wow	95	100	97	89	79	70
	boah	95	23	100	11	73	36
Threat	hey1	95	41	70	37	26	23
	hey2	90	19	55	22	26	38
Disgust	buäh	100	69	97	59	53	37
	ih	95	97	90	82	53	45
Elation	ja1	85	90	90	74	51	52
	ja2	70	44	80	40	49	68
Boredom	yawn	95	100	97	96	58	49
	hmm	85	81	86	85	70	51
Relief	sigh	100	100	93	74	46	56
	uff	100	88	90	78	47	45
Startle	int. breath	100	100	100	96	33	34
	ah	90	74	87	48	22	41
Worry	oje	100	34	87	58	62	45
	oh-oh	85	71	97	65	65	45
Contempt	pha	95	81	87	82	35	48
	tse	100	71	87	77	55	50
Anger	growl1	90	81	80	74	37	23
	growl2	80	58	70	48	32	22
Average		92	71	87	65	49	44

(German); ‘Ja, toen zei ik tegen mezelf, probeer het maar een keer ⟨⟨pause⟩⟩ en toen heb ik het gedaan!’ (Dutch); ‘Yeah, then I told myself, why don’t you try it ⟨⟨pause⟩⟩ and then I did it!’ (English translation). In both the German and the Dutch sentence, the pause was 750 ms long. The affect bursts were mixed into the sentence starting at 150 ms into the pause, without modifying the pause duration. In other words, the feedback and the second part of the speaker utterance overlapped for those affect bursts that were longer than 600 ms. All affect bursts were normalised to the same average power as the sentence into which they were embedded. In order to mask the different recording conditions between the speaker sentence and the feedback, a low-intensity white noise (at – 60 dB) was added to the resulting stimuli.

The test was carried out in a web-enabled setup, using the open source tool RatingTest. The 20 stimuli were presented in an automatically randomised order. For each stimulus, subjects answered two questions. In a forced choice setup comparable to the one used by Schröder (2003), they identified the emotion expressed by the listener from a list of 10 categories. In addition, they rated on a continuous scale the question of how well the listener’s interjection fits into the dialogue.

In the German test, 30 subjects participated (15 female; mean age: 24.1 years). And 11 of these took the test in a controlled setting in a quiet office room; the remaining subjects took part in the test via the web. In the Dutch test, 27 subjects participated via the web (5 female; mean age: 24.2 years). A separate group of 32 Dutch listeners also rated the affect bursts in isolation, in order to provide Dutch data comparable to the results in Schröder (2003).

3.2.2 Results

The first observation to make in Table 2 is that the recognition rates for affect bursts in isolation are lower for Dutch listeners than for German listeners. Differences are rather small for the vast majority of bursts; only four bursts that were highly recognised by German listeners are not recognised by Dutch listeners. The two threat bursts were badly recognised, confirming the finding in Schröder (2003) that the threat and anger categories cannot be fully distinguished. Also, Dutch listeners do not seem to make the clear distinction that Germans make between ‘boah’ (expressing admiration) and ‘buäh’ (expressing disgust), leading to a very low recognition for ‘boah’. Similarly low is the recognition of worry ‘oje’, suggesting that in both cases, the language-specific segmental form may be crucial to the emotional meaning.

Regarding the recognition in context, it can be seen from Table 2 that overall recognition rates are slightly lower than for perception in isolation. However, the distribution of recognition rates across categories is very similar to the perception in isolation. One can conclude that the role of context on emotion recognition in this case appears to be very small.

Acceptability ratings showed clear differences between the stimuli, but the pattern is not easy to interpret. One can observe (Table 2) that ratings tend to be consistent within emotion categories. Acceptability was rated very high for admiration (leaving aside the Dutch rating of the ‘boah’ burst not recognised as admiration); moderately high for boredom, worry, elation and relief; moderately low for disgust and contempt; and very low for threat, anger and startle.

Interpretation is not made easier by the inherent ambiguity of the question of ‘good fit’ that the subjects were asked to rate. It may have been interpreted by the subjects as a general appropriateness in the context, as was intended; or one might have found it strange as a reaction to the meaning of the carrier sentence; it may also have been used to indicate technical aspects such as a mismatch between the sound quality of context and burst or the timing of the burst; finally it may have been used to indicate social appropriateness in the given context, in the sense of Ekman’s *display rules*: social norms prescribed by one’s culture as to ‘who can show what emotion to whom, when’ (Ekman, 1977).

Pursuing this issue of social appropriateness, one can attempt to account for the pattern found in terms of display rules. The results can make sense if seen as a cue to display rules whose underlying logic classifies emotions in terms of their being positive or negative and the type of goal they monitor (Castelfranchi, 2000; Poggi and Germani, 2003).

The first display rule seems to point at a general bias against expressing negative emotions. More specifically, the most sanctioned emotions are those linked to goals of aggression (anger and threat), while a somewhat lower sanction holds over negative emotions linked to goals of evaluation (disgust and contempt). Moving up to higher scores, one finds worry, relief and elation, emotions linked to the goal of well-being, and then, even higher, admiration, linked to the evaluation of others. Therefore, a positive bias towards the expression of emotions may hold, first, over emotions that show a positive evaluation of the other (admiration); then positive emotions like elation and relief; and finally over negative emotions like worry. Actually, there is a common feature to elation, relief and worry when expressed after another sentence: They may all be viewed as empathic reactions to the other’s narration.

The experiments described in this section have focussed on how backchannel expressions can express the attitudes of listeners in a conversation rather than at their conversation management functions. From the experiments it appears that the listener responses can have important interpersonal functions. In the context of embodied conversational agents, the relationship between feedback and the effects on the interpersonal relationship has been looked at most closely in the context of rapport. This is discussed in the next section.

4 Agents That Build Rapport

This section presents the Rapport Agent (Gratch et al., 2006b). This agent attempts to create a sense of rapport simply by generating listening feedback based on shallow observable features of a speaker’s bodily movements and speech prosody. We discuss the results of a study that demonstrates the Rapport Agent can produce some of the beneficial social effects associated with rapport. Such agent technology has potential as a powerful and novel methodological tool for uncovering the key factors that influence rapport in face-to-face interactions. It also has potential as a training system to enhance communication skills – for example, to reduce the impact of public speaking anxiety (Pertaub et al., 2001) – or to teach students to recognise specific patterns of nonverbal feedback, such as those that might predict clinical pathologies (Bouhuys and van den Hoofdakker, 1991), those that might cause intercultural misunderstandings (Gratch et al., 2006a) or those that arise in the context of deception.

Up to now, only a few systems can condition their listening responses to features of the user's speech, though typically this feedback occurs only after an utterance is complete. For example, Neurobaby analyses speech intonation and uses the extracted features to trigger emotional displays (Tosa, 1993). More recently, Breazeal's Kismet system extracts emotional qualities in the user's speech (Breazeal and Aryananda, 2002). Whenever the speech recogniser detects a pause in the speech, the previous utterance is classified (within 1 or 2) as indicating approval, an attentional bid or a prohibition, soothing or neutral. This recognition feature is combined with Kismet's current emotional state to determine facial expression and head posture. People who interact with Kismet often produce several utterances in succession, thus this approach is sufficient to provide a convincing illusion of real-time feedback.

Only a few systems can interject meaningful nonverbal feedback during another's speech and these methods usually rely on simple acoustic cues. For example, REA will execute a head nod or paraverbal (e.g. 'mm-hum') if the user pauses in mid-utterance (Cassell et al., 1999). Also the Gandalf system produced gaze shifts, back-channel feedback in real time based on the automatic analysis of prosody and gesture input (Thórisson, 1996).

Some work has attempted to extract extra-linguistic features of a speaker's behaviour, but not for the purpose of informing listening behaviours. For example, Brand's voice puppetry work attempts to learn a mapping between acoustic features and facial configurations inciting a virtual puppet to react to the speaker's voice (Brand, 1999).

In all of the cases the feedback by the agent is produced relying on a shallow analysis of some superficial features in the speaker's speech or nonverbal expressions. The feedback that is being produced is mostly intended as showing contact, attention and engagement (Sidner and Lee, 2007), but does not contain much other content. (Jonsdottir et al., 2007, made a first timid attempt to provide more contentful feedback.) The reliance on superficial features seems to be warranted by an experience that most of us have had that it is possible to signal attention by providing feedback even if one is attending only superficially while being preoccupied with other things (Bavelas et al., 2000) – which leads Schegloff (1982) to claim that the term *signal* may not be correct.

> It is worth noting, however, that 'uh huh', 'mm hmm', 'yeah', head nods, and the like *claim* attention and/or understanding, rather than 'showing' it or 'evidencing' it.

Although the feedback produced by listening agents may be based on a shallow analysis, this is not to say that it only has effects on the quality of the process of communication. The feeling of engagement that the feedback is supposed to create will also have an effect on the interpersonal level of communication. Although there is considerable research showing the benefit of such feedback on human to human interaction, there has been almost no research on their impact on human to virtual human rapport (cf. Bailenson and Yee, 2005; Cassell and Thórisson, 1999). In the Rapport Agent, this aspect is being studied in some depth.

Rapport is a crucial factor in establishing successful relationships. Capella (1990) states rapport to be 'one of the central, if not the central, constructs necessary to understanding successful helping relationships and to explaining the development of personal relationships'. It is closely related to some other concepts from social psychology and anthropology, e.g. 'interpersonal sensitivity' (Hall and Bernieri, 2001), 'social glue' (Lakin et al., 2003), 'interactional synchrony' (Bernieri and Rosenthal, 1991), 'mutuality' (Burgoon and Hale, 1987) and empathy (Sonnby-Borgstrom et al., 2003). Tickel-Degnen and Rosenthal (1990) equate rapport with behaviours indicating positive emotions (e.g. head nods or smiles), mutual attentiveness (e.g. mutual gaze) and coordination (e.g. postural mimicry or synchronised movements).[3]

That interpersonal rapport is perceptible and is a factor in the success of goal-directed activities is well established in the field of social psychological research. Naive observers will readily make judgements concerning whether participants in dyadic interactions, viewed on video for example, have rapport with one another. A study by Grahe and Bernieri (1999) determined that nonverbal behaviours are more significant than verbal factors in making such judgements. These judgements have been found to correlate reasonably well with the self-assessments of the members of the interacting dyad (Ambady et al., 2000).

Rapport is argued to underlie social engagement (Tatar, 1997), success in teacher–student interactions (Bernieri and Rosenthal 1988), success in negotiations (Drolet and Morris, 2000), improving worker compliance (Cogger, 1982), psychotherapeutic effectiveness (Tsui and Schultz, 1985), improved test performance in classrooms (Fuchs, 1987) and improved quality of child care (Burns, 1984).

Studies have also indicated that rapport can be experimentally induced or disrupted by altering the presence or character of several nonverbal signals (e.g. Bavelas et al., 2000; Drolet and Morris, 2000). Such findings have encouraged the development of embodied conversational agents that can induce rapport through the appropriate generation of nonverbal behavior.

When it comes to creating synthetic agents that simulate human nonverbal behavior, research has focused on half of the equation. Systems emphasise the importance of nonverbal behavior in speech production. Few systems attempt the tight sense-act

[3]See also the chapter by Marinetti et al. 'Emotions in Social Interactions: Unfolding Emotional Experience' in Handbook Area at the beginning of this volume.

loops that seem to underlie rapport and, despite considerable research showing the benefit of such feedback on human to human interaction, few studies have investigated its impact in human to virtual human interaction (cf. Cassell and Thórisson, 1999; Bailenson and Yee, 2005).

4.1 Rapport Agent

The Rapport Agent (Gratch et al., 2006b) was designed at the Institute of Creative Technologies to establish a sense of rapport with a human participant in face-to-face monologs where a human participant tells a story to a silent but attentive listener. In such settings, human listeners can indicate rapport through a variety of nonverbal signals (e.g. nodding, postural mirroring). The fluid, contingent nature of nonverbal behaviour associated with rapport suggests that it could be induced by rapidly responding to a speaker's physical movements. The Rapport Agent attempts to replicate these behaviours through a real-time analysis of the speaker's voice, head motion and body posture, providing rapid nonverbal feedback. The system is inspired by findings that feelings of rapport are correlated with simple contingent behaviours between speaker and listener, including behavioural mimicry (Chartrand and Bargh, 1999) and backchannelling (e.g. nods, see Yngve, 1970). The Rapport Agent uses a vision-based tracking system and signal processing of the speech signal to detect features of the speaker and then uses a set of reactive rules to drive the listening mapping displayed in Table 1. The architecture of the system is displayed in Fig. 2.

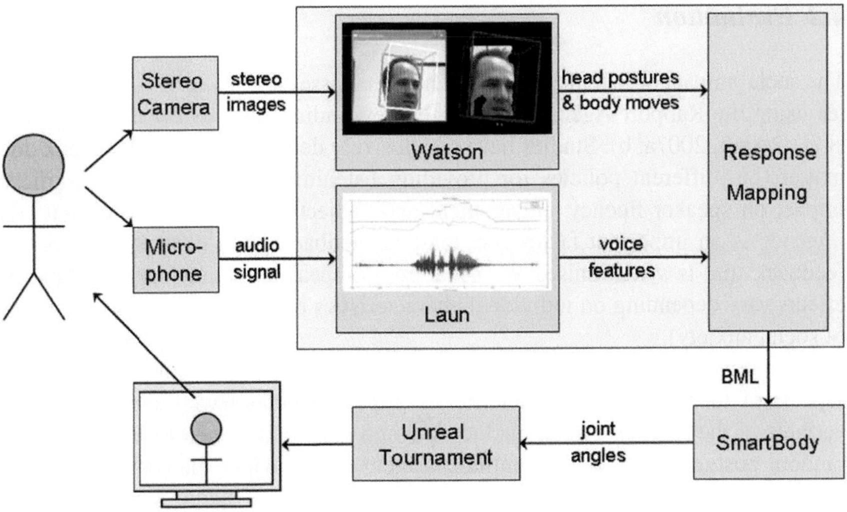

Fig. 2 Rapport Agent Architecture

To produce listening behaviours, the Rapport Agent first collects and analyses the speaker's upper-body movements and voice. For detecting features from the participants' movements, the system detects speaker's head movements. Watson (Morency et al., 2005) uses stereo video to track the participants' head position and orientation and incorporates learned motion classifiers that detect head nods and shakes from a vector of head velocities. Other features are derived from the tracking data. For example, from the head position the Rapport Agent can infer the posture of the spine given that the participant is seated in a fixed chair. Thus, the system detects head gestures (nods, shakes, rolls), posture shifts (lean left or right) and gaze direction.

Acoustic features are derived from properties of the pitch and intensity of the speech signal, using a signal processing package, Laun, developed by Mathieu Morales. Speaker pitch is approximated with the cepstrum of the speech signal (Oppenheim and Schafer, 2004) and processed every 20 ms. Audio artefacts introduced by the motion of the speaker's head are minimised by filtering out low-frequency noise. Speech intensity is derived from amplitude of the signal. Laun detects speech intensity (silent, normal, loud), range (wide, narrow) and backchannel opportunity points (derived using the approach of Ward and Tsukahara (2000).

Recognised speaker features are mapped into listening animations through a set of authorable mapping rules. These animation commands are passed to the SmartBody animation system (Kallmann and Marsella, 2005) using a standardised API (Vilhjalmsson et al., 2007). SmartBody is designed to seamlessly blend animations and procedural behaviours, particularly conversational behaviour. These animations are rendered in the Unreal Tournament™ game engine and displayed to the speaker.

4.2 Evaluation

The social impact of listening feedback has been assessed in a series of formal studies using the Rapport Agent. Some of the key findings are reviewed here (Gratch et al., 2006b, 2007a, b). Studies have conclusively demonstrated that feedback does matter (i.e. different policies for providing listening feedback have a significant impact on speaker fluency, engagement and subjective experience) and that contingency is an important factor (i.e. random feedback gives different results than feedback that is synchronised with features of speaker's behaviour), but that the effects vary depending on individual characteristics of speakers (such as their level of social anxiety).

Interactive virtual agents allow experimenters to carefully manipulate subtle aspects of the feedback and quantify its impact. Studies have contrasted several variants of the Rapport Agent, including a non-responsive agent that displays only random posture shifts, a non-contingent agent that provides the same distribution of feedback as the Rapport Agent but disrupts feedback synchrony (subjects actually see the feedback that was given to a different speaker), an avatar condition that accurately displays the actual movements of a human listener, as well as compared performance with face-to-face interaction.

All studies have involved speakers retelling a recently watched movie (either a funny Sylvester and Tweety cartoon or a serious presentation about sexual harassment in the workplace) to the agent (or a human listener).[4]

Findings show that the presence of listening feedback tends to improve listener performance along several dimensions. When compared with agents that did not provide positive listening feedback (i.e. the unresponsive agent), the Rapport Agent produced more engagement as indexed by the length of stories produced by speakers (Gratch et al., 2006b) and elicited more fluent speech, meaning speakers produced fewer filled pauses, repetitions and broken words (Gratch et al., 2006b, 2007a). One study found that the Rapport Agent could even elicit longer stories than face-to-face interaction between strangers (Gratch et al., 2007b). In general, engagement is positively correlated with the amount of positive feedback, i.e. agents or people that generated more nods tended to elicit longer stories.[5]

Findings also demonstrate that the feedback must be well timed to features of the speaker's behaviour to achieve these beneficial effects, i.e. random feedback is inadequate. When compared with the Rapport Agent or face-to-face interaction, the non-contingent agent produces significantly higher levels of speech disfluency, including far more broken words, repetitions and filled pauses (Gratch et al., 2006b, 2007a, b). This suggests that speakers were distracted by ill-timed feedback, possibly resulting in higher cognitive load. Indeed, subjects rated the non-contingent agent as highly distracting.

Finally, speaker's subjective feeling about the interaction varied with the quantity and quality of feedback, although when compared with observable behaviour (e.g. number of words and disfluencies produced), feelings depend on additional factors, such as their disposition to be anxious in social situations. For example, findings show that subjects that rated high in social anxiety were much more sensitive to non-contingent feedback, reporting higher embarrassment and lower self-perception of performance when compared with less anxious subjects. This suggests the contingency of feedback is especially critical to people who are socially anxious.

Collectively, the findings suggest that virtual agents can achieve some of the elements of rapportful interaction simply by recognising and responding to low-level features of a speaker's non-verbal behaviour. By improving the quality of such feedback, extending its scope to include more features such as gaze and facial expressions and, ultimately, by blending these low-level behaviours with the higher-level semantic understanding more commonly explored by embodied conversational agents, one may be able to realise many of the empirical benefits of rapport on learning and persuasion.

---

[4]It should be noted that interactions with virtual characters can vary depending on if subjects believe the character is an avatar (controlled by a human) or an agent (controlled by software). In the results we report here, subject were led to believe they were interacting with an avatar to assess the impact of the quality of feedback while holding other factors constant.

[5]It should be noted that listening agents that produced more head nods were also rated as more insincere, arguing for some caution when generating listening feedback.

4.3 Conclusion

In this chapter it was shown how human communication involves a complex synchronisation of actions of multiple participants that are highly connected. Each action calls forth a next one and simultaneously constitutes a reply to a previous one. It is successful or appropriate only in the context of actions that go on in parallel. How actions in human–human communication are intertwined has been studied intensively by linguists, psychologists and sociologists. Creating artificial systems that show the same proficiency in producing behaviours that are equally contingent on the behaviours of human interlocutors is a big challenge. However, it is obvious from studies such as those reported on above that when we want to create virtual agents that we would like to interact with, the agents should be able to at least pretend that they are listening to what we have to say.

Acknowledgments Many of the sections of this chapter are based on previous work of the authors that was done in collaboration with other colleagues. Therefore they wish to acknowledge the big help they got from Sue Duncan, Marion Tellier and Rieks op den Akker.

References

Allwood J (1993) Feedback in second language acquisition. In: Perdue C. (edi) Adult language acquisition. Cross linguistic perspectives. Cambridge University Press, Cambridge, NY, pp 196–232

Allwood J, Cerrato L (2003) A study of gestural feedback expressions. In: Paggio P, Jokinen K, Jonsson A. (eds) First nordic symposium on multimodal communication, Copenhagen, 23–24 September, pp 7–22

Allwood J, Nivre J, Ahlsén E (1992) On the semantics and pragmatics of linguistic feedback. Semantics 9(1):1–26

Ambady N, Bernieri FJ, Richeson JA (2000) Toward a histology of social behavior = Judgment accuracy from thin slices of the behavioral stream. Academic, San Diego, CA, pp 201–271

Austin JA (1962) How to do things with words. Oxford University Press, London

Bailenson JN, Yee N (2005) Digital chameleons = automatic assimilation of nonverbal gestures in immersive virtual environments. Psychol Sci 16:814–819

Bakhtin M (1999) The problem of speech genres. In: Jaworski A, Coupland N. (eds) The discourse reader. Routledge

Bavelas JB, Chovil N (1997) Faces in dialogue. In: Russell J, Fernandez-Dols JM, (ed) The psychology of facial expressions. Cambridge University Press, Cambridge, pp 334–346

Bavelas JB, Coates L, Johnson T (2000) Listeners as co-narrators. J Pers Soc Psychol 79(6): 941–952

Bernieri FJ, Rosenthal R (1991) Interpersonal coordination, behavior matching and interactional synchrony. In: Feldman RS, Rimé B, (ed) Fundamentals of nonverbal behavior. Cambridge University Press, Cambridge

Bevacqua E, Heylen D, Pelachaud C, Tellier M (2007) Facial feedback signals for ECAs. In: Proceedings of AISB'07: artificial and ambient intelligence, Newcastle University, Newcastle upon Tyne, UK, April 2007

Bouhuys AL, van den Hoofdakker RH (1991) The interrelatedness of observed behavior of depressed patients and of a psychiatrist. An ethological study on mutual influence. J Affect Disorders 23:63–74

Brand M (1999) Voice puppetry. In: ACM SIGGRAPH. ACM Press/Addison-Wesley, New York, NY

Breazeal C, Aryananda L (2002) Recognition of affective communicative intent in robot-directed speech. Autono Robots 12:83–104
Burgoon J, Hale J (1987) Validation and measurement of the fundamental themes of relational communication. Commun Monogr 54:19–41
Burns M (1984) Rapport and relationships. The basis of child care. J Child Care 2:47–57
Capella JN (1990) On defining conversational coordination and rapport. Psychol Inquiry 1(4): 303–305
Cassell J, Thórisson KR (1999) The power of a nod and a glance, envelope vs. emotional feedback in animated conversational agents. Int J Appl Artif Intell 13(4–5):519–538
Cassell J, Bickmore T, Billinghurst M, Campbell L, Chang K, Vilhjálmsson H, Yan H (1999) Embodiment in conversational interfaces. Rea. In: Conference on human factors in computing systems, Pittsburgh, PA, pp 520–527
Castelfranchi C (2000) Affective appraisal versus cognitive evaluation in social emotions and interactions. In: Paiva A, (ed) Affective interactions. Towards a new generation of computer interfaces. Springer, Berlin
Cauldwell R (2000) Where did the anger go? The role of context in interpreting emotion in speech. In: Proceedings of the ISCA workshop on speech and emotion, Northern Ireland, pp 127–131
Chartrand TL, Bargh JA (1999) The chameleon effect, the perception-behavior link and social interaction. J Personal Soc Psychol 76(6):893–910
Chovil N (1991) Social determinants of facial displays. J Nonverbal Behav 15:141–154
Clark H, Krych MA (2004) Speaking while monitoring addressees for understanding. J Mem Lang 50:62–81 Clark.Krych.04.pdf
Clark H, Schaefer E (1991) Contributing to discourse. Cogn Sci 13:259–294
Clark HH (1996) Using language. Cambridge University Press, Cambridge
Cogger JW (1982) Are you a skilled interviewer? Personnel J 61:840–843
Condon WS, Ogston WD (1966) Sound film analysis of normal and pathological behavior patterns. J Nerv Dis 143(4):338–347
Drolet AL, Morris MW (2000) Rapport in conflict resolution = accounting for how face-to-face contact fosters mutual cooperation in mixed-motive conflicts. Exp Soc Psych 36:26–50
Duncan S, Fiske DW (1977) Face-to-face Interaction. Erlbaum, Hillsdale NJ
Duncan SD, Niederehe G (1974) On signalling that its your turn to speak. J Exp Soc Psychol 10:234–47
Ekman P (1977) Biological and cultural contributions to body and facial movement. In: Blacking J, (ed) The anthropology of the body. Academic, London, pp 39–84
Ekman P (1979) About brows: emotional and conversational signals. In: Cranach von M, Foppa K, Lepenies W, Ploog D, (ed) Human ethology. Cambridge University Press/Editions de la Maison des Sciences de l'Homme, Cambridge, pp 169–202
Fuchs D (1987) Examiner familiarity effects on test performance, implications for training and practice. Top Early Child Spec Educ 7:90–104
Goffman E (1976) Replies and responses. Lang Soc 5(3):2257–313
Goodwin C (1984) Notes on story structure and the organization of participation. In: Atkinson MJ, Heritage J, (ed) Structures of social action. studies in conversation analysis. Cambridge University Press, Cambridge, pp 225–246
Grahe JE, Bernieri FJ (1999) The importance of nonverbal cues in judging rapport. J Nonverbal Behav 23:253–269
Gratch J, Okhmatovskaia A, Duncan S (2006a) Virtual humans for the study of rapport in cross cultural settings. In: 25th army science conference, Orlando, FL, 27–30 November 2006
Gratch J, Okhmatovskaia A, Lamothe F, Marsella S, Morales M, van der Werf R, Morency LP (2006b) Virtual rapport. In: 6th international conference on intelligent virtual agents, Springer, Berlin Marina del Rey, CA
Gratch J, Wang N, Gerten J, Fast E (2007a) Creating rapport with virtual agents. In: 7th international conference on intelligent virtual agents, Paris, France, 2007a

Gratch J, Wang N, Okhmatovskaia A, Lamothe F, Morales M, van der Werf R, Morency L-P (2007b) Can virtual humans be more engaging than real ones? In 12th international conference on human-computer interaction, Beijing, China

Grice HP (1975a) Logic and conversation. In: Cole P, Morgan JL, (ed) Syntax and semantics: Vol 3: speech acts. Academic, San Diego, CA, pp 41–58

Grice HP (1975b) Meaning. Phil Rev 66(3):377–388

Gumperz J (1982) Discourse strategies. Cambridge University Press, Cambridge, England

Hadar U, Steiner TJ, Rose CF (1985) Head movement during listening turns in conversation. J Nonverbal Behav 9(4):214–228

Hall J, Bernieri F. (ed) Interpersonal sensitivity. LEA, Mahwah, NJ

Heylen D (2007) Multimodal backchannel generation for conversational agents. In: Proceedings of the workshop on multimodal output generation (MOG 2007), University of Twente, 2007. CTIT Series, p 8192, 25–26 January 2007

Heylen D, Bevacqua E, Tellier M, Pelachaud C (2007a) Searching for prototypical facial feedback signals. In: Proceedings of the 7th international conference on intelligent virtual agents, Paris, France, 17–19 September, pp 147–153

Heylen D, Nijholt A, Poel M (2007b) Generating nonverbal signals for a sensitive artificial listener. In: Esposito A, Faunder-Zanny M, Keller E, Marinaro M. (ed) Verbal and nonverbal communication behaviours, Lecture notes in computer science. Springer, Berlin, pp 264–274

Jefferson G (1984) Notes on a systematic deployment of the acknowledgement tokens 'yeah' and 'mm hm'. Papers in Linguistics, 17:197–206

Jonsdottir GR, Gratch J, Fast E, Thórisson KR (2007) Fluid semantic back-channel feedback in dialogue: challenges and progress. In Proceedings of the 7th international conference on intelligent virtual agents, Paris, France, 17–19 September, pp 154–160

Kallmann M, Marsella S (2005) Hierarchical motion controllers for real-time autonomous virtual humans. In: Intelligent virtual agents, Kos, Greece, 2005. Springer, Berlin

Kendon A (1970) Movement coordination in social interaction: some examples described. Acta Psychol 32:100–125

Kendon A (1967) Some functions of gaze direction in social interaction. Acta Psychol 26:22–63

Kraut RE, Lewis SH, Swezey LW (1982) Listener responsiveness and the coordination of conversation. J Pers Soc Psychol 43(4):718–731

LaFrance M (1979) Nonverbal synchrony and rapport: analysis by the cross-lag panel technique. Soc Psychol Quart 42(1):66–70

Lafrance M, Ickes W (1981) Posture mirroring and interactional involvement: sex and sex typing effects. J nonverb behav 5:139–154

Lakin JL, Jefferis VA, Cheng CM, Chartrand TL (2003) Chameleon effect as social glue, evidence for the evolutionary significance of nonconscious mimicry. J Nonverb Behav 27(3):145–162

McClave EZ (2000) Linguistic functions of head movements in the context of speech. Journal of Pragmatics, 32:855–878

Morency L-P, Sidner C, Lee C, Darrell T (2005) Contextual recognition of head gestures. In: 7th international conference on multimodal interactions, Torento, Italy, 4–6 October 2005, pp 18–24

Oppenheim AV, Schafer RW (2004) From frequency to quefrency. A history of the cepstrum. IEEE Signal Process Mag September:95–106

Pelachaud C, Bilvi M (2003) Computational model of believable conversational agents. In: Huget M-P (ed) Communication in multiagent systems, vol 2650 of Lecture notes in computer science. Springer, Berlin, pp 300–317

Pertaub D-P, Slater M, Barker C (2001) An experiment on public speaking anxiety in response to three different types of virtual audience. Presence Teleoperators and Virtual Environ 11(1): 68–78

Poggi I (2007) Minds, hands, face and body. Weidler Buchverlag, Berlin

Poggi I, Germani M (2003) Emotions at work. In: International conference on human aspects of advanced manufacturing: agility and hybrid automation, Rome, Italy, 27–30 May 2003, pp 461–468

Rosenfeld HM (1987) Conversational control functions of nonverbal behavior. In: Siegman AW, Feldstein S (eds) Nonverbal behavior and communication. Lawrence Erlbaum Associates, Hillsdale NY, pp 563–601
Rosenfeld HM, Hancks M (1980) The nonverbal context of verbal listener responses. In: Key MR (ed) The relationship of verbal and nonverbal communication. Mouton, The Hague, pp 193–206
Sacks H, Schegloff EA, Jefferson G (1974) A simplest systematics for the organization of turn-taking for conversation. Language 50:696–735
Schegloff EA (1982) Discourse as interactional achievement: some uses of "uh huh" and other things that come between sentences. In: Tannen D. (ed) Analyzing discourse, text, and talk. Georgetown University Press, Washington, DC, pp 71–93
Schegloff EA, Sacks H (1973) Opening up closings. Semiotica 8:289–327
Scherer K (1994) Affect bursts. In: van Goozen SHM, van de Poll NE, Sergeant JA. (ed) Emotions: Essays on emotion theory. Lawrence Erlbaum, Hillsdale, NJ, pp 161–193
Schröder M (2003) Experimental study of affect bursts. Speech Commun Special Issue Speech a Emot 40(1–2):99–116
Schröder M, Heylen D, Poggi I (2006) Perception of non-verbal emotional listener feedback. In: Proceedings of speech prosody 2006, Dresden, Germany, 2–5 May 2006
Searle JR (1969) Speech acts: an essay in the philosophy of language. Cambridge University Press, Cambridge
Sidner CL, Lee C (2007) Attentional gestures in dialogues between people and robots. In: Nishida T (ed) Conversational informatics. Wiley, New York, NY
Sonnby-Borgstrom M, Jonsson P, Svensson O (2003) Emotional empathy as related to mimicry reactions at different levels of information processing. J Nonverb Behav 27(1):3–23
Tatar D (1997) Social and personal consequences of a preoccupied listener. PhD thesis, Department of Psychology, Stanford University, Stanford, CA
Thórisson KR (1996) Communicative Humanoids: a computational model of psycho-social dialogue skills. PhD thesis, Massachusetts Institute of Technology
Tickle-Degnen L, Rosenthal R (1990) The nature of rapport and its nonverbal correlates. Psychol Inquiry 1(4):285–293
Tosa N (1993) Neurobaby. In: SIGGRAPH-93 visual proceedings, tomorrow's realities, ACM, pp 212–213
Trouvain J (2004) Non-verbal vocalisations – the case of laughter. Paper presented at Evolution of Language: Fifth International Conference, 2004. Max Planck Institute for Evolutionary Anthropology, Leipzig, Germany, 31 March – 3 April 2004
Trouvain J, Schröder M (2004) How (not) to add laughter to synthetic speech. In: Proceedings of Workshop on affective dialogue systems, Kloster Irsee, Germany, 14–16 June 2004, pp 229–232
Tsui P, Schultz GL (1985) Failure of rapport. Why psychotherapeutic engagement fails in the treatment of Asian clients. Am J Orthopsychiatry, 55:561–569
Vilhjalmsson H, Cantelmo N, Cassell J, Chafai NE, Kipp M, Kopp S, Mancini M, Marsella S, Marshall AN, Pelachaud C, Ruttkay ZS, Thorisson KR, van Welbergen H, van der Werf R (2007) The behavior markup language, recent developments and challenges. In: International conference on intelligent virtual agents, Paris, France, 2007. Springer, Berlin
Ward N, Tsukahara W (2000) Prosodic features which cue back-channel responses in English and Japanese. J Pragmatics 23:1177–1207
Yngve VH (1970) On getting a word in edgewise. In: Papers from the sixth regional meeting of the Chicago Linguistic Society. Chicago Linguistic Society, Chicago, IL, pp 567–577

Coordinating the Generation of Signs in Multiple Modalities in an Affective Agent

Jean-Claude Martin, Laurence Devillers, Amaryllis Raouzaiou, George Caridakis, Zsófia Ruttkay, Catherine Pelachaud, Maurizio Mancini, Radek Niewiadomski, Hannes Pirker, Brigitte Krenn, Isabella Poggi, Emanuela Magno Caldognetto, Federica Cavicchio, Giorgio Merola, Alejandra García Rojas, Frédéric Vexo, Daniel Thalmann, Arjan Egges, and Nadia Magnenat-Thalmann

Abstract In order to be believable, embodied conversational agents (ECAs) must show expression of emotions in a consistent and natural looking way across modalities. The ECA has to be able to display coordinated signs of emotion during realistic emotional behaviour. Such a capability requires one to study and represent emotions and coordination of modalities during non-basic realistic human behaviour, to define languages for representing such behaviours to be displayed by the ECA, to have access to mono-modal representations such as gesture repositories. This chapter is concerned about coordinating the generation of signs in multiple modalities in such an affective agent. Designers of an affective agent need to know how it should coordinate its facial expression, speech, gestures and other modalities in view of showing emotion. This synchronisation of modalities is a main feature of emotions.

1 Introduction

As explained in Sect. 6.1.1.1, an embodied conversational agent (ECA) is a multimodal interface in which an animated character displayed on a computer screen converses with the user with several human-like modalities such as speech, gesture and facial expressions (Cassell et al., 1999; Kopp et al., 2003). Using an affective ECA (affective agent for short) in an interface is expected to lead to an intuitive and friendly human–computer interaction, for example, via the display of emotional expressions that can be useful in games or education as a means of motivating students.

In order to be believable, ECAs must show expression of emotions in a consistent and natural looking way across modalities. The ECA has to be able to display

J.-C. Martin (✉)
Computer Sciences Laboratory for Mechanics and Engineering Sciences (LIMSI), Paris, France
e-mail: Martin@limsi.fr

P. Petta et al. (eds.), *Emotion-Oriented Systems*, Cognitive Technologies,
DOI 10.1007/978-3-642-15184-2_18,

coordinated signs of emotion during realistic emotional behaviour. Such a capability requires one to study and represent emotions and coordination of modalities during non-basic realistic human behaviour, to define languages for representing such behaviours to be displayed by the ECA, to have access to mono-modal representations such as gesture repositories.

This chapter is concerned about coordinating the generation of signs in multiple modalities in such an affective agent. Designers of an affective agent need to know how it should coordinate its facial expression, speech, gestures and other modalities in view of showing emotion. This synchronisation of modalities is a main feature of emotions (Scherer and Ellring, 2007; Scherer, 2000). Yet, generating multimodal expressions of emotions raises several questions to designers such as What are the key signs required for the expression of emotions in individual modalities? What are the different ways to combine and synchronise these signs across modalities? This also raises methodological questions such as How to be sure that the signs displayed by multiple modalities will be perceived and understood sign by sign, as well as a whole multimodal expression? How to collect knowledge on the coordination of signs in a specific situation? How can multimodal emotional corpora be useful and how to collect them? How to validate models of multimodal expressions of emotions?

This chapter has connections with the following other chapters of the handbook. In relation to Chapter "Emotion: concepts and definitions", we are concerned with the multimodal expression of emergent emotions that are quite relevant for the design of affective agents and that might involve intensity, rapidity of change, brevity, event focus, appraisal elicitation and synchronisation. We will describe not only the display of basic emotions but also the display of several simultaneous emotions that we will call blends of emotions (cf. below). In relation to Chapter "Emotions in Social Interactions: Unfolding Emotional Experience", this chapter also considers the perception of emotional expression and its impact on interaction.

While Chapter "Image and Video Processing for Affective Applications" focuses on the recognition of affective states from a user, for example, while interacting with an embodied agent or not), this chapter deals with the expression and generation of coordinated signs of affects displayed by an embodied agent. Thus, when considering video corpora or vision processing techniques in the current chapter, we do it only for the sake of collecting knowledge on how the ECA should be specified. We do not consider image processing of user's behaviour.

As we will explain below, databases of multimodal emotional behaviours (Part III "Data and Databases") can be useful for the design of affective agents. While the Part VII "Usability" is concerned with the design and evaluation of affective systems without a specific focus on affective agents, this chapter is of interest for designers of multimodal agents and provides detailed explanations on how the agent needs to coordinate the modalities. We do not consider how a user would perceive these signs when using a system as the field is not yet mature enough. Rather we focus on perceptual experiments studying how subjects perceive different signs and different ways to combine them. Such experiments are required for validating models of multimodal coordination.

Finally, as to the Part IV "Emotion in Interaction" itself, we explain how designers of affective agents can collect knowledge on how to coordinate several modalities during the display of affective messages. While Chapter "Fundamentals of Agent perception and Attention modelling" deals with perception, this chapter focuses on generation. While Chapter "Generating Listening Behaviour" deals with specific communicative functions related to listening behaviour, this chapter focuses on the expression of emotions. Chapter "Representing Emotions and Related States in Technological Systems" deals with internal representations of emotions; this chapter deals with their multimodal expressions. Finally, this chapter can be seen as input for Chapter "Embodied Conversational Characters: Representation Formats for Multimodal Communicative Behaviours" on behaviour representation languages, since here we will describe the approaches and methodology to collect knowledge on how to coordinate signs, which can then be described formally using behaviour representation languages.

2 Human Multimodal Communication

In this section, we provide some background information about multimodal communication by humans that needs to be considered when designing affective agents.

Human communicative modalities include speech, hand gestures (and other bodily movements), facial expressions, gaze (and pupil dilation), posture, spatial behaviour, bodily contact, clothes (and other aspects of appearance), non-verbal vocalisations, smell (Argyle, 2004). There is a wide range of literature on non-verbal communication (Argyle, 2004; Feldman and Rim, 1991; Knapp and Hall, 2006, Siegman and Feldstein, 1985). Different coding schemes for the different non-verbal modalities have been defined in psychology and can inspire the design of ECAs (Harrigan et al., 2005).

2.1 Facial Expressions

Facial expressions have been studied by Ekman (1999, 2003; Ekman and Friesen 1975). He described how "the rapid signals (seconds or fractions of seconds) are produced by the contractions of the facial muscles, resulting in temporary changes in facial appearance, shifts in the location and shape of the facial features, and temporary wrinkles." Facial signals not only send emotional signals but also send emblematic messages (meaning of which is very specific, the nonverbal equivalent of a common word or phrase such as a head nod for "yes" and "no"). Raising the brows and holding them while keeping the rest of the face blank is an example of emblem signalling questioning. If the brow raise is done together with an upward head movement, it is an exclamation. Some facial emblems called emotional emblems are a conventional reference to a feeling. Facial signals are also used as conversational punctuators (emphasise a particular word). Thus, any movement of a given facial area might have several meanings. For example, the lowered, drawn-together brow is part of anger, but it is also an emblem (for determination, concentration and perplexity) and a punctuator (Ekman and Friesen, 1975).

2.2 Coding Scheme

The Facial Action Coding System (FACS) (Ekman and Friesen, 2002) is a physically based coding scheme and does not include behavioural interpretation. It explains how to classify facial movements as a function of the muscles which are involved. Action units (AU) are used to represent the muscular activity that produces momentary changes in facial appearance. An AU can represent the movements of one or several muscles. Symmetrically, for some facial muscles, the effects of different parts of the same muscles are detailed using several AUs. FACS also includes the description of gaze behaviours and head movements. Ekman described in detail the distinctive clues to each basic emotion (Ekman, 2003) and also depending on the intensity of the emotion and its meaning how many areas can/must be involved and how their signals are modified according to intensity. When considering three areas (brows, eyes, mouth), surprise can be shown in just two areas, with the third area uninvolved. The combination of these two-area surprise faces has a slightly different meaning:

- Eyes + brows = questioning surprise
- Eyes + mouth = astonished surprise
- Brows + mouth = dazed or less interested surprise

2.3 Complex Facial Expressions of Emotion

Real-life emotions are often complex and involve several simultaneous emotions (Ekman and Friesen, 1975; Scherer, 1998; Ekman, 2003). They may occur either as the quick succession of different emotions, the superposition of emotions, the masking of one emotion by another one, the suppression of an emotion or the overacting of an emotion. These blends of emotions produce "multiple simultaneous facial expressions" (Richmond and Croskey, 1999). Depending on the type of blending, the resulting facial expressions are not identical. A masked emotion may leak over the displayed emotion (Ekman and Friesen, 1975), while superposition of two emotions will be shown by different facial features (one emotion being shown on the upper face while another one on the lower face) (Ekman and Friesen, 1975). Ekman described in a systematic manner the blending of facial expressions for all the pairs of six acted basic emotions (happiness, sadness, surprise, fear, anger, disgust) (Ekman and Friesen, 1975). Ekman photographed models who were instructed to move particular facial muscles. He photographed three areas which are capable of independent movement: the brow/forehead; the eyes/lids and the root of the nose;

the lower face including the cheeks, mouth, most of the nose and chin. He also illustrates the context which can lead to each of these blends (for example, a woman being both angry and sad after a driver has run over her dog). He explains the complexity of these blended behaviours with regard to the timing and the subtleties of individual emotions (the families of facial expressions). A blend does not require that different facial areas show the different emotions. It can also be achieved by blending the appearance of the two emotions within each area of the face. For example, the blend anger/disgust is described by lips pressed (anger), upper lip raised (disgust), nose wrinkled (disgust).

2.4 Bodily Expression of Emotion

Several researchers experimentally studied expression of emotion in movement. Wallbott conducted a study in which actors were elicited with the following emotions: elated joy, happiness, sadness, despair, fear, terror, cold anger, hot anger, disgust, contempt, shame, guilt, pride and boredom (Wallbott, 1998). The actors had to utter meaningless sentences. Twelve drama students were asked to code body movements and postures performed by these actors. Wallbott observed distinctive patterns of movement and postural behaviour associated with some of the emotions studied. Some of these distinctive features seem to be typical of certain emotions, such as "arms crossed in front of chest" for the emotion of pride or the number of "self-manipulators" for shame. A number of movements and posture categories merely distinguish between "active" emotions (like hot anger or elated joy) and more "passive" emotions. His results indicate that movements and postural behaviour are indicative of the intensity of different emotions. Movement and postural behaviour when encoding different emotions (at least as captured with the categories used here) are to some degree specific to at least some emotions. For example, expressions of hot anger were observed to have the following discriminative features of postural behaviour: shoulder up, arms stretched out frontal and lateralised movements, pointing, opening/closing, back of hands sideways, many illustrators. Expressions of hot anger were also observed to have discriminative features of movement quality: high movement activity, expansive movements, high movement dynamics.

2.5 Multimodal Emotional Corpora

Analog video has been used for a long time for observing and manually annotating non-verbal behaviours. Since the last 10 years, several studies using digital video and computer-based annotations have been conducted in a variety of context and data (laboratory, meeting, TV material, field studies). Both analog and digital video corpora are used in two main goals: experimental studies or computational models. Compared to the social sciences experiments, digital corpus-based studies aim at producing computational models of multimodal behaviour including details of individual behaviours that can be useful for the design of affective ECAs with individual profiles. The ISLE project surveyed several corpora of human multimodal communication built before 2002 (Wegener Knudsen et al., 2002a, b). A key issue of multimodal and emotional corpora is the availability of so-called real-life data (Douglas-Cowie et al., 2003; Batliner et al., 2000). Different corpora enable to study different dimensions of emotional behaviours:

- location of recording (laboratory/TV on stage/TV in street/natural environment (e.g. at home));
- instructions staging/spontaneous/portrayed/posed;
- timescale (minutes (TV clip)/hours (meeting)/days);
- awareness of being filmed, awareness of emotion study;
- subjects (actors/experts on emotion/ordinary people);
- interactivity; monologue/multiparty/social constraints on interactivity;
- human–human or human–computer or mixed;
- number of recorded modalities, resolution and quality of collected data; intrusiveness of recording media

The reader can find more details about methodologies and tools for the collection and annotation of multimodal corpora in Martin et al. (2004, 2006), and Maybury and Martin (2002) and about emotional corpora in Devillers et al. (2006).

Videos of users interacting with the SAL system were manually annotated and used for the definition of a list of nonverbal behaviour generation rules (Lee and Marsella, 2006). The multimodal behaviour of a human storyteller was annotated in de Melo and Paiva (2006) and used for the design of an ECA in a narrative application. A parametric model for iconic gesture generation is defined by Tepper et al. (2004) following the analysis of a video corpus. Cartoon has been used as a video material for studying the role of irregularities and discontinuities modulations in expressivity (Ech et al., 2006). Recent emotional corpora aim at studying the synchronisation between the different modalities (Bänziger and Scherer, 2007).

2.6 Coordination of Multiple Modalities

In this section, we explain the different means for coordinating modalities. One main feature of multimodal communication is that there are multiple mappings between meanings and multimodal signals (Poggi 1996, 2003). Several signals can be displayed in one or several modalities for a single meaning. For example, the meaning "emphasis" of a given word can be displayed via an eyebrow raise, a head nod or both. Symmetrically, a given signal can be used for several meanings: an eyebrow raise might mean surprise, emphasis or suggest.

Due to the ambiguity of meanings of behaviour in the different modalities, multimodal behaviours must be interpreted by analysing the different modalities altogether. Several classifications have been proposed for describing how modalities may combine either in human communication or in multimodal interfaces (Knapp and Hall, 2006; Scherer, 1984; Buisine, 2005; Poggi, 2006; Martin et al., 2001; Martin and Béroule 1993; Coutaz et al., 1995; Kendon, 2004; Engle, 2000). The following notions are introduced in these classifications:

- Equivalence/substitution: one modality conveys a meaning not borne by the other modalities (while it could be conveyed by these other modalities)
- Redundancy/repetition: the same meaning is conveyed at the same time via several modalities
- Complementarity:
  - Amplification accentuation/moderation: one modality is used to amplify or attenuate the meaning provided by another modality
  - Additive: one modality adds congruent information to another modality
  - Illustration/clarification: one modality is used to illustrate/clarify the meaning conveyed by another modality
- Conflict/contradiction: the meaning transmitted on one modality is incompatible or contrasting with the one conveyed by the other modalities; this cooperation occurs when the meaning of the individual modalities seems conflicting but indeed the meaning of their combination is not and emerges from the conflicting combination of the meanings of the individual modalities.
- Independence: the meanings conveyed by different modalities are independent and should not be merged.

These combinations of modalities can be brought into play at different levels (Poggi, 2006). A signal (e.g. a pointing finger) may have a repetitive function with regard to its literal meaning, whereas its indirect meaning has an additive function (e.g. accusation).

3 Embodied Conversational Agents

In this section, we explain the importance of behaviour generation in ECAs. One makes a distinction between two parts in an ECA: its body and its mind. The body is in charge of the display of the behaviours. The mind is in charge of the reasoning, planning and management of the interaction with the user or the surrounding environment (Pelachaud et al., 2004).

3.1 Definitions and Examples

Several models have been proposed for the generation, selection and planning of multimodal behaviours in an ECA (Cassell et al., 2000; Prendinger and Ishizuka,

2004). André et al. (2000) describe the design of dialogue for agents. One approach consists in inserting gesture tags in the text to be spoken by the agent (Cassell et al., 2001). Automatic learning of these manual rules for gesture selection and timing from manual specifications is proposed in Kipp, (2006). With respect to the synchronisation of different modalities, the gestures of ECAs are often timed so that the stroke coincides with the stressed syllable of the co-occurring concept in speech (Cassell et al., 1994). The affective dimension is recognised of importance for pedagogical agents (Jacques et al., 2004). Jaques et al. (2004) describe some affective tactics that their mediating agent should bring into play to promote a positive mood in the student. Several experimental studies have evaluated the utility of such pedagogical agents for learning or to motivate students (Wonisch and Cooper, 2002; Lester et al., 1997; Moreno et al., 2001, Dehn and van Mulken, 2000).

Cosmo is a full body 3D cartoon like pedagogical agent in a learning environment for the domain of Internet packet routing (Lester et al., 2000). Given a communicative goal, the explanation planner determines the content and structure of an agent's explanations and then passes specifications of triplets (communicative act, topic, referent) to the deictic planner. The deictic planner exploits a world model, a user model, the current problem state, one focus history for speech and one focus history for gesture. Then the deictic behaviour planner selects and coordinates locomotive, gesture and speech behaviours. This deictic planning operates in three phases: (1) ambiguity appraisal, (2) gesture and locomotion planning, and (3) utterance planning and coordination (real-time composition of deictic gesture and gaze components). If the referent was mentioned in a recent utterance, only a spoken pronominal reference is generated (e.g. "it has low traffic" where "it" stands for a subnet that was pointed to by the agent in the previous utterance). The emotional planner that relies on a set of emotional behaviours has been designed so that it corresponds to variations of the most frequent pedagogical acts required in the application. The graphical design of the behaviours is based on knowledge from the literature of the animation industry.

The formation and motion characteristics of pointing by hand at objects of different sizes and locations have been investigated and used for embodied agents acting as weather reporters (Noma et al., 2000), cooperative assembly companion (Kopp et al., 2003) or explaining directions (Cassell et al., 2007). Pointing/reaching hand gesture should be preceded by gaze. The interplay of gaze, head and posture when looking at a direction has also been studied. Horizontal gaze shift greater than 25 degree, and vertical gaze shift greater than 10 degree are performed by the combination of head and gaze movement (Chopra-Khullar and Balder, 2001). These principles have been applied to visually explore the environment as well as follow moving targets by pointing or gaze. Gaze has been used to identify and follow "interesting", new and moving objects or people in a virtual environment (Hill et al., 2002), as well as in case of responsive faces reflecting emotions and attention according to the number of people appearing and moving around in a real world. See Sect. 6.2 (this handbook) for more details on such models.

3.2 Affective Agents

Greta incorporates conversational and emotional qualities (Pelachaud, Poggi et al., 2005). To determine speech accompanying nonverbal behaviours, the system relies on the taxonomy of communicative functions proposed by Isabella Poggi (2003). Communicative functions are defined by meaning, signal, expressivity. For example, looking away from the interlocutor (signal) conveys a take-turn meaning, within the communicative function "turn allocation".

3.3 Multimodal Emotional Corpora for Informing the Design of Affective Agents

Several studies rely on computer-based annotation of video-taped human behaviours to inform the design of ECAs. Such an approach has two advantages: it is experimentally grounded and thus can be more relevant for the task at hand than general rules provided by the literature. The second advantage is that it enables to model individual use of nonverbal behaviours, which can be a means to achieve animated agents with an individual profile (Kipp, 2004).

Elisabeth André addresses several issues related to the use of corpus-based approaches to behaviour modelling for embodied agents via observation, modelling, generation and perception (André, 2006). She reviews different approaches such as "understanding via learning to generate" (e.g. Wachsmuth), "analysis by observation and synthesis" (e.g. Krahmer), "Study, model, build, test cycle" (e.g. Cassell). The author also recommends to start from a model when collecting data.

3.4 Coordination of Multiple Modalities in ECAs

Redundancy was observed to provide better results than complementarity in a study investigating recall and preference after technical presentations made by an ECA (Buisine and Martin, 2007a). Yet, individual differences were observed with respect to gender and extraversion of the subjects (Buisine and Martin, 2007b). Redundancy might also seem unnatural if too frequent and involving too many modalities in a systematic way during a long presentation (Cassell et al., 1994).

Synchrony between speech and facial expression occurs at all levels of speech (Condon, 1986): phonemic segment, word, phrase or long utterance (Cassell et al., 1994). Different facial motions are characteristic of these different groups. Some of them are more adapted to the phoneme level, like an eye blink, while others act at the word level, like a frown. In the illustrative example "Do you have a blank check?", a raising eyebrow might start and end on the accented syllables "check", while a blink might start and end on the pause marking the end of the utterance. Facial expression of emphasis can match the emphasised segment, showing synchronisation at this level (a sequence of head nods can punctuate the emphasis). Moreover, some movements reflect encoding–decoding difficulties and therefore coincide with hesitations and pauses inside clauses. Many hesitation pauses are produced at the beginning of speech and correlate with avoidance of gaze (the head of the speaker turns away from the listener) as if to help the speaker to concentrate on what she is going to say (Duncan and Friske, 1977).

One example of coordination between multiple modalities occurs in the case of response to external stimuli. An embodied agent should react to the visual and acoustic stimuli from the surrounding virtual and/or real world, in order to be perceived as life like and believable. Reacting to signals from the real world contributes to "sharing the real world with the human interlocutor" experience. Such signals play a role in human nonverbal behaviour in two different ways:

(a) Invoke some behaviour, often on a biological reflex level. For example, a sudden noise triggers a blink and attracts attention; the pupil size adjusts itself to the lighting intensity; sudden intense light evokes blinking.
(b) Modify some behaviour, concerning its timing and/or formation. For example, when looking at a moving object, the eyes, the head, the upper torso or even the entire body may be turned, depending on the relative location and speed of the object being looked at. Or when approaching a train to get on, one not only generates a shortest path with respect to the other passengers and obstacles, but may speed up his locomotion when hearing the whistle of the conductor.

Coordination/intersynchrony to accomplish a common task: In case of interpersonal gestures involving physical touch, the behaviours of two embodied agents need to be simultaneously coordinated, like in the case of a handshake, a hug or kiss. Both physical and social factors play a role in the coordination strategy. Though among humans the physical contact is established by getting feedback via peripherical vision, the protocol of such gestures too involves synchrony with mutual gaze: eye contact is to be established during the handshake. The duration of the handshake, the proximity and eventual head nod or bowing move of the upper body are to not only be subtly coordinated, but also be used according to the protocol reflecting the gender, ethnicity and social relationship of the two parties.

3.5 Facial Expressions of Emotions in Affective Agents

In this section, we focus on the signs of emotions in facial expressions.

3.5.1 MPEG-4

MPEG-4 is an ISO/IEC international standard defined in 1998. It aims at overcoming all the divisions in the world of facial animation by defining a standard way to deal with synthetic faces using a model-based approach. It led to the definition of a set of animation parameters and semantic rules which can be used to drive any synthetic face compliant with MPEG-4 (Pandzic and Forchheimer, 2002). The six archetypal expressions (joy, sadness, anger, fear, disgust and surprise) can be described using MPEG-4 (Malatesta et al., 2007).

Emotions that belong to the same category can be rendered by animating the same FAPs using different intensities. For example, the emotion group fear also contains worry and terror; these two emotions can be synthesised by reducing or increasing the intensities of the employed facial animation parameter (FAP). Creating profiles for emotions that do not clearly belong to a universal category is not straightforward. Apart from estimating the range of variations for FAPs, one should first define the FAPs which are involved in the particular emotion.

Very few models of blended emotions have been developed so far for ECAs. The interpolation between facial parameters of given expressions is commonly used to compute the new expression (Pandzic and Forchheimer, 2002; Albrecht et al., 2005; Ruttkay, 2003; Duy Bui, 2004; Malatesta et al., 2007).

3.5.2 Complex Expression of Emotions

Distinguishing various types of blends of emotions in ECA systems is relevant as perceptual studies have shown that people are able to recognise facial expression of felt emotion (Ekman, 1982; Wiggers, 1982) as well as fake emotion (Ekman, 1982) from real life as well as on ECAs (Pandzic and Forchheimer, 2002). Moreover, in a study on deceiving agent, Rehm and André (2005) found that the users were able to differentiate when the agent was displaying expressions of felt emotion or expression of fake emotion. Aiming at understanding whether facial features or regions play identical roles in emotion recognition, researchers performed various perceptual tasks or studied psychological facial activity (Bassili, 1979; Gouta and Miyamoto, 2000; Constantini et al., 2005; Cacioppo et al., 1986). They found that positive emotions are mainly perceived from the expression of the lower face (e.g. smile) while negative emotion from the upper face (e.g. frown). One can conclude that reliable features for positive emotion, that is, features that convey the strongest characteristics of a positive emotion, are in the lower face.

On the other hand, the most reliable features for negative emotion are in the upper face.

Based on such findings computational models for facial expressions of blend of emotions can be defined (Ochs et al., 2005; Niewiadomski, 2007). Such models compose facial expressions from those of single emotions using fuzzy logic-based rules, for example, for assessing the similarity between facial expressions (Niewiadomski and Pelachaud, 2007).

3.6 Bodily Expression of Emotions in Affective Agents

Studies from the social sciences observed that discriminative features of emotions are found (1) in postures and (2) in movement quality (Wallbott, 1998). Thus, both should be considered when designing an affective agent. In posture, three aspects are required in posture design: achieving a set of world space constraints, finding a body shape that reflects the character's inner state and personality and making adjustments to balance that act to strengthen the pose and also maintain realism (Neff and Fiume, 2005). Experimental studies are being conducted to collect more knowledge on the posture features which can be used with a discrete or a dimensional model (Kleinsmith and Bianchi-Berthouze, 2007). Few studies have explored the combination of postures with other modalities.

3.6.1 Expressivity

The Greta ECA (Pelachaud) features a model of gesture expressivity that acts on the production of communicative gestures. The model of expressivity is based on studies of nonverbal behaviour (Wallbott, 1998; Wallbott and Scherer, 1986; Gallaher, 1992). Expressivity is described as a set of six dimensions (Hartmann et al., 2002, 2005). Each dimension acts on a characteristic of communicative gestures. Spatial extent describes how large the gesture is in space. Temporal extent describes how fast the gesture is executed. Power describes how strong the performance of the gesture is. Fluidity describes how two consecutive gestures are co-articulated one merging with the other. Repetition describes how often a gesture is repeated. Overall activity describes how many behaviour activities there are over a time span. For the face, the expressivity dimensions act mainly on the intensity of the muscular contraction and its temporal course. In order to consider individual agents, an individual expressive profile can be assigned to the agent specifying in which modalities the agent is the most expressive. The expressivity attributes can act over a whole animation, on gesture phases or on every frame (Ech et al., 2006). A corpus-based approach with acted data collected in laboratory was used to provide values for such parameters using image processing techniques (Caridakis et al., 2006). Image processing was also used for analysing bodily movement of a dancer (Volpe, 2005). Emotional reaction of affective agents: An example of a body gesture that depends on the emotional context is the reaction movement. Reactions are unconscious behaviours that improve the realism of affective agents. Based on observation experiments of real people reacting, different types of reactions were identified: avoid, face and protect. These types of reactions are associated with personality traits (Garcia-Rojas et al., 2007). To synthesise these reactions into embodied agents, a semantic model that represents the animation synthesis, agent's geometry and a description of individual agents through personality, emotion, age, gender, etc. was created. The advantage of such a semantic model is that it enables scalability for implementation of concepts.

3.7 Coordinating the Generation of Signs in Multiple Modalities in an Affective Agent

When expressing superposition of two emotions in several audio and visual modalities, how should the signs of the different emotions be split into different modalities? Some researchers study how to generate mixed emotions in ECAs (Carofiglio et al., in press, Karunaratne and Yan, 2006; Lee at al., 2006). Arfib (2006) uses time warping to align video of one emotion and vocal tempo of another emotion using a time-warping algorithm, hence artificially creating multimodal blends of emotions.

Video corpora of TV interviews enable to explore how people behave during such blended emotions (Devillers et al., 2006; Martin et al., 2005) not only by their facial expression but also by their gestures and their speech (Douglas-Cowie et al., 2005). Yet, these corpora call for means of validating subjective manual annotations of emotion.

Annotation is composed of two steps: Step 1 aims at the automatic annotation of the video with data that can be useful either for the manual annotation of the video or the specification of the agent's behaviour: pitch, intensity, etc. Step 2 involves manual annotations of the video. The word by word transcription including punctuation is achieved following the LDC norms for hesitations, breath, etc. The video is then annotated at several temporal levels (whole video, segments of the video and behaviours observed at specific moments) and at several levels of abstraction. The global behaviour observed during the whole video is annotated with communicative act, emotions and multimodal cues. The segments are annotated with emotion labels and the modalities perceived as relevant with regard to emotion. We have grounded this coding scheme in requirements collected from the parameters known as perceptually relevant for the study of emotional behaviour and the features of our emotionally rich TV interviews. Movement expressivity is annotated for torso, head, shoulders and hand gestures.

Manual steps can be followed for defining an animation of an affective agent replaying annotated multimodal emotional behaviours observed in an annotated video. Such a copy-synthesis approach enables to identify the levels of representation that are required (Martin et al., 2006) via several steps: annotation, extraction and generation. Perceptual experiments are recommended to compare the perception of the original videos and the animations of the corresponding behaviour replayed by an affective agent (Buisine et al., 2006).

Similarly, automatic mimicry approach lies in the fact that both facial and gestural aspects of the user's behaviour are analysed and processed. The mimicry consists of perception, interpretation, planning and animation of the expressions shown by the human, resulting not in an exact duplicate rather than an expressive model of the user's original behaviour (Caridakis et al., 2007).

4 Conclusions

In this section, we explained the complexity of expressions of emotions in individual modalities. Some discriminative features of specific emotions have been pointed out in facial expressions, in posture and in expressive movements. Several possible ways to combine different modalities were presented for multimodal coordination and synchronisation.

From a methodological point of view, we also described how to apply corpus-based approaches for studying how different modalities are coordinated in human communication. We explained how to specify such expressions of complex emotions in an affective agent using facial expressions and gestures.

Where the display of signs in individual modalities by an affective ECA can be seen as quite advanced (e.g. displaying expression of basic or intermediate emotions using MPEG-4), the coordination of different modalities during the expression of complex emotions was only recently brought on the forefront and deserves dedicated experimental studies.

Researchers are also paying attention to individual differences in the perception of multimodal behaviours displayed by affective agents. For example, extrovert and introvert versions of static text and static graphical display of posture were combined by Isbister and Nass (2000). It was found that users identify the intended personality trait. Furthermore, users preferred consistent verbal and nonverbal, as well as agents with a complementary extroversion compared to their own.

References

Albrecht I, Schrder M, Haber J, Seidel H-P (2005) Mixed feelings: Expression of non-basic emotions in a muscle-based talking head. J Virtual Reality Lang Speech Gesture 8(4 Special issue):201–212

André E (2006) Corpus-based approaches to behavior modeling for virtual humans: a critical review. Modeling communication with robots and virtual humans. In: Workshop of the ZiF: research group 2005/2006 "Embodied communication in humans and machines". Scientific organization: Ipke Wachsmuth (Bielefeld), Günther Knoblich (Newark)

Andr E, Rist T, van S, Mulken, Klesen M, Baldes S (2000) The automated design of believable dialogues for animated presentation teams. In: Cassell JSJ, Prevost S, Churchill E (eds) Embodied conversational agents. MIT Press, Cambridge, MA, pp 220–255

Arfib D (2006) Time warping of a piano (and other) video sequences following different emotions. In: Workshop on "subsystem synchronization and multimodal behavioral organization" held during Humaine summer school. Genova

Argyle M (2004) Bodily communication, 2nd edn. Routledge, London and Taylor and Francis, New York, NY

Bänziger T, Scherer K (2007) Using actor portrayals to systematically study mul-timodal emotion expression: the GEMEP corpus. In: 2nd international conference on affective computing and intelligent interaction (ACII 2007) Lisbon, Portugal, pp 476–487

Bassili JN (1979) Emotion recognition: the role of facial movement and the relative importance of upper and lower areas of the face. J Pers Soc Psychol 37(11):2049–2058

Batliner A, Fisher K, Huber R, Spilker J, Noth E (2000) Desperately seeking emotions or: Actors, wizards, and human beings. ISCA Workshop on speech and emotion: a conceptual framework for research Newcastle, Northern Ireland, pp 195–200

Buisine S, Abrilian S, Niewiadomski R, Martin J-C, Devillers L, Pelachaud C (2006) Perception of blended emotions: from video corpus to expressive agent. 6th International Conference on Intelligent Virtual Agents (IVA'2006). Best paper award. Springer, Marina del Rey, CA, pp 93–106

Buisine S (2005) Conception et évaluation d'Agents conversationnels multimodaux bidirectionnels. PhD Thesis. Doctorat de Psychologie Cognitive – Ergonomie, Paris V. 8 avril 2005. Direction J.-C. Martin and J.-C. Sperandio. 2005. URL http://stephanie.buisine.free.fr/. Accessed on 4 November 2010

Buisine S, Martin JC (2007a) The effects of speech-gesture cooperation in animated agents' behavior in multimedia presentations. Interact Comput 19:484–493

Buisine S, Martin J-C (2007b) The influence of personality on the perception of embodied agents' multimodal behavior. In: 3rd conference of the international society for gesture studies

Cacioppo JT, Petty RP, Losch ME, Kim HS (1986) Electromyographic activity over facial muscle regions can differentiate the valence and intensity of affective reactions. J Pers Soc Psychol 50:260–268

Caridakis G, Raouzaiou A, Karpouzis K, Kollias S (2006) Synthesizing gesture expressivity based on real sequences. In: Workshop "multimodal corpora. From multimodal behaviour theories to usable models". 5th international conference on language resources and evaluation (LREC'2006), Genova, Italy, pp 19–23

Caridakis G, Raouzaiou A, Bevacqua E, Mancini M, Karpouzis K, Malatesta L, Pelachaud C (2007) Virtual agent multimodal mimicry of humans. J Lang Res Eval (Special issue) "Multimodal Corpora". Lang Res Eval (41):367–388

Carofiglio V, de Rosis F, Grassano R (2008) Dynamic models of multiple emotion activation. In: Canamero L, Aylett R (eds) Animating expressive characters for social interactions. John Benjamins, Amsterdam, pp 123–141

Cassell J, Pelachaud C, Badler N, Steedman M, Achorn B, Becket T, Douville B, Prevost S, Stone M (1994) Animated conversation: rule-based generation of facial expression, gesture and spoken intonation for multiple conversational agents. In: ACM SIGGRAPH'94, pp 413–420

Cassell J, Bickmore T, Billinghurst M, Campbell L, Chang K, Vilhjálmsson HH, Yan H (1999) Embodiment in conversational interfaces: Rea. CHI'99 (SIGCHI conference on Human factors in computing systems: the CHI is the limit) Pittsburgh, PA, USA, pp 520–527

Cassell J, Bickmore T, Campbell L, Vilhjlmsson H, Yan H (2000) Human conversation as a system framework: designing embodied conversational agents. In: Cassell J, Sullivan J, Prevost S, Churchill E (eds) Embodied conversational agents. MIT Press, Cambridge, MA, pp 29–63

Cassell J, Vilhjálmsson H, Bickmore T (2001) BEAT: the Behavior Expression Animation Toolkit. In: 28th Annual Conference on Computer Graphics and Interactive Techniques (SIGGRAPH '01) Los Angeles, CA, pp 477–486

Cassell J, Kopp S, Tepper P, Ferriman K, Striegnitz K (2007) Trading spaces: how humans and humanoids use speech and gesture to give directions. In: Nishida T (ed) Conversational informatics. Wiley, New York, NY, pp 133–160

Chopra-Khullar S, Badler N (2001) Where to look? Automating attending behaviours of virtual human characters. In: 4th Conference on AAMAS, pp 9–23

Condon WS (1986) Communication: rhythm and structure. rhythm in psychological, linguistic and musical processes. Charles C Thomas Publisher

Constantini E, Pianesi F, Prete M (2005) Recognizing emotions in human and synthetic faces: the role of the upper and lower parts of the face. In: Intelligent User Interfaces (IUI'05) San Diego, CA, USA, pp 20–27

Coutaz J, Nigay L, Salber D, Blandford AE, May J, Young RMY (1995) Four easy pieces for assessing the usability of multimodal interaction. In: Interact'95 pp 115–120

de Melo C, Paiva A (2006) A Story about Gesticulation Expression. In: 6th International Conference on Intelligent Virtual Agents (IVA'06) Marina del Rey, CA, pp 270–281

Dehn DM, van Mulken S (2000) The impact of animated interface agents: a review of empirical research. Int J Hum Comput Stud 52:1–22

Devillers L, Cowie R, Martin J-C, Douglas-Cowie E, Abrilian S, McRorie M (2006) Real life emotions in French and English TV video clips: an integrated annotation protocol combining continuous and discrete approaches. In: 5th in-ternational conference on Language Resources and Evaluation (LREC 2006), Genoa, Italy

Devillers L, Martin J-C, Cowie R, Douglas-Cowie E, Batliner A (2006) Workshop "Corpora for research on emotion and affect". In: 5th international conference on language resources and evaluation (LREC'2006). Genova, Italy

Douglas-Cowie E, Campbell N, Cowie R, Roach P (2003) Emotional speech; Towards a new generation of databases. Speech Commun 40

Douglas-Cowie E, Devillers L, Martin J-C, Cowie R, Savvidou S, Abrilian S, Cox C (2005) Multimodal databases of everyday emotion: facing up to complexity. In: 9th European Conference on Speech communication and technology (Interspeech'2005), Lisbon, Portugal, pp 813–816

Duncan S, Fiske D (1977) Face-to-face interaction: research, methods and theory. Lawrence Erlbaum, Hillsdale, N J

Duy Bui T (2004) Creating emotions and facial expressions for embodied agents. PhD Thesis. University of Twente

Ech Chafai N, Pelachaud C, Pelé D, Breton G (2006) Gesture Expressivity Modulations in an ECA Application. In: 6th international conference on intelligent virtual agents (IVA'06) Marina del Rey, CA, pp 181–192

Ekman P (1982a) Emotion in the human face. Cambridge University Press

Ekman P, Friesen W (1982) Felt, false, miserable smiles. J Nonverb Behav 6:4

Ekman P (1999) Basic emotions. In: Dalgleish T, Power MJ (eds) Handbook of cognition & emotion. Wiley, New York, NY, pp 301–320

Ekman P (2003a) Emotions revealed. Understanding faces and feelings. Weidenfeld and Nicolson, London

Ekman P (2003b) The face revealed. Weidenfeld and Nicolson, London

Ekman P, Friesen WV (1975) Unmasking the face. A guide to recognizing emotions from facial clues. Prentice-Hall, Englewood Cliffs, NJ.

Ekman P, Friesen WC, Hager JC (2002) Facial action coding system. The Manual on CD ROM

Engle RA (2000) Toward a theory of multimodal communication: combining speech, gestures, diagrams and demonstrations in instructional explanations. PhD Thesis, Stanford University

Feldman RS, Rim B (1991) Fundamentals of nonverbal behavior. Studies in emotion and social interaction. Cambridge University Press, Cambridge

Gallaher P (1992) Individual differences in nonverbal behavior: Dimensions of style. J Pers Soc Psychol 63:133–145

Garcia-Rojas A, Vexo F, Thalmann D (2007) Semantic representation of individualized reaction movements for virtual humans. J Virtual Reality 6(1):25–32

Gouta K, Miyamoto M (2000) Emotion recognition, facial components associated with various emotions. Shinrigaku Kenkyu 71(3):211–218

Harrigan JA, Rosenthal R, Scherer K (2005) The new handbook of methods in nonverbal behavior research. Series in Affective Science. Oxford University Press, Oxford

Hartmann B, Mancini M, Pelachaud C (2002) Formational parameters and adaptive prototype instantiation for MPEG-4 compliant gesture synthesis. In: Computer animation (CA'2002) Geneva, Switzerland, pp 111–119

Hartmann B, Mancini M, Pelachaud C (2005) Implementing expressive gesture synthesis for embodied conversational agents. In: Gesture Workshop (GW'2005), Vannes, France

Hill R, Han C, van Lent M (2002) Perceptually driven cognitive mapping of urban environments. In: First international joint conference on autonomous agents and multiagent systems, Bologna, Italy

Isbister K, Nass C (2000) Consistency of personality in interactive characters: verbal cues, non-verbal cues, and user characteristics. Int J Hum Comput Stud 53:251–267

Jacques PA, Vicari RM, Pesty S, Bonneville J-F (2004) Applying affective tactics for a better learning. In: 16th European Conference on Artificial Intelligence (ECAI 2004). Valncia, Spain, IOS, Amsterdam, pp 109–113

Karunaratne S, Yan H (2006) Modelling and combining emotions, visual speech and gestures in virtual head models. Signal Process Image Commun 21:429–449

Kendon A (2004) Gesture : visible action as utterance. Cambridge University Press, Cambridge

Kipp M (2004) Gesture generation by imitation. From human behavior to computer character animation. Boca Raton, Dissertation.com Florida

Kipp M (2006) Creativity meets automation: combining nonverbal action authoring with rules and machine learning. In: 6th international conference on intelligent virtual agents (IVA'06) Marina del Rey, CA, pp 230–242
Kleinsmith A, Bianchi-Berthouze N (2007) Recognizing affective dimensions from body posture. In: 2nd international conference on affective computing and intelligent interaction (ACII 2007) Lisbon, Portugal, pp 48–58
Knapp ML, Hall JA (2006) Nonverbal communication in human interaction, 16th edition. Thomson and Wadsworth, Belmont, CA
Kopp S, Jung B, Lessmann N, Wachsmuth I (2003) Max - A multimodal assistant in virtual reality construction. KI-Künstliche Intelligenz. Vol. 4/03, pp 11–17
Lee J, Marsella S (2006) Nonverbal behavior generator for embodied conversational agents. In: 6th international conference on intelligent virtual agents (IVA'06) Marina del Rey, CA, pp 243–255
Lee B, Kao E, Soo V (2006) Feeling ambivalent: a model of mixed emotions for virtual agents. In: 6th international conference on intelligent virtual agents (IVA'06) Marina del Rey, CA, pp 329–342
Lester J, Converse S, Kahler S, Barlow T, Stone B, Bhogal R (1997) The Persona effect: affective impact of animated pedagogical Agents. In: CHI '97 Atlanta, pp 359–366
Lester JC, Towns SG, Callaway CB, Voerman JL, P, F (2000) Deictic and emotive communication in animated pedagogical agents. Embodied conversational agents. The MIT Press, Cambridge, MA
Malatesta L, Raouzaiou A, Karpouzis K, Kollias S (2007) MPEG-4 facial expression synthesis. J Pers Ubiquitous Comput 'Emerg Multimodal Interfaces' (Special issue) following the special session of the AIAI 2006 Conference. Springer, 13(1):77–83
Martin JC, Béroule D (1993) Types et buts de coopérations entre modalités. Cinquièmes Journées sur l'Ingénierie des Interfaces Homme-Machine Lyon, France, pp 17–22
Martin JC, Grimard S, Alexandri K (2001) On the annotation of the multimodal behavior and computation of cooperation between modalities. In: Workshop on "Representing, Annotating, and Evaluating Non-Verbal and Verbal Communicative Acts to Achieve Contextual Embodied Agents" in conjunction with the 5th International Conference on Autonomous Agents (AAMAS'2001) Montreal, Canada, pp 1–7
Martin J-C, den Os E, Kuhnlein P, Boves L, Paggio P, Catizone R (2004) Workshop "Multimodal corpora: models of human behaviour for the specification and evaluation of multimodal input and output interfaces". In: Association with the 4th international conference on language resources and evaluation LREC2004 URL http://multimodal-corpora.org/. Accessed on 4 November 2010. Centro Cultural de Belem, LISBON, Portugal
Martin J-C, Abrilian S, Devillers L (2005) Annotating multimodal behaviors occurring during non basic emotions. In: 1st International Conference on affective computing and intelligent interaction (ACII'2005), Beijing, China, pp 550–557
Martin J-C, Kuhnlein P, Paggio P, Stiefelhagen R, Pianesi F (2006) Workshop "Multimodal Corpora: from Multimodal Behaviour Theories to Usable Models". In: Association with the 5th international conference on language resources and evaluation (LREC2006), Genoa, Italy
Martin J-C, Niewiadomski R, Devillers L, Buisine S, Pelachaud C (2006) Multimodal complex emotions: gesture expressivity and blended facial expressions. J Humanoid Rob (Special issue). In: Pelachaud C, Canamero L (eds). Achieving human-like qualities in interactive virtual and physical humanoids. 3(3):269–291
Maybury M, Martin J-C (2002) Workshop on "Multimodal Resources and Multi-modal Systems Evaluation". In: Conference on language resources and evaluation (LREC'2002), Las Palmas, Canary Islands, Spain
Moreno R, Mayer RE, Spires HA, Lester JC (2001) The case for social agency in computer-based teaching: do students learn more deeply when they interact with animated pedagogical agents? Cognition Instr 19:177–213
Neff M, Fiume E (2005) Methods for exploring expressive stance. Graph Models. Special issue on SCA 2004. 68(2):133–157

Niewiadomski RA (2009) model of complex facial expressions in interpersonal relations for animated agents. PhD Thesis. PhD. dissertation, University of Perugia

Niewiadomski R, Pelachaud C (2007) Fuzzy similarity of facial expressions of embodied agents. In: 7th international conference on intelligent virtual agents (IVA'2007) Paris, France, pp 86–98

Noma T, Zhao L, Badler N (2000) Design of a Virtual Human Presenter. IEEE J Comput Graph Appl 20(4):79–85

Ochs M, Niewiadomski R, Pelachaud C, Sadek D (2005) Intelligent expressions of emotions. In: 1st International Conference on Affective Computing and Intelligent Interaction (ACII'2005), Springer-Verlag, Beijing, China, pp 707–714

Pandzic IS, Forchheimer R (2002) MPEG-4 facial animation. The standard, implementation and applications. Wiley and Sons, LTD

Pelachaud C (2005) Multimodal expressive embodied conversational agent. ACM Multimedia, Brave New Topics session, Singapore, 6–11 November, ACM

Pelachaud C, Braffort A, Breton G, Ech Chadai N, Gibet S, Martin J-C, Maubert S, Ochs M, Pelé D, Perrin A, Raynal M, Réveret L, Sadek D (2004) AGENTS CONVERSATIONELS : Systmes d'animation Modlisation des comportements multimodaux applications : agents pdagogiques et agents signeurs. Action Spcifique du CNRS Humain Virtuel

Poggi I (1996) Mind markers. In: 5th International pragmatics conference, Mexico City

Poggi I (2003) Mind markers. Gestures. Meaning and use. University Fernando Pessoa Press

Poggi I (2006) Social influence through face, hands, and body. In: Second Nordic Conference on Multimodality Goteborg, Sweden, pp 5–29

Poggi I, Pelachaud C, de Rosis F, Carofiglio V, De Carolis B (2005) GRETA. A Believable Embodied Conversational Agent. In: Stock O, Zancarano M (eds) Multimodal intelligent information presentation. Kluwer, Dordrecht, pp 3–26

Prendinger H, Ishizuka M (2004) Life-like characters. Tools, affective functions and applications. Springer, Berlin

Rehm M, André E (2005) Catch Me If You Can – Exploring Lying Agents in Social Settings. In: International Conference on Autonomous agents and multiagent systems (AAMAS'2005) Utrecht, the Netherlands, pp 937–944

Richmond VP, Croskey JC (1999) Non Verbal Behavior in Interpersonal relations. Allyn and Bacon

Ruttkay Z, Noot H, ten Hagen P (2003) Emotion Disc and Emotion Squares: tools to explore the facial expression face. Comput Graph Forum 22(1):49–53

Scherer KR (1984) Les fonctions des signes non verbaux dans la communication. In: Cosnier J, Brossard A (eds) La communication non verbale. Delachaux & Niestl, Paris, pp 71–100

Scherer KR (1998) Analyzing Emotion Blends. In: Proceedings of the 10th conference of the international society for research on emotions Wrzburg, Germany, pp 142–148

Scherer KR (2000) Emotion. Introduction to social psychology: a European perspective. Blackwell, Oxford

Scherer KR, Ellgring H (2007) Multimodal expression of emotion: affect programs or componential appraisal patterns? Emotion 7:1

Siegman AW, Feldstein S (1985) Multichannel integrations of nonverbal behavior. LEA–Routledge, New York, NY

Tepper P, Kopp S, Cassell J (2004) Content in context: generating language and iconic gesture without a gestionary. Workshop on balanced perception and action in ECAs at automous agents and multiagent systems (AAMAS), New York, NY, USA

Volpe G (2005) Special issue on expressive gesture in performing arts and new media. J New Music Rese Taylor and Francis 34:1

Wallbott HG (1998) Bodily expression of emotion. Eur J Soc Psychol 28:879–896

Wallbott HG, Scherer KR (1986) Cues and channels in emotion recognition. J Pers Soc Psychol 51(4):690–699

Wegener Knudsen M, Martin J-C, Dybkjr L, Berman S, Bernsen NO, Choukri K, Heid U, Kita S, Mapelli V, Pelachaud C, Poggi I, van Elswijk G, Wittenburg P (2002a) Survey of NIMM data resources, current and future user profiles, markets and user needs for NIMm resources. ISLE natural interactivity and multimodality. Working Group Deliverable D8.1

Wegener Knudsen M, Martin J-C, Dybkjr L, Machuca Ayuso M-J, Bernsen NO, Carletta J, Heid U, Kita S, Llisterri J, Pelachaud C, Poggi I, Reithinger N, van Elswijk G, Wittenburg P (2002b) Survey of multimodal annotation schemes and best practice. ISLE natural interactivity and multimodality. Working Group Deliverable D9.1

Wiggers M (1982) Jugments of facial expressions of emotion predicted from facial behavior. J Nonverb Behav 7(2):101–116

Wonisch D, Cooper G (2002) Interface agents: preferred appearance characteristics based upon context. In: Virtual conversational characters: applications, methods, and research challenges in conjunction with HF2002 and OZCHI2002 Melbourne, Australia

Representing Emotions and Related States in Technological Systems

Marc Schröder, Hannes Pirker, Myriam Lamolle, Felix Burkhardt, Christian Peter, and Enrico Zovato

Abstract In many cases when technological systems are to operate on emotions and related states, they need to represent these states. Existing representations are limited to application-specific solutions that fall short of representing the full range of concepts that have been identified as relevant in the scientific literature. The present chapter presents a broad conceptual view on the possibility to create a generic representation of emotions that can be used in many contexts and for many purposes. Potential use cases and resulting requirements are identified and compared to the scientific literature on emotions. Options for the practical realisation of an Emotion Markup Language are discussed in the light of the requirement to extend the language to different emotion concepts and vocabularies, and ontologies are investigated as a means to provide limited "mapping" mechanisms between different emotion representations.

1 Aims and Purpose

Machines that register when a user is emotional, machines that express emotions, and machines that reason about the appropriate emotion in a given situation – what used to be regarded as science fiction not long ago – is starting to become a reality. Many systems that implement one or the other aspect of this kind of behaviour are being built and show first successes in individual aspects of these complex tasks. From the point of view of applications, the modelling of emotion-related states in technological systems can be important for two reasons:

1. To enhance computer–mediated or human–machine communication. Emotions are a basic part of human communication and should therefore be taken into account, e.g. in emotional chat systems or emphatic voice boxes. This involves specification, analysis and display of emotion-related states.

M. Schröder (✉)
Deutsches Forschungszentrum für Künstliche Intelligenz, Saarbrücken, Germany
e-mail: schroed@dfki.de

P. Petta et al. (eds.), *Emotion-Oriented Systems*, Cognitive Technologies,
DOI 10.1007/978-3-642-15184-2_19,

2. To enhance systems' processing efficiency. Emotion and intelligence are strongly interconnected. The modelling of human emotions in computer processing can help to build more efficient systems, e.g. using emotional models for time-critical decision enforcement.

For early systems, it is acceptable – even inevitable – to use ad hoc representations. For example, an early emotion recognition system may use simple words such as "happy", "angry", "sad" to describe its recognition outputs; an emotion-related reasoner program may conclude that a "good event" should trigger the emotion of "joy"; and an early expressive system may mix up behavioural with emotional labels and use "happy" or "smiling" interchangeably. Such ad hoc solutions, tailor-made to the requirements of the immediate application domain, are fine as long as the various systems are not trying to communicate with each other.

However, more recently, integrated emotion-oriented computing systems are appearing, which typically consist of multiple components covering various aspects of data interpretation, reasoning, and behaviour generation. In this case, the need for a standardised way of representing emotions and related states is becoming clear: emotion-related information needs to be represented at the interfaces between system components in a way that allows one system component to make sense of the output from another component. We call such a representation an "Emotion Markup Language". This chapter describes a number of basic considerations that should be addressed by an Emotion Markup Language.

A standardised way to markup the data needed by such "emotion-oriented systems" has the potential to boost development both in academia and industry primarily because

(a) data that were annotated in a standardised way can be interchanged between systems more easily, thereby simplifying the reuse of emotional databases;
(b) the standard can be used to facilitate the creation of reusable submodules of emotion processing systems, e.g. submodules for the recognition of emotion from text, speech, or multimodal input;
(c) the use of a standardised technology in a new endeavour can help to ensure that best practice experiences from previous projects are taken into account.

The present contribution provides an overview of work in the area of representing emotions in technological systems. A number of existing markup and representation languages containing elements of emotion representation are briefly presented. However, the main focus of the chapter is on the considerations that should underlie a standardised, reusable representation. In that context, the chapter draws on recent work in the W3C Emotion Incubator Group,[1] an international endeavour investigating the prospects of defining a general-purpose emotion annotation and

[1] http://www.w3.org/2005/Incubator/emotion/

representation language.[2] We present the outcomes of the group's work regarding the kinds of application scenarios ("use cases") which would benefit from a standardised emotion representation and the requirements towards such a representation that arise from the use cases. We discuss how these requirements relate to existing scientific work in the area of emotion research. Finally, we formulate a number of alternatives for representing the emotions on a technical level, pointing out strengths and weaknesses of various design choices.

2 Related Work

Wherever emotion-related behaviour is to be analysed from user behaviour or generated in system behaviour, there is a need to represent emotional states and related information. Thorough and scientifically well-founded representations are being proposed in the context of data annotation and are described in *Part III on "Data and Databases"* (Douglas et al., this volume).

Representations aimed at being used in technological systems, on the other hand, have generally been shaped by application concerns; indeed, the investigation of appropriate means and models to represent emotion-related states in technological systems is going on since first such trials have been performed. In emotion recognition research, the preliminary state of the art – only a small number of distinct states can be reasonably recognised – has required only simple class labels to represent the emotional states. In research systems generating emotional behaviour, on the other hand, emotion representations have been built into several markup languages. For example, the Virtual Human Markup Language VHML (Gustavsson et al., 2001) was created in order to control the behaviour of animated characters (virtual agents); in addition to markup for facial animation, speech synthesis, dialogue management etc., the specification also contains a section for representing emotions. The actual representations are very simple: a set of nine emotions is encoded directly as XML elements, e.g.

Example 1. Representation of a simple emotion

```
<afraid intensity="40">
Do I have to go to the dentist?
</afraid>
```

The Affective Presentation Markup Language, APML (de Carolis et al., 2004), provides an attribute "affect" to encode an emotion category for an utterance (a "performative") or for a part of it:

---

[2]The W3C incubator group closed on 20, November 2008 with the report, "Elements of an EmotionML 1.0". The current working draft of the Emotion Markup Language is available at http://www.w3.org/TR/emotionml/ (accessed 2010-05-31).

Example 2. Affective Presentation Markup Language (APML)

<performative affect="afraid">
Do I have to go to the dentist?
</performative>

The Rich Representation Language, RRL (Krenn et al., 2002), uses an element "emotion", embedded in a dialogue act, to represent the emotion. The emotion category and its intensity can be expressed, as well as the three emotion dimensions "activation", "evaluation" and "power". In addition, there is a conceptual distinction between feeling and expressing an emotion:

Example 3. Rich Representation Language (RRL)

<dialogueAct>
...
<emotion>
<emotionExpressed type="afraid" intensity="0.3" activation="0.3" evaluation="-0.6" power="-0.3"/>
</emotion>
<entence> <text> Do I have to go to the dentist? </text>...</sentence>
</dialogueAct>

All these languages include the representation of an emotional state as one aspect in a complex representation oriented towards the generation of behaviour for an embodied conversational agent (ECA). None of the representations aim for reusability in different contexts, and none reach a representational power coming anywhere near the complexity considered to be necessary in emotion research (see, e.g. Cowie et al. this volume).

The Emotion Annotation and Representation Language, EARL (Schröder et al., 2006), was introduced as an attempt to address both issues: reusability and a representation approaching what is considered scientifically necessary. It can represent emotions alternatively in terms of categories, dimensions or appraisals; the intensity of the state can be indicated; several kinds of regulation are previewed, e.g. the simulation, suppression or amplified expression of an emotional state; complex emotions can be represented, as in situations of regulation or when more than one emotion is present. For example

Example 4. Emotion Annotation and Representation Language (EARL)

<emotion category="afraid" intensity="0.4" suppress="0.6" activation="0.3" evaluation="-0.6" power="-0.3">
Do I have to go to the dentist?
</emotion>

In view of generic use, several ways of defining the scope of the emotion were previewed in EARL: simply embedding the annotated content (as in the above example), cross-linking or the indication of start and end times. Furthermore, the actual lists of categories, dimensions and appraisals to use were designed to be flexible, giving the user a choice of using a label set appropriate for a specific use.

One major difference between EARL and ECA-related languages such as VHML, APML or RRL is the fact that EARL explicitly excludes the description of behaviour, linguistic structures, facial expressions etc. It aims to be a specialised, plug-in language to be used in combination with other languages. The advantage of this design approach is that it is easier to add emotion representation to a variety of systems. In particular, where a system already exists, it is possible to complement the existing representations with an emotion plug-in language such as EARL. For example, a multimodal dialogue system could add some emotional competence by using EARL at the interface between its processing components: the output of a facial analysis system could include a classifier's estimation of the emotions present and the respective probabilities, encoded in EARL; a subsequent dialogue manager could interpret the emotion, and generate an emotionally coloured reaction, again using an EARL representation to transmit the intended expression to the audio and visual generation components. The use of a standardised emotion representation language would increase the chance that components can be reused and/or integrated in different systems.

The difficulty faced by any standard, and especially in the context of emotions where so many different views exist on what an emotion is and how it can be defined and described, is satisfying everyone without becoming overly complex. A language will only be used if it provides what users need; at the same time, the language should not become too complicated, so that it is easy enough to understand how to use it. For that reason, it is essential to get a clear picture of what potential users require from a markup language. The following two sections describe, in some detail, how the W3C Emotion Incubator Group has attempted to answer that question.

3 Use Cases

In order to compile a new technological framework, it is good practice to start with a collection of descriptions of situations in which the need for such a framework might arise ("use cases"). So as a first step, the Emotion Incubator Group gathered together as complete a set of use cases as possible for the language, with two primary goals in mind: to gain an understanding of the many possible ways in which this language could be used, including the practical needs which have to be served, and to determine the scope of the language by defining which of the use cases would be suitable for such a language and which would not. Individual use cases were grouped into three broad categories: data annotation, emotion recognition and emotion generation. Many of the individual use case scenarios fall clearly into one of the categories; however, naturally, some cross the boundaries between categories. The types of use cases identified are summarised below.

3.1 Data Annotation

The data annotation use cases comprise a broad range of scenarios involving human annotation of the emotion contained in some data, e.g. speech samples or video clips. These scenarios show a broad range with respect to the material being annotated, the way this material is collected, the way the emotion itself is represented, and, notably, which kinds of additional information about the emotion are being annotated.

One simple case is the annotation of plain text with emotion dimensions, notably valence, as well as with emotion categories and intensities. Recent work on naturalistic multimodal emotional recordings has compiled a much richer set of annotation elements (Douglas et al., this volume) and has argued that a proper representation of these aspects is required for an adequate description of the inherent complexity in naturally occurring emotional behaviour. Examples of such additional annotations are multiple emotions that co-occur in various ways (e.g. as blended emotions, as a quick sequence, as one emotion masking another one), regulation effects such as simulation or attenuation, confidence of annotation accuracy, or the description of the annotation of one individual versus a collective annotation. Data are often recorded by actors rather then observed in naturalistic settings. Here, it may be desirable to represent the quality of the acting, in addition to the intended and possibly the perceived emotion. With respect to requirements, data annotation poses the most complex kinds of requirements towards an Emotion Markup Language, because many of the subtleties humans can perceive are far beyond the capabilities of today's technology.

3.2 Emotion Recognition

The general context of the emotion recognition use case has to do with low- and mid-level features which can be automatically detected, either offline or online, from human–human and human–machine interaction. In the case of low-level features, these can be facial features, such as action units (AUs) (Ekman and Friesen, 1978) or MPEG 4 facial action parameters (FAPs) (Tekalp and Ostermann, 2000), speech features related to prosody (Devillers et al., 2005) or language, or other, less frequently investigated modalities, such as bio-signals, like heart rate or skin conductivity (e.g. Picard et al., 2001). All of the above can be used in the context of emotion recognition to provide emotion labels or extract emotion-related cues, such as smiling, shrugging or nodding, eye gaze and head pose, etc. These features can then be transferred into higher levels of abstraction, stored for further processing or reused to synthesise expressivity in an embodied conversational agent (ECA) (Bevacqua et al., 2006).

In the case of unimodal recognition, the most prominent examples are speech and facial expressivity analysis. Regarding speech prosody and language, the CEICES data collection and processing initiative (Batliner et al., 2006) as well as exploratory extensions to automated call centres (Burkhardt et al., 2005) are the main factors that define the essential features and functionality of this use case.

Furthermore, individual modalities can be merged, either at feature or decision level, to provide multimodal recognition (Castellano et al., 2008; McIntyre and Göcke, 2008). A fusion on the feature level means that the classifier combines features from several modalities and decides on the basis of the unified features.

An alternative would the classification (or recognition) based on the outcome of several classifiers, one for each modality (decision level). Features and timing information (duration, peak, slope, etc.) from individual modalities are still present, but an integrated emotion label is also assigned to the multimedia file or stream in question. In addition to this, a confidence measure for each feature and decision assists in providing flexibility and robustness in automatic or user-assisted methods.

3.3 Emotion Generation

The use cases in the generation category were divided into three further subcategories, dealing with the simulation of modelled emotional processes, the generation of face and body gestures and the generation of emotional speech.

The first subset of generation use cases is termed "affective reasoner", to denote emotion modelling and simulation. The use cases in this category have a number of common elements that represented triggering the generation of an emotional behaviour according to a specified model or mapping. In general, emotion eliciting events are passed to an emotion generation system that maps the event to an emotion state which could then be realised as a physical representation, e.g. as gestures, speech or behavioural actions as described in the use cases of the other two subsets.

The second subset deals with the generation of automatic facial and body gestures for characters. With these use cases, the issue of the range of possible outputs from emotion generation systems becomes apparent. While all focused on generating human facial and body gestures, the possible range of systems that they connect to is large, meaning the possible mappings or output schema would be large. Both software and robotic systems are represented and as such the generated gesture information could be sent to both software-and hardware-based systems on any number of platforms. While a number of standards are available for animation that are used extensively within academia (e.g. MPEG-4 (Tekalp and Ostermann, 2000), BML (Kopp et al., 2006)), they are by no means common in industry.

The final subset is primarily focused on issues surrounding emotional speech synthesis, dialogue events and paralinguistic events. Similar to the previous subsets, this is also complicated by the wide range of possible systems to which the generating system will pass its information. There does not seem to be a widely used common standard, even though the range is not quite as diverse as with facial and body gestures. Some of these systems make use of databases of emotional responses and as such might use an emotion language as a method of storing and retrieving this information.

4 Requirements for an Emotion Markup Language

How useful is a markup language for emotions? This crucially depends on its ability to provide the representations required in a given use case.

The Emotion Incubator Group has analysed the above-mentioned use cases in order to make the implicit requirements contained in the descriptions explicit: to structure them in a way that reduces complexity and to agree on the boundaries between what should be included in the language itself, and where suitable links to other kinds of representations should be used. Given the thematic breadth of use cases, the integration of requirements from all the use cases into one coherent document was not straightforward. For example, many application domains have their own specific kinds of expressive behaviour that are important and that must be linked with the emotion representation. Also, what is called "input" in the case of recognising emotions, such as a facial expression, would be called "output" in the case of generation and vice versa. Therefore, to delimit the area to be covered by the Emotion Markup Language, and in order to find a common vocabulary that can be used across application domains, two basic principles were agreed.

1. The emotion language should not try to represent sensor data, facial expressions, etc. but define a way of interfacing with external representations of such data.
2. The use of system-centric vocabulary such as "input" and "output" should be avoided. Instead, concept names should be chosen by following the phenomena observed, such as "experiencer", "trigger" or "observable behaviour".

The resulting requirements are reported in detail in the final report of the W3C Emotion Incubator Group (Schröder et al., 2007); the key elements are summarised in the following.

4.1 Core Emotion Description

For the emotion (or emotion-related state) itself, three types of representation are envisaged, which can be used individually or in combination.

- Emotion categories (words) are symbolic shortcuts for complex, integrated states; an application using them needs to take care to define their meaning properly in the application context. The emotion markup should provide a generic mechanism to represent broad and small sets of possible emotion-related states. It should be possible to choose a set of emotion categories (a label set), because different applications need different sets of emotion labels.
- Alternatively, or in addition, emotion can be represented using a set of continuous dimensional scales, representing core elements of subjective feeling and of people's conceptualisation of emotions. As for emotion categories, it is not possible to predefine a normative set of dimensions. Instead, the language should provide a "default" set of dimensions, such as arousal, valence and power, that can be used if there are no specific application constraints, but allow the user to "plug-in" a custom set of dimensions if needed.
- As a third way to characterise emotions and related states, appraisal scales can be used, which provide details of the individual's evaluation of their environment.

Examples include novelty, goal significance or compatibility with one's standards.
- Finally, it may potentially be relevant to characterise emotions in terms of the action tendencies that accompany them, such as approach or avoidance.

Other concepts that appear to be necessary for describing the "core" of an emotion include

- the intensity of an emotion;
- the representation of multiple and complex emotions, in cases where emotions may be co-occurring (such as being sad about one thing and angry about another at the same time) or in cases of regulation;
- regulation, i.e. the complex of phenomena where an experiencer modifies the emotion itself or its expression, e.g. by masking, inhibiting, simulating
- temporal aspects of the emotion. A generic mechanism for temporal scope should allow different ways to specify temporal aspects, such as start and end times, start time and duration, timing relative to another entity (e.g. "start 2 seconds before an utterance starts and end with the second noun-phrase..."). In addition, a sampling mechanism providing values for a parameter at evenly spaced time intervals would allow for the description of continuous values such as intensity or dimensional scales as they change over time.[3]

4.2 Meta Information About Emotion Description

Three additional requirements with respect to meta-information have been elaborated:

1. information concerning the degree of acting of emotional displays; so, a mechanism should be defined to add special attributes for acted emotions such as perceived naturalness, authenticity, quality;
2. information related to confidences and probabilities of emotional annotations. The emotion markup should provide a generic attribute for representing the confidence (or, inversely, uncertainty) of any aspect of the representation. For example, such an attribute can be used to reflect the confidence of a human annotator that the particular value is as stated (e.g. a confidence of 0.8 that a subject's expression corresponds to the emotion "happiness"). More generally, it may be necessary to represent such a confidence with respect to each level of representation: intensity, degree of acting, etc.
3. and finally the modalities involved (e.g. not only face, voice, body posture or hand gestures, but also lighting, font shape).

All of this information thereby applies to each annotated emotion separately.

[3]Note that the timing of any associated behaviour, triggers etc. is covered in the section "Links to the rest of the world"....

4.3 Links to the "Rest of the World"

In order to be properly connected to the kinds of data relevant in a given application scenario, several kinds of links are required, referring to external media objects or to a position on a timeline within a media file. Start and end times are important to mark onset and offset of an emotional episode. Relevant information to link to can be of various sorts, so that a mechanism should be defined for flexibly assigning meaning to a link. A reasonable initial set of meanings for such links to the "rest of the world" should include the following: the experiencer, i.e. the person who "has" the emotion; the observable behaviour "expressing" it; the trigger, cause, or eliciting event of an emotion; and the object or target of the emotion, that is, what the emotion is about.

4.4 Global Metadata

Representing emotion, be it for annotation, detection or generation, requires the description of the context not directly related to the description of emotion per se but also the description of a more global context which is required for exploiting the representation of the emotion in a given application. Examples are data on person(s) like ID, date of birth, gender, language, personality traits, culture or level of expertise as labeller, information about the intended application (e.g. purpose of classification; application type – call centre data, online game; possibly, application name and version) and furthermore, it should be possible to specify the technical environment, for example, links to the particular camera properties, sensors used (model, configuration, specifics) or indeed any kind of environmental data. Finally, information on the social and communicative environment will be required, such as the type of collected data or the situational context in which an interaction occurs (number of people, relations, link to description of individual participants).

5 Different Descriptive Schemes for Emotions

5.1 Glancing at Theories

The collection of use cases and subsequent definition of requirements presented so far was performed in a predominantly bottom-up fashion and thus provides a strongly application-centred, engineering-driven perspective. The purpose of this section is to put this into a more theory centred perspective. A representation language should be as theory independent as possible but by no means ignorant of psychological theories. Therefore, a crosscheck to which extent components of existing psychological models of emotion mirrored in the currently collected requirements is performed.

The old Indian tale of the blind men and the elephant gained some popularity in the psychological literature as an allegory for the conceptual difficulties to come up with unified and uncontroversial descriptions of complex phenomena such as emotions. In this tale several blind men, who never have encountered an elephant before, try to come up with an understanding of the nature of this unknown object. Depending on the body part each of them touches they provide strongly diverging descriptions. An elephant seems to be best described as a rope if you hang to its tail only, is an ensemble of columns if you just touched its legs, appears as a piece of cloth if you encountered its ears, etc.

This metaphor fits nicely with the multitude of definitions and models currently available in the scientific literature on emotions, which come with a fair amount of terminological confusion added on top. Cowie et al. (this volume) give a very good overview. There are no commonly accepted answers to the questions on how to model the underlying mechanism that are causing emotions, on how to classify them, on whether to use categorical or dimensional descriptions, etc. But leaving these questions aside, there is a core set of concepts that are quite readily accepted to be essential components of emergent emotions.

One terminological issue quite relevant for the discussion in this section is the semantics of the term emotion itself, which has been used in a broad and a narrow sense.

In its narrow sense, used, e.g., by Scherer, the term refers to what is also called a prototypical emotional episode (Russell and Feldman, 1999), full-blown emotion, emergent emotion (Cowie et al., this volume): a short, intensive, clearly event triggered emotional event, of which fear when encountering a bear in the woods and fleeing in terror is the favourite example.

Especially in technological contexts there is a tendency to use the term emotion(al) in a broad sense, in its extreme for almost everything that cannot be captured as a purely cognitive aspect of human behaviour. More established but still not concisely defined terms for the range of phenomena that make up the elements of emotional life are "emotion-related states", "affective states" and "pervasive emotions". Whatever term used, there is quite some agreement that apart from emergent emotions the group of affective states includes moods, interpersonal stances, preferences/attitudes and affect disposition.

The envisaged scope of an emotion representation language clearly is concerned with emotions in the broad sense, i.e. it should be able to deal with different emotion-related states. Emergent emotions – not without reason also termed prototypical emotional episodes – can be viewed as the archetypical affective states and many emotional theories focus on emergent emotions. Empirical studies (Cowie et al., this volume; Wilhelm et al., 2004) on the other hand indicate that while there are almost no instances where people report their state as completely unemotional, examples of full-blown emergent emotions are really quite rare. As the majority of the ever present emotional life consists of moods, stances towards objects and persons and altered states of arousal, these naturally should play an increasingly prominent role in emotion-related computational applications and are thus clearly in the scope of the representation language.

5.2 Core Concepts and Their Role in the Representation Language

As stated above, although there is much disagreement on how best to theoretically model emotions, there is a certain consensus on a number of components that do play a role in the emotional life. The following list presents prominent concepts that have been used by psychologists in their quest for describing emotions. It will be evaluated whether and how these concepts are mirrored in the current list of requirements.

A general heuristic in the design of a representation language for emotions is to focus on those concepts that are observable in some way, thus hidden processes and constructs that are defined conceptually and not experientially should not be part of the representation.

Subjective component: Feelings. Feelings have not been mentioned in the requirements at all. They are not to be explicitly included in the representation for the moment being, as they are defined as internal states of the subject and are thus not accessible to observation. Applications can be envisaged where feelings might be of relevance in the future though, e.g. if self-reports are to be encoded. It should thus be kept as an open issue whether to allow for an explicit representation of feelings as a separate component in the future.

Cognitive component: Appraisals. As a reference to appraisal-related theories, the OCC model (Ortony et al., 1988), which is especially popular in the computational domain, has been brought up in the use cases. In these models emotions are elicited by a cognitive evaluation of perceived events or situations by a number of checks along different dimensions (e.g. relevance, coping potential). No choice for the exact set of appraisal conditions is to be made here. An open issue is whether models that make explicit predictions on the temporal ordering of appraisal checks (Sander et al., 2005) should be encodable to that level of detail. In general, appraisals are to be encoded in the representation language via attributing links to trigger objects. The encoding of other cognitive aspects, i.e. effects of emotions on the cognitive system (memory, perception, etc.) is to be kept as an open issue.

Physiological component: Changes in heart rate, breathing, sweating, etc. obviously are a component of emergent emotions. They also are quite interconnected with other components in this list. Feelings, e.g., can just be conceived as perception and classification of these physiological states, as in James (1984). Physiology, e.g. changes in the muscular tone, also accounts for changes of expressive features in speech (prosody, articulatory precision) or in the appearance (posture, skin colour). Physiological measures have been mentioned in the context of the use case of emotion recognition. They are to be integrated in the representation via links to externally encoded measures conceptualised as "observable behaviour".

Behavioural component: Action tendencies, such as the tendency to approach, avoid or reject, are a central concept in the work of Frijda (1986). Action tendencies can be viewed as a link between the outcome of an appraisal process and actual actions. It remains an issue of theoretical debate whether action tendencies, in contrast to actions, are among the set of actually observable concepts. Nevertheless these should be integrated in the representation language. This once again can be achieved via the link mechanism, this time an attributed link can specify an action tendency together with its object or target.

Expressive component: Expressions, which are not only most frequently studied in the face, but also as signs in the voice, posture and gesture, play a central role in the research tradition that relies on so-called basic emotions, e.g. in the work of Ekman (1992). But for the sake of the representation language, no stance needs to be taken whether expressions are produced by innate mechanisms as proposed by evolutionary models or are fundamentally communicative. Expressions are frequently referred to in the requirements. There is agreement to not encode them directly but again to make use of the linking mechanisms to observable behaviours.

Dimensional descriptions somehow cut across different components mentioned so far. There is a rich tradition of models that describe emotions in terms of a combination of typically two or three dimensions going back as far as Wundt (1905). Though there are differences in the concrete terminology, valence (positive or negative, pleasantness) and activation (arousal, energy) are rather undisputed dimensions. Dimensions, however, are not only used as global descriptors for emotions; typically, appraisals are formulated in terms of dimensions as well. Russell and Feldman (1999) use dimensions for describing feelings (core-affects in their terminology), and there are studies where dimensions are used for statistically clustering lexical terms for emotions in different languages (Mehrabian, 1995; Roesch et al., 2006). For the design of a representation language it can thus be concluded that expressive means for dimensional encodings should be made quite generally available.

Category-based descriptions, i.e. systems that make use of single terms such as anger or fear for representing emotions, are appealingly simple and are of course a prominent option in the envisaged representation. They are not without problems though. As Cowie et al. (this volume) spell out, these terms often lack concise definitions, might be too unspecific, obscure the influence of the context, etc. In a theoretical context, categorical descriptions are typically connected to theories that tie emotions to basic innate neural systems and claim that there is a small number of such basic emotions that are nicely reflected in everyday natural language terms. In the context of a representation language for computational purposes, this strong commitment to a certain psychological theory plays a less important role. Even when using simple terms like fear, it still can be conceived as just a handy abbreviation for a complex process, where a subject encountering a bear in the woods initiates the evaluation of a cascade of appraisal dimensions, experiences strong changes in their physiological states, their perceptive and cognitive capabilities, feelings of fear and develops not only an action tendency for fleeing but also performs this action, while crying out in panic. Also, another researcher might use the label fear for quite a different experience, for instance, a user realising loss of sensible data. Good use of meta-information is hence required when using category-based descriptions, accurately describing the situation in which the emotion occurred.

Figure 1 summarises the way in which these components of emotion are related to the requirements for an Emotion Markup Language as summarised in Sect. 6.4.4.

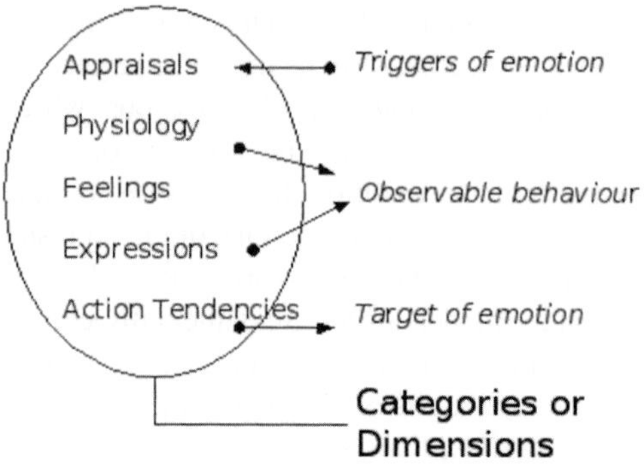

Fig. 1 Overview of how components of emotions are to be linked to external representations

5.3 Emergent Emotions vs. Other Emotion-Related States

At the start of this section it was elaborated that the scope of the representation language should not be restricted to emergent emotions even though these have received most attention so far. A positive aspect is that, although emergent emotions make up only a very small part of the emotion-related states, they nevertheless are sort of archetypes. Representations developed for emergent emotions should be usable as a basis for the encoding of other important emotion-related states such as moods and attitudes.

Scherer (2000), who systematically defines relationships between emergent emotions and other emotion-related states, supports this hypothesis. Scherer is proposing a small set of so-called design features for characterising the various kinds of affective state: intensity, duration, synchronisation, event focus, appraisal elicitation, rapidity of change and behavioural impact. For example, emergent emotions are defined as having a strong direct impact on behaviour, high intensity, being rapidly changing and short, are focussing on a triggering event and involve strong appraisal elicitation. Moods, as another kind of emotion-related states, are described using the same set of design features: they are characterised as not having a direct impact on behaviour, being less intense, changing less quickly and lasting longer and not being directly tied to an eliciting event. In this framework, different types of emotion-related states thus just arise from differences in the design features. In technological contexts, however, descriptions will rarely reach this level of detail. Therefore, the representation language should provide simple means for encoding the type of an emotion-related state explicitly.

6 Options for Realisation in XML

This section addresses the question how a generic language to annotate and represent emotions should be realised syntactically. Such a language should be (i) easily combined with other markup languages and (ii) extensible in order to adapt it to specific domains.

Several options are available. Given the fact that most recent markup languages are defined in XML (eXtensible Markup Language), it appears to be a good choice to formulate an Emotion Markup Language in XML. An alternative to this is the use of RDF (resource description framework), a formalism for representing information as subject–predicate–object expressions. RDF is particularly well suited for representing ontological structures, i.e. relations among entities. An XML formulation of RDF exists, making the two alternatives non-exclusive. Another alternative is the use of OWL (Ontology Web Language) which facilitates the representation of ontologies (cf. Sect. 8) by providing additional vocabulary along with a formal semantics. OWL adds more vocabulary for describing properties, classes of concepts and their interrelationships. It is based on XML and RDF. Without aiming to prejudge this question, the following discussion will focus on XML realisation.

6.1 Flat vs. Deep Structures

The structure of annotation in a markup language can be flat or deep. Deep structures have the advantage that the meaning of information is fully explicit in the structure; however, this adds overhead which becomes a burden especially in simple cases. Flat structures, on the other hand, can represent simple cases in a simple way, but may become difficult to read for more complex annotations.

The HUMAINE EARL (Emotion Annotation and Representation Language) (Schröder et al., 2006) is an example of a flat structure. Its design was guided by the principle that "simple cases should look simple". Indeed, the annotation of a picture with an emotion category and intensity would be as simple as the following:

Example 5. Emotion category and intensity in EARL

```
<emotion xlink:href="picture.jpg" category="contentment" intensity="0.7" />
```

When this annotation is to be enriched with additional annotations, such as the appraisal "high goal conduciveness" and the dimensional ratings "positive and passive", these are simply added to the list of attributes:

Example 6. Additional annotations in EARL

```
<emotion xlink:href="picture.jpg" category="contentment" intensity="0.7"
goal_conduciveness="0.9" valence="0.5" arousal="-0.3" />
```

In such a structure, the meaning of the various attributes is implicit – the user needs to know what they being doing, or else it is easy to mix things in an invalid way.

A deep structure could help avoid this problem, e.g. by making the status of category, dimension or appraisal of an annotation explicit, e.g.:

Example 7. Deep structure in EARL

```
<emotion>
<object xlink:href="picture.jpg" />
<category intensity="0.7"> contentment </category>
<appraisal>
<goal_conduciveness> 0.9 </goal_conduciveness>
</appraisal>
<dimensions>
<valence> 0.5 </valence>
<arousal> -0.3 </arousal>
</dimensions>
</emotion>
```

In this example, the trade-off between simplicity and explicit structures becomes clearly visible. The disambiguation of annotations is paid for by a more complicated structure.

Which alternative is better to use remains to be seen. As both alternatives have all their information embedded in one < *emotion* > tag, there is no difference regarding their capabilities for being combined with other markup languages.

7 Extensibility

The question of extensibility is a challenge with respect to the definition of any standard. In the present case, where very different emotion representations are suitable in different circumstances, and sets of values are often domain specific, the challenge is particularly marked.

The simplest solution to providing this kind of flexibility would be to leave the sets of values open. It would provide maximum flexibility, but on the other hand would make it impossible to verify that the markup is correct and meaningful.

The HUMAINE EARL specification has proposed a more controllable method for defining custom sets of values. It allows users to "plug-in" their own, tailor-made sets of emotion categories, dimensions and/or appraisals. Technically, this is achieved by splitting the EARL schema design in four parts: for any given EARL "dialect", the EARL base schema, which defines the structure of EARL documents, is complemented by three small schemas defining the set of categories, dimensions and/or appraisals to be used. In this way, the EARL is actually conceived as a family of EARL dialects – all of them sharing a common structure, but each with its own set of valid values and identified by its own XML namespace. As a "default" EARL language, a set of 48 categories, 3 dimensions and 19 appraisals were proposed. These can be used when there are no specific requirements to go beyond them.

An alternative approach may be for the markup to indicate, as an attribute value, a namespace to be used for validating substructures. This approach was used in the W3C working draft language EMMA (extensible multimodal annotation) in its < *emma* : *model* > mechanism: A namespace reference can be used to indicate the possible substructures of an < *emma* : *interpretation* > element. This mechanism appears not only to be more flexible, but also to introduce more overhead compared to the HUMAINE EARL approach.

8 Ontologies of Emotion Descriptions

In an emotion representation language, different emotion representations need to be made possible because no preferred representation has yet emerged for all types of use. Instead, the most suitable representation to use depends on the application. As a result, complex systems such as many foreseeable real-world applications will require some information about (1) the relationships between the concepts used in one description and about (2) the relationships between different descriptions. In order to enable components in complex systems to work together even though they use different emotion representations, an Emotion Markup Language should be complemented with a mapping mechanism based on ontologies of emotion descriptions.

The concepts in an emotion description are usually not independent, but are related to one another. For example, emotion words may form a hierarchy, as suggested, e.g., by prototype theories of emotions. For example, Shaver et al., (1987) classified cheerfulness, zest, contentment, pride, optimism enthrallment and relief as different kinds of joy, irritation, exasperation, rage, disgust, envy and torment as different kinds of anger.

Such structures, be they motivated by emotion theory or by application-specific requirements, may be an important complement to the representations in an Emotion Markup Language. In particular, they would allow for a mapping from a larger set of categories to a smaller set of higher level categories.

Different emotion representations (e.g. categories, dimensions and appraisals) are not independent; rather, they describe different parts of the "elephant" of the phenomenon emotion. However, from a scientific point of view, it will not always be possible to define such mappings. For example, the mapping between categories and dimensions will only work in one direction. Emotion categories, understood as short labels for complex states, can be located on emotion dimensions representing core properties; but a position in emotion dimension space is ambiguous with respect to many of the specific properties of emotion categories and can thus only be mapped to generic super-categories.

Similarly, it may be possible to define a mapping from categories to appraisals, when categories are understood as "shortcuts" for appraisal configurations. In the opposite direction, however, many appraisal combinations will not be associated with any category in an exact manner; instead, they may be "similar" to one or more categories.

The mapping mechanism required here could easily become as complex as the field of emotion theory itself, and the attempt to define such mappings could end up as an interminable discussion about theoretical notions. Pragmatically, however, a subset of possible mappings may be defined, at least in application-specific ways, and the concrete needs of applications may be a suitable guideline for the definition of a mapping mechanism, helping to avoid the pitfall of getting stuck in theoretical debate.

9 Conclusion and Outlook

This chapter has described a number of basic considerations that should be addressed by an Emotion Markup Language. We have briefly reviewed existing work, pointing out the potential benefits of a reusable standard "plug-in" representation of emotions and related states. We have described the compilation of a rich set of requirements from use cases in the W3C Emotion Incubator Group and have compared the work with scientific descriptions of emotion. This comparison has shown that many aspects studied in the literature are also relevant for the technological use cases envisaged, whereas some aspects of high scientific importance, such as feeling, play only a limited role. Finally, we have pointed out basic design choices available for the syntactic realisation of an emotion representation, notably in terms of a flat or deep structure, and have pointed out the potential use of emotion ontologies for the interoperability of components using different types of emotion descriptors.

The considerations described here are a solid basis for the development of an emotion markup specification. While various aspects may need to be simplified in view of implementability, the present collection is valuable as an outline of the actual complexity of the phenomenon and should be able to serve as a guideline for future standardisation activities in the area of representing emotions and related states.

References

Batliner A, Steidl S, Schuller B, Seppi D, Laskowski K, Vogt T, Devillers L, Vidrascu L, Amir N, Kessous L, Aharonson V (2006) Combining efforts for improving automatic classification of emotional user states. In: Erjavec T Gros J (ed) Language technologies, IS-LTC 2006 Infornacijska Druzba (Information Society), Ljubljana, Slovenia, [pp, 240–245]

Bevacqua E, Raouzaiou A, Peters C, Caridakis G, Karpouzis K, Pelachaud C, Mancini M (2006) Multimodal sensing, interpretation and copying of movements by a virtual agent. In: Proceedings of perception and interactive technologies (PIT'06), Kloster Irsee, Germany, June 19–20

Burkhardt F, van Ballegooy M, Englert R, Huber R (2005) An emotion-aware voice portal. In: Proceedings of Electronic Speech Signal Processing ESSP. Prague, Czech Republic, September 2005, pp 123–131

Castellano G, Kessous L, Caridakis G (2008). Emotion recognition through multiple modalities: face, body gesture, speech. In: Christian P, Russell B (eds) Affect and emotion in human-computer interaction. LNCS, vol 4868. Springer, Heidelberg, pp 92–103

Cowie R, Sussman N, Ben-Ze'ev A (this volume) Emotions: concepts and definitions

de Carolis B, Pelachaud C, Poggi I, Steedman M (2004) APML, a mark-up language for believable behavior generation. In: Prendinger H (ed) Life-like characters. Tools, affective functions and applications, Springer, New York, NY, pp 65–85

Devillers L, Vidrascu L, Lamel L (2005) Challenges in real-life emotion annotation and machine learning based detection. Neural Netw 18:407–422

Douglas-Cowie E, Cox C, Lowry O, Martin J-C, Devillers L, Abrilian S, Pelachaud C, Peters C (this volume). The HUMAINE Database

Ekman P, Friesen W (1978) The facial action coding system. Consulting Psychologists Press, San Francisco, CA

Ekman P (1992) An argument for basic emotions. Cogn Emot 6:169–200

Frijda N (1986) The emotions. Cambridge University Press, Cambridge

James W (1884) What is an emotion? Mind 9:188–205

Gustavsson C, Beard S, Strindlund L, Huynh Q, Wiknertz E, Marriott A, Stallo J (2001) VHML specification working draft v0.3, October 21st 2001. http://www.vhml.org/downloads/VHML/vhml.pdf. Accessed 24 Oct 2007

Kopp S, Krenn B, Marsella S, Marshall A, Pelachaud C, Pirker H, Thórisson K, Vilhjálmsson H (2006) Towards a common framework for multimodal generation in ECAs: the behavior markup language. In: Proceedings of the 6th International Conference on Intelligent Virtual Agents (IVA'06), Marina del Rey, USA, pp. 205–217

Krenn B, Pirker H, Grice M, Piwek P, Deemter K, van, Schröder M, Klesen M, Gstrein E (2002) Generation of multimodal dialogue for net environments. In: Proceedings of Konvens. Saarbrücken, Germany

McIntyre G, Göcke R (2008) The composite sensing of affect. In: Christian P, Russell B (eds) Affect and emotion in human-computer interaction. LNCS, vol 4868. Springer, Heidelberg

Mehrabian A (1995) Framework for a comprehensive description and measurement of emotional states. Genet Soc Gen Psychol Monogr, 121:339–361

Ortony A, Clore GL Collins A (1988) The cognitive structure of emotions. Cambridge University Press, New York, NY

Picard RW, Vyzas E, Healey J (2001, October) Toward machine emotional intelligence – analysis of affective physiological state. IEEE Trans Patt Anal Mach Intell 23(10):1175–1191

Roesch EB, Fontaine JB, Scherer KR, (2006) The world of emotion is two-dimensional – or is it? In: Paper presented to the HUMAINE Summer School 2006, Genoa, Italy

Russell JA, Feldman BL (1999) Core affect, prototypical emotional episodes, and other things called emotion: dissecting the elephant. J Pers Soc Psychol, 76: 805–819

Sander D, Grandjean D, Scherer KR (2005) A systems approach to appraisal mechanisms in emotion. Neural Netw 18:317–352

Scherer KR (2000) Psychological models of emotion. In: Borod J C (ed) The neuropsychology of emotion. Oxford University Press, New York, NY, (pp. 137–162)

Schröder M, Pirker H, Lamolle M (2006) First suggestions for an emotion annotation and representation language. In: Proceedings of LREC'06 Workshop on Corpora for Research on Emotion and Affect, Genoa, Italy, pp 88–92

Schröder M, Zovato E, Pirker H, Peter C, Burkhardt F (2007) W3C Emotion incubator group final report, Published online on 10 July 2007:http://www.w3.org/2005/Incubator/emotion/XGR-emotion/. Accessed 2 November 2010

Shaver P, Schwartz J, Kirson D, O'Connor C (1987) Emotion knowledge: further exploration of a prototype approach. J Pers Soc Psychol, 52:1061–1086

Tekalp M, Ostermann J (2000). Face and 2d mesh animation in MPEG-4. Image Commun J 15:387–421

Wilhelm P, Schoebi D, Perrez M (2004) Frequency estimates of emotions in everyday life from a diary method's perspective: a comment on Scherer et al.'s survey-study "Emotions in everyday life". Soc Sci Info 43(4):647–665

Wundt W (1905) Grundriss der psychologie [Fundamentals of psychology]: vol 3, 5th ed. Engelmann, Leipzig

Embodied Conversational Characters: Representation Formats for Multimodal Communicative Behaviours

Brigitte Krenn, Catherine Pelachaud, Hannes Pirker, and Christopher Peters

Abstract This contribution deals with the requirements on representation languages employed in planning and displaying communicative multimodal behaviour of embodied conversational agents (ECAs). We focus on the role of behaviour representation frameworks as part of the processing chain from intent planning to the planning and generation of multimodal communicative behaviours. On the one hand, the field is fragmented, with almost everybody working on ECAs developing their own tailor-made representations, which is amongst others reflected in the extensive references list. On the other hand, there are general aspects that need to be modelled in order to generate multimodal behaviour. Throughout the chapter, we take different perspectives on existing representation languages and outline the fundamental of a common framework.

1 Introduction

This contribution deals with the requirements on representation languages employed in planning and displaying communicative multimodal behaviour of embodied conversational agents (ECAs). The term ECA has been coined in Cassell et al. (2000) and refers to human-like virtual characters that typically engage in a face-to-face communication with the human user, employing various synchronised channels of communication such as facial expression, hand–arm gestures and body posture, as well as tone of voice and text. The embodiment of ECAs ranges from talking heads to full-bodied 2D and 3D characters. Underlying are complex AI systems that model the character's capabilities to meaningfully engage in communicative situations with the human user or other ECAs. This includes modelling of the character's self, its tasks, goals, related believes and intentions, its personality, emotions and interpersonal stances, but also modelling of the character's (perceived)

B. Krenn (✉)
Austrian Research Institute for Artificial Intelligence, Freyung 6, 1010, Vienna, Austria
e-mail: brigitte.krenn@ofai.at

P. Petta et al. (eds.), *Emotion-Oriented Systems*, Cognitive Technologies,
DOI 10.1007/978-3-642-15184-2_20,

environment, including social scenarios, communication situations and partners, in order to generate situationally adequate and believable behaviours.

ECA systems may implement a face-to-face dialogue with the user, model scenarios where humans and artificial agents interact with each other in a virtual or mixed environment or generate communicative interactions between different artificial characters in which the user actively takes part or which are displayed to the user like in a product presentation, a TV spot or a stage play. See for instance André and Rist (2000) and Nijholt (2006) where one or more virtual agents present information to the user, Krenn et al. (2002) where a virtual car seller and buyer engage in a conversation about various features of cars and Mateas and Stern (2003) where in an interactive drama the interaction of a user with two characters representing a couple on the verge of divorce influences the outcome of the couple's story. Rehm and André (2005) present GAMBL, an interactive test-bed for human–ECA interaction. Other examples of ECA implementations are the Real Estate Agent REA (Cassell et al., 1999) which implements a full perception–action loop of communication by interpreting multimodal user input and generating multimodal agent behaviour; the pedagogical agent Steve (Rickel and Johnson, 1998) which functions as a tutor in training situations; MAX (Kopp and Wachsmuth, 2004), a virtual character geared towards simulating multimodal behaviour; and Carmen (Marsella et al., 2003), a system that supports humans in emotionally critical situations such as advising parents of infant cancer patients. ECAs can adopt several roles, such as being a teacher (Johnson et al., 2005; Moreno, 2007), a doctor (De Rosis et al., 2003), a museum guide (Kopp and Wachsmuth, 2004), a real-estate agent REA (Cassell et al., 1999) or a companion (Bickmore and Cassell, 2005; Hall et al., 2006). Even though the above examples represent only a small fraction of the vast and constantly growing work on ECAs, they are well suited for illustrating the broad range of applications.

Human communicative behaviour covers a broad range of skills, including natural language generation and production, co-verbal gesture, eye gaze and facial expression. People produce such behaviours in real time with ease and in a broad range of circumstances. In order to simulate human-like multimodal communicative behaviour, advanced ECA systems need to incorporate a whole range of complex processing steps, from intent to behaviour planning to behaviour realisation including some sort of scene or story generation, multimodal natural language generation, speech synthesis, the temporal alignment of verbal and non-verbal behaviours and behaviour realisation employing particular animation libraries and engines. In the current contribution, we focus on the role of behaviour representation frameworks as part of the processing chain from intent planning to the planning and generation of multimodal communicative behaviours.

2 Background

Imagine a situation, where we want to model the following encounter between a character C and a user U: C is in a good mood, encountering the appearance of U in the system makes C particularly happy and leads C to greet U effusively. Finally, we want to see an animation including the following behaviours: C displays a neutral but friendly face, directs its attention to U, broadly smiles at U and says *Hello my friend! Good to see you after such a long time!* As regards the spoken utterance, we want to put emphasis on *hello, good* and *such.*

At the intentional level we thus have something like the following ingredients which we present in a pseudo-notation (in reality XML[1] formats are widely used):

Example 1. Intentional Level pseudo-notation

```
mood(C) = happy;
event = encounter(C,U) -->
emotion(C) = happy & communication_act(C) = greet(U).
```

At the behaviour planning level, we need to further specify the non-verbal and verbal behaviours, bring them into a temporal order and specify the relative dependencies between the communication channels involved. Employing our pseudo-notation, this might be represented as follows:

Example 2. Behaviour planning level pseudo-notation

```
c1: face(C) = neutral_friendly;
c2: gaze(target(C))= user;
c3: face(C) = broad_smile;
c4: gesture(C) = wave;
c5: utterance(C) = emph {hello} my friend! m1 emph{good} to
see you after such a emph{long} time!;
start(c1) = t0;
start(c2) = encounter(C,U);
start(c3) = encounter(C,U);
end (c3) = start(c5);
start(c4) = encounter(C,U);
end(c4) = m1.
```

The above notation defines the partial behaviours (c1 to c5) including face, gaze, gesture and utterance, puts them in a relative temporal order and specifies the following dependencies: At time t0, the beginning of the animation, C looks neutral_friendly and starts to look at the user when encountering them. At the same time, C starts a broad smile and waves. When beginning to speak C stops smiling. The wave ends after *friend* has been spoken out and before the onset of *good.*

At the behaviour realisation level, we need to specify the actual behaviours at an even greater level of detail such that the realisation engine is able to generate a sequence of integrated multimodal behaviours for playing. In particular, the concrete animations are selected, the utterance is synthesised and the timing of the partial behaviours is transformed from relative to absolute. For lip-synchronised

[1] http://www.w3.org/XML/

speech, phonemes (transcripts of consonants and vowels) are aligned with visemes (visualisations of mouth shapes which may also include the tongue). Depending on the speech synthesis component employed, also a markup of prosodic information including syllables, intonation phrases and related accents may be available. This information is necessary to synchronise eyebrow raises and accents.

As our small example has shown, there is a variety of information that needs to be modelled in order to represent multimodal communicative behaviour. Accordingly, considerable effort has been put into the development and documentation of representation formats in the last years. The overall complexity of ECA systems motivated different strands of development. On the one hand it gave rise to a number of XML-based markup languages which are aimed at providing means for human authors to easily annotate text with multimodal behaviours. On the other hand attention was geared towards the specification of representation formats for the exchange of information between sub-components of an ECA system.

The goal of this contribution is to give an overview on representation formats proposed so far and will discuss specific representational needs raised by selected sub-tasks such as modelling at different levels of processing emotional display and spoken dialogue accompanying bodily behaviours. An evaluation in terms of the general acceptance and dissemination of different representation formats will be provided. Moreover, we will present an initiative for a common architecture for ECA systems and the prospects for the future development of representation languages. The community is still investigating ways to come up with strategies of unifying the existing variety of representation formats. We expect benefits of such an endeavour only if representation formats and source codes of related processing components are made available for free to the research community.

3 Different Views on Representation Languages/Formats for Behaviour Generation of ECAs

Numerous representation languages or formats have been proposed in the literature. They include markup languages for annotating text with behaviour directives, representation languages that declaratively and to different degrees of detail model various aspects of information required at different stages of behaviour generation, and languages that incorporate procedural knowledge in their annotations. Some languages attempt to cover a broad range of information relevant for behaviour generation. Most of the representations, however, have been designed for specific applications. In order to structure the wealth of proposals, we will, in the following, offer two views on existing representations: First we provide examples for different representation formats ranging from text markup to representations that contain aspects of high-level programming languages. Second we will present examples for representation languages that attempt to cover a broad range of information versus representations that have been designed with a specific application purpose in mind.

3.1 Formats – Markup– Versus Representation– Versus Scripting Languages

Markup languages typically define sets of markups that give the non-expert user (usually a web designer) the possibility to annotate text with high-level behavioural information in order to produce pre-scripted presentations for ECAs. VHML (Beard and Reid, 2002)[2] is an example for this type of languages. It has been designed for creating interactive multimodal applications with talking heads or full-bodied ECAs. Other examples of ECA markup languages where text is annotated with high-level concepts are APML (De Carolis et al., 2004) and MPML (Zong et al., 2000).

Representation languages in contrast aim for the technically detailed annotation of theory-specific information. In this respect, the Emotion Annotation and Representation Language (EARL) addressed in Schröder et al. (2010) is more a representation than a markup language. This holds in general for all languages that become more and more detailed in modelling and describing multimodal behaviours. Thus representation languages are well suited to function as data representation formats inside a system, especially as representations at the interfaces between the individual sub-components. RRL (Rich Representation Language[3], Piwek et al., 2002) is an example for such a language that defines an XML format for representing the input and output of all the components used for realising the processing steps from intent to behaviour planning to behaviour realisation.

Scripting languages in addition also incorporate means for encoding procedural knowledge, e.g. conditional execution of behaviours such as "if event X occurs, then execute behaviour Y". Thus scripting languages are comparable to high-level programming languages. Examples in the field of ECAs are STEP and its XML variant XSTEP (Huang et al., 2003), and ABL (Mateas and Stern, 2004). The expressive power of scripting languages comes with a price though, e.g. the complexity of writing specifications in ABL comes close to programming in Java. The choice of the appropriate level of representation thus has to take different constraints into account. On the one hand, markup languages are indispensable in application development, because the application designer need not necessarily be an expert in all the fields underlying the development of ECA systems. On the other hand, representation languages are crucial in research contexts, because of the necessity to represent highly specific, low-level information. With the increased demand for truly interactive systems, the need for including at least some procedural capabilities typical for scripting languages into the representations becomes more and more of an issue, resulting in new hybrid formats. This trend is, e.g., exemplified in the evolution from MPML to MPML3D (Nischt et al., 2006). While MPML was designed as a markup language that allows the non-expert user to create pre-synchronised presentations, MPML3D has evolved to an authoring system that allows for the embedding of scripts and for the design of reactive scenarios. Summing up, while multimodal

[2] http://www.vhml.org, Virtual Human Markup Language

[3] http://www.ofai.at/research/nlu/NECA/RRL

markup languages are designed to allow non-experts create multimodal presentations easily, representation languages in the above definition are designed to ease the integration and exchange of components in multimodal systems. Rist (2004) argues though that with the advent of more and more sophisticated authoring tools in the future, simplicity eventually will become less of a design criterion for markup languages and the distinction between the different types of representation languages will become more vague.

3.2 Scope – General Purpose Versus Application-Specific

To give an assessment of different strands of endeavours, we present, in the following, attempts to develop representation formats of broad scope and contrast them with languages that have been developed to serve much more restricted purposes, either being developed in the contexts of and thus particularly geared towards certain ECA implementations or aiming at the representation of certain aspects in multimodal behaviour generation.

3.2.1 General-Purpose Initiatives

HumanML (Brooks, 2000) was an initiative hosted by the "Organization for the Advancement of Structured Information Standards" (OASIS) to come up with a mark-up language for describing virtually all properties not only of artificial characters but of human beings. It set off with the goal to provide information for human-to-human and human-to-machine communications in a machine-readable form. The language aimed to encode information related to human communicative behaviour from high level (culture, emotion) to low level (signal, kinesics) and aimed to be of relevance for such diverse areas as anthropology, medicine, business communication and virtual reality. It was planned to specify tags related to physiology, proxemics, kinesics, haptics, beliefs, intentions, emotions, etc. and aimed to provide attributes related to community, culture, context/location of the conversation, personality, thoughts and signals. The initiative came to a halt soon after proposing a rough XML scheme with place-holders for the high-level concepts which never were specified in more detail.

Virtual Human Markup Language (VHML) presents a more down-to-earth initiative for a language that should facilitate the interaction of a talking head or a virtual human (Beard and Reid, 2002). It was designed as a confederation of various relatively simple sub-languages, each of them concerned with a sub-task: dialogue management, emotion, facial animation, body animation, hypertext and speech. VHML has a hierarchical structure, i.e. elements of a lower level inherit information from the higher level. The one typical example for the hierarchical encoding is emotion tags which are inherited by all the sub-components for speech, face and gesture. The specification of VHML did not leave the draft level and was mainly geared towards the control of a talking head, for which also sample implementations were implemented. That is, the most detailed specification was available for the head and the face, while the gesture markup language only comprised a small set of six atomic behaviours (shrug, agree, disagree, concentrate, emphasise, smile).

3.2.2 Special-Purpose Applications

Languages that have been developed for specific purposes are, e.g., SiGML for sign language (Elliott et al., 2004) and MURML for the reproduction of gesture kinematics (Kranstedt and Kopp, 2002). Other examples are MPML for a presentation agent (Mori et al., 2003); RRL to represent information relevant at the interfaces of system components in a pipeline for generation of animated presentation dialogues (Piwek et al., 2002); APML (De Carolis et al., 2004), AML and CML (Arafa et al., 2002) for agent communicative behaviour; and BEAT for verbal and non-verbal synchronisation (Cassell et al., 2001). We will describe MPML, RRL and APML in greater detail directly in the following. MURML will be addressed when it comes to gesture coding (Sect. 4.3.2) and BEAT when we talk about multimodal natural language generation (Sect. 4.2.1).

Multimodal Presentation Markup Language (MPML) aims at developing a language to easily create animated agents within interactive presentations. Agents may be set up on the web and the user can interact directly with the agents. The general goal of MPML is that, unlike most other web agents for presentation applications, the presentation of information is no longer presented sequentially, but its content is generated dynamically as the conversation between the agent and the user evolves (Mori et al., 2003). Furthermore, MPML has been designed for mouse control, voice control, text-to-speech and agent's action description (Tsutsui et al., 2000). A specialised scripting language, SCRipting Emotion-based Agent Minds (SCREAM), may be interfaced with MPML (Prendinger et al., 2004). SCREAM has been designed to create emotionally and socially appropriate responses of animated agents placed in an interactive environment. SCREAM is specialised in scripting the agent's mind. SCREAM may be used within applications that compute the verbal content of the interaction between the user and the agent. The role of SCREAM is to compute the emotion that may arise during the conversation. The instantiation of the signals and their intensity for a given emotion is then computed taking into account many factors, such as the social setting of the conversation as well as the agent's mental state.

The Rich Representation Language, RRL, has been developed to manage interactive dialogue scenes between two or more virtual agents (Piwek et al., 2002). RRL is used as a link between a scene generator, a multimodal natural language generator, a speech synthesis component, a gesture assignment component and finally a media player. A scene description contains information related to the set of communicative acts and the temporal ordering of these acts. An affective reasoner is embedded in the scene description to compute the corresponding emotion that may be triggered by given acts. The emotion is defined by its type, intensity and optionally by the object that causes the emotion. A scene description is input to the multimodal natural language generator that computes the corresponding linguistic and non-linguistic forms of the communicative acts. The role of the speech synthesis and gesture assignment components is to instantiate the acoustic and visual data for a given emotion and dialogue act. This results in an XML script representing the multimodal dialogue where the verbal and non-verbal behaviour is fully specified and the integrated temporal specification of the various communication channels (speech, facial, expression, gesture) is absolute. The script forms the common basis from which the specific animation directives required to drive individual players are derived via syntactic transformation.

The Affective Presentation Markup Language (APML) is based on a taxonomy of communicative functions proposed by Isabella Poggi (Poggi et al., 2000). A communicative function is defined as a pair (meaning, signal) where the meaning corresponds to the communicative value the agent wants to communicate and the signal is the behaviour used to convey this meaning. Communicative values are differentiated into four categories namely information about the speaker's beliefs, intentions, affective state and mental state. APML tags correspond to the meaning of a given communicative function. The conversion from meaning to signals is done by looking up a library of meaning–signal pairs.

Noot et al. (2004) developed GESTYLE, a complex representation language which is based on several dictionaries. Each dictionary reflects a certain aspect of the style of a character, e.g. cultural characteristics, profession or personality, and defines the association of meaning to signals. In addition physical information such as manner of gesturing (smooth, jerky, etc.) or tiredness can be specified. To create an agent with style one needs to specify this set of parameters (e.g. an Italian extrovert professor), and the proper set of mappings between meanings and signals is then instantiated.

To summarise, as we can see from the examples in this section, all proposed languages somehow model the relation between the ECA's mental states (goals, beliefs, intentions, emotions) and their display via concrete verbal and non-verbal behaviours. The details addressed in the various languages, however, differ widely, as they depend on the particular application the ECA system is built for, and thus on the system components realised.

4 Inventory of Information Relevant in the Generation Process

As we have seen in the previous section, all the languages or representation formats presented so far model some specific mix of information required in the complex process of multimodal behaviour generation. Obviously there is a considerable overlap between different representations, but at the same time they are not directly compatible or cannot be easily adapted for the needs of individual ECA projects. Therefore in the past we have seen (re-)building of representations over and over again. In this section, we will take a step back and concentrate not so much on existing languages, but look at the kinds of information relevant at different steps of the generation process.

Figure 1 depicts one way to conceptually organise the multitude of different processing steps and associated modules into fairly high-level blocks and helps to

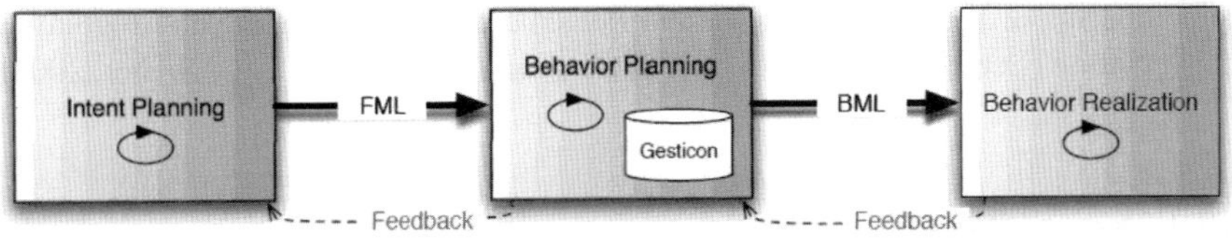

Fig. 1 Outlay of the processing pipeline for the generation of ECA behaviour as proposed by the SAIBA framework (Kopp et al., 2006)

divide the overall task into separate sub-components. This grouping into three top-level modules namely intent planning, behaviour planning and behaviour realisation has been proposed by the SAIBA framework Situation, Agent, Intention, Behaviour and Animation (Kopp et al., 2006; Vilhjálmsson et al., 2007). All three boxes have to be understood as complex systems with a variety of sub-components. One of the main guiding principles in the design of SAIBA architecture was to aim for a clear distinction between function and behaviour. FML thus stands for Functional Markup Language and BML for Behaviour Markup Language. Functional markup is to include all information regarding agent's mental, communicative and affective state that is necessary to create a link between intent and behaviour planning. It needs to provide a large spectrum of information including semantic, communicative, discursive, pragmatic and epistemic information. Behaviour markup comprises all those representations that are necessary for the realisation of behaviour. This includes textual and prosodic information, facial display, gestures and postures, eye gaze, etc. and, very importantly, it includes directives for the temporal synchronisation of behaviours. This format nevertheless aims for specifying behaviours independent of specific behaviour realisation systems, most specifically independent of concrete animation engines. In the following we are organising the discussion on information requirements and existing representation format along the lines of the SAIBA architecture.

4.1 Intent Planning

In order to be able to specify an ECA's communicative behaviours, first of all the underlying intent needs to be determined. At this stage, the basic semantic units related to the communicative event are computed. There is no reference to any physical or verbal behaviour yet. All in all, intent planning requires the computation of the mental, affective and communicative state of the agent. One possibility to implement the mental state of the agent is a BDI approach (Belief, Desire, Intention; Louis and Martinez, 2007). FML has not yet been defined in such a detail as BML has been. However, several proposals have been made for relevant concepts to be modelled during intent planning; see the collection of papers in Heylen et al. (2008). Contributions cover amongst others the specification of communicative actions (Kopp and Pfeiffer-Leßmann, 2008), different cognitive functions such as remembering or recalling (Mancini and Pelachaud, 2008), planning and regulating conversation (Lee et al., 2008), as well as emotional states (Krenn and Sieber, 2008).

4.1.1 Personality and Emotion

Personality and emotion are important aspects guiding the display of human behaviour. Emotions and emotion related states influence the way communication proceeds, its wording, voice quality, facial expression and other aspects of bodily behaviours such as posture and the dynamics of gesture. While emotion is responsible for the temporary changes in the quality of expression, personality determines the global tendencies of an individual's expression and thus may function as a means to establish coherency and consistency in the behaviour of an individual so that it becomes more predictable for the observer (Ortony, 2003). The expression of joy, for instance, in an extrovert person is overall much more pronounced than when expressed by an introvert person.

In the ECA community two approaches to personality and emotion are widely employed, the Five-Factor (OCEAN) model (Wiggins, 1996) and the OCC model (Ortony et al., 2003), respectively. This is reflected in the inventory of several representation languages for ECAs, e.g. PAR (Allbeck and Badler, 2002) makes provisions for OCC and OCEAN; MPML incorporates labels for the 22 emotions defined in OCC (Zong et al., 2000); RRL encodes the OCC labels plus their extension by (Elliott, 1992), a subset of OCEAN labels and a politeness attribute; APML defines its own set of emotion labels geared towards the communication situation (medical counselling) and the type of ECA (a talking head) used.

OCC is an example of an appraisal model. It defines emotions as positive and negative reactions to events, to other characters' or people's actions and to objects. Events are evaluated with respect to their desirability, actions according to their praiseworthiness, and objects in terms of their appeal to the agent. The subjective appraisal of a situation depends on the goals, standards and attitudes of the agent. Whereas attitudes are long-term affective states, emotions have a strong onset, but diminish over time. The latter is typically modelled via a decay function (Gebhard et al., 2003). Gebhard (2005) has presented a computational model that integrates emotion, mood and personality, standing for short-, mid- and long-term aspects of affect, respectively. As the terminology to describe human emotional life is manifold, we would like to redirect you to the opening chapter of the handbook (Cowie et al., 2010) for a discussion of concepts and definitions.

Modelling of emotion comes into play in intent and behaviour planning as well as in behaviour realisation. Whereas in intent planning, appraisal models have shown to be well suited (Ortony et al., 2003; Egges et al., 2003; Gebhard et al., 2003; Gebhard, 2005; Rank and Petta, 2005; Ochs et al., 2008; Marsella and Gratch, 2009), basic emotion categories (Ekman, 1993) are still the predominant representation when it comes to behaviour realisation. As Ekman's original research on basic emotions has focused on facial expressions, it unsurprisingly is still very influential in the field of facial animation for ECAs. Alternatively dimensional models

(Scherer, 2000) have been successfully employed for modelling emotional speech (Schröder et al., 2001; Schröder, 2004).

Personality models have been integrated in agents to model behaviour tendencies as well as intent planning (Moffatt, 1995; André et al., 2000; Johns and Silverman, 2001; Ball and Breese, 2000; Egges et al., 2004; Kshirsagar and Magnenat-Thalmann, 2002). The Five-Factor model of personality (McCrae and Costa, 1996) is used in most of the cited works. The interplay between personality and emotion has been studied. Moffatt (1995) views personality and emotion as similar states that differ in time and duration. Moreover personality ensures coherency of reactions to similar events, i.e. the emotional answers of an individual to these events are coherent through time. See also Ortony (2003).

4.1.2 Dialogue

Modelling of affect and personality not only is required for generating believable non-verbal behaviour, but also influences the agents dialogue; see Piwek (2003) for a survey. Even more importantly, automatic dialogue generation requires a representation of the domain, as it determines to a large extent what the virtual actors can talk about. Moreover behaviour, including dialogues, adheres to social conventions. Depending on the social relationship between the communication partners and the formality or informality of the situation, things are said differently and different display rules for the body behaviour apply (Walker et al., 1996; Rehm and André, 2005; De Carolis et al., 2001; Niewiadomski and Pelachaud, 2007; Ball and Breese, 2000), e.g. people would normally avoid crying in a business meeting whereas such a behaviour is much more likely in a private, intimate setting. What kinds of behaviours are socially acceptable and which ones are not strongly depends on cultural conventions. Just think about what is considered as good table manners in Europe as opposed to China. For instance, smacking and burping while eating will be considered as rude in Europe, but may be expected in China as an indicator for the positive appreciation of the food.

For modelling the communicative state, some dialogue planner is required that generates the initial version of a dialogue as a sequence of dialogue acts. A dialogue act represents an abstract communicative function, such as requesting for information, answering a question and giving feedback. Such communicative functions can be realised in many different ways depending, for example, on the personality and affective state of the actor. The structures produced by the dialogue planner represent communicative strategies that can be observed in a particular genre or domain and (partial) plans of how the communication should proceed. These plans include choice points according to which the communication differently proceeds based on the input from the outside world. This can be utterances from the communication partner(s), but also events occurring in the environment. van Deemter et al. (2008), e.g., describe the plan generation process for whole scenes of car sales dialogues enacted by two virtual characters, a customer and a seller.

This kind of presenter agents (see also André et al., 2000) realises very specific cases of dialogue where the whole dialogue is planned in one go depending on the initial settings given by the user. While the dialogue proceeds, no interference from the user is possible. Carmen's bright ideas (Marsella, 2003), FearNot! (Hall et al., 2006) and Faade (Mateas and Stern, 2003), are examples where the way how the story proceeds depends on the user's contribution to the dialogue.

4.2 Behaviour Planning

Given a particular intention and/or emotional state the agent aims to communicate, the system needs to decide which non-verbal signals will be used. The behaviour planner takes as input a given intention and/or emotional state, for instance to greet the communication partner happily, and outputs representations for the visual and acoustic signals to be generated by the behaviour realisation modules, such as a waving hand gesture, a broad smiling face and a greeting utterance for instance "Hello my friend! It is great to see you!". The behaviour planner has to instantiate the communicative acts, in our case greeting with the emotional colouring happy, which were generated during the intent planning phase and which are defined in terms of their meaning, into signals and then decide which modalities (facial expression, gaze, gesture, posture, voice, etc.) will be used to convey the particular meaning. Apart from selecting the modalities to convey meaning, proper synchronisation between the modalities is crucial. Examples of ECA systems that use such an approach are, for instance, the Greta behaviour engine (Pelachaud, 2005), BEAT (Cassell et al., 2001), SmartBody (Thiébaux et al., 2008) and MAX (Kopp and Wachsmuth, 2004). The task of behaviour planning thus at least comprises the sub-tasks multimodal planning and multimodal alignment.

If you go back to our example 2 on page 3, multimodal planning accounts for modelling the communication channels face, gaze, gesture and utterance expressed in c1 to c5, whereas multimodal alignment takes care of the timing of the expressions in the different channels relative to each other, which we have modelled with the start–end mechanism. Absolute timing is only available at the stage of behaviour realisation, when the speech is synthesised and concrete animations have been selected.

4.2.1 Multimodal Behaviour Planning

Non-verbal and verbal behaviour needs to be tightly integrated. Thus, in the most sophisticated ECA implementations, planning of non-verbal behaviour is coupled with natural language generation (NLG) leading to multimodal natural language generation (MNLG). In NLG the overall generation task is traditionally divided into two separate phases. In strategic generation it is decided "What to say", i.e. which propositions are to be expressed in a still language-independent representation. Tactical generation then deals with the "How to say", i.e. it is responsible to come up with the concrete wording.

As in NLG, the multimodal generation process is divided into a planning phase where the communicative acts are semantically outlined and a realisation phase where the behaviours that are going to be actually displayed are specified. Gestures are planned on the basis of the semantic and pragmatic content of the natural language utterances and are aligned with the respective representations of the utterance. At this stage, information on the concrete realisation of the body behaviour as well as the surface realisation of the utterances, i.e. the concrete wording, is still under-specified. The idea of intertwining gestural and syntactic structure has been proposed in different works. Cassell et al. (2000) describe a mechanism for applying the SPUD natural language generator (Stone et al., 2000) to multimodal generation. SPUD makes use of the "Lexicalized Tree Adjoining Grammar" (LTAG) formalism (Joshi et al., 1975; Schabes, 1990) and integrates the natural language grammar with motion events. Integration of gestures and syntax is particularly suitable for gestures that can express semantic content and therefore present an alternative to linguistic expression of the same content. For instance, if one wants to express that some X has a square shape one could say "X is squared", "X is of square shape", etc. However, one could as well say "X looks like this" and produce a gesture depicting a square. Gestures can also be used to express discourse functions. For instance, a question can be accompanied by an eye brow raise or a head tilt, assertions by a head nod, etc. van Deemter et al. (2008) describe an approach to MNLG which is based on typed feature structures representing deep syntactic, semantic and pragmatic content of dialogue acts and referring expressions. For the semantic representations, Discourse Representation Theory is employed (Kamp and Reyle, 1993). An extra module associates gestures and body postures with specific dialogue acts.

4.2.2 Alignment of Multimodal Behaviours

At the stage of behaviour planning signals across modalities must be aligned to each other. The most prevalent task is the temporal synchronisation between verbal and non-verbal signals, i.e. typically between eye gaze, facial expression, gestures and speech. Emphasis, e.g., is usually encoded via the synchronisation of an eye brow movement and a beat gesture with the stressed syllable of the word to be emphasised. Also speech, gaze and deictic gestures are typically synchronised. Take for example an utterance such as "give me this cake", which might be accompanied by an eye gaze and a deictic gesture towards the cake, with gaze and pointing being aligned with the phrase "this cake". This example shows that apart from the temporal alignment across channels, also the spatial alignment of movements and objects in the world needs to be handled. Getting back to the example, there is the cake as a target of pointing and gaze, and there is the addressee of the utterance who also needs to be looked at in order to establish and maintain the communication channel. In order to enable the behaviour realisation modules to take care of the actual synchronisations, the representations on the level of behaviour planning should make provisions for appropriate synchronisation points. Knowledge about the location of stressed syllables and sentence accents, e.g., is most crucial for proper synchronisation in the speech modality. For gestures usually the exact placement of the stroke phase, which is defined as the most meaningful part, is essential. In Fig. 2 the especially fine-grained inventory of synchronisation points offered by BML is depicted. The purpose of the different points is explained as follows.

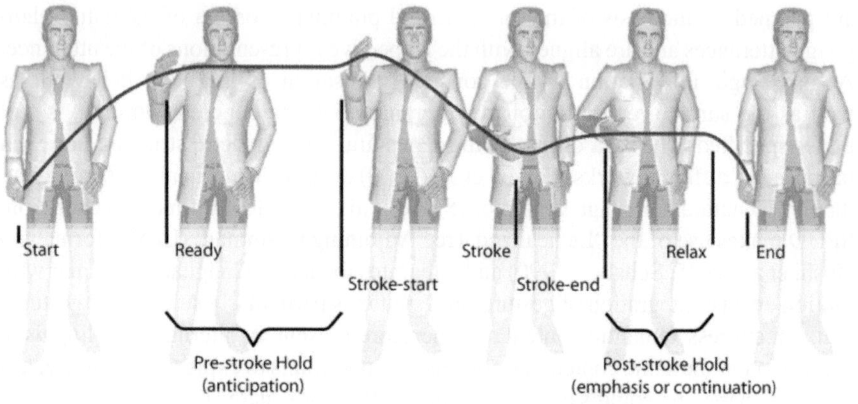

Fig. 2 The synchronisation points of a communicative behaviour (Kopp et al., 2006)

> The preparation for or visible anticipation of the behaviour occurs between start and ready, and the retraction back to neutral or previous state occurs between relax and end. The actual behaviour takes place between the ready and relax, with the most significant or semantically rich motion during the stroke phase, between stroke-start and stroke-end, with the greatest effort coinciding with the stroke point. (. . .) If no preparation or relaxation is needed, then start and ready refer to the same point in time, and relax and end refer to the same point in time. Quoted from Kopp et al. (2006)

4.3 Behaviour Realisation

Behaviour realisation is concerned with the generation of the concrete realisation of the behaviours. This component deals with tasks such as selecting the one most appropriate deictic gestures from a repository of candidate gestures that shall be finally realised by the media player, realising a very specific facial expression, generating speech on the basis of natural language text and replacing the specification of relative timing of the synchronised behaviours from the planning phase with absolute time values. Until today, it is common in ECA implementations to use speech as the guiding medium for temporal alignment. In the best case the granularity of temporal information in the speech channel goes down to the level of phoneme durations (in milliseconds), though in many applications only information on the location of word boundaries is employed. With the availability of fine-grained prosodic information, facial and gesture animation can be time-aligned to individual phonemes, accented syllables or boundaries of intonation phrases.

What is still missing is an integrated approach where not only speech defines the timing of its accompanying facial expressions and gestures, but also motor activation constrains voice quality, e.g. to lengthen the duration of a prosodic phrase and postpone the onset of its successor in order to wait for an accompanying gesture to finish.

4.3.1 Speech

Text-to-speech (TTS) systems take as input text possibly annotated with additional information. The TTS determines pronunciation and prosodic properties, such as location and type of pitch accents, prosodic phrase boundaries and duration of phonemes. Based on this information sound files are generated. For generating multimodal behaviour the TTS should provide fine-grained temporal information such as the list of phonemes, the location of phoneme boundaries to allow for the synchronisation of visemes (mouth shapes related to sounds/phonemes) or the location of accented syllables to allow for the exact temporal alignment of beat gestures. Though this information at some point is available within virtually every TTS system, there is no standard way to gain access to this information, given it is accessible at all. Unfortunately many of the commercial products do not provide interfaces to this sort of information, and only a few of the research-related systems provide access as easily as, e.g., the Mary system (Schröder and Trouvain, 2004) or the Festival TTS (Black and Taylor, 1997).

The W3C Speech Synthesis Markup Language SSML[4] has been developed to assist the generation of synthetic speech. It provides a standard way to mark up text in order to control aspects of speech such as pronunciation, volume, pitch and rate and is supported by a variety of speech synthesis systems. SSML can thus be seen as an example of a success story when it comes to the specification of a unimodal markup language. This standard, however, only provides a very rough interface both to and from the synthesis engine. For instance the only feedback mechanism is the insertion of event throwing tags in the text, which limits the temporal granularity of the feedback to the level of orthographic words. For the purpose of multimodal generation, in addition to SSML, some standardised speech synthesis output format would be highly desirable in order to make TTS's internal decisions on pronunciation, timing, pitch, accenting, phrasing, etc. accessible to other components.

As previously mentioned, speech is usually the leading modality providing the timeline for the other modalities (face, gaze, body, gesture) to align with. This, however, requires to be changed in favour of models where the timing of speech should be sensitive to restrictions from other modalities posing additional challenges for the specification of proper interfaces from and to the speech synthesis component. One example is the generation of explanations while manipulating objects, where the manipulations become the leading modality and pauses between intonation phrases are adapted accordingly. In this case, control of sub-sentential chunks has to be guaranteed (Kopp and Jung, 2000). Another possible application where a demand of increased temporal control is to be foreseen is the implementation of immediate reactivity, e.g. an ECA that is interrupting an utterance as an instant reaction to a barge-in or to some other observed user behaviour.

[4] http://www.w3.org/TR/speech-synthesis/

4.3.2 Gestures

Gestures are complex, being composed by one or a sequence of basic gesture elements, each of which describe a basic hand–arm movement trajectory. A trajectory is defined by a sequence of key points where each point corresponds to a position of the wrist in 3D space and a hand configuration. Depending on the granularity of representation, a gesture spans one or more phases, e.g. preparation, stroke, hold, retract. See also Fig. 2. Methods for encoding of gestures can be classified on a continuum ranging from purely semantic representations (related only to the meaning of the gesture) to formats which encode the form of the gesture exclusively. Most existing representation languages for computational systems are founded upon annotation systems in psychology and sign language research. In the following, we will briefly review some of the foundational works and then give an outline of scripting languages used for ECA systems. A more detailed review of existing gesture coding schemes can be found in Serenari et al. (2002). McNeill (1992) provides a semantic classification into iconic, metaphoric, deictic and beat gestures. To localise gestures, a grid-like gesture space is introduced in front of the actor. The descriptions of gesture form are holistic-imagistic though and not readily adaptable to automatic processing, because the shape of a gesture, especially in the case of iconic and some metaphoric gestures, refers to the meaning of the gesture. For example, a cup-shaped hand in certain contexts (for instance when visitors come to the house in France) carries the meaning “to offer something to the guests” as it is interpreted as representing a bowl of food. To make use of this representation for behaviour realisation, it requires the instantiation of the rough semantic classification into concrete representations, such as shape descriptions. Calbris (1990), thus, describes gestures by the meaningful form of their components morphology (segment, configuration, orientation, localisation and movement). Components may be linked to physical properties, for example, the flat vertical hand held between the speaker and the listener to mean stop symbolises the erection of a wall between the speaker and the listener as to show refusal. This act of refusal can also be done by throwing the head backward or even turning the head away. Calbris (1990) describes how gesture variants can be gathered as a class of gestures carrying one meaning. In contrast a gesture may be associated to several meanings depending on the contexts it occurs in. Stokoe (1978) introduces the concept of breaking gestural configurations down into formational parameters such as location and orientation of the wrist and hand shape. A hand shape is described by a thumb orientation and shapes of the other four fingers. The more recent HamNoSys notation framework (Prillwitz et al., 1989; Hanke, 2004), originally developed for sign languages, follows this breakdown into formational parameters and provides a dictionary of the most frequently used configurations. MURML (Kranstedt and Kopp, 2002) is an XML-based description language which has been influenced by HamNoSys. It allows for detailed control of parallel and sequential components of gestures, whereas hand shapes and facial expressions are specified by simple labels. Several parameters have been defined such as the wrist location in space, the palm and finger orientation, hand shape and the wrist trajectory. Values of these parameters create hand and arm configuration.

Such configurations must be described for each phase of a gesture (preparation, pre-stroke hold, stroke, etc). Timing information related to the duration of a gesture to the temporal constraints between body parts involved within the gesture can be provided. MURML allows for a precise description of behaviour. Hartmann et al. (2002) describe a language for ECA animation that unites features of McNeill and HamNoSys. FORM2 by Martell et al. (2003) is an annotation scheme that captures the exact configuration and orientation of the arms and hands of a gesturer. It focuses on a detailed description of the geometrical properties of gestures. A corpus of annotated video material is available.

Scripting Technology for Embodied Persona, STEP (Huang et al., 2004), works for ECAs based on H-Anim.[5] It offers a set of sensors and effectors through which agents can perceive the world they are placed in and can take appropriate actions. STEP includes two main primitive actions (turn, move) to specify body movement. The first defines the rotation to apply to a given joint of the virtual agent, while the second relates to the displacement of one effector. More complex actions can be obtained by combining these primitive actions with three operators: seq, the sequential operator; par, the parallel operator; and T, the repeated operator.

4.3.3 Facial Expression

As with the other kinds of information relevant in multimodal behaviour generation, several coding schemes to describe facial expressions have been devised too. MPEG-4, for instance, is an ISO/IEC standard which defines specifications for the animation of face and body models within an MPEG-4 terminal (Doenges et al., 1997; Ostermann, 1998; Pandzic and Forchheimer, 2002). Two sets of parameters describe and animate the 3D facial model: the facial animation parameters (FAPs) and the facial definition parameters (FDPs). FDPs define the shape of the model, FAPs define the facial actions. FAPs represent a large set of basic facial actions including head, tongue, eye and mouth movement. In combination they represent facial expressions. Facial expression may also be coded using the Facial Action Coding System (FACS) developed by Ekman et al. (Ekman and Friesen, 1978; Ekman et al., 2002). It is a framework to measure facial signals using minimal action units (AUs). With FACS facial action units can be encoded on a scale of five intensities. Behaviour changes along this scale are carefully described. Paradiso and L'Abbate (2001) have established an algebra to create facial expressions. The authors have elaborated operators that combine and manipulate facial expressions. Another definition language has been proposed by De Carolis et al. (2004) and by Paradiso and L'Abbate (2001). In their language, an expression may be defined at a high level (a facial expression is a combination of other facial expressions already pre-defined) or at a low level (a facial expression is a combination of facial parameters). The low-level facial parameters correspond to the MPEG-4 FAPs. The language is also suitable to create easily extendable facial display dictionaries.

[5] http://www.h-anim.org/

5 Towards a Common Framework for Representations in Multimodal Behaviour Generation

The non-exhaustive overview above provides an impression on the amount of effort that has up to now been invested in the design of representation formats. In the last decade we have seen the development of a significant number of ECA systems and applications. In parallel, we have experienced the publication of an almost equal number of usually XML-based markup and representation languages – cf. Arafa et al. (2003) for an impressive list of such languages. To some extent the proliferation of representation languages can also be explained with the enthusiasm about XML in its early hey-days. Employing XML came with an implicit promise of reusability and ease of application and, in addition to presenting ECA systems, the associated representation formats suddenly became a topic of interest worth publishing. Though there are identifiable similarities between existing representation languages still there are not many examples where a sharing of representation formats – not to speak of software – has taken place. Considering the work that has been put to this topic, there is a significant lack on formats that ever got reused outside their original institution or ECA project. This lack of success in terms of acceptance in the community is most obvious for VHML and especially HumanML, which were explicitly designed with the aim to become standards. For other languages reusability might not have been the primary goal in the first place, as they were designed to fit the needs of a concrete ECA application. But even if the main motivation for existing representations might not have been to trigger and sustain the development of reusable and exchangeable system components via the specification of open interface formats, there is an obvious demand for such formats. The implementation of ECA systems requires expertise in such diverse research topics as emotion modelling, behaviour planning, natural language generation, speech synthesis and computer animation. Only for some of these tasks off-the-shelf modules are available; other components are still in their early stages of research and development. Given the overall complexity of ECA systems, exchangeable sub-components that would allow for a plug-and-play approach for the system development would clearly be desirable. Due to the lack of common standards and architectures, similar functionalities need to be implemented over and over again for different systems. At a second thought the problem can also be stated the other way round: It is not so much the lack of standardised interfaces which prevents the development of common software modules, but the lack of widely used system components hinders the establishment of common representation languages. There is a mutual dependency between interface specifications and the availability of system components, but of course software modules that are both useful and usable for a broader group of developers do indeed provide a high incentive to promote the interface formats connected with these components. These considerations have led to a renewed interest in the development of common interfaces, which should then foster the development of system components that could be shared and reused among different ECA projects, thus avoiding the replication of effort. An exemplary activity in this direction is the work of the SAIBA initiative where several research groups have joined forces to come up with a commonly agreed on framework for multimodal behaviour generation. The goal for this initiative is to develop common specifications for representation languages, which are meant to be application independent and graphics model independent, and to present a clear-cut separation between information types (function versus behaviour specification). Intermediate results of this joint endeavour are documented in Kopp et al. (2006) and Vilhjálmsson et al. (2007). Experience with earlier initiatives in this direction like VHML provides strong evidence that success or failure of such a representation format is tightly coupled to the availability of software components that actually provide an immediate benefit for the system developer. This of course is a chicken or egg problem not easily solved. We thus are re-evaluating exemplary sub-topics for which we think that shared representations and jointly developed software modules could succeed in the intermediate future. It is not by chance that the work within the SAIBA initiative has by now mostly focused on issues concerning the specification of BML, i.e. on schemes that deal with the encoding of behaviours. Though still complicated enough, there is a joint understanding of the concepts necessary to describe the communicative actions of the human body. Human anatomy and the specifics and needs of existing animation techniques and speech synthesisers help to guide the development of BML. For FML, i.e. the encoding of functional categories, it is much more difficult to come up with general, application-independent representations. Among the information types affiliated to the functional domain in SAIBA, the representation of emotions is a prime candidate for the development of a joint representation format; see the chapter "Representing Emotions and Related States in Technological Systems" by Schröder et al. (2010) in this part.

One important issue in the current development of BML is the specification of non-verbal communicative behaviour. Existing schemes which are using joint-angles and segment translations such as MPEG-4 or BVH (Pandzic and Forchheimer, 2002) do provide exact and detailed physical information on body shapes. Nevertheless they are viewed as way too specific. They are lacking flexibility both in the modification of temporal and spatial properties, e.g. they run into problems of collision if body proportions are changed. Also functional information, e.g. the identification of stroke phases, which is crucial for the proper temporal alignment across modalities, is completely missing. Coming up with a common higher- level format for the representation of facial expression, gesture and posture which would function as a sort of middle layer between the specification of intentions and the formats used for actually rendering body movements is by no means a trivial task. But still the problem seems to be confined in a tractable way. There is much agreement on the overall requirements for such a representation, and the development can partly be based on experience gained with existing languages like MURML and FORM. The prospect of coming up with a representation for the physical appearance and bodily actions of an ECA that is independent of individual animation engines is obviously appealing. A strong motivation for working on such a representation for behaviours has been the idea to use it for providing repositories of communicative gestures, for which the terms gesticon and gestuary have been proposed (Krenn and Pirker, 2004), i.e. collections of gestures reusable in different

ECA systems. At the same time the immediate requirement for supplementing the representation format with concrete software that actually provides non-trivial functionalities becomes evident. In the terminology of the SAIBA architecture this would be, e.g., behaviour realisation modules which are interpreting BML representations and render them to different animation engines. As long as no components for interpreting and transforming to player-specific code are available, this intermediate representation does not provide any additional functionality and developers of ECA systems would skip it and stick to their own representations. When it comes to representation formats for speech synthesis, we are facing a special situation. Speech synthesis is the one domain where exchangeable off-the-shelf components actually do exist. Basically all these systems provide an input interface and an output interface that are universally accepted, namely text and audio files, respectively. Even if developers of ECA systems might not always be happy with the quality of the synthesiser's output, these systems deal with a clearly defined, specialised and complex task, and not many developers feel inclined to intermingle with this functionality themselves. Open issues one could think of when it comes to ECAs and speech synthesis are missing standards to specify emotion and the missing ability for incremental speech synthesis which would be desirable for the really interactive systems that would, e.g., react to interruptions by the user in mid-sentence. But there is another issue on representation formats and speech synthesis that is not so much a technical problem. In most implementations of multimodal systems, speech is the leading modality which provides the temporal grid to which the other modalities (lip movements, gestures) are synchronised. Information on the temporal locus of stressed syllables, phonemes, accents, etc. thus is crucial, but in most TTS systems this data is not made available to the user. This is not due to technical reasons, but this information is usually suppressed because of mere ignorance of an existing demand for it. The promotion of a standardised format for this kind of information, i.e. a kind of speech output format, could trigger an increased awareness of TTS suppliers for that kind of information demand by ECA developers. As stated above, not only does the fate of any representation format rely on design factors such as expressive power, flexibility and ease of application but its acceptance in a wider community also strongly depends on the availability of implementations that actually support the creation and interpretation of the proposed format. We are concluding this contribution with a short survey on possible insights which could either be gained from or shared with research domains outside the narrow ECA paradigm. Descriptive schemes used for ECAs of course have much in common with coding schemes that describe the behaviour, physical appearance or emotional states of real humans, and ECA developers have been adapting coding schemes originally designed for humans in the past. FACS (Ekman and Friesen, 1978; Ekman et al., 2002), the coding scheme for facial expressions, was developed in the context of psychological and anthropological research and was very influential for specifying codes for facial animations. HamNoSys (Hanke, 2004), a representation format for sign language, has been adapted for the encoding of gestures in ECAs. Descriptive schemes for gestures have been developed for the manual annotation of multimodal conversational data, e.g. ANVIL (Kipp, 2001) and CoGest (Trippel et al., 2004).

Demands for a common gesture description language have also been brought up in the field of automatic gesture recognition and interpretation (Kölsch and Martell, 2006). Another source for inspiration and possible synergies is the gaming industry. Yue and de Byl (2006) present a standardisation initiative for AI components used in games. They deal with problems that are related to the intent and behaviour planning and not so much to the behaviour realisation, e.g. path finding and steering in a game environment and action planning. On an abstract level these are functionalities that resemble those in the behaviour planning component in SAIBA. There are insights to be gained by the way this standardisation process is organised. One interesting aspect is that this initiative does not bother with XML formats but deals with the specification of Application Programming Interfaces (APIs), i.e. formulates their interfaces in directly implementation-related terms.

The gaming industry also provides examples on how pseudo-standards actually may emerge from the spreading of tools and vice versa. BVH (Biovision Hierarchy) is a graphics format developed for storing motion-captured data. Tools for creating key-frame animations in BVH are also emerging, e.g. Cal3D, an open source character animation library,[6] and plug-ins for exporting BHV format from widely used commercial graphics programs such as 3D Studio Max. The BVH format is also used for animating avatars in the virtual world of Second Life and is an example of how an application provides the incentive for the development of tools and thus for the proliferation of specific representation formats. It remains to be seen whether a similar momentum can be gained in the development of representation formats for ECAs in the future.

References

Allbeck J, Badler N (2002) Towards representing agent behaviours modified by personality and emotion, In: Marriott A, Pelachaud C, Rist T, Ruttkay Z, Vilhjalmsson H (eds) Embodied conversational agents: let's specify and compare Them!, workshop notes, autonomous agents and multiagent systems 2002, July 16, University of Bologna, Bologna, Italy

André E, Rist T (2000) Presenting through performing: on the use of life-like characters in knowledge-based presentation systems. In: Proceedings of the 2000 international conference on intelligent user interfaces, New Orleans, LA, USA, 9–12 January 2000

André E, Rist T, van Mulken S, Klesen M, Baldes S (2000) The automated design of believable dialogues for animated presentation teams. In: Cassell J, Sullivan J, Prevost S, Churchill E (eds) Embodied conversational agents. MIT Press, Cambridge, MA

André E, Klesen M, Gebhard P, Allen S, Rist T (2000) Integrating models of personality and emotions into life-like characters. In: Paiva A (ed) Affective interactions: towards a new generation of computer interfaces. Lecture Notes in Computer Science, Vol 1814, Springer, Berlin

Arafa Y, Kamyab K, Mamdani E, Kshirsagar S, Guye-Vuilléme A, Thalmann D (2002) Two approaches to scripting character animation. In: Marriott A, Pelachaud C, Rist T, Ruttkay Z, Vilhjálmsson H (eds) Embodied conversational agents: let's specify and compare them!, workshop notes, autonomous agents and multiagent systems 2002, July 16. University of Bologna, Bologna, Italy

[6] https://gna.org/projects/cal3d/

Arafa Y, Kamyab K , Mamdani E (2003) Character animation scripting languages: a comparison. In: Rosenschein JS et al. (eds) Proceedings of the second international joint conference on autonomous agents and multiagent systems (AAMAS 2003), July 14–18, Melbourne, Australia. ACM Press, New York, NY, pp 920–921

Ball G, Breese J (2000) Emotion and personality in a conversational agent. In: Cassell J, Sullivan J, Prevost S, Churchill E (eds) Embodied conversational agents. MIT Press, Cambridge, pp 189–219

Beard S, Reid D (2002) MetaFace and VHML: A first implementation of the virtual human markup language. In: Marriott A, Pelachaud C, Rist T, Ruttkay Z, Vilhjalmsson H (eds) Embodied conversational agents: let's specify and compare them!, workshop notes, autonomous agents and multiagent systems 2002, July 16. University of Bologna, Bologna, Italy

Bickmore T, Cassell J (2005) Social dialogue with embodied conversational agents. In: van Kuppevelt J, Dybkjaer L, Bernsen N (eds) Advances in natural, multimodal dialogue systems. Kluwer, New York, NY

Black AW, Taylor PA (1997) The festival speech synthesis system: system documentation. Technical Report HCRC/TR-83, Human Communication Research Centre, University of Edinburgh, Scotland, UK. http://www.cstr.ed.ac.uk/projects/festival/. Accessed 3 May 2010

Brooks R (ed) (2002) Human Markup Language Primary Base Specification 1.0, OASIS HumanMarkupTC. http://www.oasis-open.org/committees/download.php/60/HM.Primary-Base-Spec-1.0.html. Accessed 31 May 2010

Calbris G (1990) The semiotics of French gestures. University Press, Bloomington, IN

Cassell J, Pelachaud C, Badler N, Steedman M, Achorn B, Becket T, Douville B, Prevost S, Stone M (1994) Animated conversation: rule-based generation of facial expression, gesture and spoken intonation for multiple conversational agents. In: Proceedings of Siggraph 94, ACM SIGGRAPH, Addison Wesley, Massachu setts, pp 413–420

Cassell J, Bickmore T, Billinghurst M, Campbell L, Chang K, Vilhjálmsson H., Yan H (1999). Embodiment in conversational interfaces: rea. In: Proceedings of the CHI'99 Conference, Pittsburgh, PA, pp 520–527

Cassell J., Stone M, Yan H (2000) Coordination and context-dependence in the generation of embodied conversation. In: First international natural language generation conference (INLG'2000), June 12, Mitzpe Ramon, Israel, pp 171–178

Cassell J, Sullivan J, Prevost S, Churchill E (eds) Embodied conversational agents. MIT Press, Cambridge, MA

Cassell J, Vilhjálmsson H, Bickmore T (2001) BEAT: The behavior expression animation toolkit. In: Proceedings of SIGGRAPH '01, Los Angeles, CA, pp 477-486, August 12–17

Cowie R, Sussman N, Ben-Ze'ev A (2010) Emotions: concepts and definitions. In: this volume

De Carolis B, Pelachaud C, Poggi I, De Rosis F (2001) Behavior planning for a reflexive agent. In: Proceedings of IJCAI 2001, Oporto, Portugal, April, 2001

De Carolis B, Pelachaud C, Poggi I, Steedman M (2004) APML, a mark-up language for believable behavior generation. In: Prendinger H, Ishizuka M (eds) Life-like characters. tools, affective functions and applications, Springer, Berlin, pp 65–85

de Rosis F, Pelachaud C, Poggi I, Carofiglio V, De Carolis N (2003) From greta's mind to her face: Modeling the dynamics of affective states in a conversational embodied agent. Special Issue on "Applications of affective computing in human-computer interaction". Int J Human-Comput Stud 59(1–2): 81–118

Doenges P, Capin TK, Lavagetto F, Ostermann J, Pandzic IS, Petajan E (1997) MPEG-4: Audio/video and synthetic graphics/ audio for real-time, interactive media delivery, signal processing. Image Commun J. 9(4): 433–463

Egges A, Kshirsagar S, Magnenat-Thalmann N (2003) A model for personality and emotion simulation. In: Knowledge-based intelligent information and engineering systems. Lect Notes Comput Sci 2773/2003: 453–461

Egges A, Kshirsagar S, Magnenat-Thalmann N (2004) Generic personality and emotion simulation for conversational agents. J Visual Comput Animation 15(1): 1–13

Ekman P (1993) Facial expression of emotion. Am Psychol 48: 384–392

Ekman P, Friesen W (1978) Facial action coding system. Consulting Psychologists Press, Palo Alto, CA

Ekman P, Friesen W, Hager J (2002) Facial action coding system: the manual. A Human Face, Salt Lake City

Elliott CD (1992) The affective reasoner: a process model of emotions in a multi-agent system. Ph.D. Thesis, Northwestern University, Illinois

Elliott R, Glauert J R W, Jennings V, Kennaway J R (2004) An overview of the SiGML notation and SiGMLSigning software system. In: Streiter O, Vettori C (eds) 4th international conference on language resources and evaluation (LREC 2004), Lisbon, Portugal, May 26–28, 2004, pp 98–104

Gebhard P (2005) ALMA – a layered model of affect. In: Proceedings of the fourth international joint conference on autonomous agents and multiagent systems (AAMAS'05), Utrecht University, Utrecht July 25-29, 2005, pp 29–36

Gebhard P, Kipp M, Klesen M, Rist T (2003) Adding the emotional dimension to scripting character dialogues. In: Proceedings of the 4th international working conference on intelligent virtual agents (IVA'03), – Irsee, Germany, 15-17 September, 2003, pp 48–56

Hall L, Vala M, Hall M, Webster M, Woods S, Gordon A, Aylett R (2006) FearNot's appearance: reflecting children's expectations and perspectives. In: Gratch J, Young M, Aylett R, Ballin D, Olivier P (eds) 6th international conference on intelligent virtual agents (IVA'06), Springer, Berlin, LNAI 4133, pp 407–419

Hanke T (2004) HamNoSys; representing sign language data in language resources and language processing contexts. In: Proceedings of 4th international conference on language resources and evaluation (LREC 2004), Lisbon, Portugal, 26–28 May, 2004, pp 1–6

Hartmann B, Mancini M, Pelachaud C (2002) Formational parameters and adaptive prototype instantiation for MPEG-4 compliant gesture synthesis. In Proceedings of computer animation 2002 (CA 2002), Geneva, Switzerland, 19-21 June, 2002, p 111

Heylen D, Kopp S, Marsell S, Pelachsud C, Vilhjalmsson H (eds) (2008) Why conversational agents do what they do. Functional representations for generating conversational agent behavior, AAMAS 2008 Workshop 2, April 9, Estoril

Huang Z, Eliens A, Visser C (2003) XSTEP: a markup language for embodied agents. In: Proceedings of the 16th international conference on computer animation and social agents (CASA'2003), May 8–9, Rutgers University, New-Brunswick, NJ, USA, IEEE Computer Society, Washington, DC, pp 105–110

Huang Z, Eliens A, Visser C (2004) STEP: a scripting language for embodied agents. In: Prendinger H, Ishizuka M (eds) Life-like characters, tools, affective functions and applications, Springer, Berlin

Johns M, Silverman, BG (2001) How emotions and personality effect the utility of alternative decisions: a terrorist target selection case study. In: Tenth conference on computer generated forces and behavioral representation. SISO. Norfolk, Virginia, pp 55–64

Johnson JH, Vilhjálmsson H, Marsella S (2005) Serious games for language learning: how much game, how much AI? In: 12th international conference on artificial intelligence in education, Amsterdam, The Netherlands, 18–22 July, 2005

Joshi AK, Levy L, Takahashi M (1975) Tree adjunct grammars. J Comput Syst Sci 10: 136–163

Kamp H, Reyle U (1993) From discourse to logic. Kluwer, Dordrecht

Kendon A (1990) Conducting interaction. Cambridge University Press, Cambridge

Klesen M, Gebhard P (2004) Player markup language. Version 1.2.4, DFKI, internal document

Kölsch M, Martell C (2006) Toward a common human gesture description language, workshop on specification of mixed reality user interfaces: approaches, languages, standardization, IEEE Virtual Reality Conference (VR 06), Alexandria, VA, 25 March, 2006

Kipp, M (2001) Anvil – a Generic annotation tool for multimodal dialogue, In: Proceedings of the 7th European conference on speech communication and technology (Eurospeech), Aalborg, Denmark, 3–7 September (2001) pp 1367–1370

Kopp S, Jung B (2000) An anthropomorphic assistant for virtual assembly: max. In: Proceedings of the autonomous agents '00 workshop: communicative agents in intelligent environments, Barcelona, Spain

Kopp S, Wachsmuth I (2004) Synthesizing multimodal utterances for conversational agents. J Comput Animation Virtual Worlds 15(1): 39–52

Kopp S, Krenn B, Marsella S, Marshall A, Pelachaud C, Pirker H, Thórisson K, Vilhjálmsson H (2006) Towards a common framework for multimodal generation: the behaviour markup language. In: Gratch J et al (eds) Intelligent virtual agents 2006, LNAI 4133. Springer, Berlin, pp 205–217

Kopp S, Pfeiffer-Leßmann N (2008) Functions of speaking and acting. In: Heylen D, Kopp S, Marsell S, Pelachsud C, Vilhjalmsson H (eds) Why conversational agents do what they do. Functional representations for generating conversational agent behavior, AAMAS 2008 Workshop 2, April 9, Estoril

Kranstedt A, Kopp S, Wachsmuth I (2002) MURML: a multimodal utterance representation markup language for conversational agents. In: Marriott A, Pelachaud C, Rist T, Ruttkay Z, Vilhjalmsson H (eds) Embodied conversational agents: let's specify and compare them!, workshop notes, autonomous agents and multiagent systems 2002, University of Bologna, Bologna, Italy, 16 July, 2002

Krenn B, Pirker H (2004) Defining the gesticon: language and gesture coordination for interacting embodied agents. In: Proceedings of the AISB-2004 symposium on language, speech and gesture for expressive characters, University of Leeds, UK, March 29–April 1, 2004, pp 107–115

Krenn B, Sieber G (2008) Functional Mark-up for behaviour planning. Theory and practice. In: Heylen D, Kopp S, Marsell S, Pelachsud C, Vilhjalmsson H (eds) Why conversational agents do what they do. Functional representations for generating conversational agent behavior, AAMAS 2008 Workshop 2, April 9, Estoril

Krenn B, Grice M, Piwek P, Schröder M, Klesen M, Baumann S, Pirker H, van Deemter K, Gstrein E (2002) Generation of multi-modal dialogue for net environments. In: Proceedings of KONVENS-02, Saarbrcken, Germany, September 30–October 2 (2002)

Kshirsagar S, Magnenat-Thalmann N (2002) A multilayer personality model. In: Proceedings of 2nd international symposium on smart graphics, ACM Press, New York, NY, pp 107–115

Lee J, DeVault D, Marsella S, Traum D (2008) Thoughts on FML: behavior generation in the virtual human communication architecture. In: Heylen D, Kopp S, Marsell S, Pelachsud C, Vilhjalmsson H (eds) Why conversational agents do what they do. Functional representations for generating conversational agent behavior, AAMAS 2008 Workshop 2, 9 April 2008, Estoril

Louis V, Martinez T (2007) JADE semantics framework. In: Developing multi-agent systems with jade. Wiley, Chichester, pp 225–246

Mancini M, Pelachaud C (2008) The FML-APML language. In: Heylen D, Kopp S, Marsell S, Pelachsud C, Vilhjalmsson H (eds) Why conversational agents do what they do. Functional representations for generating conversational agent Behavior, AAMAS 2008 Workshop 2, 9 April 2008, Estoril

Marsella S (2003) Interactive pedagogical drama: Carmen's bright IDEAS assessed. In: Proceedings of the 4th international working conference on intelligent virtual agents (IVA'03), Irsee, Germany, 15-17 September, 2007, pp 1–4

Marsella, S. and Gratch, J (2009) EMA: A model of emotional dynamics. J Cogn Syst Res 10(1): 70–90

Marsella S, Johnson WL, LaBore C (2003) Interactive pedagogical drama for health interventions. In: Proceedings of the 11th international conference on artificial intelligence in education AIED 2003, Sidney, Australia, 20–24 September 2003

Martell C (2002) FORM: an extensible, kinematically-based gesture annotation scheme. In: Proceedings of ICSLP-2002, Denver, Colorado, 16–20 September, 2002, pp 353–356

Martell C, Howard P, Osborn C, Britt L, Myers K (2003) FORM2 kinematic gesture. Video recording and annotation. Linguistic Data Consortium LDC, Philadelphia, PA

Mateas M, Stern A (2003) Facade: an experiment in building a fully-realized interactive drama. In: Game Developer's Conference: Game Design Track, San Jose, CA, USA, 20–24, March 2003

Mateas M, Stern A (2004) A behaviour language: joint action and behavioural Idioms. In: Prendinger H, Ishizuka M (eds) Life-like characters. Tools, affective functions, and applications. Springer, Berlin, pp 19–38

Matheson C, Pelachaud C, de Rosis F, Rist T (2003) MagiCster: believable agents and dialogue. In: Künstliche Intelligenz, special issue on "Embodied Conversational Agents", November 2003, 4, pp 24–29

McCrae R R, Costa P T Jr. (1996) Toward a new generation of personality theories: theoretical contexts for the five-factor model. In: Wiggins SJ (ed) The five-factor model of personality: theoretical perspectives. Guilford, NY, pp 51–87

McNeill D (1992) Hand and mind – what gestures reveal about thought. The University of Chicago Press, Chicago, IL

Moffat D (1995) Personality parameters and programs. In: Lecture notes in artificial intelligence: creating personalities for synthetic actors: towards autonomous personality agents. LNCS. doi:10.1007/BFb0030565, pp 120–165

Moreno, R (2007) Animated software pedagogical agents: how do they help students construct knowledge from interactive multimedia games? In: Lowe R, Schnotz W (eds) Learning with animation. Cambridge University Press, New York, NY, pp 183–207

Mori K, Jatowt A, Ishizuka M. (2003) Enhancing conversational flexibility in multimodal interactions with embodied lifelike agents. In: Proceedings of the International conference on intelligent user interfaces (IUI 2003), Miami, Florida, 12–15 January 2003, pp 270–272

Niewiadomski R, Pelachaud C (2007) Model of facial expressions management for an embodied conversational agent. In: Proceedings of ACII 2007, September 12-14, Lisbon. LNCS. doi:10.1007/978-3-540-74889-2, pp 12–23

Nijholt A (2006) Towards the automatic generation of virtual presenter agents. In: Proceedings of InSITE 2006, June 25-28, Salford. Infor Sci 9: 97–115

Nischt M, Prendinger H, André E, Ishizuka M (2006) Creating three-dimensional animated characters: an experience report and recommendations of good practice. Upgrade: virtual environments 7(2): 35–41. http://www.upgrade-cepis.org/issues/2006/2/upgrade-vol-VII-2.pdf. (Accessed) 31 May 2010

Nischt M, Prendinger H, André E, Ishizuka (2006) MPML3D: a reactive framework for the multimodal presentation markup language. In: Proceedings of the 6th international conference on intelligent virtual agents (IVA'06), August 21-23, Marina del Rey, CA, LNCS. doi:10.1007/11821830, pp 218–229

Noot H, Ruttkay Z (2004) Gesture in style. In: Camurri A, Volpe G (eds) Gesture-based communication in human-computer interaction – GW 2003. LNCS vol 2915, Springer, Berlin, pp 471–472

Ochs M, Pelachaud C, Sadek D (2008) An empathic virtual dialog agent to improve human-machine interaction. In: Seventh international joint conference on autonomous agents and multi-agent systems, AAMAS'08, Estoril Portugal, 12–16, May 2008, pp 89–96

Ortony A (2003) On making believable emotional agents believable. In: Trappl R, Petta P, Payr S (eds) Emotions in humans and artefacts. MIT Press, Cambridge, MA pp 189–212

Ortony A, Clore GL, Collins A (1988) The cognitive structure of emotions. Cambridge University Press, Cambridge

Ostermann J (1998) Animation of synthetic faces in MPEG-4. In: Proceedings of the computer animation' 98, Philadelphia, PA, USA, 8–10 June 1998, pp 49–51

Pandzic IS, Forchheimer R (eds) (2002) MPEG4 facial animation – the standard, implementations and applications. Wiley, New York, NY

Paradiso A, L'Abbate M (2001) A model for the generation and combination of emotional expressions. In: Proceedings of the AA' 01 workshop on multimodal communication and context in embodied agents, Montreal, Canada, 29 May 2001

Pelachaud C (2005) Multimodal expressive embodied conversational agents. In: Proceedings of the 13th annual ACM international conference on multimedia. SESSION: brave new topics 2: affective multimodal human-computer interaction; Singapore, 6–11, November 2005, pp 683–689

Peltz J, Kumar Thunga R (2005) HumanML: The Vision. TheHumanMLReport-WhiteP, July 12. http://www.oasis-open.org/committees/download.php/13625/HumanMLReport-WhitePaper.pdf Accessed 31 May 2010

Piwek P (2003) An annotated bibliography of affective natural language generation. version 1.3. (version 1.0 appeared in 2002 as ITRI Technical Report ITRI-02-02, University of Brighton). http://www.itri.brighton.ac.uk/projects/neca/affect-bib.pdf Accessed 31 May 2010

Piwek P, Krenn B, Schröder M, Grice M, Baumann S, Pirker H (2002) RRL: a rich representation language for the description of agent behaviour in NECA. In: Marriott A, Pelachaud C, Rist T, Ruttkay Z, Vilhjalmsson H (eds) Embodied conversational agents: let's specify and compare them!, workshop notes, Autonomous Agents and Multiagent Systems 2002, University of Bologna, Bologna, Italy, 16 July, 2002

Poggi I, Pelachaud C, de Rosis F (2000) Eye communication in a conversational 3D synthetic agent. AI Commun 13(3): 169–182

Predinger H, Ishizuka M (eds) (2004) Life-like characters. Cognitive technologies. Springer, Berlin

Prendinger H, Saeyor S, Ishizuka M (2004) MPML and SCREAM: scripting the bodies and minds of life-like characters. In: Predinger H, Ishizuka M (eds) Life-like Characters. Cognitive technologies. Springer, Berlin, pp 213–242

Prillwitz S, Leven R, Zienert H, Hanke T, Henning J (1989) Hamburg notation system for sign languages: an introductory guide. In: International studies on sign language and communication of the deaf, vol 5. Signum Press, Hamburg, Germany

Rank S, Petta P (2005) Appraisal for a character-based story-world. In: Panayiotopoulos T et al (eds) Intelligent virtual agents, 5th international working conference, IVA 2005. Kos, Greece, September 12–14. Springer, Berlin pp 495–496

Rehm M, André E (2005) From chatterbots to natural interaction – face to face communication with embodied conversational agents. IEICE transactions on information and systems, special issue on life-like agents and communication. Oxford University Press Oxford, Oxford, UK, pp 2445–2452

Rehm M, André E (2005) Catch me if you can: exploring lying agents in social settings. In: Proceedings of the fourth international joint conference on autonomous agents and multiagent systems AAMAS '05. July 25 – 29, Utrecht, The Netherlands. ACM, New York, NY, pp 937–944

Rehm M, André E (2005) Informing the design of embodied conversational agents by analysing multimodal politeness behaviors. In: AISB symposium for conversational informatics, University of Hertfordshire, Hatfield, England, 12–15 April 2005

Rickel J, Johnson WL (1998) STEVE: a pedagogical agent for virtual reality. In: Sierra C et al (eds) Proceedings of the second international conference on autonomous agents (Agents'98), 9–13, Minneapolis/St. Paul, MN, USA. ACM Press, New York, NY, pp 332–333

Rist T (2004) Issues in the design of scripting and representation languages for life-like characters. In: Prendinger H, Ishizuka M (eds) Life-like characters. Tools, affective functions, and applications. Springer, Berlin, pp 463–468

Schabes Y (1990) Mathematical and computational aspects of lexicalized grammars. Ph.D. thesis, Computer Science Department, University of Pennsylvania

Scherer K (2000) Emotion. In: Hewstone M, Stroebe W (eds) Introduction to social psychology: a European perspective. Wiley-Blackwell, Oxford, UK, pp 151–191

Schröder M (2004) Speech and emotion research: an overview of research frameworks and a dimensional approach to emotional speech synthesis (Ph.D thesis). vol 7 of Phonus, Research Report of the Institute of Phonetics, Saarland University

Schröder M, Trouvain J (2003) The German text-to-speech synthesis system MARY: a tool for research, development and teaching. Int J Speech Tech 6: 365–377

Schröder M, Cowie R, Douglas-Cowie E, Westerdijk M, Gielen S (2001) Acoustic correlates of emotion dimensions in view of speech synthesis. In: Proceedings of Eurospeech 2001, Aalborg, Denmark, 3–7 September 2001, pp 87–90

Schröder M, Pirker H, Lamolle, Burkhardt F, Peter C, Zovato E (2010) Representing emotions and related states in technological systems. In: this volume

Searle J R (1969) Speech acts: an essay in the philosophy of language. Cambridge University Press, Cambridge

Serenari M, Dybkjaer L, Heid U, Kipp M, Reithinger N (2002) Survey of existing gesture, facial expression, and cross-modality coding schemes. IST-2000-26095 Deliverable D2.1, Project NITE

Stokoe WC (1978) Sign language structure: an outline of the communicative systems of the American deaf. Linstock Press, Silver Spring

Stone M, Bleam T, Doran C, Palmer M (2000) Lexicalized grammar and the description of motion events. In: TAG+5, Workshop on tree-adjoining grammar and related formalisms, Paris, France, 25–27 May 2000

Thiébaux M, Marsella S, Marshall AN, Kallmann M (2000) SmartBody: behavior realization for embodied conversational agents. In: Padgham L, Parkes DC, Müller J, Parsons S (eds) Proceedings of conference on autonomous agents and multi-agent systems (AAMAS08), Estoril, Portugal, 12–16 May 2008, pp 151–158

Trippel T, Gibbon D, Thies A, Milde JT, Looks K, Hell B, Gut U (2004) CoGesT: a formal transcription system for conversational gesture. In: Proceedings of 4th international conference on language resources and evaluation (LREC 2004), Lisbon, Portugal, 26–28 May 2004

Tsutsui T, Saeyor S, Ishizuka M (2000) MPML: a multimodal presentation markup language with character agent control functions. In: Proceedings of world conference on the WWW and internet, WebNet 2000, San Antonio, TX, USA, October 30–November 4

van Deemter K, Krenn B, Piwek P, Klesen M, Schröder M, Baumann S (2008) Fully generated scripted dialogue for embodied agents. Artif Intell J 172(10):1219–1244

Vilhjálmsson H, Cantelmo N, Cassell J, Chafai NE, Kipp M, Kopp S, Mancini M, Marsella S, Marshall AN, Pelachaud C, Ruttkay Z, Thórisson KR, van Welbergen H, van der Werf RJ (2007) The behavior markup language: recent developments and challenges. In: Pelachaud C et al (eds) Intelligent virtual agents. Springer, Berlin, pp 99–111

Walker M, Cahn J, Whittaker S (1996) Linguistic style improvisation for lifelike computer characters. In: Proceedings of the AAAI Workshop on AI, Alife and Entertainment. August, Portland, Oregon, USA

Wiggins J (1996) The five-factor model of personality: theoretical perspectives. The Guilford Press, New York, NY

Yue B, de Byl P (2006) The state of the art in game AI standardisation. In: Proceedings of the 2006 international conference on game research and development. December 4, Perth, Australia, ACM International conference proceedings series Vol 223. Murdoch University, Australia, pp 41–46

Zong Y, Dohi H, Ishizuka M (2000) Multimodal presentation markup language MPML with emotion expression functions attached. In: Proceedings of the international symposium on multimedia software engineering (IEEE Computer Soc), Taipei, Taiwan

Part V
Emotion in Cognition and Action

Overview of Emotion in Cognition and Action

Lola Cañamero

Abstract In humans, emotions entail distinctive and integrated ways of perceiving and assessing situations, processing information, and modulating and prioritizing actions. Our perception of the world and our action on, and interaction with it, is colored by emotions. Given the pervasiveness of emotions, in order to perceive, interact with, and accept social robots as believable partners, they need to be aware of our affective states and respond to them with appropriate affective responses tailored to ours. In addition, to be sustainable over the long term, a believable affect-aware social partner expressive and affective behavior in general must be coherent, adaptive, and sensible in the long term. This can only be achieved if the artifact is endowed with an underlying emotion architecture that processes emotion-relevant information available on the environment and generates appropriate emotional behavior adapted to that of the human. This capability is often known in the literature as emotion synthesis. Within the HUMAINE project, this issue was explored under the thematic area "Emotion in Cognition and Action." The great complexity underlying modeling the processes involved in the generation of emotion-colored behavior and cognition necessitates an inclusive and multidisciplinary approach that takes account of the main issues and challenges posed from different perspectives. This section provides some examples of the various disciplines and approaches contributing to this investigation.

1 Introduction and Background

In humans, emotions entail distinctive and integrated ways of perceiving and assessing situations, processing information, and modulating and prioritizing actions. Our perception of the world and our action on and interaction with it is colored by emotions; not only do we naturally anthropomorphize and easily attribute emotional

---

L. Cañamero (✉)
School of Computer Science, University of Hertfordshire, College Lane, Hatfield, Herts, UK
e-mail: L.Canamero@herts.ac.uk

P. Petta et al. (eds.), *Emotion-Oriented Systems*, Cognitive Technologies,
DOI 10.1007/978-3-642-15184-2_21,

states even to inanimate things around us, including different sorts of artifacts and technology (Reeves and Nass, 1996), but we also expect the objects with which we interact to respond in emotionally and socially appropriate ways to our behavior. A widespread approach commonly found, for example, in commercial toys is to display expressive manifestations of "affective states" – typically facial expressions of basic emotions such as joy, sadness, anger – as a response to simple cues in the behavior of the human-like responding to "keywords" in a rather automatic way. While good animations can prove very successful in the short term, the unavoidably repetitive behavior that such approach gives rise to soon produces lack of engagement, boredom, and finally the rejection of the artifact by the human. In order to be a believable affect-aware social partner, expressive and affective behavior in general must be coherent and sensible *in the long term*; this means not only sustained over a long period of time, but also adapted to the behavior of the human and adaptive to the dynamics of interactions. This can only be achieved if the artifact is endowed with an underlying emotion architecture that processes emotion-relevant information available on the environment and generates appropriate emotional behavior adapted to that of the human.

This capability is often known in the literature as *emotion synthesis* (Picard, 1997). Within the HUMAINE project, this issue was explored within the thematic area "Emotion in Cognition and Action," which within the organizational structure of the project was grouped under Workpackage 7, or WP7 for short.

Emotion synthesis, and the "Emotion in Cognition and Action" thematic area, makes four major contributions to the emotion-oriented systems that are the object of HUMAINE:

1. An underlying emotion system increases the believability of the artifact by providing coherence to its behavior and interactions with humans, as they respond to a common and well-defined model.
2. Mapping the role that emotions play in the coordination and synchronization of other (cognitive, behavioral, and bodily) subsystems in humans and animals, a properly grounded emotion system should play a major role in the synchronization and coordination of the overall architecture and behavior of the emotion-oriented artifact, improving its coherence not only from the point of view of the human user (believability of its observable manifestations) but also regarding the functioning and performance of the system.
3. When the underlying emotion system is elaborated taking inspiration from human emotional systems (or rather some aspects of them), it can serve as a "model" of the human user that can help to build an emotional and personality profile of the human user as a part of the "user model."
4. Artificial emotion systems that are psychologically or biologically plausible can be used as "virtual laboratories" or tools that can contribute to the understanding of human emotions, providing feedback to emotion theorists with a synthetic perspective (by building systems with parameters that are easy to manipulate and test) that complements their analytic studies.

2 Objectives, Scope, and Approach

The "Emotion in Cognition and Action" Workpackage was thus primarily concerned with the investigation of (computational) "internal" mechanisms (emotion architectures) that allow to synthesize or generate emotions and model their involvement in various aspects of cognition and action in emotion-oriented systems. Elaborating robotic and computer-based models that capture the key aspects of the effects that emotions have in the "cognitive systems" (in the broad meaning of this term, which includes behavior and more generally action) of biological systems is a complex, multi-faceted problem that poses numerous conceptual, technical, and integration challenges.

To explore such challenges, WP7 carried research aimed at laying sound foundations for addressing these issues in a multi-disciplinary framework. Our goal was to achieve a better understanding of basic issues through a shared critical reflection (rather than by developing isolated engineering projects) about key developments, open research topics, modeling and evaluation approaches, methods and tools, key research and application scenarios, regarding the involvement of emotions in cognition and action, and potential for cross-fertilization among disciplines, with a view to grounding and promoting sound research into artificial emotional systems for artifacts that must interact with humans.

Our integrative efforts in this area aimed to shed light toward the development of sound robotic and computer-based models of emotions that could have a twofold function:

(a) to improve – according to relevant objective quantitative and qualitative performance metrics related to problem solving or human – robot/human – computer interaction parameters – the behavior of emotion-oriented systems and our interactions with them; and
(b) to provide feedback to emotion theorists in order to gain further insight into their understanding of human emotions. In this respect, the contribution of such models concerns two main aspects (Cañanamero, 2005). On the one hand, they endow the observable behavior of the artifact with the features of autonomy and coherence that are required to achieve long-term interactions adapted to humans. On the other hand, they contribute toward a better understanding of human emotions by providing a synthetic approach (by building systems) that complements the analytic studies carried out in disciplines such as psychology and cognitive neuroscience.

To meet both objectives, we adopted an inclusive approach that set to analyze the rationale, scope, advantages, limits, and complementarities of the main research paradigms, methods, and tools available in the area. We termed our endeavor – or "exemplar," as it was termed within the organizational structure of the project – *Approaches to Emotion-Oriented Architectures: Assumptions, Integration Challenges and Guidelines for Future Research*. This title was chosen to stress several ideas:

- "Comparative approaches" wanted to emphasize the fact that WP7 acknowledged and welcomed the diversity of conceptual and computational models and frameworks that can be used to model emotional systems. These different approaches are not necessarily equivalent or redundant, and the task of this exemplar was to understand (in a deep sense) their scope, limits, incompatibilities, and complementary aspects.
- "Comparative approaches" also wanted to de-emphasize the idea of a "unified" design or model for an emotion-based architecture.
- The subtitle "Assumptions, integration challenges and guidelines for future research" stresses the nature of our common, principled integration effort in setting sound grounds that can guide future research in the area.

For the sake of analysis, our exemplar was divided into four main elements or sub-problems corresponding to major approaches in the conceptualization and computational modeling of emotions and their influence on cognition and action in emotion-oriented architectures: (1) emotion in "lower lever" or embodied cognition and action; (2) emotion in "higher level" or deliberative cognition and action; (3) emotion in bridging the gap between "lower-level" and "higher-level" cognition and action; and (4) emotion in social cognition and interaction.

Each sub-problem was addressed by a dedicated working group. However, such division only made sense for organizational purposes, since there are numerous interrelations across those issues and paradigms, and the working groups had regular exchanges and discussions. For this reason, we also decided not to organize the chapters included in this section strictly in terms of those sub-topics, although the selected chapters cover the four "elements." In addition to the topics represented by these chapters, many others were investigated in the workpackage. We refer the reader to the "Emotion Architecture" link of the Bibliography page of the HUMAINE portal (http://emotion-research.net) for a more comprehensive list of publications resulting from our research on "Emotion in Cognition and Action."

3 The Chapters in This Section

The chapters in this section provide a small but representative sample of the research carried out as part of the "Emotion in Cognition and Action" thematic area. They provide a broad coverage of the topics and research paradigms investigated.

"A Bottom-Up Investigation of Emotional Modulation in Competitive Scenarios," by Lola Cañamero and Orlando Avila-García, adopts the perspective of embodied artificial intelligence to study the adaptive value of emotional mechanisms underlying flee-fight behaviors. Taking inspiration from neuroscience, relevant aspects of flee-fight-related emotions are modeled in a robotic platform as patterns of (neuro-)hormonal modulation of an underlying "nervous system" – or its robotic equivalent in this case. The "message to take home" of this work is multiple. First, it illustrates the advantages that conceptualizing emotions as different patterns of (neuro-)modulation presents in terms of understanding and modeling emotional influences on (emotional "coloring" of) cognitive (e.g., perceptual, memory, attentional) and action (e.g., motor strength, expressive, and other types of behavior) capabilities. Second, this model provides very useful "guidelines" to modelers in terms of economy of design, since it permits to model different emotions using the same underlying robotic architecture (e.g., the same neural network) by simply changing the way it functions through the action of (simulated) chemicals released under different environmental pressures and circumstances – related to the physical, social, or "internal" environment. Finally, the caveat to avoid over-designing complex architectures to model complex phenomena that might in fact be produced by the (possibly complex) dynamics of interactions of very simple underlying mechanisms can also be applied to the understanding of our own emotions and their underlying mechanisms.

"Novelty Processing and Emotion: Conceptual Developments, Empirical Findings and Virtual Environments" by Didier Grandjean and Christopher Peters provides an example of collaboration between neuroscience, psychology, and virtual environment modeling. The focus of this model is on the emotional relevance of novelty detection, a "cognitive" capability closely related and highly relevant to the flee-fight system and the emotional modulation of action selection presented in the previous chapter, but involving interactions among "higher-level" and "lower-level" brain structures and functions. Two points that this chapter makes have particularly important consequences for building computational or robotic models of novelty detection. First is the fact that novelty processing is not a unitary phenomenon, but may be dissociated in different kinds of processes, not only at the conceptual level but also in terms of temporal dynamic processing and the relationship with attentional processes. The distinction among different types of novelty processing at the conceptual level permits not only the implementation of different types of novelty in virtual environments, but also clarifying different concepts of novelty useful for experimental research (perceptual, contextual, partial, and semantic novelty). Second, it is important to understand the unfolding of novelty detection in terms of timing, since it allows researchers to distinguish between the early stages of processing and later ones involving different kinds of attentional processes: These two steps of information processing are related to different levels of processing, with early stages related to exogenous attentional processes and later stages more related to endogenous attentional processes.

"Anticipation and Emotion" by Cristiano Castelfranchi and Maria Miceli analyzes in a systematic way the multiple relations between emotion and anticipation, another key aspect of emotion linked to both the attentional processes that constituted the object of chapter "Novelty Processing and Emotion: Conceptual Developments, Empirical Findings and Virtual Environments", and emotional modulation of decision making presented in chapter "A Bottom-Up Investigation of Emotional Modulation in Competitive Scenarios". This chapter discusses in ample detail two main topics: (a) the role of emotions in anticipation in terms of their eliciting either preparatory behaviors or anticipatory mental representations and (b) the role of cognitive anticipation in different kinds of emotions: some of them feelings associated with expectations (like hope or fear); some others consequences of the invalidation of expectations (like surprise, disappointment, relief, discouragement, and sense of injustice); still some others are the anticipated representation (and possibly also feeling) of future emotions. The key "message to take home" is perhaps that the relationship between emotion and anticipation is neither of overlap nor of inclusion; they are rather partially overlapping sets of phenomena. On the one hand, anticipation is not necessarily emotion-based. Non-emotional systems might be endowed with an "anticipatory" capacity not only in behavioral but also in cognitive terms. On the other hand, emotions are neither necessarily based on anticipatory representations nor necessarily anticipated, as illustrated by the robotic study presented in chapter "Novelty Processing and Emotion: Conceptual Developments, Empirical Findings and Virtual Environments".

"Cognitive Evaluations and Intuitive Appraisals: Can Emotion Models Handle Them Both?" by Fiorella de Rosis, Cristiano Castelfranchi, Peter Goldie, and Valeria Carofiglio discusses in detail the more "higher-level" and "cognitive" (in the sense of "deliberative") aspects of emotions. In a deeply multi-disciplinary dialogue, these authors first endeavor to clarify some common misunderstandings regarding the use of the terms "evaluations" and "appraisal" in various disciplines, and their relation to rational thought, to later consider the relation of emotions to beliefs and goals. Such reflection does not remain at an abstract level, since the authors then provide guidelines and elements that should be included when building deliberative computational emotion models and a brief review of main deliberative emotion systems that can be found in the literature and how they deal with the issues previously discussed. Their theoretical framework is then applied to elaborate a cognitive model of fear. The chapter concludes by pointing at some problems related to the computational implementation of cognitive emotion models, with particular attention to issues of parameter sensitivity and cognitive inconsistency.

Finally, "Socially Situated Affective Systems" by Sabine Payr and Peter Wallis not only looks at affective systems from the point of view of their embeddedness in a social environment, but also stresses their social construction in dynamic interactions among agents. For socially situated agents, the world is populated by other agents with which they interact in a "strong" sense, i.e., the action–reaction sequence typical of the physical world is replaced by mutual acting-with and acting-upon the other. This chapter first introduces conceptual framework, grounded in sociology, for their analysis of emotion as glue and regulatory system of social relationships both on the individual and on the collective level. It then focuses on language as the pre-eminent medium of social interaction and discusses, by way of examples, the functioning of conversation beyond information exchange as both norm-following and norm-building human social behavior. The chapter concludes by stressing the fact that, to design proper socially situated affective systems, more research is needed on aspects of interaction that are often overlooked since they might seem deceivingly trivial, and presents methods to uncover and analyze human–machine interaction.

References

Cañanamero L (2005) Emotion understanding from the perspective of autonomous robots research. Neural Netw 18:445–455

Picard RW (1997) Affective computing. MIT Press, Cambridge, MA

Reeves B, Nass C (1996) The media equation: how people treat computers, television, and new media like real people and places. CSLI Publications and Cambridge University Press, New York, NY

A Bottom-Up Investigation of Emotional Modulation in Competitive Scenarios

Lola Cañamero and Orlando Avila-García

Abstract In this chapter, we take an incremental, bottom-up approach to investigate plausible mechanisms underlying emotional modulation of behavior selection and their adaptive value in autonomous robots. We focus in particular on achieving adaptive behavior selection in competitive robotic scenarios through modulation of perception, drawing on the notion of biological hormones. We discuss results from testing our architectures in two different competitive robotic scenarios.

1 Introduction

One of the main problems for autonomous robots is behavior selection or "what to do next" (Maes, 1991). Motivation-based architecture (Maes, 1991; Blumberg, 1997; Spier and McFarland, 1997; Avila-Garcia et al., 2003) integrate a combination of internal and external factors to select the appropriate behavior and satisfy the robot's needs in real time. However, these architectures are not always sufficiently adaptive to rapid environmental changes. Previous work (Canamero, 1997) postulated the use of second-order mechanisms, akin to some of the functions of emotions in biological systems, that act on other elements in the architecture for improved performance in dynamic, unpredictable, and dangerous environments. In that architecture and others that have followed a similar approach, the adaptive functions of emotions are predefined by the designer. While this nowadays widespread design practice can produce efficient behavior selection, it leaves unanswered the question of which are the underlying mechanisms and how they integrate and interact with other elements to achieve adaptive behavior. In the work presented here, we take an incremental approach to investigate plausible mechanisms underlying emotional

---

L. Cañamero (✉)
School of Computer Science, University of Hertfordshire, College Lane, Hatfield, Herts, UK
e-mail: L.Canamero@herts.ac.uk

This chapter was originally published in *Affective Computing and Intelligent Interaction: Proceedings of the Second International Conference, ACII 2007, Lisbon, Portugal, September 12–14, 2007,* pp. 398–409, LNCS 4738, Springer Berlin/Heidelberg. Reprinted by permission of Springer.

P. Petta et al. (eds.), *Emotion-Oriented Systems*, Cognitive Technologies,
DOI 10.1007/978-3-642-15184-2_22,

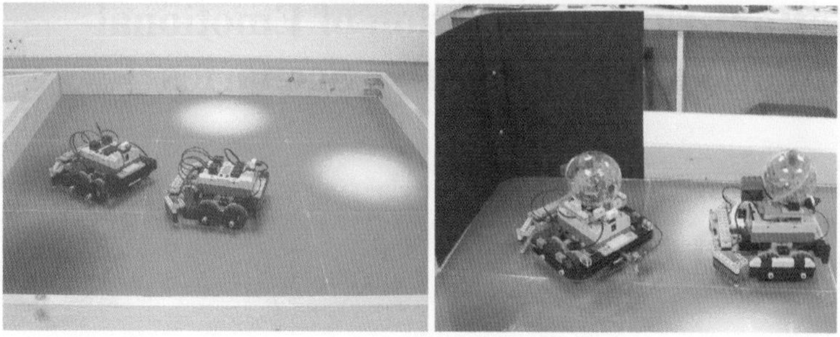

Fig. 1 Experimental setups used to carry out the studies: a competitive two-resource problem scenario (*left*), and a "prey-predator" scenario showing the prey robot inside the nest and the predator outside (*right*)

modulation of behavior selection and their adaptive value. We are particularly interested in how such modulation can achieve different functionalities from the same architecture by interacting with other elements rather than including emotions as additional components. In this chapter, we focus on discussing how behavior selection can be made adaptive (i.e., its output biased) to different environmental situations (two different competitive robotic scenarios) by modulating different sensory channels – perception of external and internal stimuli. Drawing on the notion of biological hormones, we have modeled two of the functionalities ascribed to them in order to improve the adaptation of motivation-based architectures to different problems. To achieve different functionalities from the same architecture, we have taken inspiration from neuroscience models of hormonal control (Kravitz, 1988; Levitan and Kaczmarek, 2002), in particular regarding the following ideas: (a) sensory inputs enhance the release of hormones that act at different levels of the nervous system; (b) they act as gain-setting sensitization processes that bias the output of the organism in particular directions; and (c) after modulation, the organism responds to particular sensory stimuli with an altered output appropriate to the new situation. We have tested our "hormone-like" mechanisms in two dynamic and unpredictable competitive robotic scenarios depicted in Fig. 1 and show how they improve adaptation and performance using quantitative indicators based on the notion of viability. Finally, we analyze the results in terms of interesting behavioral phenomena that emerge from the interaction of these artificial hormones with the rest of architectural elements and the environment and that resemble "emotional" behavior in biological systems confronted to similar situations.

2 Behavior Selection Architecture

Following [5], in our architecture behavior selection results from the interactions of a number of elements integrated through an artificial physiology and in interaction with the environment.

The physiology consists of (1) survival-related, homeostatically controlled essential variables and (2) hormones. *Essential variables* are abstractions representing the level of internal resources that the robot needs in order to survive. They must be kept within a range of permissible values for the robot to remain viable or "alive," thus defining a physiological space (McFarland, 1974) or viability zone (Ashby, 1952; Meyer, 1995) within which survival (continued existence) is guaranteed, whereas transgression of these boundaries leads to "death." *Hormones* can be seen as second-order control mechanisms that affect the behavior of other elements of the architecture.

Motivations are abstractions representing tendencies to act in particular ways as a function of internal and external factors (Toates, 1986). Internal factors are mainly (but not only) physiological deficits ($0 \leq d_i \leq 1$) or bodily needs – traditionally known as "drives" – that set urges to action to maintain the state of the controlled physiological variables within the viability zone. External factors are environmental stimuli, commonly termed "incentive cues" in ethology, ($0 \leq c_i \leq 1$) that allow to satisfy bodily needs through behavior execution. In our implementation, each motivation performs homeostatic control of one physiological variable. We have used the equation proposed in Avila-Garcia et al. (2003) to combine cue and physiological deficit when computing motivational intensities:

$$m_i = d_i + (d_i \times \alpha c_i) \tag{1}$$

In addition to physiological deficits (d_i) and incentive cues (c_i), this equation introduces a weighting factor ($0 \leq \alpha \leq 1$) that affects the relevance given to the external cue.

Behaviors are coarse-grained subsystems (embedding simpler actions) that implement behavioral competencies similar to Canamero, (1997) and Maes (1991). Following a classical distinction in ethology (McFarland, 1999), motivated behavior can be consummatory – "goal achieving" and needing the presence of an incentive stimulus to be executed – or appetitive – "goal-directed" search for a particular incentive stimulus. In addition to modifying the external environment, the execution of a behavior has an impact on (increases or decreases) the level of specific physiological variables. Therefore, behaviors take part in the homeostatic control to maintain the state of the physiological variables within the viability zone.

Behavior selection is performed in a continuous loop consisting of three main steps: (1) the deficit of the physiological variables (internal needs) and the intensity of the external stimuli are calculated; (2) motivational intensities are computed combining (perception of) deficits and external stimuli ponderated by the weight α, following Eq. (1); and (3) the behavior that (best) satisfies the motivation with the highest intensity is executed, modifying the physiology and possibly the position of the robot relative to external stimuli in the environment.

3 Competition for Resources

In previous work (Avila-Garcia et al., 2003) we analyzed different motivation-based behavior selection architectures within a static Two-resource problem (TRP) in which a single robot must maintain appropriate levels of two internal variables by consuming two resources available in the external environment. The TRP constitutes the minimal scenario to test behavior selection mechanisms, and it has become a standard testbed for behavior selection both in animals – see, e.g., Spier and McFarland, (1997) – and autonomous agents and robots – e.g., Avila-Garcia and Canamero (2004); Blumberg (1997); Givard et al. (2002). Its simplicity, although not devoid of problems, favors a systematic analysis of results. The particular implementation of the TRP in Avila-Garcia et al. (2003) used a Lego Mindstorms robot (see Fig. 1, left, for a similar arena, although the TRP uses only one robot), with the need to maintain temperature and energy levels by consuming heat (white gradients on the floor of the arena) and food (black gradients), respectively. The robot had two motivations: m_{cold} to increase temperature, which can be satisfied by executing the consummatory behavior b_{warmup} and m_{fatigue} to increase energy, which can be achieved by executing the consummatory behavior b_{feed}. In addition, the robot had a reflex obstacle avoidance behavior b_{avoid} and the appetitive behavior b_{search}. The execution of all behaviors affects both essential variables.[1]

To measure results in TRP, we used different performance indicators based on the notion of viability, in particular: *Life span*, defined as the time that the robot survived in each run ($\text{LS} = t_{\text{life}}/t_{\text{run}}$); *Overall comfort*, the average level of satisfaction of the physiological variables during a run ($\text{OvC} = \sum_{i=1}^{t_{\text{life}}}(1 - \overline{d_i})/t_{\text{life}}$); and *physiological balance*, the homogeneity with which physiological needs are satisfied during a run ($\text{PhB} = \sum_{i=1}^{t_{\text{life}}}(1 - \sigma^2(d_i))/t_{\text{life}}$). We also noted that, when doing behavior selection in TRP, the robot executed regular cycles of activities rather than isolated behaviors, and those activity cycles were reflected in the physiological space of the robot, as shown in Fig. 2: from the initial state, the robot would start looking for a given resource, e.g., heat (arrow noted as A in the figure), then consume it until satiated (B), then start looking for the other resource (C), consume it until satiated (D), and start all over again. The position of the cycles in the physiological can be changed: the same cycle (i.e., with the same shape and duration of each activity) would be executed closer to the ideal state, therefore preserving viability "better," or farther away from it (and therefore in a "less viable" way) depending on the value of α, the parameter that weighed the significance of external stimuli in Eq. (1), as depicted in Fig. 2 (right). The regular shape of those activity cycles reflects the fact that behavior selection in TRP was static and highly predictable.

The competitive two-resource problem[2] (CTRP) is an extension of this problem that consists in the introduction of two robots in the same environment simultaneously performing their own TRP, as depicted in Fig. 1 (left). The robots do not explicitly communicate or compete; however, the fact that they have to use the same resources to satisfy their needs introduces competition for those resources, as both

[1] At each execution cycle, b_{warmup} increases temperature by 0.3 units while decreasing energy by 0.1 units, b_{feed} increases energy by 0.3 units while decreasing temperature by 0.1 units, and b_{avoid} and b_{search} decrease each variable by 0.2 units.

[2] We refer the reader to Avila-Garcia and Canamero (2004) for an in-depth technical quantitative analysis of this scenario, while here we focus on a qualitative discussion of the adaptive value of hormonal modulation and its significance from the point of view of emotion.

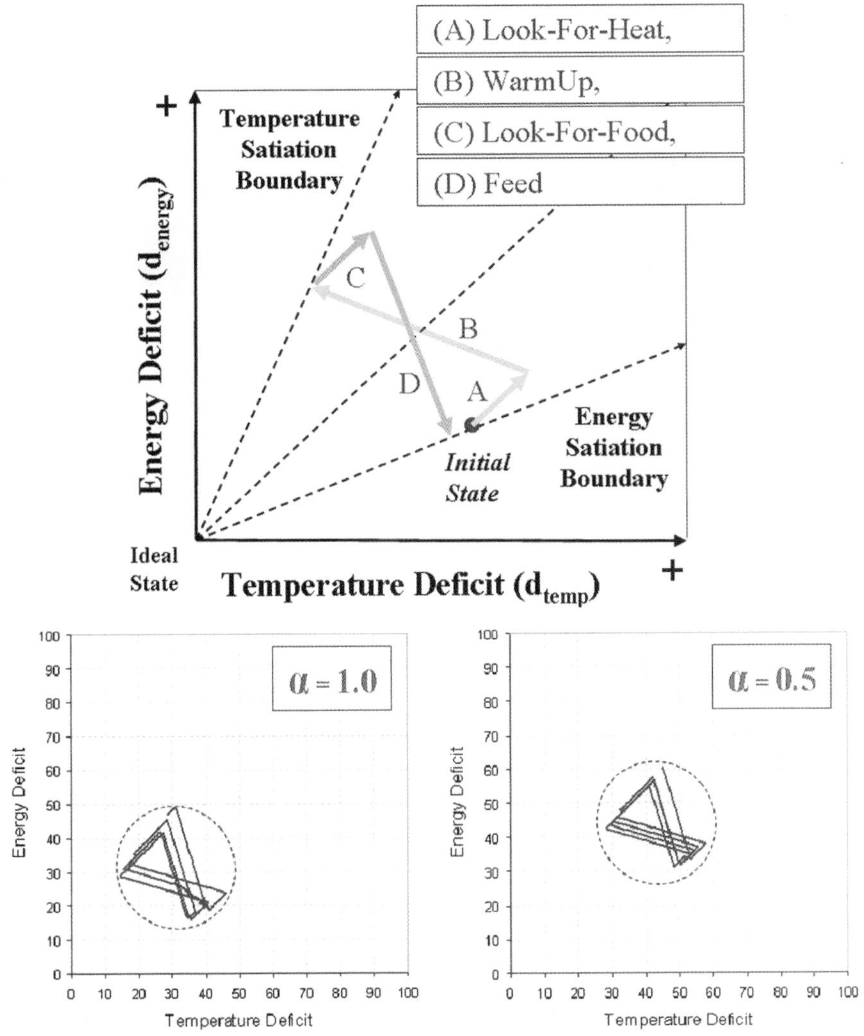

Fig. 2 Activity cycles in TRP. *Top*: cycle as reflected in the physiological space. *Bottom*: position of cycles in the physiological space as a function of α

robots might need access to the same resource at the same time. Therefore, new forms of environmental complexity – availability and accessibility of resources – appear due to the interaction between robots, breaking the predictability and symmetry of TRP. The question that needs to be examined here is to what extent the architecture used for the TRP can solve the CTRP.

Analysis showed that the new forms of complexity dramatically decrease the performance of that behavior selection architecture, as clearly reflected by the different viability indicators and the activity cycles depicted in Fig. 3. In particular, analysis

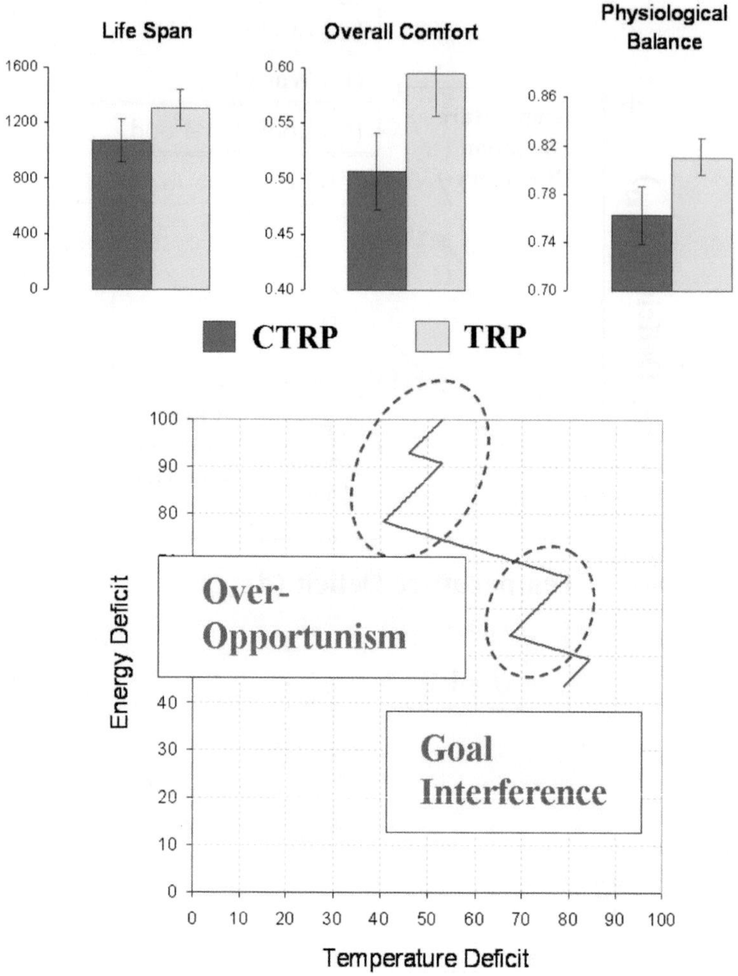

Fig. 3 Decreased performance in the CTRP, as measured by viability indicators (*top*) and activity cycles (*bottom*)

of the activity cycles shows that the cycles easily lose the regularity and symmetry they showed in TRP, as illustrated in Fig. 3 (right), and that the robot very often dies from two problems that the behavior selection mechanism used within the TRP presents when used in the CTRP. First, the robot can fall in a pathological sequence of opportunistic activities – consuming the same resource – that eventually can drive it to death due to over-opportunism. Second, when one robot is located on top of a resource – i.e., consuming it – the other robot might bump into it and push it out of the resource. This will result in the interruption of the ongoing consummatory activity and to death due to goal interference.

The next step in our incremental design approach is to analyze what needs to be added to the architecture to be able to solve those problems. A solution to the "over-opportunism" problem requires shifting attention away from less needed resources when the robot is in a high risk of death (RoD) that we define as the inverse of the distance between physiological state ($d_{\text{temper}}, d_{\text{energy}}$) and lethal boundaries. A solution to the "goal interference" problem requires that the robot in need of an occupied resource does not avoid the "intruder" as if it were a mere obstacle. Both problems can be solved by altered perception of external stimuli, i.e., by modulation of exteroception.

3.1 Modulation of Exteroception

Rather than adding more structural elements to our architecture, our solution consists in trying to achieve additional functionality from the same architecture. A single "hormone-like" modulatory mechanism can alter perception in both cases, with a twofold effect. First, by acting on the parameter α of Eq. (1) – i.e., by biasing the relevance given to external cues – the hormone reduces the perception of both incentive cues, therefore reducing opportunistic activities when there is any risk of death. Second, by cancelling the perception of obstacles s_{obstacle} (carried out using the bumper sensor), and hence the avoidance reflex behavior, when the robot is facing the competitor, the hormone potentiates the competition skills of the robot by enhancing its capacity to push the other robot out the resources and not to be interrupted. To achieve this twofold functionality, the concentration of hormone will be a function of the risk of death (RoD) and the perception of the competitor, given by $0 \leq s_{\text{competitor}} \leq 1$. Hormone concentration is computed as

$$c_g = \text{RoD} + s_{\text{competitor}} \tag{2}$$

The relation between hormone concentration and the cancellation of the perception of incentive cues and obstacles is as follows. To achieve the first functionality, the cancellation of α is directly proportional to the increment in hormone concentration, i.e., when RoD increases, α decreases: $\alpha = \min(1 - c_g, 0)$ The second functionality is obtained by cancelling the perception of s_{obstacle} – i.e., bumpers – when the competitor is in front of the robot. For this mechanism to be efficient, two conditions must be fulfilled to make a coherent pushing of the other robot. First, the robot must avoid getting engaged in fights when it has high RoD. Second, it must only bump blindly into the other robot, not against the walls of the arena. To produce that effect the cancellation of the bumpers must be at hormonal levels $c_g \simeq 1$ and $c_g \simeq 2$.

It is worth noting that the motivation-based behavior selection architecture has suffered no modification; the only difference with respect to the TRP is the fact that now one of its parameters (α, cf. Eq. 1) is modulated by the hormonal feedback mechanism.

3.2 Experiments and Results

We tested the robots for a total of 16 runs of 1200 steps (approximately 5 min) each, one step representing a loop of the behavior selection mechanism that takes 260 ms in the 16 MHz onboard microcontroller. As shown in Fig. 4, the robot with hormone-like mechanism recovers the stability and viability of activity cycles. We refer the reader to Avila-Garcia and Canamero (2004) for a detailed quantitative analysis, while we focus here on various interesting functionalities that emerged as a result of modulating the exteroception of the robot. The first functionality is to stop consuming resources when the robot detects its competitor approaching. This could be interpreted by an external observer as abandonment of a situation (waiting for the other robot at the resource) in which competing is disadvantageous. Instead, the robot will leave the resource and go straightforward toward the competitor until it reaches it; at that moment, two things can happen. If there is some level of RoD, the bumpers of the robot will not be cancelled and it will avoid the competitor, showing a behavior that an observer could interpret as "fear" after evaluating the competitor. On the contrary, if there is no RoD, the hormonal system will cancel the bumpers and the robot will push the competitor unconditionally – as if it showed some sort of "aggression" against it. If we study the whole picture as external observers, such behavioral phenomena could well be interpreted as some sort of "protection of resources."

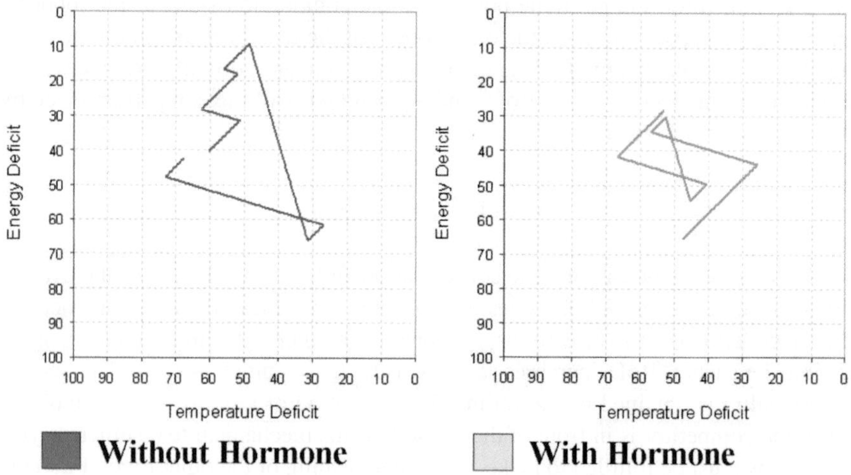

Fig. 4 Activity cycles in CTRP in unmodulated architecture (*left*) and with modulation of exteroception (*right*)

4 Prey–Predator Scenario

The previous scenario involved no active interaction between the two robots. It therefore seems natural to ask whether an active relation between the robots would introduce additional complexity, and how the previous behavior selection architecture would cope with it. We thus designed a prey–predator scenario (Fig. 1, right) that we call the Hazardous 3-Resource Problem (H3RP). In H3RP, a "predator" robot actively chases and can damage a "prey" robot by hitting a home-made contact sensor in the form of a ring. To make this interaction possible, we had to introduce new elements in the environment – a nest in one of the corners of the arena, in which the prey can "hide" and recover from damage – and in the architecture of the prey, namely (a) a third physiological variable, integrity, which is a metaphor of the essential need any organism has to keep its tissue – the boundary between the organism and its environment—intact and that is unpredictably reduced by the attacks of the predator; (b) a new motivation m_{damage} to decrease the integrity deficit; and (c) an appropriate consummatory behavior b_{recover} to satisfy the new need.

Initial experiments showed very quickly that a purely motivation-based behavior selection mechanism does not perform well within the new framework, since the prey invariably died as a consequence of predator attack (see the right graph of Fig. 6 for quantitative results of additional experiments). The main cause seemed to be the inability of the prey to react timely to the attack of the predator, which was perceived in close proximity only. In other words, the behavior selection mechanism paid low attention to the new motivation to recover integrity, even when the predator is in sight. The probability to lose integrity rises when the predator is around, therefore it would be advantageous for the prey robot to "anticipate" that loss and start "preparing in advance" to recover integrity.

In the animal world, exposure to predators triggers what has been termed "predator-induced stress" or "predator stress" for short, characterized by high levels of corticoids or "stress hormones" and a number of responses related to increased attention to and avoidance of the predator. Such reactions occur not only in the presence of a predator. Prey animals use unconditioned and conditioned predator cues to assess risk of predation, and they even seem to be able to perceive risks in the absence of such cues (Curio, 1993). An example of the latter is the phenomenon known as "risk of permanence" – maintained levels of vigilance after predator's disappearance. Risk of predation strongly influences prey decision-making (for example, when and where to feed, vigilance, or the use of nest), which in this circumstance can be considered as a mechanism to allow an animal to manage predator-induced stress (Lima, 1998). Risk of predation has been proposed to increase the animal's level of "apprehension," i.e., the reduction in attention to other activities (e.g., foraging) as a result of increasing the time spent executing defense-related activities such as vigilance or refuge use (Kavaliers and Choleris, 2001).

4.1 Modulation of Interoception

We have again applied "hormonal" modulation to our behavior selection architecture to achieve such "anticipatory" behavior, this time exploiting the temporal dynamics of hormonal decay to produce long-term modulatory effects triggered by short-term exposure to a stimulus (Levitan and Kaczmarek, 2002).

To achieve this, a simple solution consists in using one of the existing sensors of the prey robot to detect the predator from a distance. Given the morphology of the robot, this sensor must be the same as that used to locate the nest. The problem of using that sensor is that it is fixed, pointing forward. Since the predator does not pass in front of the prey very often and only does it for very brief periods, the additional stimulus (s_{predator}) will be too weak to make any difference. However, long-term hormonal modulation acts as a mechanism for predation risk assessment in the absence of predator cues. Hormone concentration makes the system more sensitive to integrity deficit after the detection of the predator. Hormonal secretion follows the detection of the stimulus s_{predator} and increases the *perceived* integrity deficit. Due to the hormone's temporal dynamics, modulation will be acting in the system long time after the predator has disappeared. Hormone concentration modifies again one of the sensory inputs of the architecture – interoceptive in this case – biasing behavior selection.

We have modeled hormonal temporal dynamics – release and dissipation – using an artificial endocrine system similar to that proposed in Neal and Timmis (2003) and described by Eqs. (3) and (4). A gland g releases hormone as a function of the intensity of the external stimulus predator (s_{predator}) at a constant releasing rate β_{g}:

$$r_{\text{g}} = \beta_{\text{g}} \cdot s_{\text{predator}} \tag{3}$$

Hormone concentration[3] suffers two opposite forces over time: it increases with the release of hormone by the gland and dissipates or decays over time at a constant rate γ_{g}:

$$c(t+1)_{\text{g}} = \max[(c(t)_{\text{g}} \cdot \gamma_{\text{g}}) + r_{\text{g}}, 100] \tag{4}$$

In this implementation, the hormone increases the perception of the integrity deficit ($d_{\text{integrity}}$), i.e., the higher the hormone concentration, the higher the reading of the $d_{\text{integrity}}$ interoceptor:

$$d_{\text{integrity}}^{\text{new}} = \max(d_{\text{integrity}} + \delta_{\text{g}} \cdot c_{\text{g}}, 1) \tag{5}$$

Factor δ_{g} determines how susceptible to hormonal modulation the interoceptor ($d_{\text{integrity}}$) is. We use $\delta_{\text{g}} = 0.005$, which implies that the level of perceived $d_{\text{integrity}}$ is increased by 0.5 when hormonal concentration is maximum ($c_{\text{g}} = 100$). In other words, although the level of integrity is at its ideal value ($d_{\text{integrity}} = 0$), the interoceptor will perceive a level of 0.5 if hormone concentration is maximum. Note that there is a constraint to avoid the level of integrity deficit to be perceived beyond the maximum possible value ($d_{\text{integrity}} = 1$).

[3]We constrained hormonal concentration to a maximum of $c_{\text{g}} = 100$ in order to keep more control on the hormone's dynamics and thus facilitate the analysis of results.

4.2 Experiments and Results

We tested the robot for 16 runs of 1600 steps each, i.e., each architecture (non-modulated and modulated) was tested for almost 2 h in H3RP.

The prey robot presented higher viability levels in terms of life span, at the cost of overall comfort, when equipped with the modulatory mechanism, as shown in Fig. 5. Long-term hormonal modulation acts as a mechanism for predation risk assessment in the absence of predator cues. It can be regarded as increasing the level of "apprehension" of the prey robot after short-term predator exposure, and this is reflected in an increment of the motivation to recover and of the execution time of recover-related (consummatory and appetitive) activities – the robot spends more time looking for the nest and recovering integrity in it – at the cost of other activities, namely feed and warm-up, as reflected in Fig. 6 (left). This increment in the execution time of recover-related activities is statistically highly significant. Another important phenomenon is the interruption of ongoing consummatory feeding or warming-up activities (Fig. 6, center). When the robot is under the effect of

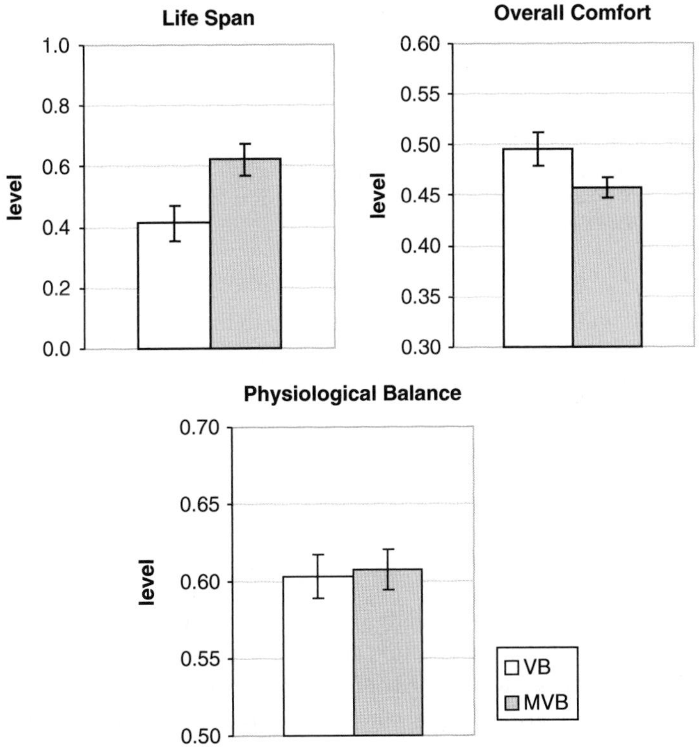

Fig. 5 Average performance of non-modulated (*light bars*) and modulated (*dark bars*) architectures in terms of life span, physiological balance, and overall comfort. *Bars* show standard error of the mean

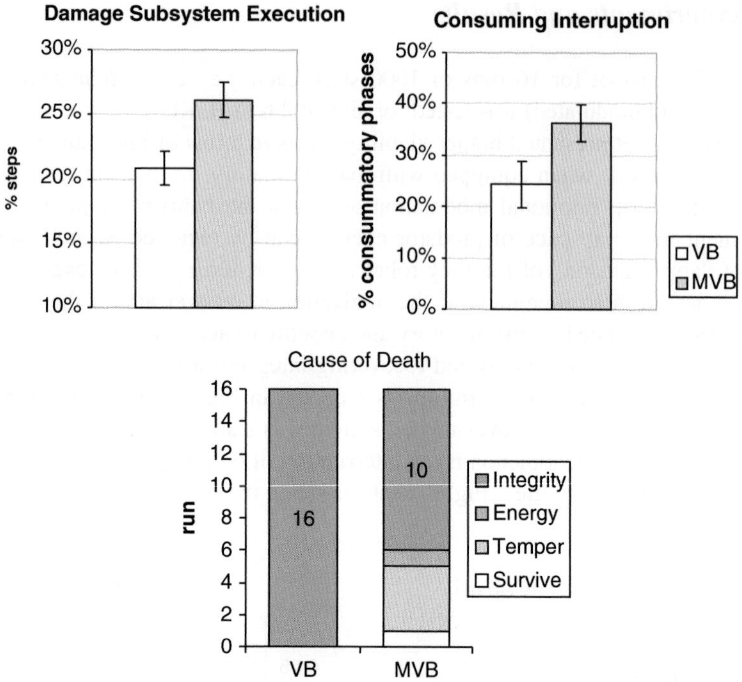

Fig. 6 Comparison between non-modulated (*bars* on the left of each graph) and modulated (*bars* on the *right*) architectures in terms of execution time of recover subsystem (*left*), average number of interruptions of consummatory feed and warmup behaviors (*center*), and causes of death in the 16 runs (*right*). *Bars* show standard mean error

the hormone it will abandon the resource and go to the nest before the motivation has been satiated. The prey robot, when equipped with the hormonal mechanism, presents statistically higher levels of interruption of ongoing feeding or warming-up activities. Finally, analysis of the causes of death (Fig. 6, right) shows substantial differences with respect to the non-modulated architecture.

5 Conclusion

We have discussed a bottom-up study of plausible mechanisms underlying emotional modulation of behavior selection and their adaptive value, in particular how such modulation applied to a motivation-based architecture can achieve different functionalities found in biological emotions, to face different emotionally relevant problems posed by different competitive scenarios. We have considered a first scenario in which obtaining resources in competition with others is the main survival-related problem, and a second scenario in which the attack of a predator constitutes the main threat. Drawing on the notion of biological hormones, we have focused on achieving adaptive behavior selection in these different competitive robotic scenarios by modulating perception of external stimuli in the first case, and of internal stimuli in the second, as depicted in Fig. 7. In addition to improving behavior selection performance and adaptation, modulation has given rise to some emergent behavioral phenomena that could be interpreted by an external observer as "emotional," such as aggressive/defensive behavior in the first, "fleeing" and "apprehension" in the second. We suggest that such modulatory mechanisms provide a more principled integration of different behavior selection elements and functions, in addition to improving the adaptation of a robot to changing environments. The type of adaptation fostered by such mechanisms is different from other mechanism such as learning or evolution, for which "past solutions" are "overwritten" by new ones.

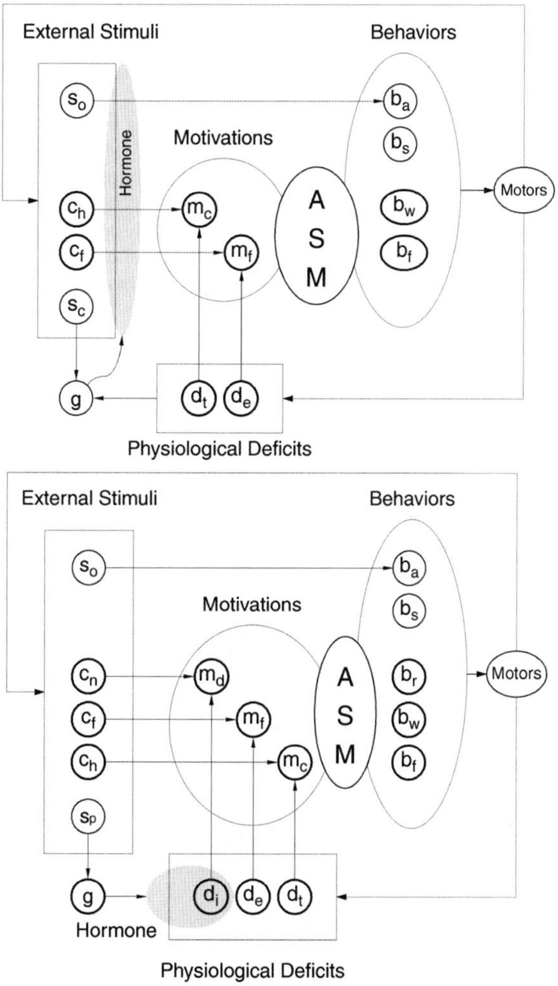

Fig. 7 Hormonal modulation of exteroception (*top*) and of interoception (*bottom*)

Current and future work includes the integration in the same architecture of both types of hormonal modulation presented here, to face a more complex prey–predator problem requiring interactions among both mechanisms. We will also continue our incremental study of plausible modulatory mechanisms underlying emotions by changing and complexifying the environment to give rise to other behavior selection problems.

Acknowledgments This research was funded partly by the European project HUMAINE (grant FP6-IST–507422 to Lola Cañamero) and partly by a University of Hertfordshire studentship to Orlando Avila-García.

References

Ashby WR (1952) Design for a brain. Chapman & Hall, London

Avila-García O, Cañamero L, te Boekhorst R (2003) Analyzing the performance of "winner-take-all" and "voting-based" action selection policies within the two-resource problem. In Proceedings of the seventh European conference in artificial life (ECAL03). Springer, Heidelberg pp 733–742

Avila-García, O. and Cañamero, L (2004). Using hormonal feedback to modulate action selection in a competitive scenario. In: Proceedings of the eighth international Conference on simulation of adaptive behavior (SAB04). MIT Press, Cambridge, MA pp 243–252

Blumberg B (1997) Old tricks, new dogs: ethology and interactive creatures. PhD Thesis. MIT Media Laboratory, Cambridge, MA (Unpublished)

Cañamero LD (1997) Modeling motivations and emotions as a basis for intelligent behavior. In: Johnson WL (ed) *Proceedings of the first international conference on autonomous agents.* ACM Press, New York, NY, pp 148–155

Curio E (1993) Proximate and developmental aspects of antipredator behavior. Adv Study Behav 22:135–238

Girard B, Cuzin V, Guillot A, Gurney KN, Prescott TJ (2002) Comparing a brain-inspired robot action selection mechanism with "winner-takes-all". In: Proceedings of the seventh international conference on simulation of adaptive behavior. MIT Press, Cambridge, MA

Kavaliers M, Choleris E (2001) Antipredator responses and defensive behavior: ecological and ethological approaches for the neurosciences. Neurosci Biobehav Rev 25:577–586

Kravitz EA (1988) Hormonal control of behavior: amines and the biasing of behavioral output in lobsters. Science 241:1175–1781

Levitan IB, Kaczmarek LK (2002) The neuron: cell and molecular biology, 3rd edn. Oxford University Press, Oxford

Lima SL (1998) Stress and decision making under the risk of predation: recent developments from behavioral, reproductive, and ecological perspectives. Adv Study Behav 27:215–290

Maes P (1991) A bottom-up mechanism for behavior selection in an artificial creature. In: Proceeding of first international conference on simulation of adaptive behavior (SAB90). MIT Press, Cambridge, MA, pp 238–246

Meyer J-A (1995) The animat approach to cognitive science. In: Roitblat HL, Meyer J-A (eds) Comparative approaches to cognitive science. MIT Press, Cambridge, MA, pp 27–44

McFarland D (ed) (1974) Motivational control systems analysis. Academic, London

McFarland D (1999) Animal behaviour, 3rd edn. Addison Wesley Longman, New York, NY

Neal M, Timmis J (2003) Timidity: a useful emotional mechanism for robot control? Informatica 27:197–204

Spier E, McFarland D (1997) Possibly optimal decision making under self-sufficiency and autonomy. J Theor Biol 189:317–331

Toates F (1986) Motivational systems. Cambridge University Press, Cambridge

Novelty Processing and Emotion: Conceptual Developments, Empirical Findings and Virtual Environments

Didier Grandjean and Christopher Peters

Abstract Novelty detection is a crucial ability of organisms to detect changes in the environment and to adapt their behaviours accordingly. In this chapter we review a conceptual framework of novelty detection informed by cognitive neuroscience and cognitive psychology. The relationship between attentional processes and novelty detection is also discussed and developed, supported by a case study highlighting methods for implementing a novelty detection capability for artificial agents in virtual environments.

1 Introduction

The growing knowledge about the functioning of human central nervous system and the possibilities for modelling such systems, at least in part, have created the potential for interdisciplinary work, for example, on the implementation of cognitive or emotional processes for artificial agents inhabiting virtual environments. The collaboration between cognitive neuroscience researchers and computer scientists working on virtual environments is important, supporting both the creation of realistic behaviours for agents inhabiting virtual environments and also allowing, through reconstruction attempts, testing of high-level behaviour implications of neuroscience results based on simple virtual analogies of real systems. In this chapter, we review the cognitive and cognitive neuroscience knowledge about novelty processing and provide an example of the first steps towards a computational implementation for supporting the dynamic interactions of virtual agents with their environment. Novelty processing is an essential ability of organisms to process novel information in order to increase their knowledge about their own environment. This mechanism allows organisms to detect changes in the environment and, with

D. Grandjean (✉)
Neuroscience of Emotion and Affective Dynamics Laboratory, Department of Psychology, University of Geneva, Geneva, Switzerland
e-mail: Didier.Grandjean@unige.ch

P. Petta et al. (eds.), *Emotion-Oriented Systems*, Cognitive Technologies,
DOI 10.1007/978-3-642-15184-2_23, © Springer-Verlag Berlin Heidelberg 2011

memory processes, to learn invariants in the environment. Even at a low level and for basic living systems, like nematodes, novelty processing is important, allowing these organisms to detect changes in their environment in order to allocate resources (i.e. energy), to avoid potentially dangerous situations or to orient towards a specific location space (i.e., respectively, to avoid some increase of salt gradient or to orient towards a source of food).

2 Novelty and Emotional Processes

The importance of novelty in emotional processes was suggested by appraisal theorists in the 1980s (Lazarus, 1991; Scherer, 1984). Novelty is not only closely related to surprise (which could be either positive or negative) but could be also determinant, at early stages of processing, in appraisal processes for several other emotions. In the context of emotion theories, the component process model (CPM) proposes emotional processes as massive synchronised changes occurring at the same time in the different sub-systems of the organism (Grandjean et al., 2008; Scherer, 2001). These systems have been described as the cognitive, peripheral, motivational, expressive and the monitor sub-systems interacting dynamically. In this chapter we will focus on a specific process, the detection of novelty, embedded into the cognitive system in the emotional process. In the context of appraisal processes, novelty detection has been proposed as the first subprocess occurring in time when an organism appraises its environment. Actually, Scherer has formulated the notion of a fixed sequence of different appraisals in a theoretical proposition in 1984 (Scherer, 1984). Based on phylogenetical, ontogenical and logic arguments (Sander et al., 2005) the main idea is that the different appraisal subprocesses have to be organised in a fixed sequence in time. Every check would compute a preliminary closure impacting on the different other components. This sequential view is compatible with further parallel processing, for example, to represent novelty at higher levels. Moreover, we have developed a series of arguments to the sequential nature of appraisal process based on cognitive neuroscience empirical studies (see also Grandjean, 2005). Recent findings based on electroencephalography (EEG) have supported this view (Grandjean and Scherer, 2008). In two EEG experiments in the visual modality, different appraisals were manipulated in order to test the sequential hypothesis; novelty was manipulated by the probability of the occurrence of stimuli, intrinsic pleasantness was operationalised by positive, negative and neutral pictures (IAPS pictures;(Lang et al., 2005)), goal relevance was manipulated by the participant's task and finally goal conduciveness was manipulated by reward–punishment contingencies of different types of visual stimuli. Based on several types of EEG analysis, ERPs, topographical analysis and frequency analysis, the results indicate (1) a first modulation of brain electrical field at 50–100 ms related to novelty manipulation and (2) a later modulation of 150 ms related to intrinsic pleasantness and later components related to goal relevance and goal conduciveness (Grandjean and Scherer, 2008).

In the following paragraphs, we present and discuss a new typology of the concept of novelty. Indeed, in the literature in cognitive psychology and cognitive neuroscience, this term has been used in our point of view to refer to many different kinds of phenomena. For the purposes of clarification, we propose the following distinctions between different types of novelty processes. We distinguish between four major kinds of novelty processes, namely perceptual novelty, partial novelty, contextual novelty and semantic novelty. We define perceptual novelty as the construction of a new representation of an object never perceived in the past by the organism and requiring a new encoding in short-term and long-term memory. This "real" novelty detection has been largely neglected in cognitive and neuro-cognitive research. Indeed, EEG measures, often used to investigate novelty processes, require a number of repetitions of different stimuli types in order to compute event-related potentials (ERPs) and perform the subsequent statistical analysis. The investigation of perceptual novelty is not easy due to the difficulty of creating a series of totally new stimuli which are not different in terms of their basic physical features, such as luminance, visual complexity, content in spatial frequency; several confounds are possible and may be attributed to novelty effects while they may in fact be better related to basic physical differences between stimuli. For example, if someone wants to test the ERPs differences between familiarity and novelty, the different stimuli used in the experimental conditions have to be similar and controlled for basic physical visual features (Delplanque et al., 2007).

Partial novelty would be involved when an organism perceives an object that looks like a previously perceived object, but which has some perceptual differences for one or several characteristics. For example, if a baby used to play with a red ball, a new violet ball will not be totally new (in terms of perceptual novelty defined just above); it is only one or several of these characteristics which are new; the categorisation processes are very important in this concept of partial novelty. The fact that organisms are able to categorise different percepts in terms of one category, as, for example 'chairs', is a crucial ability allowing the extraction of the invariant in different dimensions; in this example, a functional utility, i.e. an object on which you can sit. This partial novelty would be sensitive to the boundaries of the category and then influenced by the mental setting of the individual, for example in terms of needs or desires.

Contextual novelty refers to situations in which an object is perceived in a new context or has emerged in a given stable context. For example, a ball on the table, when such objects are usually experienced on the floor, would induce a novelty detection phenomenon. Another example is the case of a familiar environment in which something new appears: on a monitor board a red light turns on to signal something unusual taking place; in this case the board is known, the red light as well, but the timing concomitance between the board and the red light is a new event in the present context. This type of novelty has been extensively investigated in cognitive neuroscience research, particularly using EEG. The need to have several dozen trials for one given experimental condition in EEG to compute ERPs has induced a bias in cognitive neuroscience towards the study of this type of novelty. In this kind of traditional EEG experiment, called the 'oddball paradigm', a flow of background stimuli (identical stimuli repeatedly presented to the participant) is interrupted by a 'new' stimulus, a deviant one. In the auditory domain, for example, the ERP component, called mismatch negativity (MMN), corresponding to a negative electrophysiological component appears when the deviant stimuli are presented (for example, a 500 Hz pure tone in a flow of 250 Hz tone Naantanen et al., 2007). Finally, semantic novelty refers to a situation in which the relationships between the objects or the concepts are organised in a new manner and have never been perceived as such in the past. The fact that individuals are able to create a new tool or a new concept from a series of well-known objects or ideas is characteristic of this type of novelty. Notice that this process is related to an active process involving an ability to create something new from something well known. The example of the use of stones as a tool to obtain food from different animals is exactly what we refer about this semantic novelty (Savage-Rumbaugh et al., 2003). This type of novelty is strongly related to creativity and the organisms abilities in terms of high cognitive computations.

Scherer has suggested distinguishing between different kinds of subprocesses related to novelty detection. Namely, he has proposed that suddenness, familiarity and predictability are particularly important in the processing of novelty detection (Scherer, 2001). Three different features of novelty have been proposed: (i) suddenness or abruptness of onset (Tulving and Kroll, 1995; Tulving et al., 1994), often coupled with high stimulation intensity, producing an orientation response (Siddle and Lipp, 1997); (ii) familiarity with the object or event (Habib et al., 2003), generally based on schema matching; and (iii) predictability, as based on past observations of regularities and probabilities for specific events. These three processes could be involved for each type of novelty that we described above. For example, suddenness is closely related to attentional processes and the orientation of attention and could be detected automatically or be processed at a higher level; for example, with a conscious perception of suddenness and reappraisal (see also below). The evaluation of familiarity and unpredictability is obviously involved for each type of novelty. These processes are strongly related to memory and could occur at different levels of processing: reflex, schematic or conceptual (Leventhal and Scherer, 1987). As already mentioned above, in the context of cognitive neuroscience, an important corpus of studies has investigated the modulations of different neural networks related to the experimental manipulation of novelty. The ERPs technique, based on the electrical signals of the brain, is one of the methods used to address the temporal dynamic of novelty processing at the brain level. Using this technique, cognitive researchers have addressed the brain mechanisms involved in novelty processes in contrast to the processes related to stimulus repetition. Stimuli repetition is classically accompanied by a decrease of neuronal activity in cortical and subcortical regions (Hensen and Rugg, 2003; Ranganath and Rainer, 2003; Ringo, 1996). A well-known electrophysiological component, the P300, has been extensively studied in the context of novelty processing (Friedman et al., 1993, 2001). This positive component, appearing at 300 ms after the onset of the visual stimuli, has been divided into two more specific electrophysiological components: the P3a and the P3b (Comerchero and Polich, 1999; Polich, 2007; Polich, 1988). Several studies have been demonstrated that the P3a is more prominent on the anterior part of the scalp compared to the

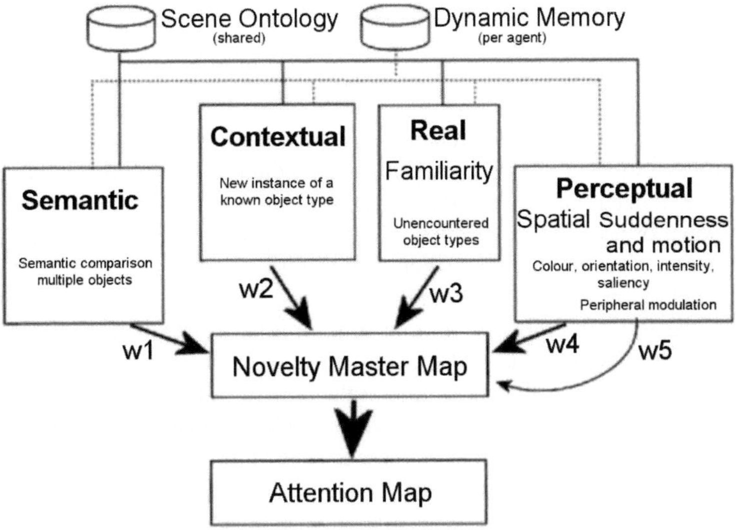

Fig. 1 Different types of novelty and their integration into a master map for guiding attention

P3b and that the former precedes in time the P3b. The shape of the P3a and its intensity are correlated to the novelty experimental manipulations while the P3b characteristics are more affected by the task performed by participants such as target detection task (Fig. 1; for a review see Ranganath and Rainer, 2003). The P3a and P3b neuronal generators are mainly located in the anterior and temporal regions. Human intracranial recordings have shown that anterior and temporal evoked potentials are compatible with the P3a or novelty P3 (Halgren et al., 1995a, b). More specifically, evoked responses have been found in the orbital, ventrolateral and dorsolateral prefrontal regions; similarly, different temporal regions have been shown being modulated by novelty, namely the medial part of the temporal region; perirhinal and posterior parahippocampal cortex, the hippocampus, the temporoparietal cortex and the cingulate gyrus. In the same way, clinical neuropsychology studies have confirmed these findings showing a decrease in novelty effects on the P3a for patients with lateral prefrontal (Knight, 1984), lateral temporoparietal or posterior medial temporal lobe lesions resulting from strokes (Knight, 1996). Functional magnetic resonance imaging (fMRI) studies have shown that the ventrolateral prefrontal cortex, cingulate gyrus and anterior insula are strongly involved in novelty processing (Knight and Nakada, 1998). Indeed, the distributed neuronal network involved in novelty processing includes areas in the lateral prefrontal cortex, orbital prefrontal, anterior insular and anterior temporal cortex, as well as temporoparietal cortex, medial temporal areas along the parahippocampal gyrus, hippocampal formation, amygdala and cingulated gyrus. A recent study investigating novelty and target processes using EEG and fMRI has shown that, despite a considerable overlap of regions activated during novelty and target processing, bilateral superior temporal and right inferior frontal areas are more activated in the novelty processing condition (Knight and Nakada, 1998).

Novelty detection induces an orientation response in the organism (Bernstein, 2002; Scherer, 2001); this orientation response is the behavioural part of the related attentional process effects and the subsequent consequences; that is, the recruitment of attentional and cognitive resources towards the new stimuli (Bernstein, 2002; Desire et al., 2004). The relationships between novelty detection and attentional processes have to be very intensive and modulated by the context of novelty appearance. Indeed, novelty detection may be related to the current goal needs of an individual or not. The modulation of the attentional processes should be different in these two extreme cases: when an organism detects something new in its environment independently and not related to these current task goals or needs, the organism has to inhibit this new information, especially if it is not relevant. In this case, new information is like a distractor and may induce inhibition processes in order to focus the organism's attention on the current task goals. The second case concerns new information appearance which could be related to the current task goals or needs of the individual; in this case the new information has to be processed deeply and involves sustained voluntary attentional processes. Indeed, several different kinds of attentional processes are recruited in novelty detection, dependent on the current task goals or needs of the individual. The unfolding of novelty processes can, indeed, include several steps which can occur at different levels of processing: (1) an early process involves the construction of a new representation; a first step of the encoding of a new stimuli; (2) the relatively automatic detection of unfamiliar, novel stimuli (indexed by the N2 for example) and the modulation of attentional processes; (3) the voluntary allocation of resources determined by the broader context in which a novel event occurs (indexed by the P3); and (4) the sustained processing of novelty, indexed by late-positive slow-wave activity (Chong et al., 2008). The first and second steps of information processing described above are mainly related to automatic detection and could take place at a low level of processing; especially at the reflexive level, eventually also at the schematic level (Leventhal and Scherer, 1987). The third and fourth steps would be more related to endogenous attentional processes involving the voluntary allocation of resources to process in greater depth the new information detected by the previous steps of information processing; see the schematic graph of this process.

As we have described above, novelty processing is not a unitary phenomenon, but may dissociate in different kinds of processes, not only at the conceptual level but also in terms of temporal dynamic processing and the relationship with attentional processes. At the conceptual level, we distinguish different types of novelty processing, allowing not only the implementation of different types of novelty in virtual environments, but also to clarify different concepts of novelty useful for experimental research (perceptual, contextual, partial and semantic novelty). In terms of timing, an understanding of the unfolding of novelty detection is important, allowing researcher to distinguish between the early stages of processing and later ones involving different kinds of attentional processes: these two steps of information processing are related to different levels of processing, with early stages related to exogenous attentional processes and later stages more related to endogenous attentional processes. Processing novelty in the environment is not independent of individual characteristics of the organism. Indeed, Cloninger et al. (1993) have proposed a psychobiological model including the concept of novelty-seeking defining the tendency for a given individual to orient and prefer novel stimuli or new situations. This temperament predisposition to search for and process preferentially novel information has been demonstrated to be related to different kinds of behaviours, like addiction and aggression (Ball, 2004). The predisposition of novelty seeking and other characteristics related to temperament and personality and their relationship with information processing should be taken into account in virtual environments to increase the richness of the possible behaviours that the avatars and agents can produce, not only in different situations but also in terms of individual differences. In the next section we describe how to apply the concepts defined above to present an outline for implementing novelty processing in virtual environments.

3 Novelty in Artificial Systems

The ability to detect novel elements of the environment is of prime importance for the survival and functioning of humans and other animals in dynamic and complex environments. As we have seen in Sect. 2, the growing knowledge about the functioning of the human central nervous system and the possibilities for the computational modelling of subsets of such systems have created the potential for interdisciplinary work, including computational implementations of cognitive and emotional processes. These implementations have been applied in many different domains, including data mining, surveillance, intrusion detection and robotics and have been based on a selection of approaches (Markou and Singh, 2009) and methodologies (Hodge and Austin, 2004). In the research literature, work relating to computational novelty detection has been referred to under a host of different names, including outlier detection and noise detection; see Hodge and Austin (2004, for overview.

An area of particular relevance to our discussion is robotics. Here, the concept of novelty has been linked to comparisons between pre-acquired memorised environment representations and current sensory data of robots in order to detect deviations from normal patterns. For example, Marsland et al. (2000) have presented an autonomous robot that senses an environment through sonar sensors and produces a novelty measure for each scan relative to the model it has learned. Metrics related to habituation are used to assign novelty values to perceptions. This concept is extended in Marsland et al. (2002), where different novelty filters are used in a variety of different contexts, allowing the context in which stimuli appear to be a factor in novelty judgments. More recently Neto and Nehmzow (2007) have combined an attention model with a novelty filter in order to process only salient locations, rather than the whole scene. This helps to improve the performance of such systems by allowing them to allocate extra processing resources to those stimuli deemed to be of higher potential relevance.

Novelty detection is particularly significant to computational agents if they are autonomous and are not granted full access to the scene database, but rather must process it through synthetic perception (see (Chapter "Fundamentals of Agent Perception and Attention Modelling")). Instead of considering real systems, such as mobile robots, we consider virtual agents embedded inside virtual environments.

4 Case Study: Novelty Detection in Virtual Environments

In this case study, we describe one application of computational novelty in more detail: generating sensible attentive behaviours for autonomous agents in virtual environments. It provides an outline of how one might construct and integrate novelty detection for real-time virtual agents operating in virtual environments. These agents have a graphical 3D appearance and interact within the confines of a 3D virtual environment. They are often humanoid in appearance, in the case of the agents described here, have a degree of autonomy and are endowed with a synthetic perception for filtering and sensing their virtual surroundings. It should be noted that the nature of novelty processing as it relates to these agents is much divorced from the real systems described earlier in this chapter: in many ways, the real-time lightweight systems described here are very different and are still extremely inflexible when compared with their real-world counterparts and also more complex computational systems. Nonetheless, the aims and functional requirements are very similar for such agents, who are autonomous to some degree, do not have full access to the world database, inhabit complex virtual environments and attempt to operate under real-time constraints with very limited processing resources.

Here we list components for supporting novelty processing, a technical framework into which novelty processing can be placed and example applications. The overall purpose of the novelty system is to contribute to a prioritisation scheme for agents, allowing them to select and process only those details of potential relevance and importance to them according to the context of the situation or task: novelty is but one factor that may contribute to this – see (Chapter "Fundamentals of Agent Perception and Attention Modelling") for possible others.

4.1 Supporting Capabilities

At least three fundamental capabilities are necessary for supporting novelty detection in computational systems:

1. An internal model of 'normality' based on past experience, e.g. from the external environment. There are numerous ways in which a model may be formed, for example as a static system with manually predefined details, or dynamic and learned based on exploratory behaviours, or blends of both.

2. A sensing system for sampling information from the environment. This may be based on different types of sensors, e.g. a simple temperature monitor, or more sophisticated camera-based sensor utilising computer vision techniques.
3. A comparison operation for comparing new incoming sensory data with the internal model of the environment, based on similarity (i.e. distance) metrics.

The existence of an internal model of normality which is subjective, based on experience and need not coincide with the real state of the environment, implied the need for some form of internal storage mechanism. The memory model (Sect. 4.1.1) fulfils this role. The state of the environment is sampled by a sensing (Sect. 4.1.2) system to provide a model of normality for the agent, against which newly encountered stimuli may be compared.

4.1.1 Memory Model

The memory system allows agents to store or create a model of the environment that can be regarded as 'normal' to them. The memory system described here (Peters, 2006) consists of two components – a static area and a dynamic object-based area. The static area of the agent's memory consists of an object ontology – it could be considered to be part of the agent's pre-defined long-term memory. Here, it is assumed that the ontology is static and no learning takes place, i.e. ontology node values do not change during the simulation, and the agent cannot add new object categories at runtime.

The dynamic area of the memory system provides a storage for object percepts that have been previously sensed by the agent, in addition to memory management descriptors. Unlike the real world, where defining the concept and nature of an object from input is a tenuous task, e.g. using vision approaches to segment a scene, in the virtual environment, the task is simplified as objects are defined by the creator of the scene when it is being constructed.

The role of the memory system and scene ontology is to store a history of familiarity with objects, such as how often they have been previously encountered and the degree to which they have been scrutinised in the dynamic area of memory, and the semantic similarities between objects in the static.

4.1.2 Synthetic Vision

A synthetic vision module samples stimuli from the environment in a snapshot manner by means of an orientable synthetic vision sensor that is locked to the gaze direction of the agent, to create an 'active vision' system – see Sect. 4.3. This sensor renders the scene from the viewpoint of the agent in two modes, allowing for two different basic types of scene representation to exist: spatial and object. By combining the full-coloured and false-coloured representations, the agent has access to a view-dependent representation of the scene, so it is capable of processing both object and spatial information from scenes (see (Chapter "Fundamentals of Agent Perception and Attention Modelling")).

4.2 Implementing Novelty Detection

The memory and synthetic perception capabilities make it possible to create a limited computational implementation of each of the four types of novelty previously described (see Fig. 1 and Peters and Grandjean 2008). As we have seen, defining novelty is a difficult prospect, as are the challenges awaiting those who attempt computational implementations. There are many ways in which items or events may appear novel, so an attempted categorisation is an imperative for supporting attempts towards the creation of computational novelty systems.

4.2.1 Real Novelty

Real novelty refers to a new object instance that has never been or rarely encountered before, although it may fall into a known category. For example, an elephant may be novel because one has never been observed before or to a lesser extent, a car may be deemed novel as this particular one (or instance) has not been observed before, despite the agent's familiarity with road vehicles.

In this case, the agent's dynamic memory system (see Sect. 4.1.1) acts as storage: An uncertainty metric is stored for each object and is a simplification used to represent the well formedness and accuracy of the agent's internal representation with respect to the actual state of the object being perceived. As the agent attends to an object, its uncertainty level is reduced in memory in order to signify the agent's increased knowledge about the object's state. Other information regarding the state of the object is also stored in memory – for example position and velocity. These allow the agent to have its own internal representation of the virtual environment that does not necessarily match its actual state. For example, if the agent is not attending to an object when its position changes, the object will maintain its old position (now incorrect) in the agent's memory.

4.2.2 Perceptual Novelty

Perceptual novelty operations are those relating to the spatial appearance of objects, without consideration of their semantics, from the point of view of the agent. For example, one of three identical cubes may appear novel if it is displayed at a different angle or under different lighting conditions to the others. This could also be regarded as a subtype of contextual novelty.

Perceptual novelty refers to the way in which spatial elements of the visible scene may contrast with others due to their perceived differences, in particular due to their suddenness and saliency. Perceptual novelty therefore deals with comparisons of perceptual features rather than with comparisons of object, scene or event semantics: for example, a small object in a group of large objects, or a red object in a group of green objects would be highlighted for novelty due to saliency, while an object moving in the periphery could be deemed novel due to its eccentricity and velocity.

We incorporate a saliency map (Itti et al., 1998) for describing this type of spatial novelty (see (Chapter "Fundamentals of Agent Perception and Attention

Modelling")), referring to it within the wider term of perceptual novelty. As has been pointed out in Itti and Baldi (2009), the saliency map can be thought of as novelty across space, while the novelty filters often described in robotics research could be considered as being novelty across time. In addition, this term also includes the concept of suddenness, which is related to perceptual influences such as the eccentricity at which a stimulus appears on the retina and its relative velocity. Perceptual novelty does not exclude contextual factors: for example, the saliency map considers spatial context in the sense of contrast of one spatial location with others, while visual similarity considers comparisons between all objects in the scene.

In addition to the general scene saliency, calculations regarding the visual similarity of objects may also be considered. A visual similarity module could be based on a perceptual grouping algorithm, such as that presented in Thorisson (1994). For a given scene rendering, this algorithm creates a graph where the nodes represent objects and the edges record the similarity between the objects with respect to their visual characteristics, such as colour, size and proximity. This process is based solely on visual input and does not utilise any semantic object information.

Suddenness may be related to the motion of objects, particularly in the periphery of vision. Eccentricity and relative velocity may also contribute to a notion of suddenness. The appearance of objects from nowhere, i.e. sudden onset, would also be one of a host of other factors which may contribute to the calculation of such a metric.

4.2.3 Partial Novelty

Partial novelty refers to when a known object type is encountered, but where that specific instance of the object has not been encountered before. This new instance of the object may contain sub-features that have not been observed previously in other instances of the same type. Partial novelty may be modelled by storing dynamic values for each object instance in the memory system.

An important concept here is habituation (Marsland et al., 2000), which relates to the reduction in behavioural response to a stimulus when it is perceived repeatedly. Unlike other forms of behavioural decrement, in the case of habituation, a change in the nature of the stimulus restores response back to its original level, a process known as dishabituation.

The memory model (Sect. 4.1.1) can be used to track objects in the environment that have been observed by the agent and the degree to which it has been familiarised with them through, e.g., repeated exposure and exploration. Objects falling within the gaze of the agent may be regarded as being attended to: as objects are attended to by the agent, the uncertainty level in memory corresponding to the object is reduced. As mentioned above, this value represents the completeness of the agent's representation of the object in memory. Thus, as the agent attends to objects, their uncertainty value decreases. This value can be linked to habituation, so that those objects with low uncertainty values are regarded as having been habituated. Changes in an object's state should lead to a corresponding increase in the uncertainty of the object, thus modelling a form of dishabituation.

4.2.4 Contextual Novelty

Contextual novelty refers to known objects occurring in unusual situations and implies a semantic comparison occurring between multiple objects. This type of novelty can also consist of matching the foreground elements in a scene with its predefined gist (Oliva and Torralba, 2001), such as an elephant in a street scene or a car in a fridge. Here, the ontology is necessary for calculating the semantic difference between objects. This operation could consist of a calculation for each object in the scene of what context it appears in now, and how many times this relationship has been previously encountered. An approach similar to that used in Oliva and Torralba (2001) may be adopted, where the scene is represented as a single entity with a vector of contextual features such as openness, naturalness. An important concept for applying this to virtual environments is that of scene partitioning, i.e. the splitting of the virtual environment into partitions. A scene 'gist' vector can be manually attributed to each zone in the environment. All objects lying within a certain zone can then be compared with the scene gist vector tagged for that zone to establish the novelty of the particular object in that scene context.

4.2.5 Semantic Novelty

Semantic novelty concerns the relationship between scene objects based on their categorisations according to the ontology. Approaches are available for computing semantics-based similarity decisions for ontologies (see, for example Chen and McLeod, 2005), so as to provide basic semantic relationships between objects present in the scene. The ontology is represented as a graph containing entities at nodes and their relationships as edges, as has been used by Navalpakkam and Itti (2002) for task relevance. Relationships may be of the following types: *is a*, *includes*, *part of*, *contains*, *similar* and *related*.

4.3 An Integrating Framework: Visual Attention

In addition to supporting components and a specification of the types of novelty, an important further consideration for constructing a system is how integration may occur between a novelty component and an agent framework. This section focuses on how all of the components can be integrated together, to form a coherent system capable of perception, decision and action. An integration concept described here is that of visual attention, where a number of models have been proposed or adapted for driving the attentive behaviours of computer agents based on visual external stimuli, to create what are referred to as 'active vision' (Aloimonos et al., 1987) for agents and robots: a more extended description can be found in (Chapter "Fundamentals of Agent Perception and Attention Modelling"). In this case, novelty detection may be one in which potentially relevant parts of the scene can be identified so that agents can adjust their attention appropriately towards new or otherwise important stimuli. The element we demonstrate here is part of the perceptual novelty category previously mentioned and consists of generating a saliency map signalling the conspicuity of constituent elements of the visual field (Itti et al., 1998).

4.3.1 Synthetic Perceptual Maps

The saliency map described in Sect. 4.2.2 is just one specific example of a number of different maps, based on the visual field of the agent, that may be used to represent the results of perceptual processing. A generalisation of these maps, referred to as synthetic perceptual maps (Peters, 2007), provides a way for different features to be represented, combined and operated on in a homogenous fashion based on the agent's synthetic perception (see Fig. 2). For the visual modality, synthetic perceptual maps are a virtual analogy of topographic retinotopic maps that represent the visual world as seen through the eyes of a viewer. They are rectangular, 2D gray scale maps corresponding to the agents field of view, where the value of a location in the map represents the strength of some particular feature or resultant operation based on the corresponding spatial location. Bottom-up saliency maps (Itti et al., 1998), task relevance maps (Navalpakkam and Itti, 2002) and the maps described here can all be viewed as instantiations of SPM's in our model. At each update of the agent's vision (see Sect. 4.1.2), two initial synthetic perceptual maps are created as inputs to the agent's perceptual pipeline. Two initial types of perceptual maps are created during each perceptual update: a full-scene map, based on a full colour rendering of the scene, and a false-colour map, based on a rendering of the scene where each object's colour is rendered uniquely without any extra operations, such as lighting or texturing, enabled. The saliency map can be viewed as a SPM created from the full-scene map. In the same way, this methodology allows for the modulation of the agents visual perception based on emotional (Adolphs, 2004), novel or even

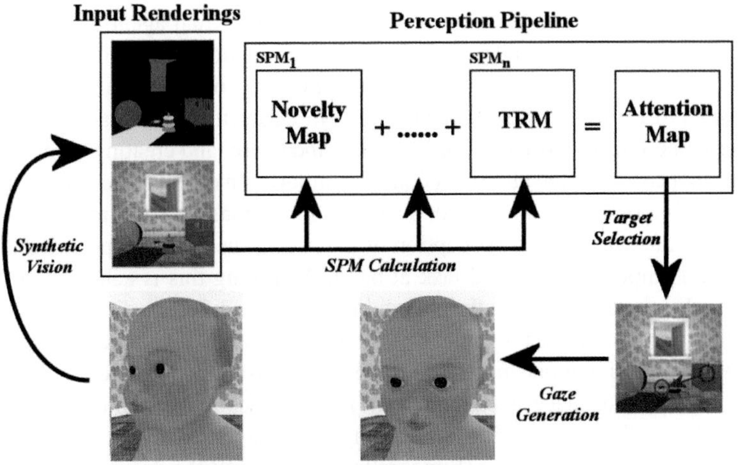

Fig. 2 Illustration of a perceptual pipeline for agents, creating a loop between perception and action, using novelty calculations to guide behaviour

memorised elements of the scene. For example, scene objects can be associated with threat values and based on an initial false-colour perceptual map, a specialised threat map can be created so that the strength in each element of the map is related to the threat value of the perceived object. This technique can also be used for establishing novelty based on objects' uncertainty values, creating a novelty master map.

4.3.2 Visual Perception Pipeline

The visual perception pipeline is a branched pipeline structure consisting of a number of different stages and partitions (see Fig. 2). Each stage in the pipeline contains a number of synthetic perceptual maps, each one the result of a processing operation on either the input maps or a map from a previous stage in the pipeline. SPM's can be combined together and the result of the perception pipeline is always a single master attention map, from which targets for the agent's gaze motions are selected.

The novelty master map can be further amalgamated with other features, the threat map previously mentioned, for example, in order to create a final master map, the attention map, which may be used for driving attentive behaviours. As with the other types of synthetic perceptual maps, the attention map is gray scale in nature; the intensities in the map represent the agent's attention to those respective parts of its visual field. An overall scene metric can also be evaluated, so that the novelty over the full scene can be used to drive explorative behaviours or have the agent engage in idle looking if the scene novelty falls below a certain threshold.

4.4 Outstanding Issues

The framework and components described here represent only one way in which a limited novelty detection system may be created for virtual agents. There are still many difficult theoretical and practical issues that remain to be addressed. A central issue involves the multitude of ways in which objects or events may be compared or considered to be novel from each other. No doubt there are many more super- and sub-categories to those that we have described here, and creating computational systems that can compare all dimensions of similarity in calculations seems an arduous, if not impossible task. Indeed, such categorisations must ultimately rest with the viewer, in terms of their perception of the scene, their prior experience, current mental and emotional state and task at hand. Assuming this is achieved, a further question then relates to how the weighting of these different dimensions should occur in order to construct a single overall novelty estimate, i.e. the novelty master map. Again, top-down biasing seems to be an important factor here, but how exact control should take place is still unclear. These issues make evaluation a difficult task. Additionally, the question of learning and adaptation of the system to new stimuli needs to be considered (Lungarella et al., 2003); the design presented here is far from flexible and is meant as one possible starting point for implementing a novelty capability. This capability can be framed as an early stage of an appraisal processing (Scherer, 2001) system for autonomous agents, by honing their attention to the relevant and important, supporting forgetting mechanisms and ultimately allowing them to better function in complex, unpredictable environments, real or virtual (Breazeal and Brooks, 2005).

Glossary

real novelty Refers to a new object instance that has never been or rarely been encountered before, although it may fall into a known object category. For example, a car may be deemed novel as this particualr instance has never been observed before.

perceptual novelty Perceptual novelty is the assembly of a new representation of an object never perceived in the past by the organism and requiring a new encoding in short-term and long-term memory.

partial novelty Partial novelty would be involved when an organism perceives an object looking like an already perceived object in the past but presenting some perceptual differences on one or several of these characteristics. For example if a baby used to play with a red ball, a new violet ball will not be totally new; its only one or several of these characteristic which are new; the categorization processes are very important in this concept of partial novelty.

contextual novelty Contextual novelty refers to the situations in which an object is perceived in a new context or emerged in a given stable context. For example a ball on the table, while such objects are usually on the floor, would induce a novelty detection phenomenon.

semantic novelty Semantic novelty refers to a situation in which the relationships between the objects or the concepts are organized in a new manner and have never been perceived such as in the past. The fact that individuals are able to create a new tool or a new concept from a series of well-known objects or ideas is characteristic of this type of novelty.

synthetic perceptual maps Gray scale topographic retinotopic maps that represent the visual world as seen through the eyes of a viewer, where the value of a location in the map is the strength of a feature or resultant operation based on the corresponding spatial location.

References

Adolphs R (2004, November) Emotional vision. Nat Neurosci 7(11):1167–1168

Aloimonos JY, Weiss I, Bandopadhay A (1987) Active vision in Proc. 1st Int. Conf. Comput. Vis., London, UK, pp 35–54

Ball SA (2004) Personality traits, disorders, and substance abuse. In: Stelmack RM (ed) On the psychobiology of personality: essays in honor of Marvin Zuckerman. Elsevier, New York, NY, pp 203–222

Bernstein AS (2002) The orienting response and stimulus significance: further comments. Biol Psychol 12(2–3):171–185

Breazeal C, Brooks R (2005) Robot emotion: a functional perspective. In: Fellous JM, Arbib MA (eds) Who needs emotions? The brain meets the Robot, Oxford University Press, NY, pp 271–310

Chen A, McLeod D (2005, May) Semantics-based similarity decisions for ontologies. In: ICEIS 2005, proceedings of the 7th international conference on enterprise information systems, Miami, USA, May

Chong H, Riis JL, McGinnis SM, Williams DM, Holcomb PJ, Daffner KR (2008) To ignore or explore: top-down modulation of novelty processing. J Cogn Neurosci 20(1):120–134

Cloninger CR, Svrakic DM, Przybeck TR (1993) A psychobiological model of temperament and character. Archiv General Psychiatry 50(12):975–990

Comerchero MD, Polich J (1999) P3a and P3b from typical auditory and visual stimuli. Clin Neurophysiol 110(1):24–30

Delplanque S, N'Diaye K, Scherer K, Grandjean D (2007) Spatial frequencies or emotional effects? A systematic measure of spatial frequencies for IAPS pictures by a discrete wavelet analysis. J Neurosci Methods 165(1):144–150

Desire L, Veissier I, Despres G, Boissy A (2004) On the Way to Assess Emotions in Animals: Do Lambs (Ovis aries) Evaluate an Event Through Its Suddenness, Novelty, or Unpredictability? J Comp Psychol 118(4):363–374

Friedman D, Cycowicz YM, Gaeta H (2001) The novelty P3: an event-related brain potential (ERP) sign of the brain's evaluation of novelty. Neurosci Biobehav Rev 25(4):355–373

Friedman D, Simpson G, Hamberger M (1993) Age-related changes in scalp topography to novel and target stimuli. Psychophysiology 30(4):383–396

Grandjean D (2005) Etude lectrophysiologique des processus cognitifs dans la gense de l'motion. University of Geneva, Geneva

Grandjean D, Sander D, Scherer KR (2008) Conscious emotional experience emerges as a function of multilevel, appraisaldriven response synchronization. Conscious Cogn 17(2):484–495

Grandjean D, Scherer K (2008) Unpacking the cognitive architecture of emotion processes. Emotion 8(3):341–351

Habib R, McIntosh AR, Wheeler MA, Tulving E (2003) Memory encoding and hippocampally-based novelty/familiarity discrimination networks. Neuropsychologia 41(3):271–279

Halgren E, Baudena P, Clarke JM, Heit G, Liegeois C, Chauvel P, Musolino A (1995a) Intracerebral potentials to rare target and distractor auditory and visual stimuli. I. Superior temporal plane and parietal lobe. Electroencephalogr Clin Neurophysiol 94(3):191–220

Halgren E, Baudena P, Clarke JM, Heit G, Marinkovic K, Devaux B, Vignal JP, Biraben A (1995b) Intracerebral potentials to rare target and distractor auditory and visual stimuli. II. Medial, lateral and posterior temporal lobe. Electroencephalogr Clin Neurophysiol 94(4): 229–250

Henson RN, Rugg MD (2003) Neural response suppression, haemodynamic repetition effects, and behavioural priming. Neuropsychologia 41(3):263–270

Hodge VJ, Austin J (2004) A survey of outlier detection methodologies. Artif Intell Rev, 22(2): 85–126

Itti L, Baldi P (2009) Bayesian surprise attracts human attention. Vis Res 49(10):1295–1306

Itti L, Koch C, Niebur E (1998 November) A model of saliency-based visual attention for rapid scene analysis. IEEE Trans Pattern Anal Mach Intell (PAMI) 20(11):1254–1259

Knight RT (1996) Contribution of human hippocampal region to novelty detection. Nature, 383(6597):256–259

Knight RT (1984) Decreased response to novel stimuli after prefrontal lesions in man. Electroencephalogr Clin Neurophysiol 59(1):9–20 (1984)

Knight RT, Nakada T (1998) Cortico-limbic circuits and novelty: a review of EEG and blood flow data. Rev Neurosci 9(1):57–70 (1998)

Lang PJ, Bradley MM, Cuthbert BN (2005) International affective picture system (IAPS): affective ratings of pictures and instruction manual. Technical Report A-6. University of Floridao, Gainesville, FL

Lazarus R (1991) Emotion and adaptation. Oxford University Press New York, NY

Leventhal H, Scherer K (1987) The relationship of emotion to cognition: a functional approach to a semantic controversy. Cogn Emot 1(1):3–28

Lungarella M, Metta G, Pfeifer R, Sandini G (2003) Developmental robotics: a survey. Connection Sci 15(4):151–190

Markou M, Singh S (2003) Novelty detection: a review, part 2: neural network based approaches. Signal Process 83(12):2499–2521

Marsland S, Nehmzow U, Shapiro J (2002) Environment-specific novelty detection. In: From animals to animats, Proceedings of 7th international conference on simulation of adaptive behaviour, Edinburgh

Marsland S, Nehmzow U, Shapiro J (2000 January) Detecting novel features of an environment using habituation. In: Proceedings of simulation of adaptive behaviour. MIT Press, Cambridge, MA, pp 189–198

Naatanen R, Paavilainen P, Rinne T, Alho K (2007) The mismatch negativity (MMN) in basic research of central auditory processing: a review. Clin Neurophysiol 118(12):2544–2590

Navalpakkam V, Itti L (2002, November) A goal oriented attention guidance model. Lect Notes Comput Sci 2525:453–461

Neto HV, Nehmzow U (2007) Visual novelty detection with automatic scale selection. Robot Auton Syst 55(9):711–719

Neto HV (2006) Visual novelty detection for autonomous inspection robots. PhD thesis, University of Essex, Colchester, UK

Oliva A, Torralba A (2001, May) Modeling the shape of the scene: A holistic representation of the spatial envelope. Int J Comput Vis 42(3):145–175, May

Peters C (2007) Designing an emotional and attentive virtual infant. In: Proceedings of the 2nd international conference on affective computing and intelligent interaction (ACII), Lisbon, Portugal, September 12–14, 2007. Lect Notes Comput Sci (LNCS) 4738:386–397

Peters C (2006, September) Designing synthetic memory systems for supporting autonomous embodied agent behaviour. In: Proceedings of the 15th international symposium on robot and human interactive communication (RO-MAN), University of Hertfordshire, Hatfield, UK, pp 14–19

Peters C, Grandjean D (2008, May) A visual novelty detection component for virtual agents. In: Paletta L (ed) Proceedings of the fifth international workshop on attention and performance in computational vision (WAPCV), Santorini, Greece, pp 289–300

Polich J (1988) Bifurcated P300 peaks: P3a and P3b revisited? J Clin Neurophysiol 5(3):287–294

Polich J (2007) Updating P300: an integrative theory of P3a and P3b. Clin Neurophysiol 118(10):2128–2148

Ranganath C, Rainer G (2003) Neural mechanisms for detecting and remembering novel events. Nat Rev Neurosci 4(3):193–202

Ringo JL (1996) Stimulus specific adaptation in inferior temporal and medial temporal cortex of the monkey. Behav Brain Res 76(1–2):191–197

Sander D, Grandjean D, Scherer KR (2005) A systems approach to appraisal mechanisms in emotion. Neural Netw 18(4):317–352

Savage-Rumbaugh ES, Toth N, Schick K (2003) Kanzi Learns to Knap Stone Tools. In: Washburn DA (ed) Primate perspectives on behavior and cognition. American Psychological Association Washington, DC, pp 279–291

Scherer K (2001) Appraisal processes in emotion: theory, methods, research, chapter Appraisal considered as a process of multilevel sequential checking. Oxford University Press, New York, NY, pp 92–120

Scherer KR (1984) On the nature and function of emotion. A component process approach. In: Scherer KR, Ekman P (eds) Approaches to emotion. Erlbaum, Hillsdale, pp 293–317

Siddle DAT, Lipp OV (1997) Orienting, habituation, and information processing: The effects of omission, the role of expectancy, and the problem of dishabituation. In: Lang PJ, Simons RF, Balaban MT (eds) Attention and orienting: sensory and motivational processes. Lawrence Erlbaum Associates, Mahwah, NJ, pp 23–40

Thorisson K (1994) Simulated perceptual grouping: an application to human computer interaction. In: Proceedings of the 16th annual conference of cognitive science society, Atlanta GA, pp 876–881

Tulving E, Kroll N (1995) Novelty assessment in the brain and long-term memory encoding. Psychonomic Bull Rev 2(3):387–390

Tulving E, Markowitsch HJ, Kapur S, Habib R, Houle S (1994) Novelty encoding networks in the human brain: positron emission tomography data. Neuroreport 5(18):2525–2528

Cognitive Evaluations and Intuitive Appraisals: Can Emotion Models Handle Them Both?

Fiorella de Rosis, Cristiano Castelfranchi, Peter Goldie, and Valeria Carofiglio

Abstract This chapter deals with the complex relationships between cognitive representations and processes (not reduced to "epistemic" representations but including the motivational ones: goals) and emotions. It adopts a belief–desire–intention paradigm (the explicit account of mental representations and of their "reading" in interaction), but psychologically and computationally sophisticated: for example, by a "dual-process" theory, distinguishing the "intuitive thinking" from the "deliberative thinking," or by a probabilistic approach to the beliefs–goals network. This representation of the mental background of the emotion is also necessary for accounting for emotional interaction, which is based on mind reading, not just on emotional expressions.

We examine the following:

The relationships between emotions and "evaluations," stressing the intrinsic appraisal function of affective reactions to events and thoughts but also disentangling the "cognitive evaluations" (evaluative judgments and beliefs) from the felt, intuitive "appraisal," related to somatic markers, and attraction/repulsion, pleasure/displeasure reactions.

The relationships between emotions and goals: How emotions are related to goals (caused by the possible frustration or realization of monitored goals), while goals can be activated by emotions (impulses).

The relationships between emotions and beliefs: How they influence each other. Emotions influence thinking and believing, while complex emotions (like guilt or envy) are triggered by and based on a specific structure of beliefs, and we can manipulate them by adding or changing some belief.

C. Castelfranchi (✉)
Institute of Cognitive Sciences and Technologies, National Research Council, Rome, Italy; Department of Communication Sciences, University of Siena, Siena, Italy
e-mail: cristiano.castelfranchi@istc.cnr.it

Fiorella de Rosis is deceased.

P. Petta et al. (eds.), *Emotion-Oriented Systems*, Cognitive Technologies,
DOI 10.1007/978-3-642-15184-2_24,

In particular, the expectation-based emotions are examined in the perspective of a formal and computational model, with a special focus on different kinds of "fear" (merely reactive, anticipation-based; counterfactual). We try to explain common features in different emotions, oscillation, and uncertainty. We discuss some computational model of emotional minds. We model fear not only formally but also within the computational emotional mind model: activation-based processes within a beliefs–goals network, also with a detailed analysis of uncertainty and of parameter sensitivity. We also deal with the crucial issue of "cognitive inconsistency" and of the real function of emotional reactions like escaping in fear or like helping the other in guilt feeling: reducing the subjective malaise or others functions like learning? In this perspective we characterize, for example, the "misleading emotions" and the "recalcitrant emotions," and in general the important dynamics of possible emotional conflicts. In sum, formal and computational sophisticated models of human emotions and their dynamics are an important research direction.

1 Introduction

Affective interaction is a sub-kind of social interaction, which entails an interaction between minds and a reaction to the mind behind a perceived behavior.[1] A given behavior can have completely different meanings, and we actually react to these meanings rather than directly to the perceived behavior. B is giving one dollar to A: is she "paying" A? Or "lending" him one dollar, or "giving him back" this money, or "offering" it, or "handing it out"? Or is she giving A one dollar so that he can buy something for her? and so on. An advantage of the belief–desire–intention (BDI) paradigm – in a broad sense, not as a specific AI architecture – with emotions and affective interaction is the fact that an agent A should not necessarily respond to an emotion of another agent B as a global expressive input, and with an emotional reaction in its turn.

- First, A can *perceive, recognize* (read) something *behind* the emotion: the beliefs, desires, intentions, and feelings of B, which motivate or constitute it (Carofiglio and de Rosis, 2005).
- Second, rather than with an emotional, "mirrored" reaction, A can *react* to the emotion of B with a belief about her mental state, and perhaps without any specific emotion toward this. For example, rather than reacting with pity to B's display of pain or anger, A may think "She clearly *believes* that John has taken her dog" or "She clearly has the *goal* to be approved/esteemed by them!".
- Third, A can *specifically* react to those beliefs, desires, intentions, or feelings. For example, if B shows some pain, A can react either empathically or not by saying:

[1] To Fiorella de Rosis, *in memoriam*. Not only to our affective link with her, but to her intelligence, innovation in this domain, and to her promotion of this chapter.

Are you suffering a lot? (to show a sympathetic attitude) or *You are wrong; it is not as you believe* (to try to stop an unmotivated emotion) or *It is not fair; you should be angry, not sorrow!* (to attempt to change B's attitude from that of a victim to an active, angry, indignant one) or *You shouldn't care of these stupid things* (to reduce B's emotion by instilling a bit of indifference, superiority in her), etc. In those cases, what triggers in A an emotion toward B or his own emotion are precisely A's beliefs about B's attitudes, that is, about beliefs, desires, intentions, and feelings he ascribes to B.

In interacting with other (human or artificial) agents, we have a causal theory and a mental model of their emotion (including their psychological processes) and we recognize and interpret this emotion not in terms of what we see but in terms of what we cannot see: their mind.

2 Cognitive Evaluations vs. Intuitive Appraisals

In the psychological literature on emotions, there is a systematic confusion between two kinds of "evaluation":

- A *declarative or explicit* form of evaluation, which contains a judgment of a means–end link, frequently supported by some reason for this judgment, relative to some "quality" or standard satisfaction. This reason-based evaluation can be discussed and explained based on arguments, as well as the goal of having/using the well-evaluated entity (which is the declarative equivalent of "attraction"). See the "motto" (in Aristotelian spirit): *it is pleasant/we like it, because it is good/beautiful.*
- A *non-rational* (but adaptive) evaluation, not based on justifiable reasons (James, 1884; Izard, 1993); this is a mere "appraisal" based on associative learning and memory.

Castelfranchi (2000) proposed to distinguish between *appraisal* – the unconscious or automatic, implicit, intuitive orientation toward what is good and what is bad for the organism – and *evaluation* – the cognitive judgments relative to what is good or bad for someone (and why). Appraisal is therefore an associated, conditioned somatic response; it has a central component and involves pleasure/displeasure, attraction/repulsion; it is a preliminary, central, and preparatory part of motor response. This response can be merely central or more complete, involving overt motor or muscle responses or somatic emotional reactions. It is automatic and frequently unconscious; it is a way of "feeling" something, thanks to its somatic (although central) nature; it gives valence to the stimulus by making it attractive or repulsive, good or bad, pleasant or disagreeable; it has intentionality, as the association/activation makes nice or bad, fearful or attractive what we feel about the stimulus. When it is a response just to the stimulus, appraisal is very fast and primary; it anticipates high-level processing of the stimulus – like meaning retrieval – and even its recognition – which can be subliminal. However, associative, conditioned, automatic responses to high-level representations may occur as well: to beliefs, to hypothetical scenarios and decisions (Damasio, 1995), to mental images, to goals, etc.

We propose to change our usual view of cognitive "layers," where association and conditioning are relative to only stimuli and behaviors, not to cognitive mental representations. While not all emotions suppose or imply a cognitive evaluation of the circumstances, any emotion as a response implies an appraisal in the above-mentioned sense; it implies the elicitation of a central affective response involving pleasure/displeasure, attraction/repulsion, and central somatic markers if not peripheral reactions and sensations. This is what gives emotions their "felt" character.

Evaluation and appraisal can also derive from each other. Merely affective reactions toward some event can be verbalized and translated into declarative appreciations. The opposite path – from a cold evaluation to a hot appraisal – is also possible, especially for personal, active, important goals, and in particular for felt kinds of goals like needs and desires. The appraisal of an event produces a feedback on the beliefs and confirms them: *Since I'm afraid, it should be dangerous. I was right!*

Evaluations do not necessarily imply emotions; not any belief about the goodness or the badness of something necessarily implies or induces an emotion or an attraction/rejection with regard to that something. Cold evaluations also exist. Evaluations are likely to have emotional consequences if they are about our own goals and if these goals are currently active and are important. On the other hand, emotions do not necessarily imply evaluations; attraction or rejection might be considered per se as forms of evaluation of the attractive or the repulsive object; we view attraction and rejection as pre-cognitive implicit evaluations that we call "appraisal."

To summarize, mind should be incorporated into emotion models with a structured and detailed representation of beliefs (evaluations, expectations, etc.) and goals (intentions, needs, desires), by going deep into the understanding of their specific relationships with "emotional feelings," that is, with the body appraisal and response.

2.1 Relationship Between Emotions and Goals

A general consensus exists on the hypothesis that emotions are a biological device aimed at monitoring the state of reaching or threatening our most important goals (Oatley and Johnson-Laird, 1987). In Lazarus's primary appraisal, the relevance of a given situation to the individual's relevant goals is assessed (Lazarus, 1991). At the same time, emotions activate goals and plans that are functional to re-establishing or preserving the well-being of the individual, challenged by the events that produced them (secondary appraisal; Lazarus, 1991). There is a strong relationship, then, between goals and emotions: goals, at the same time, *cause* emotions and *are caused by* emotions. They cause emotions since, if an important goal is achieved or threatened, an emotion is triggered; emotions are therefore a feedback device that monitors the reaching or the threatening of our high-level goals. At the same time, emotions activate goals and action plans that are functional to re-establishing or preserving the well-being of the individual that was challenged by the events that produced them (Poggi, 2005). As personality traits may be viewed in terms of weights people put on different goals, the strength of the relationship between goals and emotions also depends on personality traits. These considerations suggest the following criteria for categorizing emotions (Poggi, 2005):

- *The goals the emotion monitors*: For instance, *Preserving self from – immediate or future – bad* may activate distress and fear; *achieving the – immediate or future – good of self* may activate joy and hope. *Dominating others* may activate envy. *Acquiring knowledge and competence* activates the cognitive emotion of curiosity. *Altruistic emotions* like guilt or compassion involve the goal of "defending, protecting, helping others." *Image emotions* like gratification involve the goal of "being evaluated positively by others"; *self-image emotions* like pride involve the goal of "evaluating oneself positively", and so on.
- *The level of importance of the monitored goal:* this may be linked to the type of goal but is also an expression of some personality traits (e.g., neuroticism).
- *The probability of threatening/achieving the monitored goal*: some emotions (joy, sadness) are felt when the achievement or thwarting is certain, others (hope, fear) when this is only likely.
- *The time in which the monitored goal is threatened/achieved*: some emotions (joy, sadness) are felt only after goal achievement or thwarting, others (enthusiasm) during or before goal pursuit.
- *The relevance of the monitored goal with respect to the situation*: if a situation is not relevant to any individual's goal, then it cannot trigger any emotion in the individual.
- *The relationship between the monitored goal and the situation in which an individual feels an emotion*: this is linked to the emotion valence, but also to its intensity. Some emotions belong to the same "family" but differ in their intensity (disappointment, annoyance, anger, fury).

Another less relevant relation between emotions and goals is that – given that emotions are pleasant or unpleasant experiences – to feel or not to feel an emotional state (for example, excitement, amusement; or fear, boredom, ...) can "per se" become a goal for the individual. We do not necessarily always avoid negative emotions (like fear or disgust); there are movies specialized for these experiences. There are even individuals or situations where the goal is "not to feel emotions at all," "not to be moved or disturbed"; or whose goal is to feel some emotion, to be aroused, just in order to "feel alive."

2.2 Relationship Between Emotions and Beliefs

Emotions are biologically adaptive mechanisms that result from evaluation of one's own relationship with the environment. However, not all appraisal variables are translated into beliefs; some of them influence emotion activation without passing through cognitive processing, while others do. These cognitive appraisals are beliefs about the importance of the event, its expectedness, the responsible agent, the degree to which the event can be controlled, its causes, and its likely consequences, in particular (as we said) their effect on goal achievement or threatening. Different versions of this assumption can be found in various cognitive appraisal theories of emotion (Arnold, 1960; Lazarus, 1991; Ortony et al., 1988; Scherer et al., 2004). In individual emotions, cognitive appraisals are first-order beliefs, while in social emotions, second-order beliefs occur as well.

Beliefs therefore influence activation of emotions and are influenced by emotions in their turn (Frijda and Mesquita, 2000). They may trigger emotions either directly (as emotion antecedents) or through the activation of a goal (Miceli et al., 2006). On the other hand, psychologists agree in claiming that emotions, in their turn, may give rise to new beliefs or may strengthen or weaken existing beliefs. Classical examples are jealousy, which strengthens perception of malicious behaviors, or fear, which tends to increase perception of the probability or the amount of danger of the feared event (we will come back again on this issue in Sect. 6). As events that elicit emotions may fix beliefs at the same time, this effect may strengthen the cyclical process of belief holding–emotion activation–belief revision; when you hear on the TV about a plane that crashed, your fear strengthens your beliefs about the risks of flying. These beliefs may be temporary and last for as long as the emotion lasts. But they may become more permanent as a result of a "rumination and amplification" effect; in this case, for instance, feeling fearful after watching the TV may bring one to detect more fearful stimuli in the news, which increase fear in their turn.

Finally, emotions may influence beliefs indirectly, by influencing thinking; they are presumed to play the role of making human behavior more effective, by activating goals, and orienting thinking toward finding a quick solution to achieving them. They hence influence information selection by being biased toward beliefs that support emotional aims.

In complex emotions, like guilt, jealousy, or envy, a specific set of beliefs is necessary for triggering the emotion as interpretation, evaluation of the event or as its prediction. Beliefs characterize, in this case, the mental attitude of that emotion; they persist during the emotion and, if they are invalidated, the emotion is stopped, reduced, or changed. They are explicit, that is, not necessarily conscious or under attention; they are subject to possible "argumentation," for example, from a friend of ours, or of manipulation from ourselves, in order to reduce the emotion or change it.

To protect oneself from pity or guilt one can, for example, introduce a belief that the harm and the possible suffering are "fair" and merited, for example, *The guy deserved that punishment, it is his fault*. Or – to transform envy – one may think that the other's enviable condition "is unfair, is an unmotivated privilege," that "it is a case of injustice"; this belief will transform the despicable, hidden, and passive "envy" into a noble and open "sense of injustice," "indignation" (Miceli and Castelfranchi, 2007).

To some authors (Feldman-Barrett et al., 2005; Miceli and Cactelfranchi, 1997), for typical and complex human emotions, even a belief of "recognition" or "categorization" of that mental and feeling state as such an emotion is necessary for specifically and fully feeling that emotion; this is culturally specific, and we have learnt to discriminate it from similar emotions also, thanks to their cultural label.

3 How Can We Apply These Theories to Build Emotion Models?

In trying to formalize a computationally tractable model of emotion, we have to deal with a very complex reality. Individuals can experience several emotions at the same time, each with a different intensity, which is due to the importance of the goal, the likelihood of the impact of the perceived event on that goal, the level of surprise about the event, the measure of how likely it is that the event will occur, the level of involvement in the situation that leads to emotion triggering. Once activated, emotions can decay over time, with a trend which depends on the emotion and its intensity; duration is also influenced by personality traits and social aspects of interaction.

3.1 Aspects to Include in a Computational Emotion Model

a. *Appraisal components of emotions*: How is it that two persons report feeling different emotions in the same situation? Clearly, the main source of difference is due to the different structure of beliefs and goals of the two individuals: the goals they want to achieve, the weights they assign to achieving them, and the structure of links between beliefs and goals. Differences among experienced emotions may be due to (1) the link between *situation* characteristics (which event – and consequences – which social aspects – role of individuals involved in the situation – which individual's personality traits) and emotion components (appraisal variables and other internal dispositions -beliefs and goals) and (2) the mutual links between emotion components themselves (van Reekum and Scherer, 1997).
b. *Emotional dispositions*: A second major source of difference is in individual emotional dispositions. An emotional disposition is a disposition to have certain kinds of occurrent thoughts and feelings toward a certain kind of thing – that is, toward a *focus* (Helm, 2001; Goldie, 2004). So John's fear of cats is an emotional disposition, with cats as the focus. The presence of this disposition will explain why John feels fear on seeing a cat or on being told that the house he is going to for dinner is full of cats; whereas other people will not feel fear in these situations. This is not a personality trait (it is too focused for that), nor is it obviously a

"different structure of beliefs and goals" – at least if it is the latter, then only derivatively so, in the sense that John will have a goal of avoiding cats – but only because he is afraid of them (in the dispositional sense).

c. *Influence of personality factors*: Individual differences can alter experienced emotions. Some personality traits may be viewed in terms of the general "propensity to feel emotions" (Poggi and Pelachaud, 1998; Plutchik, 1980). Picard (1997) calls this subset of personality traits "temperament", while other authors relate them directly to one of the factors in the "big-five" model (Mc Crae and John, 1992), for instance, neuroticism. These traits imply a lower threshold in emotion feeling (Ortony et al., 1988): a "shy" person is keener to feel "shame," especially in front of unknown people; a "proud" person attributes a high weight to his goal of self-esteem, etc. A personality trait (proud) is therefore related to attaching a higher weight to a particular goal (self-esteem, autonomy) and since that goal is important to that kind of person, the person will feel the corresponding emotion (pride or shame) with a higher intensity.
d. *Emotion intensity and decay/duration*: In the OCC theory (Ortony et al., 1988), desirability, praiseworthiness, and appealingness are key appraisal variables that affect the emotional reaction to a given situation, as well as their intensity; they are therefore called *local intensity variables*. For example, the intensity of prospect-based and confirmation emotions is affected by the *likelihood* that the prospected event will happen; the *effort* (resources needed to make the prospected event happen or to prevent its happening); and the *realization* (degree to which the prospected event actually happens). The intensity of fortune-of-others emotions is affected by presumed *desirability* for the other, *liking* (the individual appraising the situation has a positive or a negative attitude toward the other) and *deservingness* (how much the individual appraising the situation believes that the other deserved what happened to him). Some *global intensity variables* also affect the intensity of emotions: *sense of reality* (how much the emotion-inducing situation is real); *proximity* (how much the individual feels psychologically close to the emotion-inducing situation); *unexpectedness* (how much the individual is surprised by the situation); and *arousal* (how much excited the individual was before the stimulus).
e. *Emotion mixing and oscillation: partially overlapping emotions*: Emotions may co-occur or mutually influence each other; this can be due to overlapping of their cognitive backgrounds. In a sense, for example, *guilt* feeling toward a victim B "contains" *pity*, as they have the same constituents: (i) the belief that B received a serious harm/loss; (ii) the idea that (because of this) he is suffering, will, or might suffer; (iii) an empathic disposition and feeling based on this perceived or imagined suffering; and (iv) the idea that such a harm and suffering is not fair, is not a right punishment for B and his conduct. This cognitive pattern elicits a goal, an impulse to care about B, to worry about his condition, to succor and help him if possible. However, pity is the empathic and charitable feeling of the bystander who is not responsible for such an unfair suffering. On the contrary, guilt is the affective attitude of those who feel responsible for that harm; so the

helping impulse not only reduces pity but also alleviates the sense of irresponsibility and guilt. Within the broader frame of guilt, the same generous impulse, based on the same beliefs, changes its "flavor"; in a sense, therefore, guilt toward a victim contains "pity," but in another sense – in the scope of a larger gestalt – it is no longer just pity.

Other examples of partial overlapping are *guilt* and *shame*, which overlap in important constituents though being focused on different goals. While what matters in shame is the other's opinion of us, our social "image," what matters in guilt is the violation of a moral standard. In both cases, however, there are negative evaluations about B (since being a morally bad person is a negative judgment) as well as self-evaluations. This is why shame and guilt can coexist, or why we can oscillate from one to the other if we care about social esteem and the other's possible judgment. This does not mean that when there is shame, there must be guilt; shame is not necessarily a "moral" emotion; it can be an "aesthetic" one (Sabini and Silver, 1982; Castelfranchi and Poggi, 1990), for example, concerning our physical aspect and not implying any responsibility at all. But in all cases, we do not correspond to some evaluative standard, and we get negative judgments. In shame, I have some inferiority, I lack something, I am defective; in guilt, I did something bad, I am a bad person. The two emotions can be transformed into each other also because of my condition; my lacking something can harm or make suffer somebody or – vice versa – my bad behavior can be due to some defect or to my being in an inferior position.

In considering the problem of emotion mixing, Picard (1997) proposed two metaphors: in a *microwave oven* metaphor, juxtaposition of two or more states may occur even if, at any time instant, the individual experiences only one emotion; emotions do not truly co-occur, but the affective state switches among them in time; this is, for instance, the case of love–hate. In the *tub-of-water* metaphor, the kind of mixing allows the states to mingle and form a new state; this is, for instance, the case of feeling wary, a mixture of interest and fear. Different emotions may coexist because an event produced several of them at the same time or because a new emotion is triggered, while the previous ones did not yet decay completely. Picard evokes the *generative mechanism* as the key factor for distinguishing between emotions that may *coexist* and emotions that *switch* from each other over time. She suggests that co-existence may be due, first of all, to differences in these generative mechanisms; but it may be due, as well, to differences in time decay among emotions that were generated by the same mechanism at two distinct time instants. This idea of generative mechanism is very close to Castelfranchi's idea of overlapping cognitive background.

f. *Role of uncertainty*: In some emotion-modeling systems, emotion intensity is measured in terms of a mathematical function combining the previously mentioned variables (Prendinger et al., 2002; Elliott and Siegle, 1993) or of logic rules (Turrini et al., 2007). An alternative approach adopts a probabilistic representation of the relationships among the variables involved in emotion triggering and display. These are usually dynamic models that represent the generative mechanism of emotions, the intensity with which they are triggered, the time decay of this intensity, and the transition through various emotional states. Due to the presence of various sources of uncertainty, *dynamic belief networks* (DBNs) are a good formalism to achieve this aim (Nicholson and Brady, 1994); in the next section we will focus our attention on some remarkable examples of this kind.

3.2 Main Experiences

In a first attempt in this domain (Ball and Breese, 2000), a simple model is proposed in which emotional state and personality traits were characterized by discrete values along a small number of dimensions (valence and arousal, dominance and friendliness). These internal states are treated as unobservable variables in a Bayesian network model. Model dependencies are established according to experimentally demonstrated causal relations among these unobservable variables and observable quantities (expressions of emotion and personality) such as word choice, facial expression, and speech. EM (Reilly, 1996) simulates the emotion decay over time for a specific set of emotions, according to the goal that generated them. A specific intensity threshold is defined, for the point at which each emotion is triggered. Affective reasoner (Elliott and Siegle, 1993) is a typical multi-agent model. Each agent uses a representation of both itself and the interlocutor's mind, evaluates events according to the OCC theory, and simulates a social interactive behavior by using its knowledge to infer the other's mental state. Emile (Gratch and Marsella, 2001) extends affective reasoner by defining emotion triggering in terms of plan representation; the intensity of emotions is strictly correlated to the probability of a plan being executed, which is responsible for agents' goal achievement. Conati proposed a model based on dynamic decision networks to represent the emotional state induced by educational games and how these states are displayed (Conati, 2002). The emotions represented in this model (reproach, shame, and joy) belong to the OCC classification; some personality traits are assumed to affect the student's goals. The grain of representation is not very fine, as among the various attitudes that may influence emotion activation (first- and second-order beliefs and goals, values, etc.), only goals are considered in the model. Subsequently, the same group (Conati and McLaren, 2005) described how they refined their model by adding new emotions and learning parameters from a data set collected from real users. In Prendinger and Ishizuka (2005), an artificial agent empathizes with a relaxed, joyful, or frustrated user by interpreting physiological signals with the aid of a probabilistic decision network; these networks include representation of events, agent's choices, and utility functions. In emotional mind, Carofiglio et al. (2008) advocate a fine-grained cognitive structure in which the appraisal of what happens in the environment is represented in terms of its effects on the agent's system of beliefs and goals, to activate one or more *individual emotions* in the OCC classification. We will demonstrate the representation power of this modeling method in Sect. 4.2 by considering, in particular, the emotion of fear.

4 The Case of Expectation-Based Emotions

We will now focus our discussion on the category of "expectation-based emotions." In our view, *expectation* is not a mere forecast (a belief about the future) (Castelfranchi and Lorini, 2003); it is a complex and unitary mental object including three components:

- A *forecast* (more or less certain, strong) about a future state or event Ev by agent A: (Bel A $\Diamond$Ev)[2];
- An *expecting* activity or disposition: A has the *goal of coming to know* whether his prediction was correct, whether the event actually happens:

  $$\text{Goal A (Know} - \text{whether A Ev)}.$$

  This is the basic meaning of "expecting" and of the derived term "expectation." This is actually a specific kind of "epistemic goal" and activity: an activity aimed at acquiring knowledge, the goal to "know whether."
- But why has A the goal to know whether her prediction is correct or not? Because the event is "relevant" for A, because A is concerned by Ev. That is, a goal of A is "into question" because of the – possible – event. Thus, the third ingredient of a full expectation is A's goal about Ev: (Goal A ¬Ev):

$$\text{Overall}: (\text{Bel A } \Diamond\text{Ev}) \wedge (\text{Goal A (Know - whether A Ev)}) \wedge (\text{Goal A } \neg\text{Ev}) \quad (1)$$

This formalization explains why currently *computers can make forecasts but cannot have expectations*; they are not "concerned" and expecting for. A "forecast" or "prediction" is the result of a previous process of "predicting," "forecasting," while "expecting" is an activity or disposition which follows the formulation of the forecast and makes it an "expectation." To model the activity of monitoring, of knowledge acquisition that is typical of anticipation, computers should be modeled so as to become "involved" about event evolution; they should be endowed with the ability, the interest, and the curiosity to know how things will evolve. We will come back to this point in Sect. 6.

There are various kinds of expectations (positive, negative, neutral, and ambivalent) which can be defined relatively to A's goals:

- *Positive expectation* (goal conformable): (Bel A $\Diamond$Ev) ∧ (Goal A Ev)
  A expects Ev to be eventually true, and wants it
- *Negative expectation* (goal opposite): (Bel A $\Diamond$Ev) ∧ (Goal A ¬Ev)
  A expects Ev to be eventually true, and does not want it
- *Neutral expectation:* (Bel A $\Diamond$Ev) ∧ ¬(Goal A Ev) ∧ ¬(Goal A ¬Ev)
  X expects Ev to be eventually true, and is explicitly indifferent about this

[2] where $\Diamond$ denotes "eventually," "in a more or less near – and possible – future."

- *Ambivalent expectation:* (Bel A $\Diamond$Ev) $\wedge$ (Goal A Ev) $\wedge$ (Goal A $\neg$Ev)
  X expects Ev to be eventually true, and on one hand he wants it while on the other hand he doesn't. For instance, he may desire something and fear it at the same time.

4.1 The Example of Fear

The general function of fear is to avoid a (represented or not represented) harm, threatening event. There are, in our view, different kinds of fear; not all of them are "anticipatory" emotions in a strict sense.

(i) A first form of fear (the reaction of startle and fright) can be due to a mere, very fast stimulus before any imagination or realization of possible dangers, even before the full recognition of the triggering stimulus itself. This form of fear (that we might call *stimulus-based* or *merely reactive*) is anticipatory only functionally, that is, the function of the induced reaction is oriented to the future, to a possible impending danger. However, it is not cognitively anticipatory, i.e., based on a prediction, a belief (an image) about the future event. "Anticipation" means that our behavior is *coordinated with a future event* (cit), but this event is not necessarily explicitly represented in our mind or in the control system governing our behavior. "Prediction" is the explicit representation of a future event. Only some forms of anticipatory behaviors are based on prediction and on the explicit mental representation of a future event.

(ii) The most typical form of fear is *prediction based*; it is activated by a prediction contrary to our goals, a threatening prediction, or a perceived dangerous event Ev. A predicts/expects – on the basis of her reasoning (inferences) or of mere associative learning and activation – a future bad event. "Bad" means contrary to her goals. This can be explicitly appraised, by A's judgment, with an explicit representation of the goal, or can be implicitly appraised as "bad" (contrary to some goal) by the associated affective reaction or evocation of a "somatic marker."

Notice that in this case, A does not perceive P from the environment; the danger is not there, it is imagined or inferred, it is coming. She is reacting to her mind, to a self-provided stimulus.

(iii) A third form of fear is in a sense not anticipatory, not relative to the future but to the past! This is a post hoc fear about something which did not actually happen, and that A does not expect to happen. This is a *counterfactual fear* due to the mere idea (inference, imagination) of what *might have happened* (but did not happen in fact). This fear (like the anticipatory one) gives rise to "relief" for the escaped harm; in this case, however, that harm was not expected before. Given the unexpected event Ev, I realize (imagine) that something very bad might have happened, and I feel fear in front of such an idea (image), although I know that it did not happen and I do not expect that it will happen.

Also in this case it is the mere endogenously produced mental representation of something, which is not there, which terrifies me; but it is not a predicted event.

The first two forms of fear share their "anticipatory" character, while the second and the third one share their "imagination-based" ("hallucinatory") character. Of course, evolutionarily speaking, the third form of fear also has a function, and this is oriented to the future, but not as a short-term prediction and preparation/reaction. Its main function is learning, that is, remembering that those situations can be dangerous. Although nothing actually happened, I know that something might have happened, and by learning from my imagination like from my actual experience, I will evoke my mental bad experience (the merely imagined harm) like a real harm and will become prepared and cautious in those circumstances.

The three forms of fear may coexist: reactive fear can be due to a "false alarm," which triggers counterfactual thinking or expectation of incoming dangers, or the idea of what might have happened can make us cautious and fearful about what might have happened. In some cases, they may even contradict each other; a merely reactive fear may be felt, while events contrary to own goals are not (yet) predicted; if something in the periphery of our field of vision moves suddenly, we jump back or get gooseflesh before or without forecasting any future danger.

4.2 A Cognitive Model of Fear

To formalize how fear may be activated, we consider its prototypical form, the *prediction-based* one. Its cognitive skeleton is a *negative expectation*:

A worries about Ev, she doesn't like Ev, desires that Ev will not happen, would like to avoid Ev; however, she predicts and expects that Ev will happen:

$$(\text{Bel A } \Diamond\text{Ev}) \wedge (\text{Goal A } \neg\text{Ev}) \tag{2}$$

or

A worries about ¬Ev, she likes Ev, desires that Ev will happen, would like Ev; however, she predicts and expects that unfortunately Ev will not happen:

$$(\text{Bel A } \neg\Diamond\text{Ev}) \wedge (\text{Goal A Ev}) \tag{3}$$

Fear occurs, as well, in ambivalent expectations, that is, when an event is taken into account as a possible outcome, it is desired and feared at the same time and one can focus on one side or the other, or anxiously oscillate from each other.

4.3 Fear in Emotional Mind

We now describe how *cognitive models of prediction-based emotion activation* are built in *emotional mind* (de Rosis et al., 2003; Carofiglio and de Rosis, 2005; Carofiglio et al., 2008) and how such an emotion may mix up with other kinds of emotion.

In this model, emotions are activated by the belief that a particular important goal may be achieved or threatened; therefore, our simulation is focused on the change in the belief about the achievement (or threatening) of the goals of agent A over time. In this monitoring system, the cognitive state of A is modeled at the time instants $\{T_1, T_2, \ldots, T_i, \ldots\}$. Events occurring in the time interval (T_i, T_{i+1}) are observed to construct a probabilistic model of the agent's mental state at time T_{i+1}, with the emotions that are eventually triggered by these events. The general structure of our model includes the following static components:

- M(T_i): the agent's mind at time T_i, with its beliefs and goals;
- Ev(T_i, T_{i+1}): an event occurring at (T_i, T_{i+1}), with its causes and consequences;
- M(T_{i+1}): the agent's mind at time T_{i+1};
- Em-feel(T_{i+1}): a particular emotion activated, in the agent, at time T_{i+1}.

M(T_{i+1}) depends on M(T_i) and Ev(T_i, T_{i+1}). The feeling of emotion depends on both M(T_i) and M(T_{i+1}). We calculate the intensity of emotions as a function of two parameters: (1) the uncertainty in the agent's beliefs about the world and, in particular, about the possibility that some important goal is achieved or threatened and (2) the utility assigned to this goal.

Figure 1 shows an example of application of this model. Here, the agent A is a mother M observing her little child C who just learnt to bicycle. This entails two consequences: the child may be at risk (Ev = AtRisk-C) and he is growing up (Ev' = GrowsUp-C). Ev is a future, possible, and negative event, while Ev' is a positive and present one. Thus, "fear" and "happy-for" may be activated at the same time, because of the two events co-occurring or just imagined by M; the child is growing up (Bel M GrowsUp-C) but might be at risk in the more or less near future (Bel M ◊AtRisk-C). As M is in a warm relationship with her child (WarmRelationship M C), the "social" emotion of happy-for is triggered after believing that C is growing up, that this is what C also desires [Bel M (Goal C GrowsUp-C)], and hence that the goal of "getting the good of others" is achieved (Bel M Ach-GoodOf-C). The intensity of this emotion depends on how much this probability varies when evidence about the mentioned desirable event is propagated in the network. It depends, as well, on the weight M attaches to achieving that goal which is, in its turn, a function of the agent's personality (in the example, how much the mother is attached to her child). The figure shows that, at the same time, fear may be activated in M by the belief that C might fall down by cycling because of his low experience: (Bel M ◊AtRisk-C) and (Bel M UnDesirable(AtRisk-C)). This may threaten the goal of self-preservation (Bel M ◊Thr-GoodOf-Self).

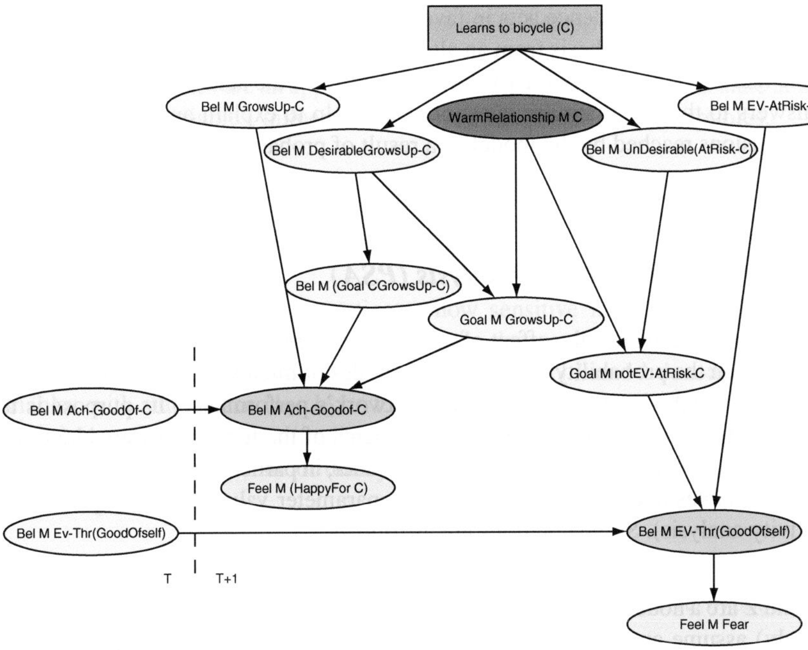

Fig. 1 An example of model for activation of fear and happy-for according to the "tub-of-water" metaphor

5 Problems in Building Emotion Models as DBNs

The problem of how to estimate parameters when building probabilistic models is still a matter of discussion. In particular, in probabilistic emotion-triggering models, the following questions are raised: What is the probability of an event? How much threatening for my life is this event? How important is it, to me, to avoid this risk? and others. Similar considerations apply to defining variables involved in the recognition process. BN parameters can be estimated by learning them from a corpus of data (frequentist approach) or according to subjective experience or common sense (neo-Bayesian approach). To validate probabilistic emotion activation and expression models, we advocate the need to investigate two different issues:

1. *Robustness of the model*: How sensitive are the results of a belief update (evidence propagation) to variations in the values of the parameters in the model (parameter sensitivity analysis)?
2. *Predictive value of the model*: Does the dynamic behavior of the model reflect what an external, independent (and competent) source would expect? Evidence sensitivity analysis may help. This may, for instance, give answers to questions like what are the minimum and maximum beliefs produced by observing

the alternative hypothesis (activation of hope), the discrimination requires to compare, in terms of Bayesian likelihood ratio, the impacts of all subsets ε' of a set of evidence ε.

6 An Open Problem: Cognitive Inconsistency

Let us introduce this topic with a premise about the difference between positive and negative emotions. While positive emotions are usually enjoyed as long as possible, and one tries to rehearse them so that they persist, with negative emotions one is induced, by the malaise or suffering, to bring an end "here and now" to that mental state or at least to make it less severe, not just to avoid the malaise and its eliciting situation in the future. So, in general a negative emotion activates a behavioral or mental activity of coping, some distraction or defense and reduction, some way out. What is remarkable is that frequently, if not normally, the activated impulses tend to alleviate or "give vent" to the painful psychological state; so escaping will reduce the fear; relieving the other will alleviate the felt pity or guilt for a victim, confessing will stop the guilt feeling, and aggression will give vent to the rage. The function of the impulse is not to reduce the subjective malaise; it is a specific function of that emotion, which is explained by the goal this emotion monitors. The function of suffering reduction is to learn and improve reaction in similar circumstances and to strengthen the induction of that behavior. This is one of the possible reasons on cognitive–emotional inconsistency, which we consider in this final section.

6.1 Consistency in Computational Models of Emotions

Irrespective of the formalism adopted and of the grain size in knowledge representation, computational models of emotions are built on the hypothesis of *consistency* among the variables included in the model: variables that represent appraisal of the environment (occurred, occurring or future events), cognitive aspects that translate these variables into "attitudes" (first- and second-order beliefs, goals, values, etc.), and emotions that are presumably activated. Various psychological studies are, however, seriously discussing this consistency hypothesis. We will refer to two of them, which give a slightly different interpretation of this phenomenon.

Goldie denotes with *misleading emotions* the emotions that are not useful in picking up saliences in the environment and enabling quick and effective action, as (on the contrary) emotions would be presumed to do (Goldie, 2009). To Brady, *recalcitrant emotions* are those emotions which involve a conflict between an emotional perception and an evaluative belief; the subject perceives his situation to be thus-and-so while believing that it is not thus-and-so (Brady, 2007). For instance, a recalcitrant bout of fear is one where someone is afraid of something, despite believing that it poses little or no danger; he may believe both that flying is dangerous and that it is perfectly safe or (in the mother–child example we introduced in Sect. 4) that running with a bike is safe while believing that it is not. Interpretation of the reasons of this contradiction differs in the two authors. Goldie follows the "dual-process" theory (Sloman, 1996), according to which our perception and response to emotional situations would be processed through two routes: (i) a *fast and frugal* route (also called "intuitive thinking"), which involves imagination, operates fast, and uses limited resources and speed of processing and (ii) a *more complex, slower route* (also called "deliberative thinking"), whose function would be to operate as a check or balance on intuitive thinking. The dual-process theory acknowledges the possibility that the two routes do not work in perfect agreement; therefore, it may happen that some emotions resulting from intuitive thinking *mislead us* and that deliberative thinking does not succeed in correcting them. Goldie's hypothesis is that intuitive thinking would be performed through some *heuristics* that were built after environmental situations humans had to face in their past history have changed. Changes in environmental conditions (that he calls *environmental mismatches*) would then be responsible for producing misleading emotions, which conflict with deliberative thinking, and that this complex and slower route is not able to detect and correct. Rather than accepting the hypothesis of recalcitrant emotions as "irrational," Brady proposes a positive role for this contradiction, as a means to facilitate the processing of emotional stimuli; even if our deliberative thinking recognizes the felt emotional state as "unreasonable," our emotional system would ensure that our attention remains fixed on the dangerous objects and events, thus checking them and facilitating a more *accurate* representation of the danger (or the insult, the loss, for emotions different from fear).

The excessive and recalcitrant emotion of fear in the mother in our example is functional to her checking carefully that her child does not adopt a dangerous attitude in cycling. This is very close to the kind of *counterfactual fear* that we considered in Sect. 4, due to the mere idea (inference, imagination) of what *might have happened* (but did not happen in fact). Its function is oriented to the future, not only as a short-term prediction and preparation/reaction but also to remember that those situations can be dangerous. Nothing actually happened but M, by learning from her imagination, evokes the merely imagined harm like a real harm and becomes prepared and cautious.

This contradiction between emotional and cognitive states is one of the cases of *cognitive dissonance* that was originally described by Festinger (1957). To this author, "cognitions" are elements of knowledge, such as beliefs, attitudes, values, and feelings (about oneself, others, or the environment); dissonance may occur among any of these attitudes. His definition of dissonance is quite strong, as cognitions are said to be "dissonant with one another when they are *inconsistent* with each other or *contradict* one another," therefore, a logical view of contradiction. In Brady and in Goldie, on the contrary, "weak" contradictions may also occur, as they may involve (again, for instance, in the case of fear) the estimation of a "degree of dangerousness" that influences a "degree of inconsistency" or incongruence with other attitudes.

6.2 Some Examples

We found several typical examples of weak contradiction between cognition and negative emotions in the ISEAR Databank.[3] To elaborate on the examples we considered in the previous sections, we will cite and reason, here, on two cases of fear.

Example 1:

I was coming home from a relative's place and it was about 9.30/10 P.M. I felt slightly apprehensive when I got off the bus and started walking towards my place. I was confident that nothing would happen to me, yet there was this slight feeling of fear.

Example 2:

If I walk alone in the night, It might happen that I will be attacked. I feel fearful, even though I don't believe that it is likely that I will be attacked

Example 1 describes a real event occurred: a "slight" fear was felt, in contradiction to a belief that the situation was *not* dangerous ("nothing would happen to me"). In Example 2, the situation is hypothetical (or was possibly experienced in the past); this time, fear is felt in front of a *low-risk* danger (it is not likely that I will be attacked).

Festinger's theory, subsequently elaborated by Harmon Jones (2000), was focused on the study of the (negative) emotions that result from becoming aware of a state of cognitive dissonance and on how these negative effects can be reduced. We evoke this theory in the context of this chapter in order to highlight that influential psychological theories exist, according to which the human mind cannot be assumed to be internally consistent. And this problem should not be ignored in building computational models of emotion activation, irrespective of the formalism employed. In our view, the following *alternatives* then arise in building these models: (a) *don't make room for the possibility of conflict:* but then an important part of emotional life is eliminated; (b) *make room for the possibility of conflict:* but then the problem should be considered of how to emulate such a kind of representational state.

The second alternative might be implemented by representing the dual process with two separate models, one for intuitive and one for reflective thinking (as envisaged in Cañamero, 2005). In this case, the cognitive component should represent the ability to correct errors introduced by fast and frugal intuitive algorithms; however, it should leave space, at the same time, for occurrence of "misleading" or "recalcitrant" emotions and should deal with them. This is a quite demanding and, to our knowledge, still not tackled challenge.

---

[3]In the 1990s, a large group of psychologists collected data in a project directed by Klaus R. Scherer and Harald Wallbott (Geneva University). Student respondents, both psychologists and non-psychologists, were asked to report situations in which they had experienced all of seven major emotions (joy, fear, anger, sadness, disgust, shame, and guilt). The final data set thus contained reports on these emotions by close to 3,000 respondents in 37 countries on all five continents. ISEAR is available from the page http://www.affective-sciences.org/researchmaterial (last accessed on 14 November 2010).

Emotion models that deal with uncertainty, however, do enable representing cognitive–emotional contradictions, although in a quite simplified way. Let us consider again Example 2: *If I walk alone in the night, it might happen that I will be attacked. I feel fearful, even though I don't believe that it is likely that I will be attacked.* If we denote with A the agent on which we are reasoning, with a the action A is performing, with Ev an event, with t_h, t_k two time instants, and with $\Rightarrow?$ an "uncertain implication," the activation of fear may be represented as follows:

$$\{\text{Bel A}[\text{Does}\,(\text{A}, a, t_h) \Rightarrow?\ \Diamond\text{Happens}\,(\text{Ev}, \text{A}, t_k)]\ (\text{with } t_k > t_h) \text{ and }\\ \text{Unpleasant}\,(\text{Ev})\} \Rightarrow?\ \text{Fearful}\,(\text{A1}, t_h)$$

If A believes that performing the action a at time t_h might produce the event Ev at a subsequent time t_k, and Ev is "unpleasant," then A may feel fear.

Here, the likelihood of the unpleasant event Ev we are considering ("to be attacked") is, in fact, related to the conditional likelihood that this event will occur in a given situation ("walking alone in the night"); the same is true for the likelihood that dangerous consequences will occur, because of being attacked. If the first likelihood (of being attacked by walking alone in the night) is low, the likelihood of dangerous consequences will be low as well. In principle then, I should not feel fear; however, the intensity of this feeling depends on how much dangerous are the consequences of being attacked and how much importance I give to my self-preservation. If I know that I am risking my life or I tend to adopt a wise attitude (because of my personality or my past experiences), even a low likelihood will make me feel fearful. Hence, my contradictory state. Contradiction may be further increased by at least two factors:

a. *Overestimating conditional likelihoods*: As Kahneman et al. (1982) pointed out, in subjectively estimating probabilities, humans apply some "quick-and-dirty" heuristics which are usually very effective but may lead them to some bias in particular situations. *How dark is the place in which I'm walking*, *whether there is some unpleasant noise*, or *whether I'm nervous*, are examples of factors that may bias this estimate.
b. *Overweighting of the losses due to the negative event:* The "risk aversion" effect may bring the subject to overweight the losses and to be distressed by this perspective.

This kind of bias can be represented by trying to emulate the way humans make inferences about unknown aspects of the environment. Uncertainty in computational models of emotion activation will be represented, in this case, with some algorithm capable of making "near-optimal inferences" with limited knowledge and in a fast way, like those proposed by Gigerenzer and Goldstein (1996). Alternatively, as in the Bayesian network representation we considered in this chapter, they can be built on probability theory. In this case, models will have to include consideration of context variables that might bring the subject to over- or underestimate conditional likelihoods and losses or gains due (respectively) to negative or positive events. This is another reason for including context in emotion activation models, a direction we followed in our dynamic, uncertain models.

7 Acknowledgments

This work was financed, in part, by HUMAINE, the European Human–Machine Interaction Network on Emotions (EC Contract 507422).

References

Arnold MB (1960) Emotion and personality. Columbia University Press, New York, NY

Ball G, Breese J (2000) Emotion and personality in a conversational agent. In: Prevost S, Cassell, J , Sullivan, J, Churchill, E (eds) Embodied conversational agents. The MIT Press, Cambridge, MA, pp 89–219

Brady M (2007) Recalcitrant emotions and visual illusions. Am Philos Q 44(3):273–284

Cañamero L (Coordinator) (2005) Emotion in cognition and action. Deliverable D7d of the HUMAINE NoE

Carofiglio V, de Rosis F (2005) In favour of cognitive models of emotions. In: Proceedings of the Joint Symposium on Virtual Social Agents, 12–15 April 2005, University of Hertfordshire, Hatfield, UK, AISB'05: Social Intelligence and Interaction in Animals, Robots, and Agents, published by SSAISB (The Society for the Study of Artificial Intelligence and the Simulation of Behaviour) Press, Hove, East Sussex, UK, pp 171–176

Carofiglio V, de Rosis F, Grassano R (2008) Dynamic models of mixed emotion activation. In: Canamero L, Aylett R (eds) Animating expressive characters for social interactions, vol 74, iss 8. John Benjamins, Amsterdam, pp 123–141

Castelfranchi C (2000) Affective appraisal vs cognitive evaluation in social emotions and interactions. In: Paiva A (ed) Affective interactions. Towards a new generation of computer interfaces, LNAI 1814. Springer, Heidelberg, pp 76–106

Castelfranchi C, Lorini E (2003) Cognitive anatomy and functions of expectations. In: Proceedings of IJCAI'03 workshop on cognitive modeling of agents and multi-agent interactions, Acapulco, Mexico, 9–11 Aug 2003

Castelfranchi C, Poggi I (1990) Blushing as a discourse: was Darwin wrong? In: Crozier R (ed) Shyness and embarrassment: perspective from social psychology. Cambridge University Press, New York, NY

Conati C (2002) Probabilistic assessment of user's emotions in educational games. Appl Artif Intell [Special Issue on 'Merging Cognition and Affect in HCI'] 16:555–575

Conati C, MacLaren H (2005) Data-driven refinement of a probabilistic model of user affect. In: Ardissono L, Brna P, Mitrovic A (eds) User modeling 2005, LNAI 3538. Springer, Berlin, pp 40–49

Coupé VMH, Van der Gaag LC (2002) Properties of sensitivity analysis of Bayesian belief networks. Ann Math Artif Intell 36(4):323–356

Damasio A (1995) Descartes' error. Emotion, reason, and the human brain. Penguin Books, East Rutherford, NJ

de Rosis F, Pelachaud C, Poggi I, De Carolis N, Carofiglio V (2003) From Greta's mind to her face: modelling the dynamics of affective states in a conversational embodied agent. Int J Hum Comput Stud 59(1/2):81–118

Elliott C, Siegle G (1993) Variables influencing the intensity of simulated affective states. In: Reasoning about mental states – formal theories and applications. Papers from the 1993 AAAI spring symposium [Technical report SS-93-05]. AAAI Press, Menlo Park, CA, pp 58–67

Feldman-Barrett L, Niedenthal PM, Winkielman P (eds) (2005) Emotion and consciousness. Guilford Publication, Boston, MA

Festinger L (1957) A theory of cognitive dissonance. Stanford University Press, Stanford, CA

Frijda NH, Mesquita B (2000) The influence of emotions on beliefs. In: Frijda NH, Manstead ASR, Bem S (eds) Emotions and beliefs: how feelings influence thoughts. Cambridge University Press, Cambridge, MA

Gigerenzer G, Goldstein DG (1996) Reasoning the fast and frugal way: models of bounded rationality. Psychol Rev 103(4):650–669
Goldie P (2004) On personality. Routledge, London
Goldie P (2009) Misleading emotions. In: Brun DG, Doguolu U, Kuenzle D (eds) Epistemology and emotions. Ashgate Publishing, Aldershot
Gratch J, Marsella S (2001) Tears and fears: modelling emotions and emotional behaviors in synthetic agents. In: Proceedings of the 5th international conference on autonomous agents. ACM Press, New York, NY, pp 278–285
Harmon Jones E (2000) A cognitive dissonance theory perspective on the role of emotion in the maintenance and change of beliefs and attitudes. In: Frijda NH, Manstead ASR, Bem S (eds) Emotions and beliefs, how feelings influence thoughts. Cambridge University Press, Cambridge, MA, pp 185, 211
Helm B (2001) Emotional reason: deliberation, motivation, and the nature of value. Cambridge University Press, Cambridge, MA
James W (1884) What is an emotion? Mind 9:188–205
Jensen FV (2001) Bayesian networks and decision graph. Springer, New York, NY
Kahneman D, Slovic P, Tversky A (1982) Judgment under uncertainty, heuristics and biases. Cambridge University Press, Cambridge, MA
Izard CE (1993) Four systems for emotion activation: cognitive and non cognitive processes. Psychol Rev 100(1):68–90
Lazarus RS (1991) Emotion and adaptation. Oxford University Press, New York, NY
Mc Crae R, John OP (1992) An introduction to the five-factor model and its applications. J Pers 60:175–215
Miceli M, Castelfranchi C (1997) Basic principles of psychic suffering: a preliminary account. Theory Psychol 7:769–798
Miceli M, Castelfranchi C (2007) The envious mind. Cogn Emot 21:449–479
Miceli M, de Rosis F, Poggi I(2006) Emotional and non emotional persuasion. Appl Artif Intell Int J 20(10):849–879
Nicholson AE, Brady JM (1994) Dynamic belief networks for discrete monitoring. IEEE Trans Syst Men Cybern 24(11):1593–1610
Oatley K, Johnson-Laird PN (1987) Towards a cognitive theory of emotions. Cogn Emot 13:29–50
Ortony A, Clore GL, Collins A (1988) The cognitive structure of emotions. Cambridge University Press, Cambridge, MA
Picard RW (1997) Affective computing. The MIT Press, Cambridge, MA
Plutchik R (1980) A general psycho-evolutionary theory of emotion. In: Plutchik R, Kellerman H (eds) Emotion: theory, research, and experiences, vol 1: Theories of emotion. Academic, New York, NY, pp 3–33
Poggi I (2005) The goals of persuasion. Pragmat Cogn 13(2):297–336
Poggi I, Pelachaud C (1998) Performative faces. Speech Commun 26:5–21
Prendinger H, Descamps S, Ishizuka M (2002) Scripting affective communication with life-like characters. Appl Artif Intell [Special Issue on 'Merging Cognition and Affect in HCI'] 16(7–8):519–553
Prendinger H, Ishizuka M (2005) The empathic companion. A character-based interface that addresses users' affective states. AAI 19:267–285
Reilly N (1996) Believable social and emotional agents. Ph.D. dissertation, School of Computer Science, Carnegie Mellon University, Pittsburgh, PA
Sabini J, Silver M (1982) Moralities of everyday life. Oxford University Press, New York, NY
Scherer KR, Wranik T, Sangsue J, Tran V, Scherer U (2004) Emotions in everyday life: probability of occurrence, risk factors, appraisal and reaction pattern. Soc Sci Inf 43(4):499–570
Sloman SA (1996) The empirical case for two systems of reasoning. Psychol Bull 119(1):3–22
Turrini P, Meyer J, Castelfranchi C (2007) Controlling emotions by changing friends. In: Dastani MM, Bordini R (eds) Proceedings of the 5th European workshop on multi-agent systems (EUMAS-2007), Hammamet, Tunisia, pp 482–496
van Reekum CM, Scherer KR (1997) Levels of processing in emotion-antecedent appraisal. In: Matthews G (ed) Cognitive science perspectives on personality and emotion. Elsevier, Amsterdam, pp 259–300

Anticipation and Emotion

Cristiano Castelfranchi and Maria Miceli

Abstract This work tries to provide a systematic outline of the manifold relations between emotion and anticipatory activity. We first address the route from emotion to anticipation, which implies considering the anticipatory function of emotion in a twofold sense. On the one hand, emotions may mediate the relationship between a stimulus and a response, by triggering anticipatory behaviors which are not based on cognitive representations of future states or events (preparatory emotions). On the other hand, emotions may accomplish the function of signaling underlying mental states (premonitory emotions), that is, the fact of experiencing a certain emotion may induce some anticipatory belief. Then we address the route from anticipation to emotion, by considering those emotional states which are elicited by anticipatory representations (expectation-based emotions). Whereas in premonitory emotions, the latter induce some expectation, in expectation-based emotions, the causal relationship is reversed; the expectation of a certain event elicits an emotional response. Here we are in the domain of cognitive appraisal proper, with the sole restriction that the appraisal regards future events. Moreover, the route from anticipation to emotion also accounts for those emotions – such as disappointment and relief – which are elicited by the invalidation of expectations (invalidation-based emotions). Finally, we discuss a third kind of interaction between emotion and anticipatory activity, that is, the anticipation of future emotions. Emotions are here the object of anticipatory representations, rather a response to them. Two kinds of expected emotions are identified, "cold" expectations versus "hot" expectations of emotions (which include some anticipated feeling), and their role in decision making is discussed.

C. Castelfranchi (✉)
Institute of Cognitive Sciences and Technologies, National Research Council, Rome 00185, Italy; Department of Communication Sciences, University of Siena, Siena, Italy
e-mail: cristiano.castelfranchi@istc.cnr.it

P. Petta et al. (eds.), *Emotion-Oriented Systems*, Cognitive Technologies,
DOI 10.1007/978-3-642-15184-2_25,

1 Introduction

The relationship between emotion and anticipation is manifold and very strict. To start with, one of the functions ascribed to emotions is precisely that of anticipating events, especially when they are relevant to central concerns and the well-being of the organism. In fact, the negative bias against emotions, and more generally against "irrational" responses – traditionally viewed as contrary to utility and disruptive for both rational thinking and effective behavior[1] – has been practically reversed in the last decades. The functional value of emotions has been widely acknowledged, and their anticipatory, interpretive, and evaluative features have been especially emphasized (e.g., Frijda, 1986; Oatley and Jenkins, 1996; Parrott and Schulkin, 1993; Smith and Lazarus, 1990).

Turning to anticipatory representations of the world, it is worth emphasizing that their use is epistemic, that is, for predicting the future. Anticipatory representations can tell us not only how the world will be but also how the world should be, or better how the organism would like it to be. They can have a motivational, axiological, or deontic nature. In fact, these representations can be used as goals and drive the behavior of the organism. Whereas a merely *adaptive* organism may tend to adjust its epistemic representations (knowledge, beliefs) to the world, to make them as accurate as possible, a goal-directed system tries to adjust the external world to its endogenous representations (which also paves the way for both hallucinations and utopias) (Castelfranchi, 2005). A goal-directed system tries to change the world (through its actions) and make it as close as possible to its mental picture. Thus, any goal-directed system is necessarily anticipatory, since it is driven by the representation of the goal state and activated by the latter's mismatch with the current state of the world (e.g., Miller et al., 1960; Rosenblueth et al., 1968).

Emotions show a relationship not only with epistemic anticipatory representations but also with goals. Emotions *monitor* and *signal* goal pursuit, achievement, and failure (e.g., Frijda, 1986; Gordon, 1987). Moreover, they *generate* goals (e.g., Frijda, 1987, 1993); once an emotion has signaled the achievement or the failure of a certain goal, generally an action tendency or an actual behavioral response is elicited, which implies the activation of some goal (of either the approach or the avoidance type). For instance, the emotion of fear not only signals the presence of a possible danger but also generates the goal to avoid it. Finally emotions may *translate* into goals (Miceli and Castelfranchi, 2002a; Miceli et al., 2006), that is, agents may perform (or avoid performing) an action *in order* (not) *to feel a certain emotion*: I may give you a present to feel the joy of making you happy or do my own duty not to feel guilty. In behavioristic terms, emotions are often (positive or

[1] Just to provide some example, consider Norman's (1981) "action slips," that is, the mistakes which typically occur in task execution when intentional behavior is "disturbed" by unconscious processes, which divert one's attention from the task itself. In the same vein, consider Elster's (1985, p. 379) remark about emotions, which, when involved in action, would "tend to overwhelm or subvert rational processes."

negative) reinforcements, favoring either the reoccurrence or the extinction of certain behaviors. Hence the important role emotions play in learning: a given action can be performed (or avoided) not only on the grounds of the agent's expectations about its outcome and evaluations of its costs but also in order to feel (or not to feel) the associated expected emotions.

In this work, we will try to provide a systematic, albeit schematic, outline of the reciprocal relations between emotion and anticipation. We will first address the route from emotion to anticipation, then the reverse one, from anticipation to emotion. Finally we will consider a third class of interaction between emotion and anticipatory activity, consisting of the anticipation of future emotions.

2 From Emotion to Anticipation

From the perspective of biological evolution (e.g., Tooby and Cosmides, 1990), emotions are psychological mechanisms that evolved to solve adaptive problems – such as escaping dangers and predators, finding food, shelter and protection, finding mates, being accepted and appreciated by one's conspecifics – and thus surviving and delivering one's genes to one's offspring. Therefore, emotions *generate* goals and behaviors our ancestors had to pursue in order to answer recurrent ecological demands. Of course, the instrumental relation between such emotion-generated goals and their functions was far from being represented in our forefathers' minds.

Addressing the route from emotion to anticipation implies considering the anticipatory function of emotion. This function may be accomplished in two basic ways. On the one hand, emotions may mediate the relationship between a stimulus and a response, by triggering anticipatory behaviors which are *not* based on cognitive representations of future events. We may call this a "preparatory" function. On the other hand, emotions may accomplish the "premonitory" function of signaling underlying mental states, that is, the fact of experiencing a certain emotion (e.g., anxiety) induces some anticipatory belief (e.g., about some impending danger). Let us now better specify either the *preparatory* or the *premonitory* roles of emotions.

2.1 Preparatory Emotions

Not every anticipatory behavior is based on explicit cognitive representations of future events, that is, on predictions. Many instances may exist of "implicit" or merely *behavioral anticipation* or *preparation.* This occurs whenever some stimulus that is precursory to a forthcoming event is associated with a certain behavior, which has been selected to react to the forthcoming event (*preparatory behavior*). For example, the jumping of a grasshopper at a rustle is not only a simple reaction to the noise itself but also (functionally) "meant" to avoid possible predators (Fig. 1).

Often, the relationship between the precursory stimulus and the behavioral anticipation is mediated by emotions. A classical example is offered by fear, whose

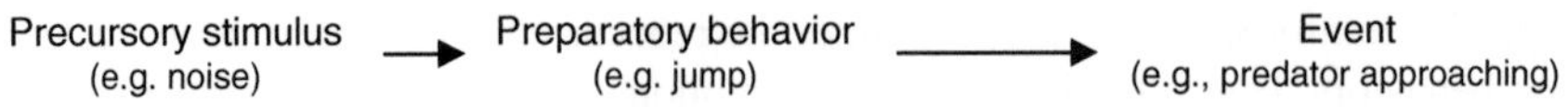

Fig. 1 Behavioral anticipation

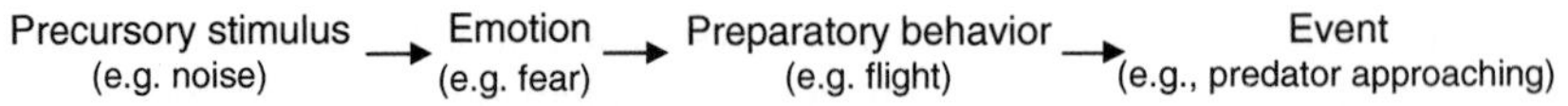

Fig. 2 Behavioral anticipation mediated by emotion

implied bodily activation is preparatory for a flight behavior. That is, the precursory stimulus elicits an internal emotional response, and the latter activates the anticipatory behavior, which is preparatory for the forthcoming event (Fig. 2).

There may be a variant of the preceding process, where no observable behavior is present. In such cases the precursory stimulus elicits an emotion which is itself preparatory for the upcoming event. For example the fear elicited by a noise may just activate a state of vigilance and alertness, without any overt behavior, and this very state is the preparatory response to the upcoming event.

The basic question about this kind of phenomena is: Why did living systems evolve from an S⇒R mechanism to the S⇒E⇒R mechanism, with an emotional (internal) mediating response? If the function of the E response is just to elicit a certain behavior, why not have the simpler direct S⇒R association? A possible answer is that the internal response E is also likely to play other roles besides that of eliciting a certain behavior. Mediating the relationships between Ss and Rs might allow that (a) various stimuli Ss elicit the same internal reaction that in turn elicits R and also, more important, (b) E may play a role in *learning*; it is a "reinforcement" since it can be pleasant or painful. Finally, E remains associated with S in memory and is automatically retrieved – in a very fast and automatic way – when S is perceived, thus representing *an implicit evaluation* of S (based on the past experience) (Miceli and Castelfranchi, 2000a).

2.2 Premonitory Emotions

Emotions may be *signals* of underlying mental states that account for and justify them. That is, emotions accomplish an informative function by providing some insight into oneself and one's relationship with the environment (e.g., Lazarus, 1991; Schwarz, 1990). In particular, as proposed by Oatley and Johnson-Laird's (1987) communicative theory of emotions, they accomplish the function to help the cognitive system manage multiple goals in an uncertain world, by communicating that some part of a plan requires the system's attention.

Indeed we may realize and evaluate what is going on in a given situation, not before but after we experience some emotion. So, I can feel anger or fear, and then realize that something has happened that makes me angry (someone has harmed me) or afraid (something is threatening me). To be sure, my interpretation may be incorrect, due to the ambiguity or the vagueness of the emotional arousal combined with the interpretive bias favored by a given context (e.g., Schachter, 1964). In any case, what we want to emphasize here is that one's emotions call for some interpretation and they demand some mini-theory about the reasons why one experiences them.

By now, it is widely acknowledged that so-called gut feelings may help a cognitive system to form rational beliefs and make rational decisions. A paradigmatic example is offered by the neuropsychological studies conducted by Damasio and his collaborators on the role of emotion in evaluation and decision making (e.g., Bechara et al., 1997). Elster (1996, pp. 1393–1394), though still maintaining that emotions may often interfere with rationality, points out:

> First, many pieces of information that we possess are not consciously acknowledged. Secondly, the cognitive basis of the emotions includes unconscious knowledge. If those premises are true, we can use our emotional reactions as cues to our unconscious assessment of a situation. Suppose you meet a person who makes you feel vaguely uncomfortable. Although you are unable to formulate a belief about the person that would justify that emotion, you can infer from the emotion that you must have some such belief. That belief, in turn, may serve as a premise for action . . ., e.g. for a decision not to have anything more to do with the person.

This view is akin to Schwarz and Clore's (1988) informative functions approach, according to which one's feelings are used as a source of information and a basis for judgment on a given target, by asking oneself "how one feels" about it (see also de Sousa, 1987).

Often, the very fact of experiencing a certain emotion or being in a certain mood (even independent of external stimuli) elicits some *anticipatory* belief about a future state or event. For instance, my experiencing anxiety makes me suppose some impending danger; conversely, my cheerful mood this morning can induce me to feel that today is going to be a nice day. These are cases in which emotional states induce or favor cognitive expectations about the future (Fig. 3).

In common usage, *expectation* is an ambiguous word. Sometimes it coincides with hope (or fear), sometimes with forecast, and sometimes it implies both. A simple forecast or *prediction* can be defined as a belief that a certain future event *p* is (more or less) probable, and it involves no necessary personal concern or goal about *p*. In contrast, by *expectation* here we mean an internally represented wish or goal about a future event together with the belief that the (un)desired outcome is possible or (more or less) probable. In other words, an expectation is a prediction the subject is personally concerned about.

Typically, people in a happy mood are likely to harbor positive expectations about desired outcomes, whereas people in a sad mood tend to have negative expectations (e.g., Johnson and Tversky, 1983; Nygren et al., 1996). However, the impact

(Precursory stimulus) ⟶ Emotion ⟶ Expectation

Fig. 3 From emotion to expectation

of emotions on expectations is not limited to the mere correspondence between the respective valences of emotions (or moods) and expectations. Emotions of the same valence may in fact differ in terms of their appraisals (e.g., Lazarus, 1991; Smith and Ellsworth, 1985), and the specific content of the latter impacts on the individual's expectations and consequent behavioral orientations. For instance, as shown by Lerner and Keltner (2000), two negative emotions, fear and anger, affect expectations (in the case at hand, judgments of risk) in opposite ways: fear favors pessimistic expectations, whereas anger favors optimistic expectations.

3 From Anticipation to Emotion

Addressing the route from anticipation to emotion implies considering those emotional states which are elicited either by anticipatory representations or by their assumed invalidation. Also here we can find two classes of emotion: *expectation-based* and *invalidation-based* emotions.

3.1 Expectation-Based Emotions

The expectation of *p* (that is, the prediction of a certain event *p* coupled with the goal that *p* or not-*p*) is likely to elicit some emotion. For instance, if I expect failure at an exam, I will feel sadness and helplessness, or apprehension and anxiety; conversely, if I expect success, I will feel hope, joy, or pride. Here we are in the domain of cognitive appraisal proper (e.g., Lazarus, 1991), with the sole restriction that the appraisal regards future events.

Whereas in premonitory emotions the latter induces some expectation, here the causal relationship between expectation and emotion is reversed. It is the expectation of a certain event (positive or negative, depending on its congruency with one's own goals) that elicits an emotional response and a consequent behavior. For instance, suppose the following scenario: A child hears in the night a series of noises and recognizes them as those made by his father coming back home, usually drunk; this perceptual stimulus, or better its recognition and evaluation, elicits the negative expectation (grounded on the child's previous experience in similar circumstances) that his father will thrash him; the expectation induces the emotion of fear, which in turn activates a preparatory behavior, such as curling up in bed, awaiting the thrashing (Fig. 4).

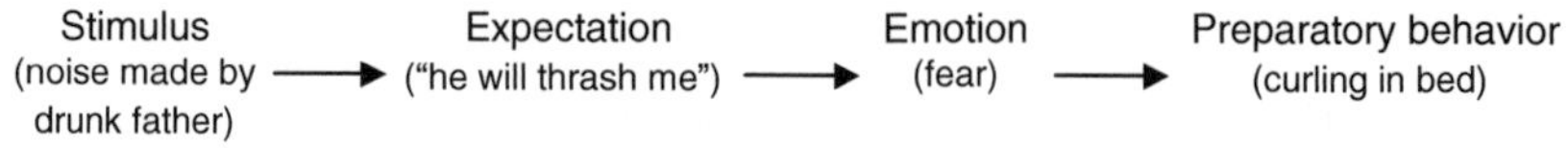

Fig. 4 From expectation to emotion

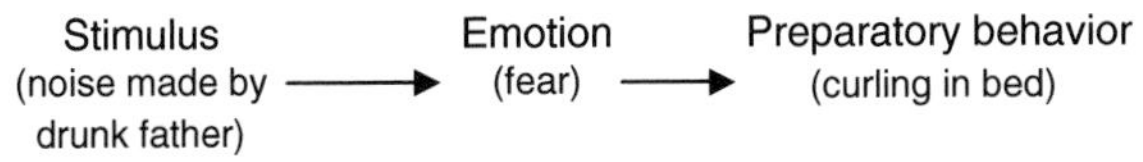

Fig. 5 From an expectation-based emotion back to a preparatory one

However, some interesting cases may occur in which an originally expectation-based emotion turns into a preparatory or even a premonitory one. Let us sketch how.

Through habituation the process above may undergo a "short circuit," where there is no longer any explicit expectation, that is, the emotion comes to be directly triggered by the stimulus (Fig. 5).

In this way, an expectation-based emotion can turn into a preparatory one. Moreover, the process may undergo some generalization. For instance, similar noises in similar contexts may directly trigger, independent of any explicit expectation, the same emotion (fear) which was originally based on an explicit expectation and now has turned into a preparatory one.

Interestingly enough, such a preparatory emotion, if consciously felt and reasoned upon ("Why am I feeling this terror? Why am I feeling this urge to hide?"), may favor in some cases the retrieval of the forgotten expectation. In other cases, it can generate a generic expectation of threat: "Since I am scared, there must be some impending danger somewhere." In this way, the preparatory emotion turns into a premonitory one, where, as already pointed out, a stimulus elicits an emotion, the emotion induces an expectation, and the latter in turn a preparatory behavior (see Fig. 3).

3.2 Invalidation-Based Emotions

Suppose that at time t_1 I have a certain (positive or negative) expectation and that at time t_2 my expectation is invalidated. The fact of having had an expectation and its being invalidated is likely to produce some emotion (Fig. 6).

If the expectation was positive – that is, my goal was congruent with my prediction – I would experience disappointment. For instance, if I both predict and want that John comes and sees me (or I both predict and want that he does not come) and I find my expectation invalidated by actual facts, I will be disappointed. Conversely, if my expectation was a negative one – that is I wanted *p* and predicted not-*p*, or vice versa – and it is invalidated, I would experience relief.

Fig. 6 Invalidation-based emotion

3.2.1 Disappointment and Relief

A number of relevant remarks are worth making in this connection. First of all, the anticipatory belief and its invalidation play a crucial role in disappointment and relief. In fact, these emotions *cannot be elicited without anticipatory beliefs*. Mere goal fulfillment or frustration, if devoid of any specific prediction (e.g., I want John to come and see me, but I do not make any particular forecast on this matter), can of course elicit some emotion (either pleasant, such as joy, or unpleasant, such as sadness). But I cannot feel (cognitive) relief unless I predicted some goal thwarting that does not come true. In the same vein, disappointment proper can arise only if my goal was accompanied by a (more or less certain) prediction about its fulfillment, and this prediction has been invalidated. As shown by Zeelenberg et al. (2000), in fact, disappointment is experienced when a chosen option turns out to be worse than expected.

Given the cognitive ingredients we postulate in these "invalidation-based" emotions, we assume that the intensity of the emotion is a function of its components. In particular:

- *the more (subjectively) certain the prediction, the more intense the disappointment or the relief* and
- *the more (subjectively) important (valuable) the goal, the more intense the disappointment or the relief.*

However, the impact of expectation invalidation on the emotional system is not limited to such feelings as disappointment and relief. On the negative side, at least a couple of other feelings are worth mentioning: discouragement and sense of injustice.

3.2.2 Discouragement

We view discouragement as a special kind of disappointment. As already pointed out, the latter implies a process of transition or transformation of a positive expectation into a negative one. A disappointed expectation is in fact a positive expectation (with varying degrees of certainty) that *becomes* negative (with varying degrees of certainty). Also in discouragement there is a transformation of a positive expectation into a negative one. In particular, discouragement implies a transition from a situation where one has the "courage," that is, one feels to manage it, to a situation where one loses heart and feels not to manage it, that is, one comes to despair of achieving some goal after having expected a positive outcome.

However, a discouragement is something more specific than a simple disappointment. Discouragement implies disappointment, whereas there may be disappointment without discouragement. Suppose yesterday I expected to have a sunny weather today; if today my expectation is invalidated, I get disappointed, but not discouraged. In this case, there is nothing to be discouraged about. In fact, one may be disappointed about mere goals, whereas, for being discouraged, there should necessarily be some intention (that is, some goal *chosen for pursuit*) implied (Miceli and Castelfranchi, 2000b). One is discouraged from pursuing some intention (because one's positive expectations have been disappointed). Going back to the previous example, discouragement might come into play if the expected sunny weather was considered an enabling condition for pursuing the intention of, say, taking a trip. In such a case, I would be discouraged with regard to that intention, while I am just disappointed relatively to the goal of having a sunny weather.

Moreover, discouragement shows another important difference from mere disappointment. In discouragement the focus of attention is on one's *lack of* (either internal or external) *power* to achieve a certain intention *p*, whereas disappointment is, so to say, *unmarked with regard to power* (Miceli and Castelfranchi, 2000b). Though both imply a transition from a positive expectation to a negative one, in the case of discouragement the positive expectation consisted of a belief of the type "I can manage it," while in the case of mere disappointment, it could just be "*p* will happen." This is quite in line with Weiner et al.'s (1979, p. 1216) view of disappointment as "independent of attributions but dependent on outcomes" (see also Zeelenberg et al., 2000).

3.2.3 Sense of Injustice

Sense of injustice is also a likely response to invalidated positive expectations. The stronger the positive expectation (that is, the more certain its implied prediction and the more important its implied goal), the more its invalidation subjectively looks like an ill-treatment, as if one were suffering something *unfair*. In fact, anger is a common reaction to a violated positive expectation (Averill, 1982; Burgoon, 1993; Levitt, 1991; Shaver et al., 1987), as well as to perceived unfairness (Fehr and Baldwin, 1996; Fitness and Fletcher, 1993; Shaver et al., 1987). The assumed violation is accompanied by a sense of rebellion and refusal of facts (actually, they "shouldn't have gone" as they did). What I expected resembles what I was entitled to obtain. I feel I did not *deserve* what has happened (Miceli and Castelfranchi, 2002b). This feeling of injustice is somewhat metaphorical in that no explicit subjective equation of "expected" with "deserved" is necessarily implied. There is just a sort of implicit and analogical overlap of the two concepts. The reason for this implicit overlap lies in a special normative component typical of positive expectations which is absent from the other kinds of anticipatory representations of the future. Positive expectations in fact do not simply consist in "predictions plus goals"; they also imply a normative component, which results from the translation of the epistemic normativity typical of predictions into a deontic normativity: What, in probabilistic terms, "should" happen, and I want to happen, turns into what I feel entitled to obtain.

But why should the epistemic "norm" be made equal to the deontic one? Because *a positive expectation favors an "as if" state of mind, according to which the desired state is viewed as (almost) realized*, and the individual feels already allowed to enjoy its satisfaction. Therefore the realization of the goal is represented as something to be *maintained* rather than acquired. Because a maintenance goal (as opposed to an acquisition one) is likely to be viewed as grounded on some supposed right (a sort of usucaption), people feel entitled to obtain what they expect (Miceli and Castelfranchi, 2002b). In other words, the relationship between maintenance goals and positive expectations can account for the ease of translation of the epistemic norm into the deontic one.

Another, more general, reason for such a translation lies in the common tendency to turn mere implications into equivalences. Because perceived rights create positive expectations, we are also likely to surreptitiously assume that positive expectations create some right! As often happens in everyday reasoning, conditionals "invite" the biconditional interpretation (e.g., Geiss and Zwicky, 1971; Oaksford and Stenning, 1992; Wason and Johnson-Laird, 1972), and simple implications ("if *p* then *q*," that is, "if there is a right, there is a positive expectation") are turned into reciprocal ones ("if *p* then *q*" *and* "if *q* then *p*"), i.e., equivalences. As a consequence, "if there is a positive expectation, there is a right."

3.2.4 Invalidated Negative Expectations: Surprise and Relief

The sense of injustice that is typical of disconfirmed positive expectations does not seem to be experienced when *negative* expectations are disconfirmed. The reason for this difference lies in our view in the absence of a normative component in negative expectations. Actually, when I want *p* but I predict not-*p* (or vice versa), I do not set any deontic norm that *p* or not-*p* ought to happen. I just believe, on the grounds of my experience or previous knowledge, that not-*p* is likely to happen, whereas I would prefer the opposite. When my negative expectation is disappointed, of course I will be *surprised*. But my surprise will take on a positive color, because my goal *p* has been fulfilled. I will neither protest nor look for somebody's responsibility, nor feel I have been treated unfairly. Rather, I will feel *relieved*, because, contrary to my prediction, my desire is fulfilled. Actually, an unexpected happy ending typically elicits such feelings as surprise and relief. The latter will be all the greater the more important is the goal fulfilled, and the more unexpected its fulfillment. Relief in fact implies a more or less explicit comparison between the anticipated distress and the actual positive situation.

Thus, a *normative component is implied only in positive expectations*. This amounts to saying that the normative component results from the *joint* force of predictions and goals. If predictions and goals are congruent with each other, then *p* "ought" to occur. If they do not converge (I predict not-*p* and I want *p*, or vice versa), no normative component will be implied. A negative expectation, when invalidated, is just disappointed, whereas a positive expectation, when invalidated, is "violated."

3.2.5 Expectation Validation and Emotions

We have pointed to the relationship between expectation invalidation and emotion. But, what about expectation *validation*? Does any specific emotion depend on the validation of one's own expectations? We do not suppose any remarkable *qualitative* difference in emotion elicitation between a case in which a mere goal (without prediction) is fulfilled or thwarted, and a case in which an expectation (either positive or negative) is validated. To be more precise, we assume that the possible difference lies in the *intensity* of the emotions experienced rather than in their *quality*. As a general rule, we suggest that, if compared with the emotions elicited by mere goal (without prediction) fulfillment or thwarting, *those emotions which are elicited by validated* (either positive or negative) *expectations should be lower in intensity*. That is, the pre-existing prediction plays the role of "watering down" the (positive or negative) emotion associated with the destiny of the goal. In fact, expected outcomes have lower emotional impact compared to unexpected ones. As expected, negative outcomes are less painful than unexpected ones, so expected positive outcomes are less elating than surprising ones (e.g., Mellers et al., 1997; Miceli and Castelfranchi, 2002b).

3.2.6 Prediction Invalidation and Emotions

So far, we have considered the emotional responses associated with the invalidation of *expectations* proper, that is, predictions *plus* goals. However, not only expectation proper but also mere *prediction* invalidation (that is a disconfirmed forecast that *p* devoid of any goal that *p* or not-*p*) may elicit some emotion. This is, again, the case of *surprise*.

As just remarked, mere predictions do not imply any personal concern about *p*, in the sense that *p*'s occurrence does not affect any of the person's goals. For instance, my prediction that next Wednesday John will visit Mary (because this happens each Wednesday) may have nothing to do with my goals: I have no interest in the fact by itself. In this sense, a prediction is a "cold," or better neutral, belief that "probable *p*." However, if this neutral *p* does not occur, we are likely to experience a surprise which contains a certain degree of distress as if we *wanted p* to become true. The more certain the prediction, that is, the more *p*'s assumed probability is close to 100%, the more the surprise turns into a bewilderment that is tinged with a negative connotation. But, if we do not have the goal that *p* by itself, what is the "goal" implied in a prediction?

People have a need for prediction, that is, they need to know what causes will come into play to produce what effects (whether beneficial or harmful). The need for prediction implies both a need to anticipate future events and the consequent need to find such anticipations validated by facts. This is Bandura's (1982) *predictability*, i.e., the cognitive component of *self-efficacy* (as distinct from the other component, *controllability*, i.e., the need to exert power over events). However, we assume that the need for prediction is *not* a goal proper, that is, it is *not* a regulatory state represented (consciously or unconsciously) in the person's mind, but a *meta-goal*, that is, a *regulatory principle* concerning one's mental functioning (Miceli and Castelfranchi, 1997). Consider belief consistency. In a sense, we "want" to maintain consistent beliefs. In fact, if a contradiction is detected, we try to eliminate it. However, the mind has this "goal" as a function. It is not necessary to express these finalistic effects as represented goals on the basis of which the mind reasons and plans. It is sufficient to conceive these principles as *procedures*, which are implemented when a contradiction is detected. If they are unsuccessful, a form of cognitive distress is likely to be experienced.[2] In the same vein, the mind's architecture includes the meta-goal to make predictions and to find those predictions validated by the evidence.[3] Finally, the meta-goal to find one's predictions validated implies the further meta-goal that *p* happens (since according to one's beliefs, it should happen). This can account for the surprise experienced and its likely negative connotation, which is stronger the more certain the prediction, and comes close to a sense of bewilderment, because the world is less predictable than expected. This view can also account for the tendency to behave in accordance with one's predictions in those cases when one's behavior can affect the likelihood of the predicted event (see Sherman, 1980).

4 Expected Emotions

A third general case of interaction between expectations and emotions is offered by explicit representations of *future states which coincide with emotions*. In other words, emotions are here the *object* of anticipatory representations, rather a reaction to them: "If I do *a*, I will feel guilty" (or happy, ashamed, relieved, and so on). Two kinds of expected emotions can be identified: "cold" expectations and "hot" expectations of emotions or, better, *expected and non-pre-felt emotions* versus *expected and pre-felt emotions*. The latter include some anticipated feeling. In both cases, expected emotions may play a remarkable role in the decision-making process: Expecting possible emotions as a consequence of one's candidate decisions affect the latter, changing one's preferences about the given options.

Although this aspect has been already acknowledged by some authors (e.g., Frijda, 1986), contemporary research has mainly focused on the motivational, decisional, and behavioral effects of past or current emotional experiences (e.g., Frijda, 1986; Lazarus, 1991; Schwarz and Bohner, 1996). Only more recently, the role played by the anticipation of future emotions in decision making and behavior has started to be systematically addressed. In particular, decision theorists have started to modify the traditional expected-utility theory so as to account for the role played in decision by anticipated emotions such as anticipated pleasure or pain, disappointment or regret (e.g., Bell, 1985; Loomes, 1987; Mellers and McGraw, 2001). However, as already remarked by some authors (e.g., March, 1978; Schwarz, 2000),

[2]Quite in line with the basic assumptions of dissonance theory (e.g., Festinger, 1957), inconsistency in fact produces an unpleasant psychological state (Carlsmith and Aronson, 1963; Cooper and Fazio, 1984; Fazio and Cooper, 1983), and the consequent attempt to eliminate it by restoring consistency.

[3]Of course, this does not rule out the possibility of humans also having some internal goals concerning their predictive activity (as well as the consistency of their beliefs), just as they can translate any biological or social function into an internal goal of their own.

anticipated regret and disappointment are not the sole feelings that may affect decisions. More general models have recently been proposed (Parker et al., 1995; Perugini and Bagozzi, 2001; Richard et al., 1995) which build upon Ajzen's (1991) theory of planned behavior and try to widen it by introducing, among the other things, a variety of anticipated emotions as determinants of purposive behavior.

4.1 Expected and Non-pre-felt Emotions

By "expected and non-pre-felt emotions" we mean those emotions the agent predicts to feel as a consequence of a candidate decision, but the agent is not actually feeling here and now. The main point to be remarked is that a "not-yet-felt" but *expected* emotion can enter the overall evaluation of which goals are worth pursuing, adding a new way of linking emotions to decision making. Thus, expecting to feel an emotion is sufficient for changing the decision process or its results, although the agent does not have to feel that emotion either at the time of the expectation or later. That is, the prediction may be wrong, and the agent may happen to experience something different (see, for instance, Kahneman and Snell, 1990). However, what matters for decision making is the expected emotion.

Expected emotions belong to the set of tools an agent can use for discriminating among different choices. Thus, evaluating which choice leads to the best outcome includes the associated emotions one would like to feel, or at least those one would be more able to stand (in the case of choices implying unpleasant emotions). In other words, while anticipating some future course of action, the agent is also likely to anticipate that he or she would feel some particular emotion; this (positive or negative) expected emotion induces the goal (not) to feel it, and this goal enters the decision-making process with a given value, possibly modifying the value of the available options.

It is worth pointing out that the valence of an anticipated emotion is not necessarily consistent with the perceived value of a certain outcome. That is, the affective consequences of a goal attainment (success) are not necessarily (or not only) positive; in the same vein, the affective consequences of goal failure are not necessarily (or not only) negative. Whereas decision theorists mainly focus on such aspects as the expected "pleasure" (elation, satisfaction) associated with a success or the expected "pain" (disappointment, regret, sense of guilt) associated with a failure, we wish to stress that some expected negative emotion may be associated with a goal attainment, and vice versa, some positive emotion with a goal failure. This implies that expected emotions may affect decision making very heavily, going even *against* the choice of highly desired outcomes. For instance, while considering how to obtain an advancement at work (a goal attainment I value very positively) and anticipating some way for cheating a colleague of mine (which is instrumental to my goal), I expect to feel guilty; this expectation can induce the goal not to feel guilty, to such a point that I give up the option of cheating and possibly even the goal of obtaining the advancement, if there are no other means available (Fig. 7).

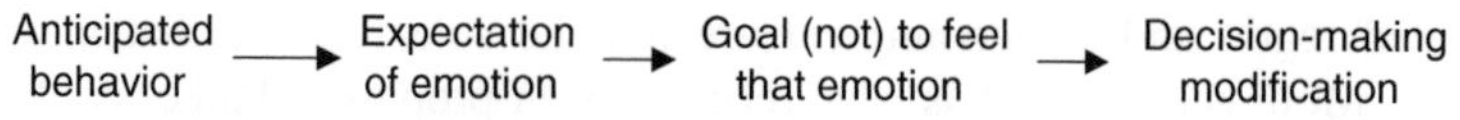

Fig. 7 Expected and non-pre-felt emotion and decision making

4.2 Expected and Pre-felt Emotions

The expected emotion can also induce an anticipated feeling. While anticipating a possible behavior and its context, I am in fact likely to "self-empathize," so to say, or "foretaste" the emotion I expect to feel, at least to some degree of intensity (if not to the same degree as when the anticipated situation is actualized). For instance, going back to the previous example, I may feel guilty at the prospect of cheating my colleague, that is, I may "hallucinatorily" experience what I (believe I) would feel if I cheat my colleague. In such cases, the impact of the expected emotion on decision making is probably stronger than that of expected but non-pre-felt emotions. In fact, here the mere cognitive expectation about some emotional reaction is reinforced by its "foretaste" (Fig. 8).

Sometimes, the expectation that one will feel a certain emotion e_2 may elicit an emotion e_1 which is different from the expected one. For instance, at time t_1 I may feel fear at the prospect of feeling guilty at time t_2. In such cases, the emotion experienced at time t_1, rather than being a foretaste of the expected emotion, is an expectation-based emotion (see above) in the strictest sense. In comparison with the expectation-based emotions we have already considered, here the difference lies in a further specification; the expectation concerns an emotion (the emotion I will or would feel at time t_2). Thus, an expectation of emotion may either favor the foretaste (pre-feeling) of the emotion expected or elicit some other emotion (a sort of "meta-emotion") *about* the expected emotion. And in any case, such feelings are likely to impact on the decision-making process.

A hybrid case, which we might call "expectation-elicited emotion," is the following: An expected event ("the boss will fire me") elicits an emotion about it (say, anger), and this emotion "tells" me what I will probably feel when the event happens, that is, the experienced emotion is the evidence on which I ground my expected emotion (Fig. 9).

This is an interesting case which testifies to the complexity of the relationships between anticipation and emotion. On the one hand, in fact, it resembles the process implied in expectation-based emotions in that here also an emotion is elicited by an

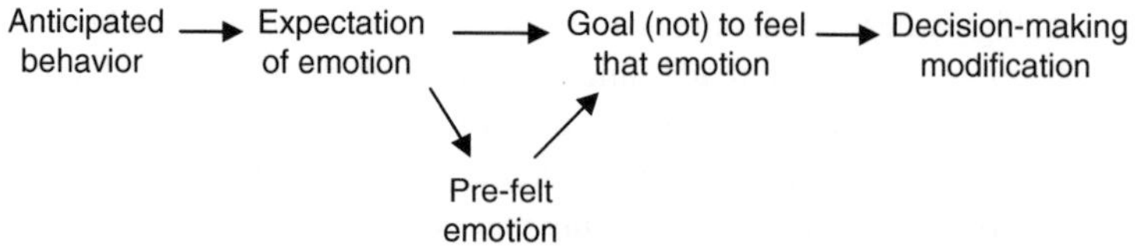

Fig. 8 Expected pre-felt emotion and decision making

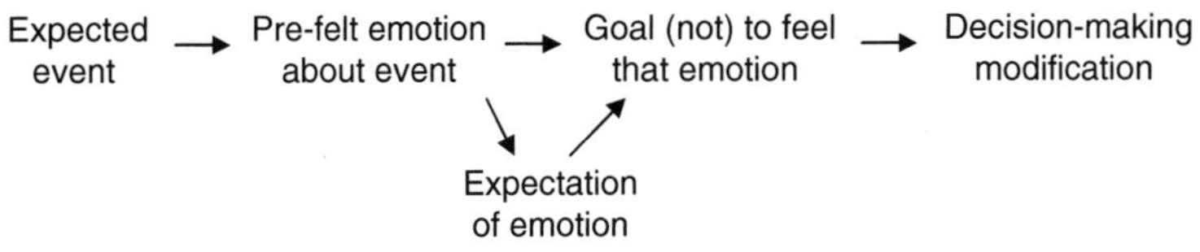

Fig. 9 Expectation-elicited emotion and decision making

expectation; whereas in the expectation-based process the emotion is experienced as regarding a *future* event (e.g., I *now* feel hope, fear, disappointment, discouragement at the *prospect* of a certain outcome), an expectation-elicited emotion implies a sort of *actualization* of the future event: while imagining my boss firing me, I feel (an amount of) the anger I will/would feel at that point in the future, when/if that event happens. On the other hand, expectation-elicited emotions are akin to expected and pre-felt emotions in that in both cases, pre-feeling the expected emotion impacts on the agent's decision making. Whereas in the expected and pre-felt emotions an expectation *about* an emotion favors its foretasting, in the expectation-elicited emotions it is the other way around: foretasting favors the expectation that I will feel that emotion (because I am pre-feeling it now), that is, foretasting plays a *premonitory* role.

5 Concluding Remarks

We have tried to analyze and systematize (a) the role of emotions in anticipation in terms of their eliciting either preparatory behaviors or anticipatory mental representations and (b) the role of cognitive anticipation in different kinds of emotions: some of them are feelings associated with expectations (like hope or fear); some are consequences of the invalidation of expectations (like surprise, disappointment, relief, discouragement, and sense of injustice); some are the anticipated representation (and possibly also feeling) of future emotions.

It is worth specifying here that the relationship between emotion and anticipation is neither of overlap nor of inclusion. We view emotion and anticipation as *partially* overlapping sets of phenomena. On the one side, in fact, anticipation is not *necessarily* emotion based. Non-emotional systems might be endowed with an "anticipatory" capacity not only in behavioral but also in cognitive terms, that is, in principle they might build internal representations of future events (predictions and expectations). On the other side, emotions are neither necessarily based on anticipatory representations nor necessarily anticipated.

However, as we have tried to show, the relationships between emotion and anticipation are manifold and very strict. Emotions produce anticipation (either anticipatory behaviors or anticipatory mental representations) and anticipation produces emotions. In particular, the capacity for anticipatory representations allows to experience "new" emotions. That is, some kinds of emotions cannot be experienced by systems devoid of anticipatory representations: hope, anxiety, and disappointment belong to this class of emotions. Moreover, the capacity for anticipatory representations creates the possibility for expected emotions, as well as emotions *about* expected emotions (for example, fear of possible shame).

Acknowledgments Preparation of this work was in part supported by HUMAINE (European Project IST-507422) and by MindRACES (European Project IST-511931).

References

Ajzen I (1991) The theory of planned behavior. Organ Behav Hum Decis Process 50:179–211
Averill J (1982) Anger and aggression: an essay on emotion. Springer, New York, NY
Bandura A (1982) Self-efficacy mechanism in human agency. Am Psychol 37:122–147
Bechara A, Damasio H, Tranel D, Damasio AR (1997) Deciding advantageously before knowing the advantageous strategy. Science 275:1293–1295
Bell DE (1985) Disappointment in decision making under uncertainty. Oper Res 33:1–27
Burgoon JK (1993) Interpersonal expectations, expectancy violations, and emotional communication. J Lang Soc Psychol 12:30–48
Carlsmith JM, Aronson E (1963) Some hedonic consequences of the confirmation and disconfirmation of expectancies. J Abnorm Soc Psychol 66:151–156
Castelfranchi C (2005) Mind as an anticipatory device: for a theory of expectations. In: De Gregorio M, Di Maio V, Frucci M, Musio C (eds) Brain, vision, and artificial intelligence. Springer, Berlin, pp 258–276
Cooper J, Fazio RH (1984) A new look at dissonance theory. In: Berkowitz L (ed) Advances in experimental social psychology, vol 17. Academic, San Diego, CA, pp 229–266
De Sousa R (1987) The rationality of emotion. MIT Press, Cambridge, MA
Elster J (1985) Sadder but wiser? Rationality and the emotions. Soc Sci Inf 24:375–406
Elster J (1996) Rationality and the emotions. Econ J 106:1386–1397
Fazio RH, Cooper J (1983) Arousal in the dissonance process. In: Cacioppo JT, Petty RE (eds) Social psychophysiology: a sourcebook. Guilford, New York, NY, pp 122–152
Fehr B, Baldwin M (1996) Prototype and script analyses of lay people's knowledge of anger. In: Fletcher GJO, Fitness J (eds) Knowledge structures in close relationships: a social psychological analysis. Erlbaum, Hillsdale, NJ, pp 219–246
Festinger LA (1957) A theory of cognitive dissonance. Row & Peterson, Evanston, IL
Fitness J, Fletcher GJO (1993) Love, hate, anger, and jealousy in close relationships: a prototype and cognitive appraisal analysis. J Pers Soc Psychol 65:942–958
Frijda NH (1986) The emotions. Cambridge University Press, New York, NY
Frijda NH (1987) Emotion, cognitive structure, and action tendency. Cogn Emot 1:115–143
Frijda NH (1993) Moods, emotion episodes, and emotions. In: Lewis M, Haviland JM (eds) Handbook of emotions. Guilford Press, New York, NY, pp 381–403
Geiss MC, Zwicky AM (1971) On invited inferences. Linguist Inq 2:561–566
Gordon RM (1987) The structure of emotion. Cambridge University Press, Cambridge, MA
Johnson E, Tversky A (1983) Affect, generalization, and the perception of risk. J Pers Soc Psychol 45:20–31
Kahneman D, Snell J (1990) Predicting utility. In: Hogarth R (ed) Insights in decision making. University of Chicago Press, Chicago, IL, pp 295–310
Lazarus RS (1991) Emotion and adaptation. Oxford University Press, New York, NY
Lerner JS, Keltner D (2000) Beyond valence: toward a model of emotion-specific influences on judgment and choice. Cogn Emot 14:473–493
Levitt MJ (1991) Attachment and close relationships: a life span perspective. In: Gerwitz JL, Kurtines WF (eds) Intersections with attachment. Erlbaum, Hillsdale, NJ, pp 183–206

Loomes G (1987) Testing for regret and disappointment in choice under uncertainty. Econ J 97:118–129
March J (1978) Bounded rationality, ambiguity and the engineering of choice. Bell J Econ 9: 587–608
Mellers BA, McGraw AP (2001) Anticipated emotions as guides to choice. Curr Dir Psychol Sci 10:210–214
Mellers BA, Schwartz A, Ho K, Ritov I (1997) Decision affect theory: emotional reactions to the outcomes of risky options. Psychol Sci 8:423–429
Miceli M, Castelfranchi C (1997) Basic principles of psychic suffering: a preliminary account. Theory Psychol 7:769–798
Miceli M, Castelfranchi C (2000a) The role of evaluation in cognition and social interaction. In: Dautenhahn K (ed) Human cognition and agent technology. Benjamins, Amsterdam, pp 225–261
Miceli M, Castelfranchi C (2000b) Nature and mechanisms of loss of motivation. Rev Gen Psychol 4:238–263
Miceli M, Castelfranchi C (2002a) Emozioni. In: Castelfranchi C, Mancini F, Miceli M (eds) Fondamenti di cognitivismo clinico. Bollati Boringhieri, Torino, pp 96–129
Miceli M, Castelfranchi C (2002b) The mind and the future: the (negative) power of expectations. Theory Psychol 12:335–366
Miceli M, de Rosis F, Poggi I (2006) Emotional and non emotional persuasion. Appl Artif Intell 20:849–879
Miller GA, Galanter E, Pribram KH (1960) Plans and the structure of behavior. Holt, New York, NY
Norman DA (1981) Categorization of action slips. Psychol Rev 88:1–15
Nygren TE, Isen AM, Taylor PJ, Dulin J (1996) The influence of positive affect on the decision rule in risk situations. Organ Behav Hum Dec Process 66:59–72
Oaksford M, Stenning K (1992) Reasoning with conditionals containing negated constituents. J Exp Psychol Learn Mem Cogn 18:835–854
Oatley K, Jenkins JM (1996) Understanding emotions. Blackwell, Oxford
Oatley K, Johnson-Laird PN (1987) Towards a cognitive theory of emotions. Cogn Emot 1:29–50
Parker D, Manstead ASR, Stradling SG (1995) Extending the theory of planned behaviour: the role of personal norm. Br J Soc Psychol 34:127–137
Parrott WG, Schulkin J (1993) Neuropsychology and the cognitive nature of the emotions. Cogn Emot 7:43–59
Perugini M, Bagozzi RP (2001) The role of desires and anticipated emotions in goal-directed behaviours: broadening and deepening the theory of planned behaviour. Br J Soc Psychol 40:79–98
Richard R, van der Pligt J, de Vries N (1995) Anticipated affective reactions and prevention of AIDS. Br J Soc Psychol 34:9–21
Rosenblueth A, Wiener N, Bigelow J (1968) Behavior, purpose, and teleology. In: Buckley W (ed) Modern systems research for the behavioral scientist. Aldine, Chicago, IL, pp 221–225
Schachter S (1964) The interaction of cognitive and physiological determinants of emotional state. In: Berkowitz L (ed) Advances in experimental social psychology, vol 1. Academic, New York, NY, pp 49–80
Schwarz N (1990) Feelings as information: informational and motivational functions of affective states. In: Higgins ET, Sorrentino RM (eds) Handbook of motivation and cognition: foundations of social behavior, vol 2. Guilford Press, New York, NY, pp 527–561
Schwarz N (2000) Emotion, cognition, and decision making. Cogn Emot 14:433–440
Schwarz N, Bohner G (1996) Feelings and their motivational implications: moods and the action sequence. In: Gollwitzer PM, Bargh JA (eds) The psychology of action. Guilford Press, New York, NY, pp 119–145
Schwarz N, Clore GL (1988) How do I feel about it? Informative functions of affective states. In: Fiedler K, Forgas J (eds) Affect, cognition and social behavior. Hogrefe, Toronto, pp 44–62

Shaver P, Schwartz J, Kirson D, O'Connor C (1987) Emotion knowledge: further explorations of a prototype approach. J Pers Soc Psychol 52:1061–1086
Sherman SJ (1980) On the self-erasing nature of errors of prediction. J Pers Soc Psychol 39: 211–221
Smith CA, Ellsworth PC (1985) Patterns of cognitive appraisal in emotion. J Pers Soc Psychol 48:813–838
Smith CA, Lazarus RS (1990) Emotion and adaptation. In: Pervin L (ed) Handbook of personality: theory and research. Guilford Press, New York, NY, pp 609–637
Tooby J, Cosmides L (1990) The past explains the present: emotional adaptations and the structure of ancestral environment. Ethol Sociobiol 11:375–424
Wason PC, Johnson-Laird PN (1972) Psychology of reasoning: structure and content. Harvard University Press, Cambridge, MA
Weiner B, Russell D, Lerman D (1979) The cognition–emotion process in achievement-related contexts. J Pers Soc Psychol 37:1211–1220
Zeelenberg M, van Dijk WW, Manstead ASR, van der Pligt J (2000) On bad decisions and disconfirmed expectancies: the psychology of regret and disappointment. Cogn Emot 14:521–541

Socially Situated Affective Systems

Sabine Payr and Peter Wallis

Abstract For a socially situated agent, the world is populated by other agents with which it interacts in the true sense of the word, in that the action–reaction sequence typical of the physical world is replaced by mutual acting-with and acting-upon the other. This chapter first introduces the sociological bottom line of emotion as the glue and regulatory system of social relationships both at the individual and at the collective level. It then focuses on language as the pre-eminent medium of social interaction and discusses, by way of examples, the functioning of conversation beyond information exchange as both norm-following and norm-building human social behaviour. We conclude that, for the design of socially situated affective systems, more research on the often overlooked because seemingly trivial aspects of interaction is needed, and we present methods for the uncovering and analysis of the workings of human–machine interaction.

1 Introduction

Emotion research and modelling have been mainly inspired by psychological and brain research. Consequently, the focus has been on modelling personality and emotions as individual properties and experiences. In this section, and by contrast, the focus will be on emotions as the primary social glue. Consider the following description of primate interaction:

> To be groomed by a monkey is to experience primordial emotions: the initial frisson of uncertainty in an untested relationship, the gradual surrender to another's avid fingers flickering expertly across bare skin, the light pinching and picking and nibbling of flesh as hands of discovery move in surprise from one freckle to another newly discovered mole. The momentarily disconcerting pain of pinched skin gives way imperceptibly to a soothing sense of pleasure, creeping warmly outwards from the centre of attention. You begin to relax into the sheer intensity of the business, ceding deliciously to the ebb and flow of the

---

S. Payr (✉)
Austrian Research Institute for Artificial Intelligence, Vienna, Austria
e-mail: Sabine.Payr@ofai.at

P. Petta et al. (eds.), *Emotion-Oriented Systems*, Cognitive Technologies,
DOI 10.1007/978-3-642-15184-2_26, © Springer-Verlag Berlin Heidelberg 2011

> neural signals that spin their fleeting way from periphery to brain, pitter-pattering their light drumming on the mind's consciousness somewhere in the deep cores of being (Dunbar, 1996).

What is described here is a deep emotional experience, but the point is that being groomed is social intercourse. He goes on:

> To recognize what this simple gesture signals in the social world of monkeys and apes, you need to know the intimate details of those involved: who is friends with whom, who dominates and who is subordinate, who owes a favour in return for one granted the week before, who has remembered a past slight. The very complexity of the social whirl creates those ambiguities we are so familiar with from our own lives. From an evolutionary perspective, emotions are one means by which the individual is integrated into the social whirl of cooperative human action [ibid.].

Social grooming has gone out of fashion in the human species, maybe for scarcity of hair and, consequently, fleas. This chapter argues that emotional experiences have ever since remained the foundation of the complex 'social whirl' in societies of primates to which we humans belong. We have only changed the methods to supply them.

The chapter proceeds as follows: Sect. 2 outlines a crude picture of how sociology conceives the emotional regulation of the relationship between the individual and society, and presents some experimental ECAs that are designed with a social relationship model in mind. Section 3 then turns to the main substitute of social grooming and other primate social interaction, which is language. Case studies are used to illustrate some aspects of what is socially done with language, which allows us to shed light on problems of acceptance of dialogue systems and requirements for socially situated systems. Section 4 deals with the methods of studying social interaction in human–human and human–machine interactions.

2 Modelling Social Identities

For the agent, the social world is populated not only by physical entities but also by other agents. Social agents are physical entities (or their virtual representation), but that does not characterize them. What is specific about them is their agentness, that is, their autonomy and capacity of intentional action. The action–reaction sequence typical for the physical world here becomes real *inter*action, i.e. the mutual acting-upon and acting-with agents. Social interaction serves both the management of social relations and the socialization of new members. In the latter case, social interaction modifies the actions and reactions of the involved agents and modifies the social agents themselves. The process of change is most evident in children's development, where from very early onwards biological development is complemented and superseded by socialization (Mead, 1934). Vygotsky (1978) used the term internalization for the gradual transformation of interpersonal processes (interactions) into intrapersonal ones, through which the Other is (finally) generalized and taken into the Self (Mead), so that the agent is no more a 'naked self' but a social person embedded in her culture, norms, and roles, and continually linked to society. The transformation, however, is never complete and never stops during the lifetime of the social agent. The social world is rich and dynamic, and the agent continuously adapts to it and adapts it to herself.

The core statement of the sociology of emotions is that emotions are the driving force of social life (Turner and Stets, 2005). At the root of any form of society is the social bond between individuals inside the group. Evolutionary theory provides arguments for this literally radical view; life in a group (or pack, or herd, or tribe) must have offered survival advantages to our primate ancestors. The capacity to bond emotionally hence probably was genetically selected for. The ability to bond socially is as basic as other emotion-based capabilities, e.g. quick reaction to danger. This fundamental need to belong (cf. Fearon, 2004; Scheff, 1990; Kemper, 1990) makes people acutely sensitive to the degree to which they are being accepted by other people (Leary, 2000).

2.1 Sociological Models of Human Relationships

The human tribe, however, has never been a peer group. It consists of individuals of different age, sex and increasingly diversified talents, capabilities, functions, and status. Social bonding correspondingly is the relationship not only among equals but also among the stronger and the weaker, the adult and the child, the leader and the follower – in short, among the more powerful and the less powerful. Each relationship has both aspects: that of closeness/distance (love, friendship) and that of up/down (dominance). Pairs of related notions for these two dimensions turn up repeatedly in the sociology of emotions, for example:

- Integration and regulation (Durkheim, 1893, 1897)
- Status (accord) and power (Kemper, 1990)
- Positive and negative face (Goffman, 1959; Brown and Levinson, 1987)[1]

There are also three-dimensional models:

- Evaluation, potency, activity (EPA) (Heise, 2002)
- Pleasure, arousal, dominance (PAD) (Mehrabian, 1995)

The latter models are based on Osgood's work (Osgood et al., 1957) and are used, e.g., in questionnaires to elicit affective meanings and ratings. The concepts used are therefore more general and applied to social relationships in the proper sense. They can be seen as combinations of the two sociological dimensions of power

[1] Positive face is the desire for appreciation and approval by others (integration) and negative face is the desire not to be imposed on by others (control, power). The terminology creates the wrong impression of a single dimension of 'face'.

(which becomes potency/dominance) and integration (evaluation/pleasure) with the psychological dimension of arousal. At first sight, it seems far-fetched to put integration and pleasure in one basket, but if we remember that integration stands here for the 'warmth', closeness, familiarity of, and commitment in social relationships, the connection becomes obvious.

The focus of emotion research and modelling on supposedly 'natural' and hence universal, permanent characteristics and experiences of persons can be contrasted with those of sociologists, in which people have multiple changing identities. Sociologists see the complex of self-conceptions as a continuum from temporary roles to principle-level identities (Burke, 1991; Tsushima and Burke, 1999) that incorporate socio-cultural values and beliefs and constitute a moral identity. Identity is not innate but a response to the people around you.

For emotion sociologists in the tradition of symbolic interactionism, individuals have and develop self-conceptions that consist of enduring meanings and emotions about a person and become a significant influence on how individuals present themselves to others (personality). Individuals also reveal more context-specific and situational identities, i.e. conceptions of themselves for specific spheres of activity in complex social structures. Individuals also become cognizant of the expectations for the roles being played and the identities presented by others. Emotional dynamics typically revolves around the processes of confirmation or disconfirmation of situational identities (Turner and Stets, 2005).

Following from what has been said above, a socially situated system in the narrow sense is a system which incorporates a model of social relationships which is used for both evaluating information from and controlling behaviour in the social world. The following examples do this in part or in specific domains.

2.2 Examples of Socially Situated Systems

2.2.1 Categorization of Emotions: Max

The multimodal conversational agent Max is equipped with a system for the simulation and visualization of the emotional state over time (Becker et al., 2004; Kopp et al., 2005; Kopp, 2006). The system combines mood with short-term emotional dynamics and, in addition, models boredom. A three-dimensional space for the categorization of emotions is used, namely pleasure, arousal, and dominance (PAD, see above). However, the PAD scheme is used only for categorizing and labelling emotions, while the model of emotional dynamics, combining emotions and moods, remains two dimensional (valence and arousal). The values of the dynamic component are mapped onto the dimensions of pleasure and arousal in the categorization component. There is no value for dominance to be mapped, but 'the BDI interpreter of the cognitive architecture ... is capable of controlling the state of dominance in an adequate way' (Becker et al., 2004).

The emotion system interacts with the cognitive architecture, which supplies valence values on the basis of the user's utterance. The emotion system in turn supplies the cognitive architecture of Max with both cognitive (emotion category and intensity) and non-cognitive (mood valence, PAD values, and degree of boredom) information. Non-cognitive information is used in the generation of non-verbal behaviour (eye blink, breathing, speech pitch, and rate), whereas emotion categories are used by the dialogue manager. The categorical output of the emotion system is incessantly asserted as belief of the agent so that it influences plan selection and facial expression, but can also be verbalized.

Max has been given a social identity implicitly. 'The perception module delivers a positive impulse each time a user has been visually perceived'. This corresponds well to the friendly, submissive identity (found in many ECAs) that is the cultural script according to which receptionists, customer service agents, and the like are (normatively) expected to behave. The zero point of equilibrium towards which the emotion system gravitates can therefore be understood as the structural sentiment of this identity and the ultimate goal as one of verification of this identity.

2.2.2 Emotion Regulation: SCREAM

SCREAM (scripting emotion-based agent minds) allows to specify a character's mental make-up and endow it with emotion and personality (Prendinger and Ishizuka, 2003). The level of detail may vary from (personality) traits to full awareness of the social interaction situation. Following the OCC model, emotion types are classes of labelled eliciting conditions with varying intensity. Since a reasonably interesting agent will have a multitude of mental states, more than one emotion is typically triggered in an interaction. The emotion resolution and maintenance modules determine the most dominant emotion and handle the decay process, respectively. The process that decides whether an emotion is expressed or suppressed is called emotion regulation. Regulatory parameters can be social threats for the agent (role, distance, power) and those that refer to its capability of (self-)control (personality, interlocutor personality, linguistic style). The personality model has only the two dimensions of extroversion and agreeableness.

The virtual character 'Genie', based on this architecture and implemented in an interactive Black Jack game, is assumed as rather agreeable and extrovert, socially close to the user, and also (initially) slightly likes the user. Its goals are that the user wins (with low intensity) and that the user follows its advice (with high intensity). A test run with the user never following the agent's advice has the result that the positive emotions of the agent decay slowly and that, finally, it slightly dislikes the user and gloats over the user's losses. It seems that this result is considered 'natural' and consistent by users.

2.2.3 Trust and Responsibility: Intermediary Agent

The intermediary agent (IA) was developed as an embodied character in an educational simulation game whose aim is to raise awareness of and develop competencies in interpersonal collaboration (Martinez-Miranda et al., 2008). The IA is the interface between the user (learner) and an underlying management simulation game.

The learner, in the role of a manager, has to implement actions in the company to introduce changes but, in this scenario, can act only through the IA. The task is to collaborate with the IA in change management, while the underlying goal of the learning experience is to recognize traps and challenges in interpersonal collaboration.

The attitude of the IA towards collaboration is dynamic. The players influence the degree of 'collaborativeness' through their actions and decisions, mostly through their choice of utterances and communication style. The following main variables modulate the IA's behaviour:

- Responsibility for the final decision in launching an initiative in the underlying simulated company
- Trust change tendency: the agent's tendency to modify trust towards the player, influenced by the players' communication style and compliance with the IA's suggestions.
- Trust level: updated after every initiative, based on the IA's rating of the outcome and the assigned responsibility.

Collaborativeness of the IA is directly proportional to the current trust value and is reflected in the types of actions suggested to the players (i.e. the presumed positive/negative outcome), the tendency to comply with players' requests, and the amount of information reported back to the players.

Responsibility and trust are easily recognizable as varieties of the power/integration dimensions. In the IA, this social model is complemented by a largely independent simple personality model which currently only represents the agreeableness trait from the big-five model. This separation allows for scenarios with, e.g., a friendly but uncollaborative agent or an unfriendly but highly collaborative agent.

2.2.4 Social Attributions and Reasoning: EMA

The focus of social reasoning in EMA is on the attribution of social causality and responsibility (Gratch and Marsella, 2004; Mao and Gratch, 2005, 2006). The computational model can automatically derive judgements that underlie attributions of responsibility and blame. Two important sources of evidence contribute to the inferences of key variables. One source is the causal evidence about the actions and effects of the observed agents. The other is the observations of the actions performed by the observed agents, including both physical and communicative acts (e.g. in a conversational dialogue). The inference process acquires beliefs from communicative events (i.e. dialogue inference) and from the causal information about the observed action execution (i.e. causal inference). The social inference module takes the observed communicative events and executed actions as inputs. Causal information and social information are also important inputs. Causal information includes an action theory and a plan library. Social information specifies the power relationship of roles. The inference process first applies dialogue inference and then causal inference. Both inferences make use of the common-sense heuristics and derive beliefs about the variable values. The values are then served as inputs for the algorithm, which determines responsibility and assigns certain blame or credit to the responsible agents.

In line with the goal of the module to attribute credit and blame, the focus of social role knowledge is on power relationships. They are important to infer beliefs about (the degree of) coercion that is at stake in an action. Coercion is assessed through both social obligation and (un)willingness of the actor. Other factors contributing to the attribution of blame or credit are agency, intentions, and foreknowledge.

The scope of this social reasoning module is specifically to attribute responsibility for the outcome of actions. Given the application background in military training, the 'society' can indeed be represented by knowledge about power relationships, i.e. about who can command and who has to obey.

3 Emotion in Language

The experimental systems presented here are approaches to design ECAs that take as a starting point the key insight of the sociology of emotions, namely that the social agent (human or artificial) is intrinsically emotional and that relationships among such agents are regulated by emotion.

The most common method for humans to interact is by talking. This section looks into how social relationships are built, regulated, and ended through language, and how emotion comes into it. To be clear, 'emotional language' is not some special language (e.g. emotion terms) or some special language use (e.g. expression of emotion) but is omnipresent in each and every verbal event or product, be it a phone call or a research paper. Emotions are fundamental for and pervasive in social interaction. Our focus will be on conversation, as this is the type of interaction that ECAs, robots, and voice-based systems aim at.

Commonly, human expressivity is seen as an axis. At one end of the scale, there are the signs that we 'give off' (Goffman, 1959) unintentionally, e.g. the facial expression of an emotion, and at the other end, there is the part of language that analytic philosophers are concerned with, namely language (*langue*) as a medium for information transfer and for formulating propositions (including speech acts). Language 'in use' however is not located at one end of the scale but covers nearly all of it as a continuum from noises and fillers up to the textbooks' complete, grammatically well-formed sentences (which are the exception in conversation). So does gesture; it goes from beat gestures to fully symbolic emblematic gestures. Social interaction uses all the media – language, prosody, gesture, gaze, posture, distance – at the same time, inextricably layered and interwoven.

For researchers in natural language understanding, the dominant background assumption has been what Mel'cuk and his colleagues have called meaning-text theory (Mel'cuk, 1981). By this model of language, sounds are produced by a speaker and the conversational partner does speech to text, morphology, normalizes the syntax, composes the sentence semantics from lexical meaning, adds in some pragmatics and world knowledge, and produces the meaning of that which is said (Allen, 1987). Having done that, there is some fairly trivial processing that determines the response, and the process is reversed from (a representation of) meaning back through some text to speech software and into the conversational partner's head. In the field of natural language processing, the 'interesting' thing about language has been the mechanism for mapping from text to meaning, and back again. This 'conduit metaphor' (Reddy, 1993) has had its critics, but the notion of dialogue being about information transfer is still the dominant approach, with emotion being given the role of 'colouring' or of emerging and fading episodes.

An alternate view is that an understanding of emotion is key to 'real natural language processing' (Cowie and Schröder, 2005). Indeed those interested in conversational artefacts (ECAs, call systems, robots, embedded speech interfaces, etc.) have emphasized the emotional and social side of being a human interacting with a computer that can or might do things other than process facts. But there is still a long way to go to fully integrate dialogue into this picture of interaction. As in other domains, emotion has not been in the focus of attention in the study of human discourse until a few years ago. Methods and results need to be revisited in the light of our insights into the essential role and function of emotion.

3.1 Making Sense in Conversation

Looking at ideas from outside computational linguistics, there are some recurring themes about what people actually do with language. Some components of natural language might be amenable to a model of language use based on signs and signifieds, but language in use depends on the hearer's common-sense reasoning about the world and about the motivations and beliefs of the speaker. What makes this difficult to model on a computer is that, it turns out, 'no detail is too small to ignore, but not every detail is relevant' to the interpretation process (Hutchby and Wooffitt, 1998).

People commit to a conversation in much the same way as one commits to riding a bicycle; there is a process to go through to get started and another process to stop. In between, one is committed to an interaction. In conversation, these processes are normative and so, like in riding a bike, one risks something unpleasant if the processes are not followed. In conversation it is worth noting that the interaction is only vaguely a series of turns in language in use. Turn critical units are not marked by pauses, rather we humans barge in, interrupt each other, talk over the top, and leave gaps. Timing, like every other detail of a speaker's actions, is interpreted if it is not 'normal'. The fine-tuning of turn-taking, and therefore the perceptiveness of out-of-tune turns, has been shown to be based on rhythmic entrainment and alignment well below the verbal level (Suzuki et al., 2003) which presupposes that participants commit emotionally to the joint performance of a conversation.

Having agreed to have a conversation, the participants are willing to work hard to figure out why their conversational partner 'said that, in that way, right then'

(Hymes, cf Eggins and Slade, 1997). This happens however against a background of normative behaviour. Much of language in use is ritualized – people return greetings and goodbyes, they answer questions and apologize, all within the normal formula for such interactions. Using the terminology of Seedhouse (2004), these utterances go 'seen but unnoticed'. However, there are countless variants of these rituals, and participants use them with considerable flexibility to solve the interaction problem at hand, like: Who may start a first or any following topic? How to signal that one wants to continue or discontinue a topic? How to establish that everything has been said and done, and that no new topic will follow? Every detail of a routine may be a constitutive part of the joint achievement, that is, the conversation (Schegloff, 1986).

Participants orient their behaviour and their expectations along a norm. Whatever departs from it is noticed, and they will work hard to 'account for' the other's behaviour. As human listeners, we are willing to put considerable effort into figuring out why the speaker said, what he or she said, and how it was said. The process exploits not only the hearer's common-sense reasoning about the world but also their understanding of other peoples' motivations and beliefs. By this model then, the interesting part of language in use often involves the hearer's theory of mind (often referred to as a ToM) to interpret why the speaker said what they did. Our model of other peoples' minds is sophisticated (Dennett, 1987). According to this model, we believe that other people act intentionally and in their own interest, and we understand what they do on the grounds of our assumption about their interests and intentions (intentional stance). This is so obvious that the reader is probably wondering why it needs saying – it is just common sense to us. The point is that computers do not have common sense and need to be told how to think this way. Note that it is the hearer who uses a ToM; it is the hearer who must be able to provide an account of the speaker's actions. Note also that the hearer must be committed to the conversation for this to work, and so the notion of sequential relevance relies on conversational partners being engaged in the dialogue as discussed above.

Participants in a conversation produce utterances that go 'seen but unnoticed' or that are 'noticed and accounted for'. If there is no accounting, then, according to Seedhouse, the speaker 'risks sanction'. Consider this conversation with a conversational agent from an early Loebner Prize competition:

PROGRAM[09:03:47]	G'day mate.
JUDGE 02[09:03:55]	G;day. How's it going?
PROGRAM[09:04:11]	My mother once said "How it is going is a mystery." She was a bit nuts.
JUDGE 02[09:04:22]	Uh-huh.
PROGRAM[09:04:31]	Are you still learning to speak, my friend?
JUDGE 02[09:04:52]	NO, no. But your anecdote took me a bit by surprise.
JUDGE 02[09:04:52]	How was your morning?
PROGRAM[09:04:59]	Oh well, that's a pity!
JUDGE 02[09:05:15]	I'm sorry; what's a pity?
PROGRAM[09:05:20]	Apologies accepted.

JUDGE 02[09:05:35]	I feel compelled to insult you, and we've only been talking a minute.
PROGRAM[09:05:49]	We are bloody well talking, aren't we?
JUDGE 02[09:06:14]	Oh baby. We sure are. So here are some keywords for you: shit mother baseball
	– conversation continues –

What is happening in this conversation? Judge 2 feels 'compelled to insult' the system because Judge 2 cannot 'account for' the system's sporadic behaviour. The sanction – when the system refuses to play along even after an explicit warning – is brought on by frustration and manifests in swearing, despite the fact the judge knows others are 'looking over his or her shoulder'.

This extract also illustrates nicely that conversation is per se emotional. If it works smoothly and the participants are in tune, the emotion tends to go unnoticed. Human emotional awareness (the 'sociometer'; Leary, 2000) is more sensitive with regard to negative emotions, which are in line with a conception of social–relational emotions as 'interpersonal wake-up calls' (Miller, 2004) that alert humans to situations threatening the social bond.

3.2 Beyond Information Exchange

One strategy to avoid a conversational system getting involved in the social–emotional business of interaction has been to have it stick to 'pure' information that takes the form of simple statements. The problem with this approach is that if machines use language – even if their purpose is only to inform, book taxis, or transfer money – they operate within a context in which humans perceive social effect. The erroneous assumption at its bottom is that there is such a thing as an emotionally and socially neutral utterance, an assumption grounded in the tradition of the reduced, non-natural language of ordinary language philosophy and even much of linguistics.

The most popular way to tell a conversational system what to say is to sit down and start writing responses to things a human might say. Tools for doing this include AIML (Artificial Intelligence Mark-up Language), initially used to write dialogue instructions for the ALICE chatbot and subsequently used on a large number of Web-based ECAs. In a similar vein, VoiceXML is a way of specifying conceptual web pages that can be navigated using spoken language, usually over telephone. There is also a long tradition of academics hand-editing state transition diagrams that specify what a system should say. Alternatively, discourse-level grammars are written that specify how utterances follow each other in (usually) two-person conversations. All of these approaches rely on the introspection of the author. The underlying assumption is that if a caller wants, for example, a taxi and the information the caller gives about place and time is understood by the system, then the task of booking the taxi is simply one of filling in the slots to make a search and accomplish the booking. This assumption starts out from the needs of the system and treats the call as information input.

Table 1 The 10 most frequent DAs in the Switchboard corpus (Jurafsky et al., 1997)

SWBD-DAMSL tag	Example	Count	Percent
Statement – non-opinion	Me, I'm in the legal department	72,824	36
Acknowledge (backchannel)	Uh-huh	37,096	19
Statement – opinion	I think it's great	25,197	13
Agree/accept	That's exactly it	10,820	5
Abandoned or turn–exit	So, –	10,569	5
Appreciation	I can imagine	4,624	2
Yes–no question	Do you have any special training?	4,624	2
Non-verbal	[Laughter], [Throat clearing]	3,548	2
Yes answers	Yes	2,934	1
Conventional closing	Well, it's been nice talking to you	2,486	1

Quantitative analysis of marked-up dialogues seems to confirm this view of conversation as information exchange. The following table gives the 10 commonest classes of dialogue moves (out of 42) as found in the Switchboard corpus (Jurafsky et al., 1997) (Table 1).

It is true that the 'non-opinion statement', i.e. an utterance that gives information, is by far the most frequent category. What is ignored here are the multiple things that are done with simple statements beyond and besides giving information. The very property of being simple (as opposed to, e.g., mitigated, elaborated, embedded, etc.) is perceived as an intentional choice made by the speaker and interpreted in its relational consequences by the hearer against the background of (assumed) other possible ways of accomplishing the same.

3.3 Mixed Initiative

Mixed initiative has been in the focus of research on conversational interaction between users and intelligent virtual agents for some years now. It has been defined as 'a flexible interaction strategy, where each agent can contribute to the task what it does best' (Allen, 1999). This definition has to be kept in mind against more recent uses of the concept, where it is applied in a much broader sense to human–machine dialogues whether they deserve the name or not. Evidence for mixed initiative has been gathered from human–human dialogues in experimental settings (e.g. Fischer, 2006; Byron and Stoia, 2005; Byron and Fosler-Lussier, 2006; Ferguson et al., 1996; Chu-Carroll and Brown, 1997) and tested in several human–machine environments (Brindöpke et al., 1995). Allen (1999) has set mixed initiative against 'system control', exemplified by automated call centres that achieve not much more than form-filling dialogues (Fairclough, 2001), to the dissatisfaction or even annoyance of numerous users.

Both system control and mixed initiative thus defined remain concepts that are firmly embedded in the paradigm of dialogue as information exchange (see above). It assumes that agents in a conversation are (exclusively) interested in getting and/or giving information. A study of initiative in human–human task-oriented dialogue

(Payr, 2007) has shown that initiative is not given and taken on such rational grounds (which, by the way, would entail that conversationalists have complete knowledge of the information the other has or needs). Rather, situational roles, together with speaker preferences, influence who dominates a dialogue, irrespective of the optimal distribution of initiative. Initiative is hardly ever negotiated explicitly, but certain signals and strategies are used and understood to make clear whether initiative is taken, kept, or given away. This is different from human–machine dialogues even where an effort is made to give initiative to the user. The following excerpt is taken from the DARPA Communicator corpus (Wallis, 2008) of human–machine conversations with state-of-the-art spoken language understanding systems.

Sys: please remember to speak after the tone. if you get confused at any point you can say start over to cancel your current itinerary. are you calling from a cellular phone?
000 Usr: n?o
Sys: what are your travel plans? (1)
001 Usr: i'd like t' go from _duluth::_(0.3)minnesota(0.4)to chicago illinoi_s_
Sys: what month and day do you wish to depart toulouse?(2)
(from dialogue 60941_XX_XX_04_20010608_062510.xml)

The system utterance marked with (1) is an open question. An open question is in most cases understood as passing the initiative to the other. The callers should feel free to state his/her plans in whatever way they wish. This is the polite way of opening such service calls by human agents (e.g. 'How can I help you?'). It is polite in Brown and Levinson's (1987) sense of preserving the caller's negative face, i.e. the wish for autonomy and control, by leaving open all options. However, just imitating this move in automated call systems is already misleading, because such systems invariably take back the initiative after this opening, and they do so without signalling it (as in the line marked (2) in the extract). It also has to be noted that for all current ways of dialogue act classification, both questions would be of the same type (wh- or open), while there is a big difference in the genuinely open question in (1) and the slot-filling question in (2) where the user is prompted not only for a specific type of information but also for its sequence and form.

As a contrast, here is an extract from a related call to a human agent. It was made by one of the authors, written down from memory, and is used here in translation:

Agent: <name of service>. What can I do for you?
Caller: Hello. My customer number is 098135, and I want to buy tickets for a trip from Vienna to Innsbruck. Now what do you need to know first? (1)
Agent: So . . . I'm speaking with Mrs. Payr . . . Sabine?
Caller: Yes.
Agent: Okay. You will be going from Vienna . . .?
Caller: Westbahnhof . . . to Innsbruck.
Agent: right, to Innsbruck Main Station. What day?
Caller: Saturday, April 22. (2)
Agent: Caller: Yes.
[call continues]

The opening question passes the initiative to the caller and confirms 'caller hegemony' (Hutchby, 2001). In (1), the caller herself then uses an open question to signal both that the caller is aware of holding the initiative and that she is ready to give it up, in order to let the agent proceed in his/her usual sequence of slot filling. With this move, the agent is also given license to do a form-filling dialogue, i.e. the caller signals that she will not take the following form-filling dialogue as a face threat.

3.4 Politeness and Social Knowledge

In the late 1990s, one of the authors (P.W.) worked on a project to implement a BDI architecture in an interactive database assistant called FRANCO (Wallis et al., 2001). The agent should be able to negotiate between the understanding of the human and the ontology of the database. To model this in BDI, knowledge about how people make decisions in dialogue was needed. The method used to elicit this knowledge was CTA, cognitive task analysis, a method explicitly intended to capture the folk reasoning of a community of practice by recording expects in activity and interviewing them about their actions.

The project team set up a Wizard-of-Oz style experiment in which one of the office staff (K.T.) provided a telephone interface to the car pool-booking system. Both sides of the phone calls of staff to book one of the departmental cars were recorded and transcribed. CTA was then used to elicit world knowledge that K.T. used to provide her service. The result, in a nutshell, was: K.T. did not need to know very much about cars and driving, destinations, and routes. What she spent most effort on was 'being polite':

– call continues –
KT: right. and your extension number?
IW: 12345
KT: 12345 and where are you going to be going?
IW: ah the, it's called the UWB facility (1)
KT: UWB
IW: yeah
KT: facility
IW: which is on the raaf base(.) and also be going to store (2)
KT: okay and do you know where the keys are for the car?
– call continues –

At the point in the transcript marked with (1), the caller introduced the 'UWB facility'. The question was: Why did she repeat 'UWB' as her response? Her answer was that she did not know where the UWB facility was and she asked in this way because 'well, [any alternative] wouldn't be polite'.

In a variant of the experiment, the team had some of the younger research staff ring up K.T. with inappropriate requests, e.g. to book a bus for a trip to the local public bar. When the caller mentioned what he wanted the bus for, K.T. said 'Oh' and left a pause but did not pursue the issue. When asked about the event, she explained that it was 'not her place' to do anything about it. K.T. has knowledge of her role in the organization and in the world, of what is appropriate for her to say, and what not in a given situation. Her understanding of politeness is not based on specific utterances or, indeed, on the specific individual at the other end of the line. It is her general knowledge of institutions and her identity in them that provides a framework for deciding what is appropriate and what callers (in their roles, respectively) will probably expect and accept.

Although designers of conversational systems try to account for the institutional application background, design of and for social role taking and management remains an issue for further research. It is especially in such contexts that, for example, what is intended as a 'simple statement' by designers can easily end up being understood as dominating or rude by human users.

4 The Study of Social Interaction

If we concentrate, in this section, on qualitative methods for studying social interaction in humans and machines, this is not because we do not attribute relevance and importance to quantitative studies. Quantitative analysis has its firm place in the canon of empirical methods when it comes to testing hypotheses; categorization of dialogue acts, statistical analysis of disfluencies, measurement of utterance length, and many other methods in this paradigm presuppose that we have a sound theory of what the counted and measured talk phenomena mean, i.e. what the speakers who used them intended to achieve through them. Some of these phenomena have been the subject of countless in-depth studies, e.g. in the conversation analysis tradition, down to considerable detail. The recurrent conclusion that is drawn by authors, however, is that of multi-functionality (polysemy) of the studied word, filler, or utterance, depending on the context of use. So even if a quantitative study reveals significant differences in the occurrence of a certain speech phenomenon between human–human and human–agent talk, researchers have to answer the difficult question of what the speakers used it for. To formulate hypotheses that are worth testing, it is necessary to gain in-depth knowledge about the phenomenon in question, be it through the literature or be it through own – qualitative – analyses, and best through both, because what is established knowledge about human–human interaction cannot be automatically transferred to human–machine dialogues.

4.1 Collecting Data of Human–Machine Interaction

There exist already huge corpora of human–machine dialogues, e.g. the Switchboard, TRAINS or DARPA Communicator corpus. Text-based systems are particularly easy to use also for data gathering, because the data are in a format that is ready for use in analysis. Even in speech-based systems there is already one side of the dialogue (the machine's) that is readily available and does not need transcription.

Data collection becomes more demanding where the non-verbal part of interactions is of interest, as it will be especially where the research focuses on emotion in conversation. Video recordings are then necessary, and their transcription, annotation, and analysis are time consuming.

The bigger problem, however, is the kind of data that are collected. Most conversational systems, and especially robots, are in the stage of research and have not left the lab yet. Their current incapacity to socially 'survive' in open environments is avoided by experimental settings where test subjects interact with them on given tasks, in a controlled environment, and for short periods of time. There are interesting data to collect even in such circumstances, but their relevance for real-life applications of systems, e.g. as assistive robots in the home environment or as multi-functional long-term companions, is limited.

There are, as yet, only a handful of studies of relationships with embodied conversational artefacts 'in the wild', especially when long-term use (i.e. for any duration beyond wearing-off of the novelty effect) is an issue (cf. Bickmore et al., 2005; Bickmoire and Picard, 2005). The European project SERA (Social Engagement with Robots and Agents) has set out to, among other things, collect video data of older people having a companion-like robotic user interface set up in their homes for about 10 days (Payr et al., 2009). Although speech recognition in such environments is currently still an unsolved problem, the interaction modalities elicit interesting social behaviours that can be studied in their development over time.

From a methodological viewpoint, data collection in human–machine interaction cannot evade the problem of a priori hypotheses, i.e. the dialogue has to be designed first on the grounds of literature, common sense, and experiential knowledge researchers have as members of the speech community. In doing this, they necessarily make assumptions which can be, and often are, wrong. However, data do not become useless because of them as long as they can be made explicit.

4.2 Methods for Analysing Human–Machine Interaction

4.2.1 Ethnomethodology and Conversation Analysis

CA in its original form (cf. ten Have, 1999; Hutchby and Wooffitt, 1998; Seedhouse, 2004; Schegloff, 1986; Schegloff and Sacks, 1973) took recordings of actual telephone conversations, transcribed them, and then analysed all and only that which could be found 'in the text'. It is and has been closely associated with ethnomethodological studies which attempt to explain the behaviour of people from their own point of view. Explanations are couched in terms that make sense to the members of the community being studied.

As members of the community of practice, we have 'direct access' to what is done with language. Researchers may have theories about face-threatening acts, and these theories may be good science with explanatory power, but they are separate from the everyday recognition that 'well, saying that would be rude'. The skill required to formulate theories requires scientific training; the skill required to recognize when something is rude is a necessary part of being a member of that society. Members have fairly reliable interpretations of what is being done with any particular utterance, because that is what the utterance is made for. As a member of the community, the researcher knows when someone is rude, but as a scientist interested in language, she should be wary of her explanations of how and why it is rude.

An issue for those studying common-sense reasoning in their own community is the problem of 'noticing'. Garfinkel, who invented ethnomethodology, used what he called breaching experiments (Garfinkel, 1967), in which the researcher acts out of character in everyday settings. The effect is to highlight the everyday assumptions that enable a community of practice to work. Conversation analysts have developed a methodology that assists in the task of noticing, generally said by asking for every detail of language use: 'why this, in this way, right here?' (see Sect. 3).

4.2.2 Interaction Analysis

Interaction analysis is an interdisciplinary method for the empirical investigation of the interaction of human beings with each other and with objects in their environment. It investigates human activities such as talk, non-verbal interaction, and the use of artefacts and technologies, identifying routine practices and problems and the resources for their solution. Its roots lie in ethnography (especially participant observation), sociolinguistics, ethnomethodology, conversation analysis, kinesics, proxemics, and ethology. An introduction and overview can be found online (Jordan and Henderson, 1995). The use and analysis of video recordings is considered vital in advancing the study of interaction so that (in contrast to conversation analysis) rich non-verbal data are collected and included in the study, allowing, e.g. the study of gaze and listener feedback (Bavelas et al., 2000, 2002).

4.2.3 Critical Discourse Analysis (CDA)

A key concept that CDA adds to the analysis of discourse is ideology, i.e. the projection of particular practices as universal and common sensical. Common-sense assumptions are implicit in the conventions according to which people interact linguistically and of which people are generally not consciously aware. These assumptions are called ideologies and are closely related to issues of power. From a CDA perspective, both researchers and interactants remain docilely inside the boundaries of existing power relationships if they take common sense as unquestionable 'natural' resource. But social structures determine not only social practice but also a product of them. Discourse determines and reproduces social structure, where 'reproduction' must not be seen solely as an uncreative repetition and conservation, but also as the locus of change and struggle over social structures.

There are no genuine CDA methods if it is not intensive questioning of texts in order to lay bare underlying (shared) assumptions. The usefulness of CDA then lies mainly in raising awareness to the workings of ideology and power relationships in discourse (Fairclough, 2001), and to help the analysts to solve the problem of noticing.

4.2.4 Interactional Sociolinguistics

Interactional sociolinguistics is concerned with how speakers signal and interpret meaning in social interaction. The term and the perspective are grounded in the work of John Gumperz (1982a, b), who blended insights and tools from anthropology, linguistics, pragmatics, and conversation analysis into an interpretive framework for analysing such meanings. Interactional sociolinguistics attempts to bridge the gulf between empirical communicative forms – e.g., words, prosody, register shifts – and what speakers and listeners take themselves to be doing with these forms. Methodologically, it relies on close discourse analysis of audio- or video-recorded interaction.

4.2.5 Appraisal Framework

Appraisal framework, based on systemic functional linguistics, is an approach to analysing the language of evaluation and stance. Systemic functional grammar is concerned primarily with the choices that are made available to speakers of a language by their grammatical systems. These choices are assumed to be meaningful and relate speakers' intentions to the concrete forms of a language. Meanings are in systemic functional grammar divided into three broad areas, called metafunctions: the ideational, the interpersonal, and the textual. The ideational is grammar for representing the world (propositional content). The interpersonal is grammar for enacting social relationships (interaction, society, culture). Finally, the textual is grammar for binding linguistic elements together into broader texts (rhetorical structure). The appraisal framework is concerned with the interpersonal metafunction and distinguishes three interacting domains: attitude, engagement, and graduation. *Attitude* is concerned with our feelings, emotional reactions (*affect*), judgements of behaviour (*judgement*), and evaluation of things (*appreciation*). *Engagement* deals with the play of voices around opinions in discourse, and *graduation* attends to grading phenomena whereby feelings are amplified and categories blurred. The appraisal framework is thus an attempt to provide the linguistic tools for studying emotion in social interaction, see Eggins and Slade (1997) with a focus on oral conversation and Martin and White (2005) with a focus on written language. The approach originates in the field of language teaching so that its application to analysis, especially of conversation, is secondary and as yet not sufficiently elaborated.

Acknowledgements This work was financed, in part, by HUMAINE, the European Human–Machine Interaction Network on Emotions (EC FP6 Contract 507422) and SERA, Social Engagement with Robots and Agents (FP7 contract no. 231868). The Austrian Research Institute for Artificial Intelligence is supported by the Austrian Federal Ministry for Science and Research and the Austrian Federal Ministry for Transport, Innovation and Technology.

References

Allen J (1987) Natural language understanding. The Benjamin/Cummings Publishing, Menlo Park, CA

Allen JF (1999) Mixed-initiative interaction. IEEE Intell Syst 6:14–16

Bavelas JB, Linda C, Trudy J (2000) Listeners as co-narrators. J Pers Soc Psychol 79:941–952

Bavelas JB, Linda C, Trudy J (2002) Listener responses as a collaborative process: the role of gaze. J Commun 52:566–580

Becker C, Kopp S, Wachsmuth I (2004) Simulating the emotion dynamics of a multimodal conversational agent. In: André E (ed) Proceedings of the ADS 2004. Springer, Heidelberg, pp 154–165

Bickmore T, Caruso L, Clough-Gorr K, Heeren T (2005) "It's just like you talk to a friend": relational agents for older adults. Interact Comput 17(6):711–735

Bickmore T, Picard RW (2005) Establishing and maintaining long-term human–computer relationships. ACM Trans Comput Hum Interact (ToCHI) 59(1):21–30

Brindöpke C, Häger J, Johanntokrax M, Phade A, Schwalbe M, Wrede B (1995) Darf ich Dich Marvin nennen? Instruktionsdialoge in einem WoZ-Szenario. Szenario-Design und Asuwertung. In: SFB-Report "Situierte künstliche Kommunikatoren" 95/16, Universität Bielefeld, Bielefeld

Brown P, Levinson SC (1987) Some universals in language usage. Cambridge University Press, Cambridge, MA

Burke PJ (1991) Identity processes and social stress. Am Sociol Rev 56:836–849

Byron DK, Fosler-Lussier E (2006) The OSU Quake 2004 corpus of two-party situated problem-solving dialogs. In: Proceedings of the 15th language resources and evaluation conference (LREC'06), Genoa/Italy, pp 395–400

Byron DK, Stoia L (2005) An analysis of proximity markers in collaborative dialogs. In: Proceedings of the 41st annual meeting of the Chicago Linguistic Society, Chicago Linguistic Society, Chicago, IL

Chu-Carroll J, Brown MK (1997) Tracking initiative in collaborative dialogue interactions. In: Proceedings of the 35th annual meeting of the Association for Computational Linguistics (ACL/EACL–97), Madrid, pp 262–270

Cowie R, Schröder M (2005) Piecing together the emotion jigsaw. In: Bengio S, Bourlard H (eds) MLMI 2004. LNCS 3361. Springer, Berlin, pp 305–317

De Angeli A (2006) On verbal abuse towards chatterbots. In: De Angeli A, Brahnam S, Wallis P, Dix A (eds) CHI 2006 workshop: misuse and abuse of interactive technologies. http://www.agentabuse.org/papers.htm. Last visited on 4 November 2010

Dennett DC (1987) The intentional stance. The MIT Press, Cambridge, MA

Dunbar R (1996) Grooming, gossip, and the evolution of language. Harvard University Press, Cambridge, MA

Durkheim E (1893) The division of labor in society. The Free Press reprint 1997

Durkheim E (1897) Suicide. The Free Press reprint 1997

Eggins S, Slade D (1997) Analysing casual conversation. Cassell, London

Fairclough N (2001) Language and power, 2nd edn. Longman, Harlow

Fearon DS (2004) The bond threat sequence: discourse evidence for the systematic interdependence of shame and social relationships. In: Tiedens LZ, Leach CW (eds) The social life of emotions. Cambridge University Press, Cambridge, pp 64–86

Ferguson G, Allen JF, Miller B (1996) TRAINS-95: towards a mixed-initiative planning assistant. In: Drabble B (ed) Proceedings of the 3rd conference on artificial intelligence planning systems (AIPS-96). Edinburgh, Scotland, May 29–31. The AAAI Press, Edinburgh, pp 70–77

Fischer K (2006) What computer talk is and isn't. AQ-Verlag, Saarbrücken

Garfinkel H (1967) Studies in ethnomethodology. Polity (reprint), Cambridge, MA

Goffman E (1959) The presentation of self in everyday life. Doubleday, New York, NY

Gratch J, Marsella SC (2004) A domain-independent framework for modeling emotion. Cogn Syst Res 5:269–306

Gumperz JJ (1982a) Discourse strategies. Cambridge University Press, Cambridge, UK; New York

Gumperz JJ (1982b) Language and social identity. Cambridge University Press, Cambridge [Cambridgeshire], UK; New York

Heise DR (2002) Understanding social interaction with affect control theory. In: Berger J, Zelditch M (eds) New directions in contemporary sociological theory. Rowman and Littlefield, Boulder, CO, pp 17–40

Hutchby I (2001) Conversation and technology. From the telephone to the Internet. Cambridge, Polity Press.

Hutchby I, Wooffitt R (1998) Conversation analysis: principles, practices, and applications. Polity Press, Cambridge, MA

Jordan B, Henderson A (1995) Interaction analysis: foundations and practice. Learn Sci 4(1): 39–103

Jurafsky D, Shriberg E, Biasca D (1997) Switchboard SWBD-DAMSL labeling project coder's manual. Technical report 97–02, University of Colorado, Institute of Cognitive Science

Kemper TD (1990) Social relations and emotions: a structural approach. In: Kemper TD (ed) Research agendas in the sociology of emotions. State University of New York Press, New York, NY, pp 207–236

Kopp S (2006) How people talk to a virtual human. Conversations from a real-world application. In: Fischer K (ed) How people talk to computers, robots, and other artificial communication partners. Report Series of the Transregional Collaborative Research Center SFB/TR 8 Spatial Cognition: Universität Bremen/Universität Freiburg, SFB/TR 8 spatial cognition, pp 101–111

Kopp S, Gesellensetter L, Krämer N, Wachsmuth I (2005) A conversational agent as museum guide – design and evaluation of a real-world application. In: Panayiotopoulos T, Gratch J, Aylett R, Ballin D, Olivier P, Rist T (eds), Intelligent virtual agents, 5th international working conference, IVA 2005, Kos, Greece, September 12–14, Proceedings, LNCS 3661. Berlin, Springer, pp 329–343

Leary MR (2000) Affect, cognition, and the social emotions. In: Forgas JP (ed) Feeling and thinking. Cambridge University Press, Cambridge, UK, pp 331–356

Mao W, Gratch J (2005) Social causality and responsibility: modeling and evaluation. In: Panayiotopoulos T, Gratch J, Aylett R, Ballin D, Olivier P, Rist T (eds), Intelligent virtual agents, 5th international working conference, IVA 2005, Kos, Greece, September 12–14, Proceedings, LNCS 3661. Berlin, Springer, pp 191–204

Mao W, Gratch J (2006) Evaluating a computational model of social causality and responsibility. AAMAS 2006:985–992

Martin JR, White PRR (2005) The language of evaluation. Appraisal in English. Palgrave Macmillan, New York, NY

Martinez-Miranda J, Jung B, Payr S, Petta P (2008) The intermediary agent's brain: supporting learning to collaborate at the inter-personal level. In: Padgham L, et al (eds) 7th international joint conference on autonomous agents and multiagent systems (AAMAS 2008). Estoril, Portugal, May 12–16, 2008, International foundation for autonomous agents and multiagent systems, IFAAMAS, vol. 3, pp 1277–1280

Mead GH (1934) Mind, self, and society. Chicago University Press, Chicago, IL

Mehrabian A (1995) Framework for a comprehensive description and measurement of emotional states. Genet Soc Gen Psychol Monogr 121:339–361

Mel'cuk I (1981) Meaning-text models: a recent trend in soviet linguistics. Ann Rev Anthropol 10:27–62

Miller RS (2004) Emotion as adaptive interpersonal communication: the case of embarrassment. In: Tiedens LZ, Leach CW (eds) The social life of emotions. Cambridge University Press, Cambridge, UK, pp 87–104

Osgood CE, Suci GJ, Tannenbaum PH (1957) The measurement of meaning. University of Illinois Press, Urbana, IL

Payr S (2007) So let's see: taking and keeping the initiative in collaborative dialogues. In: Pelachaud C, Martin J-C, André E, Chollet G, Karpouzis K, Pelé D (eds) Intelligent virtual agents. Proceedings of the IVA'07, Paris. Springer, Heidelberg, pp 175–182

Payr S, Wallis P, Cunningham S, Hawley M (2009) Research on social engagement with a rabbitic user interface. In: Tscheligi M, de Ruyter B, Soldatos J, Meschtscherjakov A, Buiza C, Streitz N, MirlacherSalzburg T (eds) Roots for the future of ambient intelligence. Adjunct proceedings, 3rd European conference on ambient intelligence (AmI09), ICT&S Center

Prendinger H, Ishizuka M (2003 Aug) Designing and evaluating animated agents as social actors. IEICE Trans Inf Syst E86-D(8):1378–1385

Reddy MJ (1993) The conduit metaphor: a case of frame conflict in our language about language. In: Ortony A (ed) Metaphor and thought. Cambridge University Press, Cambridge, MA, pp 164–201

Scheff T (1990) Microsociology: discourse, emotion, and social structure. University of Chicago Press, Chicago, IL

Schegloff EA (1986) The routine as achievement. Hum Stud 9(2–3):111–151

Schegloff EA, Sacks H (1973) Opening up closings. Semiotica 8(4):289–327

Seedhouse P (2004) The interactional architecture of the language classroom: a conversation analysis perspective. Blackwell, Malden, MA

Suzuki N, Takeuchi Y, Ishii K, Okada M (2003) Effects of echoic mimicry using hummed sounds on human–computer interaction. Speech Commun 40:559–573

ten Have P (1999) Doing conversation analysis. Sage, London

Tsushima T, Burke PJ (1999) Levels, agency, and control in the parent identity. Soc Psychol Q 62:173–189

Turner JH, Stets JE (2005) The sociology of emotions. Cambridge University Press, Cambridge, MA

Vygotsky LS (1978) Mind in society. Harvard UP, Cambridge, MA

Wallis P (2008) Revisiting the DARPA communicator data using conversation analysis. Interact Stud 9(3):434–457

Wallis P, Mitchard H, Dea DO', Das J (2001) Dialogue modelling for a conversational agent. In: Stumptner M, Corbett D, Brooks M (eds) AI2001: advances in artificial intelligence. Proceedings of the 14th Australian joint conference on artificial intelligence. Springer, Heidelberg, pp 532–544

Part VI
Persuasion and Communication

Editorial: "Persuasion and Communication"

Massimo Zancanaro

Abstract The chapters in this part elaborate the aspect of computers as social actors and investigate the more ambitious topic of intelligent systems that have contextual goals to pursue and aim at inducing the user to change attitudes or behavior. The potential for applications of persuasive systems is huge in fields like health, business, safety, and education. Computer-induced human change is also going to have a very important impact on society for conflict management and resolution, personal coaching and training, group facilitation, shaping of economical behaviors, adoption of sustainable lifestyles, and so on.

Since the beginning of the 1990s, the field study called *captology* – which stands for computers as persuasive technologies (Fogg, 2003) – investigates how interactive systems, such as Web sites or mobile phones, can be purposefully designed with the aim of changing users' attitudes and behaviors. The persuasive nature of a computer application usually does not reside with the object itself: for a computer program to be classified as "persuasive," the context of creation, distribution, and adoption has to be taken into account (Fogg, 1999). At a first glance, one can ascribe the study of persuasion by means of computers to the general field of media studies. Yet, as discussed in the seminal book by Reeves and Nass (2001), human beings tend to attribute to computer properties and characteristics that are typically human. Interactive computer applications therefore are closer to persuasive actors than to persuasive media even when this is not intended by their designers.

The chapters in this part elaborate the aspect of computers as social actors and investigate the more ambitious topic of intelligent systems that have contextual goals to pursue and aim at inducing the user to change attitudes or behavior. This long-term objective requires a full understanding of the cognitive mechanisms of persuasion and how they relate to other aspects of human mind, like emotions and their behavioral manifestations.

M. Zancanaro (✉)
Fondazione Bruno Kessler-Irst, Trento, Italy
e-mail: Zancana@fbk.eu

P. Petta et al. (eds.), *Emotion-Oriented Systems*, Cognitive Technologies,
DOI 10.1007/978-3-642-15184-2_27,

The foundational nature of the research described in this part resides in the observation that the capability of affecting other beings is paramount for autonomous agents; there are no prospects for the whole endeavor of pervasive and ubiquitous computing unless machines are made capable of affecting their users.

Autonomy implies the capability of making and pursuing ethically informed choices that have individual and societal values as their driving forces (Guerini et al., 2005). Indeed, as this kind of computer applications becomes more complex and more common in everyday life, the need for them to be designed ethically is becoming more compelling. The challenge in the long run is to create ethically aware agents: able to reason about the ethicality of their actions and possibly correct their behavior accordingly.

Nevertheless, the potential for applications of persuasive systems is huge in fields like health, business, safety, and education. Computer-induced human change is also going to have a very important impact on society for conflict management and resolution, personal coaching and training, group facilitation, shaping of economical behaviors, adoption of sustainable lifestyles, and so on (Stock et al., 2006).

Scholars have not yet agreed on a unified definition of persuasion. Many theories and perspectives on the topic have been proposed from fields such as rhetoric, psychology, sociology, and marketing. All these different points of view, although they can bewilder the occasional reader, do eventually contribute to a better understanding of this powerful tool to foster changes in society. Similarly, in the contributions of this chapter, the authors take and discuss their own definitions. Rather than enforcing a reductive compromise, we decided to keep the richness of the diverse points of view. Indeed, the chapters of this part provide a unique opportunity to cast a look across the field of automatic persuasion and all together they present a grand picture of the studies in this field.

The first chapter, by Miceli et al., models persuasion in the general context of social influence. In particular, the chapter relays on the specific role played by emotions. In this approach, the point of view of the persuader is taken and an architecture is described for an autonomous agent that deliberately plans to communicate so as to induce a recipient to process the conveyed information in the intended way. Persuasion is related to two main forms of emotional strategies: those that appeal to expected emotions and those that try to persuade through arousal of emotions. Even when emotional persuasion strategies are applied, a careful, context-sensitive, and rational planning of the strategy to apply is required on the side of the persuader.

The second chapter, by Guerini et al., proposes a view on automatic persuasion that emphasizes verbal communication. Natural language generation (NLG), either written or spoken, is introduced as a central aspect of automatic persuasion. Extending NLG to deal with persuasive messages implies advancing the state of the art on both strategic and tactical levels since the effectiveness of the generated message can be enhanced by appropriate content selection, text planning, and linguistic choices. The section also provides a reasoned overview of many systems addressing verbal persuasion issues for intelligent user interfaces, including several that go beyond language and generate multimodal messages.

Indeed, only a small percentage of communication involves words; the focus of the third contribution to this part, by André et al., is concerned with a very specific and crucial aspect of multimodality, namely nonverbal behavior in face-to-face communication. The authors discuss the fundamental factors to be considered when constructing persuasive agents in the form of embodied virtual agents, in particular, the need to appear credible, trustworthy, confident and non-threatening. The chapter is concluded by a rich list of recommendations for nonverbal behaviors to be considered when constructing this kind of persuasive systems.

Finally, the last chapter, by Strapparava et al., discusses a peculiar yet important topic in this field. This section surveys the basic theories of humor and presents the main contributions made in the field of computational verbal humor, including applications for automatic humor generation and humor recognition. From an application point of view, the use of humor in persuasive messages has great potential, because it helps getting and keeping people's attention and helps remembering.

References

Fogg BJ (2003) Persuasive technology using computers to change what we think and do. Morgan-Kaufmann, Amsterdam

Guerini M, Stock O (2005) Toward ethical persuasive agents. In: Proceedings of the IJCAI workshop on computational models of natural argument, Edinburgh, Jul 2005

Reeves R, Nass C (2001) The media equation. Cambridge University Press, New York, NY

Stock O, Guerini M, Zancanaro M (2006) Interface design and persuasive intelligent user interfaces. In: Bagnara S, Crampton Smith G (eds) The foundations of interaction design. Lawrence Erlbaum, Hillsdale

Emotion in Persuasion from a Persuader's Perspective: A True Marriage Between Cognition and Affect

Maria Miceli, Fiorella de Rosis, and Isabella Poggi

Abstract We will first try to place persuasion in the general context of social influence, suggest a definition of persuasion, and discuss its implications in terms of the basic principles of any persuasive attempt. Our model takes the persuader's perspective, thus focusing on their theory of the recipient's mind, and their planning for influencing the recipient. We will address persuasion strategies, focusing on the distinction between emotional and non-emotional ones. Once the basic relationships existing between emotions and goals, which are at the foundation of emotional persuasion, are outlined, we will present two general kinds of emotional strategies, *persuasion through appeal to expected emotions* and *persuasion through arousal of emotions*, and illustrate the typical features of each kind. Special attention will be paid to *persuasion through arousal of emotions*, to some problems it raises, and in particular – by focusing on the arousal of two "germane" emotions, envy and emulation – to the analysis of the persuader's reasoning and planning implied by this strategy. Finally, we will briefly compare our model with the dual-process theories of persuasion and provide some concluding remarks on the specificity of our approach, as well as on possible directions of research on persuasion.

1 Introduction

Persuasion is a form of social influence, a topic addressed by many disciplines and approaches: marketing and advertising, law, linguistics and rhetoric, social psychology and communication studies, politics, public relations, human–computer interaction, and the so-called "captology" (e.g., Fogg, 1999), that is, the use of computers as persuasive technologies. Social influence, concerning the production of any kind of change of others' beliefs, goals, or behavior, comprises a much broader

M. Miceli (✉)
Institute of Cognitive Sciences and Technologies, CNR, Rome, Italy
e-mail: maria.miceli@istc.cnr.it

Fiorella de Rosis is deceased

P. Petta et al. (eds.), *Emotion-Oriented Systems*, Cognitive Technologies,
DOI 10.1007/978-3-642-15184-2_28,

class of phenomena than mere persuasion. Even considering only the *intentional* change of others' attitudes, we can find different forms of intentional influence.

To start with, one may want to change others' goals or behaviors without acting upon their beliefs. Mere reinforcement may work in such cases: by rewarding or punishing others' behaviors, one may influence the establishment or extinction of such behaviors. This kind of influence is very far from persuasion, one of the reasons (albeit not the only reason) being precisely that it does not necessarily imply the medium of others' beliefs. In this regard, consider a nice example reported by (Rhoads, 1997), about a psychology professor who was teaching his students the basics of behavioristic psychology, of rewards and punishments:

> The story goes that his students decided to test the professor's theories ... Whenever the professor lectured from the right side of the room, the students became distracted and noisy. When he moved to the left side of the room, they listened with rapt attention. By selectively rewarding the professor's location in the room, the class was able to shape the professor's lecturing behavior to the point that he would enter the room and walk immediately to the left wall, leaning against it during the entire lecture. When the students finally revealed their prank ... the professor denied any influence over his behavior, and insisted that lecturing while leaning against the left wall was simply his preferred style!

It might be argued that mere rewards and punishments (those devoid of any anticipation through the influencer's promise of reward or threat of punishment) may often induce some (conscious or unconscious) beliefs in the rewarded or punished people (e.g., "If I do *x*, dad will give me a thrashing") which are likely to impact on their subsequent behavior. However, it still remains that the provider of mere rewards or punishments does not act directly on such beliefs. Thus, so far we might draw the conclusion that for being considered a form of persuasion, social influence should directly act upon the target's beliefs.

On the other hand, one may want to change others' beliefs regardless of the latter's impact on their goals and behaviors. More precisely, though a potential impact of beliefs on goals is generally unavoidable, one may be uninterested in such effects. For instance, I may want you to believe that "American Beauty" is the best movie made in the last 20 years (because, say, I wish to achieve consistency between your judgments and mine) independent of the possible impact of such belief on your goals and behaviors. Such forms of influence are aimed at directly modifying others' beliefs, but we would hardly see them as kinds of persuasion proper, as long as the belief change is not meant to be instrumental to changing the others' goals. As pointed out by Guerini et al. (2003) "while argumentation is concerned with the goal of making the receiver believe a certain proposition ..., persuasion is concerned with the goal of making the receiver perform a certain action." Argumentation may be (and often is) implied in persuasion, but *for persuasion proper to apply, belief change should be aimed at goal change*.

Actually, one generally wants to change others' beliefs *in order to* change their goals, or their importance, and induce a consequent behavior. Persuasion is placed in this area. However, again, one's intention to change others' beliefs in order to change others' goals or their importance (and induce a consequent behavior) is still insufficient to qualify "persuasion." Consider one's provision of some perceptual input (or, more generally, of some physical conditions) to another when such provision is precisely aimed at changing the other's beliefs as a means for changing their goals. For instance, suppose that one sets fire to a room, or simply produces some smoke in the room, in order to induce another's belief that a fire is breaking out, in order to make the other get out of the room. We doubt that this kind of influencing should be considered a case of persuasion. Conversely, if the influencer says something like "Don't you see the smoke? Better we get out!," this would more likely resemble a case of persuasion. In other words, *communication* appears to be a necessary ingredient of persuasion. We endorse here a strict notion of communication, according to which the sender of a (verbal or non-verbal) message should have a *communicative goal*. By communicative goal we mean the goal of making someone believe not only a given proposition *p* but also *one's own goal of making them believe p* (Castelfranchi and Parisi, 1980). In fact, one may want to make another believe something without making them assume one's own goal of making them believe it (like in the case of the mere provision of perceptual input).[1]

So far then, we have stated that for social influence to be a form of persuasion it should be intended to (a) act upon others' beliefs (b) in order to change their goals and behaviors and (c) this should be accomplished through communication. However, such forms of influence as orders, threats, and promises seem to satisfy the three requirements above, and still one would hardly see them as forms of persuasion. The reason is that they induce *compliance without agreement*. In fact, persuasion typically involves the addressee's conviction of the *intrinsical* relevance of the conveyed beliefs to their own goals and behaviors. Conversely, orders, threats, and promises impose an extrinsical relationship between beliefs and goals and between means-goals and end-goals, through the medium of the influencer's will: "I want you to do *p*; if you don't do *p*, I will prevent you from obtaining *q*"; "If you do *p*, I will allow you to get *q*." By itself, "doing *p*" is not a means for "getting *q*": it is only through the influencer's power over the addressee that the relationship is extrinsically or "artificially" established (see Castelfranchi and Guerini, 2007). Conversely, for persuasion to occur, the change of the addressee's goals and behaviors should be a "free" change, independent of the influencer's exercising power over them (Poggi, 2005). More precisely, we might say a *minimal* condition for persuasion to apply is that the persuader should want that the recipient intends to do a certain action *not only* because P wants R to do so (Miceli et al., 2006). This implies that the *ad baculum* argument (see Walton, 1996a) is outside the realm of persuasion to the extent to which it involves the exercise of power or force by P over R.

As pointed out by Pratkanis and Aronson (1991), Western societies prefer persuasion more than other societies do. We are more interested in changing others' minds than mere behaviors and in doing so through others' agreement, rather than mere

[1]It is worth specifying that communication is by no means restricted to verbal messages. In fact, there are innumerable instances of non-verbal communication. For example, by ostensively taking an umbrella while looking at you before going outdoors, I am making you believe not only that it is raining (or going to rain), but also my *goal* of making you believe it. We view such cases as instances of communication proper.

compliance. This preference stems both from our ethical and democratic values of freedom and from utilitaristic and pragmatic considerations. In fact, persuasion proves to be far more effective than coercion, especially in the long run. Coercion requires a good deal of power over the addressee (to obtain compliance, one should be *able to punish* non-compliance), as well as constant monitoring of their behavior (non-compliance is very likely if there is insufficient control). Conversely, once persuasion has been accomplished, the "instilled" goal or behavior is much more likely to last, regardless of any further intervention from the persuader. Thus, to the basic requirements of persuasion we have identified so far – (a) intended modification of others' beliefs (b) in order to change their goals and behaviors (c) performed through communication – we should add (d) without coercion.

A further relevant issue is that of manipulation. Manipulation might be defined as P's acting on R's beliefs through communication in order to induce R to conceive intentions or perform actions that are in fact instrumental to P's own goals, while pretending to assume (and making R assume) they are in R's interest. Unlike what happens in coercion, however, here R should believe (according to P's plan) that her beliefs and goals are "freely" changing. That is, R should intend to do a certain action independent of P's exercising his power over her. (From now on, we will refer to P as a he and to R as a she.) P is in fact not imposing any extrinsical relationship between R's beliefs and goals, and between R's means-goals and end-goals, through the medium of his will. He is just concealing that R's intention and behavior is (also) instrumental to some goal of his own. Three basic features characterize manipulation: (a) P's *exploitation* of R, as if she were an object or means; in fact, P takes R's goals into account not as ends, but only as means (hence the etymological meaning of "manipulating") to his own goals and interests; (b) P's *deception* about the end-goal of his persuasive plan: deception is somehow necessarily required by exploitation in that, should R believe that her feelings and goals are cared about and looked for by P only as means to his own goals, R would resist P's influencing; (c) *unfairness*, which is implied by both exploitation and deception.

According to some authors, e.g., Burnell and Reeve (1984), manipulation is not a form of persuasion, in that the latter should be limited to those cases in which P "acts in good faith," that is, in R's interest, without taking advantage of R in view of P's own interests and without any deceptive intent. We see this notion of persuasion as too narrow and prefer to talk of either manipulative or non-manipulative persuasion (see below).

In what follows, we will first introduce our definition of persuasion, implying the constituting features we have discussed so far. We will then outline the implications of our definition in terms of the basic principles of any persuasive attempt. Further, we will address persuasion strategies. These are one of the main concerns of any approach that considers the persuader's perspective, starting from rhetoric up to human–computer interaction and persuasive technologies. The latter, however, while acknowledging the transferability of theories and methods already developed in other fields (such as psychology or communication studies) into captology (e.g., Fogg, 1999), seem primarily concerned with finding some useful device which might increase the effectiveness of a persuasive message, such as for instance providing the persuasive message at just the time when R has to make a decision

(Intille et al., 2003) or informing R that her behavior will be monitored or tracked, thus influencing (by the way, without actually *persuading*) R to comply with some norm (King and Tester, 1999). However, finding devices of this sort is different from understanding and implementing P's theory of R's mind, as well as P's consequent reasoning and planning. This lack of theoretical grounding in our view becomes especially apparent when emotional persuasion is called into play. In fact, when distinguishing rational persuasion from emotional persuasion, some intuitive dimensions come to mind, such as the *cold* versus *warm* one, which in turn implicitly suggest some easy "recipes" for emotional persuasion. Whereas cold persuasion is associated with the provision of information about serious matters and goals (like health, justice, public policy), and a formal and impersonal communication style, warm persuasion typically refers to "futile" goals, like attractiveness or popularity, and uses a more personalized and informal style. As a consequence, for appealing to emotions in persuasion it might seem enough to grab R's attention with some surprising novelty, resort to an easygoing way of communicating, and enrich the message with a number of "catchy" adjectives. For example, while a cold advertisement of a new car would appeal to such aspects as its cost or safety, a typically warm one "might depict the car as fun, comfortable, and possibly as sexy and exciting" (Rosselli et al., 1995, p. 165).

However, emotional persuasion is far more complex than that. As we will try to show, it implies an accurate and rational planning, as well as a careful consideration of both its advantages and shortcomings or risks. Our attempt to model P's reasoning and planning to induce an intentional state in R by appealing to her emotions might provide some useful suggestions to user modeling research carried on within HCI or conversational systems.

Once we outline the basic relationships existing between emotions and goals, which are at the foundation of emotional persuasion, we will present two general kinds of emotional strategies, *persuasion through appeal to expected emotions* and *persuasion through arousal of emotions*, and illustrate the typical features of each kind. We will then focus on *persuasion through arousal of emotions*, discuss some issues and problems it raises, and in particular analyze the persuader's reasoning and planning implied by this strategy, focusing on the arousal of two "germane" emotions, envy and emulation. Finally, we will briefly compare our model with the dual-process theories of persuasion – the elaboration likelihood model (Petty and Cacioppo, 1981, 1986) and the heuristic–systematic model (Chaiken, 1980, 1987) – and provide some concluding remarks on the specificity of our approach, as well as on possible directions of research on persuasion.

2 A Definition of Persuasion

By persuasion we mean "an agent P's intention to modify, through communication, R's beliefs or their strength, as a means for P's superordinate goal to have R freely generate, activate or increase the strength of a certain goal q and, as a consequence, to generate an intention p instrumental to q, and possibly to have R pursue p; but

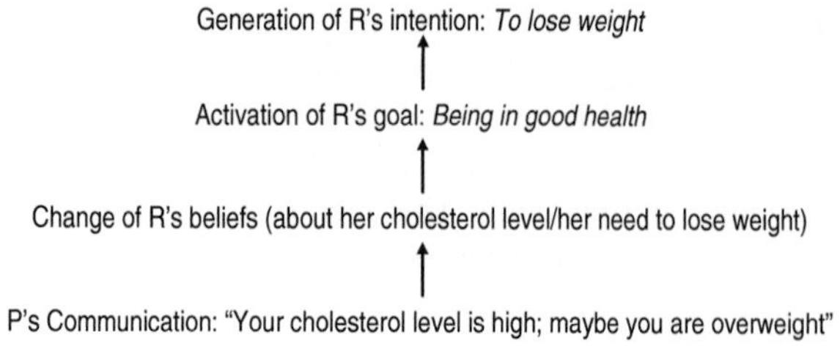

Fig. 1 An example of persuasive attempt

the minimal condition is that R has that intention" (Miceli et al., 2006). To provide a very simple example (see Fig. 1), consider P's message, "Your cholesterol level is high; maybe you are overweight," aimed at modifying R's beliefs (about R's cholesterol level and consequent need to lose weight), as a means for making R freely activate goal *q* of being in good health and, as a consequence, to favor R's generation of the intention *p* to lose weight, instrumental to *q*. It is worth specifying that what is sketched in Fig. 1 is just P's persuasive plan, not necessarily its effect on R's mind.

As is apparent from the definition above, our model takes the persuader's perspective, focusing on his theory of R's mind and his planning for influencing R, that is, for changing R's mental state so as to make her intend to do a certain action or plan. We also circumscribe the notion of persuasion in relation to such criteria as *accidental* versus *intentional* (intentionality is necessary), *communicative* versus *non-communicative* (communication is necessary), and *coercive* versus *non-coercive* persuasion (non-coercion is necessary).

With regard to a criterion such as *success* versus mere *attempt* at persuasion, being interested in P's planning aimed at persuasion, we do not view its actual success as a necessary requirement for its being a persuasive planned intention. The latter remains a persuasive intention irrespective of its effects (which may depend on a variety of factors, including contextual or accidental causes). Thus, by persuasion we mean a persuasive intention and attempt rather than a successful persuasion.

With regard to the *manipulative* versus *non-manipulative criterion*, as already pointed out, we view the requirement of non-manipulation as too restrictive. Our definition is general enough to include both manipulative and non-manipulative persuasion, depending on the content of the ultimate goal of P's persuasive attempt. Such an ultimate goal (*w*) may either coincide or not with R's goal *q* (which motivates R's intention *p*). Going back to the example above, P's ultimate goal may be either R's good health or some other goal of his own. In fact though *p* (losing weight) is instrumental to *q*, it may be meant by P to favor, at the same time, the achievement of his own goal *w* (say, having a thinner, more attractive wife to show to his friends). This instrumental relationship represents some kind of manipulation, that is, of P's "unfair" use of the persuasive message in order to achieve his own goals. In this case, to effect his purpose, P will have to conceal this relationship from R, by making her believe that *q* corresponds, as well, to his own final goal. Thus, just as our notion of persuasion encompasses both manipulative and non-manipulative cases, it also includes both *sincere* and *deceitful* persuasion. In any case P, in attempting to persuade R, is trying to make her believe that a relation holds between p and q, whether P himself believes this relation to hold or not and whether the arguments P uses to convince R are shared by him or not.

In our definition we distinguish *goal* from *intention*. Our notion of *goal* is very basic, in terms of regulatory state of a system, that is, a representation that the system tries (through its actions) to liken the world to (e.g., Miller et al., 1960; Rosenblueth et al., 1968). This "goal" is actually a complex family, including wishes, needs, and intentions. In a world where resources are bounded, not any goal is *chosen for being pursued* (Bell and Huang, 1997; Castelfranchi, 1996; Haddadi and Sundermeyer, 1996). This choice depends on a variety of criteria, including the perceived importance of the goals, their feasibility, and the amount of resources required to accomplish them.

An *intention* is a special kind of goal, which mediates the relationship between mental attitudes and behavior (Ajzen, 1985; Fishbein and Ajzen, 1975). It is a goal endowed with the following properties: it is conscious; consistent with both the agents' beliefs about its possible achievement and their other intentions; chosen, i.e., implying a decision to pursue it; and planned for. So, an intention is always about some action or plan. The decision to pursue the goal implies the agent's *commitment* to it (e.g., Cohen and Levesque, 1990). However, also an intention is not necessarily pursued. If a goal is *chosen for pursuit* and some *planning* is being done for it, this goal is already an intention, namely, a "future-directed intention," rather than an "intentional action" (Bratman, 1987).

In our definition we also distinguish goal *generation* from goal *activation*. A goal is *active* when it is included in the agent's "goal balance" (Castelfranchi, 1990), that is, when the agent starts to assess its importance and feasibility through comparison with other candidate goals, in view of its possible translation into an intention. An active goal is not yet an intention; it may *become* so if that goal is finally chosen for pursuit. An *inactive* goal of R (that is, a goal currently not included in her goal balance) can be *activated* by P when, in various possible ways, P makes the goal enter into R's goal balance. By contrast, a *generated* goal is a new goal, i.e., a regulatory state that comes to be newly represented in an agent's mind. Goals are generated as means for pre-existing goals (Conte and Castelfranchi, 1995). The means–end relationship between a generated goal and a pre-existing one may be either internally represented (that is, planned by R) or external to R's mind. For instance, the goal to have sex is functional to reproduction, but at the psychological level R might want to have sex just for its own sake, regardless of its superordinate function. Also an intention may be *generated* as a means for a pre-existing goal, on condition that this goal is *active* in R's mind.

3 Basic Principles of Persuasion

First of all, as our definition implies, *P should have a* (more or less explicit) *theory of R's mind.* This includes a general, "universal" theory of mind, that is, a theory of the basic relations between mental attitudes, and between the latter and actual behavior. For instance, P should know that one's having goal *q* and assuming that *p* is a means for *q* increases the likelihood that one pursues *p*; or that one's being angry at another increases the likelihood of one's trying to hurt that other; and so on. It includes as well a general theory of "personality" and its possible impact on attitude change. For instance, P might consider that self-esteem is likely to affect one's receptivity to some form of persuasion (see, e.g., McGuire, 1969). Finally, P should also have a more specific theory of his individual target's characteristics and dispositions, that is, R's personality, hence R's typical goals and values, as well as R's specific goals and beliefs which are currently active in the situation at hand.

Secondly, a general principle of any form of persuasion is in our view that of *goal "hooking"* (Poggi, 2005). That is, *in order to have R intend p, P should "hook" p to some other goal q that (P believes) R already has of her own.* Any persuasive attempt implies P's acting on some of R's pre-existing goal while (more or less explicitly) suggesting a means–end relationship between the intention he wants to induce in R and that pre-existing goal of R. The principle of goal hooking reminds to some extent of the notion of "Socratic effect" (McGuire, 1960). The latter is grounded on cognitive consistency and posits that interrelated beliefs become more consistent if they are made *salient* to the individual in close temporal proximity. Therefore, if applied to persuasion, the Socratic effect would imply that P has nothing to do but make "salient" what is *already* believed by R. As stated by (McGuire, 1960), "The postulate of cognitive consistency suggests that persuasion could be effected by the quite different technique of eliciting the persuasive material from the person's own cognitive repertory, rather than presenting it from outside" (p. 345). We do not totally endorse such a view, in that we believe that some "persuasive material" may be presented from "outside" as well. (For instance, we do not assume that R should necessarily know that *p* is a means for *q*.) However, we assume that persuasion cannot apply if P does not make "salient" some pre-existing goal of R's. We also assume that P's attempts at increasing the consistency of R's system of beliefs and goals are often instrumental to indicating a means–end relationship between some pre-existing goal *q* of R and the intention *p* he wants to induce in R (thus "hooking" *p* to *q*). Such a means–end relationship may or may not be already known by R, but in any case it should be derivable from her pre-existing beliefs.

Finally, any persuasive attempt implies *P's* (more or less explicit) *goal to show his unselfish concern*, i.e., to show that the end-result of persuasion is in the interest of R (both when P is acting in his own interest and when he is genuinely acting in R's interest). If R (rightly or wrongly) supposes that P is motivated by a selfish concern, she is likely to harbor the suspicion of being manipulated, which would in turn hamper R's free activation or generation of the candidate goals P wants to induce in R. In other words, P's goal to show his unselfish concern is aimed at preventing R's *resistance* to persuasion. Actually, resistance may depend on a variety of causes, for instance, R's experiencing negative affect and attributing it to the content or the source of the persuasive message (Cacioppo and Petty, 1979; Zuwerink and Devine, 1996) or the mere fact of being exposed to attitude-incongruent information (Frey, 1986; Gilbert, 1993). However, the suspicion of manipulation is in our view a powerful instigator of resistance: if P appears to "use" an alleged interest of R as a mask for his own interest, there is indeed reason to question the truth value of the persuasive message itself. As a consequence, the suspicion of manipulation may favor R's counter-arguing, which in turn impacts on metacognition, increasing the certainty of R's original attitudes (e.g., Tormala and Petty, 2002, 2004) which P was trying to change. A special form of resistance is *reactance*, which is merely concerned with the perceived threat to one's freedom, and typically implies the attempt at restoring such freedom through a behavior which is the opposite of the persuasive message (e.g., Brehm, 1966; Wicklund, 1974). Since the point of reactance is to contrast any threat to R's self-determination, reactance can occur even when R is convinced that p (the intention that P is suggesting her to pursue) is a means for her goal q, because in any case accepting P's suggestion thwarts R's goal of deciding by herself, which may be much more important to her than any possible q. But, a fortiori, if R suspects that P is trying to influence her in view of some personal advantage, R's need for self-determination and control is particularly threatened, and (independent of the content of the message) R is highly motivated to show that P has no influence on her.

However, the difficulties implied by R's concern for her freedom and possible suspiciousness toward manipulation are mitigated by a certain degree of "gullibility" which is favored by the typical tendency to overestimate one's own control over one's attitudes and behavior (e.g., Fischhoff, 1994; Kelley, 1971; Langer, 1975; Taylor and Brown, 1988).

4 Emotional Versus Non-emotional Persuasion

Innumerable possible typologies of influencing and persuasion strategies have been suggested (e.g., Kellerman and Cole, 1994; Levine and Wheeless, 1990; Marwell and Schmitt, 1967; Mulholland, 1994). However, taxonomies and typologies are not always founded on solid theoretical grounds. The poor interest in theorizing typical of some influence domains, for instance advertising, is one of the reasons for such theoretical weakness. In fact, clever advertising may identify and employ effective strategies without being interested in either analyzing *why* they work or their possible conceptual and functional relationships.

We personally have a taste for general classes of strategies, which concern the mental mechanisms and processes implied, rather than (or before) focusing on the specific content of persuasive messages, the kinds of goal to "hook," or the positive versus negative valence of the goal. Representative examples of general classes of strategies are the well-known dual-process theories of persuasion: the elaboration likelihood model (Petty and Cacioppo, 1981, 1986) and the heuristic–systematic model (Chaiken, 1980, 1987). Although these models present some important differences, they share a number of basic assumptions: (a) a "least effort" principle, according to which people tend to process information superficially unless they are motivated otherwise; (b) a "capability" principle, according to which, for engaging in a systematic, effortful elaboration of the information received, people should be endowed with sufficient cognitive skills, need for cognition, background knowledge, and time; otherwise, they are likely to process information superficially; (c) the existence of two general modes of thinking, corresponding to two general routes to persuasion – the "central" or systematic one, characterized by careful elaboration and evaluation of the message and the quality of its arguments; and the "peripheral" or heuristic one, characterized by the use of cues (for instance, the attractiveness or supposed expertise of the communicator) or heuristics, that is, simple decision rules (for instance, "scarcity implies preciousness") to judge the validity of a message; (d) the differential impact of persuasion variables depending on the matching between the persuasion "tool" (arguments or cues) used by P and the thinking mode used by R.

We propose two general classes of persuasion strategies: emotional and non-emotional (which, as we will discuss later on, present connections with, as well as differences from, such dual-process theories of persuasion). Our definition of persuasion is general enough to cover both emotional and non-emotional persuasion. Actually, we view emotional persuasion as a sub-case of general persuasion. By emotional persuasion we mean a *persuasive intention which appeals to R's emotions*. That is, its specificity lies in the *means* used by P while trying to generate, activate, etc., R's goals (Miceli et al., 2006). This may happen through the medium of R's emotions in a twofold sense: either through the actual elicitation of some emotion in R (*persuasion through arousal of emotions*) or by appealing to R's expected emotions, that is, to R's beliefs and goals *about* her emotions (*persuasion through appeal to expected emotions*) (see also O'Keefe, 2002).

Before describing these two kinds of emotional persuasion, however, we have to discuss more generally why appealing to emotions is functional to persuasion and outline the basic relationships existing between emotions and goals, which we view at the foundation of emotional persuasion strategies.

The importance of appealing to emotions for persuasion has been acknowledged since the most ancient times. As Aristotle (1991) already argued, persuasion relies on the interplay of three basic ingredients: the persuader's credibility and trustworthiness, especially his moral character (*ethos*); a logical and well-reasoned argument (*logos*); and the feelings of the audience (*pathos*). From Aristotle on, it has been widely acknowledged that persuasion is likely to appeal to both the informational and the emotional sides. Attitudes are complex constructs composed of action tendencies, a complex of beliefs, and emotional states associated with, or aroused by, the object of the attitude (Fishbein and Ajzen, 1975). Modifying an attitude implies modifying its three components.

In particular, emotional responses are characterized by a special strength and immediacy. Under certain conditions, the emotional component seems to hold a sort of primacy over the informational one. For instance, inconsistencies between affective and cognitive components are more likely to be resolved in favor of the affective components, that is, by changes in cognition rather than affect (Jorgensen, 1998).

The reason for the particular strength and immediacy of the impact of emotions on attitude change lies in our view in their strict and manifold relationship with goals. In fact, three distinct relations may hold between emotions and goals: emotions *signal* goal pursuit, achievement, and failure; they *generate* goals; and they may *translate* into goals (Miceli and Castelfranchi, 2002; Miceli et al., 2006)

4.1 Relationships Between Emotions and Goals

First, emotions *signal* the (possible) achievement or thwarting of goals (e.g., Frijda, 1986; Gordon, 1987): the experiences of fear, anxiety, shame, guilt, surprise, joy, and so on, all work as signals of the destiny of our goals, thus accomplishing an informative function about our relationship with the environment (e.g., Damasio, 1994; Keltner and Ekman, 2000; Lazarus, 1991; LeDoux, 1996; Schwarz, 1990).

Second, once an emotion has signaled the destiny of some goal, a behavioral response is likely to follow, which implies the generation of some other goal. For instance, once fear has signaled the presence of a possible danger, it produces the goal to avoid it; A's envy toward B, besides signaling that A's goal of "not being inferior to B" has been thwarted, generates A's goal that B suffers some harm (Miceli and Castelfranchi, 2007). This generative relationship between emotions and goals is at the foundation of what we call *persuasion through arousal of emotions*.

It is worth pointing out that the goal generated by an emotion (for instance, in the case of envy, the goal that B is damaged) is often *instrumental* to the goal whose achievement or thwarting is signaled by the same emotion (not being inferior to B). However, the means–end connections of the goals generated by emotions are not necessarily represented in a person's mind (in envy, for instance, A wishes B's harm as an end in itself, not as a means for not being less than B.) Here a particular perspective on psychological mechanisms and processes is implied: the functional and evolutionary one, which is concerned with the psychological mechanisms that evolved to solve adaptive problems (such as escaping dangers; finding food, shelter, and protection; finding mates; being accepted and appreciated by one's conspecifics) and thus surviving and delivering one's genes to one's own offspring. From the perspective of evolutionary psychology (e.g., Tooby and Cosmides, 1990), emotions generate goals our ancestors had to pursue in order to answer such ecological demands. And, of course, the instrumental relation between such emotion-generated goals and their functions was far from being represented in our forefathers' minds.

Finally, emotions "become" goals. More precisely, the anticipation that a certain emotion will (not) be felt may give rise to the goal of (not) feeling it. As a consequence, agents may perform (or avoid performing) an action *in order* (not) *to feel a certain emotion*: I may give you a present to feel the joy of making you happy or do my own duty so as not to feel guilty. In behavioristic terms, emotions are often (positive or negative) reinforcements. Hence the important role they play in learning: a given action can be performed not only on the grounds of one's expectations about its outcome and evaluations of its costs, but also in order (not) to feel the associated emotions. Expected emotions play a remarkable role in decision making. An expected emotion may induce the goal (not) to feel it, which may enter the decision-making process with a given value, possibly modifying the value of the available options. For instance, while considering to cheat my colleague to obtain a promotion at work, I expect to feel guilty; this expectation induces the goal not to feel guilty, which impacts on my decision making to such a point that I give up the option of cheating.

This kind of relationship between emotions and goals is, as we are going to see, at the foundation of persuasion through *appeal to expected emotions*.

4.2 Persuasion Through Appeal to Expected Emotions

In persuasion through appeal to expected emotions, *P's intention to modify R's beliefs or their strength is a means for P's super-goal to activate, or increase the strength of, R's goal of (not) feeling a certain emotion and to induce in R an intention instrumental to this goal*. For instance (see Fig. 2), P's saying to R "If you are kind to John, you will not feel guilty" is meant to activate R's goal *q* of "not feeling guilty," while suggesting the intention *p* of "being kind to John" as a means for *q*.

An appeal to expected emotions is structurally indistinguishable from any other "argument from consequences" or, in our terms, "intention generation by goal activation." The only difference resides in the *content* of the activated goal: in the appeal to expected emotions, this content is precisely that of "feeling" a certain emotion rather than having a certain state of the world true. There is no structural difference between "If you are kind to John, you will not feel guilty" and, for instance, "If you are kind to John, you will obtain an advancement at work." In fact, *persuasion through appeal to expected emotions* is a form of rational and *argumentative* persuasion, in that it applies typical rules of reasoning about means–end relationships, with the sole specification that the ends concern a special class of goals: the goals to feel (or not to feel) certain emotions. By contrast, as we are going to see, *persuasion through arousal of emotions* contains an a-rational component as long as the aroused emotion generates a certain goal independent of any reasoning.

Decision theorists have started to modify the traditional expected-utility theory so as to account for the role played in decision by anticipated emotions, such as anticipated pleasure or pain, disappointment, or regret (Bell, 1985; Loomes, 1987;

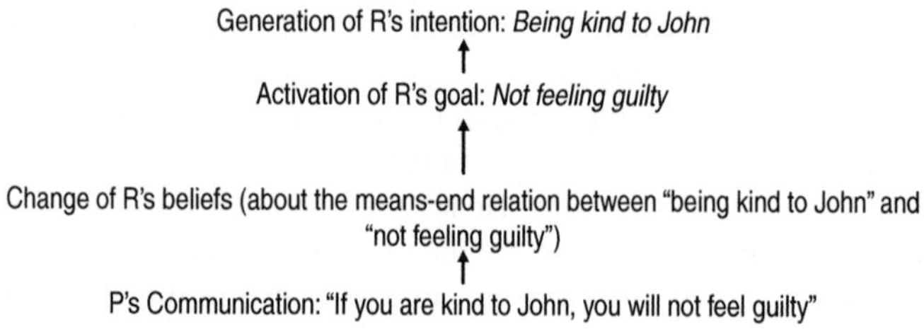

Fig. 2 An example of persuasion through appeal to expected emotions

Mellers and McGraw, 2001). Anticipated emotions seem to work quite well as predictors of intentional behavior (e.g., Parker et al., 1995; Richard et al., 1996; Zeelenberg and Beattie, 1997). Therefore, persuasion through appeal to expected emotions offers a valuable means to affect others' decision making by anticipating the possible emotional consequences of their behavior. Of course, P's knowledge of the basic components of emotions and their interrelations, as well as of R's dispositions and personality, is a crucial requirement for the applicability of this strategy.

4.3 Persuasion Through Arousal of Emotions

So far the impact of felt emotions on persuasion has been addressed in psychology by mainly focusing on the differential influence of positive and negative moods on attitude change (Eagly and Chaiken, 1993; Petty et al., 2001). As emphasized by DeSteno et al. (2004), this represents a gross oversimplification of the emotional experience and its influence on persuasion, because it disregards the different appraisals implied by discrete emotions (Lazarus, 1991; Smith and Ellsworth, 1985), even when they are of the same valence, as well as their potential impact on attitude change. For instance, two negative emotions, fear and anger, are likely to affect judgment and behavioral orientations in opposite ways (Lerner and Keltner, 2000).

Unlike the mainstream orientation, DeSteno et al. (2004) address the impact of discrete emotions on persuasion. However, they are interested in the role played by *incidental* emotions, that is "emotions with no direct connection to the message topic" (p. 44), induced in the participants *before* exposure to a persuasive message. In contrast, we are interested in emotion arousal *through* a persuasive message.

More relevant to our perspective is the study of so-called "fear appeals" (Gleicher and Petty, 1992; Janis, 1967; Leventhal, 1970; McGuire, 1969; Rogers, 1983; Sutton, 1982), which are intended to induce a specific emotion, fear. However, much research on fear appeals presupposes a questionable equation between "fear" and "threat," which makes it hard to distinguish persuasion through fear arousal from any form of "threatening" argumentation (see Witte, 1992). In other words, where is *fear* in such appeals? What is the difference between a "fear appeal" and an "argument from negative consequences" (Walton, 1996b)? We are not claiming that, for a fear appeal to occur, fear should be *actually aroused* in the addressee. (This may or may not happen depending on a variety of factors, including accidental circumstances.) We are just claiming that, for a fear appeal to occur, the arousal of fear should be explicitly *planned* by a persuader. This implies a constitutive difference between the planning typical of persuasion through emotional arousal and that typical of other strategies.

We can now provide our definition of *persuasion through arousal of emotions:* in persuasion through arousal of emotions, *P's intention to modify R's beliefs or their strength is a means for P's superordinate goal (super-goal) to arouse an emotion in R, which in turn is a means for P's further super-goal to generate a goal in R, and then an intention instrumental to it*. For instance (see Fig. 3), supposing P's saying

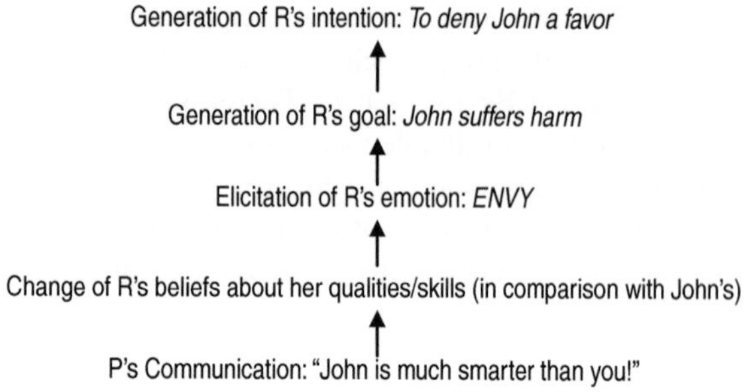

Fig. 3 An example of persuasion through emotional arousal

to R "John is much smarter than you" is meant to provoke R's envy, this feeling should generate R's goal that John suffers some harm and induce, as a means for it, her intention to deny John a favor.

What is the difference between inducing goals (and then persuading) through mere beliefs and inducing goals through *emotion-arousing* beliefs? Beliefs cannot generate goals by themselves alone. A belief can only *activate* a pre-existing goal. It is the pre-existing goal which, in interaction with the belief, can *generate* a sub-goal. Suppose I learn that tomorrow there will be shortage of water. This belief will activate my pre-existing goal "to have water," which in turn will generate my goal "to stock up on water" as a means for it. The cognitive activation of goals is in fact strictly related to the typical planning and reasoning procedures about goals, means, and enabling conditions. By contrast, if a belief arouses an emotion, the latter can directly *generate* a goal, independent of any planning and reasoning, i.e., independent of any *represented* means–end relation between the generated goal and some other pre-existing goal. Supposing the belief that "John is more intelligent than I am" arouses my envy toward John, this emotion is able to generate by itself the goal that "John suffers some harm." True, as already noted, such a goal is *functional* to a more general goal of "not being less than John," but this means–end relation is not (necessarily) represented in my mind, and in any case it is not the *motive* why I want that John suffers some harm. If I envy John, I want this for its own sake (because of my envious ill will against John), not as a means for not being less intelligent than John. Thus, unlike the purely cognitive activation, the emotional triggering of goals is a form of *direct generation* of goals.[2]

[2]It might seem that, precisely for this reason, the principle of goal hooking does not apply to persuasion through arousal of emotion. In fact, if the motivating goal (in our example, the goal that John suffers some harm) of R's behavior is directly *generated* by the emotion aroused (in our example, envy), one might conclude that P is not acting on a *pre-existing* goal of R. Strictly speaking, this conclusion is correct. However, it is also true that such goal is "hardwired" in the emotion itself. So, by arousing the emotion, P is planning a more sophisticated goal hooking: that between the goal hardwired in the emotion (and in this sense, "pre-existing") and a possible intention *p* (for instance, in our example, the intention to damage John).

The advantages offered by such strategy stem from the immediate motivating force of emotions. A goal which is generated independent of any planning and reasoning is less likely to undergo scrutiny or evaluation of its value as a means for other goals, as well as of its costs or side effects. In our example, R would feel (according to P's planning) this "urge" that John suffer some harm, regardless of any careful evaluation of its instrumental relationship, or of its interference, with other goals and would profit by any opportunity to satisfy it (through some intention *p* such as that of denying John a favor, which P, on his part, would be ready to suggest!).

However, the strategy also presents a number of possible drawbacks. To start with, emotions, when unpleasant, may favor some form of resistance. As shown for instance by research on fear appeals (e.g., Hovland et al., 1953; Janis and Feshbach, 1953; Janis and Mann, 1977), defensive avoidance – implying the receivers' inattention to the fear-arousing message or their further suppression of any thought regarding the threat – is a common reaction. The experience of a negative emotion may in fact foster emotion control processes (which in the case of fear are very likely when the receiver has low self-esteem, lacks coping skills, and is very anxious; see Rosen et al., 1982).

Moreover, persuasion through arousal of (either positive or negative) emotions may be perceived as particularly *unfair* by R, if she detects or suspects that P is "playing" with her emotions. In fact, persuading through emotional arousal is like playing a game by violating its rules, or using uneven weapons. Since it triggers goals in a more compelling and uncontrollable way than plain reasoning, it may be seen as unfair and manipulative in itself. What we have already observed about *reactance* is particularly relevant to this kind of persuasive strategy. Emotions are viewed and experienced as subjective, spontaneous, and endogenously produced reactions. Moreover, they are perceived as hardly controllable by the experiencing person. If R suspects that P is trying to influence her through emotional arousal, she perceives a very serious threat to her freedom, because P is "using" her spontaneous and hardly controllable feelings in view of some strategic end. (Even when R considers such end to be in her own interest, she may be disappointed by P's strategy, because P has resorted to a means – R's emotions – which is unlikely to be under R's conscious control, and threatens her "freedom to feel.")

Further, persuasion through emotional arousal is a risky strategy because often there is no one-to-one relationship between emotions and goals. More precisely, the goal generated by a certain emotion is often so general and high level that it can "instantiate" a variety of more specific goals, depending on contextual as well as personality factors. For instance, suppose P tries to arouse R's shame about her shape so as to generate R's goal of "saving face" and induce, as a means for it, R's intention *p* to lose weight. The goal of saving one's face, once generated, may actually favor either one's active attempt at obtaining more positive evaluations of oneself from others (and in this case intention *p* to lose weight would be a suitable means) or one's attempt at avoiding exposure to others' evaluation (by avoiding social interaction). Therefore, P's persuasive plan that aims to induce *p* in R may actually obtain quite a different outcome.

Finally, it is often hard to identify the differences between "germane" emotions (e.g., anger versus indignation; envy versus emulation). This bears important consequences in persuasion through emotional arousal. A persuasive message aimed to arouse, say, emulation may happen to arouse envy. These emotions (as we are going to argue) share many components, but are likely to generate *different* goals: whereas emulation induces a "self-enhancement" goal, envy induces an "other-diminution" goal. Such considerations point to the crucial role played in persuasion through emotional arousal by P's knowledge of the basic components of emotions and their interrelations, as well as of R's dispositions and personality.

5 An Example of the Persuader's Theory of Emotions: Emotions Arising from Sense of Inferiority

As Marcus Tullius Cicero already argued, emotional persuasion requires "a thorough acquaintance with all the emotions with which nature has endowed the human race" (Cicero, *De Oratore*). Let us suppose that P wants to arouse either envy or emulation in R, in order to have R generate, respectively, an "other-diminution" goal or a "self-enhancement" one and then some intention *p* instrumental to the generated goal. First of all, P should have a more or less explicit theory of such emotions. This theory may provide the building blocks for a computational model of P's reasoning aimed at assessing the emotions that are likely to be aroused in a given context, together with their presumable consequences. In this model two agents are involved: E (either the "Envier" or the "Emulator") and A (the "Advantaged Other"), and the following components of E's state of mind are represented:

1. *E's beliefs about the state of the world* φ: (Bel E φ), where φ may denote
   - the ownership of an (abstract or concrete) object or quality x by an agent (be it E or A): Has(E x) or its negation;
   - E's ability to come to hold it in a more or less near future: CanEvHave(E x)or its negation;
   - some relationship between E and A, either generic: Similar(A E) or pertinent to x: InferiorTo(E A x);
   - the responsibility about the specific relationship occurring between E and A, related to x: CauseOfInferiority(E A x).
2. *E's desires or goals about the state of the world* ψ*:* (Goal E ψ), where ψ may denote
   - again, the ownership of x by E: Has(E x);
   - a relationship between E and A, either related to x: ¬InferiorTo(E A x) or generic: EqualTo(E A).

3. *E's feeling of an emotional state:* (Feel P ε), where ε may denote, for the emotions we are considering, the following states:
   - being hopeful about x: Hopeful(E x);
   - being hopeless about x: Hopeless(E x);
   - being hostile toward A: Hostile(E A);
   - suffering because of a sense of inferiority toward A: SufferingInferiority (E A).

Let us now sketch a basic cognitive "anatomy" of envy and emulation, respectively. (For a more detailed treatment of these emotions, see Miceli and Castelfranchi, 2007.)

5.1 Basic Cognitive Components of Envy

According to a mild notion of envy, agent E (the "Envier") envies another agent A (the "Advantaged other") if

(a) E wants something (any type of object or state of affairs, from material goods to spiritual gifts, from social positions to psychological states) and
(b) E believes that A already has that "good," whereas
(c) E does not have it and suffers from this lack.

In addition, there is some good reason to suppose that huge differences in well-being are less likely to induce envy than minimal ones, which are typical of peers. We typically envy those who are "close" to us in terms of time, space, age, and reputation (Aristotle, 1991); in general, "overwhelming and astounding inequality ... arouses far less envy than minimal inequality" (Schoeck, 1969, p. 62). Accordingly, Ben-Ze'ev (1992) believes that *slight* differences in social status might foster envy and describes an egalitarian society such as the kibbutz in which the frequency and intensity of envious comparisons seem likely to increase. The social comparison literature (Festinger, 1954; Goethals and Darley, 1977; Wheeler et al., 1997; Wood, 1989), stressing the tendency to compare oneself with *similar* others, lends indirect support to this view. Therefore, to the above-mentioned basic ingredients, one might add (d) a "perceived similarity" component.

So far, then, our cognitive anatomy of envy implies the following components:

$$\text{Goal E Has}\,(\text{E} \cdot x)\,;\ \text{Bel E}\,\neg\text{Has}\,(\text{E} \cdot x)\,;\ \text{Bel E Has}\,(\text{A} \cdot x)\,;\ \text{Bel E Similar}\,(\text{A} \cdot \text{E})\,. \tag{1}$$

However, the previous analysis seems still insufficient to account for envy proper. In fact, it may account for the existence of a mere unfulfilled goal or desire of E's, complemented by E's comparison with A. In other words, mere greed or craving for a certain "good," might account for this state of affairs, with the specification that the craving has been instigated, favored, or increased by the sight of that good in a similar other. This is not yet envy proper.

An important feature of envy is E's wish that A *loses* the desired good (Parrott and Smith, 1993). However, also this wish by itself is not sufficient to account for envy. Suppose that A's losing x increases the likelihood of having it for E: this is what happens in mere *competition* for scarce resources, where one's success implies another's failure. By contrast, envy may arise independent of such conditions: One may envy another his new car, for instance, even though there are innumerable identical specimens in existence. More importantly, E's wish that A loses the envied good often does not increase the likelihood of having it for E. And still, E is very likely to harbor such a wish.

This feature is very informative about the real object of envy, because it indicates that what characterizes envy is not E's mere lack of x, but E's perceived *inferiority* to A. Any disadvantage between two people can in fact be filled by either the disadvantaged party's acquiring the lacking good or the advantaged party's losing it. This is why E harbors the wish that A loses x (especially when the former option is viewed as unlikely). Thus, the specific good (or goal) is often a mere opportunity for social comparison. It is not the mere fact of not having x that matters to E, but the fact of being inferior to A. The real object of envy is not a given x, but superiority, or better non-inferiority. Thus, as a consequence of her comparison with A,

(e) E comes to believe she is inferior to him. (From now on, we will refer to E as a she and to A as a he.) Such a belief, coupled with
(f) E's opposite goal of not being inferior to A (which, as a consequence, has been thwarted), in turn implies that
(g) E suffers because of her perceived inferiority to A, that is, some emotional consequence is added to the mere belief of inferiority. Therefore, we have to add such components of envy to the previous ones:

$$\begin{aligned}&\text{Bel E InferiorTo}\,(\text{E} \cdot \text{A}x)\,;\ \text{Goal E}\ \neg\text{InferiorTo}\,(\text{E} \cdot \text{A}x)\,;\\&\text{Feel E SufferingInferiority}\,(\text{E} \cdot \text{A})\,.\end{aligned} \tag{2}$$

Are all the "premises" above sufficient to account for E's envy toward A? Sense of inferiority is no doubt necessary for envy to arise, but still insufficient to account for the "full-blown" feeling of envy. If E's sense of inferiority limits itself to provoke discontent, depressive feelings, and a lowered self-esteem in E, without any special negative feeling toward A, we do not have envy proper. Actually, there is a weak notion of envy which consists essentially in the ingredients above. This is called "good" or "benign" envy, as distinct from "malicious" envy (Farrell, 1989; Neu, 1980; Roberts, 1991; Taylor, 1988; Young, 1987), and can be readily confessed (e.g., Heider, 1958). A declaration of such envy conveys a sort of appreciation of the person and their achievements: "I envy you" is better translated into "I would like to be in your shoes because I regard what you have as valuable." By contrast, "malicious" envy, or envy proper, implies something more: while orienting (consciously or unconsciously) her attention toward A,

(h) E assumes A to be the *cause* of her inferiority ("if it weren't for his superiority, I would not be inferior") and
(i) E feels ill will or hostility toward A.

To avoid misunderstanding, it should be stressed that this mental scenario does not imply E's experiencing a sense of injustice, because the latter should entail E's attribution of *responsibility* proper to, and blame on, A, which is not necessarily implied here. In fact, A can be perceived as the mere *cause* of E's suffering. One might ask why a hostile or angry reaction should result from one's mere focusing on the advantaged other as the cause of one's own inferiority. The elicitation of anger is here simply related to external attributions for one's perceived failure (in envy, inferiority). In fact, the perceived controllability or the intentionality of the outcome is not strictly a necessary antecedent of an angry reaction. What can be viewed as necessary are the motivational relevance of the outcome, its inconsistency with the person's goals, and the "other-accountability" of the outcome (where accountability does *not* necessarily overlap with responsibility proper and blameworthiness; see Smith and Lazarus, 1993). Thus, we add the following components of envy to the previous ones:

$$\text{Bel E Cause Of Inferiority } (A \cdot E \cdot x)\text{ ; Feel E Hostile } (E \cdot A)\,. \tag{3}$$

Such ill will in turn implies E's goal that A not achieve (some of) his goals (not necessarily limited to the original having x). Though sometimes E may confine her aggressive goal to A's losing x, this is not necessarily the case. More generally, what E wishes is A's disadvantage, which can be instantiated in a variety of possible failures of A's goals (also depending on opportunities offered by contextual factors).

The previous analysis is still incomplete. A further necessary ingredient of true envy seems to be E's negative expectations about overcoming her inferiority. If such expectations are positive, envy is likely to be weakened and possibly change into emulation (see later on), whereas envy will persist in the case of negative expectations. In fact, ill will is sustained by one's helplessness and hopelessness. In other words, if E is hopeless about overcoming her inferiority (because she believes she is unable to obtain x or restore the power balance in some other way), she is left with her ill will against A. Conversely, if E were confident to overcome her inferiority, she would put her efforts in trying to meet A's standard. Research on the different effects of upward social comparison (e.g., Buunk et al., 1990; Lockwood and Kunda, 1997; Taylor et al., 1996; Testa and Major, 1990) in fact shows that if the standard's position is perceived as unattainable, the comparer perceives himself or herself as hopelessly inferior to the standard and is likely to experience discouragement and self-deflation, and possibly envy. Conversely, if the standard's position is perceived as attainable, the comparer is likely to experience self-enhancement, and inspiration and encouragement to strive for it. Therefore, we should add these further components, namely,

(j) E's belief that she cannot obtain *x* eventually and
(k) E's consequent feeling of hopelessness:

$$\text{Bel E } \neg\text{CanEvHave } (\text{E} \cdot x)\,;\text{Feel E Hopeless } (\text{E} \cdot x)\,. \tag{4}$$

At this point, from both E's hopelessness and her ill will against A, we can derive E's aggressive goal that A suffer some harm, that is that (some of) A's goals are thwarted, which we might generically represent as follows:

$$\text{Goal E } \neg\text{Has } \ (\text{A} \cdot y) \tag{5}$$

provided that E believes having *y* to be one of A's goals (and *y* might even coincide with *x*, as it happens in some instances).

Therefore our final anatomy of envy implies the components described in (1), (2), (3), (4), and (5), whose interrelations are illustrated in Fig. 4.

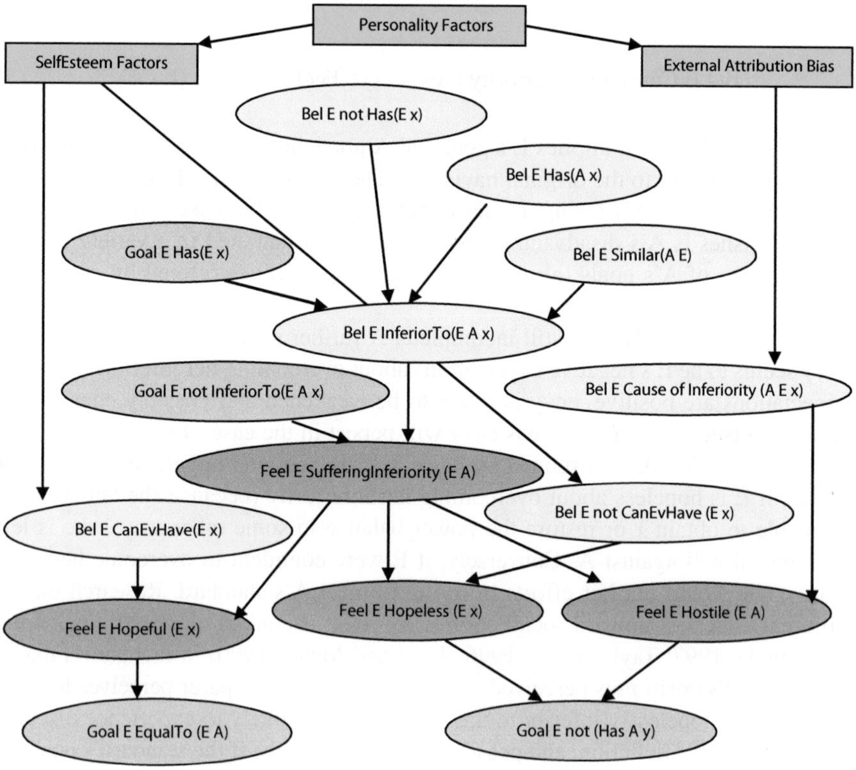

Fig. 4 Tracing of the reasoning steps followed by the persuader

5.2 Basic Cognitive Components of Emulation

Emulation shares with envy the unfavorable comparison and the related sense of inferiority. In other words, the components in (1) and (2) are common to the two emotions. However, emulation implies some remarkable distinguishing features. First of all, in emulation there is no hostility toward the advantaged party. The latter is not perceived as a cause of one's own inferiority, but as an example to follow (and possibly surpass), and one's present disadvantage is viewed as a challenge rather than a threat for one's self-esteem. This perception of one's own disadvantage as contingent and surmountable can be traced back to one's own efficacy beliefs. That is, in emulation one feels that the lacking good is within one's own reach; thus one feels capable of overcoming the present disadvantage. Therefore, we can add the following distinguishing components to the previous ones:

$$\text{Bel E CanEvHave}\,(\text{E} \cdot x); \text{Feel E Hopeful}\,(\text{E} \cdot x)\,.$$

In fact, whereas envy implies E's helplessness and hopelessness about the possibility of overcoming her disadvantage – which favors and sustains her ill will against A – in emulation one's efficacy beliefs, together with the suffering implied in the (contingent) sense of inferiority experienced, provide the driving force for the emulative motivation, which we might generically represent as a goal to surmount one's disadvantage, thus reaching (at least) equality with A:

$$\text{Goal E EqualTo}\,(\text{E} \cdot \text{A})\,.$$

Upward social comparison is likely to exert positive effects on one's self-view when the disadvantaged party believes they are able to reduce the discrepancy: "The realization that one is currently less successful than another may lose its sting if it is accompanied by the belief that one will attain comparable success in the future" (Lockwood and Kunda, 1997, p. 93). The advantaged person's success comes to play, as already observed, a self-enhancing and "inspiring" role, also providing useful suggestions about strategies to learn with the view of obtaining a similar success.

An important social–psychological problem is "understanding why some individuals can use unflattering social comparisons as a basis for more constructive, emulative impulses, whereas others seem overcome by destructive, hateful feelings" (Smith, 1991, p. 96). We believe that the comparison between envy and emulation bears important implications for persuasive and educational strategies. When attempts are made to favor high achievement motivation through social comparison and competition, special care should be taken that what is instigated will be emulation rather than envy. In fact, whereas emulation implies the goal to improve and enhance oneself, envy may favor, through ill will, the goal or wish that the envied be harmed. Envy tries to achieve equality through the other's *diminution* rather than one's own enhancement. The method of self-protection typical of envy "is that of *undercutting* the other person. If we redouble our own efforts because we are shamed by a rival's attainments, we are not considered to be envious ... we are ... indulging in virtuous emulation" (Silver and Sabini, 1978, p. 108). As remarked by Farber (2000), "by demeaning the envied one and aggrandizing the envier, envy attempts to redress inequality without the risk of intervening effort or development. In this way, envy opposes change, enforces the status quo, and is inimical to learning" (p. 242). For inducing an emulative motivation through a painful comparison with the better off, it is necessary to favor the efficacy beliefs of the target of one's persuasive strategies. Even when starting with envy, one might change it into its noble sister, emulation, by revising one's own efficacy beliefs.

5.3 Reasoning on the Receiver's Mental State, in Envy and Emulation

As already noted while discussing the basic principles of persuasion (Sect. 3), being endowed with some general knowledge of the emotions to be aroused is just one of the prerequisites P should satisfy if he aims to persuade R through emotional arousal. In fact P should also be able to apply such a general knowledge to the specific receiver of his persuasive message, which implies reasoning on R's mental state and situation, and being able to evaluate whether a certain emotion is likely to be aroused in R. In applying his general model of emotion elicitation to R, the latter will take the role of E. As "germane" emotions are at stake, P should make sure to arouse in R the "right" one – the emotion which will generate the goal P is interested in, that is, the goal which might act as a motivation for the specific intention P wants to induce in R. The knowledge of P about R will include second-order beliefs about

(a) R's goals and wishes, possibly those important to her (e.g., P believes that x is desirable for R);
(b) R's power comparison with a third person A: A's possessing/obtaining x versus R's lacking it;
(c) R's beliefs and feelings about this power comparison and its consequences, with special reference to R's comparative self-evaluation (R's sense of inferiority to A) and self-confidence about her capability to obtain x or overcome her inferiority in some other way;
(d) R's possible feelings toward A (either hostility against the cause of her inferiority or mere consideration of A as an "inspiring" example to follow and possibly surpass).

As one can see, here some knowledge about R's personality is at stake, which is very likely to exert a remarkable impact on the success of P's persuasive attempts. Such knowledge can be either derived by P's personal relationship with R, and thus be fairly detailed and analytical, or based on P's recognition in R of a few basic traits, from which it is reasonable to infer a number of more specific attitudes and dispositions. Let us consider Fig. 4 again: if, for instance, R shows low self-esteem,

P can reasonably infer a strong tendency to social comparison on R's part, as well as R's overreliance on social comparison to assess her self-worth (Wayment and Taylor, 1995). The more R's low self-esteem is seen as global and enduring, the more it is reasonable to infer a sense of inferiority (stemming from the disadvantageous comparison) which is associated with helplessness and hopelessness (Abramson et al., 1978; Epstein, 1992). From R's helpless and hopeless sense of inferiority, P can reasonably expect R's resort to self-serving biases, in particular the attributional bias, according to which one's failures are likely attributed to external causes whereas one's successes are attributed to internal causes (e.g., Miller and Ross, 1975). Finally, from R's external attribution for her failure (inferiority) P can reasonably expect R's consequent hostility against the cause of such failure, if that "cause" happens to be another agent. Conversely, from R's high self-esteem, especially when stable, it is reasonable to infer R's sense of self-efficacy and hope to overcome her inferiority, because her global feeling of self-worth is unlikely to be affected by contingent negative outcomes or disadvantageous comparisons (e.g., Kernis, 2005).

The persuader's beliefs about R's mental state and dispositions are second-order beliefs, mainly constituted by hypotheses on R's mind, and therefore they are typically uncertain and incomplete. They can be organized in an oriented graph, whose root-nodes regard P's beliefs about R's beliefs and goals in a given context, whereas the leaf-nodes regard the aroused emotions and the goals generated by such emotions. The intermediate nodes represent the intermediate cognitive steps in the process of emotion arousal. The graph in Fig. 4 can be viewed as a sketchy representation of the cognitive elements conducive to the emotions of envy and emulation, and of their interrelations. It shows that the common root of both envy and emulation is suffering because of a sense of inferiority arising from a disadvantageous social comparison, whereas their differences stem from a sense of self-efficacy, and consequent hope or hopelessness to overcome one's inferiority, and from attributional dispositions with regard to the cause of one's inferiority.[3] The possible impact in the whole process of personality features, namely, self-esteem factors and the external attribution bias[4], is also generically indicated. This graph can be viewed as a second-order model that P employs as a tool to reason on R's mind and decide whether to try or not to arouse either emotion. In this case, specific knowledge about R's characteristics and dispositions is introduced by P in the form of "evidence" about some

[3]Not all the steps of the mental process are represented in this figure; in particular, when several cognitive items contribute to "produce" another item, we do not specify the causal relationships between the producing items. For instance, while representing the activation of the sense of inferiority, we omit the step of "comparison between E and A." We believe such simplifications to be reasonable, considering that we wish to employ this model to simulate the process of persuasion strategy selection; however, we are aware that it might reduce the possibility of repairing failed strategies.

[4]It is worth specifying that, strictly speaking, the external attribution bias, being common to the majority of people, cannot be considered a personality trait. However, self-defensive people, and particularly those with a "projective" personality, are indeed characterized by a much greater proneness to such a self-serving bias (e.g., Kaney and Bentall, 1992).

nodes (their truth value). Propagation of this evidence in the graph simulates a predictive kind of reasoning aimed at assessing the emotion that will likely be aroused in the given context, with its presumable consequences.

6 Emotional–Non-emotional Versus Central–Peripheral (or Systematic–Heuristic) Strategies

As already mentioned, there are both similarities and differences between our emotional–non-emotional dimension and the central–peripheral or systematic–heuristic ones. To start from a basic similarity, our *persuasion through emotional arousal* is a clear case of a strategy which acts on R's peripheral or heuristic mode of thinking, because of the non-reasoned and "automatic" quality of the goal generated by the emotion aroused.

However, *our persuasion through appeal to expected emotions*, being a particular form of argumentative persuasion, shows no peripheral or heuristic connotation: in fact, it acts on R's central or systematic mode. This amounts to saying that, in our view, the possible analogy between "emotional," on the one hand, and "peripheral" or "heuristic," on the other hand, is limited to *persuasion through emotional arousal*.

Moreover, the overlap between "emotional" and "peripheral" or "heuristic" is not complete for a second reason: not every peripheral route necessarily involves the arousal of emotions. The dual-process theories of persuasion consider many *non*-emotional "cues" people use for judging the validity of persuasive messages: for instance, the expensiveness of a resource as a cue of its value; or social consensus on some opinion as a cue of its validity. Actually, such cues are in a sense akin to typical "rational" arguments. Compare, for instance, social consensus as a cue of the validity of a certain choice with the "argument from popular practice" (Walton, 1996b) which assumes precisely that, if a large majority of people do something, they probably believe that doing this is right; and if something is generally considered as right, doing it corresponds to a prudent course of action. The only (crucial) difference between such cues and true arguments is that the former are schematic, non-effortful, automatic, etc. They constitute pre-fabricated, "frozen" heuristics which are mindlessly applied to a given context.

Finally, according to the elaboration likelihood model (Petty and Cacioppo, 1986; Petty et al., 1987), the two "routes" to persuasion are mutually exclusive, that is, they cannot be followed at the same time. However, the heuristic–systematic model (Chaiken et al., 1989; Chaiken and Maheswaran, 1994) assumes that, under certain conditions (for instance, when systematic processing does not contradict the judgmental implications of heuristic processing), they can co-occur. Actually, we view the mingling and intertwining of emotional and non-emotional strategies *in the same* persuasive attempt as possible, and even likely. In particular, we view it as possible to rapidly shift from one "route" to the other.

Consider persuasion through arousal of emotion. We have stated that, once an emotion has been aroused in R, it generates a goal, regardless of any reasoning and planning on R's part. However, what *precedes* emotional arousal, as well as what *follows* goal generation, can be object of reasoning and planning.

Let us start from what might precede the emotional arousal. We suppose that P can arouse R's emotion through argumentation. For instance, to arouse R's envy toward A, P can start arguing in favor of the value for R of a certain resource or condition *x* (say, "good shape"), which R lacks; P can then compare R with A, explain that A is on advantage over R because of *x*, and say (and even show) that R is unable to obtain *x*, and reach A's standing: "Yesterday I met Jane: she is really in good shape! It's unbelievable: at the primary school, you were quite akin, but now ... It's a pity, but it's practically impossible to fill the gap." All of this (and much more) can be done in quite a systematic mode, through sound and convincing arguments, and R may come up to be persuaded of her helpless inferiority to A. Also, P can try to instigate R's ill will against A by stressing how bothering it is to see that some people (like Jane) enjoy beauty and health, whereas others (like R) are treated so badly by Mother Nature. If P succeeds in his endeavor, R will envy A, and this feeling will generate (without any reasoning and planning) the goal that A suffers some harm.

Now let us see what follows goal generation by first considering R's mind. Though the goal that A suffers some harm has been generated regardless of any evaluation of its instrumental relationship (or its possible interference) with other goals, this by no means prevents R from reasoning and planning in view of its achievement. That is, the goal is not represented as *instrumental* to other goals, but, as any terminal goal, it may be the end-goal of a plan, which can be worked out even with sophisticated reasoning and cunning. In our example, once this "urge" that A suffer some harm is felt by R, the latter may be ready to start such reasoning (about, say, A's goals, and enabling conditions) and planning (to thwart some of A's goals).

At this point, P may use again some argumentative tools to suggest some suitable means for R's planning. We should in fact remember that P is not interested in R's generated goal per se, but rather he wants to act on it as a *motivator of a specific intention* he wants to induce in R. Thus, P is interested in R's intention to do a specific action or plan. P may either share R's goal that A suffer some harm or have some other goal of his own, which R's behavior can (contribute to) satisfy. Let us suppose that P's end-goal is to receive R's help to prepare for an exam (because R is a very clever student), and this goal is at present quite difficult to achieve because A (Jane), who is R's customary studying companion, does not like P to join them (or so P assumes). Thus, P has some good reason to suggest the following means for R's goal to damage A: to deprive A of R's company and help in studying. This may be accomplished with some malicious insinuation, like "But it is unlikely that Jane is able to get ahead in everything! She too should have some weak point ... I bet that if she were not studying with you, she would meet with serious difficulties at school ... ."

7 Concluding Remarks

We have presented a model of persuasion which takes the persuader's perspective and focuses on P's theory of the recipient's mind and on P's planning for influencing R. Rather than directing our analysis on how information is actually processed by the recipient, we have addressed how the persuader consciously plans to communicate so as to induce the recipient to process the conveyed information. In particular, in describing what we see as the two main forms of emotional persuasion (persuasion through appeal to expected emotions and persuasion through arousal of emotions), we have tried to show that, even when emotional persuasion strategies are applied, this requires a careful, context-sensitive, and rational planning of the strategy to apply, on the side of the persuader.

Reasoning is the first step of any persuasion attempt, or any form of "practical argumentation," to use Walton's terminology (e.g., Walton, 1990). In the kind of planning we have described, the persuader builds a model of the recipient's mental state that is based on his general theory of mind and personality: a theory of how emotion arousal is affected by beliefs, goals, and personality traits and how, once aroused, emotions may in turn influence various aspects of R's mental state. The persuader uses this general model in a "what-if" reasoning mode, to predict the possible, emotional and non-emotional, consequences of a given communication. Specific knowledge about the recipient (her characteristics and dispositions) is introduced in the model as "evidence" available, and the consequences of propagating this evidence in the model are "observed." Of course, this kind of reasoning is presumptive and plausible; we did not describe in this chapter how we propose to deal with the various forms of uncertainty that may influence its results: for this particular aspect, please refer to Carofiglio et al. (2008).

Reasoning is only the first step of the persuasion task (Walton, 1990). Once a strategy has been selected, the persuader has to translate it into a good "persuading text." In fact, on the one hand, a text may fail to be persuasive as a result of the weakness or erroneousness of the selected persuasion strategy: typically, because the persuader's hypotheses about the receiver were not correct. In our example of emotions arising from sense of inferiority, self-esteem factors and attribution biases are the starting assumptions that make the difference between inducing envy or emulation: hence, an attempt to activate a goal of emulation by, e.g., saying "Yesterday I met Jane, she is really in good shape!" may fail just because P's assumptions about R's self-esteem were wrong. On the other hand, however, even a persuasion attempt which is grounded on P's "correct" reasoning may fail because its translation into a text was improper. If compared with the richness of human persuasion messages, the examples we included in this chapter suggest how difficult it is to generate a "good" text: the strategy must be instantiated into a "discourse plan" in which the items to mention, their presentation order, and the rhetorical relations among them have to be carefully established. The plan has then to be translated into a natural language message, implying a phase of "surface generation," in which careful choice of the syntax and wording of sentences must be applied (Mazzotta et al., 2007).

Some suggestions on how this task may be accomplished are already available: a variety of "argumentation schemes" have been proposed that formalize the structure of an argumentation text in terms of premises and conclusions (Walton, 1996b). Other studies (Kibble, 2006) have proposed a set of rhetorical relations that may be employed to strengthen internal consistency in these texts. However, these schemes mirror the prevailing attention toward forms of "legal" or "rational" argumentation and persuasion. In our view, much work has still to be done, and much reflection on the role of rhetorical relations is needed, to extend this list to the two forms of emotional persuasion we consider in this chapter.

Particular attention should be paid to the problem of which part of the reasoning process can (or should) be omitted from the persuasion text. For example, not all the nodes in the reasoning diagram displayed in Fig. 4 should enter the discourse plan. To start with, some of them may be omitted for the sake of simplicity and understandability of the message: for instance, sentences representing some of the intermediate nodes in this diagram. This reflects the classical view of enthymemes as "propositions not explicitly stated in the text of discourse, even though it may be clear enough that the speaker was relying on it, or including it, as part of the argument" (Walton, 2001): typically, common knowledge, known positions of the speaker, etc. In those cases, the speaker assumes that the receiver will likely fill those gaps, and that this will increase the intelligibility and strength of the persuasion message.

However, things are probably more complex than that, especially in emotional persuasion. In this case, as we said, accepting a suggestion is not the direct consequence of *accepting all the premises of the reasoning* followed by the persuader. As pointed out by Weaver (1967), "the missing proposition of an enthymeme is sometimes suppressed because the maker of an argument knows that, if we look carefully at his premises, we may question or reject some of them." Weaver goes on observing that much advertising, as well as a considerable part of political argumentation, is presented in the form of enthymemes for just this purpose. Other authors (Walton and Reed, 2005) also acknowledge that dialectical factors are involved in the use of enthymemes. In particular, in the context of a critical discussion, they observe that an arguer will try to *use premises that the audience accepts*. This leaves room to the possibility that the arguer also tries to *select*, among the available premises, the most "agreeable" ones and conversely tries to conceal the less "agreeable" ones, especially if weak or questionable in themselves.

We view persuasion through arousal of emotions as one of the exemplary cases in which enthymemes take the lion's share. They play a substantial role precisely because here some of the elements of the reasoning process are not only likely to be omitted, but *should* be omitted from the argumentation message, to avoid failing of the persuasive attempt. As we said in Sect. 3, the suspicion of manipulation is a powerful instigator of resistance to persuasion by the recipient, which may favor her counter-arguing and a final strengthening of her original attitudes. Persuasion through arousal of emotions may be perceived as such a form of manipulation, and therefore as *unfair* by R, if she detects or suspects that P is "playing with her emotions." Generation of an argumentation text in which this form of reasoning is applied by P should carefully take this risk into account, and argumentation schemes should be defined accordingly. We would suggest the role of enthymemes, and related problems, as one of the objects of future research on emotional persuasion.

References

Abramson LY, Seligman MEP, Teasdale JD (1978) Learned helplessness in humans: critique and reformulation. J Abnorm Psychol 87:49–74

Ajzen I (1985) From intentions to actions: A theory of planned behaviour. In: Kuhl J, Beckmann J (eds) Action control: from cognition to behavior., Springer, Berlin

Aristotle (1991) On rhetoric. A theory of civic discourse (trans: Kennedy GA). Oxford University Press, New York, NY

Bell DE (1985) Disappointment in decision making under uncertainty. Oper Res 33:1–27

Bell J, Huang Z (1997) Dynamic goal hierarchies. In: Bell J, Huang Z, Parsons S (eds) Proceedings of the second workshop on practical reasoning and rationality, University of Manchester, Manchester, UK

Ben-Ze'ev A (1992) Envy and inequality. J Philos 89:551–581

Bratman ME (1987) Intentions, plans, and practical reason. Harvard University Press, Cambridge, MA

Brehm JW (1966) A theory of psychological reactance. Academic, San Diego, CA

Burnell P, Reeve A (1984) Persuasion as a political concept. Br J Polit Sci 14:393–410

Buunk BP, Collins RL et al (1990) The affective consequences of social comparison: either direction has its ups and downs. J Pers Soc Psychol 59:1238–1249

Cacioppo JT, Petty RE (1979) Effects of message repetition and position on cognitive response, recall, and persuasion. J Pers Soc Psychol 37:97–109

Carofiglio V, de Rosis F, Grassano R (2008) Dynamic models of mixed emotion activation. In: Canamero D, Aylett R (eds) Animating expressive characters for social interactions. John Benjamins, Amsterdam

Castelfranchi C (1990) Social power: A missed point in DAI, MA and HCI. In: Demazeau Y, Mueller JP (eds) Decentralized AI. Elsevier, North-Holland

Castelfranchi C (1996) Reasons: belief support and goal dynamics. Mathw Soft Comput 3:233–247

Castelfranchi C, Guerini M (2007) Is it a promise or a threat? Pragmat Cogn 15:277–311

Castelfranchi C, Parisi D (1980) Linguaggio, Conoscenze e scopi. Il mulino. Bologna, Italy

Chaiken S (1980) Heuristic versus systematic information processing and the use of source versus message cues in persuasion. J Pers Soc Psychol 39:52–766

Chaiken S (1987) The heuristic model of persuasion. In: Zanna MP, Olson JM, Herman CP (eds) Social influence: the Ontario symposium, vol 5. Erlbaum, Hillsdale, NJ

Chaiken S, Liberman A, Eagly AH (1989) Heuristic and systematic information processing within and beyond the persuasion context. In: Uleman JS, Bargh JA (eds) Unintended thought. Guilford, New York, NY

Chaiken S, Maheswaran D (1994) Heuristic processing can bias systematic processing: effects of source credibility, argument ambiguity, and task importance on attitude judgment. J Pers Soc Psychol 66:460–473

Cohen PR, Levesque HJ (1990) Intention is choice with commitment. Artif Intell 42:213–261

Conte R, Castelfranchi C (1995) Cognitive and social action. University College London, London

Damasio AR (1994) Descartes' error. Avon Books, New York, NY

DeSteno D, Petty RE, Rucker DD et al (2004) Discrete emotions and persuasion: the role of emotion-induced expectancies. J Pers Soc Psychol 86:43–56

Eagly AH, Chaiken S (1993) The psychology of attitudes. Harcourt Brace Jovanovich, Fort Worth, TX

Epstein S (1992) Coping ability, negative self-evaluation, and overgeneralization: experiment and theory. J Pers Soc Psychol 62:826–836
Farber LH (2000) The ways of the will. Basic Books, New York, NY (Original work published 1966)
Farrell D (1989) Of jealousy and envy. In: Graham G, LaFollette H (eds) Person to person. Temple University Press, Philadelphia, PA
Festinger LA (1954) Theory of social comparison processes. Hum Relat 7:117–140
Fischhoff B (1994) What forecasts (seem to) mean. Int J Forecast 10:387–403
Fishbein M, Ajzen I (1975) Belief, attitude, intention, and behavior: an introduction to theory and research. Addison-Wesley, Reading, MA
Fogg BJ (1999) Persuasive technologies. Commun ACM 42:26–29
Frey D (1986) Recent research on selective exposure to information. In: Berkowitz L (ed) Advances in experimental social psychology, vol 19. Academic, San Diego, CA
Frijda NH (1986) The emotions. Cambridge University Press, New York, NY
Gilbert DT (1993) The assent man: mental representation and the control of belief. In: Wegner DM, Pennebaker JW (eds) Handbook of mental control. Prentice Hall, Englewood Cliffs, NJ
Gleicher F, Petty RE (1992) Expectations of reassurance influence the nature of fear-stimulated attitude change. J Exp Soc Psychol 28:86–100
Goethals GR, Darley JM (1977) Social comparison theory: an attributional approach. In: Suls JM, Miller RL (eds) Social comparison processes: theoretical and empirical perspectives. Hemisphere, Washington, DC
Gordon RM (1987) The structure of emotion. Cambridge University Press, Cambridge, UK
Guerini M, Stock O, Zancanaro M (2003) Persuasion models for intelligent interfaces. In: Carenini G, Grasso F, Reed C (eds) Workshop on computational models of natural arguments. International joint conference on artificial intelligence (IJCAI'03), Acapulco, Mexico
Haddadi A, Sundermeyer K (1996) Belief-desire-intention agent architectures. In: O'Hare GMP, Jennings NR (eds) Foundations of distributed artificial intelligence. Wiley, London
Heider F (1958) The psychology of interpersonal relations. Wiley, New York, NY
Hovland CI, Janis IL, Kelley HH (1953) Communication and persuasion: psychological studies of opinion change. Yale University Press, New Haven, CT
Intille SS, Kukla C, Farzanfar R, Bakr W (2003) Just-in-Time technology to encourage incremental dietary behavior change. In: Proceedings of the AMIA 2003 Symposium. http://web.media.mit.edu/~intille/papers-files/IntilleKuklaFarzanfarBakr03.pdf. Accessed 30 Mar 2007
Janis IL (1967) Effects of fear arousal on attitude change: recent developments in theory and experimental research. In: Berkowitz L (ed) Advances in experimental social psychology, vol 3. Academic, San Diego, CA
Janis IL, Feshbach S (1953) Effects of fear-arousing communications. J Abnorm Soc Psychol 48:78–92
Janis IL, Mann L (1977) Decision-making: a psychological analysis of compact, choice, and commitment. Free Press, New York, NY
Jorgensen PF (1998) Affect, persuasion, and communication processes. In: Anderson PA, Guerrero LK (eds) Handbook of communication and emotion. Academic Press, San Diego, CA
Kaney S, Bentall RP (1992) Persecutory delusions and the self-serving bias. J Nerv Ment Dis 180:773–780
Kellerman K, Cole T (1994) Classifying compliance gaining messages: taxonomic disorder and strategic confusion. Commun Theory 4:3–60
Kelley HH (1971) Attribution in social interaction. General Learning Press, Morristown, NJ
Keltner D, Ekman P (2000) Facial expression of emotion. In: Lewis M, Haviland-Jones JM (eds) Handbook of emotions, 2nd edn. Guilford, New York, NY
Kernis MH (2005) Measuring self-esteem in context: the importance of stability of self-esteem in psychological functioning. J Pers 73:1569–1605

Kibble R (2006) Dialectical text planning. In: Grasso F, Kibble R, Reed C (eds) Proceedings of 6th workshop on computational models of natural argumentation, Riva del Garda, Italy

King P, Tester J (1999) The landscape of persuasive technologies. Commun ACM 42:31–38

Langer EJ (1975) The illusion of control. J Pers Soc Psychol 32:311–328

Lazarus RS (1991) Emotion and adaptation. Oxford University Press, New York, NY

LeDoux JE (1996) The emotional brain. Simon & Shuster, New York, NY

Lerner JS, Keltner D (2000) Beyond valence: toward a model of emotion-specific influences on judgment and choice. Cogn Emot 14:473–493

Leventhal H (1970) Findings and theory in the study of fear communications. In: Berkowitz L (ed) Advances in experimental social psychology, vol 5. Academic, San Diego, CA

Levine TR, Wheeless LR (1990) Cross-situational consistency and use/nonuse tendencies in compliance-gaining tactic selection. South Commun J 56:1–11

Lockwood P, Kunda Z (1997) Superstars and me: predicting the impact of role models on the self. J Pers Soc Psychol 73:91–103

Loomes G (1987) Testing for regret and disappointment in choice under uncertainty. Econ J 97:118–129

Marwell G, Schmitt DR (1967) Dimensions of compliance-gaining behavior: an empirical analysis. Sociom 30:350–364

Mazzotta I, de Rosis F, Carofiglio V (2007) PORTIA: a user-adapted persuasion system in the healthy eating domain. IEEE Intell Syst 22:42–51

McGuire WJ (1960) Cognitive consistency and attitude change. J Abnorm Soc Psychol 60:345–353

McGuire WJ (1969) The nature of attitudes and attitude change. In: Lindzey G, Aronson E (eds) The handbook of social psychology, vol 3. Addison-Wesley, Reading, MA

Mellers BA, McGraw AP (2001) Anticipated emotions as guides to choice. Curr Dir Psychol Sci 10:210–214

Miceli M, Castelfranchi C (2002) Emozioni. In: Castelfranchi C, Mancini F, Miceli M (eds) Fondamenti di cognitivismo clinico. Bollati Boringhieri, Torino

Miceli M, Castelfranchi C (2007) The envious mind. Cogn Emot 21:449–479

Miceli M, de Rosis F, Poggi I (2006) Emotional and non emotional persuasion. Appl Artif Intell 20:1–31

Miller GA, Galanter E, Pribram KH (1960) Plans and the structure of behavior. Holt, New York, NY

Miller DT, Ross M (1975) Self-serving biases in the attribution of causality: fact or fiction? Psychol Bull 82:213–225

Mulholland J (1994) Handbook of persuasive tactics: a practical language guide. Routledge, London

Neu J (1980) Jealous thoughts. In: Rorty AO (ed) Explaining emotions. University of California Press, Berkeley, CA

O'Keefe DJ (2002) Persuasion: theory and research. Sage, Thousand Oaks, CA

Parker D, Manstead ASR, Stradling SG (1995) Extending the theory of planned behavior: the role of personal norms. Br J Soc Psychol 34:127–137

Parrott WG, Smith RH (1993) Distinguishing the experiences of envy and jealousy. J Pers Soc Psychol 64:906–920

Petty RE, Cacioppo JT (1981) Attitudes and persuasion: classic and contemporary approaches. Brown, Dubuque, IA

Petty RE, Cacioppo JT (1986) The elaboration likelihood model of persuasion. In: Berkowitz L (ed) Advances in experimental social psychology, vol 19. Academic, San Diego, CA

Petty RE, Cacioppo JT, Kasmer JA, Haugtvedt CP (1987) A reply to Stiff and Boster. Commun Monogr 54:257–263

Petty RE, DeSteno D, Rucker DD (2001) The role of affect in attitude change. In: Forgas JP (ed) Handbook of affect and social cognition. Erlbaum, Mahwah, NJ

Poggi I (2005) A goal and belief model of persuasion. Pragmat Cogn 13:297–336

Pratkanis AR, Aronson E (1991) Age of propaganda: the everyday use and abuse of persuasion. Freeman, New York, NY
Rhoads K (1997) Ethics of influence & persuasion. In: Working Psychology. Introduction to influence. The Working Psychology Web site http://www.workingpsychology.com/ethics.html. Accessed 30 May 2010
Richard R, Van der Pligt J, De Vries N (1996) Anticipated affect and behavioral choice. Basic Appl Soc Psychol 18:111–129
Roberts R (1991) What is wrong with wicked feelings? Am Philos Q 28:13–24
Rogers RW (1983) Cognitive and physiological processes in fear appeals and attitude change: A revised theory of protection motivation. In: Cacioppo JT, Petty RE (eds) Social psychophysiology: a source book. Guilford, New York, NY
Rosen TJ, Terry NS, Leventhal H (1982) The role of esteem and coping in response to a threat communication. J Res Pers 16:90–107
Rosenblueth A, Wiener N, Bigelow J (1968) Behavior, purpose and teleology. In: Buckley W (ed) Modern system research for the behavioral scientist. Aldine, Chicago, IL
Rosselli F, Skelly J, Mackie DM (1995) Processing rational and emotional messages: the cognitive and affective mediation of persuasion. J Exp Soc Psychol 31:163–190
Schoeck H (1969) Envy: a theory of social behaviour. Harcourt, Brace & World, New York, NY
Schwarz N (1990) Feelings as information: informational and motivational functions of affective states. In: Higgins ET, Sorrentino RM (eds) Handbook of motivation and cognition: foundations of social behavior, vol 2. Guilford, New York, NY
Silver M, Sabini J (1978) The perception of envy. Soc Psychol 41:105–117
Smith RH (1991) Envy and the sense of injustice. In: Salovey P (ed) The psychology of jealousy and envy. Guilford, New York, NY
Smith CA, Ellsworth PC (1985) Patterns of cognitive appraisal in emotion. J Pers Soc Psychol 48:813–838
Smith CA, Lazarus RS (1993) Appraisal components, core relational themes, and the emotions. Cogn Emot 7:233–269
Sutton SR (1982) Fear-arousing communications: A critical examination of theory and research. In: Eiser JR (ed) Social psychology and behavioral medicine. Wiley, New York, NY
Taylor G (1988) Envy and jealousy: emotions and vices. Midwest Studies Philos 13:233–249
Taylor SE, Brown JD (1988) Illusion and well-being: a social psychological perspective on mental health. Psychol Bull 103:193–210
Taylor SE, Wayment HA, Carrillo M (1996) Social comparison, self-regulation, and motivation. In: Sorrentino RM, Higgins ET (eds) Handbook of motivation and cognition. Guilford, New York, NY
Testa M, Major B (1990) The impact of social comparison after failure: the moderating effects of perceived control. Basic Appl Soc Psychol 44:672–682
Tooby J, Cosmides L (1990) The past explains the present: emotional adaptations and the structure of ancestral environment. Ethol Sociobiol 11:375–424
Tormala ZL, Petty RE (2002) What doesn't kill me makes me stronger: the effects of resisting persuasion on attitude certainty. J Pers Soc Psychol 83:1298–1313
Tormala ZL, Petty RE (2004) Resistance to persuasion and attitude certainty: the moderating role of elaboration. Pers Soc Psychol Bull 30:1446–1457
Walton D (1990) What is reasoning? What is an argument? J Philos 87:399–419
Walton D (1996a) Practical reasoning and the structure of fear appeal arguments. Philos Rhetor 29:301–313
Walton D (1996b) Argumentation schemes for presumptive reasoning. Erlbaum, Mahwah, NJ
Walton D (2001) Enthymemes, common knowledge and plausible inference. Philos Rhetor 34:93–112
Walton D, Reed CA (2005) Argumentation schemes and enthymemes. Synth 145:339–370
Wayment HA, Taylor SE (1995) Self-evaluation processes: motives, information use, and self-esteem. J Pers 63:729–757

Weaver RM (1967) A rhetoric and handbook. Holt, Rinehart and Winston, New York, NY
Wheeler L, Martin R, Suls J (1997) The proxy model of social comparison for self-assessment of ability. Pers Soc Psychol Rev 1:54–61
Wicklund RA (1974) Freedom and reactance. Wiley and Sons, New York, NY
Witte K (1992) Putting the fear back into fear appeals: the extended parallel process model. Commun Monogr 59:329–349
Wood JV (1989) Theory and research concerning social comparisons of personal attributes. Psychol Bull 106:231–248
Young R (1987) Egalitarianism and envy. Philos Stud 52:261–276
Zeelenberg M, Beattie J (1997) Consequences of regret aversion 2: additional evidence for effects of feedback on decision making. Organ Behav Hum Decis Process 72:63–78
Zuwerink JR, Devine PG (1996) Attitude importance and resistance to persuasion: it's not just the thought that counts. J Pers Soc Psychol 70:931–944

Approaches to Verbal Persuasion in Intelligent User Interfaces

Marco Guerini, Oliviero Stock, Massimo Zancanaro, Daniel J. O'Keefe, Irene Mazzotta, Fiorella de Rosis, Isabella Poggi, Meiyii Y. Lim, and Ruth Aylett

1 Introduction

People tend to treat computers as social actors, even if these are usually designed as mere tools. This forces computers to play a social role without having the social skills to be successful (Reeves and Nass, 1996).

Persuasion is likely to become a hot topic for intelligent interfaces (Stock et al., 2006). As opposed to traditional scenarios of intelligent user interfaces (hereafter IUI), future intelligent systems may have contextual goals to pursue that aims at inducing the user, or in general, the audience, to perform a specific action in the real world.

These systems will have to take into account the social environment, exploit the situational context, and enhance emotional aspects in communication. Scenarios that can be envisaged include dynamic advertisement, preventive medicine, social action, and edutainment. In all these scenarios what really matters is not just the content, but the overall impact of the communication.

In this chapter, we will consider natural language generation (NLG) of persuasive messages (either written or spoken), as the central aspect of communication. NLG is the branch of natural language processing that deals with the automatic production of texts (Reiter and Dale, 2000).

The aim is to provide an overview of theories and systems that have the capability of reasoning about the effectiveness of the message, as well as about the high-level goals and content. We will also consider those aspects of multimodal realization that are strictly connected to NLG (e.g. markup languages, see the final chapter in Part IV).

M. Guerini (✉)
Fondazione Bruno Kessler-Irst, Povo, Trento, Italy
e-mail: guerini@fbk.eu

Fiorella de Rosis is deceased.

P. Petta et al. (eds.), *Emotion-Oriented Systems*, Cognitive Technologies,
DOI 10.1007/978-3-642-15184-2_29,

This chapter is structured as follow: Sect. 2 introduces the basis about persuasive communication (theories, main dimensions included in the phenomena, and related concepts). Sect. 3 describes the connection between NLG and persuasion, along with the main areas of research and challenges. Finally, Sect. 4 presents a review of persuasive systems.

2 Persuasive Communication and Related Concepts

According to Perelman and Olbrechts-Tyteca (1969) persuasion is a skill that human beings use – in communication – in order to make their partners perform certain actions or collaborate in various activities (a similar definition has been proposed by Moulin et al., 2002). Historically, in the human sciences various definitions similar to Perelman and Olbrechts-Tyteca's have been proposed. Most of them have a common core, addressing

methodologies aiming at changing, by means of communication, the mental state of the receiver.

We refer to the chapter by M. Miceli et al., this volume for a thorough analysis. In these definitions there are three elements at play that are strictly interleaved: the persuader, the persuadee, and the message. All these elements are necessary in explaining how the persuasive process occurs, although some definitions may focus on one element more than on the others, as below:

- Those focusing mainly on message structure – for example, how an argument can be structured to be effective (Toulmin, 1958; Perelman and Olbrechts-Tyteca, 1969; Walton, 1996).
- Those that address the persuader side and the cognitive processes that take place in their mind for generating effective arguments – e.g., the selection of the "heuristics" through which the receiver may be persuaded (e.g., Chaiken, 1980, or Cialdini, 1993).
- Those that focus on the persuadee side trying to understand on which cognitive processes persuasive messages hinge on. Petty and Cacioppo (1986), for example, focus on the inferential routes that the receiver uses to process the message (central or peripheral, i.e., rational or "affective").
- Others that broadly refer to the beliefs and goals of both persuader and persuadee and how they are related (Poggi, 2005).

2.1 Dimensions of Verbal Persuasion

There are several dimensions of persuasion along which broad areas of study can be classified. To build persuasive systems it is necessary to individuate systematically these dimensions of persuasion at play and the kind of interaction we want to address:

1. Action vs. behavior and attitude inducement:

   a. Behavior inducement – Changing, in a stable and persistent manner, the way an agent acts, for example, in response to certain events or state of affairs in the world. This is meant to be a long-term effect.
   b. Attitude inducement – Changing, in a stable and persistent manner, the way an agent evaluates events, state of affairs, or objects. This effect is also long term.
   c. Action inducement – Changing a particular planned action of an agent. This effect is short term.

2. Argumentation specific vs. fully persuasive:

   There are situations in which a simple argumentative approach may suffice in creating effective messages (e.g., when persuadee "resistance" to changing is low or the performing of the required action is more a matter of knowledge rather than a matter of will). In other situations mixed approaches are needed.

3. Audience specific vs. universal

   a. One of the definitions of persuasion given by Perelman and Olbrechts-Tyteca (1969) claims that what characterizes persuasion is it being audience specific, namely its capacity of adapting the topic to the specific listeners.
   b. On the other hand, Cialdini (1993) takes the opposite position: all the strategies he analyzes are meant to be universal (since they use cognitive patterns of the receiver which are common to everybody).

4. Monological vs. dialogical

   a. Perelman's analysis of persuasion, since concerned with rhetoric (how to create effective discourses), is more involved with monological interaction.
   b. Cialdini's analysis of persuasion, since concerned mainly with selling scenarios, is focused on dialogical interactions (e.g., *foot in the door* strategy, *door in the face* strategy).

5. Domain specific vs. universal

   a. Some strategies are typically domain specific (like *fake discount* strategy to sell more)
   b. Other strategies are universal, that is, applicable to every situation (like *fear appeals* strategy with impressionable persons)

6. Sole language vs. multimodality

   This distinction is relevant especially with the coming of the new media and for IUI, it is a crucial field of research.

2.2 Elements Addressed by Persuasion

Persuasion mechanisms include the following four aspects:

1. The cognitive state of the participants – The beliefs and goals of both persuader and persuadee.
2. Their social relations – Social power, shared goals, and so on.
3. Their emotional state – Both persuader's and persuadee's.
4. The context in which the interaction takes place.

A brief description of these aspects follows:

Beliefs and goals of both the participants about the domain of the interaction are prerequisite for persuasive interaction, because persuasion is a communication leading to belief adoption, with the overall goal of inducing an action by the user by modifying their preexistent goals (see the Chapter by M. Micel; et al., this volume).

Social relations exist between persuadee and persuader (that can play the role of a museum guide, a car seller, etc.) and between persuadee and other relevant persons such as experts and parents.

Emotional elements can enhance or lower message effectiveness. Gmytrasiewicz and Lisetti (2001) propose a useful framework on how the emotions *felt* by an agent can change their own behavior. Still, for persuasive purposes, the focus should be posed on how emotional elements (either on persuader or persuadee side) can be *used* to increase or diminish the persuasiveness of a message. There are four dimensions to be considered:

1. The current emotional state of the persuadee – How it affects the strategy selection of the persuader.
2. The emotional state expressed by the persuader – What emotion the persuader must display to maximize the persuasive force of the message.
3. The emotional state possibly produced in the persuadee by the message – The induced emotional state may not be desirable, and it may be necessary to take it into consideration for subsequent interactions.
4. The current emotional state of the persuader – How it affects their strategy selection.

It is still a matter of debate whether the current emotional state of the persuader should actually be taken into consideration in persuasive interfaces. The two main standpoints are as follows:

1. A perfect persuasive agent should be emotion neutral; they just have to display the most effective emotion for the current persuasive goal.
2. For a persuader, to feel emotions is a good way to handle unpredicted situations and a resource for responding to the persuadee's moves.

2.3 Related Concepts

Social influence Affecting or changing how someone behaves or thinks (by changing their mental state). Social influence is a superset of persuasion since the aforementioned core definition of persuasion (see the previous section) restricts the field of coverage by making reference to both the concepts of "aim" and "communication." The term "aim" indicates that persuasion is an intentional process: there is persuasion only when there is the intention to produce a change in the mental state of the receiver (this is not the case, for example, of unintended induction of emulation phenomena). The term "communication" rules out those effects of (social) influence that, for example, are caused by mere exposition to repeated stimuli.

Negotiation = Broadly speaking, negotiation is an interactive process between two or more parties trying to influence each other to achieve goals which they cannot (or prefer not to) achieve on their own; these goals may conflict or depend upon one another. Negotiation is thus a form of alternative dispute resolution. Even though persuasion can be a resource for negotiation, in negotiation participation is voluntary (whereas in persuasion, it can be unwilled), and the structure is intrinsically dialogical (whereas in persuasion, it can also be monological).

Argumentation = Intuitively, when talking about the relation between argumentation and persuasion, a dichotomy between these two concepts is put forward. The former is seen as a process that involves "rational elements," while the latter uses "arational elements" like emotions. In our view, however, argumentation is a resource for persuasion because

1. Planning of persuasive messages involves a "rational" activity, even when emotion inducement is employed as a means to increase the persuasion strength of a message. On the other hand, the way persuasion is performed (items selected, their order of presentation, their "surface" formulation) also depends on the emotional state of the persuader.
2. Argumentation is concerned with the goal of making the receiver believe a certain proposition (influence their mental state) and, apart from coercion, the only way to make someone do something (persuasion) is to change their beliefs (Castelfranchi, 1996).

Since persuasion includes "arational" elements as well, it is a "superset" of argumentation, but this does not rule out the fact that there is a role for emotion within argumentation (Miceli et al., 2006): through arousal of emotions (see Rhetoric) or through appeal to expected emotions. In classical argumentation, though, these problems are not addressed since emotional argumentation is often considered as some sort of "deceptive" argumentation (Grasso et al., 2000).

In our view a better distinction between argumentation and persuasion can be drawn considering their different foci of attention: while the former is focused on the *correctness* of the message (it being a *valid* argument) the latter is more concerned with its *effectiveness*. The point is that an argument can be valid but not effective or, on the contrary, can be effective but not valid (as an example see the discussion about the goat problem explanation by Horacek, 2006).

Natural argumentation = The recent area of natural argumentation tries to bridge argumentation and persuasion by focusing, for example, on the problem of the adequacy – effectiveness – of the message (Fiedler and Horacek, 2002). Even in professional settings, such as juridical argumentation, extra-rational elements can play a major role (Lodder, 1999). Recent works have studied applications of natural argumentation (Walton and Reed, 2002; Das, 2002); argumentation-based text generation has been proposed by Zukerman and colleagues (2000), relying on a Bayesian approach. Negotiation has also been widely investigated and modeled in a computational framework; see, for instance, Kraus et al. (1998) and Parson and Jennings (1996). A thorough survey on the area can be found in the work by Reed and Grasso (2007).

Coercion = using force to "persuade" someone to do something they are not willing to do. Obviously coercion falls out of the definition of persuasion.

Rhetoric = the study of how language can be used effectively. This area of studies concerns the linguistic means of persuasion (one of the main means, but not the only one).

Irony = It refers to the practice of saying one thing while meaning another. Irony occurs when a word or phrase has a surface meaning, but another contradictory meaning beneath the surface. Irony is a widely used rhetorical artifice, especially in advertisement. For a more detailed description, please see the last chapter in this Part.

3 Natural Language Generation for Persuasion

3.1 What Is NLG

At the root of the theme of intelligent information presentation we can consider several scientific areas, but at least NLG is fundamental. NLG is the branch of natural language processing that deals with the automatic production of texts in human languages, often starting from non-linguistic input (Reiter and Dale, 2000). Normally the field is described as the investigation of communicative goals, the dynamic choice of what to say, the planning of the overall rhetorical structure of the text, and the actual realization of sentences on the basis of grammar and lexicon. The three-stages model proposed in Reiter and Dale (2000) is usually taken as a reference:

1. *Document planning* that decides the content and the structure of the message to be generated (sometimes called *strategic planning*).
2. *Microplanning* that decides how information structure should be expressed linguistically, involving mainly lexical choice.

3. *Surface realization* that generates the final output according to the decisions of the previous stages and according to, for example, grammatical and anaphoric constraints (sometimes called *tactical planning*).

NLG approaches can be roughly divided into two areas (Reiter et al., 2003b):

1. Knowledge-based approaches (for an overview see Scott et al., 1991)
2. Statistically based approaches (for an overview see Jurafsky and Martin, 2000).

While at the beginning of NLG the main approaches have been knowledge based, in recent years, statistical approaches are becoming widely used (especially not only for tactical planning – e.g., by using N-grams (Langkilde and Knight, 1998) – but also for strategic planning (Duboue and McKeown, 2003)). Other approaches such as the one proposed by Radev and McKeown (1997) rely on automatic acquisition of sentences mapped on their functional description, to overcome the problem of simple canned texts extraction. More recently, Guerini et al. (2008a) investigated an approach for persuasive natural language processing based on the analysis of a specific corpus of political speeches tagged with audience reactions.

3.2 NLG and Monological Persuasion

Most systems and approaches in NLG are based on descriptive tasks, focusing on texts which realize a single, often informative, communicative goal, as opposed to persuasive NLG where the communicative goal is usually surmounted by reasoning about the persuadee's behavior modification.

Persuasive features can have an impact on both strategic and tactical levels since the effectiveness of a message can be enhanced by appropriate content selection, text planning, and linguistic choices.

1. In strategic planning, for example, a widely used reference theory is the one proposed by Mann and Thompson. This theory, called Rhetorical Structure Theory (RST) (Mann and Thompson, 1987) – formalized by Marcu (1997) – puts forward the idea that the structure of many texts is a tree built recursively starting from atomic constituents (e.g., clauses) connected through particular relations. These relations, called rhetorical relations (RRs), account for the structure and content ordering of the text. In almost every relation a text span plays a major role: this is often referred to as "nucleus" (as opposed to "satellite" that plays an ancillary role). There have been various attempts and debates about the possible use of RRs in persuasive message generation, e.g., Marcu (1996), Reed and Long (1998), Kibble (2006), and Guerini et al. (2004).
2. In tactical planning, for example, the lexical choice of what to say can be – persuasively – driven by reasoning on the different emotional impact that words can have in conveying a given meaning: techniques which allow the speaker to

slant a text (e.g., selecting “kick the bucket” instead of “die”) or stress particular concepts (e.g., by means of repetition) are needed. For a general discussion about this point see chapter 4 of Hovy (1988); for a specific survey on affective lexicalization components see de Rosis and Grasso (2000) or Piwek (2002). An interesting approach has been proposed by Fleiscman and Hovy (2002).

Multimodal generation addresses similar problems as NLG, but goes “a step further” since the message is communicated across more than one modality, and persuasive features in this case can have a deep impact.

3.3 NLG and Dialogical Persuasion

Instead of being seen as a predefined, integrated set of propositions, in a dialogical perspective argumentation is seen as a sequence of moves in which two parties (persuader and persuadee) are reasoning together on some subjects. The dialogue may be more or less symmetrical, as far as the initiative in persuasion and argumentation is concerned: the role of persuader and persuadee may be fixed, or alternate, during interaction.

Dialogic persuasion is not restricted to situations in which two parties are trying to resolve a conflict of opinion or attempt to influence each other’s behavior. Some argumentative exchanges may occur in almost any kind of context: one of the most recent examples is the case of Online Dispute Resolution, in which an arbitration environment supports communication and discussion in web-based groups (Vreeswijk and Lodder, 2005; Walton and Godden, 2005). In Walton’s New Dialectic (Walton, 1998), six basic types of dialogues are proposed:

- Persuasion: the goal is to resolve a conflict of opinion;
- Inquiry: the goal is to find or verify evidence and proving or disproving a hypothesis.
- Negotiation: the starting point is a conflict of interests; the goal is to find a compromise between what each participant wants;
- Information seeking: the goal is to acquire or give information;
- Deliberation: the goal is to decide the best available course of action;
- Eristic: the goal is to reveal the basis of a conflict.

In some of these dialogues, persuasion and argumentation are central, while in others they may enter only in some phases. On the other hand, during an argumentation sequence, there can be “dialectical shifts” of context from one type of dialogue to another. This makes the modeling task particularly difficult.

While monological persuasion is characterized by the three stages of planning (see Sect. 3.1) in “pure” persuasion dialogues, the sequence of exchanges includes some typical phases, and forms of reasoning, by the persuader:

- Make a proposal: after reasoning on the persuadee's mind, propose an action or a claim by giving reasons as grounds for supporting the proposal;
- Observe the persuadee's reaction: what does they say or express differently;
- Classify it (a request of justification, an objection, with or without counter-argumentation, a refusal, etc.);
- Reason (again) on the persuadee's mind to interpret the persuadee's reaction by placing it into her presumed set of attitudes: this requires a belief–desire–intention model of mind and reasoning (Wooldridge, 2002), eventually enhanced with emotions in a BDI&E model (Carofiglio and de Rosis, 2005);
- Justify it or defend the own proposal if possible; retract it if needed, find an alternative, and relate new argumentation to the previous one.

A proposal may be criticized by the persuadee in several ways (e.g., by questioning the goal premises, by attacking them with counter-arguments, by undercutting the inferential link between premises and conclusion (Walton, 2006a)), and the persuader must be able to respond appropriately to all the situations.

The complexity of Walton's distinction between "reasoning and argumentation" (Walton, 1990) – that is, between a phase of reasoning on the persuadee's mind to select an appropriate strategy and a phase of translation into a coherent message – increases when argumentation becomes dialogic. At every dialogue step, the persuader must decide on which part of their reasoning to make explicit in generating the argument and which one to hide or to postpone. In addition, a refined ability to "interpret and reason on the persuadee's reaction" must be added to the system.

This new reasoning ability becomes quite complex when context, personality, and emotional factors are considered: research about consumers' behaviors and attitudes contributed considerably in increasing knowledge in this domain. For example, not only the message features but also the persuadee's perceived features, like "credibility," "likability," "attractiveness," (O'Keefe, 2002), must be considered, as well as the persuadee's evaluation of the effectiveness and appropriateness of the persuasion tactics (Friestad and Wright, 1994).

On the other hand, receivers may be vulnerable to blatant persuasion attempts (Bosmans and Warlop, 2005). They may also be biased toward a persuasive attempt, being skeptical, defensive, or hostile, either in general or toward a particular persuader (Ahluwalia, 2000). This kind of "resistance" can include different mixtures of rational and emotional components (Coutinho and Sagarin, 2006).

More in general, evaluation of persuasion attempts by persuadees may be influenced by affective factors. Some of these factors are stable (like personality traits), others are more or less transient. For instance, positive mood seems to reduce systematic processing of information, whereas negative mood enhances it (Hullett, 2005); positive feelings lead to more positive evaluation of information received, while the opposite seems to hold for negative feelings (Petty et al., 1991). If a persuasion move aims at influencing the persuadee's attitude, it has been demonstrated that the persuader's attitudes are influenced, in turn, by the success or failure of their persuasion attempts: this is known as a referral-backfire effect (Geyskens et al., 2006).

Although theoretical aspects of dialogic persuasion have been extensively investigated in the philosophical and the marketing studies domains, examples of dialogical persuasion prototype systems are few and quite recent.

3.4 Other Aspects of Verbal Persuasion: Storytelling

A recent topic of research in verbal persuasion deals with narration and storytelling.

Story represents a fundamental structuring of human experience, both individual and collective (Young, 2001). According to the constructivist theory, people are not passive recipients of their experience but active constructors of their own reality through mental activity (Piaget, 1972). Hence, story can be viewed as a process of sense making via narrative organization, the process which Aylett (2000) termed as the storification process. It is a specific mechanism through which the real world can be created in the imagination of the receivers.

Story has been widely used in subjects where temporal complexity and social interaction are significant issues. These include language development, literature study, history, and social science. Personal, social, and health education, in which the pedagogical aims involve attitudes and behaviors and not just knowledge, has also applied role-play widely. Social interaction has been used in educational role-play as the stimulus for challenging and changing existing beliefs (Piaget, 1972) and can result in significant behavioral changes (Lewin, 1951). Surprisingly, there is not much research on persuasion up to date that attempts to use story as a source of persuasive message, considering its high significance for social and emotional learning (Davison and Arthur, 2003; Henriksen 2004).

The nature of story as persuasive message can vary depending on the information that the persuader has about the persuadee. If the persuader has a complete model of the persuadee's current emotional and cognitive states, a full story can be generated using natural language approaches discussed in the previous sections. If this information is not available, improvisational storytelling techniques (Ibanez, 2004; Lim, 2007) may be adopted, whereby the persuader shapes the story dynamically based on the persuadee's feedback as in dialogical situations.

Since a story molds and reflects human experiences, emotions and personality are important elements that contribute to its structure and content. According to Nass et al. (1995), in human–computer interaction, people prefer computer agents that align to them and they tend to rate these agents as more helpful and more intelligent. This implies that by assigning the story with a personality that conforms to the persuadee's personality, an improved persuasive effect may be achieved.

A particular viewpoint can also be slipped into the persuadee more easily if the persuader can induce the persuadee's mental attitude and mold a message or story to suit the persuadee's feeling. This is due to the fact that people tend to believe rumors or arguments that are consonant to their emotional attitude (Frijda et al., 2000), that is, we pay attention to what interests us. Emotions can also generate functional beliefs that help to achieve emotional goals such as increasing one's sense of competence, making one feel better, or getting rid of dissonance as demonstrated by Harmon-Jones (2000). So, a persuader can persuade the persuadee to look at a particular subject or event from its own point of view and experiences, challenging or changing their existing beliefs, and structuring their mental picture of the natural world by invoking empathy through stories. For more arguments on how emotions can lead to belief changes in society, please refer to Lim (2007).

From the above discussion, the use of story as persuasive message seems to be a promising direction for further study.

3.5 Meta-analytic Research for Persuasive NLG

There already exists a substantial body of social-scientific theory and research concerning persuasive communication. This work offers the possibility of informing the development of NLG persuasion systems, by providing empirical evidence concerning (for example) how specific message variations affect persuasive outcomes. However, the task of obtaining dependable generalizations about factors influencing persuasive effectiveness has been hampered in two ways. First, on many research questions, relatively little research evidence is in hand. No single study of a particular message variation can provide good evidence for broad generalizations about the effects of that factor. It would obviously be unwise to design NLG persuasion systems based on poorly evidenced conclusions. Second, even when a substantial number of studies have been conducted on a given question, relatively little attention has been devoted to research organization and synthesis in this area. However, in recent years, a number of meta-analytic reviews of persuasion effects research have appeared, including reviews of the persuasiveness of one-sided and two-sided messages (O'Keefe, 1999a), fear appeals (Witte and Allen, 2000), guilt appeals (O'Keefe, 2000), the use of metaphor (Sopory and Dillard, 2002), conclusion explicitness (O'Keefe, 1997), and gain-framed and loss-framed appeals (O'Keefe and Jensen, 2006).

Considered broadly, these meta-analytic reviews point to two general conclusions about how message variations can influence persuasive success (O'Keefe, 1999b). First, the average effect size (the magnitude of difference that it makes to persuasiveness to use one kind of appeal as opposed to another) is usually relatively small; expressed as a correlation, average effect sizes rarely exceed 0.20 and commonly are in the neighborhood of 0.10. Second, for any given factor, effect sizes are usually quite variable from one study to another. This finding obviously underscores the dangers of relying on any single study's results (in trying to identify dependable generalizations), but it also emphasizes that the implementation (in an NLG persuasion system) of any specific message design principle is likely to itself produce variable effects from one instance (one application) to another. Taken together, these two conclusions imply that there is no "magic bullet" for persuasion – no message variation that consistently (in every instance) produces large increases in persuasiveness. Even so, NLG persuasion systems should be designed in ways that make use of the accumulated knowledge about how to maximize the likelihood of success of persuasive messages.

3.6 Challenges

In this section a list of the main challenges that are to be taken into consideration for building effective persuasive IUI is presented.

1. *Porting of strategies*. There is a rich repertoire of persuasive strategies coming from human sciences (in addition to those already mentioned) that can be used for persuasive and multimodal NLG. Most of them address specific aspects of persuasion, for example, guilty feeling induction (Miceli, 1992), use of promises and threats (Castelfranchi and Guerini, 2007), ordering of positive or negative arguments to create proper frames in persuadee (Prospect Theory (Kahneman et al., 1982)). This collection is not structured: social, emotional, and cognitive aspects interact with each other. There is the need for a porting and structuring of such concepts. The *media equation* supports this view (Reeves and Nass, 1996).
2. *Knowledge representation*. To simulate natural argumentation and (emotional) persuasion, it is necessary to define new methods for representing knowledge, for reasoning on it, and for generating natural language and multimodal messages (both in monological and dialogical situations). Starting from the continuum which characterizes the various (emotional and non-emotional) persuasion modes, a framework which tries to unify the various items of this continuum must be given (see, for example, Guerini et al. 2003). Investigating the various (emotional and non-emotional) persuasion and argumentation strategies (like those in Prakken et al. 2003 Toulmin 1958) and proposing a method to formalize them, by representing the various sources of uncertainty and incomplete knowledge they may include, is just the first step. Related aspects, fundamental for dialogue interactions, like critical questions, counter-arguments, must be taken into consideration.
3. *Measures needed in persuasion.* To handle the problem of uncertainty, to model the concept of effectiveness of a message, and to foster the process of choosing the best strategy to be used at every interaction, it is necessary to furnish models of measurement of the strength of persuasive strategies and of other related concepts such as argumentation strength (Sillince and Minors, 1991), probative weight (Walton, 2000), dialectical relevance (Walton, 1999), and impact (Zukerman et al., 1999; Zuckerman, 2001). Various methods have been proposed for representing uncertainty in this domain: see for instance, BIAS (Zukerman et al., 1999; Zuckerman, 2001) and Carofiglio (2004). Starting from these concepts more complex measures can be created, like emotional impact of a message that has been explored in Carofiglio and de Rosis (2003).
4. *User modeling*. To produce *effective* communication, tailored messages are needed (regardless of the universal-or audience-specific nature of the used persuasive strategies). Detailed *user models* are then necessary for tailoring the message. These models can have different degrees of complexity depending on

the kind of interaction modeled (static user model for monological interaction may suffice, dynamic user models are instead necessary for dialogical situations).
5. Integrated models of emotion manipulation and beliefs and goal induction are necessary; they can use BDI&E (belief, desire, intention, and emotions) approaches to model how intentions and commitments are produced (and induced). These models can be used not only to describe the characteristics of persuadee (and how these characteristics are affected by a persuasive message) so to select the "best" persuasive move the system can make, in a given situation; they can also be used to describe the process of persuasion itself (i.e., the planning of a persuasive message).
6. *Multimodality*. The realization of a persuasion message requires the expression in a communication language. In most approaches natural language is the main modality, but it can be combined with music, kinetic typography, ECAs, and so on. For instance, a talking head may express the mood of the message originator, a music theme may emphasize a given emotional aspect, or simply, a relevant image can be combined with the produced text. Multimodality poses lots of challenges: for example, the question whether an ECA should be credible or realistic (it can be argued that, with children, cartoon-style ECAs are more credible – and persuasively effective – than realistic ones).
7. *Evaluation*. Persuasive systems need to be evaluated. That is, it is not sufficient that they are theoretically sound: they also have to be proven effective with real users. Evaluation is not straightforward at all: it is necessary to point out carefully all the variables that can affect the effectiveness of the system and how they can correlate (context of use, scenario of the interaction, typology of the user, required task, persuasive strategies at hand, and so on). Specific evaluation methodologies have to be defined.
8. *Indirect aspects*. "Indirect aspects" like attention and memorization can affect the effectiveness of persuasive messages. For example, if the attention of the user is low, or there are key concepts persuader wants to stress, then persuadee's attention has to be focused or enhanced by using various means. Among them we consider of high importance the use of irony or affectively "colored" terms. Similar considerations can be made about memorization.
9. *Ethical reasoning*. Finally, ethical issues must be addressed. As artificial agents are becoming more complex and common in our everyday life, the need for an ethical design of such agents is becoming more compelling, especially if, as in the present case, the focus is on persuasiveness. A set of principled guidelines for design and implementation of ethical persuasive agents is necessary (the seminal work of Berdichevsky and Neuenschwander (1999) goes in this direction). Future challenges will address meta-planning models of ethical reasoning, see, for example, Guerini and Stock (2005).

These challenges are strictly interconnected; modeling decisions of one aspect often have consequences on other aspects.

4 Persuasive Systems

In the following section we provide an overview of various theories and systems addressing verbal persuasion issues for IUI. These systems have been classified according to the most salient characteristics they address (monological, argumentative and dialogical, emotional, multimodal aspects). The fact that a system appears in one group does not rule out the fact that it can also address aspects of the other groups.

4.1 Monological Persuasion Systems

The area of health communication is one of the first where the potentials of persuasive features for NLG were investigated. In particular, the focus was posed on tailoring messages (for a detailed overview on NLG tailoring systems for health communication, see Kukafka, 2005). Here we will mention just a couple of these systems, then we will focus on general purpose (not domain-specific) systems.

- STOP is one of the best known systems for behavior inducement that exploited persuasion (Reiter et al., 2003a). STOP is (mainly) an NLG system aiming at inducing users to stop smoking. It produces a tailored smoking-cessation letter, based on the user's response in a questionnaire. The NLG process follows the classical three-stage pipeline as described in Reiter and Dale (2000). The tailoring process is done by dividing smokers into categories and then applying category-specific schema to generate the letter. The main limitation is that this approach is strongly domain specific and all the reasoning is based on expert-oriented knowledge acquisition for the clinical smoking domain (Reiter et al., 2003b). This renders the system unportable.
- *Migraine* (Carenini et al., 1994) is a natural language generation system for producing information sheets for migraine patients. It consists of three main components: (a) an interactive history-taking module that collects information from patients; (b) an intelligent explanation module that generates explanations tailored to individual patient and responds to follow-up questions; and (c) an interaction manager that presents the interactive information sheet on the screen and manages the subsequent interaction with the patient. *Migraine* is different from STOP in that it adds interactivity to the interaction with the patients, furnishing some limited form of "dialogical" interaction (in a question answer fashion, by clicking on portions of the text and by selecting available questions from menus).
- *Promoter* (Guerini et al., 2007; Guerini, 2006) is a prototype that uses strategies gathered from different persuasive theories and subsumed in a general planning framework for multimodal message generation. The planning framework consists of a taxonomy of strategies (rules that have some applicability conditions based on the social, emotional, and cognitive context of interaction) and a meta-reasoning module (used for content selection, ordering, and modification

to create complex messages). *Promoter*, by means of selection theorems, also accounts for the interaction between persuasion and rhetorical relations selection.

- Another persuasive NLG system is the one presented by Reed and colleagues (Reed et al., 1996, 1997; Reed and Long, 1997) that uses two modules, argument structure (AS) and eloquence generation (EG), connected to the LOLITA system (Smith et al., 1994) for natural language realization. In this approach the system starts with a generic goal of making the user believe a certain proposition. The AS module produces a logical form of the argument, employing various logical and fallacy operators. Then the EG level modifies this structure by using heuristics to render the message more persuasive. The main limitation is that the system generates an "argumentative" text and then modifies it to make it (more) persuasive. But persuasion is a phenomenon that drives text planning, not simply a "modifier" of the process.
- DIPLOMAT (Kraus and Lehmann, 1995) is a negotiating automated agent built for playing the *Diplomacy* game with human users. The ability to negotiate requires different persuasive skills. In particular this agent is able to perform persuasive actions like threatening, promising, giving explanations and in general it is designed to convince other players. The architecture of DIPLOMAT is developed on a multi-agent platform. Tested with human players, DIPLOMAT outperformed them, and this is a notable result given the complexity and uncertainty of the selected scenario. Despite the success of the system, its usefulness for the present survey is limited because (1) it was developed for a task – negotiation – which is related, but not coincident, to persuasion; (2) the architecture is domain dependent; (3) it uses a negotiation language that is an English-like logical language, no real NLG process is done. Players must previously learn this language to interact with the system.
- Lim et al. (2005) present an attempt to bind persuasion and storytelling. The architecture of an empathic tour guide system (a context-aware mobile system that includes an "intelligent empathic guide with attitude") is described. It consists of two virtual agents each possessing a contrasting personality, presenting users with different versions of the story of the same event or place. An emergent empathic model with personality is proposed as a mechanism for action selection and affective processing. The guide creates personalized communication applying improvisational storytelling techniques to persuade the user to think in the way it thinks; by invoking empathy, the guide makes the user see an event in a deeper sense.

4.2 Argumentative and Dialogical Persuasion systems

The theory of argumentation dialogues originates from research about expert systems, in which an advice-giving system was built to suggest appropriate therapies in a given situation (Buchanan and Shortliffe, 1984). A key function was to support their suggestion with explanations and clarifications after requests from the user, including critiques to the suggested plan: they therefore set the framework for subsequent developments of criticizing argumentation attempts (Walton, 2006b). In the multi-agent system domain, this kind of dialogues was subsequently employed, by agents, to distribute and contract roles and tasks (negotiation); several proposals about the language to employ and how an artificial agent's mind could be simulated were made (Wooldridge, 2002; Rao and Georgeff, 1991). This representation was seen as the object of "second-order reasoning" applied in planning a dialogue move and interpreting the agent's reaction. Other computational models of dialogical persuasion (e.g., Grasso et al., 2000; Walton and Reed, 2002; Zukerman et al., 1999) were built on the seminal work developed by linguists, philosophers, and cognitive psychologists (e.g., Toulmin (1958)).

- NAG (nice argument generator) is a precursor of argumentation systems (Zukerman et al., 2000). It is concerned with the abstract form of the unfolding of the argument (e.g., *reductio ad absurdum, inference to the best explanation, reasoning by cases*), and the structure of persuasive messages is limited to one inference schema per move. The system is based on Bayesian networks (BNs): arguments are represented as networks of nodes (representing propositions) and links (representing inferences). BNs have been chosen in order to represent normatively correct reasoning under uncertainty. The system also includes a module, aimed at interpreting the persuadees' reaction according to the system's knowledge of their presumed set of beliefs, and a generation component (Zukerman and George, 2005). Although the two components have not yet been integrated into a dialogic argumentation prototype system, they set some of the principles that guide their development.
- By seeing monologues as "inner dialogues" Kibble (2006) studied the kinds of communicative acts that are employed in persuasion dialogues (in particular, in challenges and clarification requests) and how they may be represented in rhetorical structures.
- In ASD (argumentation scheme dialogue), Reed and Walton (2007) use the language of formal dialectics to define a dialectical system in terms of locution rules (statements, withdrawals, questions, challenges, and critical attacks), commitment rules (effects of locution rules on the two interlocutor's knowledge), and dialogue rules (sequencing of communicative acts).
- *Magtalo* (Multi-agent argumentation, logic, and opinion) is a prototype environment for debate. It supports flexible intuitive interaction with data in complex debate domains to facilitate understanding, assimilation, and structured knowledge elicitation, which enable the expansion of domain resources (Reed and Wells, 2007).
- The PORTIA (Mazzotta et al., 2007) prototype is based on Miceli et al.'s theory of emotional persuasion (Miceli et al., 2006). It implements Walton's idea of separation between a "reasoning" and an "argumentation" phase (Walton, 1990) by representing with Bayesian networks the uncertainty inherent in this form of reasoning. Argumentation schemes associated with Bayesian networks are chained back to translate the selected strategy into recipient-adapted messages. Answers

to the user reactions to persuasion attempts are produced after reasoning on the same knowledge base.

- ARGUER (Restificar et al., 1999a, b) focuses on methods, based on argumentation schemata, to detect *attack* or *support* relations among user's and system moves – utterances – during a dialogue. The approach is quite simplified and the dialogue is seen as a simple process of alternate *attack* and *support* utterances between the system and the user (called sometimes *ping-pong effect*).
- The system proposed by Andrè et al. (2000) makes an unusual exploitation of dialogical interactions. It generates presentations that are delivered to the user by teams of communicative agents interacting with each other, rather than addressing the user directly using a single conversational human-like agent, as if it were a face-to-face conversation. In this case the (argumentative) communication is simulated and the user experiences an indirect interaction by "overhearing" the dialogue. The dialogical model is not dynamic, there is a planner that generates a script played by the agents. This approach allows to explore the use of multiple personalities, to express multiple points of view, and to highlight relevant arguments and counter-arguments that in a direct interaction with the user could be missed.
- The system proposed in Andrè et al. (2004) and Rehm and Andrè (2005a) focuses and implements specific aspects and tactics of dialogical and emotional communication, such as politeness strategies (both for verbal (Andrè et al., 2004) and non-verbal (Rehm and Andrè, 2005a) behavior of dialogical agents). In Andrè et al. (2004) a hierarchical selection process for politeness behavior is presented (based on the theory proposed in Brown and Levinson 1987) claiming that attention should be devoted not only to the adaptation of the content to be conveyed but also to stylistic variations in order to improve the users' affective response by mitigating face threats resulting from dialogue acts.
- ARAUCARIA (Reed and Rowe, 2004), an XML-based tool for analyzing and diagramming arguments, focuses on argumentation presumptive schemes (schemes that are defeasible in their nature). The attention is posed on schemes refined with critical questions (Walton and Reed, 2002) to detect or prevent possible user's counter-moves. The model underlying ARAUCARIA is more refined than the ARGUER one but it can be used only for analysis purposes.

The study of these theories/systems enlightened the limits of applying a purely logical reasoning to real domains and the need, on the one side, of considering uncertainty (Zuckerman et al., 2001) and, on the other side, of introducing argumentation schemes more refined than logical *modus ponens* (Walton, 2000).

4.3 Emotional Persuasion Systems

Since emotional reasoning is usually performed in order to modify/increase the impact of the message, affective NLG is strictly connected to persuasive NLG. An annotated bibliography on affective NLG can be found in Piwek (2002). There are also many computational models of emotion dynamics based on cognitive theories like the one proposed by Ortony et al. (1988). Yet, it is not clear how much of these computational models have been implemented and how much of them are persuasion driven: their focus is on *believability*, for a natural communication with the user (see DeCarolis et al., 2001) or for simulation purposes, rather than on emotions use for an *effective* communication.

- Elliott (1994) presents the affective reasoner multiagent platform for simulating simple emotional "reactions" among groups of agents, depending on personality traits. However, this system does not address persuasion directly since it focuses on emotions dynamics in complex social environments rather than emotions induction for persuasive interactions.
- Carofiglio and de Rosis (2003) focus on emotions as a core element for affective message generation. The implemented model is more complex and more "persuasion oriented" than Elliott's. They use a dynamic belief network for modeling activations of emotional states during dialogical interactions. This model of emotional activation is inserted in an argumentation framework. The main limitations of their approach are the use of only one persuasive strategy at a time – system move – and the fact that they stick to an argumentative view of dialogs: they do not consider the problem of the interaction among different strategies, central in building *complex* persuasive messages.
- de Rosis and Grasso (2000) also focus on affective language generation. Although multimodality aspects are missing, the technological aspects for a "richer" NLG production are addressed, leaving aside emotion simulation needs. Their model uses plan operators – for text structuring – enriched with applicability conditions depending on the user's emotional traits, combined with rule-based heuristics for revising both strategic and tactic planning.
- In Rehm and Andrè (2005b) the focus is on the use of emotional display for particular communicative situations such as social lies and deception. In their work, the authors try to understand how users will react if the information conveyed non-verbally exhibits clues that are not consistent with the verbal part of an agent's action. The virtual agent masks "real" emotions with fake emotions that are inconsistent with the content of the message (based on Ekman's theory (1992). It has been tested both in monological and undialogical scenarios (for the latter, the GAMBLE system is used (Rehm and Wissner, 2005)).
- *Valentino* (Guerini et al., 2008b) is a tool for modifying existing textual expressions toward more positively or negatively valenced versions as an element of a persuasive system. For instance, a strategic planner may decide to intervene on a draft text with the goal of "coloring" it emotionally. When applied to a text, the changes invoked by a strategic level may be uniformly negative or positive; they can smooth all emotional peaks; or they can be introduced in combination with deeper rhetorical structure analysis, resulting in different types of changes for key parts of the texts.

In general we can say that one of the most recent subjects of interest in this trend of research concerns widening the persuasion modes from considering "rational" or "cognitive" arguments to appealing to values and emotional states (Sillince and Minors, 1991; Grasso et al., 2000; Guerini et al., 2003; Poggi, 2005).

4.4 Multimodal Realization Aspects of Persuasion

There is a wealth of research on multimodal aspects of IUI. Many "realizers" can be potentially used in combination with language for emotion displaying/induction. For example

- Embodied conversational agents (ECAs)
- Kinetic typography
- Music
- Use of images

ECAs are synthetic characters (usually human like) endowed with a physical appearance able to display dynamic expressive behavior. Most research has mainly focused on the perceptual interface side.[2] Planning of autonomous behavior led by internal beliefs, desires, and intentions is an open topic of research. Essential elements of ECAs are (1) a *3D model* of the animated agent and (2) a *representation language* for directing the 3D model (usually XML based).

Three dimensional models can be roughly divided in two groups:

1. Just face, e.g., Greta, Xface
2. Face plus body, e.g., EMOTE, PPP Persona

There is a wealth of different representation languages that have different features, partially overlapping, partially differing. These features include emotions expression, deictics specification, RR labeling, personality definition, performative type selection (request, inform, ask). Some of the tasks that these languages can accomplish are

1. speaking ECA (i.e., face), e.g., APML, RRL, SMIL-Agent
2. moving ECA (i.e., face plus body), e.g., XSTEP, MURML
3. interacting ECA, e.g., PML, ABL

Flexibly persuasive ECAs have a large potential as they allow for a richer communication with the user (for an example of a commercial application, see Ach and Morel 2007):

---

[2] For a survey on existing ECAs and empirical research on their impact see Prendinger and Ishizuka (2003) and Dehn and vanMulken (2000). For further details, please see the final chapter in Part IV.

1. They are "natural" (e.g., gaze can be used for displaying the focus of attention and for turn taking, gestures can be used for deictic references and for contributing to communicative contents). On the importance of eye gaze in ECAs to improve the quality of communication see, for example, the study of Garau et al. (2001).
2. They are more involving (they may include display of emotion and expression of personality). On behavior and emotion displaying there is a plethora of studies, see, for example, Allbeck and Badler (2002) and Poggi et al. (2001). Taking into account also the natural predisposition of humans to treat anthropomorphic agents as human peers, the possibility of ECAs to persuade them – by leveraging social responses – is crucial and critical (for example, making resort to their realism for believability).

Nevertheless the relation between realism and believability on the one side and effectiveness on the other is not totally obvious. In certain situations it is reasonable to use cartoon-like (non-realistic) characters. They can be effective because (a) they do not generate over-attribution that could lead to frustration, (b) they are more suitable for particular kinds of audience, for example, children. When talking about realism and effectiveness we are not just referring to a question of appearance but also of behavior; cartoon-like characters often display exaggerated emotions, obtaining the desired effect. In any case, we should point out that ECAs are not the only resource in persuasive interfaces.

Other resources for emotion empathy and induction are the use of music (Scherer, 1995), the use of kinetic typography (Forlizzi et al., 2003), and so on.

Kinetic typography refers to the art and the technique of expression with *animated* text. Text animation allows adding further dimensions of expressivity to the text and provides it with the ability to display emotions and to capture and direct attention. KT has demonstrated ability to add significant appeal to texts, allowing some of the qualities normally found in film and in the spoken word to be added to static texts. Kinetic typography has been widely and successfully used in film (e.g., opening credits) as well as in television and advertising (e.g., TV commercials, web banners). There are several key areas in which kinetic typography has been particularly successful (Ford et al., 1997; Ishizaki, 1998). These include

- Expression of affective content,
- Creation of characters, and
- Capture or direct attention.

If used in an appropriate way, automated kinetic typography can enhance the emotional impact of the content conveyed by persuasive messages (e.g., hopping words can be used to emphasize a happy message; in Guerini et al., 2007, an attempt to automatically render persuasive messages tagged in APML representation language is presented).

5 Conclusions

In this chapter, we have proposed a view that emphasizes verbal communication capabilities in intelligent user interfaces. Intelligent information presentation systems must take into account the specifics about the user, such as needs, interests, and knowledge; in particular, we think that the emotion dimension and the personality dimension must have a part in individual-oriented and context-aware communication systems. For this purpose, we have discussed persuasive interfaces in particular. These are interfaces that aim at inducing the user – or in general the audience – to perform some actions in the real world. Interfaces of this kind must take into account the "social environment," exploit the situational context, and value emotional aspects in communication. Modeling persuasion mechanisms and performing flexible and context-dependent persuasive actions are much more ambitious than what most current approaches to persuasive technologies aim at (see "captology" (Fogg, 2002)). As opposed to hardwired persuasive features, in this chapter we focused on those systems that have reasoning capabilities able to provide flexible persuasive communication with their users.

References

Ach L, Morel B (2007) Intelligent Virtual Agents, chapter Avatars contributions to commercial applications with living actor technology, Springer, Berlin, pp 411–412

Ahluwalia R (2000) Examination of psychological processes underlying resistance to persuasion. J Consumer Res 27(2):217–232

Allbeck J, Badler N (2002) Toward representing agent behaviours modified by personality and emotion. In: Proceedings of the 1st international joint conference on autonomous agents and multi-Agent systems, Bologna, Italy

Andrè E, Rist T, van Mulken S, Klesen M, Baldes S (2000) The automated design of believable dialogues for animated presentation teams. The MIT Press, Cambridge, MA, pp 250–255

Andrè E, Rehm M, Minker W, Buhler D (2004) Endowing spoken language dialogue systems with emotional intelligence. In: André E, Dybkjaer L, Minker W, Heisterkamp P (eds) Affective dialogue systems. Springer, Berlin, pp 178–187

Aylett RS (2000) Emergent narrative, social immersion and "storification". In: Proceedings of narrative interaction for learning environments, Edinburgh

Berdichevsky D, Neuenschwander E (1999) Toward an ethics of persuasive technology. Commun ACM Arch, 42(5):51–58

Bosmans A, Warlop L (2005) How vulnerable are consumers to blatant persuasion attempts? Technical report, DTEW Research Report 0573, K.U. Leuven, 2005

Brown P, Levinson SC (1987) Politeness Some universals in language use. Cambridge University Press, Cambridge

Buchanan BG, Shortliffe EH (1984) Rule-based expert systems. Addison-Wesley

Carenini G, Mittal V, Moore J (1994) Generating patient specific interactive explanations. In: Proceedings of 18th symposium on computer applications in medical care (SCAMC '94). McGraw-Hill

Carofiglio V (2004) Modelling argumentation with belief networks. In: Proceedings of the ECAI workshop on computational models of natural argument, Valencia, Spain

Carofiglio V, de Rosis F (2003) Combining logical with emotional reasoning in natural argumentation. In: Proceedings of the UM'03 workshop on affect, Pittsburgh, PA, USA

Carofiglio V, de Rosis F (2005) In favour of cognitive models of emotions. In: Proceedings of the AISB workshop on 'mind minding agents', 2005, Hatfield, England

Castelfranchi C (1996) Reasons: beliefs structure and goal dynamics. Mathw Soft Comput, 3(2):233–247

Castelfranchi C, Guerini M (2007) Is it a promise or a threat? Pragmat Cogn Journal, 15(2): 277–311

Chaiken S (1980) Heuristic vs. systematic information processing and the use of source vs message cues in persuasion. J Pers Soc Psychol, 39:752–766

Cialdini RB (1993) Influence. The psychology of persuasion. William Morrow & Company, New York, NY

Coutinho S, Sagarin BJ (2006) Resistance to persuasion through inductive reasoning. Stud Learn Eval Innovat Dev 3(2):57–65

Das S (2002) Logic of probabilistic argument. In: Proceedings of the ECAI 2002 workshop on computational model of natural argument, Lyon, pp 9–18

Davison J, Arthur J (2003) Active Citizenship and the development of social literacy: a case for experiential learning. Citizenship and Teacher Education, Canterbury

de Rosis F, Grasso F (2000) Affective natural language generation. In Paiva A (ed) Affective interactions. Springer Lecture Notes in Artificial Intelligence, vol 1814, pp 204–218

DeCarolis B, de Rosis F, Carofiglio V, Pelachaud C (2001) Interactive information presentation by an embodied animated agent. In: International workshop on information presentation and natural multimodal dialogue, Verona, Italy

Dehn DM, van Mulken S (2000) The impact of animated interface research: a review of empirical research. Int J Hum Comput Stud 52:1–22

Duboue P, McKeown K (2003) Statistical acquisition of content selection rules for natural language generation. In: Proceedings of EMNLP-03, 8th conference on empirical methods in natural language processing, Sapporo, Japan, pp 121–128

Ekman P (1992) Telling lies – clues to deceit in the marketplace, politics, and marriage. Norton and Co

Elliott (1994) Multi-media communication with emotion driven 'believable agents'. In: AAAI technical report for the spring symposium on believeable agents, Stanford University, pp 16–20

Fiedler A, Horacek H (2002) Argumentation within deductive reasoning. In: Proceedings of the ECAI 2002 workshop on computational model of natural argument, Lyon, pp 55–64

Fleischman M, Hovy E (2002) Towards emotional variation in speech-based natural language generation. In: Proceedings of the second international natural language generation conference (INLG02), Arden Conference Center, Harriman, NY, USA, pp 57–64

Fogg BJ (2002) Persuasive technology: using computers to change what we think and do. Morgan Kaufmann

Ford S, Forlizzi J, Ishizaki S (1997) Kinetic typography: issues in time-based presentation of text. In: CHI97 conference extended abstracts, Atlanta, Georgia, pp 269–270

Forlizzi J, Lee J, Hudson SE (2003) The kinedit system: affective messages using dynamic texts. In: CHI2003 conference proceedings, Fort Lauderdale, FL, pp 377–384

Friestad M, Wright P (1994) The persuasion knowledge model: how people cope with persuasion attempts. J Consumer Res 21(1):1–31

Frijda NH, Manstead ASR, Bem S (eds) (2000) Emotions and beliefs. chapter The inuence of emotions on beliefs, Cambridge University Press, Cambridge, pp 45–77

Garau M, Slater M, Bee S, Sasse MA (2001) The impact of eye gaze on communication using humanoid avatars. In: Proceedings of the SIG-CHI conference on Human factors in computing systems, Seattle, Washington, pp 309–316

Geyskens K, Dewitte S, Millet K (2006) Stimulating referral behavior may backfire for men: the effect of referral failure on susceptibility to persuasion. Technical report, KUL Working Paper No. OR 0609, 2006

Gmytrasiewicz PJ, Lisetti CL (2001) Emotions and personality in agent design and modeling. In: Proceedings of the user modeling 2001 workshop on attitude, personality and emotions in user adapted interaction, Sonthofen, Germany

Grasso F, Cawsey A, Jones R (2000) Dialectical argumentation to solve conflicts in advice giving: a case study in the promotion of healthy nutrition. Int J Hum Comput Stud 53(6):1077–1115

Guerini M (2006) Persuasion models for multimodal message generation. PhD thesis, University of Trento

Guerini M, Stock O (2005) Toward ethical persuasive agents. In: Proceedings of the IJCAI workshop on computational models of natural argument, Edimburh

Guerini M, Stock O, Zancanaro M (2003) Persuasion models for intelligent interfaces. In: Proceedings of the IJCAI workshop on computational models of natural argument, Acapulco, Mexico

Guerini M, Stock O, Zancanaro M (2007) A taxonomy of strategies for multimodal persuasive message generation. Appl Artif Intell J 21(2):99–136

Guerini M, Strapparava C, Stock O (2008a) Corps: a corpus of tagged political speeches for persuasive communication processing. J Info Tech Polit 5(1):19–32

Guerini M, Strapparava C, Stock O (2008b) Valentino: a tool for valence shifting of natural language texts. In: Proceedings of LREC2008, Marrakech, Marocco

Guerini M, Stock O, Zancanaro M (2004, August) Persuasive strategies and rhetorical relation selection. In: Proceedings of the ECAI workshop on computational models of natural argument, Valencia, Spain

Harmon-Jones E (2000) Emotions and beliefs. In: Frijda NH, Manstead ASR, Bem S (eds) A cognitive dissonance theory perspective on the role of emotion in the maintenance and change of beliefs and attitudes, Cambridge University Press, Cambridge, pp 185–211

Henriksen T (2004) Beyond role and play – tools, toys and theory for harnessing the imagination. chapter On the transmutation of educational role-play: a critical reframing to the role-play to meet the educational demands, Ropecon Ry, Helsinki, pp 107–130

Horacek H (2006, September) Argument understanding and argument choice – a case study. In: Proceedings of the ECAI workshop on computational models of natural argument, Riva del Garda, Italy

Hovy E (1988) Generating natural language under pragmatic constraints. Lawrence Erlbaum Associates, Hillsdale, NJ

Hullett CR (2005) The impact of mood on persuasion: a meta-analysis. Commun Res, 32(4): 423–442

Ibanez J (2004) An intelligent guide for virtual environments with fuzzy queries and flexible management of stories. PhD thesis, Departamento de Ingenieria de la Informacion y las Communicaciones, Universidad de Murcia, Murcia, Spain

Ishizaki S (1998) On kinetic typography. Statements, the Newsletter for the American Center of Design, 12(1):7–9

Jurafsky D, Martin JH (2000) Speech and languageprocessing: an introduction to natural language processing, speech recognition, and computational linguistics. Prentice-Hall, NJ

Kahneman D, Slovic P, Tversky A. (eds) Judgment under Uncertainty:heuristics and biases. Cambridge University Press, New York, NY

Kibble R (2006) Dialectical text planning. In: Proceedings of ECAI workshop on computational modelling of natural argumentation, Riva del Garda, Italy

Kraus S, Lehmann D (1995) Designing and building a negotiating automated agent. Comput Intell 11(1):132–171

Kraus S, Sycara K, Evenchik A (1998) Reaching agreements trough argumentation: a logic model and implementation. Artif Intell J, 104:1–69

Kukafka R (2005) Consumer health informatics: informing consumers and improving health care. In: Lewis D, Eysenbach G, Kukafka R, Stavri PZ, Jimison H (eds) Tailored health communication, Springer, New York, NY, pp 22–33

Langkilde I, Knight K (1998) The practical value of n-grams in derivation. In: Hovy E. (ed) Proceedings of the ninth international workshop on natural language generation', New Brunswick, NJ. Association for Computational Linguistics, pp 248–255

Lewin K (1951) Field theory in social science. Harper and Row, New York, NY

Lim MY, Aylett R, Jones C (2005) Empathic interaction in a virtual guide. In: Proceedings of virtual social agents joint symposium, University of Hertfordshire, UK, AISB Symposia, pp 122–129

Lim MY (2007) An intelligent guide with attitude. chapter Emotions, behaviours and belief regulation. School of Mathematical and Computer Sciences, Heriot-Watt University, Edinburgh

Lodder AR (1999) DiaLaw: on legal justification and dialogical models of argumentation. Kluwer

Mann W, Thompson S (1987) Rhetorical structure theory: a theory of text organization. Ablex Publishing Corporation

Marcu D (1996) The conceptual and linguistic facets of per-suasive arguments. In: Proceedings of the ECAI workshop, gaps and bridges: new directions in planning and natural language generation, Budapest, Hungary, 12 August 1996, pp 43–46

Marcu D (1997) The rhetorical parsing, summarization, and generation of natural language text. University of Toronto

Mazzotta I, Novielli N, Silvestri E, de Rosis F (2007) 'O francesca, ma che sei grulla?' Emotions and irony in persuasion dialogues. In: Proceedings of the 10th Conference of AI*IA – Special Track on 'AI for Expressive Media'. AI*IA 2007: Artificial Intelligence and Human-Oriented Computing, Springer LNCS 4733/2007, pp 602–613

Mazzotta I, de Rosis F, Carofiglio V (2007) Portia: a user-adapted persuasion system in the healthy eating domain. IEEE Intelligent Systems, Special Issue on Argumentation Technology., In press

Miceli M (1992) How to make someone feel guilt: Strategies of guilt inducement and their goals. J Theory Soc Behav 22(1):81–104

Miceli M, de Rosis F, Poggi I (2006) Emotional and non-emotional persuasion. Appl Artif Intell 20:849–879

Moulin B, Irandoust H, Belanger M, Desordes G (2002) Explanation and argumentation capabilities: towards the creation of more persuasive agents. Artif Intell Rev 17:169–222

Nass C, Moon Y, Fogg B, Reeves B (1995) Can computer personalities be human personalities? Int J Human-Comput Stud 43:223–239

O'Keefe DJ (1997) Standpoint explicitness and persuasive effect: a meta-analytic review of the effects of varying conclusion articulation in persuasive messages. Argument Advocacy 34:1–12

O'Keefe DJ (1999a) How to handle opposing arguments in persuasive messages: a meta-analytic review of the effects of one-sided and two-sided messages. Communi Yearbook 22:209–249

O'Keefe DJ (1999b) Variability of persuasive message effects: meta-analytic evidence and implications. Document Design 1:87–97

O'Keefe DJ (2000) Guilt and social influence. Communi Yearbook 23:67–101

O'Keefe DJ (2002) Persuasion: theory and research (2nd ed). Sage, Thousand Oaks, CA

O'Keefe DJ, Jensen JD (2006) The advantages of compliance or the disadvantages of noncompliance? a meta-analytic review of the relative persuasive effectiveness of gain-framed and loss-framed messages. Commun Yearbook 30:1–43

Ortony A, Clore GL, Collins A (1988) The cognitive structure of emotions. Cambridge University Press, Cambridge

Parson S, Jennings NR (1996) Negotiation trough argumentation-a preliminary report. In: Proceedings of the international conference on multi-agent system, Kyoto, pp 267–274

Perelman C, Olbrechts-Tyteca L (1969) The new rhetoric: a treatise on argumentation. Notre Dame Press

Petty RE, Cacioppo JT (1986) The elaboration likelihood model of persuasion. Adv Exp Soc Psychol 19:123–205

Petty RE, Gleicher F, Baker SM (1991) Emotion and social judgements, chapter Multiple roles for affect in persuasion. Pergamon Press

Piaget J (1972) The principles of genetic epistemology. Routledge & Keegan Paul, London

Piwek P (2002) An annotated bibliography of affective natural language generation. ITRI ITRI-02-02, University of Brighton

Poggi I (2005) The goals of persuasion. Pragmat Cogn 13(2):297–336

Poggi I, Pelachaud C, De Carolis BN (2001) To display or not to display? towards the architecture of a reflexive agent. In: Proceedings of the 2nd workshop on attitude, "Personality and emotions in user-adapted interaction". User modeling 2001, Sonthofen, Germany

Prakken H, Reed C, Walton D (2003) Argumentation schemes and generalizations in reasoning about evidence. In: ICAIL 2003, pp 32–41

Prendinger H, Ishizuka M. (eds). Life-like characters: tools, affective functions and applications. Springer, Heidelberg

Radev D, McKeown K (1997) Building a generation knowledge source using internet-accessible newswire. In: Proceedings of the 5th conference on applied natural language processing, Washington, DC, pp 221–228

Rao AS, Georgeff MP (1991) Modeling rational agens within a bdi architecture. In: Allen J, Fikes R, Sandewall R. (eds) Proceedings of the 2nd international conference on principles of knowledgerepresentation and reasoning. Morgan Kaufman, pp 473–484

Reed CA, Grasso F (2007) Recent advances in computational models of argument. Int J Intell Syst, 22(1):1–15

Reed CA, Long DP (1997) Ordering and focusing in an architecture for persuasive discourse planning. In: Proceedings of the 6th European workshop on natural language generation (EWNLG97), Duisburg, Germany

Reed CA, Long DP (1998) Generating the structure of argument. In: Proceedings of the 17th international conference on computational linguistics and 36th annual meeting of the association for computational linguistics (COLING-ACL98), Montreal, Canada, pp 1091–1097

Reed CA, Rowe GWA (2004) Araucaria: Software for argument analysis, diagramming and representation. Int J AI Tools 14(3–4):961–980

Reed CA, Walton D (2007) Argumentation schemes in dialogue. In: Hansen HV, Tindale CW, Johnson RH, Blair JA (eds) Dissensus and the search for common ground (Proceedings of OSSA 2007), Windsor, ON

Reed CA, Wells S (2007) Dialogical argument as an interface to complex debates. IEEE Intell Syst 22(6):60–65

Reed CA, Long DP, Fox M (1996) An architecture for argumentative dialogue planning. In: Practical reasoning: proceedings of the first international conference on formal and applied practical reasoning (FAPR96). Springer, Berlin, pp 555–566

Reed CA, Long DP, Fox M, Garagnani M (1997) Persuasion as a form of inter-agent negotiation. In: Multi-agent systems: methodologies and applications: selected papers from the proceedings of the 2nd australian workshop on distributed AI, pp 120–136, Cairns, Australia, 1997. Springer, Berlin

Reeves B, Nass C (1996) The media equation. Cambridge University Press

Rehm M, Andrè E (2005a) Informing the design of embodied conversational agents by analysing multimodal politeness behaviors in human-human communication. In: Proceedings of the AISB 2005 symposium on conversational informatics for supporting social intelligence and interaction, Hatfield, England

Rehm M, Andrè E (2005b) Catch me if you can – exploring lying agents in social settings. In: Proceedings of the international conference on autonomous agents and multiagent systems, Utrecht (Olanda), pp 937–944

Rehm M, Wissner M (2005) Gamble – a multiuser game with an embodied conversational agent. In: Entertainment computing – ICEC 2005: 4th international conference, New York, pp 180–191 2005. Springer, Berlin

Reiter E, Dale R (2000) Building natural language generation systems. Cambridge University Press, Cambridge

Reiter E, Robertson R, Osman L (2003a) Lesson from a failure: generating tailored smoking cessation letters. Artif Intell 144:41–58

Reiter E, Sripada S, Robertson R (2003b) Acquiring correct knowledge for natural language generation. J Artif Intell Res 18:491–516

Restificar AC, Ali SS, McRoy SW (1999a) Arguer: using argument schemas for argument detection and rebuttal in dialogs. In Proceedings of the seventh international conference on user modelling (UM-99), Banff, Canada, pp 315–317

Restificar AC, Ali SS, McRoy SW (1999b) Argument detection and rebuttal in dialogs. In: Proceedings of the twenty first annual meeting of the cognitive science society (Cogsci-99), Vancouver, Canada

Scherer KR (1995) Expression of emotion in music. J Voice 9(3):235–248

Scott AC, Clayton JE, Gibson EL (1991) A practical guide to knowledge acquisition. Addison-Wesley

Sillince JAA, Minors RH (1991) What makes a strong argument? emotions, highly-placed values and role playing. Commun Cogn 24(3&4):281–298

Smith MH, Garigliano R, Morgan RC (1994) Generation in the lolita system: an engineering approach. In: Proceedings of the 7th international workshop on natural language generation, Kennebunkport, Maine, pp 241–244

Sopory P, Dillard JP (2002) The persuasive effects of metaphor: a meta-analysis. Human Commun Res 28:382–419

Stock O, Guerini M, Zancanaro M (2006) Interface design and persuasive intelligent user interfaces. chapter The foundations of interaction design. Lawrence Erlbaum, Hillsdale, NJ

Toulmin S (1958) The use of arguments. Cambridge University Press, Cambridge, MA

Vreeswijk G, Lodder AR (2005) Gearbi: towards an online arbitration environment based on the design principles simplicity, awareness, orientation, and timeliness. Artif Intell Law 13(2): 297–321

Walton DN (1996) Argumentation Schemes for presumptive reasoning. Lawrence Erlbaum Associates, Mahwah, NJ

Walton DN (1998) The new dialectic. University of Toronto Press

Walton DN (1999) Dialectical relevance in persuasion dialogue. Inf Logic 19(2–3):119–143

Walton DN (2000) Syntheses 2000. chapter The place of dialogue theory in logic, computer science and communication Studies. Number 123. Kluwer, The Netherlands, pp 327–346

Walton DN (2006a) Examination dialogue: an argumentation framework for critically questioning an expert opinion. J Pragmat 38:745–777

Walton DN (2006b) How to make and defend aproposal in a deliberation dialogue. Artif Intell Law 14(3):177–239

Walton DN, Godden DM (2005) Persuasion dialogue in online dispute resolution. Artif Intell Law 13(2):273–295

Walton DN, Reed CA (2002) Argumentation schemes and defeasible inferences. In: Proceedings of the ECAI 2002 workshop on computational model of natural argument, Lyon

Walton DN (1990) What is reasoning? what is argument? J Philos 87:399–419

Witte K, Allen M (2000) A meta-analysis of fear appeals: implications for effective public health programs. Health Educ Behav, 27:591–615

Wooldridge M (2002) An introduction to multiagent systems. Wiley, Chichester

Young K (2001) The neurology of narrative. SubStance – Issue 94/95 30(2):72–84

Zuckerman I (2001) An integrated approach for generating arguments and rebuttals and understanding rejoinders. In: UM01 proceedings – the eighth international conference on user modeling, Sonthofen, Germany, pp 84–94

Zuckerman I, Jinah N, McConachy R, George S (2001) Recognizing intentions from rejoinders in a bayesian interactive argumentation system. In PRICAI2000, Melbourne, Australia

Zukerman I, George S (2005) A probabilistic approach for argument interpretation. User Model User-Adapt Int 15(1):5–53

Zukerman I, McConachy R, Korb K, Pickett D (1999) Explanatory interaction with a bayesian argumentation system. In: Proceedings of the sixteenth international joint conference on artificial intelligence, Stockholm, pp 1294–1299

Zukerman I, McConachy R, Korb K (2000) Using argumentation strategies in automated argument generation. In: Proceedings of the 1st international natural language generation conference, Mitzpe Ramon, Israel, pp 55–62

Non-verbal Persuasion and Communication in an Affective Agent

Elisabeth André, Elisabetta Bevacqua, Dirk Heylen, Radoslaw Niewiadomski, Catherine Pelachaud, Christopher Peters, Isabella Poggi, and Matthias Rehm

Abstract This chapter deals with the communication of persuasion. Only a small percentage of communication involves words: as the old saying goes, "it's not what you say, it's how you say it". While this likely underestimates the importance of good verbal persuasion techniques, it is accurate in underlining the critical role of non-verbal behaviour during face-to-face communication. In this chapter we restrict the discussion to body language. We also consider embodied virtual agents. As is the case with humans, there are a number of fundamental factors to be considered when constructing persuasive agents. In particular, one who wishes to persuade must appear credible, trustworthy, confident, and non-threatening. Knowing how not to behave is also a vital basis for effective persuasion. This includes resolving task constraints or other factors with the social perception considerations. These social virtual agents face many of the same problems as humans have in controlling and expressing themselves in an appropriate manner so as to establish and maintain persuasive interaction. All along the chapter, much of our discussion will handle concepts applicable both to agent and to human behaviour.

1 Introduction

Persuasion is a way to influence other people, that is, to make them do actions, pursue goals, that they would have not otherwise. But it differs from other ways to influence for three reasons: (1) it does not imply the use of force, but makes an appeal to the persuadee's free choice; (2) it claims that the goal/action proposed by the persuader is in the interest of the persuadee in that it is a subgoal to the persuadee's goals, and (3) it aims at influencing through communication: by letting the persuadee know that the persuader wants him to do so. Persuasion necessarily

E. André (✉)
University of Augsburg, Augsburg, Germany
e-mail: Andre@informatik.uni-augsburg.de

P. Petta et al. (eds.), *Emotion-Oriented Systems*, Cognitive Technologies,
DOI 10.1007/978-3-642-15184-2_30,

goes through communication: in part through argumentation, that is, words and sentences; but only a small percentage of communication involves words: as the old saying goes, "it's not what you say, it's how you say it". While this likely underestimates the importance of good verbal persuasion techniques (see the chapter by M. Guerint et al., this volume), it is accurate in underlining the critical role of non-verbal behaviour during face-to-face communication: For no matter how correct the argument or message, it is unlikely to be listened to unless the non-verbal language is congruent, communicating credibility, confidence, and trustworthiness. For example, in experiments concerning the expression of attitudes and feelings, Mehrabian (1971) found that when there was incongruence in verbal and non-verbal messages being communicated, the relative importance of messages was 7% based on verbal liking, 38% on vocal liking, and 55% on facial liking: receivers tended to predominantly favour the non-verbal aspects, in contrast to the literal meaning of the words, during the communication of attitudes and feelings.

While the environment, haptics, state of the persuadee, and appearance of persuader are also of importance, here we restrict the discussion to body language. Body language is certainly not a new topic in the domain of scientific study (Darwin, 1872), although its use in practical persuasive situations has increasingly been the focus of studies in marketing, political campaigning, and courtroom scenarios (Bernstein et al., 1994) and is becoming popularised through the publication of numerous easily accessible books (Pease and Pease, 2006). A relatively new and novel domain for the evaluation and application of body language research is that of embodied virtual agents. If embodied agents resemble humans in appearance, then they can take advantage of these extra modalities to enrich and smooth the human–computer interaction process. These social virtual agents therefore face many of the same problems as humans have in controlling and expressing themselves in an appropriate manner so as to establish and maintain persuasive interaction – as such, much of our discussion will handle concepts applicable both to agent and to human behaviour.

As is the case with humans, there are a number of fundamental factors to be considered when constructing persuasive agents. In particular, one who wishes to persuade must appear credible, trustworthy, confident, and non-threatening. Knowing how not to behave is also a vital basis for effective persuasion. This includes resolving task constraints or other factors with the social perception considerations. For example, not looking at people due to consulting one's notes may give the impression of dishonesty; slouching due to tiredness could be misinterpreted as a sign of a lack of interest. Good behaviour does not apply solely to the speaker of course. For example, the background non-verbal behaviours made when one does not hold the floor can also have a large effect on credibility ratings (Seiter et al., 2006).

In the next section of this chapter we report on several studies on the role of non-verbal behaviours in persuasion. While Sect. 3 presents studies from the standpoint of the speaker, Sect. 4 examines the effects of persuasive non-verbal behaviours on listener's perception of speaker. Sect. 5 continues to look at the listener. It addresses the issues of backchannel signals emitted by the listener as a way of assessing the effect of persuasion on the listener. In the last section of this chapter we present works on virtual agents endowed with social capabilities, in particular the capacities of controlling and managing their facial behaviours and gestures.

2 Persuasion and Emotion

To persuade the other to do something we must convince him that this is the right thing to do; and we can do so, as Aristotle put it, through logos (rational arguments), ethos (the orator's character), and pathos (the audience's emotions): in other words, to persuade we must (1) provide good reasons to do what we propose, (2) win the interlocutor's trust, and (3) induce or evoke emotions. The first aspect, logos, is the object of the science of argumentation, which studies the possible arguments and counter-arguments that can be made explicit through verbal language; the second refers to the need for the persuader to show an image of a person who is both (a) competent in the topic dealt with (competence), in such a way that his arguments are taken at face value, and (b) one who really wants to make the interests of the persuadee (benevolence), so that the persuadee does not fear to be deceived and to get harm from doing what the persuader proposes. In this work we do not go deep into the issue of argumentation, nor in the aspect of the persuader's ethos. Here we focus on the role of emotions in persuasion. Why and how can emotions be an instrument of persuasion? Emotions are linked to goals at least in three ways:

1. Emotions monitor goals: The function of emotions is to monitor humans' adaptive goals, since an emotion is felt any time an adaptively important goal is, or is likely to be, fulfilled or thwarted (Castelfranchi, 2000; Darwin, 1872; Frijda, 1986). Fear monitors the goals of survival and safety, anger the goal of justice, shame the goals of image and self-image.
2. Emotions activate goals: They have a high motivating power in that they trigger goals. Each emotion, along with feelings, physiological arousal, expressive pattern, includes a readiness to action, that is, a goal of high priority that presses to be urgently fulfilled. Fear activates fight or flight, anger triggers aggression, and compassion triggers help.
3. Emotions can become goals: Pleasant emotions generate a goal of feeling them again, while unpleasant emotions a goal of not feeling them next time. If I felt proud because I made my home assignment well, I'll have the goal to study hard again, if I felt ashamed after having aggressed a friend I'll have the goal of not doing so anymore. Thus, emotions result in a learning mechanism.

As we mentioned, persuasion is a way to influence people, that is, to lead them to pursue some goals, and to induce the persuadee to do so the persuader must convince him that the proposed goal is a goal of high value, also because it is a means for other very valuable goals. But to activate a goal in a person not only through rational but also through an emotional mechanism is not the same. Suppose the same course is taught by two teachers: one simply teaches me the discipline, but the other also gives me enthusiasm for that discipline; I will probably choose the latter, because attending the former course only fulfils my cognitive goal, while this also fulfils my affective goals. In the same vein, in persuasion attaching the proposed goal only to rational goals of the persuadee is not as effective as attaching it also to emotional goals. Since emotions are linked to biologically adaptive goals, they add a higher value to other goals.

3 The Persuasive Body

The importance of body behaviour in persuasive discourse has been stressed ever since ancient rhetoric. Cicero in "*De oratore*" and Quintilian in "*Institutiones Oratoriae*" showed how gesture, face, gaze, and posture are an important part of "Actio" (discourse delivery). Especially gestures were studied, due to their capacity of summoning, promising, exhorting, inciting, prohibiting, approving, and to their ease in expressing emotions, showing attitudes, indicating objects of the orator's thought. Quintilian's work about gestures is partly guided by normative intents – he often stresses what gestures should not be used by an orator, while they are – and just because they are – typical of comic actors; but he does so on the basis of a deep and detailed knowledge of the gestures' forms and meanings. For every gesture he also tells us in what segment of the rhetorical structure of discourse it can be used, which means that also gestures, as words, are subject to rules, and their distribution is determined by context. Moreover, from his description one can see that particular combinations of movements with the same hand shape quite precisely convey the meaning of specific speech acts – for example, we "lower [our hands] in apology or supplication (. . .) or raise them in adoration, or stretch them out in demonstration or invocation" (15) – or express emotions: "we sometimes clench the hand and press it to our heart when we are expressing regret or anger" (104). Finally, Quintilian acknowledges that sometimes through gestures we may induce persuasive effects: "Slapping the thigh (. . .) is becoming as a mark of indignation, while it also excites the audience" (123).

3.1 Persuasion and Gesture

In present times, while a huge quantity of studies address the use of gesture and other bodily signals in everyday conversation, only part of them is devoted to analysing them in persuasive discourse. Some overview some aspects of the body's relevance in political communication (Atkinson, 1984) or focus on the synchronisation of gestures with pauses and intonation and other rhetorical devices, frequently used to quell the applause (Bull, 1986). Others investigate the audience's physiological, cognitive, and emotional reactions to the politicians' facial expression and other vocal and bodily behaviours (Bucy and Bradley, 2004; Frey, 2000). Some recent works finally provide detailed morphological and semantic descriptions of gestures and on this basis give pertinent insights about the relation between gesture and persuasive discourse (Tournier, 2003; Kendon, 2004; Streeck 2008).

In her book "*The Gestural Expression of a Politician's Thought*" (Tournier, 2003), Calbris analyses the political discourses by Lionel Jospin from July 1997 through April 1998 – his first months as a prime minister. Through an insightful analysis of the metaphors exploited by his manual behaviour, she demonstrates how Jospin's gestures or aspects of their execution – for example, the shape of the hand or even which hand is used, right or left – can express abstract notions like effort, objective, decision, balance, priority, private or public stance. At the same time, though, they fulfil discursive functions: they can delimit or stress, enumerate or explicate the topics of discourse.

Kendon (2004) analyses gestures in different cultures and different types of interaction, by people that tell about their past life or comment on everyday life events, sometimes also with an argumentative intent, and distinguishes three main functions of co-verbal gestures: a referential function, of conveying parts of the propositional content of an utterance; an interactive function, of helping the turn-taking management; and a pragmatic function; within this, a gesture has a performative function if it clarifies the type of speech act that is being performed; a modal function if it alters the interpretation of the utterance, e.g. through negation or intensification; and a parsing function, marking the syntactic or textual structure of a sentence or discourse. Then Kendon analyses some gesture families, by singling out, for each specific hand shape and orientation, typical contexts of use and finding unifying semantic themes. Some of the gestures he analyses can well have a persuasive use: for example, the "ring" gestures that bear a meaning of "making precise" or "clarifying" are used every time this clarification is important "in gaining the agreement, the conviction or the understanding of the interlocutor" in Italian culture (p. 241).

3.2 Persuasion, Body Language, and Politics

Streeck (2008) analyses the gestural behaviour of the Democratic candidates during the political campaign of 2004 in the USA. They do not use many different gestures, as to hand shape and movement pattern, and the gestures they use are very rarely iconic: partly because, as anticipated by Quintilian, iconics look too much a popular style of gesturing and partly because their function in political discourse is not to convey referential but mainly pragmatic information. Yet, among the candidates' gestures with pragmatic functions, Streeck maintains they mainly fulfil a parsing function, more than a performative one, and they do not unequivocally indicate which speech act is being performed, since they do not imply a fixed form–function relationship. For example, Streeck doubts that the "ring" always has a meaning of precision or that the "power grip" of moving the fist always conveys an assertion of power. In maintaining that the candidates' gestures mainly have a parsing function,

Streeck (2008) further argues that the tempo of a gesture, more than other aspects of it, is a clear cue to the discourse structure: for example, the alternation of rapid and slow beats, or whether the stroke combines with the peak syllable or with all stressable syllables, distinguishes between background and foreground information.

Poggi and Pelachaud (2008) analysed fragments of three pre-electoral debates through an annotation scheme that described gestures as to their global meaning and to the meaning borne by their expressivity parameters and classified the meanings as to their persuasive import. Then they computed the quantity of persuasive gestures in the two candidates and their percentage of distribution across the various persuasive strategies. The persuasive strategies adopted are somewhat different between the two politicians. They differ as to the proportion of pathos and logos, for both the majority of gestures pursue an ethos strategy and both tend to project more an image of competence than one of benevolence, but for one of the two candidates the preference for the image of competence is much higher than that of benevolence. Moreover, the persuasive strategies are not always conveyed by the gesture shape, but in some cases only by the gesture expressivity or simply by an indirect meaning of the gesture. Pathos is contained more typically (exclusively, for one candidate) in the expressivity of gesture, while ethos in both orators is more often conveyed by the gesture as a whole. Further, the pathos strategy comes out only at the indirect level in both politicians, while they differ in their level of indirectness. From the analysis it resulted that the differences how much each persuasive strategy is exploited and how it is conveyed (through whole gesture, or expressivity, direct or indirect meaning) are congruent with either the context of the specific debate under analysis, or the general political strategy of the candidates, or finally with his general communicative style.

Poggi and Vincze (2008) analysed the gaze behaviour of two orators, Romano Prodi and Ségolène Royal, in political debates and interviews. The study analysed the general persuasive structure of some fragments in terms of hierarchy of goals and the items of gaze through an annotation scheme that analysed their direct and indirect persuasive import. Prodi's gaze was found to use a competence strategy much more frequently than other strategies and more than Royal (67 versus 27%), while he never used a benevolence strategy. Royal, instead, used the logos gaze most frequently (54%). Also in this study it was found that the pattern of gaze persuasive strategies of each orator was congruent with the politicians' global political strategy and with the context of the specific debate. This seems to show that during persuasion all body modalities coherently cooperate to one and the same persuasive goal.

3.3 Quality of Persuasive Body Behaviours

In a recent study, Poggi and Pelachaud (2008) wondered if it is possible to single out some gesture that one could define as "persuasive gestures". Actually, there could be some – very rare – gestures that one could call persuasive: for example, a gesture of incitation. Yet, it is more frequent that persuasiveness in a gesture does not dwell in the global meaning of the gesture per se, but rather in the expressivity of gesture (Hartmann et al., 2002): its spatial extent, temporal extent, fluidity, power, and repetition. Later, Poggi and Vincze (2008) put the same question about the existence of "persuasive gaze" (as described in the previous section).

But what does it mean that a gesture or a gaze is "persuasive"? According to Poggi and Pelachaud (2008), a gesture is persuasive when either the global meaning of the gesture or the meaning of some parts or aspects of it conveys some of the semantic contents that are contained in the persuasive structure of a discourse. Poggi and Pelachaud (2008) and Poggi and Vincze (2008) adopt the model of persuasion of Poggi (2005), according to which persuasion is an act of social influence brought about by communication, through which a persuader A aims to convince a persuadee B (that is, to make him believe with a high level of certainty) that a goal GA proposed by A is a subgoal to some goal GB that B already has; thus, A must convince B that GA is a goal of high value, possibly of higher value than alternative goals GC or GD, and must do so by exploiting the three Aristotelian strategies of logos (rational argumentation), pathos (induce emotions in B), and ethos (look credible and reliable to B). More specifically, A must show that they are certain of what they are saying (certainty), that they propose goal GA not out of their own concern but in order to achieve the goals of B (benevolence), and that they have good capacities of action and planning to pursue their own goals and to advise goals to others (competence).

Therefore, the gesture and gaze items that have a persuasive import are those that convey the following types of information:

1. Importance. The signals conveying the meaning "important" mention the high value of a proposed goal, thus trying to convince the persuadee to pursue it. This meaning is typically contained in some performative gestures, like incitation and request for attention, or other gestures like Kendon's (2004) "grappolo" ("finger bunch") that convey a notion of importance as their very meaning; but "important" is also the core meaning of beats, since every beat stresses a part of a sentence or discourse, hence communicating "this is the important part of the discourse I want you to pay attention to and to understand". Finally, this can be the meaning of some particular aspects of the expressivity of gesture, for instance irregularity or discontinuity in movement (Chafai et al., 2007).
2. Certainty. To induce certainty in the persuadee, the persuader must show self-confidence and be certain about what he is saying. This is why gestures that convey high certainty, like the "ring" mentioned by Kendon (2004) and Streeck (2008), may be persuasive. Yet, since persuading can mean either to convince to believe something or to convince to do something, the gestures that convey a high degree of certainty generally persuade to believe and only indirectly persuade to do something.
3. Evaluation. To express a positive evaluation of some object or event implies that it is a useful means to some goal; thus, to bring about that event or to obtain that object becomes desirable, a goal to be pursued. In a marketplace, to convince someone to buy a food, a "cheek screw" (rotating the tip of the index finger on

cheek), that means "good", "tasty", made by an Italian grocer, would be a good example of persuasive gesture. Of course, we cannot find an example like this in our fragments, due to obvious reason of social register. However, as we see below, a persuader, to pursue an ethos strategy, can make gestures that induce a positive evaluation of him.

4. Sender's benevolence. In persuasion not only the evaluation of the means to achieve goals is important, but also the evaluation of the persuader: the sender's ethos. If I am benevolent to you, you can trust me, so if I tell you that a goal is worthwhile you should pursue it. A gesture driven by the ethos strategy of showing one's moral reliability is, for example, putting one's hand on one's breast, which means "I am noble, I am fair". This gesture is quite frequent in various corpora of political communication (see, for example, Poggi and Pelachaud, 2008; Serenari, 2003).
5. Sender's competence. If I am an expert in the field I am talking about, if I am intelligent, efficient, you might join me and pursue the goals I propose. For example, in a pre-electoral debate a candidate, in talking of quite technical things concerning taxes, rotates his right hand curve open, with palm to left, twice. This gesture means that he is passing over these technicalities, possibly difficult for the audience; but at the same time the relaxed appearance of his movement lets them infer that he is smart because he is talking of such difficult things easily, unconstrained. This provides an image of competence in what he is talking about.
6. Emotion. If I express an emotion, and this is transferred to you through contagion (Poggi, 2004), since emotions trigger goals, a goal will be activated in you, thus implementing a pathos strategy. Another candidate says, "I cannot pretend to act in a country different from what it is", referring to his country, Italy. The movement of his forearm shows low spatial extent and fluidity (it is short and jerky) and high power and velocity, thus conveying an emotion load that aims to transmit a sense of pride of being Italian and thus to elicit the Italians' desire to vote for him. Another candidate in France observes that her opponent's politics towards the unemployed is somewhat punitive. While saying "sanctionner les chômeurs" (punish the unemployed) she raises the internal parts of eyebrows, thus performing a gaze of sadness. Thus she exhibits an emotion and aims to transmit it to the audience: a use of the pathos strategy.

4 The Effects of Persuasion

So far we have seen what persuasion is on the part of the persuader; now we can wonder about the effects of body behaviour on persuasion and about how persuasive effects can be assessed during interaction.

As to what are the effects of gestures on persuasion, several authors make the hypothesis that a persuasive discourse is more or less effective depending on the type of gestures used. Typically, as Henley (1977), Burgoon et al. (1990), and Carli et al. (1995) maintain, self-adaptors – the gestures of touching one's body – seem to have a negative effect on persuasion. However, presumably it is not so much the type of gesture – whether symbolic, iconic, self-adaptor, object adaptor, or the like – that makes the difference in persuasiveness. Our hypothesis is, instead, that it is not the type of gesture that takes persuasiveness away from discourse, but the gesture's meaning. Why, for example, are self-adaptors typically self-defeating in persuasive gesture? Probably because what a self-adaptor lets you infer is that the orator is not very self-confident, that he needs to reassure himself; and this lowers trust in the persuader. One case in which the very meaning of a gesture had a negative effect on persuasion concerns an American candidate, Howard Dean, who used to make a single gesture, "index up", or "finger wag", which seemed to have a subtly self-defeating effect. According to Streeck (2007), this gesture displayed the speaker's claim that what he was saying was important and instructive; but since Dean was enacting this "hierarchical act" in permanence, he might have given the impression he was presenting himself as one of "superior knowledge", thus spoiling, with a body behaviour somehow contemptuous towards the audience, the ascendancy credited to him by his early textual presence. Also in this case, therefore, what has negative effects on persuasion is not simply the type of gesture used, but its specific meaning.

5 Assessing the Effect of Persuasion: The Role of Backchannel

The latter issue to investigate is, How can the persuader monitor if and how much his persuasive effort has been effective? To this goal, the interlocutor's feedback must be taken into account.

When conversing, we try to get our interlocutors to engage with us in a joint project. We utter words, so that they will be heard, listened to, understood and that the other person engages in the proper acts that we are soliciting: attending, understanding, and reacting appropriately; answering our questions; believing our statements; taking up our orders; etc.

Conversational actions are undertaken to engage the addressees in taking actions in turn. When advising people, we hope people take the advice to heart. When cheering them up, we hope they feel happy for a while. When we argue in favour of a certain proposition, we hope that the other will become convinced of what we believe to be true.

In persuasive or argumentative conversations, just like with any action, it is important for speakers to check to what degree these actions are successful on all levels. In the case of conversations these checks involve monitoring the interlocutor to see his or her uptake of the joint projects proposed by the speaker. Through typical behaviours involving gaze, facial expressions, head movements, posture, and vocal backchannels listeners show that they are engaged in the conversation, paying attention, showing they are interested in what is being said, and, in the case of persuasive dialogue, whether they are starting to get convinced or not by the arguments. The interaction between speaker and audience is claimed to be essential in argumentative discourse according to van Eemeren and Grootendorst (1984).

> In ordinary conversational situations some speech acts by speakers are specifically calculated to elicit from listeners certain verbal (and possibly also non-verbal) responses in which they indicate understanding and (in particular) acceptance. In our view this applies pre-eminently to the argumentation advanced during a discussion or debate. This means that to a certain extent arguments in debates are designed to achieve precisely defined verbally externalized illocutionary and perlocutionary effects that are immediately related to the speech acts performed. [p. 24]

A particular type of response to speech is performed through backchannels. These were originally characterised by Yngve (1970) as follows:

> [B]oth the person who has the turn and his partner are simultaneously engaged in both speaking and listening. This is because of the existence of what I call the back channel, over which the person who has the turn receives short messages such as "yes" and "uh-huh" without relinquishing the turn. The partner, of course, is not only listening, but speaking occasionally as he sends the short messages in the back channel. The back channel appears to be very important in providing the monitoring of the quality of communication.

As a subset of feedback expressions, these backchannel signals in this view are short vocal messages that have an important structural function (monitoring the quality of the communication). Kendon (1967) talks about a similar category of expressions which he terms accompaniment signals: "short utterances that the listener produces as an accompaniment to a speaker, when the speaker is speaking at length". He divides these into two functional groups: attention signals (in which one appears to signal no more than that one is attending) and assenting signals that express "point granted" or "agreement". In general listener responses can serve several feedback functions. Allwood et al. (1992) distinguish four kinds: contact, perception, understanding, and attitude.

1. Contact: signals that show whether the interlocutor is willing and able to continue the interaction
2. Perception: behaviours that indicate whether the interlocutor is willing and able to perceive the message
3. Understanding: actions that display whether the interlocutor is willing and able to understand the message
4. Attitude: reactions that tell whether the interlocutor is willing and able to react and (adequately) respond to the message, specifically whether he/she accepts or rejects it.

The attitude-feedback functions, of which the assenting signals discussed by Kendon (1967) form an important subset, are interesting for the study of persuasive discourse as they might indicate the success achieved by the person who is trying to persuade.

In real-life examples it is not always easy to say which function is served by a particular feedback expression and many will combine several functions. Within the annotations of the AMI corpus (Jaimes et al., 2007), a difference was made between different types of speech acts that were mainly or exclusively used as listener responses: *backchannels* that acknowledge reception,

CommentAboutUnderstanding that indicates understanding or problems with understanding, and *assessments*. The latter are defined as follows:

> An **assessment** is any comment that expresses an evaluation, however tentative or incomplete, of something that the group is discussing. [...] There are many different kinds of assessment; they include, among other things, accepting an offer, expressing agreement/disagreement or any opinion about some information that's been given, expressing uncertainty as to whether a suggestion is a good idea or not, evaluating actions by members of the group, such as drawings. [...] It can be very short, like "yeah" and "ok".

Assessments are clearly used for the attitudinal reactions, where the speaker expresses his stance towards what is said, either acceptance or rejection.[1] It thus appears that there is much polysemy[2] in a verbal backchannel signal that can only in part be resolved by the context. In Sect. 6.1 we take a look at work on non-verbal backchannels that are important in a persuasive context for an ECA.

6 Towards Persuasive Embodied Conversational Agents

Embodied conversational agents, ECAs, are virtual human-like entities, with the ability to communicate with users and/or virtual agents. They are endowed with verbal and non-verbal communicative means, such as speech, voice intonation, gesture, body posture, facial expression, gaze. Human users have a tendency to apply the same principles when communicating with other humans or with ECAs (Reeves and Nass, 1996). ECAs are often used as dialogue partners. They can be used as personal assistant, web presenter, companion, tutor, and so on. As pointed out in the previous sections, persuasion may be relevant in each of these applications.

Several models of ECAs have been implemented. While few models have been elaborated for persuasive ECAs, several works have been undertaken to endow ECAs with communicative qualities that allow them to be persuasive in given situations. In the following we present a state of the art of such models. As described above, we have highlighted three qualities that are important in persuasion, namely gesture and expressive gestures, persuasion and emotion, and finally persuasion and backchannel. We now present the existing models in relation to these three qualities.

---

[1] A question looked at by Heylen and op den Akker (2007) was whether it is possible to distinguish an utterance containing "yeah" which expresses a stance (of partial agreement, i.e. an assessment in terms of the AMI annotation scheme) from an utterance that is simply meant as a backchannel. They achieved correct classification only for 60% of the cases when not taking into account the speech act of the previous utterance and 80% if they did.

[2] We avoid the use of the term ambiguity here as the distinction between the various categories is not strict and acknowledgements can easily shift into an assessment.

6.1 ECA – Persuasion and Gesture

Not much work has been done so far in implementing gesture features for persuasion in ECA. We report works done in modelling gesture style and expressive gestures that could be used in persuasive strategies as described in Sect. 3.

Ruttkay and her colleagues (2003) proposed a behaviour representation to encompass styles. An ECA is described over a large set of dimensions ranging from its culture and profession to its emotional and physical state. All these dimensions affect the way an ECA moves and gesticulates. EMOTE (Costa et al., 2000) implements the Laban annotation scheme for dance to change, through a set of parameters, the way a gesture looks depending on values such as the strength of the gesture and its tempo. EMOTE works as a post-filter after a gesture animation has been computed and adds expressivity to its final animation.

A model of behaviour expressivity using a set of six parameters that act as modulation of behaviour animation has been implemented (Hartmann et al., 2005). When applied at these different levels, expressivity may convey different functions in the discourse context: it can attract the attention (Chafai et al., 2007), persuade the addressee (Poggi and Pelachaud, 2008), and indicate emotional state (Martin et al., 2005). In these studies they were interested in understanding and modelling how emotion as well as persuasion can be conveyed qualitatively, in particular through gesture expressivity.

Two-dimensional cartoons were manually annotated to understand how animators used characters' movement expressivity to call for the attention of the spectators. Two types of modulation of behaviour expressivity (irregularities and discontinuities) were found to be used by the animators. They were integrated in an ECA system (Chafai et al., 2007). An evaluation study was conducted to see if these modulations played a role in attracting the user's attention when conversing with the agent. The results of the evaluation confirm that expressivity specified at gesture phase level may play a specific function; namely, here, it can act to attract the attention of the interlocutor at precise moments of the dialogue.

Expressivity parameters were extracted either manually (Devillers et al., 2005; Martin et al., 2005) or automatically (Caridakis et al., 2006) over a whole sequence and played back by an agent. The purpose of these studies was to understand which elements of behaviours, in particular behaviour expressivity, play a role for perceiving an emotional state.

6.1.1 Persuasion and Backchannel

In a series of studies Bevacqua et al., (2007) and Heylen et al., (2007), looked at the perception of facial expressions of attitudes that a conversational agent displayed when listening. They looked at a particular subset of expressions that might be relevant to use in persuasive dialogues, where the agent as listener could indicate its understanding or acceptance of the utterances of the speaker.

In a perception test, subjects were asked to watch the virtual character Greta (Bevacqua et al., 2007a) displaying a combination of facial signals and head movements. The following behaviours were presented in the 14 video clips shown in a first experiment (Bevacqua et al., 2007b):

1. a single head nod (N)
2. a head nod with a smile (NS)
3. a head nod and a raise of the eyebrows (NRE)
4. a head shake (HS)
5. a head shake and a frown (SF)
6. a head shake, a frown, and a tension in the lips that tighten and get thinner (SFT)
7. a frown and a tension in the lips that tighten and get thinner (FT)
8. a raise of the left eyebrows (RLE)
9. the eyes roll in the head (ER)
10. a head tilt to the left with sad eyebrows (TSE)
11. a head tilt to the left and a frown (TF)
12. a head tilt to the right and a raising of the eyebrows (TRE)
13. a head tilt to the right, gaze turns down to the right (TG)
14. eyes wide open (EWO)

The displays are prototypically associated with a number of meanings in a conversational setting as listener responses that are either used as conventional signals or as extensions of the meanings these displays have in, for instance, emotional expressions.

The subjects had to associate one or more meanings to these signals. They could choose from a close list of meanings. In particular, we have investigated how well these expressions matched various classes of performative displays and epistemic and affective states: agreement, disagreement, acceptance, rejection, interest and boredom, believe, disbelieve, understanding, and lack of understanding. These are assumed to be important attitudinal reactions in a persuasive dialogue. One could assume that signals containing nods and smiles would be mostly associated with the positive states whereas shakes, frowns and lip tension would be associated with negative states (Bassili, 1979). Table 1 shows for each behaviour the meaning that was most often assigned to it. In some cases there were two meanings that were equally often chosen.

This study (Heylen et al., 2007) shows that for most attitudes fairly clear prototypical visual expressions can be found. A one-to-many mapping can be established between a facial signal and a backchannel meaning. The results of the study are in line with existing literature. Nods, for example, are mostly used in the contexts of acknowledgements or agreements or as a positive answer substituting or coinciding with the utterance "yes" or an equivalent. Shakes, on the other hand, have a negative meaning (Kendon, 2003). Head tilts often suggest interest. Smiles occur a lot in conversations. They mostly show some positive attitude, except in cases such as a sarcastic grin. Eyebrow movements have a range of meanings. One of the meanings that Ekman (1979) points out is one where it can function as an agreement response indicating that the listener is attending but also understands and does not disagree

Table 1 Meanings associated with head behaviours

Nod	Acceptance
Nod smile	Liking
Nod right eyebrow	Agreement
Raise left eyebrow	Disbelieve
Tilt sad eyebrows	Not understand
Eye roll	Disbelieve
Tilt, gaze down	Boredom
Shake, frown, tense lips	Disagreement/rejection
Frown, tense lips	Disagreement
Head shake	Rejection
Shake, frown	Disagreement, rejection
Tilt, frown	Disagreement, disbelieve
Tilt right, raise brows	Believe (is this correct)
Eyes wide open	Disbelieve

with what is being said. Raising of the eyebrows together with a head nod or an agreement word is a typical agreement listener response. However, raised eyebrows can also figure in displays of surprise. Also a frown can function as an expression of perplexity but is also associated with expressions of anger and disgust. The perplexity that is often being expressed by the frown can function as a response to the speaker indicating lack of understanding or indicating that one thinks what is being said does not make sense.

6.2 Persuasion and Emotion

In persuasion one has to convince with logos (rational argumentation), ethos (the persuader's credibility and reliability), and pathos (the appeal to the emotions of the interlocutor). In both trying to be convincing and transmitting emotions can one need to exhibit a fake personality and/or fake emotions? In human–human communication, people often try to hide their real emotions because the social situation requires it. Typical example are excuses, such as "I would love to join you, but . . .". Emotions are the number-one topic that people lie about and studies show that up to 30% of social interaction longer than 10 min contains such deceptions (DePaulo et al., 1996). Endowing technical systems like embodied conversational agents with the ability to detect, represent, generate, and/or show emotions, it is thus indispensable to investigate the crucial questions how to handle false emotional expressions from the user and how and when to create false emotional expressions in the ECA.

6.2.1 Persuasion and Deception

Various attempts have been made to create synthetic agents that deliberately oppress or express a certain emotion. De Rosis and colleagues (2003) as well as Prendinger and colleagues (2001) have developed agents that are able to control their emotional displays if the social situation requires it. For instance, if the social distance between an agent and its conversational partner is high, Prendinger's agent would not show anger to the full extent. The virtual tutor COSMO developed by Lester and colleagues (2000) intentionally portrays emotions with the goal of motivating students and thus increasing the learning effect.

Earlier approaches start from the assumption that the agent is able to perfectly hide emotions if the social or pedagogical situation requires it. However, humans are not always capable of completely concealing their true emotions. For instance, masking smiles cannot entirely override the muscular programme of the original emotion because not every facial muscle can be consciously controlled. As a consequence, such a mask will always include segments of one's felt emotion. The question arises as to how to handle situations in which the agent decides to display an emotion which is in conflict with its internal appraisal processes. In some situations, it might be desirable to employ agents that perfectly convey "wrong" emotions with the aim to convince the interlocutor. Consider a sales agent on the web that has to advertise a product of minor quality. If it does not succeed in concealing its negative attitude towards the product, a decrease of the sales might be the consequence. On the other hand, agents in social settings may come across as little believable or cold if they are always able to perfectly hide their true emotions. In addition, the display of mixed emotions may even lead to a positive response from the interlocutor. For instance, students may feel sympathy towards a virtual teacher that desperately tries to hide its negative emotions caused by the students' bad performance. Last but not least, the emulation of deceptive behaviours may enrich our interactions with synthetic agents – especially in game-like environments.

Following these considerations, Rehm and André (2005a) focused on synthetic agents that may express emotions that are in conflict with their appraisal processes. Unlike earlier work, they modelled situations in which the agent fails to entirely conceal its "felt" emotions. They developed an agent whose behaviours may reflect potential conflicts between "felt" and deliberately expressed emotions. Their work concentrated on facial expressions of deception which have been profoundly researched in the psychological literature. According to Ekman and colleagues (1988), there are at least four ways in which facial expressions may vary if they accompany lies and deceptions: micro-expressions, masks, timing, and asymmetry.

1. Micro-expressions: A false emotion is displayed, but the felt emotion is unconsciously expressed for a fraction of a second. The detection of such micro-expressions is possible for a trained observer.
2. Masks: The felt emotion (e.g. disgust) is masked by a noncorresponding facial expression, in general by a smile. Because we are not able to control all of our facial muscles, such a masking smile is in some way deficient. Thus, it reveals at least in part the original emotion.
3. Timing: Facial expressions accompanying true emotions do not last for a very long time. Thus, the longer an expression lasts the more likely it is that it is accompanying a lie. A special case seems to be surprise, where elongated on- and offset times are a good indicator of a false emotion.

4. Asymmetry: Voluntarily created facial expressions like they occur during lying and deceiving tend to be displayed in an asymmetrical way, i.e. there is more activity on one side of the face than on the other.

To model the non-verbal behaviour, they employed the Greta agent system developed by Pelachaud and colleagues (2007). For example, Fig. 1 shows the facial displays for genuine joy, genuine disgust, and disgust faked by joy. In the case of natural joy (left screenshot), the corners of the mouth and cheeks are moved up symmetrically. In the case of natural disgust (middle screenshot), the eyebrows are contracted and the upper lip is moved up. To fake disgust by joy (right screenshot), the eye movements of disgust and the mouth movements of joy have been combined. Different degrees of masking are combined with different degrees of asymmetry of the facial displays resulting in 32 possible facial expressions.

A computational model of complex facial expression was developed by Niewiadomski and Pelachaud (2007a, b). It is also based on Ekman's work (1975, 2003). Complex facial expressions are obtained using a face partitioning approach where facial expression is defined by a set of eight facial areas: forehead/eyebrow, upper eyelid, eyes, lower eyelid, cheek, nose, lip (movement), and lip tension. In particular, in a case of complex facial expressions different emotions can be expressed on different areas of the face. The complex facial expressions are composed of the facial areas of input expressions using a set of fuzzy rules. Different rules have been implemented for the superposition of two felt emotions (see Fig. 2) and for the masking of a felt emotion by a fake (i.e. non-felt) emotion (see Fig. 3). The rules make use of the results showing that expressions of felt emotions are signalled by characteristic (also called reliable) features (such as the crow feet for felt happiness) (1975, 2003) and that positive emotions are mainly perceived from the lower part of the face (smile for happiness) and negative emotions from the upper face (such as frown for anger) (Bassili, 1979).

An example of such a rule for the superposition of two felt emotions (sadness and anger) is the following the more one of the input expressions is (similar to) anger and the other input expression is (similar to) sadness, the more certain is that the final expression contains brows, upper eyelids, eyes, and the lower eyelids of the first expression and the mouth area rest of the second (see Fig. 2). The output of this module is a facial expression composed of parts of the facial expression of one emotion and parts of the facial expression of the other emotion. They refer to this final expression as a complex expression as it is composed of two expressions and thus shows two emotions.

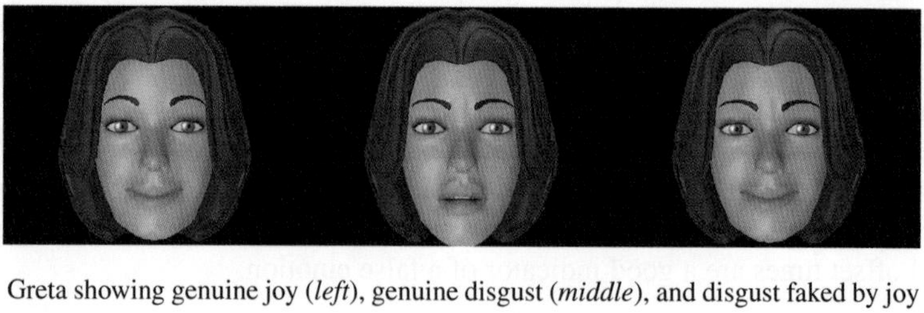

Fig. 1 Greta showing genuine joy (*left*), genuine disgust (*middle*), and disgust faked by joy (*right*)

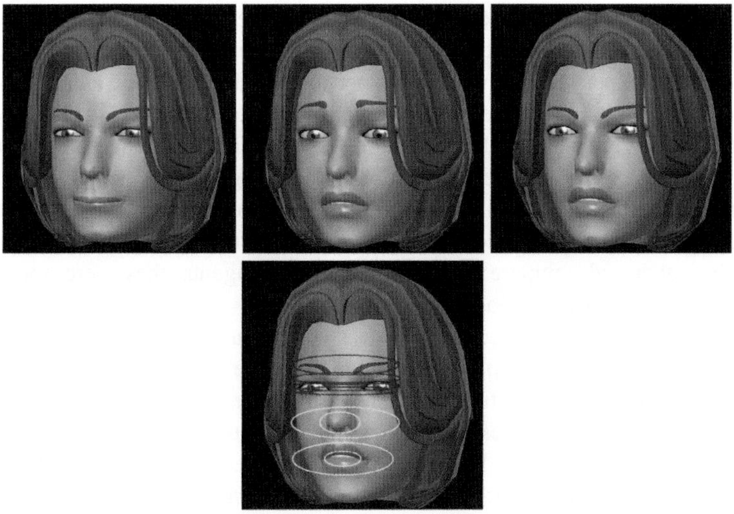

Fig. 2 Superposition of anger and sadness. From the *left* to *right*: anger (**a**), sadness (**b**), superposition of anger and sadness (**c**) superposition of anger and sadness with significant areas marked (**d**)

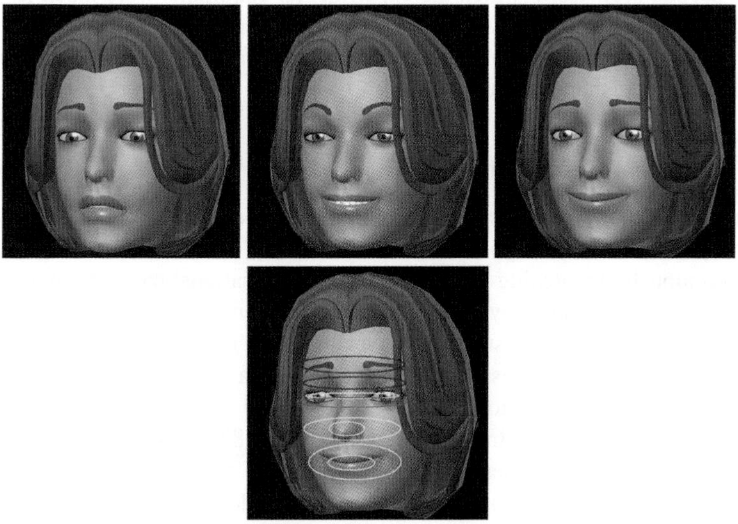

Fig. 3 Sadness masked by happiness. From the left to right: sadness (**a**), happiness (**b**), sadness masked by happiness (**c**) sadness masked by happiness with significant areas marked (**d**)

Figure 3 presents the agent displaying an expression of sadness, that is, masked by a fake happiness. According to these rules facial areas of forehead/eyebrows and upper eyelids cover the features of felt sadness (red circles in Fig. 3d) that leak over the mask of a joy (yellow circles in Fig. 3d). As a consequence, they can be observed in a final expression.

To find out how deceptive clues are subjectively perceived by a human user and to what extent users are able to correctly interpret them, Rehm and André (2005a) conducted two experiments. In the first experiment, a deceptive and a non-deceptive version of the Greta agent presented movie reviews to the users. That is, user did not interact with the agent and information was provided by the agent in terms of a monologue. The study revealed that the non-deceiving agent is perceived as being more reliable, trustable, convincing, credible, and more certain about what it said. Although people reacted to facial clues of deceit when they had the opportunity to carefully watch and compare different instances of agents, they were not able to name the reasons for these reactions.

In the second experiment, the agent was serving as a game partner in a game of dice. To win the game, players had to lie to the other players and to catch liars. In the game, the agent tried to mislead the other players by portraying facial expressions that did not correspond to her actual emotional state. For instance, she expressed false joy to make her game partners believe that she achieved a high score. Nevertheless, the agent did not lie in a perfect manner, but still revealed her deceptive behaviours by subtle facial cues. The first study indicated that even subtle expressions of deception may have an unfavourable impact on the user's perception of the agent. The results of the first study could, however, not be confirmed for the second scenario in which the experimental conditions were much less controlled. Most likely the players were too much engaged in the game to pay attention to the agent's deceptive cues. In a more natural and engaging face-to-face situation, subjects tend to disregard deceptive clues.

6.2.2 Persuasion and Politeness

If you want to persuade someone through inducing positive emotions in him/her a good way to do so may be politeness. When humans interact with each other, they risk continuously threatening the face of their conversational partners, for example, by showing disapproval or by putting the other person under pressure. To mitigate such face threats, humans usually rely on politeness tactics. For instance, instead of formulating a direct request "Solve the equation", a teacher might suggest "Why don't we solve the equation together?"

In their seminal work, Brown and Levinson (1987) analyse verbal strategies of politeness. Positive politeness seeks approval of the addressee, for example, by making him a compliment before a criticism is delivered. Negative politeness emphasises the hearer's freedom of action, for example, by formulating a suggestion instead of a request. Off-record statements are vague and the addressee has to infer the exact meaning of the speaker leaving him in the position to misunderstand the speaker and thus to not feel offended. Previous work has concentrated for the most part on verbal means to mitigate face threats. An exception is an empirical study by Trees and Manusov (1998) who found that non-verbal behaviours, such as pleasant facial expressions and more direct body orientation, may help to mitigate face threats evoked by criticism. Bavelas et al., (1995) provide a classification of gestures, some of which can be directly mapped onto Brown and Levinson's strategies of politeness. Shared information gestures mark material that is part of the interlocutor's common ground. Citing gestures refer to previous contributions of the addressee and aim at conveying the impression that the interlocutors share a common opinion. Elliptical gestures mark incomplete information that the addressee should augment for himself or herself and may take on a similar function as off-record strategies. Seeking agreement gestures directly correspond to Brown and Levinson's approval-oriented strategies. Turn open gestures can be regarded as attempts to satisfy the addressee's desire for autonomy. Linguistic means to deliver face-threatening acts have partly become part of the grammar and Bavelas classification of gestures suggests that there might be similar principled and standardised connection between non-verbal means of communication and politeness strategies.

To shed light on the question of how face threats are mitigated by non-verbal means, Rehm and André (2005b) conducted a corpus study with human speakers. This multimodal corpus consists of staged interactions of inherently face-threatening situations. They devised a scenario that forced the participants to use their (unconscious) knowledge of politeness strategies by confronting them with an inherently face-threatening situation. To code politeness strategies, they followed Walker et al.'s (1997) categorisation of direct, approval-oriented, autonomy-oriented, and off-record strategies.

The results of their corpus analysis indicate that gestures are indeed used to strengthen the effect of verbal acts of politeness (Rehm and André, 2005b). In particular, vagueness as a means of politeness is reflected not only by verbal utterances but also by gestures. Iconic and deictic gestures were overwhelmingly used in more direct criticism while there was a high frequency of metaphoric gestures in off-record strategies. Obviously, the subjects did not attempt at compensating for the vagueness of their speech by using more concrete gestures.

Walker et al. (1997) have presented one of the first approaches to implement politeness strategies based on the theory by Brown and Levinson (1987) as a means to more flexible dialogue control. They describe a selection mechanism that is based on the variables power, social distance, and ranking of speech act. Johnson et al. (2004) investigated the potential benefits of politeness in a tutoring system. Examining the interactions between a real tutor and his students, they came up with a set of templates each of which is annotated according to the amount of redress that tactic gives to the learner's face. Rehm and André (2005b) equipped the Greta agent with politeness behaviours based on the corpus study presented above.

Niewiadomski and Pelachaud (2007b) analysed the same video-corpus of Rehm and André (2005b) in order to find relations between politeness strategies and facial behaviour. They considered four types of facial displays: expression of the true emotional state, inhibited, masked, and fake expression. They analysed the frequency of occurrence for each of them and found that different types of facial expressions were not evenly distributed along the different strategies of politeness. They used this information to build their model of facial deceptive behaviour management in interpersonal relations. In their model, they considered three variables to encompass the characteristics of the interaction (Brown and Levinson, 1987) and the emotional state of the displayer, namely social distance, social power, and emotion valence. Based on the value of these variables they defined which facial management is the most appropriate, e.g. when a felt emotion should be masked or inhibited.

7 Recommendations to Build a Persuasive ECA

To conclude this chapter, we provide recommendations for non-verbal behaviours to be considered when constructing a persuasive ECA.

7.1 Gaze Behaviour

Eye contact is a powerful tool for persuasive purposes. It can be used to acknowledge presence, show interest, provide feelings of importance, and help convey an attitude that the persuader likes the persuadee, something also influenced by large pupil size. In courtroom situations, making eye contact with jurors has been referred to as being the most important type of communication possible (Bernstein et al., 1994). Speakers who engage in eye contact are perceived as being more credible (Hemsley and Doob, 1978) than those who continually avert their gaze, the latter appearing untrustworthy or unconfident. However, it is important to establish a balance between appearing confident by engaging in sufficient eye contact, while at the same time avoiding to be perceived as overpowering or threatening by maintaining eye contact for too long. This is a difficult proposition in practice, as it must account for many factors related to the context of the interaction, the persuader and the persuadee, such as culture and gender. At a higher level, gaze strategies such as logos gaze, competence, and benevolence strategies (see Sect. 2) can be adopted to support persuasion through rational arguments, establishing trust, and invoking emotions (Poggi and Vincze, 2008). In these situations, a key factor should be to ensure congruence between gaze and other modalities in order to provide coherent persuasive behaviour consistent with a single strategy. The behavioural realism of the gaze behaviour is also a factor for consideration, as virtual humans higher in behavioural realism have been found to be more influential (Guadagno et al., 2007).

7.2 Facial Expressions

Facial expressions convey a large amount of non-verbal information: a smile or a frown made at a certain moment can speak volumes. In general, facial expressions should appear to be relaxed in order to help convey a confident but non-threatening demeanour: tension should not be perceptible in the brows, jaw, mouth, or shoulders so that the persuader appears to be in control of the situation at all times.

7.3 Showing Attention and Interest

Related to eye contact and facial expression is the importance of showing interest in what the other has to say. For example, as described in Levine (2006), based on his experiences working on the documentary *Thin Blue Line*, interviewer Errol Morris highlighted the importance of appearing to be engrossed and show interest in what his guests had to say at all costs. In some situations, Morris even suggested not actually becoming too engrossed in the story, so one could concentrate totally on providing signals of interest, which he regarded as being of the utmost importance for eliciting disclosure of facts from interviewees.

7.4 Hands and Arms

Great speakers use hand gestures more than on average and gestures can give the listener confidence in the speaker as well as help the speaker maintain their turn. It is particularly important that hand gestures keep out of the personal space of the other and especially away from their face. The arms should reinforce the impression of openness on the part of the agent by being uncrossed. As with facial expressions, arm movements should not appear to be tense and should be open and expansive. Humans have a natural tendency to protect vulnerable organs especially in threatening or stressful situations – not doing so can make us feel exposed and vulnerable, but can project a courageous posture, reflecting self-confidence. Furthermore, hands should ideally be visible rather than placed in pockets: Open palms are a good way to express honesty and trustworthiness. Self-touching behaviours are to be avoided, mainly because they may convey a need for self-reassurance.

7.5 General Posture and Barriers

As with the other factors, it is important to project an open, honest, cooperative posture, as this can be perceived to reflect a psychological openness. In contrast, a poor, deflated posture is associated with a lack of confidence or a lack of interest. The orientation of the body is of importance here too: as the term "cold shoulder" suggests, when we feel uncomfortable with a person of situation, we may tend to orientate our body sideways, showing aversion of timidity. Orienting towards the audience, and not allowing obstructions between them and the persuader, is of importance in establishing a direct connection with them. Arms, obstacles, and instruments may be used to create barriers, severing this connection and portraying a defensive posture – thus, arms should not be placed at rest in front of the abdomen or chest.

References

Allwood J, Nivre J, Ahlsén E (1992) On the semantics and pragmatics of linguistic feedback. J Semantics 9(1):1–26

Atkinson M (1984) Our masters' voices. The language and body language of politics. Methuen, London

Bassili JN (1979) Emotion recognition: the role of facial movement and the relative importance of upper and lower areas of the face. J Pers Soc Psychol 37(11):2049–2058

Bavelas JB, Chovil N, Coates L, Roe L (1995) Gestures specialized for dialogue. Pers Soc Psychol Bull 21(4):394–405

Bernstein LE, Demorest ME, Eberhardt SP (1994) A computational approach to analyzing sentential speech perception: phoneme-to-phoneme stimulus-response alignment. J Acoust Soc Am 95(6):3617–3622

Bevacqua E, Heylen D, Pelachaud C, Tellier M (2007a, April) Facial feedback signals for ecas. In: Proceedings of AISB'07: artificial and ambient intelligence, Newcastle University, Newcastle upon Tyne, UK

Bevacqua E, Mancini M, Niewiadomski R, Pelachaud C (2007b) An expressive eca showing complex emotions. In: Proceedings of the AISB annual convention, Newcastle, UK, pp 208–216

Brown B, Levinson SC (1987) Politeness: some universals on language usage. Cambridge University Press, Cambridge

Bucy EP, Bradley SD (2004) Presidential expressions and viewer emotion: Counterempathic responses to televised leader displays. Soc Sci Inf 43:59–94

Bull PE (1986) The use of hand gesture in political speechers: some case studies. J Lang Soc Psychol 5:102–118

Burgoon JK, Birk T, Pfau M (1990) Nonverbal behaviors, persuasion, and credibility. Hum Commun Res 17:140–169

Caridakis G, Raouzaiou A, Karpouzis K, Kollias S (2006, May) Synthesizing gesture expressivity based on real sequences. In: Workshop on multimodal corpora: from multimodal behaviour theories to usable models, LREC 2006 Conference, Genoa, Italy, 24–26 May 2006

Carli LL, LaFleur SL, Loeber CC (1995) Nonverbal behavior, gender, and influence. J Pers Soc Psychol 68:1030–1041

Castelfranchi C (2000) Affective appraisal versus cognitive evaluation in social emotions and interactions. In: Paiva AM (ed) Affective interactions: towards a new generation of computer interfaces. Springer, Heidelberg, pp 76–106

Chafai NE, Pelachaud C, Pelé D (2007) A case study of gesture expressivity breaks. Int J Lang Resour Eval, special issue on Multimodal Corpora Modell Hum Multimodal Behav 41(3–4):341–365

Costa M, Zhao L, Badler NI (2000) The EMOTE model for effort and shape. In: Proceedings of the 27th annual Conference on Computer Graphics and Interactive Techniques SIGGRAPH '00, New Orleans, ACM Press, New York, NY, pp 173–182

Darwin C (1872) The expression of the emotions in man and animals. John Murray, London

DePaulo BM, Kashy DA, Kirkendol SE, Wyer MM, Epstein JA (1996) Lying in everyday live. J Pers Soc Psychol 70(5):979–995

deRosis F, Pelachaud C, Poggi I, Carofiglio V, De Carolis B (2003) From Greta's mind to her face: modelling the dynamics of affective states in a conversational embodied agent. Int J Hum Comput Stud 59(1–2):81–118

Devillers L, Abrilian S, Martin J-C (2005) Representing real life emotions in audiovisual data with non basic emotional patterns and context features. In: First international conference on affective computing and intelligent interaction (ACII'2005), Beijing, China, pp 519–526

Ekman P (1975) Unmasking the face. A guide to recognizing emotions from facial clues. Prentice-Hall, Inc., Englewood Cliffs, NJ

Ekman P (1979) About brows: emotional and conversational signals, pages 169–202. Cambridge University Press/Editions de la Maison des Sciences de l'Homme

Ekman P (2003) The face revealed. Weidenfeld & Nicolson, London

Ekman P, Friesen W, O'Sullivan M (1988) Smiles when lying. J Personal Soc Psychol 54(3): 414–420

Frey S (2000) Die Macht des Bildes. Der Einfluss der nonverbalen Kommunikation auf Kultur und Politik. Huber, Bern

Frijda N (1986) The emotions. Cambridge University Press

Guadagno RE, Blascovich J, Bailenson JN, McCall C (2007) Virtual humans and persuasion: the effects of agency and behavioral realism. Media Psychol 10:1–22

Hartmann B, Mancini M, Pelachaud C (2002) Formational parameters and adaptive prototype instantiation for MPEG-4 compliant gesture synthesis. In: Computer animation'02, Geneva, Switzerland, 2002. IEEE Computer Society Press

Hartmann B, Mancini M, Pelachaud C (2005) Implementing expressive gesture synthesis for embodied conversational agents. In: Gesture in human-computer interaction and simulation, 6th international gesture workshop, GW 2005, Berder Island, pp 188–199

Hemsley GD, Doob AN (1978) The effect of looking behaviour on perceptions of a communicator's credibility. J Appl Soc Psychol 8:136–144

Henley NM (1977) Body politics: power, sex, and nonverbal behavior. Prentice-Hall, Englewood Cliffs, NJ

Heylen D, Bevacqua E, Tellier M, Pelachaud C (2007) Searching for prototypical facial feedback signals. In: Pelachaud C, Martin J-C, André E, Chollet G, Karpouzis K, Pelé D (eds) Proceedings of the 7th international conference on intelligent virtual agents (IVA), vol 4722 of Lecture Notes in Computer Science, pp 147–153. Springer, Heidelberg

Heylen D, op den Akker R (2007) Computing backchannel distributions in multi-party conversations. In: Cassell J, Heylen D (eds) Proceedings of the ACL workshop on embodied language processing, vol W07-19, Prague, Czech Republic, 2007, pp 17–24. Association of Computational Linguistics

Poggi I (2004) Emotions from mind to mind. In: Proceedings of the workshop W13: empathic agents. third international conference on autonomous agent systems, AAMAS 2004, New York, NY, USA, pp 208–216

Jaimes A, Bourlard H, Renals S, Carletta J (2007) Recording, indexing, summarizing, and accessing meeting videos: an overview of the ami project. In: Fourth international conference on image analysis and processing (ICIAP 2007) and workshop on visual and multimedia digital libraries (VMDL 2007), Modena, pp 59–64

Johnson WL, Rizzo P, Bosma W, Kole S, Ghijsen M, van Welbergen H (2004) Generating socially appropriate tutorial dialog. In: ISCA workshop on affective dialogue systems, Modena, pp 254–264

Kendon A (1967) Some functions of gaze direction in social interaction. Acta Psychol 26:22–63

Kendon A (2003) Some uses of head shake. Gesture 2:147–182

Kendon A (2004) Gesture : visible action as utterance. Cambridge University Press Cambridge

Lester JC, Towns SG, Callaway CB, Voerman JL, FitzGerald PJ (2000) Deictic and emotive communication in animated pedagogical agents. In: Prevost S, Cassell J, Sullivan J, Churchill E (eds) Embodied conversational characters. MITpress, Cambridge, MA, pp 123–154

Levine R (2006) The power of persuasion. chapter Whom do we trust? Experts, honesty and likability. Wiley

Martin J, Abrilian S, Devillers L (2005) Annotating multimodal behaviors occurring during non basic emotions. In: Proceedings of the first international conference on affective computing and intelligent interaction (ACII), Pkin, Chine

Martin J-C, Abrilian S, Devillers L (2005) Annotating multimodal behaviors occurring during non basic emotions. In: 1st international conference on affective computing & intelligent interaction (ACII'2005), Beijing, China, 2005 October 22–24

Maurice Tournier (2003) Geneviève Calbris. L'Expression gestuelle de la pense d'un homme politique. CNRS Communication

Mehrabian A (1971) Silent messages. Wadsworth Publishing Company

Niewiadomski R, Pelachaud C (2007a) Fuzzy similarity of facial expressions of embodied agents. In: Pelachaud C, Martin J-C, André E, Chollet G, Karpouzis K, Pelé D (eds) Proceedings of the 7th international conference on intelligent virtual agents (IVA). Lecture Notes in Computer Science, vol 4722. Springer, Heidelberg, pp 86–98

Niewiadomski R, Pelachaud C (2007b) Model of facial expressions management for an embodied conversational agent. In: Paiva A, Prada R, Picard RW (eds) Second international conference

on affective computing and intelligent interaction (ACII). Lecture Notes in Computer Science, vol 4738. Springer, Heidelberg, pp 12–23
Pease B, Pease A (2006) The definitive book of body language. Bantam Books
Poggi I (2005) The goals of persuasion. Pragmatics Cogn 13:298–335
Poggi I, Pelachaud C (2008) Persuasion and the expressivity of gestures in humans and machines. In: Wachsmuth I, Lenzen M, Knoblich G (eds) Embodied communication in humans and machines Oxford University Press, Oxford, pp 391–424
Poggi I, Vincze L (2008) Persuasive gaze in political discourse. In: Proceedings of the AISB annual convention, Aberdeen, Scotland, UK
Prendinger H, Ishizuka M (2001) Social role awareness in animated agents. In: Proceedings of the fifth international conference on autonomous agents, Montreal, Quebec, Canada, pp 270–277
Quintilianus MF (1920) Institutio Oratoria (trans: Butler HE). Loeb Classical Library. Harvard University Press, Cambridge
Reeves B, Nass C (1996) The media equation: how people treat computers, television, and new media like real people and places. Cambridge University Press, Cambridge
Rehm M, André E (2005a) Catch me if you can – exploring lying agents in social settings. In: Dignum F, Dignum V, Koenig S, Kraus S, Singh MP, Wooldridge M (eds) the proceedings of international joint conference on autonomous agents and multi-agent systems (AAMAS), Utrecht, The Netherlands, 2005, pp 937–944. ACM
Rehm M, André E (2005b) More than just a friendly phrase: multimodal aspects of polite behavior in agents. In: Nishida T (ed) Conversational informatics: an engineering approach. Wiley, Chichester, pp 69–84
Ruttkay Z, van Moppes V, Noot H (2003) The jovial, the reserved and the robot. In: Proceedings of the AAMAS03 Ws on embodied conversational characters as individuals, Melbourne, Australia
Seiter JS, Kinzer HJ, Weger H Jr (2006, April) Background behavior in live debates: the effects of the implicit ad hominem fallacy. Commun Rep 19(1):57
Serenari M (2003) Examples from the Berlin dictionary of everyday gestures. In: Rector M, Poggi I, Trigo N (eds) Gestures. Meaning and use. Edicoes Universidade Fernando Pessoa
Streeck J (2008) A case study of the democratic presidential candidates during the 2004 primary campaign. Res Lang Soc Interact 41:154–186
Trees AR, Manusov V (1998) Managing face concerns in criticism. Hum Commun Res 24(4): 564–583
van Eemeren FH, Grootendorst R (1984) Speech acts in argumentative discussions. Walter de Gruyter and Foris, Berlin
Walker MA, Cahn JE, Whittaker SJ (1997) Improvising linguistic style: Social and affective bases for agent personality. In: First international conference on autonomous agents, Marina del Rey, CA, USA, pp 96–105
Yngve VH (1970) On getting a word in edgewise. In: Papers from the sixth regional meeting of the Chicago Linguistic Society. Chicago Linguistic Society, Chicago; pp 567–577
Watson K, Mattews B, Allman J (2007) Brain activation during sight gags and language-dependent humor. Cerebral Cortex 17(2)
Yuill N (1997) A funny thing happened on the way on the classroom: jokes, riddles and metalinguistic awareness in understanding and improving poor comprehension in children. In: Disabilities: processes and intervention. Erlbaum, Mahwah, NJ

Computational Humour

Carlo Strapparava, Oliviero Stock, and Rada Mihalcea

Abstract Computational humour is a challenge with connections and implications in many artificial intelligence areas, including natural language processing, intelligent human–computer interaction, and reasoning, as well as in other fields such as cognitive science, linguistics, and psychology. Of particular interest is its connection to emotions. In this chapter we overview the basic theories of humour and present the main contributions made in the field of computational verbal humour, including applications for automatic humour generation and humour recognition.

1 Introduction

The interaction between humans and computers needs to evolve beyond usability and productivity. There is an agreement in the field of human–computer interaction that the future stands in themes such as entertainment, fun, emotions, aesthetic pleasure, motivation, attention, engagement. Humour is an essential element in communication: it is strictly related to the themes mentioned above, and arguably humans cannot survive without it. While it is generally considered as merely a way to induce amusement, humour provides an important way to influence the mental state of people to improve their activity. Even though humour is a complex capability to reproduce, it is realistic to model some types of humour production and to aim at implementing this capability in computational systems.

Humour is a powerful generator of emotions. As such, it has an impact on people's psychological state, directs their attention, influences the processes of memorization and of decision-making, and creates desires and emotions. Actually, emotions are an extraordinary instrument for motivation and persuasion because those who are capable of transmitting and evoking them have the power to influence other people's opinions and behaviour. Humour, therefore, allows for conscious and

C. Strapparava (✉)
Fondazione Bruno Kessler-Irst, Povo, Trento, Italy
e-mail: strappa@fbk.eu

P. Petta et al. (eds.), *Emotion-Oriented Systems*, Cognitive Technologies,
DOI 10.1007/978-3-642-15184-2_31,

constructive use of the affective states generated by it. Affective induction through verbal language is particularly interesting; and humour is one of the most effective ways of achieving it. Purposeful use of humourous techniques enables us to induce positive emotions and mood and to exploit their cognitive and behavioural effects. For example, the persuasive effect of humour and emotions is well known and widely employed in advertising. Advertisements have to be both short and meaningful, to be able to convey information and emotions at the same time.

Humour acts not only upon emotions but also on human beliefs. A joke plays on the beliefs and expectations of the hearer. By infringing on them, it causes surprise and then hilarity. Jesting with beliefs and opinions, humour induces irony and accustoms people not to take themselves too seriously. Sometimes simple wit can sweep away a negative outlook that places limits on people's desires and abilities. Wit can help people overcome self-concern and pessimism that often prevents them from pursuing more ambitious goals and objectives.

Humour encourages creativity as well. The change of perspective caused by humourous situations induces new ways of interpreting the same event. By stripping away clichés and commonplaces, and stressing their inconsistency, people become more open to new ideas and points of view. Creativity redraws the space of possibilities and delivers unexpected solutions to problems. Actually, creative stimuli constitute one of the most effective impulses for human activity. Machines equipped with humourous capabilities will be able to play an active role in inducing users' emotions and beliefs and in providing motivational support.

There are many practical settings where computational humour adds value. Among them there are business world applications (such as advertisement, e-commerce), general computer-mediated communication and human–computer interaction, increase in the friendliness of natural language interfaces, educational and edutainment systems.

There are also important prospects for humour in automatic information presentation. In the Web age, presentations will become more and more flexible and personalized and will require humour contributions for electronic commerce developments (e.g. product promotion, getting selective attention, help in memorizing names) more or less as it happened in the world of advertisement within the old broadcast communication.

In this chapter we focus mainly on "verbal humour", which is the most tangible and perhaps the most widely researched form of humour. Although other forms of humour (e.g. visual or situational) have also received attention from the research community, we concentrate our work and consequently this survey on the linguistic expressions of humour.

The chapter is structured as follows. Section 2 briefly surveys the main theories of humour in philosophy, psychology, and linguistics. Sect. 3 summarizes the main research attempts in computational humour. In Sect. 4 we illustrate an example of humour generation system (HAHAcronym), while in Sect. 5 we describe an approach to deal with humour recognition. Finally, Sect. 6 concludes the chapter with some prospects on this field.

2 Background in Humour Research

In this section, we summarize the main theories of humour that emerged from philosophical and modern psychological research and survey the past and present developments in the fields of theoretical and computational linguistics. We also briefly overview related research work in the fields of psychology, sociology, and neuroscience.

2.1 Theories of Humour

There are three main theories of humour, which emerged primarily from philosophical studies and research in psychology.

2.1.1 Incongruity Theory

The incongruity theory suggests that humour is due to the mixing of two disparate interpretation frames in one statement. One of the earliest references to an incongruity theory of humour is perhaps due to Aristotle (350 BC) who found that the contrast between expectation and actual outcome is often a source of humour. He is also making a distinction between surprise and incongruity, where the later is presumed to have a resolution that was initially hidden from the audience. The incongruity theory has also found a supporter in Schopenhauer (1819), who emphasizes the element of surprise by suggesting that "the greater and more unexpected [...] the incongruity is, the more violent will be [the] laughter". The incongruity theory has been formalized as a necessary condition for humour and used as a basis for the Semantic Script-based Theory of Humour (SSTH) (Raskin, 1985) and later on the General Theory of Verbal Humour (GTVH) (Attardo and Raskin, 1991).

2.1.2 Superiority Theory

The superiority theory argues that humour is a form of expressing the superiority of one over another. As suggested by Hobbes (1840), laughter is "nothing else but sudden glory" triggered by a feeling of superiority with respect to others or with respect to ourselves in a previous moment. A closely related theory is the one supported by Solomon (2002), who suggests that humour is due to feelings of inferiority, which led to the so-called inferiority theory. Although the superiority and inferiority theories of humour have been typically perceived as diametrically opposed, they are in fact intimately related, as the "superior"/"inferior" distinctions are often due to a different point of view. In fact, it can be argued that laughter is triggered by our feelings of superiority with respect to others or ourselves in a previous moment, which are equivalent to feelings of inferiority felt by others or by ourselves in a past moment.

2.1.3 Relief Theory

The third major theory is the *relief theory*, which suggests that humour is a form of bypassing certain censors that prevent us from having "prohibited thoughts". Eluding these censors results in a release of the energy inhibited by these censors and consequently the feeling of relief. One of the strongest supporters of the relief theory is Freud (1905), who draws a connection between jokes and the unconscious, and Spencer (1860), who suggests that laughter is a form of "nervous energy". Some of these ideas have been later embraced by Minsky in his theory of humour (Minsky, 1980), to which he adds a cognitive element that attempts to explain the "faulty logic" that is typically encountered in jokes, which is normally suppressed in order to avoid "cognitive harm".

2.2 Linguistics Research on Humour

A significant fraction of the research on humour that has been carried out to date has concentrated on the linguistic characteristics of humour. Among the linguistic theories, the most influential is perhaps the General Theory of Verbal Humour (GTVH) (Attardo and Raskin, 1991), which is an extension of the earlier Semantic Script-based Theory of Humour (SSTH) (Raskin, 1985).

2.2.1 Semantic Script-Based Theory of Humour

SSTH is based on the representation of jokes as *script opposition*, which is an idea closely related to the incongruity resolution theory. Briefly, SSTH defines the structure of a joke as consisting of a *set-up* and a *punchline*. The set-up has at least two possible interpretations out of which only one is obvious, and consequently the humourous effect is created by the punchline which triggers the second less obvious interpretation in a surprising way.

The central hypothesis in SSTH is that a text is humourous if the following two conditions are satisfied. First, the humourous text has to be compatible with at least two different interpretations (scripts). And second, the two interpretations have to be opposed to each other. For instance, the following example taken from Raskin (1985) illustrates this theory: "The first thing that strikes a stranger in New York is a big car." The set-up has two possible interpretations: strike as in "impress" or "hit", which are opposed to each other ("impress" being a positive action and "hit" triggering negative feelings). The first interpretation is more obvious and thus initially preferred. However, the punchline "by a big car" will change the preference to the second interpretation, which generates the humourous effect.

According to SSTH, the opposition between scripts is binary and can fall into one of the following three generic types: actual/non-actual, normal/abnormal, possible/impossible, which in turn can be broken down into more specific oppositions, such as positive/negative or good/bad.

2.2.2 General Theory of Verbal Humour

Following SSTH, the GTVH (Attardo and Raskin, 1991) extends the script opposition theory and adds other possible knowledge resources for a humourous text. While SSTH was primarily focused on semantics, the GTVH is more general and includes other areas in linguistics such as pragmatics and style. GTVH defines six main knowledge resources that can be organized on six levels from concrete (low level) to abstract (high level).

- *Script opposition*, which is a knowledge source based on the main idea of SSTH of opposing interpretations that are both compatible with the text.
- *Logical mechanism*, which provides a possible resolution mechanism for the incongruity between scripts.
- *Situation*, which defines the context of the joke in terms of location, participants, and others.
- *Target*, which is the person or group of persons that are targeted by the joke.
- *Narrative strategy*, which defines the style of the joke, i.e. whether it is a dialogue, a riddle, or a simple narrative.
- *Language*, which defines the "surface" of the joke in terms of linguistic aspects such as lexicon, morphology, syntax, semantics.

For example, Attardo and Raskin (1991) exemplify the knowledge resources using the following joke: "How many Poles does it take to screw in a light bulb? Five. One to hold the light bulb and four to turn the table he is standing on." The script opposition is formed between the expected normal behaviour of a person when screwing in a light bulb and the "dumb" resolution proposed by the punchline; the logical mechanism is that of "reversal" of a normal behaviour; the situation is "bulb changing"; the target of the joke are the "Poles"; and finally, the narrative structure is a "riddle" (Ritchie, 2003).

An interesting experiment centred around the GTVH theory is reported by Ruch et al. (1993), where three jokes are transformed into variants that differed from the original joke in one of the GTVH parameters. A group of 500 subjects were asked to rate the similarity between each of the variants and the original joke on a scale of 1–4. The findings indicate that higher similarity is observed for those variants that differ in a low-level parameter in the GTVH hierarchy, thus suggesting that the higher level parameters such as script opposition and logic mechanism are more humour related (Ritchie, 2003).

While GTVH is perhaps the most extensive linguistic theory of humour that has been proposed to date, it has been criticized by Ritchie (2003) as lacking theoretical grounds. Ritchie raises doubts about the falsifiability of the GTVH and about the lack of systematic examples where some of the GTVH knowledge resources are missing, thus resulting in a lack of humourous effect, along with humourous examples that include the missing knowledge resources.

2.2.3 Related Work in Linguistics

Besides the SSTH and the GTVH theories, other research work in linguistics has focused mainly on the analysis of the lexical devices used in humourous text. The syntactic ambiguity often encountered in humour is analysed by Hetzron (1991), who describes the structure of jokes and punchlines and analyses the logical devices found in verbal humour. Oaks (1994) proposes an interesting account on syntactic ambiguity in humour and identifies several ambiguity "enablers". He focuses mainly on part-of-speech ambiguity and identifies verbs, articles, and other parts-of-speech that can introduce ambiguity in language (e.g. *bite* that can be either a verb or a noun).

The lexical and syntactic ambiguity as a source of humour is also studied by Bucaria (2004), who analyses the linguistic ambiguity in newspaper headlines. She identifies three main types of ambiguity: lexical (e.g. "Actor sent to jail for not finishing *sentence*."), syntactic (e.g. "Eye *drops* off shelf"), and phonological (e.g. "Is there a ring of debris around *Uranus*"). She also identifies two additional schemata for humourous ambiguity, including the disjunctor/connector model (e.g. "New study on *obesity* looks for *larger* group.") and the double ambiguity model (e.g. "Farmer *Bill dies* in house."). In an analysis of 135 headlines, the lexical and syntactic forms of ambiguity were found to be dominant (71 lexical and 63 syntactic), covering a significant fraction of the corpus, and thus providing support for the incongruity theory of humour.

2.3 Multidisciplinary Research on Humour

In addition to the research work in linguistics humour has been also studied in other areas, e.g. sociology, neuroscience, and last but not least recent efforts in computational linguistics.

2.3.1 Sociology

In sociology, humour has been frequently associated with studies concerned with patterns of communication in different groups. For instance, Duncan (1984) shows that cohesive and non-cohesive work groups have different humour patterns, suggesting a correlation between the type of humour practiced in a group and the structure of the group.

Studies have also investigated the association between gender and humour, by analysing the type and role of humour for female, male, and mixed groups. Hay (1995) used a taxonomy of humour in a gender-oriented analysis, which revealed the preference of women for observational humour and the tendency of male groups for insults and role play. Interestingly, a correlation was also observed between the gender of these groups and the function of humour; women groups used humour primarily as a social element, whereas men groups often used it as a means for increasing status. Finally, Hay's study also reported on the association between gender and humour topics, suggesting that women use more frequently humour on topics involving people, while men joke more about politics, computers, and work;

this observation correlates with recent conclusions drawn in corpus-based gender studies (Liu and Mihalcea, 2007).

Another aspect of interest in sociology is the relation between culture and humour. Work in this area has highlighted the relation between cultural background and humour appreciation, showing that the set of values and norms of a culture largely determine the content and style of humour (Hertzler, 1970). Focused studies have highlighted differences between various cultures, as for instance the study reported by Nevo (1984), which shows how Arab and Jewish communities developed a different sense of humour explained by their diverse background and different social status.

2.3.2 Psychology

Humour research in psychology has been mainly concerned with the correlation between humour and individual development. There are several studies that considered the cognitive aspects of humour and the role that humour can play in infants and children development. For instance, it has been found that humour has an important role in improving text comprehension (Yuill, 1997).

Other studies have been concerned with the relation between personality profiles and sense of humour. Along these lines, it has been suggested that extroversion and neuroticism can be predicted from humour perception (Mobbs et al., 2005). Similarly, humour was found to be related to other personality characteristics such as simplicity–complexity, intelligence, or mood (Ruch, 1998).

2.3.3 Neuroscience

In recent years, given the advances made in brain imaging techniques (fMRI or MEG), researchers have started to investigate the brain activity observed during humour detection and comprehension. Recent research findings suggest that the left and the right hemispheres are both involved in humour appreciation, which is an effect that has been observed in verbal humour as well as visual humour (Bartolo et al., 2006). Moreover, studies have also observed the activation of the amygdala and midbrain regions (also known as the "pleasure centre"), which is probably due to the pleasurable effect created by humour (Watson et al., 2007).

It is also worth noting the study reported in Mobbs et al., (2005), which shows connections between gender, personality (i.e. extroversion and neuroticism), and humour appreciation, observed using brain imaging techniques. Such associations have been typically identified through surveys conducted in psychological studies, and the study reported in Mobbs et al., (2005) confirms these previous findings by identifying patterns of brain activity occurring during humour comprehension.

3 Computational Humour: State of the Art

While humour is relatively well studied in fields such as linguistics (Attardo, 1994) and psychology (Freud, 1905; Ruch, 2002), to date only a limited number of research contributions have been made towards the construction of computational humour prototypes. Most of the computational approaches to date on style classification have focused on the categorization of more traditional literature genres, such as fiction, sci-tech, legal, and others (Kessler et al., 1997), and much less on creative writings such as humour.

The most systematic effort in this area is perhaps Ritchie's book on the linguistic analysis of jokes, which brings together research on linguistic theories and artificial intelligence. In addition to a comprehensive overview of the main research contributions in humour, Ritchie is also proposing a classification of jokes into propositional and linguistic and suggests a structural description of the jokes (Ritchie, 2003).

There are two main research directions in computational humour: (1) *humour generation*, which attempts to build computational models to generate humourous text, and (2) *humour recognition*, which deals with the problem of identifying humour in natural language.

3.1 Humour Generation

One of the first attempts in humour generation is the work described by Binsted and Ritchie (1997), where a formal model of semantic and syntactic regularities was devised, underlying some of the simplest types of puns (*punning riddles*). The model was then exploited in a system called JAPE that was able to automatically generate amusing puns. A punning riddle is a question–answer riddle that uses phonological ambiguity. The three main strategies used to create phonological ambiguity are syllable substitution, word substitution, and metathesis. Their system generates punning riddles from a fixed linguistic model of pun schemata. An example: "What do you call a murderer with fiber?" *A cereal killer*.

Tinholt and Nijholt (2007) describe a first attempt at automatically generating jokes based on cross-reference ambiguity. The idea is that when a given cross-reference ambiguity results in script opposition it is possible to generate a punchline based on this ambiguity. An example of dialogue is "User: Did you know that the cops arrested the demonstrators because they were violent?" "System: The cops were violent? Or the demonstrators? :)"

Another humour generation project is the HAHAcronym project (Stock and Strapparava, 2003), whose goal was to develop a system able to automatically generate humourous versions of existing acronyms or to produce a new amusing acronym constrained to be a valid vocabulary word, starting with concepts provided by the user. The comic effect was achieved mainly by exploiting incongruity (e.g. finding a religious variation for a technical acronym). We describe in detail this system in Sect. 4.

3.2 Humour Recognition

There are only a few studies addressing the problem of humour recognition. The study reported in Taylor and Mazlack (2004) is devoted to the problem of humour comprehension, focusing on a restricted type of wordplays, namely the

"Knock-Knock" jokes. The goal of the study was to evaluate to what extent wordplay can be automatically identified in "Knock-Knock" jokes and if such jokes can be reliably recognized from other non-humourous text. The algorithm was based on automatically extracted structural patterns and on heuristics heavily based on the peculiar structure of this particular type of jokes. While the generic wordplay recognition gave satisfactory results (67% accuracy), the identification of wordplays that had a humourous effect turned out to be significantly more difficult (12% accuracy).

In our own previous work (Mihalcea and Strapparava, 2005b; Mihalcea and Pulman, 2007), humour recognition was formulated as a text classification task, and machine learning algorithms were run on large collections of humourous texts (oneliners or humourous news articles). Both content and stylistic features were evaluated, including n-gram models, alliteration, antonymy, and adult slang, with performance figures significantly higher than a priori known baselines. We will describe in detail the used methodology in Sect. 5.

Another humour-recognition study was reported by Purandare and Litman (2006), where the recognition experiments were performed using both content features and spoken dialogue prosody features (tempo, energy, and pitch). The experiments were run on dialogues from the TV series "Friends", with significant improvements observed over the baseline. They also reported a gender study, with the improvement obtained for humour recognition in male dialogues being higher than the one obtained for female dialogues, suggesting perhaps that the humourous features are more prominent for males than for females.

4 Humour Generation: HAHAcronym

HAHAcronym was the first European project devoted to computational humour.[1] The main goal of HAHAcronym was the realization of an acronym ironic re-analyser and generator as a proof of concept in a focalized but non-restricted context. In the first case the system makes fun of existing acronyms; in the second case, starting from concepts provided by the user, it produces new acronyms, constrained to be words of the given language. And, of course, they have to be funny.

The realization of this system was proposed to the European Commission as a project that we would be able to develop in a short period of time (less than a year), that would be meaningful and well demonstrable, that could be evaluated along some pre-decided criteria, and that was conducive to a subsequent development in a direction of potential applicative interest. So for us it was essential that

1. the work could have many components of a larger system, simplified for the current setting;
2. we could reuse and adapt existing relevant linguistic resources;
3. some simple strategies for humour effects could be experimented.

[1]EU project IST-2000-30039 (partners: ITC-irst and University of Twente), part of the Future Emerging Technologies section of the Fifth European Framework Program.

One of the purposes of the project was to show that using "standard" resources (with some extensions and modifications) and suitable linguistic theories of humour (i.e. developing specific algorithms that implement or elaborate theories), it is possible to implement a working prototype.

4.1 Resources

In order to realize the HAHAcronym prototype (see Fig. 1), we refined existing resources and developed general tools useful for humourous systems. A fundamental tool is an incongruity detector/generator that makes the system able to detect semantic mismatches between word meaning and sentence meaning (i.e. in our case the acronym and its context). For all tools, particular attention was put on reusability.

The starting point consisted in making use of some standard resources, such as WORDNET DOMAINS (Magnini et al., 2002) (an extension of the well-known English WORDNET) and standard parsing techniques.

Wordnet. WORDNET is a thesaurus for the English language inspired by psycholinguistics principles and developed at the Princeton University by George Miller (Fellbaum, 1998). Lemmata (about 130,000 for version 1.6) are organized in synonym classes (about 100,000 *synsets*). A synset contains all the words by means of which it is possible to express a particular meaning: for example, the synset `knight, horse` describes the sense of "horse" as a chessman. The main relations present in WORDNET are *synonymy*, *antonymy*, *hyperonymy–hyponymy*, *meronymy–holonymy*, *entailment*, *troponymy*.

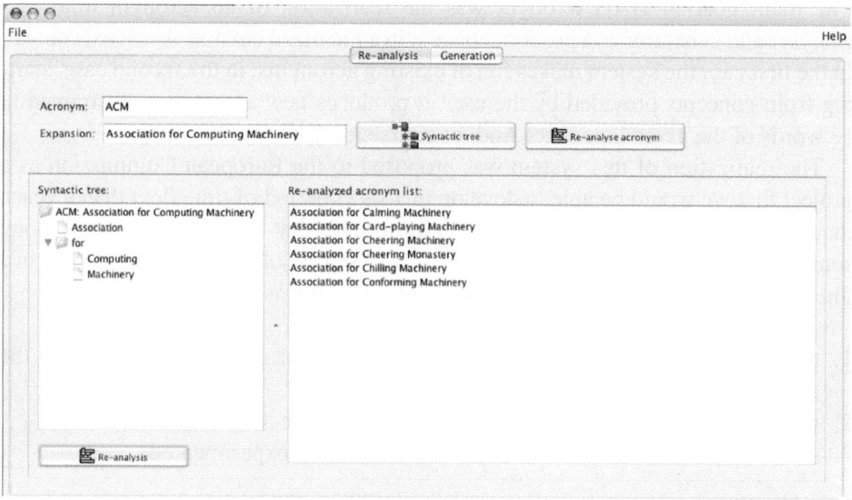

Fig. 1 A screenshot of a reanalysis in HAHAcronym

Wordnet Domains. Domains have been used both in linguistics (i.e. Semantic Fields) and in lexicography (i.e. Subject Field Codes) to mark technical usages of words. Although this is useful information for sense discrimination, in dictionaries it is typically used for a small portion of the lexicon. WORDNET DOMAINS[2] is an attempt to extend the coverage of domain labels within an already existing lexical database, WORDNET. The synsets have been annotated with at least one domain label, selected from a set of about 200 labels hierarchically organized.

The 250 domain labels are organized in a hierarchy (exploiting Dewey Decimal Classification), where each level is made up of codes of the same degree of specificity: for example, the second level includes domain labels such as BOTANY, LINGUISTICS, HISTORY, SPORT, and RELIGION, while at the third level we can find specialization such as AMERICAN_HISTORY, GRAMMAR, PHONETICS, and TENNIS.

Opposition of Semantic Fields. On the basis of well-recognized properties of humour accounted for in many theories (e.g. incongruity, semantic field opposition, apparent contradiction, absurdity) an independent structure of domain opposition was modelled, such as RELIGION vs. TECHNOLOGY, SEX vs. RELIGION. Opposition is exploited as a basic resource for the incongruity generator.

Adjectives and Antonymy Relations. Adjectives play an important role in modifying and generating funny acronyms. WORDNET divides adjectives into two categories. *Descriptive adjectives* (e.g. `big, beautiful, interesting, possible, married`) constitute by far the largest category. The second category is called simply *relational adjectives* because they are related by derivation to nouns (i.e. `electrical` in `electrical engineering` is related to the noun `electricity`). To relational adjectives, strictly dependent on noun meanings, it is often possible to apply similar strategies as those exploited for nouns. Their semantic organization, though, is entirely different from that of the other major categories. In fact it is not clear what it would mean to say that one adjective "is a kind of" (ISA) some other adjective. The basic semantic relation among descriptive adjectives is antonymy. WORDNET proposes also that this kind of adjectives is organized in clusters of synsets associated by semantic similarity with a focal adjective. Figure 2 shows clusters of adjectives around the direct antonyms *fast/slow*.

Exploiting the Hierarchy. It is possible to exploit the network of lexical and semantic relations built in WORDNET to make simple ontological reasoning. For example, if a noun or an adjective has a geographic location meaning, the pertaining country and continent can be inferred.

Rhymes. The HAHAcronym prototype takes into account word rhymes and the rhythm of the acronym expansion. To cope with this aspect the CMU Pronouncing Dictionary[3] was organized with a suitable indexing. The CMU Pronouncing Dictionary is a machine-readable pronunciation dictionary for North American English that contains over 125,000 words and their transcriptions.

---

[2]It is freely available for research purposes at http://wndomains.itc.it (visited 30 May 2010).

[3]Available at http://www.speech.cs.cmu.edu/cgi-bin/cmudict (visited 30 May 2010).

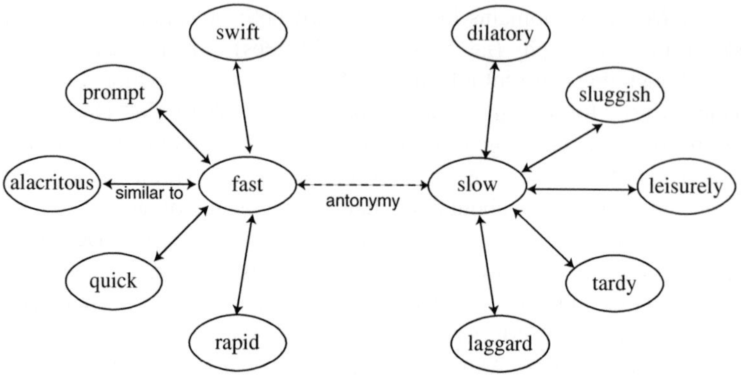

Fig. 2 An example of adjective clusters linked by antonymy relation

Parser, Grammar and Morphological Analyser. Word sequences that are at the basis of acronyms are subject to a well-defined grammar, simpler than a complete noun phrase grammar, but complex enough to require a nontrivial analyser. A well-established nondeterministic parsing technique was adopted. As far as the dictionary is concerned, the full WORDNET lexicon was used, integrated with an ad hoc morphological analyser. Also for the generation part the grammar is exploited as the source for syntactic constraints. All the components are implemented in Common Lisp augmented with nondeterministic constructs.

Other Resources. An "a-semantic" or "slanting" dictionary is a collection of hyperbolic/attractive adjective/adverbs. This is a last resource, that sometimes can be useful in the generation of new acronyms. In fact a slanting writing refers to that type of writing that springs from our conscious or subconscious choice of words and images. We may load our description of a specific situation with vivid, connotative words and figures of speech. Some examples are *abnormally*, *abstrusely*, *adorably*, *exceptionally*, *exorbitantly*, *exponentially*, *extraordinarily*, *voraciously*, *weirdly*, *wonderfully*. This resource is handmade, using various dictionaries as information sources.

Other lexical resources are a euphemism dictionary, a proper noun dictionary, lists of typical foreign words commonly used in the language with some strong connotation.

4.2 Reanalysis and Generation

To get an ironic or "profaning" reanalysis of a given acronym, the system follows various steps and strategies. The main elements of the algorithm can be schematized as follows:

- acronym parsing and construction of a logical form
- choice of what to keep unchanged (typically the head of the highest ranking NP) and what to modify (e.g. the adjectives)

- look up for possible substitutions
- exploitation of semantic field oppositions
- granting phonological analogy: while keeping the constraint on the initial letters of the words, the overall rhyme and rhythm should be preserved (the modified acronym should sound similar to the original as much as possible)
- exploitation of WORDNET antonymy clustering for adjectives
- use of the slanting dictionary as a last resource

Figures 3 and 4 show a sketch of the HAHAcronym system architecture.

HAHAcronym, making fun of existing acronyms, amounts to an ironical rewriting, desecrating them with some unexpectedly contrasting, but otherwise consistently sounding expansion.

As far as acronym generation is concerned, the problem is more complex. To make the task more attractive – and difficult – we constrain resulting acronyms to be words of the dictionary (APPLE is good, IBM is not). The system takes in input concepts (actually synsets, possibly resulting from some other process, for instance sentence interpretation) and some minimal structural indication, such as the semantic head. The primary strategy of the system is to consider words that are in ironic relation with the input concepts as potential acronyms. By definition acronyms have

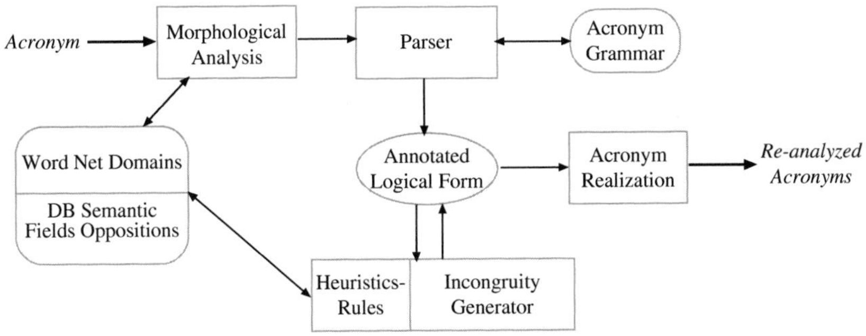

Fig. 3 The HAHAcronym system architecture

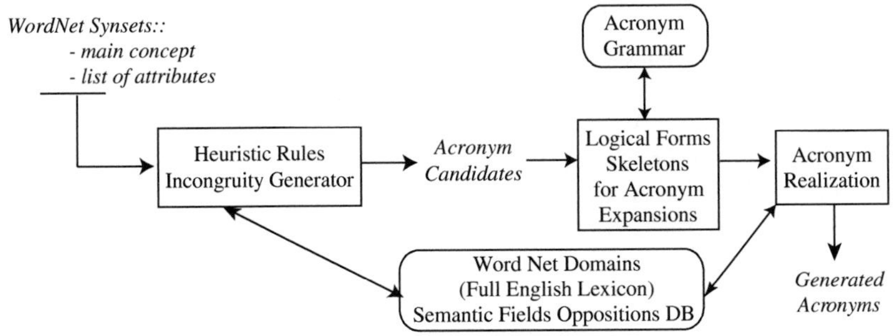

Fig. 4 Acronyms generation in the HAHAcronym system

to satisfy constraints – to include the initial letters of lexical realization, granting that the sequence of initials to satisfy the overall acronym syntax. Ironic reasoning comes mainly at the level of acronym choice and in the selection of the fillers of the *open slots* in the acronym.

For example, giving as input "fast" and "CPU", we get static, torpid, dormant. The complete synset for "CPU" is `{processor#3, CPU#1, central_proces-sing_unit#1, mainframe#2}`; so we can use a synonym of "CPU" in the acronym expansion. The same happens for "fast". Once we have an acronym proposal, a syntactic skeleton has to be filled to get a correct noun phrase. For example, given in input "fast" and "CPU", the system selects `TORPID` and proposes as syntactic skeletons

<adv>$_T$*<adj>*$_O$ `Rapid Processor`*<prep><adj>*$_I$*<noun>*$_D$

or

<adj>$_T$*<adj>*$_O$ `Rapid Processor`*<prep><noun>*$_I$*<noun>*$_D$

where "rapid" and "processor" are synonyms, respectively, of "fast" and "CPU" and the notation `<Part_of_Speech>`$_{Letter}$ means a word of that particular part_of_speech with *Letter* as initial. Then the system fills this syntactic skeleton with strategies similar to those described for reanalysis.

4.3 Examples

Here below some examples of acronym reanalysis are reported. As far as semantic field opposition is concerned we have slightly tuned the system towards the domains FOOD, RELIGION, and SEX. We report the original acronym, the reanalysis, and some comments about the strategies followed by the system.

`ACM` – Association for Computing Machinery
→ `Association for Confusing Machinery`
`FBI` – Federal Bureau of Investigation
→ `Fantastic Bureau of Intimidation`

The system keeps all the main heads and works on the adjectives and the PP head, preserving the rhyme and/or using the a-semantic dictionary.

`CRT` – Cathodic Ray Tube
→ `Catholic Ray Tube`
`ESA` – European Space Agency
→ `Epicurean Space Agency`
`PDA` – Personal Digital Assistant
→ `Penitential Demoniacal Assistant`
→ `Prenuptial Devotional Assistant`
`MIT` – Massachusetts Institute of Technology
→ `Mythical Institute of Theology`

Some re-analyses are RELIGION oriented. Note the rhymes.

As far as generation from scratch is concerned, a main concept and some attributes (in terms of synsets) are given as input to the system. Here below we report some examples of acronym generation.

Main concept: *processor* (in the sense of CPU);
Attribute: *fast*

OPEN – On-line Processor for Effervescent Net
PIQUE – Processor for Immobile Quick Uncertain Experimentation
TORPID – Traitorously Outstandingly Rusty Processor for Inadvertent Data_processing
UTMOST – Unsettled Transcendental Mainframe for Off-line Secured TCP/IP

We note that the system tries to keep all the expansions of the acronym coherent in the same semantic field of the main concept (COMPUTER_SCIENCE). At the same time, whenever possible, it exploits some incongruity in the lexical choices.

4.4 Evaluation

Testing the humourous quality of texts is not an easy task. There have been relevant studies though, such as those in Ruch (1996). For HAHAcronym, a simpler case, an evaluation was conducted under the supervision of Salvatore Attardo at Youngstown University, Ohio. Both reanalysis and generation have been tested according to criteria of success stated in advance and in agreement with the European Commission, at the beginning of the project.

The participants in the evaluation were 40 students. They were all native speakers of English. The students were not told that the acronyms had been computer generated.

No record was kept of which student had given which set of answers (the answers were strictly anonymous). No demographic data were collected. However, generally speaking, the group was homogeneous for age (traditional students, between the ages of 19 and 24) and mixed for gender and race.

The students were divided into two groups. The first group of 20 was presented the reanalysis and generation data. We tested about 80 reanalysed and 80 generated acronyms (over twice as many as required by the agreement with the European Commission). Both the reanalysis module and the generation module were found to be successful according to the criteria spelled out in the assessment protocol (see Table 1).

The acronyms reanalysis module showed roughly 70% of acronyms having a score of 55 or higher (out of a possible 100 points), while the acronym generation module showed roughly 53% of acronyms having a score of 55 or higher. The thresholds for success established in the protocol were 60 and 45%, respectively.

Table 1 Evaluation results

Acronyms	Scored > 55%	Success thresholds (%)
Generation	52.87	45
Reanalysis	69.81	60
Random	7.69	

One could think that a random selection of fillers could be often funny as well. A special run of the system was performed with lexical reasoning and heuristics disabled, while only the syntactical constraints were operational. If the syntactical rules had been disabled as well, the output would have been gibberish and it would be not fairly comparable with normal HAHAcronym production. This set of acronyms was presented to a different group of 20 students. The result was that less than 8% of the acronyms passed the 55 points score test; we conclude that the output of HAHAcronym is significantly better than random production of reanalysis.

A curiosity that may be worth mentioning: HAHAcronym participated in a contest about (human) production of best acronyms, organized by RAI, the Italian National Broadcasting Service. The system won a jury's special prize.

5 Humour Recognition: One-Liners Recognition

Previous work in computational humour has focused mainly on the task of humour generation (Stock and Strapparava, 2003; Binsted and Ritchie, 1997), and very few attempts have been made to develop systems for automatic humour recognition (Taylor and Mazlack, 2004; Mihalcea and Strapparava, 2005b). This is not surprising, since, from a computational perspective, humour recognition appears to be significantly more subtle and difficult than humour generation.

In this section, we describe experiments concerned with the application of computational approaches to the recognition of verbally expressed humour. In particular, we investigate whether automatic classification techniques represent a viable approach to distinguish between humourous and non-humourous text, and we bring empirical evidence in support of this hypothesis through experiments performed on very large data sets.

Since a deep comprehension of humour in all of its aspects is probably too ambitious and beyond the existing computational capabilities, we chose to restrict our investigation to the type of humour found in *one-liners*. A one-liner is a short sentence with comic effects and an interesting linguistic structure: simple syntax, deliberate use of rhetoric devices (e.g. alliteration, rhyme), and frequent use of creative language constructions meant to attract the readers' attention. While longer jokes can have a relatively complex narrative structure, a one-liner must produce the humourous effect "in one shot", with very few words. These characteristics make this type of humour particularly suitable for use in an automatic learning setting, as the humour-producing features are guaranteed to be present in the first (and only) sentence.

We attempt to formulate the humour-recognition problem as a traditional classification task and feed positive (humourous) and negative (non-humourous) examples to an automatic classifier. The humourous data set consists of one-liners collected from the Web using an automatic bootstrapping process. The non-humourous data are selected such that it is structurally and stylistically similar to the one-liners. Specifically, we use four different negative data sets: (1) Reuters news titles; (2) proverbs; (3) sentences from the British National Corpus (BNC); (4) commonsense statements from the Open Mind Common Sense (OMCS) corpus. The classification results are encouraging, with accuracy figures ranging from 79.15% (One-liners/BNC) to 96.95% (one-liners/Reuters). Regardless of the non-humourous data set playing the role of negative examples, the performance of the automatically learned humour recognizer is always significantly better than a priori known baselines.

5.1 Humourous and Non-humourous Data Sets

To test our hypothesis that automatic classification techniques represent a viable approach to humour recognition, we needed in the first place a data set consisting of humourous (positive) and non-humourous (negative) examples. Such data sets can be used to automatically *learn* computational models for humour recognition and at the same time *evaluate* the performance of such models.

Humourous data. While there are plenty of non-humourous data that can play the role of negative examples, it is significantly harder to build a very large and at the same time sufficiently "clean" data set of humourous examples. We use a dually constrained Web-based bootstrapping process to collect a very large set of one-liners. Starting with a short *seed* set consisting of a few one-liners manually identified, the algorithm automatically identifies a list of webpages that include at least one of the seed one-liners, via a simple search performed with a Web search engine. Next, the webpages found in this way are HTML parsed, and additional one-liners are automatically identified and added to the seed set. The process is repeated several times, until enough one-liners are collected. As with any other bootstrapping algorithm, an important aspect is represented by the set of constraints used to steer the process and prevent as much as possible the addition of noisy entries. Our algorithm uses (1) a *thematic* constraint applied to the theme of each webpage, via a list of keywords that have to appear in the URL of the webpage, and (2) a *structural* constraint, exploiting HTML annotations indicating text of similar genre (e.g. lists, adjacent paragraphs)

Two iterations of the bootstrapping process, started with a small seed set of 10 one-liners, resulted in a large set of about 24,000 one-liners. After removing the duplicates using a measure of string similarity based on the longest common subsequence, we are left with a final set of 16,000 one-liners, which are used in the

Table 2 Sample examples of one-liners, Reuters titles, proverbs, OMC and BNC sentences

One-liners	
Take my advice; I don't use it anyway. I get enough exercise just pushing my luck. Beauty is in the eye of the beer holder.	
Reuters titles	*Proverbs*
Trocadero expects tripling of revenues. Silver fixes at two-month high, but gold lags. Oil prices slip as refiners shop for bargains.	Creativity is more important than knowledge. Beauty is in the eye of the beholder. I believe no tales from an enemy's tongue.
OMCS sentences	*BNC sentences*
Humans generally want to eat at least once a day. A file is used for keeping documents. A present is a gift, something you give to someone.	They were like spirits, and I loved them. I wonder if there is some contradiction here. The train arrives three minutes early.

humour-recognition experiments. A more detailed description of the Web-based bootstrapping process is available in Mihalcea and Strapparava (2005a). The one-liners humour style is illustrated in Table 2, which shows three examples of such one-sentence jokes.

Non-humourous data. To construct the set of negative examples required by the humour-recognition models, we tried to identify collections of sentences that were non-humourous, but similar in structure and composition to the one-liners. We do not want the automatic classifiers to learn to distinguish between humourous and non-humourous examples based simply on text length or obvious vocabulary differences. Instead, we seek to enforce the classifiers to identify humour-specific features, by supplying them with negative examples similar in most of their aspects to the positive examples, but different in their comic effect.

We tested four different sets of negative examples, with three examples from each data set illustrated in Table 2. All non-humourous examples are enforced to follow the same length restriction as the one-liners, i.e. one sentence with an average length of 10–15 words.

1. *Reuters* titles, extracted from news articles published in the Reuters newswire over a period of 1 year (20 August 1996–19 August 1997) (Lewis et al., 2004). The titles consist of short sentences with simple syntax and are often phrased to catch the readers' attention (an effect similar to the one rendered by the one-liners).
2. *Proverbs* extracted from an online proverb collection. Proverbs are sayings that transmit, usually in one short sentence, important facts or experiences that are considered true by many people. Their property's of being condensed, but memorable sayings make them very similar to the one-liners. In fact, some one-liners attempt to reproduce proverbs, with a comic effect, as in e.g. *"Beauty is in the eye of the beer holder"*, derived from *"Beauty is in the eye of the beholder"*.

3. *British National Corpus (BNC)* sentences, extracted from BNC – a balanced corpus covering different styles, genres, and domains. The sentences were selected such that they were similar in content with the one-liners: we used an information retrieval system implementing a vectorial model to identify the BNC sentence most similar to each of the 16,000 one-liners.[4] Unlike the Reuters titles or the proverbs, the BNC sentences have typically no added creativity. However, we decided to add this set of negative examples to our experimental setting, in order to observe the level of difficulty of a humour-recognition task when performed with respect to simple text.
4. *Open Mind Common Sense (OMCS)* sentences. OMCS is a collection of about 800,000 commonsense assertions in English as contributed by volunteers over the Web. It consists mostly of simple single sentences, which tend to be explanations and assertions similar to glosses of a dictionary, but phrased in a more common language. For example, the collection includes such assertions as "keys are used to unlock doors" and "pressing a typewriter key makes a letter". Since the comic effect of jokes is often based on statements that break our commonsensical understanding of the world, we believe that such commonsense sentences can make an interesting collection of "negative" examples for humour recognition. For details on the OMCS data and how it has been collected, see Singh (2002). From this repository we use the first 16,000 sentences.[5]

To summarize, the humour recognition experiments rely on data sets consisting of humourous (positive) and non-humourous (negative) examples. The positive examples consist of 16,000 one-liners automatically collected using a Web-based bootstrapping process. The negative examples are drawn from (1) Reuters titles; (2) proverbs; (3) BNC sentences; and (4) OMCS sentences.

5.2 Features for Automatic Humour Recognition

We experiment with automatic classification techniques using (a) heuristics based on humour-specific stylistic features (alliteration, antonymy, slang); (b) content-based features, within a learning framework formulated as a typical text classification task; and (c) combined stylistic and content-based features, integrated in a stacked machine learning framework.

---

[4]The sentence most similar to a one-liner is identified by running the one-liner against an index built for all BNC sentences with a length of 10–15 words. We use a *tf.idf* weighting scheme and a cosine similarity measure, as implemented in the Smart system (`ftp.cs.cornell.edu/pub/smart`, visited 30 May 2010).

[5]The first sentences in this corpus are considered to be "cleaner", as they were contributed by trusted users (Push Singh, p.c.).

5.2.1 Humour-Specific Stylistic Features

Linguistic theories of humour (e.g. Attardo, 1994) have suggested many *stylistic features* that characterize humourous texts. We tried to identify a set of features that were both significant and feasible to implement using existing machine-readable resources. Specifically, we focus on alliteration, antonymy, and adult slang, previously suggested as potentially good indicators of humour (Ruch, 2002; Bucaria, 2004).

Alliteration. Some studies on humour appreciation (Ruch, 2002) show that structural and phonetic properties of jokes are at least as important as their content. In fact one-liners often rely on the reader awareness of attention-catching sounds, through linguistic phenomena such as alliteration, word repetition, and rhyme, which produce a comic effect even if the jokes are not necessarily meant to be read aloud. Note that similar rhetorical devices play an important role in wordplay jokes and are often used in newspaper headlines and in advertisement. The following one-liners are examples of jokes that include alliteration chains:

Veni, Vidi, Visa: I came, I saw, I did a little shopping.
Infants don't enjoy infancy like adults do adultery.

To extract this feature, we identify and count the number of alliteration/rhyme chains in each example in our data set. The chains are automatically extracted using an index created on top of the CMU Pronuncing Dictionary.

Antonymy. Humour often relies on some type of incongruity, opposition, or other forms of apparent contradiction. While an accurate identification of all these properties is probably difficult to accomplish, it is relatively easy to identify the presence of *antonyms* in a sentence. For instance, the comic effect produced by the following one-liners is partly due to the presence of antonyms:

A clean desk is a sign of a cluttered desk drawer.
Always try to be modest and be proud of it!

The lexical resource we use to identify antonyms is WORDNET (Miller, 1995), and in particular the *antonymy* relation among nouns, verbs, adjectives, and adverbs. For adjectives we also consider an indirect antonymy via the *similar-to* relation among adjective synsets. Despite the relatively large number of *antonymy* relations defined in WORDNET, its coverage is far from complete, and thus the *antonymy* feature cannot always be identified. A deeper semantic analysis of the text, such as word sense or domain disambiguation, could probably help in detecting other types of semantic opposition, and we plan to exploit these techniques in future work.

Adult Slang. Humour based on adult slang is very popular. Therefore, a possible feature for humour recognition is the detection of sexual-oriented lexicon in the sentence. The following represent examples of one-liners that include such slang:

The sex was so good that even the neighbors had a cigarette.
Artificial Insemination: procreation without recreation.

To form a lexicon required for the identification of this feature, we extract from WORDNET DOMAINS[6] all the synsets labelled with the domain SEXUALITY. The list is further processed by removing all words with high polysemy (≥ 4). Next, we check for the presence of the words in this lexicon in each sentence in the corpus and annotate them accordingly. Note that, as in the case of antonymy, WORDNET coverage is not complete, and the *adult slang* feature cannot always be identified. Finally, in some cases, all three features (alliteration, antonymy, adult slang) are present in the same sentence, as for instance the following one-liner:

> *Behind every great$_{al}$ man$_{ant}$ is a great$_{al}$ woman$_{ant}$, and behind every great$_{al}$ woman$_{ant}$ is some guy staring at her behind$_{sl}$!*

5.2.2 Content-Based Learning

In addition to stylistic features, we also experimented with *content-based features*, through experiments where the humour-recognition task is formulated as a traditional text classification problem. Specifically, we compare results obtained with two frequently used text classifiers, Naïve Bayes and Support Vector Machines, selected based on their performance in previously reported work and for their diversity of learning methodologies.

5.3 Experimental Results

Several experiments were conducted to gain insights into various aspects related to an automatic humour-recognition task: classification accuracy using stylistic and content-based features, learning rates, impact of the type of negative data, impact of the classification methodology. All evaluations are performed using stratified 10-fold cross validations, for accurate estimates. The baseline for all the experiments is 50%, which represents the classification accuracy obtained if a label of "humourous" (or "non-humourous") would be assigned by default to all the examples in the data set.

5.3.1 Heuristics Using Humour-Specific Features

In a first set of experiments, we evaluated the classification accuracy using stylistic humour-specific features: alliteration, antonymy, and adult slang. These are numerical features that act as heuristics, and the only parameter required for their application is a threshold indicating the minimum value admitted for a statement to be classified as humourous (or non-humourous). These thresholds are learned automatically using a decision tree applied on a small subset of humourous/non-humourous

[6] WORDNET DOMAINS assigns each synset in WORDNET with one or more "domain" labels, such as SPORT, MEDICINE, ECONOMY. See http://wndomains.itc.it.

Table 3 Humour-recognition accuracy using alliteration, antonymy, and adult slang

	One-liners			
Heuristic	Reuters (%)	BNC (%)	Proverbs (%)	OMCS (%)
Alliteration	74.31	59.34	53.30	55.57
Antonymy	55.65	51.40	50.51	51.84
Adult slang	52.74	52.39	50.74	51.34
All	76.73	60.63	53.71	56.16

examples (1000 examples). The evaluation is performed on the remaining 15,000 examples, with results shown in Table 3.[7]

Considering the fact that these features represent *stylistic* indicators, the style of Reuters titles turns out to be the most different with respect to one-liners, while the style of proverbs is the most similar. Note that for all data sets the alliteration feature appears to be the most useful indicator of humour, which is in agreement with previous linguistic findings (Ruch, 2002).

5.3.2 Text Classification with Content Features

The second set of experiments was concerned with the evaluation of content-based features for humour recognition. Table 4 shows results obtained using the four different sets of negative examples, with the Naïve Bayes and SVM classifiers.

Once again, the content of Reuters titles appears to be the most different with respect to one-liners, while the BNC sentences represent the most similar data set. This suggests that joke content tends to be very similar to regular text, although a reasonably accurate distinction can still be made using text classification techniques. Interestingly, proverbs can be distinguished from one-liners using content-based features, which indicates that despite their stylistic similarity (see Table 3), proverbs and one-liners deal with different topics.

Table 4 Humour-recognition accuracy using Naïve Bayes and SVM text classifiers

	One-liners			
Classifier	Reuters (%)	BNC (%)	Proverbs (%)	OMCS (%)
Naïve Bayes	96.67	73.22	84.81	82.39
SVM	96.09	77.51	84.48	81.86

5.3.3 Combining Stylistic and Content Features

Encouraged by the results obtained in the first two experiments, we designed a third experiment that attempts to jointly exploit stylistic and content features for

[7]We also experimented with decision trees learned from a larger number of examples, but the results were similar, which confirms our hypothesis that these features are heuristics, rather than learnable properties that improve their accuracy with additional training data.

Table 5 Humour-recognition accuracy for combined learning based on stylistic and content features

One-liners			
Reuters	BNC	Proverbs	OMCS
96.95%	79.15%	84.82%	82.37%

humour recognition. The feature combination is performed using a stacked learner, which takes the output of the text classifier, joins it with the three humour-specific features (alliteration, antonymy, adult slang), and feeds the newly created feature vectors to a machine learning tool. Given the relatively large gap between the performance achieved with content-based features (text classification) and stylistic features (humour-specific heuristics), we decided to do the meta-learning using a rule-based learner, so that low-performance features are not eliminated in favour of the more accurate ones. We use the Timbl memory-based learner (Daelemans et al., 2001) and evaluate the classification using a stratified 10-fold cross-validation. Table 5 shows the results obtained for the four data sets.

Combining classifiers results in a statistically significant improvement ($p < 0.0005$, paired t-test) with respect to the best individual classifier for the One-liners/Reuters and one-liners/BNC data sets, with relative error rate reductions of 8.9 and 7.3%, respectively. No improvement is observed for the one-liners/proverbs and one-liners/OMCS data sets, which is not surprising since, as shown in Table 3, proverbs and commonsense statements cannot be clearly differentiated using stylistic features from the one-liners, and thus the addition of these features to content-based features is not likely to result in an improvement.

The experimental results prove that computational approaches can be successfully used for the task of humour recognition. An analysis of the results shows that the humourous effect can be identified in a large fraction of the jokes in our data set using surface features such as alliteration, word-based antonymy, or specific vocabulary. Moreover, we also identify cases where our current automatic methods fail, which require more sophisticated techniques such as recognition of irony, detection of incongruity that goes beyond word antonymy, or commonsense knowledge. Finally, an analysis of the most discriminative content-based features identified during the process of automatic classification helps us point out some of the most predominant semantic classes specific to humourous text, which could be turned into useful features for future studies of humour generation.

6 Prospects for Computational Humour

Humour is an important mechanism for communicating new ideas and for changing perspectives. On the cognitive side humour has two very important properties: it helps getting and keeping people's attention and it helps remembering. Type and rhythm of humour may vary and the time involved in building the humourous effect may be varied in different cases: sometimes there is a context, as in joke telling, which allows you to expect since the beginning the humourous climax, even if it occurs after a long while. Other times the effect is obtained in almost no time, with one perceptive act. This is the case of static visual humour, of ironic pictures, when some well-established convention is reversed through the combination with an evocative surprising utterance. Many advertisement-oriented expressions have this property. The role of variation of a known expression seems to be of high importance and studies have also shown the positive impact on the audience of forms of incongruity in the resulting expressions.

As for memorization, it is a common experience to connect in our memory new knowledge with humourous remarks or events. In foreign language acquisition, it sometimes happens that involuntary humourous situations are created because of so-called false friends words that sound similar in the two languages, but have a very different meaning. The "false friends" acknowledgment is conducive to remembering the correct use of the word. Similarly, as shown experimentally, a good humourous expression has exceptionally good recall quality not only per se but also for product type and brand. For a large number of verbal expressions what it takes is the ability to perform *optimal innovation* (Giora, 2002) of existing material, with a humourous connotation. When the novelty is in a complementary relation to salience (familiarity), it is "optimal" in the sense that it has an aesthetics value and "induces the most pleasing effect". Therefore the simultaneous presence of novelty and familiarity makes the message potentially surprising, because this combination allows the recipient's mind to oscillate between what is known and what is different from usual.

A good strategy is to start from well-known expressions that are firm points for the audience and to creatively connect them to the concept or element we intend to promote. We should then be able to perform variations either in the external context, in case the material is ambiguous and the audience can be lured to a different interpretation, or, more often, within the expression itself, changing some material of the expression appropriately, while still preserving full recognizability of the original expression. For instance a good advertising expression for a soft drink is *"Thirst come, thirst served"*, an obvious alteration of a known expression.

In most fields of AI the difficulties of reasoning on deep world knowledge have been recognized for a long while. There is a clear problem in scaling up between toy experiments and meaningful, large-scale applications. The more so in an area such as humour, by many called "AI-complete", where good quality expressions require subtle understanding of situations and are normally the privilege of talented individuals. A goal of computational humour should be to produce general mechanisms limited to the humourous revisitation of verbal expressions, but meant to work in unrestricted domains. To this end, we consider the use of affective terms a critical aspect in communication and in particular for humour. Valence (positive or negative polarity) of a term and its intensity (the level of arousal it provides) are fundamental factors for persuasion and also for humourous communication. Making fun of biased expression or alluding to related "coloured" concepts plays an important role for humourous revisitation of existing expressions.

From an application point of view we think the world of advertisement has a great potential for the adoption of computational humour. Perception of humour in promotional messages produces higher attention and in general a better recall than non-humourous advertisement of the product category, of the specific brand, and of the advertisement itself (Perry et al., 1997).

The future of advertisement will include three important themes: (1) reduction in time to market and extension of possible occasions for advertisement; (2) more attention to the wearing out of the message and for the need for planning variants and connected messages across time and space; (3) contextual personalization, on the basis of audience profile and perhaps information about the situation. All three cases call for a strong role for computer-based intelligent technology for producing novel appropriate advertisements. We believe that computational humour will help produce those kinds of messages that have been so successful in the "slow" or non personalized situation we have lived in.

References

Aristotle. Rhetoric. 350 BC

Attardo S (1994) Linguistic theories of humor. Mouton de Gruyter, Berlin

Attardo S, Raskin V (1991) Script theory revis(it)ed: joke similarity and joke representation model. Humour 4(3):293–347

Bartolo A, Benuzzi F, Nocetti L, Baraldi P, Nichelli P (2006) Humor comprehension and appreciation: an fmri study. J Cogn Neurosci, 18(11):1789–1798

Binsted K, Ritchie G (1997) Computational rules for punning riddles. Humor 10(1):25–76

Bucaria C (2004) Lexical and syntactic ambiguity as a source of humor. Humor 17(3):279–309

Daelemans W, Zavrel J, van der Sloot K, van den Bosch A (2001) Timbl: Tilburg memory based learner, version 4.0, reference guide. Technical report, University of Antwerp

Duncan W (1984) Perceived humor and social network patterns in a sample of task-oriented groups: a reexamination of prior research. Hum Relat 37(11):895–907

Fellbaum C (1998) WordNet. An electronic lexical database. The MIT Press, Cambridge, MA

Freud S (1905) Der Witz und Seine Beziehung zum Unbewussten. Deutike, Leipzig

Giora R (2002) Optimal innovation and pleasure. In: Proceedings of the April fools day workshop on computational humour (TWLT20), Trento, Italy

Hay J (1995) Gender and humour: beyond a joke. Master's thesis, Victoria University of Wellington

Hertzler J (1970) Laughter: a social scientific analysis. Exposition Press, New York, NY

Hetzron R (1991) On the structure of punchlines. Int J Humor Res 4(1):61–108

Hobbes T (1840) Human nature in english works. Molesworth

Kessler B, Nunberg G, Schuetze H (1997) Automatic detection of text genre. In: Proceedings of the 35th annual meeting of the association for computational linguistics (ACL97), Madrid

Lewis D, Yang Y, Rose T, Li F (2004, December) RCV1: A new benchmark collection for text categorization research. J Mach Learn Res 5:361–397

Liu H, Mihalcea R (2007) Of men, women, and computers: data-driven gender modeling for improved user interfaces. In: Proceedings of international conference on weblogs and social media, Boulder, CO

Magnini B, Strapparava C, Pezzulo G, Gliozzo A (2002) The role of domain information in word sense disambiguation. J Nat Lang Eng 8(4):359–373

Mihalcea R, Pulman S (2007) Characterizing humour: an exploration of features in humorous texts. In: Proceedings of the conference on intelligent text processing and computational linguistics, Mexico City

Mihalcea R, Strapparava C (2005a) Bootstrapping for fun: web-based construction of large data sets for humor recognition. In: Proceedings of the workshop on negotiation, behaviour and language (FINEXIN 2005), Ottawa, Canada

Mihalcea R, Strapparava C (2005b, October) Making computers laugh: investigations in automatic humor recognition. In: Proceedings of the joint conference on human language technology/empirical methods in natural language processing (HLT/EMNLP), Vancouver

Miller G (1995) Wordnet: a lexical database. Commun ACM 38(11):39–41

Minsky M (1980) Jokes and the logic of the cognitive unconscious. Technical report, MIT Artificial Intelligence Laboratory, 1980. AI memo 603

Mobbs D, Hagan C, Azim E, Menon C, Reiss A (2005) Personality predicts activity in reward and emotional regions associated with humor. Proc Nat Acad Sci, USA, 102, 16502–16506

Nevo O (1984) Appreciation and production of humor as an expression of aggression. J Cross-Cult Psychol 15(2):181–198

Oaks D (1994) Creating structural ambiguities in humor: getting english grammar to cooperate. Int J Humor Res 7(4):377–402

Perry S, Jenzowsky S, King C, Yi H, Hester J, Gartenschlaeger J (1997) Humorous programs as a vehicle of humorous commercials. J Commun 47(1):20–39

Purandare A, Litman D (2006) Humor: Prosody analysis and automatic recognition for F*R*I*E*N*D*S*. In: Proceedings of the 2006 conference on empirical methods in natural language processing, Sydney, Australia

Raskin V (1985) Semantic mechanisms of humor. Reidel, Dordrecht

Ritchie G (2003). The linguistic analysis of jokes. Routledge, London

Ruch W, Attardo S, Raskin V (1993) Toward an empirical verification of the general theory of verbal humor. Int J Humor Res 6(2):123–136

Ruch W (1996) Special issue: measurement approaches to the sense of humor. Humor 9(3/4)

Ruch W (1998) The sense of humor: explorations of a personality characteristic. Mouton de Gruyte, Berlin

Ruch W (2002) Computers with a personality? lessons to be learned from studies of the psychology of humor. In: Proceedings of the April fools day workshop on computational humour (TWLT20), Trento, Italy

Schopenhauer A (1819) The World as Will and Idea. Kessinger Publishing, Whitefish, MT

Singh P (2002) The public acquisition of commonsense knowledge. In: Proceedings of AAAI spring symposium: acquiring (and using) linguistic (and world) knowledge for information access., Palo Alto, CA

Solomon R (2002) Are the three stooges funny? Soitainly! (or When is it OK to laugh?). In: Rudinow J, Graybosch A, (eds) Ethics and values in the information age. Wadsworth, New York, NY

Spencer H (1860) The physiology of laughter. Macmillan's Magazine 1, 395–402

Stock O, Strapparava C (2003, August) Getting serious about the development of computational humour. In: Proceedings of the 8th international joint conference on artificial intelligence (IJCAI-03), Acapulco, Mexico

Stock O, Strapparava C, Nijholt A (eds) (2002) Proceedings of the April fools day workshop on computational humour (TWLT20), Trento, Italy

Taylor J, Mazlack L (2004) Computationally recognizing wordplay in jokes. In: Proceedings of 26th annual meeting of the cognitive science society, Chicago, August 2004

Tinholt HW, Nijholt A (2007) Computational humour: utilizing cross-reference ambiguity for conversational jokes. In: Masulli F, Mitra S, Pasi G, (eds) 7th international workshop on fuzzy logic and applications (WILF 2007), Lecture Notes in Artificial Intelligence, Berlin, 2007. Springer, Berlin, pp 477–483

Watson K, Matthews B, Allman J (2007) Brain activation during sight gags and language-dependent humor. Cereb cortex 17(2):314–324

Yuill N (1997) A funny thing happened on the way to the classroom: Jokes, riddles and metalinguistic awareness in understanding and improving poor comprehension in children. In Disabilities: Processes and Intervention. Erlbaum, Hillsdale, NJ, pp 193–220

Part VII
Usability

Editorial: "Usability"

Jarmo Laaksolahti and Kia Höök

Abstract This editorial provides a brief introduction and overview of the following 4 chapters dedicated to the design and evaluation of systems for affective interaction. In such systems affect must be a consideration from the start rather than an afterthought or they are likely to fail due to mistakes in the design. Hence one of the purposes of this part of the handbook is to accommodate affective interaction into a user-centred design loop, thereby increasing the chances of arriving at successful designs. Each chapter corresponds to a step in a generic UCD process and discusses the theoretical and practical underpinnings of activities typically found in the step, lists challenges related to bringing affective interaction into the loop, suggests methods that have proven to be useful in each step, and where appropriate gives examples of systems for affective interaction that have been designed.

Emotion is everywhere. It awaits us at work as well as home and in private as well as public, providing texture to our days, making them different from one another. It sits on our shoulders influencing our decisions, making us rational rather than the opposite (Damásio, 1994). Emotion permeates all of time. It greets us in the morning as we wake up pondering whether it is to be a good or a bad day and it tucks us in at night as we summarise our experiences of the day and prepare for the next. Emotion is a constant companion in our lives.

From the preceding part and chapters it should be clear that emotion as a research topic spans a wide range of fields. This part takes a step back from the topic of emotion in itself and instead attempts to look at the role of emotions in systems that we build. What is their purpose? How should they manifest themselves? Our starting point is that all systems evoke emotions, whether they intend to or not. Embracing that fact allows us to tap into the power of emotions and use them as a resource in designing systems rather than viewing them as obstacles. However, adding emotional interaction to a system must not be an afterthought but be a part of the design process from the start or it will surely fail. The chapters in this part take

J. Laaksolahti (✉)
Department of Computer and Systems Sciences, Stockholm University/KTH, Kista, Sweden
e-mail: jarmo@sics.se

P. Petta et al. (eds.), *Emotion-Oriented Systems*, Cognitive Technologies,
DOI 10.1007/978-3-642-15184-2_32,

seriously the design of emotional interaction and the place of systems providing it in our lives.

Over the years, a number of models with accompanying methods for developing systems have been proposed, each coloured by the field (and time) from which it emerged. A common denominator is that they all view design as consisting of several activities ranging from idea generation, through sketching and prototyping, to evaluation. This chapter is mainly concerned with design and development from a human–computer interaction (HCI) perspective and is structured around a model of user-centred design (UCD). A user-centred design process will aim to involve end-users in all stages of the design process. By bringing in users repeatedly we stand a better chance of steering the development towards applications that people really want to live with, as part of their life. We avoid creating systems that really work only for one person: the designer/researcher himself/herself.

However, in order to accommodate emotions into a UCD design cycle, we need to broaden the range of goals that we are aiming for. Usability traditionally focuses on goals such as effectiveness, efficiency, safety, utility, learnability, and memorability. These objective usability goals sometimes contrast with *user experience* goals, which cover subjective qualities such as being fun, rewarding, motivating, satisfying, enjoyable, and helpful. Usability goals and user experience goals often stand in complex relationships, involving trade-offs such as safety vs. fun or efficiency vs. enjoyability (Preece et al., 2002). Introducing emotion thus raises many new dimensions for research on usability to address. It becomes, for instance, a serious issue whether users feel a system is 'sympathetic' or morally acceptable, whether it engages them emotionally. Furthermore, emotions address directly inherently adaptive faculties of humans, posing challenges for methods of user studies and artefact design.

The aim of this part then is to explore how design for emotional interaction with systems can be accomplished through a UCD process that is sensitive to user experience goals in addition to more traditional usability goals. The Area is divided into four chapters each roughly corresponding to one step in the UCD life cycle as suggested by the ISO 13407 standard (ISO 13407:1999, 1999).

As each design case is in a sense unique, there is no single correct way of working through such a cycle. Therefore each part will present a selection of methods that have proven to be useful in designing emotional interaction rather than committing to a single set. In that way, readers are presented with a choice of methods that they can use to carve their own path through the design process. Note also that the set of methods presented is by no means exhaustive but represents a selection that the authors themselves have worked with or are familiar with.

The first chapter, *The Design and Evaluation Process*, gives an introduction to the design and evaluation of affective interaction. Starting with an introductory part based on the ISO standard, it gives a general outline of the UCD process, followed by a historical overview of how evaluation practices have evolved from an activity performed by engineers to evaluate mostly technical aspects of a system, to include, as evaluation was performed by professionals from other fields, an increasing range of psychological, social, and experiential factors. However, design and evaluation as performed by researchers does not always take into consideration the challenges faced by developers in commercial environments. Therefore an overview of some of the differences between academic and commercial design and evaluation practices is presented, stressing the need for academia to reach out, and adapt, their models and practices to suit real-world situations. These threads come together in the conclusion of the first part, which introduces what has been called the third wave of human computer interaction (HCI) (Bødker, 2006). This new wave of HCI recognises the importance of experiential qualities in designing systems and brings those qualities to the front, assigning them and traditional usability goals equal importance.

The second chapter, *Understanding Users and Their Situation*, dives into the business of designing emotional interaction in earnest. The first step in a development process is to set the stage for what to design and how that should be realised. In terms of user-centred design, this includes developing a sense of who will be using the system, where it is to be used, and what it should be used for. The chapter starts by reviewing the reasons for, and importance of, methods to know users as well as some of the major challenges, e.g. that emotions are essentially subjective and embodied, presented by emotional interaction in this respect. A selection of methods, referred to as empathic design methods, with the power to at least to some extent overcome such challenges, are presented.

Designs for affective interaction are solely affected not only by how, and how well, one gets to know target users but also by designers' own view of emotions and values designed for. The next chapter reviews the *informational* and *interactional* views of emotion and how choosing one or the other can affect a design process. Being the newcomer on the scene, understanding of the interactional view is also supported by some design values and principles that are firmly rooted in it. Design always involves taking the difficult step of inventing something new that will make sense to users based on what is known about them and their situation. The chapter, *Generating Ideas and Building Prototypes*, starts by discussing the challenge of framing the problem in such a way that the design process can move forward and prototypes can be constructed. Prototypes are important in the design process as they are the main vehicles through which ideas are tested and solutions are verified. Generating ideas for prototypes is a highly creative task that can be supported by a number of methods. A selection of brainstorming methods that can be used for this, as well as other ends in the design process, are introduced. However, as emotions are experienced physically as well as cognitively, it may sometimes be necessary to do more than think about emotions. Bodystorming is presented as a way of refining ideas using low-fidelity prototypes to physically act out the emotional interaction within the design team. After idea generation comes prototyping. A number of methods are discussed that enable the process to proceed in a stepwise fashion beginning with low-fidelity prototypes such as paper prototypes, before moving on to more complex implementations. At every step the design and functionality of prototypes must be evaluated to make sure that the design is headed in the right direction. The purpose of such evaluation is not to learn how a completed system is received but rather to gain feedback to the design process. Sometimes it may even be beneficial to let humans take on some of the functionality of a system to try it out in Wizard-of-Oz study.

The final chapter, *Evaluation of Affective Interactive Applications*, covers the challenging task of finding out how the prototypes we build are received. The chapter presents two strands of evaluation methods. The first is concerned with what may be called more traditional evaluation methods: Is the system usable for the purpose it was designed for? The methods belonging to this strand are grounded in "traditional" fields that have influenced HCI such as psychology. The selection of methods covered here includes psychometric methods, physiological measures, and think aloud methods. The second strand of evaluation methods is more experience focused: Does the system provide for the kind of emotional experience that it aimed to do? Methods belonging to this second strand are often grounded in fields outside the traditional ones, such as art and design. Both strands have their strengths and weaknesses; which strand of methods to use ultimately depends on the purpose of the evaluation. The choice of evaluation method, as well as the choice of path through the design process as a whole, is different for each project. Here we have presented a variety of methods that make it possible to carve many different paths.

References

Bødker S (2006) When second wave HCI meets third wave challenges. In: Mørch A, Morgan K, Bratteteig T, Ghosh G, Svanaes D (eds) Proceedings of the 4th Nordic conference on human–computer interaction: changing roles (NordiCHI 2006), vol 189. ACM Press, New York, NY, pp 1–8

Damásio AR (1994) Descartes' error: emotion, reason and the human brain. Grosset/Putnam, New York, NY

ISO/IEC 13407:1999 (1999) Human-centred design processes for interactive systems. International Organization for Standardization, Geneva

Preece J, Rogers Y, Sharp H (2002) Interaction design: beyond human–computer interaction. Wiley, New York, NY

The Design and Evaluation Process

Joseph Jofish Kaye, Jarmo Laaksolahti, Kia Höök, and Katherine Isbister

Abstract The purpose of this chapter is to describe the design and evaluation process in the light of affective interaction. With a starting point in user-centred design we will explore what additional problems or opportunities become important when designing for affective interaction with computer systems. This chapter also provides a historical background to HCI ending with what is sometimes named the third wave of HCI – that is, designing for aesthetic, emotional experiences with and through technology.

1 User-Centred Design

Software design has traditionally been viewed as a linear process in which development proceeds stepwise from an application description, through a well-defined requirement specification and system design based on the specification, to a finished product. However, this model of software development (also known as the waterfall model, see e.g. Sommerville, 1992) has a number of problems associated with it. For instance, its linear structure does not allow for changing requirement specifications or system designs once they have been decided upon. Furthermore the requirement gathering and system design activities of the waterfall model seldom include representatives of the users that in the end will be using the system, meaning that their needs and wants are not taken into account. Instead the product, once finished, is imposed upon the users who are expected to adapt to the new situation and accommodate the system into their situation.

In contrast, user-centred design (UCD) is an iterative approach to systems development that puts users and questions regarding their needs, wants, and usage situations in focus. User-centred design is not a method per se but rather a design

J.J. Kaye (✉)
Nokia Research Center, Palo Alto, CA, USA
e-mail: jofish.kaye@nokia.com

P. Petta et al. (eds.), *Emotion-Oriented Systems*, Cognitive Technologies,
DOI 10.1007/978-3-642-15184-2_33,

philosophy that can have many incarnations, all sharing the same underlying principles (Preece et al., 1994):

- To make user issues central in the design process
- To involve users in the design process
- To carry out early testing and evaluation
- To do design iteratively

UCD recognizes the fact that users and usage contexts differ and that, consequently, systems should be designed with that in mind. Rather than forcing users to adapt to accommodate a system, the system is designed from the start to suit its users. The basic idea that systems should be designed based on what users need and want is a notion that permeates UCD methods. In practice this means paying extensive attention to users' needs, wants, and limitations in each step of the design process. It requires designers to analyse and foresee how users are likely to use a system and to test results in real-world tests with actual users. Such testing is necessary as it is often difficult for designers to understand what experiences users of their design go through.

The basic principles of UCD have been standardized in the ISO standards, 'ISO 13407:1999 Human-centred design process', and developed in 'ISO TR 18529: Ergonomics – Ergonomics of human–system interaction – Human-centred lifecycle process description'. These standards aim to provide guidance for achieving quality in use by incorporating UCD activities throughout the life cycle of interactive computer systems.

Figure 1 illustrates the four human-centred design activities that ISO 13407 suggests should take place during a system development project. First studies are performed to gain insight about contexts in which the system will be used and requirements in terms of users' needs, wants, and practices. This step may also include specifying organizational requirements that are of importance in a workplace. After analysing the results of the studies, conclusions and design ideas drawn from it are used to guide and fuel the design work. Eventually a design prototype is produced, which is tested with actual users to assess how well it meets the design intentions, which in turn are based on users' needs and wants. The produced prototypes can have various forms including scenarios and sketches showing broad functionality, paper- or screen-based mock-ups that simulate aspects of functionality, to fully working prototypes that represent the full functionality of the system. This development cycle is iterated several times during the design process allowing misconceptions to be clarified – and improvements based on user feedback to be incorporated into the design along the way.

While ISO 13407 describes the general outline of a UCD development cycle, there is great diversity in how the principles of UCD have been implemented. Consequently UCD processes differ along a number of dimensions, such as level of user involvement and control in the design process or methods used in the different steps. Methods such as participatory design advocate user control and involvement in the whole design process and often incorporate end-users into the design

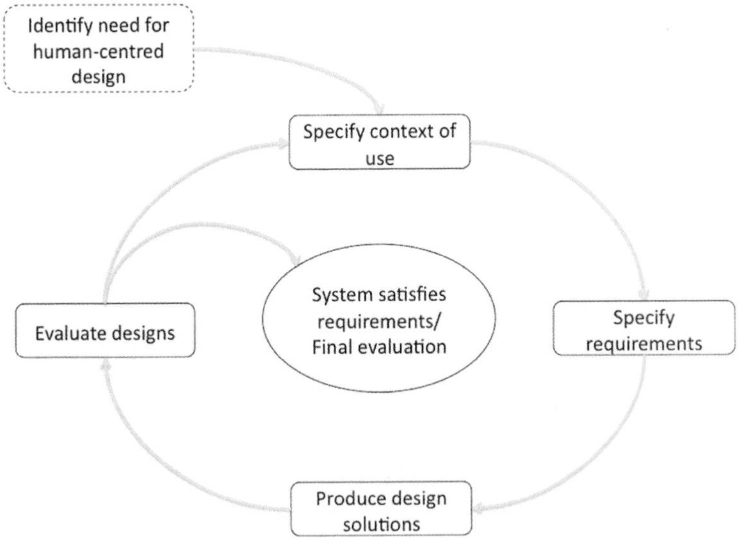

Fig. 1 ISO 13407 Human-centred design process: design activities

team (Ehn, 1992), while other methods are content with bringing in end-users as test subjects during the development cycle. Some methods use ethnography as a means to intimately understand the end-users situation, while others are content with using market research as a starting point. The 'right' combination of methods to use is ultimately determined by the characteristics and available resources for each development project.

With the increasing number of applications, such as computer games, whose primary purpose is to provide experiences rather than make us more efficient at tasks, UCD is faced with new challenges. Instead of task-completion time or productivity the important aspects for such applications are the experiences they provide and the emotions they evoke. While traditional HCI tools are well equipped for answering the first type of questions, they are less well equipped for answering the second type of questions. As of yet we do not know how to capture subjective experiences and affective involvement in such a way that it can provide feedback into the design process. Within the research community, this has led to an increasing interest in finding methods for designing and evaluating non-task-related aspects of application usage (Blythe et al., 2004; McCarthy and Wright, 2004). A UCD perspective and development philosophy is a good starting point as such but it will have to be extended to include methods that can specifically deal with designing for qualities such as flow, pleasure, fun, excitement, or fear while at the same time keeping users involved in the process.

The bulk of this chapter is concerned with such methods and will describe them and situate them in the general UCD cycle above. But before we present any methods, we will look at a historical perspective on design and evaluation from a human–computer interaction view in Sect. 2.2 to understand its origins and future directions, and have a look at issues related to design and evaluation outside research in Sect. 2.3 and finally discuss the particular challenges associated with designing for affective interaction in Sect. 2.4.

2 Evaluation: A Historical Perspective

The methods and approaches of evaluation in HCI have changed along with the field of computing itself. In particular, as different sets of users have different limiting factors on their ability to accomplish their tasks using computers, they require different forms of evaluation necessary to recognize and evaluate where those limitations occur. By tracing the history of these varying forms of evaluation in the field, we can understand how the current approaches to evaluation in HCI have come to be and further understand prospects for the future evaluation in the field, and in particular the role of affect in that evaluation.

2.1 Evaluation by Engineers

Computers were originally experienced as fundamentally mechanical and electrical machines. Vacuum tubes broke on a regular basis, and evaluation in the sense that it is used today in HCI did not exist – the closest in the published papers of the time is 'reliability' – because the biggest factor in determining how well a computer worked was how long it would do so. For example, John von Neumann brings up reliability of vacuum tubes and memory storage seven times in his draft report on the EDVAC in 1945 (von Neumann, 1945). This is an excellent example of a limiting factor on users accomplishing their tasks: limitations like the difficulty of entering commands were dwarfed by the simple fact that the computer itself would frequently cease to function after half an hour of continuous use.

2.2 Evaluation by Computer Scientists

By the 1950s, evaluation of computers was being done by computer scientists. The users were themselves programmers, although they were now freed of the necessity of understanding how the hardware itself worked, and had a level of abstraction provided for them by programming languages, assemblers, compilers, and debuggers. 'Reliability' remained an important metric: Holt (1960) uses the term to refer to how ready for use a particular piece of hardware or software is: '... some sort of reliability (or usability, or state-of-completeness) code.' But at this stage, we start to see a meaning of 'evaluation' that we can recognize; it means 'testing' or 'appropriateness for task'. The first use of 'evaluation' in this sense is Israel's (1957) paper, in which the focus is the development of a set of pre-packaged input data that can be fed to a program as part of a testing regime. Similarly, in talking of evaluation, Calingaert (1967) defines throughput, turnaround, and availability as fundamental measures of performance. Lucas (1971) refers to three major purposes for evaluating the hardware and software performance of computer systems: selection evaluation (deciding what to buy), performance projection (figuring out how fast it will run once you have bought it), and performance monitoring (figuring out how fast it runs now you own it).

These notions of evaluation emphasized the speed of the computer. This is not evaluation in the way it is commonly used in HCI, because by the late 1970s the computer was no longer necessarily the slowest part of the interface; the computer started spending a lot of time waiting for the human rather than the other way around. But at this earlier stage, the limiting factor was the speed of the machine. What is important about this stage is that it was formative in the way that computer scientists saw evaluation, and this approach was influential for many years thereafter. The evaluation is in MIPS (million instructions per second), megahertz, and megabytes, terms that computer scientists understood as giving them information about appropriateness for and speed of task execution.

A key example of this interface paradigm can be found in the influential conference paper by Royce (1970), in which he discusses the waterfall method of organizing large software programming tasks. He discusses the importance of testing software and the role of the *customer*, but the customer is entirely a business entity; there is absolutely no notion of *users*. His notion of evaluation is that it is entirely concerned with the task of programming, and one that harkens back to the engineering notion of reliability: 'Most errors are of an obvious nature that can easily be spotted by visual inspection [of the code].' Once again, the computer is the turf of engineers, of mathematicians, and of these new breeds of programmers and computer scientists, and they are the ones doing evaluation.

2.3 Evaluation by Experimental Psychologists

The situation changes around the end of the 1970s and beginning of the 1980s; probably the earliest example is the 1976 SIGGRAPH Workshop on User-Oriented Design of Interactive Graphics Systems (Treu, 1977), although there was little work published in this vein again until the beginning of the 1980s. At this point, the first signs of what we might recognize as 'end-users' show up. These are people who are using the computer as a tool to solve problems. They no longer understand what the computer is doing in the same manner as engineers and programmers do, and they do not care. They are still mainly government and corporate users, and they are still doing single, focused tasks. The speed of execution of these tasks is becoming the metric by which systems are measured rather than the speed of the calculations that enable that execution. But there was a change in the air of computing; the human was becoming an essential part of the human–computer equation, and the focus in the discourse of the uses of computing was shifting from a discussion based purely around the technology itself to a discussion about the role of the computer and the human working together. As such, experimental psychologists and cognitive scientists were starting to enter the field of human–computer interaction, bringing with them an approach to evaluation and a set of problems they are particularly equipped to solve, based on experimental and psychophysical approaches to human–computer interaction. So in publications of the time, we see an emphasis on, for example, the ergonomics of the keyboard, the legibility of different colours of text, and the role of timing in input and output. Grudin (1990) writes:

> Perceptual issues such as print legibility and motor issues . . . arose in designing displays, keyboards and other input devices . . . [new interface developments] created opportunities for cognitive psychologists to contribute in such areas as motor learning, concept formation, semantic memory and action. In a sense, this marks the emergence of the distinct discipline of human-computer interaction.

So what of evaluation at this point? Experimental psychology and cognitive science bring with them an experimental approach towards the evaluation of systems; they are used to test users in laboratory experiments to determine psychophysical parameters, such as response times to stimuli and the like. However, they are entering a field that already has computer scientists and their notions of evaluation as the dominant paradigm, and the end result is a notion of evaluation that combines the system focus of computer science with the experimental approach of cognitive science. The limiting factor becomes the ways and speed in which users can instruct the computer to do what they want.

2.4 Evaluation by Usability Professionals

As the 1980s progress, we see the advent of the personal computer and the development of something that starts to look like a coherent field of human–computer interaction. The key difference to previous stages is that the computer itself was no longer the site for action, but rather the interaction itself. Shackel (1997) writes, 'In the beginning, the computer was so costly that it had to be kept gainfully occupied for every second; people were almost slaves to feed it' (p. 977). Shneiderman (1979) describes the relationship between computer and user as one of symbiosis with implications of human and computer on an equal footing, but by this stage the human is, at least ideally, no longer the lesser player in the relationship.

In 1981, the personal computer (PC) enters the stage, bringing with it a change from the mainframe and minicomputer concept of one computer for many users to initially one computer for one user and the foundation for one-to-many human-to-computer relationships of ubiquitous computing and the many-to-many relationships of computer-supported collaborative work. To the corporation, the PC meant that each employee could have their own computer sitting on their desk, for their exclusive use: a one-to-one person-to-computer ratio for the first time. There was also increase in the number of computers in the schools and in the homes, encouraging the development of the notion of a computer for discretionary use rather than use required by a job. In general, this profusion of computing power had a number of effects on the way the computer was used, who was using it, and what they were using it for. For example, the level of expertise that was necessary to have access to computing power decreased significantly. No longer could the end-user be expected to go through a training course or walk themselves through three-ring binders of instructions; it became necessary to start to think of delivering software that was easy to use and accessible by lay people. The limiting factor in using a computer to accomplish a task started to be the difficulty to the human of making sense of the computer. The human interface rose to the top of issues that the industry needed to address to make their product more successful in the marketplace.

In 1984, the release of the Macintosh computer was accompanied by a book, the Apple Human Interface Guidelines, which set out the specifications for the human-focused graphical user interface, which defined the 'look and feel' of the Mac interface. This was necessary for two reasons. Firstly, the graphical user interface was a novel approach to human–computer interaction; software developers were simply not used to designing for it. It was not possible to just release development software and assume that developers would do the right thing with it. Secondly, the Macintosh project was set up in such a way so as to encourage the development of third-party software for the platform. External developers were a key part of how the new platform was going to be used by people everyday.[1] Apple made the Macintosh operating system 'look and feel' and 'ease of use' key parts of the way that the computer was presented to the world.

2.5 *Evaluation for the Social*

As the personal computer became rapidly more popular, they entered the workplace in droves. But when computers were placed into office settings, to peoples' surprise, it did not produce the increases in efficiency that were thought would be naturally brought about by such clearly advanced technologies. Many people using many computers would be, one would have thought, a simple matter of multiplying the effects of one person using one computer. But in practice the difference between many people-to-many computers and one person-to-one computer turned out to be as big a difference as that between one person-to-one computer and many people-to-one computer. Software specifically written to support existing ways of getting work done turned out to be hard to use and got in the way of accomplishing tasks rather than facilitating it. How, then, to solve the new problems produced by a situation in which many people in the workplace were using computers to do their own work, and yet desperately needed to communicate with others in the course of doing so?

Attempts to answer this question resulted in the development of what has become a separate field of computer-supported collaborative work, or CSCW. In 1984,

[1] See, e.g., http://www.folklore.org/StoryView.py?project=Macintosh&story=Inside_Macintosh.txt (visited 2010-05-30).

Paul Cashman and Irene Grief organized an interdisciplinary workshop to discuss the development of computer-related tools for the workplace, which was followed by the first full conference, CSCW'86, 2 years later (Greif et al., 1987). Previous efforts as part of the 'office automation' movement had hit somewhat of a dead end with a focus on building technical tools rather than understanding the environment in which they were placed (Grudin, 2005). Schmidt (1991, p. 10) uses the following quotation from Barber et al. (1983) to summarize the problem:

> In all these systems information is treated as something on which office actions operate producing information that is passed on for further actions or is stored in repositories for later retrieval. These types of systems are suitable for describing office work that is structured around actions (e.g. sending a message, approving, filing); where the sequence of activities is the same except for minor variations and few exceptions[...] These systems do not deal well with unanticipated conditions (Barber et al., 1983, p. 562).

So how did this new field, with its emphasis on the social in the workplace, deal with the problem of evaluation? In his thesis, Martin Ramage spends an entire chapter summarizing the various methods used in CSCW (1998). He divides CSCW evaluation into four types. The first, 'effects of a co-operative system in an organization', is concerned with examining the effects of a system once it has been introduced. The second, 'formative evaluation of CSCW technology', is evaluation that happens in the course of building technical systems, with the aim of developing them further and making them more useful and appropriate for the users – a notion very similar to the notions of evaluation we have been discussing in HCI. The third, 'conceptual development', is the kind of evaluation done only in a research context to evaluate the concepts that underlie the system and determine if they might be applicable. The fourth is 'what software should I buy?' Each of these categories is clearly a form of evaluation, but each requires a very different approach to the situation; as such, Ramage proceeds to synthesize a method of evaluation he refers to as systematic evaluation for stakeholder learning, or SESL, which involves multiple stakeholder evaluations of the socio-technical system, emphasizing the role of evaluation as a collaborative opportunity for learning within the organization.

2.6 Evaluation of Experience

The most recent approach to evaluation in HCI involves the evaluation of HCI focused on experience rather than tasks. Evaluations of experience-focused HCI are typically open-ended and encourage interpretation rather than eliminating ambiguity (Sengers and Gaver, 2006). Even more so than previous methods, they encourage looking at the human lived experience of the technology, rather than any kind of emphasis on the computer itself, and the researchers in this area draw from approaches that encourage this perspective, including ethnography, game design, industrial design, cultural theory, as well as art and aesthetic practices. Users are people going about their lives, generally outside of the workplace, communicating with friends, playing games, expressing and enjoying themselves. The field of experience-focused HCI has a strong theoretical grounding, with works such as McCarthy and Wright's *Technology as Experience* (2004) that leverages the work of pragmatist philosophers Dewey and Bakhtin, and Dourish's (2001) *Where the Action* that takes a phenomenological approach to HCI, recognizing the ways in which HCI engages with twentieth-century dissatisfaction with the adequacy of Descartian approaches to knowing and understanding. This is a field of research that is actively under development and has not developed agreed-upon approaches for evaluation, but there is early work that suggests likely approaches and demonstrates the emphasis on the role of design that is at the heart of approaches to experience-focused HCI.

For example, at CHI 2006, Isbister and colleagues (2006) presented an 'affective evaluation tool', taken to be focused on a nonverbal, body-based approach that encourages UCD values and be practical, portable, flexible, cross-culturally valid, and fun to use. The tool centres around the use of eight hand-held white clay figurines that are manipulated by the user to express emotions non-verbally in the course of interacting with a technological system. Similarly, Joseph Kaye et al.'s work with intimate objects emphasizes open-ended responses to cultural probe-influenced questionnaires about users' interactions with the technology and the social systems the technology is intended to influence (Kaye et al., 2005; Kaye, 2006). William Gaver references the role that commentary and discourse can play in other cultural situations by appropriating media such as documentary film and journalism for the evaluation of technological systems (Gaver, 2007). In contrast to the rest of these examples, in an excellent example of relating ideas from a new intellectual approach to existing work in the field, Charlotte Wiberg has shown how it is necessary and possible to modify traditional usability metrics to evaluate the experience of using websites designed for entertainment (Wiberg, 2005).

The emphasis shown here on open-ended and qualitative evaluations that engage both user and designer in interpretation of their affective, lived experience with the technology seems likely to be a hallmark of experience-focused evaluation and demonstrates the importance of the material covered in this report.

3 Design and Evaluation in the Real World

After the brief review of the history of evaluation in human–computer interaction, we will next contrast design and evaluation practices in product development and research: What are key differences between the two perspectives?

Researchers and practicing commercial designers, which we are calling 'real world', do tend to approach design and evaluation differently, probably due to a mixture of differences in end product goals and in the working cultures that they inhabit. We can readily see the tensions and challenges of communicating between these worlds in its evolution and iteration at the CHI conferences in recent years, which made a dedicated effort to reach out to more 'real-world' practitioners with practical tutorials and more opportunities to exhibit commercial design methods and results.[2] But how do 'real-world' and research design practices differ in detail? Let us break down this issue into the activities of design as outlined in the previous section (cf. Fig. 1).

3.1 Problem Definition and Time Frame

When defining a problem and a time frame, 'real-world' designers carry out their work within the constraints of commerce: of shipping a purchased product to end-users. Most real-world design efforts come with an extensive set of constraints within which the designer must work. It includes available budget for the project, a set time frame for releasing the product, market constraints in terms of what is considered acceptable for the target user group, and resource constraints in terms of team expertise and equipment available. Real-world project teams typically have members whose roles are assigned according to established areas of expertise (e.g. visual design, interaction design, usability testing, and project management). Most real-world teams are comprised of individuals who have demonstrated expertise and prior experience with the roles they are assigned. Team members are often asked to join in producing and committing to reasonable time estimates for the completion of projects.

Designers in research contexts also have constraints, but of a different sort. Time frames are linked to conference calendar cycles as well as the 'publish or perish' imperative of the tenure track in academia, and internal reporting cycles of research labs. Budgets are a by-product of institutional constraints (e.g. industry lab budgets, or academic grants and available infrastructure and resources) and are often not as clearly demarcated on a project-by-project basis as commercial design budgets. Team members may or may not have prior experience in a given role, depending upon the researchers' backgrounds and the experience level of student or intern participants. Often in academic contexts, projects have a secondary aim of providing apprenticeship to budding researchers who will be growing their skills as they work, which also shifts the nature of the project management role. Defining a problem in the research context proceeds in quite a different manner than in the commercial world. A researcher's aim is not simply a polished end product but also (and sometimes instead) some sort of methodological and/or theoretical contribution to the ongoing research dialog – what Zimmerman et al. call 'transformative design' (Zimmerman et al., 2007). This makes a big difference in the choices that researchers make as the process proceeds, as will be outlined in the following.

3.2 Involvement of Users

Regarding user involvement, 'real-world' designers usually work on very tight timelines and also with a high degree of dependency for success upon the satisfaction

[2] See, for example, the call to the design community for CHI 2007: http://www.chi2007.org/community/design.php (visited 2010-05-30).

of end-users with what they do. The product of their efforts must satisfy a predetermined economically viable user group, and efforts will be made on most larger scale efforts to get the designs in front of some subset of that market to ensure that the team is on the right track. Involving users is typically done at the early conceptual state, with focus groups or other methods for gleaning market acceptability of a product concept, and then at mid-to-late stages of product development, with usability testing. Real-world designers can call upon the services of in-house or consultancy-based experts to recruit members of the target user group, for fees that are often substantial. If the project is being completed for a larger company with existing customers, the designer can probably draw upon the user base as well as aggregated data about use patterns. Typically user involvement is limited to focus groups and usability testing sessions. So one could say that the accuracy of the population selection can be quite high for real-world designers, but the range of involvement is rather restricted. There are of course exceptions, as documented in papers about participatory design efforts (e.g. Puri, Byrne et al., 2004; Kristensen et al., 2006). In recent years, some commercial design firms have begun to use and to market their use of a wider range of tactics for involving users.[3]

Research-based designers often use a sample of convenience (college students or fellow researchers) due to a lack of budget for and attention paid to the importance of targeting representative populations. It is also the case that a researcher may not be exactly sure, yet, what population will be best served by a new blue-sky idea for an interface. Thus, one could say that research designers tend to involve a less varied and 'accurate' set of users but also that this may not be the main point of soliciting feedback, at least in some cases. Research designers feel much more free to try out experimental methods for eliciting feedback and participation from end-users (e.g. cultural probes, Gaver et al., 2004, or the sensual evaluation instrument, Isbister et al., 2006). These methods may provide the designer with a wider range of input from users, offering richer fodder for design. However, such input must also be interpreted with care, as the methods often will not have been subjected to many applications and rigorous discussion among designers as to their efficacy in supporting design practice.

4 Brainstorming

As explained, 'real-world' designers usually work on a very tight time frame, and idea generation must take place rapidly. Some design consultancies have developed processes for rapid and effective generation of the maximum number of best quality ideas. These firms may work in collaboration with their client companies and can also pull in designers working on other projects for consultancy.

[3] For example, IDEO's method cards (http://www.ideo.com/work/item/method-cards/; visited 2010-05-30).

Researchers are typically able to use longer time frames in the ideation phase that brings them to a final project concept, interweaving this work with completion of other projects. As with user involvement, researchers often feel freer to use more experimental methods such as bodystorming (Oulasvirta et al., 2003) to obtain a wide range of concepts for their designs. They may consider the evolution of new techniques and tools for brainstorming to be a part of their research (e.g. Bastéa-Forte and Yen, 2007).

5 Prototyping

'Real-world' designers use prototypes to serve several purposes: to help the team converge upon a working solution, to engage users in order to get corrective/confirmatory feedback, and to showcase for internal stakeholders (e.g. managers) to get approval and provide reassurance. The third purpose is quite important for commercial prototypes and can sometimes lead to substantial time spent polishing and carefully delineating what will be shown. Since real-world designers are working under time pressure, they will avoid throw-away efforts wherever possible – prototyping in commercial contexts may therefore be more conservative compared to the latitude often offered in research contexts. Some commercial designers make use of paper prototypes, wire frames, and other techniques that mitigate the early effort spent on polishing prototypes that leads to more design inertia.

Researchers vary widely in their use of prototypes – in many cases a prototype (rather than a polished product) is the end result of a research cycle. Most research prototypes are not nearly as robust and polished as commercial counterparts; however they are likely to be far more adventurous in terms of core concepts that are manifested in the design itself, for example exploring the benefits of new input modalities well before they are commercially viable (e.g. eye tracking – see Merten and Conati, 2006). Research prototypes usually do not have to stand up to hard use by outsiders or naysayers, which may lead to less being learned through exposing prototypes to users than in commercial contexts. However, there is also active research into prototyping as an activity in itself – feeding back valuable data to those in commercial practice about how and when to use low-fidelity versus high-fidelity prototypes, for example (e.g. Lim et al., 2006).

5.1 Testing with Users

'Real-world' designers, as mentioned in Sect. 3.2, typically engage in early focus group-style testing and mid-to-late-stage usability testing, with an emphasis on making sure that they test with exemplars of the target user group. Both focus group and usability testing are often done by outside companies or by internal departments that specialize in these processes, working from the belief that the designers themselves may have biases that get in the way of being able to truly test and critique their work, even though some commercial designers have questioned this bifurcated mode of working.[4] Commercial developers typically also allocate a significant amount of time to late-stage 'bug-fixing' testing of their releases.

Researchers may employ similar methods, though more likely with a convenience sample rather than a representative sample of users. As for the previous activities, they also spend time innovating user testing methods as well (as mentioned in Sect. 3.2).

5.2 Polishing and Sharing Results

A 'real-world' designer's efforts usually result in a product released to the public, which must stand up to unmonitored use. Thus, a great deal of time and money in the development cycle is spent polishing the product and ensuring that it will be reliable and able to withstand repeated use. Real-world designers sometimes write papers or deliver talks at design conferences (such as CHI, the ACM Conference on Human Factors in Computing, and DIS, Designing Interactive Systems Conference, conference series), but they may not be at liberty to share innovative details of their process if their company considers them a business advantage. In the end, commercial designers gauge their success by whether the product satisfies end-users and is successful in the marketplace.

As mentioned above, researchers often stop with a reasonably polished prototype, which may not need to be usable in an unattended fashion and may only be viewed by other relatively sympathetic and context-sharing experts. Researchers' products thus rarely achieve the level of polish and robustness of commercial designs. However, researchers must be prepared to explain and present the larger value of their work – exploring/revealing new terrain in terms of research contribution. A research prototype is meant to 'demonstrate a research contribution' (Zimmerman et al., 2007, p. 495); so publication and dissemination is an important part of the design process for researchers.

Researchers sometimes use final prototypes to help engage the community in a critique of existing practice: 'Design researchers engaged in critical design create artefacts intended to be carefully crafted questions. These artefacts stimulate discourse around a topic by challenging the status quo and by placing the design researcher in the role of a critic. The Drift Table offers a well known example of critical design in HCI, where the design of an interactive table that has no intended task for users to perform raises the issue of the community's possibly too narrow focus on successful completion of tasks as a core metric of evaluation and product success' (Zimmerman et al., 2007, p. 496, with reference to Gaver et al., 2004).

[4] See, e.g., Lane Becker (2004) 90% of all usability testing is useless (http://www.adaptivepath.com/publications/essays/archives/000328.php; visited 2010-05-30).

6 Third-Wave HCI and Evaluation Challenges

In this chapter, we have tried to provide a historical background to HCI in order to arrive at a better understanding of the challenges that affective interaction poses to design and evaluation methods. In our historical background, we have argued that the overall challenge has moved from early concerns with reliability and efficiency of the machine itself to subsequent concern with people as a part of the complete system of human and machine, to finally seeing people as collaborators in workplaces in the CSCW field. With the advent of seeing computers also as tools for play, ludology and experience design have emerged as important design goals. And it is here that we find affective interaction and its role.

We have also painted a particular picture through our focus on a UCD cycle, ending with the 'third wave of HCI' perspective. When user experience is dealt with from this stance, it becomes a problem of *interaction*. Emotion is seen as constructed in the dialogue between user and system – not as something that exists a priori in the user – and the design challenge is to design for such experiences to happen. As we shall discuss in subsequent chapters, this is a that stance differs from the more AI-oriented perspective that is still more directed at looking at the problem as a usability problem. The traditional AI view is that it should be possible to recognize some aspects of the users' emotions and then improve the efficiency of the system from this perspective.

In the following chapters, we will explore the influences of these approaches to understanding the nature of human computer interaction, and their implications for the involvement of users in the design cycle, for the kinds of systems that we aim to design, and for the evaluation metrics appropriate for such study.

References

Barber GR, de Jong P, Hewitt C (1983) Semantic support for work in organizations. In: Mason REA (ed) Information processing 83. Proceedings of the IFIP 9th world computer congress, Paris, 19–23 Sept 1983. North-Holland, Amsterdam, pp 561–566 (cited in Schmidt 1991)

Bastéa-Forte M, Yen C (2007) Encouraging contribution to shared sketches in brainstorming meetings. In CHI '07 extended abstracts on human factors in computing systems (San Jose, CA, USA, April 28–May 03, 2007). CHI '07. ACM, New York, NY, 2267–2272. DOI= http://doi.acm.org/10.1145/1240866.1240992

Blythe MA, Overbeeke K et al (eds) (2004) Funology: from usability to enjoyment. Kluwer, Boston, MA

Calingaert P (1967) System performance evaluation: survey and appraisal. Commun ACM 10(1):12–18

Dourish P (2001) Where the action is. MIT Press, Cambridge, MA

Ehn P (1992) Scandinavian design: on participation and skill. In: Adler PS, Winograd TA (eds), Usability: turning technologies into tools. Oxford University Press, New York, NY, pp 96–132

Gaver WW (2007) Cultural commentators: non-native interpretations as resources for polyphonic assessment. Int J Hum Comput Stud 65(4):292–305

Gaver WW, Boucher A et al (2004) Cultural probes and the value of uncertainty. Interactions 11(5):53–56

Gaver WW, Bowers J, Boucher A, Gellerson H, Pennington S, Schmidt A, Steed A, Villars N, Walker B (2004) The drift table: designing for ludic engagement. In CHI '04 Extended Abstracts on Human Factors in Computing Systems (Vienna, Austria, April 24–29, 2004). CHI '04. ACM, New York, NY, 885–900. DOI= http://doi.acm.org/10.1145/985921.985947

Greif I, Curtis B et al (1987) Computer-supported cooperative work (panel): is this REALLY a new field of research? In: Proceedings of the SIGCHI/GI conference on human factors in computing systems and graphics interface. ACM Press, Toronto, ON

Grudin J (1990) The computer reaches out: the historical continuity of interface design. In Proc. CHI '90, p264. ACM, New York, NY, 261–268. DOI= http://doi.acm.org/10.1145/97243.97284

Grudin J (1994) Computer-supported cooperative work: history and focus. IEEE Comput 27(5): 19–26

Grudin J (2005) Three faces of human computer interaction. IEEE Ann Comput 27(4):46–62

Holt A (1960) Over all computation control and labelling. Commun ACM 3(11):614–615

Isbister K, Höök K, Sharp M, Laaksolahti J (2006) The sensual evaluation instrument: developing an affective evaluation tool. In: Grinter R, Rodder T, Aoki P, Cutrell E, Jeffries R, Olson G (eds) Proceedings of the SIGCHI conference on human factors in computing systems, Montréal, QC. ACM Press, New York, NY, pp 1163–1172

ISO/IEC 13407:1999 (1999) Human-centred design processes for interactive systems. International Organization for Standardization, Geneva

Israel DR (1957) Simulation techniques for the test and evaluation of real-time computer programs. J ACM 4(3):354–361

Kaye JJ (2006) I just clicked to say I love you: rich evaluations of minimal communication. In CHI '06 Extended Abstracts on Human Factors in Computing Systems (Montréal, Québec, Canada, April 22–27, 2006). CHI '06. ACM, New York, NY, 363–368. DOI= http://doi.acm.org/10.1145/1125451.1125530

Kaye JJ, Levitt MK, Nevins J, Golden J, Schmidt V (2005) Communicating intimacy one bit at a time. In CHI '05 Extended Abstracts on Human Factors in Computing Systems (Portland, OR, USA, April 02–07, 2005). CHI '05. ACM, New York, NY, 1529–1532. DOI= http://doi.acm.org/10.1145/1056808.1056958

Kristensen M, Kyng M et al (2006) Participatory design in emergency medical service: designing for future practice. In: Proceedings of the SIGCHI conference on human factors in computing systems, ACM, Montréal, QC

Lim Y-K, Pangam A et al (2006) Comparative analysis of high- and low-fidelity prototypes for more valid usability evaluations of mobile devices. In: Proceedings of the 4th Nordic conference on human–computer interaction: changing roles, ACM, Oslo, Norway

Lucas H (1971) Performance evaluation and monitoring. ACM Comput Surv 3(3):79–91

McCarthy J, Wright PC (2004) Technology as experience. MIT Press, Cambridge, MA

Merten C, Conati C (2006) Eye-tracking to model and adapt to user meta-cognition in intelligent learning environments. In: Proceedings of the 11th international conference on intelligent user interfaces. ACM, Sydney

Oulasvirta A, Kurvinen E et al (2003) Understanding contexts by being there: case studies in bodystorming. Pers Ubiquit Comput 7(2):125–134

Preece J, Rogers Y et al (1994) Human–computer interaction. Addison-Wesley, Reading, MA

Puri SK, Byrne E, Nhampossa JL, Quraishi ZB (2004) Contextuality of participation in IS design: a developing country perspective. In Proceedings of the Eighth Conference on Participatory Design: Artful integration: interweaving Media, Materials and Practices – Volume 1 (Toronto, Ontario, Canada, July 27–31, 2004). PDC 04. ACM, New York, NY, 42–52. DOI= http://doi.acm.org/10.1145/1011870.1011876

Ramage M (1998) The learning way: evaluating co-operative systems. Lancaster University, Lancaster, UK

Royce WW (1970) Managing the development of large software systems: concepts and techniques. In: Proceedings of IEEE WESTCON, Los Angeles, CA

Schmidt K (1991) Riding a tiger, or computer supported cooperative work. In: Bannon L, Robinson M, Schmidt K (eds) Proceedings of the 2nd European conference on computer-supported cooperative work (ECSCW '91), 25–27 Sept 1991, Amsterdam, The Netherlands. Kluwer, Norwell, MA, pp 1–16

Sengers P, Gaver B (2006) Staying open to interpretation: engaging multiple meanings in design and evaluation. In: Proceedings of the 6th conference on designing interactive systems, ACM, University Park, PA

Shackel B (1997) Human–computer interaction – whence and whither? J Am Soc Inf Sci 48(11):970–986

Shneiderman B (1979 Dec) Human factors experiments in designing interactive systems. IEEE Computer, Los Alamitos, CA, pp 9–19

Sommerville I (1992) Software engineering. Addison-Wesley, Redwood City, CA

Treu S (ed) (1977) User-oriented design of interactive graphics systems. In: Based on ACM/SIGGRAPH workshop conducted, 14–15 Oct 1976, Pittsburgh, PA. ACM Press, New York, NY

von Neumann J (1945) First draft of a report on the EDVAC, contract No. W-670-ORD-492, Moore School of Electrical Engineering, University of Pennsylvania, Philadelphia, PA. Springer, pp 383–392

Wiberg C (2005) Affective computing vs. usability? Insights of using traditional usability evaluation methods. In: CHI 2005 workshop on innovative approaches to evaluating affective systems, Vienna, Austria

Zimmerman J, Forlizzi J Evenson S (2007) Research through design as a method for interaction design research in HCI. In: Proceedings of the SIGCHI conference on human factors in computing systems, San Jose, CA. ACM Press, New York, NY, pp 493–502

Understanding Users and Their Situation

Ylva Fernaeus, Katherine Isbister, Kia Höök, Jarmo Laaksolahti, and Petra Sundström

Abstract The first step in any design process is to set the stage for what to design and how that should be realised. In terms of user-centred design, this includes to develop a sense of who will be using the system, where it is intended to be used, and what it should be used for. In this chapter we provide an overview of this part of the development process, and its place in the design cycle, and some orienting design challenges that are specific to affective interaction. Thereafter we present a variety of methods that designers may want to consider in actual design work. We end by providing a set of examples from previous and ongoing research in the field, which could also work as inspirations or guiding sources in the early stages in a user-centred design process.

1 Why Do We Need Methods for Getting to Know the Users?

A user-centred design process often starts off with contextual explorations, where the designers attempt to get familiar with the settings and activities in which the system is meant to be used. This is useful to get inspirations both for design concepts and for testing and discussing initial ideas before having to start building anything.

This step roughly corresponds to the two first activities in the user-centred design cycle described in the previous chapter: specifying the context and specifying user requirements and organisational requirements. According to the ISO 13407 standard this includes finding out who the intended users are in terms of preferences, education, and experience among other things; specifying the purpose of the system, the activities that should be supported, and the actions that users are meant to perform; finding out details about the environment in which the system will be used, e.g. what kind of hardware is present and whether it is a public or private environment. Other

Y. Fernaeus (✉)
SICS, Stockholm, Sweden
e-mail: Ylva@dsv.su.se

P. Petta et al. (eds.), *Emotion-Oriented Systems*, Cognitive Technologies,
DOI 10.1007/978-3-642-15184-2_34,

factors that need to be specified include the intended performance of the system in economic and operational terms, communication needs between users, allocation of user's tasks, and last but not least the specific requirements of the user interface.

Contextual explorations are commonly what starts a design process but they are useful throughout the whole process. Many of the methods that we present can be used both in studies prior to the development process and along the way for evaluating and understanding the experience of how users make sense of prototypes and final systems.

Apart from exploring the context of use, projects in the field of affective interaction often start off with rather extensive literature studies in domains that are seen as relevant for the specific project at hand. Studies of film and music theory as well as comic strip design have for instance been of much value for the development of interactive drama, while for other application domains it has been considered more appropriate to make explorations into the studies of linguistics, typography, advertisement, semiotics, colour theory, animation, sociology, biology, and of course psychology. In terms of design practice, results from such studies could work not only for inspiration and generation of new ideas, but also for guidance when selecting between different solutions. By getting familiar with theories of aesthetic experiences, researchers have for instance come up with a range of dimensions such as agency, immersion, and transformation, of relevance for the design and evaluation of experiences with technology (Laaksolahti, 2008).

Recently there has been a renewed interest in design methods where the designers themselves take on a more active role in assessing the qualities of what is being built (e.g. Sengers et al., 2005). As designers get more experience in making and using affective interaction systems, approaches grounded in prior experience of the design team may in fact be very successful in reaching acceptable results. It is, however, important to be aware of the methods' strengths and weaknesses; it can, for instance, be extremely difficult as a designer to imagine (and welcome) user actions and interpretations that go beyond the intended use scenarios.

The importance of knowing their target group is well known amongst designers, and most designers are conscious of the need to design for end users. However, because of a range of constraints in development projects, designers often feel that it is difficult to really do this properly. Instead they get to base their work on data that may be considered less expensive, such as market research findings. However, the aggregated reports of market research do not always capture what people find most important, and moreover, what people *tell* researchers does not always tally with what they actually *do*, when they are studied for real (e.g. Suchman, 1987). So, relying entirely on such sources does in practice often mean that the design still gets heavily biased by prior assumptions made in the design team, which may in fact not at all be in line with the goals and interests of the users.

Recent research suggests that the most effective way of approaching the details of the user experience – and in the end create better and more acceptable technology – is by actually getting to *know* target users, through direct engagement. Through such engagement, designers more likely get to develop a sense of what the users may experience, what in the literature is referred to as 'empathic' approaches to design. However, how they should actually go about in this task depends largely on the characteristics of the project at hand, such as its project goals, time frame, budget, as well as the already existing relationships between the design team and their intended user group.

2 Some Orienting Design Challenges

Different designers naturally enter a design process with quite different ideas of what the overall result should be and what methods would be most appropriate. So apart from directly approaching the users in their design efforts, the design team may also need to agree on what kind of interaction they are envisioning. A set of orienting design aims are then useful to keep the team focused and able to select which data pertains to the situation at hand and what specific methods to use.

When designing for affective interaction there is a range of specific challenges, or themes, which could be especially useful to orient the team in such respects. These themes could be summarised as three core challenges.

The first challenge has to do with emotions as essentially subjective, difficult to capture, and changing over time. This is important to consider in the design process, especially in trying to capture how users interpret the interaction and how the emotions that they experience may be understood. It is also important to acknowledge that affective interaction often allows users to build their own interpretation out of the signs and signals emitted by the system. It may therefore be a substantial design challenge to bridge the known qualities of the software and hardware with the interpretational complexity of users' subjective experiences. An important goal is therefore to find methods that may address the user experience as an open-ended, subjective, interpretative process.

The second challenge has to do with emotions that are regarded as essentially embodied. From Damásio (1994) we learn the importance of seeing emotion as experienced and as the effects of a close intertwinement of body and mind. Hence a lived affective experience requires careful tuning of parameters that influence a users physical interaction with the system, such as balancing the timing between user actions and system responses. Affective interaction systems are therefore often designed not only to 'read' off the body and display representations, but also to encourage users to act – make gestures, new postures, etc. At the same time, we have seen a strong development of tangible interfaces, new game consoles interaction devices, and various body sensors, which can potentially tap into the bodily aspects of emotional experiences. Therefore, our design efforts should be directed towards methods that allow physical context and bodily aspects to be more directly involved and analysed.

The third challenge has to do with the intended use setting, which with affective systems often is personal and private in character. This means that while directly engaging with potential users in the intended context is ideal, this is sometimes difficult – or even impossible – to achieve for practical reasons. It might for instance be difficult to observe people in their homes for longer periods of time or to engage with people that are suffering from illness. This can be useful to have in mind when selecting the method of inquiry, since methods that place the user and usage at core are a key component in achieving the design goals of affective applications.

This means that all methods that we propose in one way or the other aim at capturing *subjective* experiences, *bodily and physical* practices, in ways that *do not intrude or disturb too much of the users' personal life*.

3 Selection of Methods

The methods presented below all bear relevance to the specific problem of 'getting to know' the users and their experiences, sometimes in the literature referred to as *empathic* design approaches, which we see as especially useful in the design of affective interaction. There is a range of further methods that we do not discuss here, but that could also be very useful in this respect. This includes various kinds of surveys, interviews, focus groups, and brainstorming sessions together with end users. The reason we exclude them here is our focus on methods that have been argued to be specifically useful for affective interaction in recent research.

3.1 Ethnography and Bodystorming

If the environment where the technology is meant to be used is specified, it is useful to perform some form of analysis of that specific environment (e.g. at a library, in the subway station, or in an office). Ethnography is a set of methods for exploring practices in an existing use context, where the ethnographer participates in people's lives for an extended period of time. By watching what happens, listening to what is being said, and asking questions, this kind of exploration may shed light on the issues that are the focus of the research. Johnson (2000, p. 111) defines ethnography as "a descriptive account of social life and culture in a particular social system based on detailed observations of what people actually do".

An ethnographic investigation typically ends with a narrative description of the user group and the activities they engage in. In such a narrative, all aspects of how users conceptualise their world and their activities are taken into account. Their tools, their cultural habits, their conceptualisations of the activity, and the way they experience the activity are all described. This provides for a rich, thick description of the situation at hand, allowing the designers to imagine how their newly designed artefact will be made part of the activity by the intended user group.

Ethnography can be especially useful to the design of affective or experience-oriented systems as the approach requires that the observer has to partly become a participant to really understand it. Or, as put by McCarthy and Wright (2004), "It tries to understand what it feels like to be the other by simultaneously observing and participating in the life of the other, writing or 'translating' that bodily, intellectual, and emotional in reflexive fieldnotes, and finally writing an analysis in a genre that gives expression to the other's experience." A good ethnographer will create an embodied, intellectual account of the activities people engage in and will try to communicate that experience to others. From this basis, the designer can more emphatically create applications that will be meaningful to the targeted user groups.

Bodystorming is a light-weight variant of ethnography that is becoming increasingly popular in design projects (Oulasvirta et al., 2003). The principle is to visit the environment where the system is meant to be used, and to use that experience as an input to design decisions. By putting oneself in the actual use setting, and by interacting with and observing users in the intended contexts, if even for a short while, several aspects that are not available otherwise become salient and possible to try out, both by physical action and by asking the people in place.

3.2 Cultural Probes and Technology Probes

Cultural probes (Gaver et al., 1999) method is used to let a selected group of users reflect on their everyday experiences, for instance, of certain media content and other situations related to the perception and interpretation of technology (already existing or one that is planned to be built). The original intention of this method was to gain inspirational knowledge from people without disturbing them by entering their lives as with common ethnography. The probes let users communicate with designers on their own terms.

When using this method participants are typically given a range of materials, such as a diary, a disposable camera, and postcards, together with a set of tasks and questions, that make participants reflect over some aspect of their life. The method has become very popular within the field of human–computer interaction (HCI), and variants of the method have made use of digital equipment, using, for example, mobile phones and text messages to prompt participants with questions (Hulkko et al., 2004). The probes given to users shall preferably allow them to make their own decisions of how to use and treat them. In turn designers are allowed to regard some materials as more important than others in their process.

Technology probes were proposed by Hutchinson and colleagues (Hutchinson et al., 2003). Their idea was to place partly unfinished technology in people's homes and then study how participants made sense of it. Kaye and colleagues have made use of a combination of the cultural and technology probes as a method for exploring how to design for intimacy (Kaye et al., 2005). Their aim was to make users reflect more on their relationships and communication of intimacy and thereby gain insights into the role that minimalistic communication channels can play in helping to keep partners close. They gave users a tool, the VIO system, serving as a technical probe, and at the same time cultural probe material that they could use to document their experiences with this system.

Recently critical voices have been raised against the extensive usage and the alterations to the cultural probes method (Boehner et al., 2007). The method as Gaver and colleagues created was aimed to collect information about users and their everyday practices without any preconceptions: information that could inspire future design areas, not a method that would capture any statistical data about users. With that in mind, the method's costs can be fairly low, depending on the material given to users; as little as a diary and a few postcards can result in rich data. People are often surprisingly willing to share quite intimate, personal aspects of their lives when confronted with a probe.

3.3 In Situ Informants

The aim of an 'in situ' exploration is to move beyond the laboratory environment where often simplistic scenarios are used to get users emotionally involved. The in situ method is used to enter and explore the subjective and distributed experiences of use, as well as how emotional experiences with technology unfold in everyday practice. The method is a combination of three well-established user-centred design methods: experience clips (Isomursu et al., 2007) and the cultural and technical probes methods (see above).

Using the in situ method, subjects are provided with packages including a technical probe, for instance in the form of a research prototype, together with some cultural probes to document their experiences with the particular prototype (Sundström et al., 2007). The packages also include material to give to a close friend who will give a more outside perspective on the user's activities. As the aim of the in situ method is to look for how emotional experiences unfold in everyday practice, the spectator needs to be someone who has a deeper understanding of the personal expressions and the body language for the specific user. Emotional body language can be highly individual and hard to interpret unless you know someone well, even though you are a professional.

The study method helps enter and explore the subjective and distributed experiences of use, as well as how emotional communication unfolds in everyday practice. When the in situ method was applied to the evaluation of the eMoto system (Fagerberg/Sundström et al., 2004), it pointed to the importance of supporting the sometimes fragile communication rhythm that friendships require – expressing memories of the past, sharing the present, and planning for the future. In the study, it turned out that emotions are not singular states that exist within one person alone, but can permeate the total situation, changing and drifting as a process between the two friends communicating. This particular study also provided insights into the under-estimated but still important physical, sensual aspects of emotional communication.

3.4 Laban's Movement Analysis

Laban analysis is a method used in dance theory to describe the physical dimensions of different movements. Rudolf Laban was a famous dance choreographer, movement analyser, and inventor of a language for describing the physical dimensions of different movements (Davies, 2001). Laban's theory is oftentimes referred to as Laban Movement Analysis (LMA) and is composed of five concepts: body, space, effort, shape, and relationship.

Shape can be described in terms of movement in three different planes: the table plane (horizontal), the door plane (vertical), and the wheel plane, which describes sagittal movements. Horizontal moments can be somewhere in-between spreading and enclosing, vertical movements are presented on a scale from rising to descending, and sagittal movements go between advancing and retiring (Fig. 1).

Effort comprises four motion factors: space, weight, time, and flow. Each motion factor is a continuum between two extremes. Figure 2 shows the graphical representation Laban uses to express effort.

Sundström et al. (2007) describe how the two components of effort and shape can help in understanding the underlying dimensions of affective body behaviours in interaction design. Effort and shape were chosen because they were found to characterise best the emotional expressions contained in gestures: shape describes the changing forms that the body makes in space, while effort involves the felt qualities of the movement and the inner attitude towards use of energy (Zhao, 2001).

Figure 3 shows still pictures of an actor acting out nine different emotions: excitement, anger, surprise-afraid, sulkiness, surprise-interested, pride, satisfaction,

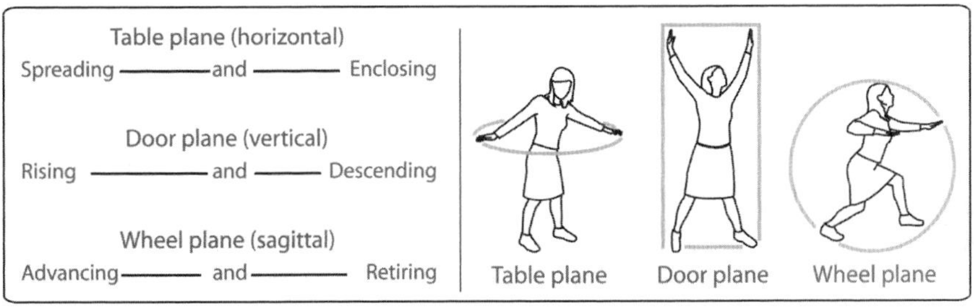

Fig. 1 Terminology to describe shape, adapted from Davies (2001)

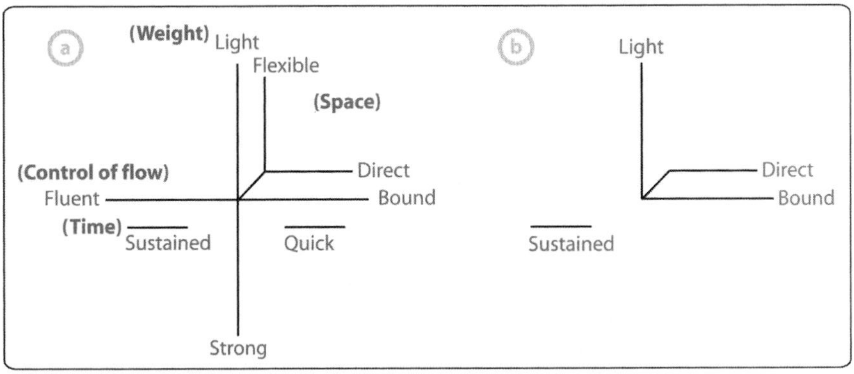

Fig. 2 (**a**) Laban's effort graph, (**b**) an example effort graph of inserting a light bulb (Laban and Lawrence)

Fig. 3 Example body movements for nine different emotions: excitement, anger, surprise-afraid, sulkiness, surprise-interested, pride, satisfaction, being in love, and sadness

sadness, and being in love. It is important to note that even though we here show still pictures, the movements analysed were the overall picture of a longer sequence of movements.

Using Laban's analysis of shape and effort on these nine examples of emotional body language the shape of these movements can be expressed in the following terms:

- Excitement – extremely spreading, rising, and advancing movements
- Anger – somewhat spreading, rising, and advancing movements
- Surprise-afraid – enclosing, somewhat descending, and retiring movements
- Sulkiness – enclosing, somewhat rising, and retiring movements
- Surprise-interested – somewhat spreading, neutral in the vertical plane, and advancing movements
- Pride – somewhat spreading, rising, and somewhat advancing movements
- Satisfaction – neutral in all planes of movements

- Sadness – enclosing, descending, and retiring movements
- Being in love – somewhat spreading, somewhat rising, and somewhat advancing movements

The analysis by Sundström et al. shows that emotions more regarded as negative are more enclosing and not so strong in their expressions. A notable exception is anger which is more like extreme excitement in terms of effort and shape. More positive emotions were found to be more advancing and outgoing in their movements than more negative emotions. Similar results were obtained by Paiva et al. (2003).

In spite of its long history, the LMA method is not established in HCI. The costs of using this method are low but the amount of work required to analyse the material is extensive. Because of the subjective nature of the analysis, it is advisable to have multiple persons analyse the same material. Even though the method could be seen as more appropriate for concrete design in the later design process, as with the probes, it can be modified to be used more generally to verify dynamics of certain movements and physical postures in interaction, as well as for evaluation purposes.

It is important to remember that this method does not lend itself easily to tabulating emotional expressions with one-to-one mappings of movements to emotions; but it can help to identify underlying characteristics of emotional body movements.

4 Learning from Others

Systems have to be built in domain-specific ways, and there is no single set of design guidelines that can be applied to all design problems that may arise. But if the goal is to design for users' personal and emotional engagement with technology, there are a number of experiences from others that could be valuable to look at and potentially brought as input to the project hand.

How knowledge can be gained and shared among designers has been increasingly discussed in HCI. Based on the theories such as that of Schön (1983), the experiences of specific cases/situations by professional practitioners are not regarded as 'mysterious', but as amenable to be actively and consciously reflected upon. Through such a reflective process, practitioners gain knowledge that can be shared and used as resource in new design situations.

More and more work in the field of affective interaction has reported on systems and studies that could work as suggestions for how to build up emotional engagement with new technology. Applying such knowledge onto the specific case at hand is not and cannot be a mechanistic process, but can inform designers developing new designs. In this section, we list a few such experiences and theories that have been developed and which can serve as guiding resources in design processes.

4.1 Informational or Interactional View

In affective interaction, two perspectives on the overall interaction have crystallised: the *informational view* and the *interactional view*. The choice of one of these perspectives in a design process will have an overarching influence on the design of the final product as well as the methods chosen for formative and evaluative processes.

In artificial intelligence, the approach to affective computing often focuses on what one might call the 'informatics of affect', in which emotions are treated as discrete units of information. Emotions are then analysed, classified, discretised, and formulated as units, the purpose of which is to inform cognition or be communicated. The often used integrative cognitive theory of emotion by Ortony et al. (1988), for example, defines emotions in terms of a set of basic types, based on the cognition or reasoning that people usually connect with them. All applications based on this 'OCC model' design emotion as a phenomenon that can be measured, isolated, and then used as a basis for how to make a system respond. This approach often means that the user is only in indirect control of what emotions are displayed or interpreted. The overall design goal is to make the system more intelligently and efficiently help users to achieve their goals – be it learning, communication, or some other design goal. While defining, classifying, creating logical structure for emotions can be useful in certain cases, there may be goals that cannot be achieved with this approach. Such goals include experience-oriented goals, ludology (engagement in play), more complex communication between friends and partners, and conscious reflections on your own emotional and behavioural processes. For these goals, the informational mindset is in danger of missing a fundamental point: affect is not just a formal, computational construct, but also a human, rich, complex, and ill-defined experience. Rationalising it may be necessary to make it computable, but an affective system that truly inspires and incorporates human emotion must include a broader cultural perspective, in which the elusive and non-rational character of emotion does not need to be explained away (Sengers et al., 2002). From this perspective, computation may be used, not to acquire and reason about user's emotional states, but rather to create experiences of affect during interaction.

The interactional view accordingly sees emotions as processes spread over people and situations – constructed in a moment-to-moment fashion (Sengers et al., 2002; Höök et al., 2008). Designs that are built from this perspective assume that the meaning of an emotional process is created by people and that affective interactive systems should be such that users are encouraged to negotiate these meanings themselves. It is not something a designer can cater for entirely, but instead, is completed, *lived*, by the experiencing person. In short, an interactional approach adheres to the following principles in designing for affect:

1. The interactional approach recognises affect as an embodied social, bodily, and cultural product.
2. The interactional approach relies on and supports interpretive flexibility.
3. The interactional approach is non-reductionist.
4. The interactional approach supports an expanded range of communication acts.

5. The interactional approach focuses on people using systems to experience and understand emotions.
6. The interactional approach focuses on designing systems that stimulate reflection on and awareness of affect.

4.2 The Affective Loop

In what has been referred to as the affective loop (Sundström et al., 2007) successive physical interactions with affective systems may invoke further reactions and emotional experiences. This means that the system does not try to infer users' emotional states, but instead to involve them in an interactive process. Users may then choose as to how and to what extent they should engage in, and in what ways to make the interaction unfold. In this scenario, the system is just staging the scene for activity. The affective loop could be understood as an interaction process where

- the user expresses their emotions through some physical interaction involving the body, for example, through gestures or manipulations of an artefact;
- the system (or another user through the system) then responds through generating affective expression, using, for example, colours, animations, and haptics;
- this in turn affects the user, making the user respond and step-by-step feel more and more involved with the system.

In a rich interactive system, these three activities take place not as a strict sequence, but largely in parallel. This loop can be in real time with immediate feedback, as in systems such as eMoto (Fagerberg/Sundström et al., 2004), SenToy (Paiva et al., 2003), or certain physical computer games, or a long-term process that invokes a deeper reflection and more lasting change, as in the use of the Affective Diary (Lindström et al., 2006).

4.3 Designing for Ambiguity

Most designers would probably see ambiguity as a dilemma for design. Gaver and colleagues, however, look upon it as "a resource for design that can be used to encourage close personal engagement" (Gaver et al., 2003). They argue that in an ambiguous situation people are forced to make a personal interpretation of what is happening. Moreover, the everyday world is inherently ambiguous and most things will have multiple meanings depending upon how we see them, which has been commonly and successfully exploited in arts, where ambiguous meanings are found to contribute to the aesthetic experience.

Based on these ideas, Gaver and colleagues created a range of systems where the meaning of the technological artefact could be interpreted in multiple ways. Their goal was to be evocative rather than didactic, and mysterious rather than obvious. It should be noted that ambiguous design does not mean fuzzy or inconsistent design – rather that it may give rise to multiple interpretations. Through this attitude, they broke with the tradition in HCI to rely entirely on understanding and 'intuitivity' as the basis for interaction. Instead, their focus was on the interpretative relationship between people and artefacts.

4.4 Familiar and Open Surfaces

From a user perspective, a system for affective interaction is not only a tool for completing a task, but an open surface that can be (actively or passively) filled with content. Thereby, the user takes an active part in shaping its functionality. Such systems may allow the functionality and content to drift over time and be flexible enough to allow for different norms, practices, and behaviours to arise.

What an 'open' design may entail is to leave certain surfaces in the interface available to users so that they can fill them with their own meaning and patterns of behaviour. But such surfaces might be very hard to understand unless there are elements in them that make them familiar (Höök, 2006). Finding the right degree of openness when it comes to possible interpretations of the design or possibilities to shape the system is a tough challenge for designers. A too 'closed' design leaves little room for interpretation and appropriation while a too 'open' design runs the risk of not allowing for meaningful interaction and interpretation.

For instance, there are systems that capture users' facial expressions or body postures in real time and represent emotional states using avatars or robot behaviours. Such systems run the risk of overly constraining the possibilities for the user to read more into the picture than what is there. If the smiley or avatar looks happy, that might make the user think that the other person is happy – curtailing any richer complexity of emotion interpretation. A more abstract design that builds on the dynamics of emotions as experienced by our physical bodies is adopted with the eMoto system. Here, colours, shapes, and animations attempt to mirror users' physical gestures addressing their inner experience of emotions (Fagerberg/Sundström et al., 2004).

5 Challenges to Designing for Affective Interaction

At the beginning of this chapter, we outlined three challenges for affective interaction when it comes to the first step in the design process where we aim to set the stage for what to design and how that should be realised. The first challenge concerned the difficulty in capturing emotional experiences as they are subjective and unique experiences – something we addressed through a range of design concepts, such as the interactional perspective on emotion, familiar and open surfaces in the design, and the affective loop concept. We also addressed it through bringing in design methods such as ethnography, bodystorming, and cultural and technical probes.

The second challenge concerned the embodied nature of emotions, where we need to address not only its cognitive aspects, but also its social and bodily parts. Here, the affective loop design concept is an interesting example of how interaction can be a persuasive bodily and cognitive process. In terms of methods, Laban analysis, and possibly bodystorming, can be useful. But more research is needed to find relevant design concepts and design methods to properly address how to design for this kind of involvement.

The third challenge has to do with the intended use setting, which with affective systems is often personal and private in character. Methods such as cultural and technical probes, in situ informants, and in some cases ethnography might help us get closer to sensitive settings without infringing on people's lives or privacy.

References

Boehner K, Vertesi J et al (2007) How HCI interprets the probes. In: Proceedings of the SIGCHI conference on Human factors in computing systems, ACM, San Jose, CA

Damásio AR (1994) Descartes' error: emotion, reason and the human brain. Grosset/Putnam, New York, NY

Davies E (2001) Beyond dance: Laban's legacy of movements analysis. Brechin Books, London

Fagerberg/Sundström P, Ståhl A, Höök K (2004) EMoto – emotionally engaging interaction. J Pers Ubiquit Comput (Special Issue on Tangible Interfaces in Perspective, Springer) 8(5):377–381

Gaver WW, Beaver J, Benford S (2003) Ambiguity as a resource for design. In: Proceedings of the conference on Human factors in computing systems (CHI'03), Ft. Lauderdale, FL

Gaver WW, Dunne T, Pacenti E (1999) Design: cultural probes. Interactions 6(1):21–29

Höök K (2006) Designing familiar open surfaces. In: Proceedings of the 4th Nordic conference on Human-computer interaction changing roles – NordiCHI '06 (presented at the 4th Nordic conference, Oslo, Norway, 2006), pp 242–251, http://portal.acm.org/citation.cfm?id=1182475.1182501&coll=DL&dl=GUIDE&CFID=112946399&CFTOKEN=73214820

Höök K, Ståhl A, Sundström P, Laaksolahti J (2008) In: Interactional Empowerment, To be presented at ACM SIGCHI conference Computer-Human Interaction (CHI2008), ACM Press, Florence

Hulkko S et al (2004) Mobile probes. In: Proceedings of the third Nordic conference on Human-computer interaction. ACM, Tampere

Hutchinson H, Mackay W, Westerlund B, Bederson BB, Druin A, Plaisant C, Beaudouin-Lafon M, Conversy S, Evans H, Hansen H, Roussel N, Eiderbäck B (2003) Technology probes: inspiring design for and with families. In: Proceedings of the SIGCHI conference on Human factors in computing systems, ACM, NY, pp 17–24

Isomursu M, Tähti M, Väinämö S, Kuutti K (2007) Experimental evaluation of five methods for collecting emotions in field settings with mobile applications. Int J Hum Comput Stud 65(4):404–418

Johnson AG (2000) The Blackwell dictionary of sociology, 2nd edn. Blackwell, Oxford

Kaye JJ, Levitt MK, Nevins J, Golden J, Schmidt V (2005) Communicating intimacy one bit at a time. In: CHI '05 extended abstracts on Human factors in computing systems, ACM, Portland, OR

Laaksolahti J (2008) Plot, spectacle, and experience: contributions to the design and evaluation of interactive storytelling. Unpublished PhD-thesis from Department of Computer and Systems Science, Stockholm University, Sweden, Feb 2008

Lindström M, Ståhl A, Höök K, Sundström P, Laaksolathi J, Combetto M, Taylor A, Bresin R (2006) Affective diary – designing for bodily expressiveness and self-reflection. In: Extended abstract CHI'06, ACM Press, Montréal, QC

McCarthy J, Wright P (2004) Technology as experience. The MIT Press, Cambridge, MA

Ortony A, Clore GL, Collins A (1988) The cognitive structure of emotions. Cambridge University Press, Cambridge, MA

Oulasvirta A, Kurvinen E, Kankainen T (2003) Understanding contexts by being there: case studies in bodystorming. Pers Ubiquit Comput 7(2):125–134

Paiva A, Costa M, Chaves R, Piedade M, Mourao D, Sobral D, Höök K, Andersson G, Bullock A (2003) SenTOY: an affective sympathetic interface. Int J Hum Comput Stud 59(1):227–235

Schön DA (1983) The reflective practitioner. Basic Books, New York, NY

Sengers P, Boehner K et al (2005) Evaluating affector: co-interpreting what "Works". In: CHI 2005 Workshop on innovative approaches to evaluating affective systems, Portland, Oregon

Sengers P, Liesendahl R, Magar W, Seibert C, Müller B, Joachims T, Geng W, Mårtensson P, Höök K (2002) The enigmatics of affect. In: Conference on Designing Interactive Systems (DIS), London, England, Jun 2002

Suchman L (1987) Plans and situated actions: the problem of human-machine communication. Cambridge University Press, New York, NY

Sundström P, Ståhl A, Höök K (2007) In situ informants exploring an emotional mobile messaging system in their everyday practice. IJHCS 65(4):388–403

Zhao L (2001) Synthesis and acquisition of Laban movement analysis qualitative parameters for communicative gestures. CIS, University of Pennsylvania, Philadelphia, PA

Generating Ideas and Building Prototypes

Katherine Isbister, Kia Höök, Petra Sundström, and Jarmo Laaksolahti

Abstract Design always involves the difficult step from seeing users and their activities to inventing something new that will make sense to them. In this chapter we turn to framing of the problem in such a way that the design process can start and the first prototypes can be constructed. Following a prototype-driven approach, we first provide a discussion of how to frame a problem, drawing on information gathered by methods presented in the previous chapter. We then show not only how to generate ideas for prototypes that would aid to validate a potential solution to that problem, but also methods to actually build and validate such prototypes. Finally, we discuss specific challenges related to affective interaction. The intention pursued with a prototype-driven approach is not to design a product, but a research vehicle for exploring a specific research idea. However, for one to say something of how successful a solution has been, a scenario for such prototype needs to be as realistic as possible, almost as if one was to design a product.

1 Framing the Problem

Once researchers have gathered information in the various ways described in the previous chapter, they must integrate all this information into a helpful framing of the problem space for themselves and all those who will collaborate on the project. Some of the information may be easy to parameterise and can provide specific guidelines for how to proceed (e.g. the Laban analysis of gesture, and using this to guide the design of a gestural interface to a mobile phone application – see Sundström et al., 2007). However, the rich, varied, and even contradictory information that can emerge from techniques such as the cultural probes requires careful and creative synthesis to properly inform the design process. Designers have evolved a range of techniques that can be used to preserve the richness of this sort of user

---

K. Isbister (✉)
Center for Computer Games Research, IT University of Copenhagen, Copenhagen, Denmark
e-mail: KIsbister@itu.dk

P. Petta et al. (eds.), *Emotion-Oriented Systems*, Cognitive Technologies,
DOI 10.1007/978-3-642-15184-2_35,

data and to incorporate it into design thinking. Some examples are mood boards, personas (Cooper, 1999), and user scenarios.

Mood boards[1] are groups of images and inspirations that designers collect to help them to envision the mood that they hope to create in the target user community through the product itself. This technique can also be used to cluster artefacts that users have contributed through cultural probes or other means that help remind the designer of the feelings that users have around the activity space for which they are designing (Fig. 1). All those who participate in the project can gather around and discuss these clusters of artefacts from users and can keep coming back to them as the team brainstorms, seeking confirmation or contradiction of emerging patterns and re-engaging the material for further inspiration.

Personas are another tactic for aggregating and framing the problem based upon a rich set of user data. Designers use the information at hand – whether gleaned from surveys, interviews, aggregate data available about the user community from prior products, and/or direct observation of target users – and create imaginary people who are composites of key features of the user community's wishes and constraints. In commercial contexts, these personas may be carefully weighted and tied to specific sub-demographics in the target user group. Inter-related personas can be constructed that can help the designer to get at how communities of users will share a product – e.g. the initial user as well as friends and family who share the application.

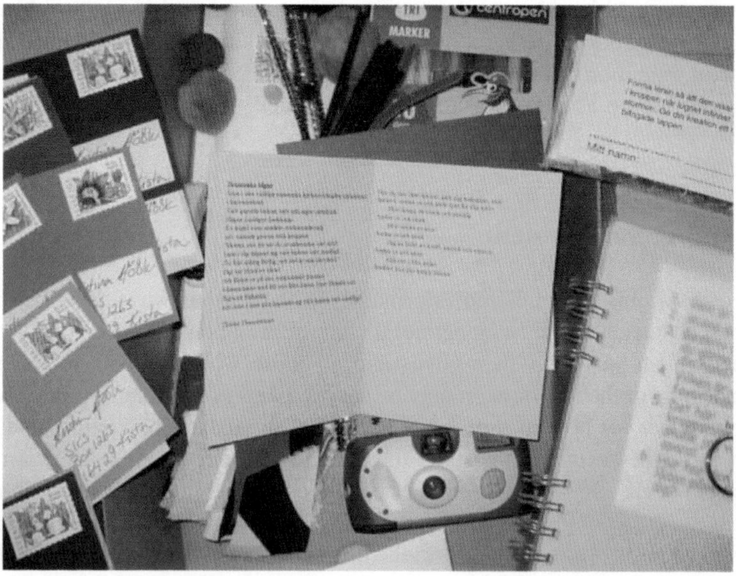

Fig. 1 An example of cultural probe materials for creating a moodboard

[1] http://www.lifeclever.com/5-reasons-to-design-with-mood-boards (accessed May 30, 2010)

There is a rich body of information in the commercial design community about using personas.[2]

A bit further into the process of idea generation user scenarios can help to even closely specify a design. User scenarios insert sample users (such as the personas described above) into carefully constructed walk-through descriptions of use of the system that will be designed. When considering in greater detail how a system should be designed, such scenarios allow designers to better imagine how users would conceptualize and use it. User scenarios can begin at a pretty high level of abstraction (e.g. 'Jane would like to rent a video online quickly, to be viewed tonight at home') to very detailed sub-areas of the target interface. For an excellent summary of a range of scenario tactics, see Benyon et al. (2005, Chapter 8).

2 Idea Generation Methods

As a next step in a prototype-driven design process one needs to generate actual ideas for prototypes that will aid in validating a potential solution to the now framed problem. Idea generation method is probably mostly thought of as regular brainstorming where a design team sits down and simply starts to talk of great new ideas, but there are in fact a range of methods for making this next step, potentially seen as mountain high, much less magic than what it most often is experienced as, looking at someone else's project.

2.1 Brainstorming Methods

Brainstorming can be used at all stages in a design process. One can brainstorm around methods for evaluation, research ideas in general, interaction models, etc. The list is endless and the various ways to set up brainstorming sessions are probably even more numerous. As examples of what a brainstorming session can look like, we present two well-established methods from product design: Random Words and Six Thinking Hats (De Bono, 1985). These methods are very much idea driven and therefore very useful when pursuing a prototype-driven approach. Even when designing a research prototype for exploring an idea rather than a product, it is important to stay in the realm of the possible to be able to assess the quality of different designs. When having a less successful scenario or prototype, one's users will most likely be concerned with the problem areas more than the overall design idea.

Random Words is used to come up with novel, inspiring, thought-provoking combinations of words from specified categories, for example, emotions, techniques for sensing emotions, and places. To start with, a group of words under each category is required, for example, 'angry', 'sad', 'happy', for the emotion category. These are

[2]See http://www.boxesandarrows.com/search?q=personas (accessed May 30, 2010)

either provided by the session leader or collected as a start-up activity for the brainstorming session. The words are written on separate pieces of paper, such as post-it notes, and placed upside down in three piles. Starting the brainstorming, the first note in each pile is turned and shown to all participants. The idea is then to brainstorm for a few minutes around what an application using that specific combination of words could be, before going on to the next combination of words. Random Words is used to come up with a range of ideas. The method is very low in cost, including time requirements. As always in brainstorming activities, it is important not to be afraid of any bad or outrageous ideas: it may well be that ultimately parts of a few of such 'bad ideas' contribute and together form a really good one. The aim is to have the mind under stress go in new directions, directions where it usually does not go. The combinations of three randomly picked words shall set up a helpful framework when having to be creative.

The same holds for the Six Thinking Hats method which is more for evaluating and developing already existing application ideas, ideas that perhaps originated from the Random Words method. In the Six Thinking Hats method, each idea is reflected upon from five different viewpoints represented by five differently coloured hats that the participants 'put on': facts and information (white hat), optimism (yellow hat), opinions and thinking (red hat), cautiousness (black hat), and creativity (green hat). These viewpoints represent five of the different hats. The last hat (blue hat) is given to the person who regulates the process. The hats can be represented by a coloured slip of paper placed in front of each member of the brainstorming design team. The participants take turns with the hats and have to act within the limitations of their current viewpoint: the wearer of the yellow hat is only allowed to be optimistic, the green hat has to be creative, etc. While wearing the white hat, one has to stick to facts and information, such as 'Bluetooth technology does not work over distances exceeding 6 m.' Factual knowledge in a new design idea can pose a formidable challenge that can be tackled either by granting Internet access to the person carrying the white hat or by informing all participants beforehand of the ideas to be discussed, so they can prepare themselves. Similar considerations hold for the person wearing the red hat, who has to be up to date on people's opinions and thinking, for example, 'Women in their forties generally think new technology is hard to learn.' Another way to work the group around these issues is to allow people to lie and make up stories and facts on the fly, e.g. 'Women in their forties generally like blue things', an approach that works surprisingly well.

2.2 Bodystorming

As emotions are not only a cognitive, but also a physical experience, a good way of testing ideas for functionality is to actually act out the interaction idea physically together in the design team. This can be done before the system even exists through so-called bodystorming techniques (Oulasvirta et al., 2003), essentially a simple way to act out a scenario as in a role playing game or an improv theatre. As characterised in Rodriguez et al. (2006, p. 964),

> Unlike brainstorming, bodystorming is the transformation of abstract ideas and concepts into physical experiences. Fun and tactile, this approach allows us to investigate different qualities that an idea may have when applied in a physical setting. It enables rapid iteration and development of ideas and relationships through a dynamic, continuous and creative process of trial and error.

In bodystorming, you typically brainstorm in situ, that is, in the location that will typically be the place where the system is aimed to be placed and used. If you are designing for a train, you spend time on the train, brainstorming together with your team, and any ideas that come up, you act out there and then on the train. As a consequence, you quickly get a grasp on how well your system will interact with and be integrated with all the other aspects of the environment for its usage. For example, assume that the aim is to design a mobile messaging system, where users should be allowed to express themselves physically when sending messages to their friends. Acting out the kinds of gestures one comes up with in the various settings where mobiles are typically used will quickly lead to the discovery that large gestures with the phone will feel silly when in public spaces. However, for some such applications it may ultimately turn out that interaction patterns that had been thought to be too extreme are in fact the ones pushing the community forward (Sundström et al., 2007).

3 Prototyping Methods

To actually build the prototypes may be yet another mountain to climb. In many cases one needs to work with new and challenging interaction techniques and not only software but also hardware that needs to be adjusted to the software and also perhaps fitted into a suitable package. To come up with a good software design can be difficult enough, but to also have it working with custom hardware requires constant validation and redesign. This holds especially when working with emotionally engaging and emotionally involving prototypes. Therefore, we discuss not only methods for final evaluation but also validation methods for all steps of a system design process, from paper sketch to digital technology.

3.1 Paper Prototyping

Paper prototyping is a method used for early usability testing in the design process once the appearance of the future prototype has been identified but before any actual code is written. The method is also well suited for workstation and laptop applications, but mainly used when designing for smaller mobile displays (Rettig, 1994). The idea is to draw all potential screen displays on pieces of paper and let a user navigate them. The analysis of the user interactions informs the adaptation of the screen displays. It is important that all buttons, interactive areas, and help texts are represented, to make the experience for the user as close as possible to the experience with an actual physical device. The main aim is to locate areas where the user runs into interaction difficulties, e.g. due to misunderstandings of feedback or other signals. While a paper prototype cannot fully replace the real experience with a working prototype, it is still a very rewarding method and a valuable design step to take.

3.2 Staged Lived Experiences

To take the paper prototyping method closer to the experience with a functional prototype, it can be combined with aspects of the bodystorming method through the creation of a *staged lived experience* (Iacucci et al., 2002). By letting the user experience the paper system in the environment where the real prototype is to be used, it is possible to approach and evaluate the actual experience rather than focusing exclusively on the usability of the user interface. While the paper prototyping method is less well suited for more tangible and alternative interaction models, the staged lived experience method can cover also these kinds of systems. By using just parts of a future system, such as a biosensor bracelet or a camera, it is possible to improve users' understanding of what that future system actually is going to be like and also of how it is, e.g., going to feel to use and wear it in public. Exposing users to such experiences also facilitates their participation in focus groups and other more informed activities. Not only users but also designers themselves can get inspirational experiences from such experiments. The main idea with the staged lived experience method is to play, pretend, and experience bits and pieces of a future system out in the wild and not in the laboratory environment where it usually is hard to experience the everyday practice.

The paper prototyping in particular is an extremely cheap method. The costs of the staged lived experience method depend on the probes, but for single-user systems the user study can be set up with one user at a time and also for multi-user systems there are ways to play and pretend that help contain costs. However, both methods are rather time consuming. For paper prototyping, the whole system needs to be thought through in detail and then sketched on pieces of paper. To set up a staged lived experience is of course even more laborious than to create a pretend setting in the laboratory but most often worth every second of it. Staged lived experience is a valuable method at all stages of a design process; in contrast, it can be argued that for final evaluation a laboratory setting is no longer acceptable, especially for affective interaction systems where a laboratory environment works more with created and staged emotions than emotions that occur in real life practices.

In their review of experiences gathered with the deployment of the related method of experience prototyping in a number of real design projects, Buchenau and Fulton-Suri (2000) illustrate how it contributed to developing an understanding of essential factors of an experience by simulating important aspects of the whole or parts of the relationships between people, places, and objects as they unfold over time; to the exploration and evaluation of ideas, providing inspiration, confirmation, or rejection of ideas based upon the quality of experience engendered; by producing answers and feedback to designers' questions about proposed solutions in terms of 'what would it feel like if ...?'; and in communicating issues and ideas, enabling direct engagement in a proposed new experience and thereby providing common ground for establishing a shared point of view.

3.3 Wizard of Oz

Designers of interactive technology often face what is best described as a chicken and egg problem: in order to design the technology they need to know something about how it will work when it is finished. By using an iterative design process where designs are repeatedly evaluated against established goals an understanding of how a system will work is gradually assembled. A number of different methods can be used during the design process to construct such an understanding. Which ones to use will depend on the particular project at hand. A method that has proved to be particularly useful when designing for very complex interaction technologies, or when entering new domains such as affective interaction, is the Wizard of Oz (WoZ) method.

The name of the method refers to the wizard in L. Frank Baum's novel *The wonderful Wizard of Oz*, who manually operated complex machinery from behind a curtain to appear more powerful. Within the human–computer interface research, the name has come to designate an iterative design method in which a human (the wizard) simulates the behaviour a computer system under development would have if it was fully functional. WoZ studies are usually performed in laboratory settings where the wizard operating the system and the participant testing it are in different rooms to maintain the illusion of a fully functional system. Sometimes participants are informed about the system status, i.e. that it is a simulated system with a human acting behind the scene, and sometimes they are not, in order to encourage natural behaviours. If participants are not made aware that the system is simulated beforehand, for ethical reasons it is important to ask for their informed consent after completion of the study, offering them a chance to withdraw their data from the study.

Using WoZ can be particularly helpful in circumstances where interpreting user input is a difficult task. In a traditional point-and-click interface one can be fairly certain that when users click a button, that is what they intended to do although the effects of pressing the button may not always be what they intended. However, when working with interfaces that include modalities such as natural language, gestures, postures, and emotions, interpreting a user's intentions or state of mind is not always as straightforward. For instance, does frowning mean that the user is annoyed or merely focused? Or does gesturing in a certain direction mean 'look there' or 'go there'?

Many natural language applications have been developed using the WoZ method. Participants behave as if communicating with a computer system using natural language in text or speech, while in reality it is the wizard who interprets and responds to their input (see Dahlbäck et al., 1993). The purpose of such studies is in general to observe the use and effectiveness of a proposed user interface rather than measuring the quality of an entire system. For example, a WoZ study may reveal a great deal about how participants would interact with a speech-based ticket booking service (e.g. what they say or what the steps of the process are) but might say nothing about the effectiveness of the natural language algorithms that would be needed for the interaction to take place. Such details are generally considered to be beyond the scope of the study. The functionality provided by the wizard may sometimes be implemented in later versions of the system but is sometimes very futuristic, far beyond the capabilities of current technology. The cost of performing WoZ studies can vary significantly depending on the system being evaluated and how (and what) data is recorded and analysed. Often special tools or systems with a special wizard backend need to be constructed in order to perform the studies, videotaped sessions may be more costly to analyse than questionnaires, etc. However, in relation to the benefits that can be reaped from performing the studies, and compared to the cost of iterating development of a fully functional system, it is often worth the extra investment.

The WoZ method has been used in several projects within the domain of affective interaction. For instance, WoZ has been used to evaluate tangible emotional interaction interfaces. Paiva et al. (2002) developed an interface to a computer game based on a tangible doll device called SenToy using the Wizard of Oz method. Their goal was to use the sensor-equipped SenToy to let players express a limited set of emotions, e.g. by jumping with the doll, shaking it, or positioning its limbs in certain configurations. These emotions would in turn control the behaviour of the player-controlled character in a fantasy game setting shown on a screen. The question that the design team faced was how players would use the SenToy doll to express the intended emotions. Based on the available literature about human behaviour, hypotheses were formed about movement patterns that were likely to be used for each emotion. These patterns, however, needed to be validated, as there could have been differences in how players expressed themselves using a doll compared to how they would using their own body (cf. Dahlbäck et al., 1993). At this point a WoZ study was performed in which players used a collection of dolls without sensors to express emotions which were mirrored by a character shown on a computer screen. The on-screen character was controlled by a wizard who sat in the same room as the player, watching their behaviour. Whenever the player performed an action with their doll that matched one of the patterns hypothesised to match an emotion according to the literature research results, the wizard would push a button to make the on-screen character show that emotion. The study provided information about which of the hypothesised patterns matched how players actually expressed emotions. In cases of insufficient correspondence, the players' actual actions with the doll suggested other patterns to look out for instead. In addition the study informed the design team about desirable qualities the doll itself should possess, such as being soft enough to bend easily and big enough to let players easily perform movements with it. A functional SenToy interface was developed and tested based on these results.

Another area where WoZ studies have been helpful is in the development of embodied conversational agents (ECAs). Such agents appear human-like and attempt to interact with users as another human being would, e.g. through conversing with them, using facial expressions and body language, displaying emotions and showing empathy. However, as noted by Dahlbäck et al. (1993), interacting with a human-like system is not the same thing as interacting with a human. Hence questions regarding interaction style including who should take initiative in a dialogue (agent or user), which emotions should be displayed by the agent, and which user expressions an agent should recognise as being emotionally charged remain largely unanswered (Cavalluzi et al., 2005; de Rosis et al., 2005). To address such issues, de Rosis et al. (2005) performed a WoZ study that investigated the forms of empathy that can be induced by ECAs in the context of promoting appropriate eating habits. The study was conducted using a WoZ tool developed specifically for the purpose of evaluating aspects of user–agent communication. The tool allows experimenters to alter various aspects of the experimental setup including physical aspects of the agent, its expressivity, and the set of dialogue moves that are available to the wizard. Thus the tool is flexible enough to handle other usage contexts as well. The study was performed iteratively aiming to gradually design a conversational agent in the chosen domain, with a particular emphasis on inducing empathy (in the broad sense of 'entering into a warm social relationship') in the user. To this effect, six rounds of WoZ tests were performed that gradually shaped the agents' personality and expressiveness. During the study, parameters such as the agents' interaction style (warm vs. cold), use of more natural sounding speech generation using different text-to-speech systems, and use of social small-talk to draw the user into a relation were varied to study their effect on the interaction. While the study did not yield any conclusive results regarding the effect of the above-mentioned parameters, it suggested that subjects were disappointed when receiving a 'cold' reply to an attempt to establish a friendly relationship. This in turn points to the need for ECAs to recognise the various forms of social contact making that humans routinely engage in.

3.4 Sensual Evaluation Instrument

The sensual evaluation instrument (SEI) is a tool for gathering affective feedback from users about a system that is a work in progress. It is a self-report measure that uses small, sculpted objects (see Fig. 2). Instead of offering verbal descriptions of how they are feeling, users indicate with the objects how they are feeling as they engage with the system prototype. This allows the designers to gain rich, nuanced feedback from users, which has not been forced into pre-conceived categories of response (e.g. 'happy' or 'sad'). Each user can create one's own taxonomy of meaning and strategies for conveying emotion through arraying the objects, gesturing with them, stacking, and the like. SEI sessions should be videotaped to review the feedback in more detail. It is also best to engage participants in a post-use discussion to elicit verbal feedback on how they used the tool and their own descriptions of personal taxonomies and use patterns that emerged for them.

Fig. 2 The sensual evaluation instrument objects

Fig. 3 A SEI session participant using multiple objects in an array that he kept close to his computer

In initial testing of the SEI, users demonstrated a wide range of usage strategies (see Figs. 3, 4, and 5).

The SEI was designed to allow for flexible, yet informative self-report of affect. The designers (see Isbister et al., 2006) worked closely with a sculptor who crafted biomorphic shapes meant to evoke a range of affective states. Preliminary research in both the USA and Sweden suggests that there are consistent emergent dimensions along which users tend to array the objects (see Isbister et al., 2007). For example, more spiky and sharp objects tend to be used to convey negative emotions. So the SEI provides some grounding common dimensions for feedback, while allowing

Fig. 4 A SEI session participant who stacked two objects

Fig. 5 A SEI session participant who held the objects in his hand and gesticulated wildly with them

for rich variance in individual expression of affect through the establishment of individual taxonomies and use patterns.

SEI has been used to evaluate three different interactive stories/games (Laaksolahti et al., 2009). The study aimed to identify the dramatic moments in the games and whether people did feel immersed. The SEI-based evaluation captured some important aspects of the emotional experiences of the interactive stories. As could be seen in the in-depth descriptions provided, participants could talk about their SEI objects and explain what emotions they portrayed in different situations. Through its purposefully ambiguous design the SEI objects are open to interpretation. In the study the objects seem ambiguous enough to accommodate a variety of emotions and shades of emotional experiences. The strength of the SEI evaluation was how it could pinpoint emotional experiences and allow for many shades of emotions. The weakness was that it only gave us hints on the local emotional experiences – not on the dramatic development of the whole game. This is something that the repertory grid technique (see below) could give a better grip on.

3.5 Repertory Grid Technique

When dealing with reports of subjective experiences a common problem is that either subjects are allowed to express themselves freely, possibly rendering large amounts of qualitative data that it is difficult to structure and compare across subjects, or the evaluators will set the boundaries for what can be expressed by asking a set of predefined questions decided by the experimental leader in a questionnaire, interview, or the like. In contrast, the basic idea behind the repertory grid technique is to elicit a set of personal constructs (or dimensions) from each participant, which are then used to evaluate the objects being studied.

The repertory grid technique (RGT) is based on Kelly's personal construct theory (Kelly, 1955). It is a tool that was designed by Kelly to gain access to a person's system of constructs by asking the person to compare and contrast 'relevant examples'. Kelly originally used the tool for investigating interpersonal relationships by having people classify a selection of persons that were important to them along a set of constructs describing relationships that were elicited. The method has later been used for many other purposes including knowledge modelling and management, construction of expert systems, and lately for capturing subjective experiential aspects of a person's interaction with various forms of technology (Fällman and Waterworth, 2005; Laaksolahti, 2008). Fällman and Waterworth also provide a good introduction to the method's underpinnings and use in the context of evaluation of artefacts.

Constructs are elicited by comparing elements with each other in various ways and extracting their similarities and dissimilarities. Constructs are usually bi-polar, taking on values between two extremes. For instance, we can judge people along dimensions such as tall–short or light–heavy. Typically, the subject is presented with three objects to be compared and has to tell which pair of objects is similar and which object is the outlier. The quality employed to separate the three objects has then to be used as a scale along which all three objects have to be assessed. This process continues until the subject cannot identify any further discriminative qualities of the objects according to his subjective experience. This process can then be used to identify experiential qualities of objects such as cars or mobile telephones.

Laaksolahti (2008) used the RGT method alongside in the SEI study (see previous section). The aim was to assess how well users become *immersed* in the stories, whether they feel they can influence it (*agency*), and to what level it allows users to *transform* themselves into the role they are playing in the interactive story. Laaksolahti's evaluation is also focused on the so-called *dramatic arc* of users' experience of the story. That is, did the story capture their interest, create a greater and greater tension, until the climax was reached and the story was completed? Or was the interactive narrative failing to produce a story-like experience? All of these concepts are very hard, elusive, qualities to evaluate. Very few structured user study methods are able to address them.

Laaksolahti modified RGT in order to capture the dynamic experience of an interactive story/game. Instead of comparing games with one another, subjects got to compare snippets of video-recordings from when they played one game, i.e. different parts of their experience. One subject expressed his experience of one of the games with the following constructs: *boring–entertaining, unengaging–engaging, mundane in a negative sense–exciting, follow the story–explorative*, and finally, *demanding–relaxed*. By following how this subject graded different snippets of video of his play the dynamics of his dramatic experience (rising and falling), his sense of being involved or not in the game, as well as his experienced ability to influence the outcome of each scene could be traced.

RGT thus allows getting at users' own subjective experiences of using interactive systems as the interaction unfolds, through their own concepts and words. Thereby it can provide vital feedback to the designer about what parts of the prototype system work and which parts need to be modified to cater to the intended kinds of experiences.

3.6 Critical Design Practice

Critical technical practice describes an approach to developing solutions to technical problems, which includes taking a core premise on which a field is founded and reversing it. It then proposes building a technology based on that reversed premise, which can contribute to the field in a novel and interesting way (Agre, 1997). Agre's key example is the notion of disembodiment that underlies classical artificial intelligence. By contrast, he proposes building fundamentally embodied agents; this notion is, e.g., at the heart of much of Rodney Brooks' early work at MIT's AI Lab (Brooks, 1986).

Critical technical practice also includes a level of reflective awareness of the discipline one is engaged in, including the field's sociological and cultural context, the philosophies it espouses at an unconscious level, and the field's key metaphors or analogies. Several designers of interactive systems have used critical technical practice as a tool to generate innovative and critically relevant systems (Sengers, 1999). For example, Simon Penny's notion of 'reflexive engineering' integrates robotics with an artist's sense of design and play. His robot Petit Mal is chaotic, whimsical, and clumsy: un-robot-like conduct that encourages the audience to generate theories as to the origin of this unusual behaviour, encouraging the public to become aware of and to consider their own notions of agency (Penny, 1997). Similarly, Gaver and colleagues (2003) propose inverting HCI's traditional goals of 'usefulness and usability' and explore the possibility of designing for rich experiences, with the potential to be intriguing, mysterious, and delightful.

Critical technical practice does not advocate the replacement of a field with one founded upon its inverse; rather, it proposes that such conceptual changes can bring insight into, awareness to, and novel contributions to a discipline. When we approach affective interaction, it may be very useful and important to use a critical design practice perspective, as it is far too easy to fall into various pitfalls where we assume that we have a good grip on emotions and emotional interaction. Contributions from a critical perspective on affective computing are, e.g., included in the chapter on the interactional approach to affective interaction.

4 Challenges Related to Affective Interaction

As can be understood from the methods we picked for this chapter, the design and prototyping phase of a project involving affective interaction and experiences has to meet a range of challenges. If we pick a method that allows for a laboratory environment, we face the challenge of making it realistic enough that our subjects get into the mood, emotion, or situated experience that we aim for. If, on the other hand, some of the prototyping involves users 'out in the wild', we will have problems with how the researcher/designer can study the situation, as most such settings do not allow the experimental leader to follow their subjects around in their daily lives.

On a general level, we might also feel troubled by the fact that people differ: we all have our own personal expressions and unique experiences. Most of the methods described in this chapter do not even attempt to generalise a larger user group. Unless we are very careful in choosing end-user groups and representatives of that group, we run the risk of getting irrelevant feedback from only a few of the participants.

References

Agre P (1997) Computation and human experience. Cambridge University Press, Cambridge, MA

Benyon D, Turner P, Turner S (2005) Designing interactive systems: people, activities, contexts, technologies. Addison-Wesley, Edinburgh

Brooks R (1986 Apr) A robust layered control system for a mobile robot. IEEE J Rob Autom RA-2:14–23

Buchenau M, Fulton-Suri J (2000) Experience prototyping. Proceedings of the conference on designing interactive systems processes, practices, methods, and techniques – DIS '00 (presented at the conference, New York City, New York, United States, 2000), pp 424–433, http://portal.acm.org/citation.cfm?id=347642.347802

Cavalluzi A et al (2005) A persona is not a person: designing dialogs with ECAs after wizard of Oz studies. In: HUMAINE WP6 workshop on "Interaction and Communication", Paris

Cooper A (1999) The inmates are running the asylum. Macmillan Publishing Co., Inc.

Dahlbäck N, Jönsson A, Ahrenberg L (1993) Wizard of Oz studies: why and how. In: Gray WD, Hefley WE, Murray D (eds) Proceedings of the 1st international conference on Intelligent user interfaces, Orlando, FL. ACM Press, New York, NY, pp 193–200

De Bono E (1985) Six thinking hats. Little, Brown and Company, New York, NY

de Rosis F et al (2005) Can embodied conversational agents induce empathy in users? In: Proceedings of the joint symposium on virtual social agents. In the scope of AISB'05, April 2005, University of Hertfordshire, Hatfield, England, pp 65–72

Fällman D, Waterworth J (2005). Dealing with user experience and affective evaluation in HCI design: a repertory grid approach. In: Workshop on Evaluating Affective Interfaces (CHI 2005), Portland, Oregon

Gaver W, Beaver J, Benford S (2003) Ambiguity as a resource for design. In: Proceedings of CHI'03, ACM Press, Fort Lauderdale, FL

Iacucci G et al (2002) Imagining and experiencing in design, the role of performances. In: Proceedings of the 2nd Nordic conference on Human-computer interaction, ACM, Aarhus

Isbister K et al (2006) The sensual evaluation instrument: developing an affective evaluation tool. In: Proceedings of the SIGCHI conference on Human Factors in computing systems. ACM, Montréal, QC, pp 1163–1172

Isbister K et al (2007) The sensual evaluation instrument: developing a trans-cultural self-report measure of affect. Int J Hum Comput Stud 65(4):315–328

Kelly G (1955) The psychology of personal constructs. Routledge, London

Laaksolahti J (2008) Plot, spectacle, and experience: contributions to the design and evaluation of interactive storytelling, PhD thesis (Stockholm University/Royal Institute of Technology, Department of Computer- and Systems Sciences, 2008)

Laaksolahti J, Isbister K, Höök K (2009) Using the sensual evaluation instrument. Digit Creativity 20(3):165–175

Oulasvirta A, Kurvinen E, Kankainen T (2003) Understanding contexts by being there: case studies in bodystorming. Pers Ubiquit Comput 7(2):125–134

Paiva A et al (2002) SenToy in FantasyA: designing an affective sympathetic interface to a computer game. J Pers Ubiquit Comput 6(5–6):378–389

Penny S (1997) Embodied cultural agents at the intersection of robotics, cognitive science, and interactive art. In: Dautenhahn K (ed) Socially intelligent agents: papers from the 1997 Fall Symposium, AAAI Press, Menlo Park, CA, pp 103–105. Technical Report FS-97-02

Rettig M (1994) Prototyping for tiny fingers. Commun ACM 37(4):21–27

Rodriguez J, Diehl JC, Christiaans H (2006) Gaining insight into unfamiliar contexts: a design toolbox as input for using role-play techniques. Interact Comput 18(5):956–976

Sengers P (1999) Practices for machine culture: a case study of integrating cultural theory and artificial intelligence. Surfaces VIII:1999

Sundström P, Ståhl A, Höök K (2007) In situ informants exploring an emotional mobile messaging system in their everyday practice. IJHCS 65(4):388–403

Evaluation of Affective Interactive Applications

Kia Höök, Katherine Isbister, Steve Westerman, Peter Gardner, Ed Sutherland, Asimina Vasalou, Petra Sundström, Joseph Jofish Kaye, and Jarmo Laaksolahti

Abstract Methods are developed for different audiences and purposes. HCI researchers develop methods to shape the future through pure, applied and blue sky research – as is still the case with most affective interactive applications. Unsurprisingly, practitioners will be more concerned that the methods they use not only are tractable but produce better and more innovative results in terms of the systems they ultimately release into the world. Researchers, on the other hand, may have other concerns, such as the novelty of their techniques. Up until recently, most HCI methods (both for researchers and practitioners) were developed for work applications and desktop situations. They focused on efficiency, learnability, transparency, control and other work-related values. They were developed in response to a theoretical orientation which viewed the user as an information processing system not so dissimilar to the computer itself. But now that HCI is concerned with technologies that enter all aspects of life, our methods have begun to change and will need to continue to change. In keeping with our changing conception of what a "user" is and a wider concern with their experience of use of new technologies, a key challenge will be to develop and expand methods for analyzing not just what people *do* with the technology but how it makes them *feel*, and not just how people understand technology but how they make sense of it as part of their lives. Methods must be concerned, not only with issues of usefulness and usability, but also with issues of aesthetics, expression, and emotion. In addition we need to focus on evaluating technology not just in the short term under controlled conditions but also in the longer term and in broader social and cultural contexts. In this section, we will therefore provide two strands of evaluation methods. The first concerns what we might see as more traditional usability evaluation: is my system usable for the purpose it was designed for? The second strand tries to get at what we have named "third wave of HCI" in the previous chapters: does my system provide for the kind of (emotional) experience that it aimed to do?

K. Höök (✉)
Department of Computer and Systems Sciences, Stockholm University/KTH, Kista, Sweden
e-mail: Kia@dsv.su.se

P. Petta et al. (eds.), *Emotion-Oriented Systems*, Cognitive Technologies,
DOI 10.1007/978-3-642-15184-2_36,

1 Evaluating Prototypes: Why?

How you evaluate something depends on your goals. For instance, are we looking for replicable 'official data' or design data? In both cases, we believe there is a value placed upon rigor – but rigor of a different ilk. If we return to the question of aims, we can see that a designer may require a rich and even divergent and unresolved set of perspectives in order to envision and create the most successful system for engaging and transforming users in their experiences. In contrast, someone required to judge a system's efficacy in eliciting particular emotions predictably requires a reliable instrument of a different sort. The latter rightly asks tough questions of the former about how valid and extensible their measures can truly be, whereas the former can point to limitations and even flaws in fundamental models and assumptions that might otherwise be overlooked. We see these approaches as complementary, and we feel the juxtaposition of stances and aims in this chapter helps to keep all of us 'honest' in our pursuit of powerful and effective affective evaluation methods.

On a practical level, Höök (2004) argues that it is necessary to divide user studies into two different levels: the first obvious challenge for affective interfaces is to find ways of checking whether the expressed emotions are understood by users and whether the system can interpret user emotions correctly. It might be that a design of an affective interactive character is perfectly valid and well-suited to the overall goal of the system, but the facial emotional expressions of the character are hard to interpret. Thus the overall design still fails. Or the other way round, the emotional expressions might be easily understood by the user, but the design still does not achieve its overall goal of entertaining or aiding the user.

Thus once the interpretation loop is bootstrapped and working, the second, even more challenging goal for evaluation of affective interfaces, is whether the overall usage scenarios are achieving their purpose of being, e.g. engaging, fun, believable or creating a relationship with the user, and how much of this can be attributed to the emotion modelling and expression. These two levels of evaluation will not necessarily be dividable into two different user studies or two different phases in the design process – instead they should be viewed as two levels of interpretation of what may be going on when a system fails to achieve its goals. What we are looking for are ways of disentangling the bad design choices from the interesting interpretative experiences end-users have with affective systems that in many cases cannot be controlled (as they are attempting to adapt the users' emotional states and thereby changes over time) or understood in a narrow sense (as they are oftentimes portraying interesting narrative or character-based dramas).

2 Qualitative or Quantitative?

A wide variety of methods are available to the usability practitioner and effective assessment procedures require that several methods are deployed as part of the software development process. Methods have different strengths and weaknesses.

Different methods are able to contribute different information to the assessment/ development process. A particularly important (and familiar) distinction can be drawn between qualitative and quantitative methods. Qualitative methods are concerned with the experience of the individual. They are particularly useful for determining what can go wrong/right with a specific design. Part of their value lies in the lack of constraints placed on the user. In contrast, quantitative methods are particularly useful when considering expected levels of performance when a system is used. They provide important information on the success (or otherwise) of software revisions and can be used to set target standards to be achieved (see the ISO 9241 standard on Ergonomics of Human System Interaction). Understanding the performance of user–computer systems from a quantitative/probabilistic point of view is essential for any form of cost–benefit analysis, as can be applied to software development. For example, it would be poor use of resources to spend substantial amounts of development time chasing a source of interface confusion that was identified by a single user only and that produced no costly detrimental outcome. Nevertheless, even when using a quantitative approach to assessment, it is important to realize that consideration of means is necessary but not sufficient. Also important are data relating to the range of scores: maxima, minima, and variability. For example, a system that performs well on average may still be unacceptable if a minority of users are encountering severe difficulties. These may be masked if only means are examined. Data relating to variability between individuals can be very useful in determining whether systems should have adaptive capabilities. A potential disadvantage of quantitative methods is that in producing circumstances in which reliable, replicable measurements can be obtained, the nature of the users' interaction can be constrained. Assessment typically focuses on specific aspects of users' performance. This can be contrasted with qualitative methods that permit greater flexibility. In summary, it is important that usability assessment is not 'paradigm bound' and that a range of complementary methods are selected as part of each system evaluation.

Within HCI, formal user studies (quantitative–scientific) are the gold standard for evaluating computational systems. But the aim in the affective interaction systems might not be best captured using formal user studies as these rarely are able to capture end-user experience (in a broader sense). We believe that informality and open-ended interpretation of users experience are key here, as done in the more ethnographically inspired parts of HCI. This approach is similar to how artwork is evaluated through art critics and informal encounters between the artist and the audience. This will not render results that are independent of time and culture – but the point is that no user evaluation studies are independent of time and culture anyway.

Informality can, e g, be observed in the HCI literature on evaluation of art-influenced speculative design. For example, the Presence project was evaluated informally by describing the designers' experience in installing the system and observing user interaction (Gaver, 2001). In the evaluation and design process of affector, Sengers and colleagues took this process even further by making the designer a subject in the evaluation study and allowing her to design and re-design the system as the evaluation process proceeded over a 6-month period (Sengers et al., 2005).

Anecdotal evidence, informal chats between users and system builders, tiny study sizes, forms structured to influence user interpretation, no discussion or analysis of results: this may sound like a to-do list for bad evaluation. But since the goal is to aid the process of improving the design until the end-user experience and the system interaction harmonize, we may sometimes prefer a rich, narrative and singular understanding over a simpler but rigorous and generalizable understanding (Höök et al., 2003).

3 Evaluation Methods from the Usability Field

In this part of the handbook we provide an overview and some examples of different approaches to the assessment of system usability. These methods differ in their generality of focus. Some are relevant for assessment of the affective responses of the user, whereas others are more general in their focus and consider users' appraisals of systems.

3.1 Context-Generic Psychometric Assessments of the User's Affective State

Psychometric principles can be applied to data of various types. Here we focus exclusively on self-report data and the term 'psychometrics' is used only to refer to this. Self-report psychometric tests typically comprise a number of items (questions) to which the test taker must respond. There is a variety of context-generic psychometric measures that can be used to assess users' affective responses when interacting with computing systems. Perhaps the more applicable of these for the present purposes are concerned with broad affective constructs and are typically based on two- or three-dimensional factorial representations of affect.

The best way of describing dimensions of affective space is the subject of continuing discussion. For example, the PANAS (Watson et al., 1988) provides assessment of the two dimensions of positive and negative affect. It requires that respondents provide ratings for each of 20 items on five-point Likert scales. The UWIST MACL (Matthews et al., 1990) is another questionnaire measure, developed on the basis of factor analysis, but produces the three scales of 'energetic arousal', 'tense arousal' and 'hedonic tone'. Mehrabian and Russell (1974) also prefer a three-dimensional description of affect, but these are 'pleasure', 'arousal' and 'dominance'. These dimensions are taken as the basis for the Self-Assessment Manikin (Lang, 1995; Hodes et al., 1985). This assessment is very quick to complete and less language dependent than is the PANAS or the UWIST MACL. It uses a pictorial approach in which a series of five cartoon-like images represent different points on each of three affect scales (five images for each). Brevity in completion is also a feature of the

Affect Grid (Russell et al., 1989), which requires a single response. Respondents signal their affective state, with reference to the two key dimensions of pleasure and arousal (cf. Russell, 2003) by positioning a cross in a 9 × 9 grid. The more central the cross, the weaker their affective experience. Brevity in assessment has some major advantages. It enables assessment with minimized disruption of associated task performance and can be more easily accommodated in repeated measure research designs.

Although these examples indicate disparity between assessments of affective dimensions, Russell (2003) suggests that this may be more apparent than real, with it being possible to accommodate each description in the circumplex of affect, albeit with different factorial orientations. Nevertheless, labelling presents a potential difficulty when considering these measures in the context of users' affective responses to computing systems. There is no clear 'mapping' between key affective constructs described in the computer literature (e.g. frustration or engagement) and the dimensions of assessment. However, a relatively new method of psychometric assessment the Geneva Appraisal Wheel (GAW: see Scherer, 2005) may hold some promise in this regard. It is also related to a circumplex model of affect, with emotions located at the periphery being more intense than those at the centre, and dimensions that bear comparison with those of Mehrabian and Russell (1974): pleasure and perceived control. However, it is more differentiated in its description of affective states than other dimensional models and incorporates a number of affective constructs that are central to much of the work that has been reported in the computer science/HCI literature.

3.2 Assessments of User Affect That Are Specific to HCI

There are a number of psychometric assessments that have been developed specifically for the purpose of assessing users' affective responses to computer-based systems. Assessment of users' satisfaction with computer-based systems has long been regarded as an important index of usability (see ISO 9241). Efforts to standardize measurement began over 20 years ago (see, e.g., Bailey and Pearson, 1983; Doll and Torkzadeh, 1988). However, there is still no clear resolution (Hornbæk, 2006). It can be argued that satisfaction is a somewhat neutral affective construct lacking in clarity (see, e.g., Edwardson, 1998) and tending to be most influential when it is absent. Perhaps higher targets should be set when considering users' affective responses to computer-based systems. Another affective computing-related construct that has been the subject of psychometric study for many years is 'computer anxiety' (Kay, 1993; Lloyd and Gressard, 1984; Dambrot et al., 1985). Initially, the primary value of assessing the constructs of satisfaction and anxiety was held to be in promoting the widespread uptake of computers in all sectors of the population. Computer anxiety was seen to be a barrier to this goal.

More recently there have been efforts to extend the view of users' affective experiences and the weight given to this construct in assessments of usability (e.g. Blythe and Wright, 2003; Dillon, 2001; Hassenzahl and Tractinsky, 2006). This may have been driven, at least in part, by the increasing availability of computer technology/software and its increasing power and sophistication. We come to expect more from the computer systems with which we interact. Consistent with this broader view, psychometric scales have been developed to assess constructs such as 'flow' (Huang, 2003; Webster et al., 1993) and 'fun' (Blythe et al., 2003; Blythe and Hassenzahl, 2003). The spread of ecommerce has led to the concept of trust being promoted up the research agenda, with a particular concern with identifying interface properties that will promote trust in the user (e.g. Basso et al., 2001; Flavian et al., 2006; Kim and Benbasat, 2006; Kim and Moon, 1998; Sillence et al., 2006; van der Heijden et al., 2003).

Recent studies have used factor analytic techniques to develop scales that distinguish ergonomic from hedonic responses to software (e.g. Hassenzahl et al., 2000; Hassenzahl, 2004; Huang, 2003, 2005; see also Voss et al., 2003). Although convergent validity should be established empirically, generally there seems to be some consistency in results. However, factors are named differently, which is unfortunate for the progression of this area of study. Moreover, divergent validity is an important area for study. It seems that relatively strong correlations remain between these constructs (see also Tractinsky et al., 2000). Factor studies of more diverse aspects of aesthetic response have also been conducted (e.g. Lavie and Tractinsky, 2004; Park et al., 2004). However, again, more work is needed to establish convergent and divergent validity. The danger, as Hornbæk (2006) points to with regard to user satisfaction, is that a diverse research literature will accumulate in which differences in approach mask similarities in constructs and the scope for application of results is diminished. It is suggested that the existing 'generic' literature on the measurement of emotion should be used as more of a 'touchstone' for studies of computer-related affective response to encourage consistency and comparability.

3.3 Psychometric Assessments of Usability

A number of questionnaire measures of system usability have been developed. These have not been specifically designed to address issues of affective computing, and integrated assessments of affective response tend to be rather limited. The usability principles set out in ISO 9241 form the basis for several of these measures (see, e.g., IsoMetrics: Gediga et al., 1999) and, consistent with this, affect tends to be considered as a uni-dimensional construct relating primarily to user satisfaction. However, there is a good deal of variation between questionnaires on other important details. These include the specific scales included and the total number of items. For this reason, tests vary in the types of assessment to which they are best suited (i.e. in selecting a test, it is necessary to consider the type of software to be studied, the nature of the user, the nature of the task to be performed and the context of use).

For example, the software usability measurement instrument (Kirakowski and Corbett, 1993) is a commercially available questionnaire based on ISO 9241 principles. User responses take the form of 'agree', 'disagree' or 'undecided' to 50 items, with aggregation of responses producing a 'general usability' scale in addition to five more specific scales: efficiency, affect, helpfulness, control and learnability (see Kirakowski, 1994). In contrast, the After Scenario Questionnaire (ASQ; Lewis, 1991) is an extremely brief measure, comprising just three items, that assesses users' satisfaction with: (i) ease of completion; (ii) time taken and (iii) the availability of support information. As discussed, there are costs and benefits associated with both long and short questionnaires and the selection of a test requires that these are weighed. In recognition of effects of scale/test length, some usability questionnaires have alternate forms allowing long or short versions to be administered depending on the nature of the study (see, e.g., Questionnaire for User Interaction Satisfaction; Chin et al., 1988; IsoMetrics: Gediga et al., 1999). Factors to consider when selecting which length of test would be appropriate for a given usability testing application include the motivation of respondents and the number of repeated measurements that are required as part of the testing protocol.

Usability questionnaires also differ in their specificity. Some include items that are specific to particular interface components that will not be present in all software [e.g. in the QUIS (Harper and Norman, 1993), one item asks users to evaluate the use of 'blinking']. Although such specific questions are capable of identifying highly important and reliable patterns of user response, it is worth evaluating the relevance of items such as this for inclusion in test sessions. Respondents rapidly become bored if faced with many items that do not apply to the specific scenario they have experienced. Finally, it is worth bearing in mind that users' evaluations of software will be constrained by the format of quantitative questionnaires. For this reason, the inclusion of some open-ended items can be very valuable as a means of identifying key usability issues that may not have occurred to the investigators. Many of the available usability questionnaires have such components (see, e.g., QUIS).

3.4 Think Aloud

The think aloud protocol originated from cognitive psychology (Ericsson and Simon, 1984). One of its applications was to understand the processes underlying problem solving. Once trained in thinking aloud, users were 'left alone' with minimal, if no interference, to verbalize their actions. The think aloud protocol has since been integrated in usability testing and is one of the most applied methods in the field. The reason for its wide appeal is that it gives insights on the 'why' behind a particular usability problem, while the data obtained is easy to analyze. Additionally, the method itself is easy to learn and to administer, thus it is cost effective.

At the onset of the usability session, users are trained on how to think aloud while performing a task by "speaking what they are doing on-screen". Following this, they are given a number of tasks to complete, one at a time. The facilitator is in the room with the user to make sure the tasks are completed, while note takers or other recording devices, e.g. video cameras, are used to capture the session. Note takers can be present in the room but also depending on the set-up they may view the session from an adjacent location (Rubin, 1994).

Usability testing, when compared to experiments in psychology, poses a unique set of requirements. A prominent example is the interference of technical errors. Faulty equipment or a faulty application interface requires technical interventions. Furthermore, the participant may believe the task has been accomplished when in reality it has not. Also, the participant may sidestep the task and find new ways for achieving it that are out of the scope of the practitioner's interest (Boren and Ramey, 2000). These issues compromise the facilitator's passive role as formulated in the original think aloud protocol (Ericsson and Simon, 1984) and invite new revisions (e.g. Wright et al., 1989; Boren and Ramey, 2000) of the think aloud that address the requirements of usability testing.

A recent approach of this kind is the "speech-genre" think aloud protocol proposed by Boren and Ramey (2000). This approach acknowledges the presence of the facilitator in the room. It clearly defines the user as the speaker, giving information about the interface, and the facilitator as the listener, who is there to learn while partaking little in the verbal protocol. After these roles are instituted by the facilitator, the usability session begins. While the session is taking place, the facilitator avoids interventions but carefully reaffirms his/her presence and active listening role with neutral speech tokens given at regular intervals, e.g. "yes", "hmm", "uh hum". This approach acknowledges the presence of the critical scenarios outlined above, e.g. technical problems, by introducing provisions in which the facilitator can exit the role of the listener and lead the discussion (Boren and Ramey, 2000).

Think aloud utterances are verbal data that have to be interpreted by the usability team. Traditionally, in usability testing, these utterances are treated as data that reveal the cause of a usability problem. As concerns affective interactions, usability practitioners are no longer interested in the revelation of a usability problem but in obtaining knowledge on users' affective experience. With this new aim in mind, the data derived from thinking aloud can be treated in two different ways.

Firstly, vocal utterances or facial expressions given off during the user's interaction with the system can be collected as objective measures for inferring emotion without interrupting the user's experience. Emotions can be extrapolated by a collection of vocal and/or facial parameters (e.g. Banziger and Scherer, 2005). By taking this approach, emotion labels can be associated with particular on-screen events. Automatic vocal analysis, as well as facial analysis, techniques exist for emotion recognition (e.g. Ioannou et al., 2005). Automatic inferences of this kind can lead to swift and thus cost-efficient analysis. However, these intelligent judgments are limited to a user's expression only and thus the context of interaction is not accounted for. Take for example a user who masks his dislike with irony. At face value, irony manifests with expressive signals of joy. An additional approach to remedy this problem is to analyze vocal and facial expression manually with the use of trained coders. This approach addresses possible concerns with context but introduces hours of manual labour (Lazzaro, 2004). It is therefore not a cost-effective approach for the field of usability.

Secondly, think aloud data can be analyzed with respect to the verbal content that is being expressed. In affective interactions, users have been found to extend the think aloud; whereas users are initially trained to verbalize their on-screen actions only, the affective nature of the application primes them to additionally describe their emotive responses (Vasalou and Bänziger, 2006). Content analysis or other qualitative analyses methods can assist in identifying patterns in the data. These patterns can subsequently expose users' experience with the application.

The think aloud protocol and the controlled lab setting it involves impose an important limitation that should be considered when choosing to use this method. When setting up the usability session, the researcher has to carefully reconstruct conditions that emulate users' natural environment. At the same time, the tasks given should represent the type of tasks users will want to achieve when using the application in the real world. If these design decisions are taken without careful consideration, the results obtained via this method may not be representative of users' real-world usage.

3.5 Physiological Measures

Human factors practitioners have been using physiological measures since the late 1970s. A range of measures have been examined as possible indices of operator workload in the context of complex systems (e.g. air traffic control). These include measures of ANS (autonomic nervous system) activity, such as galvanic skin response (GSR), measures relating to heart rate (HR), pupil diameter and measures of respiration; and also measures of cortical activity, including electroencephalograms (EEGs) and evoked potentials (ERPs). The emphasis has been on the detection of stress states with a view to being able to design systems that avoid operator overload. Promising measures include heart rate variability (HRV), the P300 component of evoked potentials and pupil diameter (see O'Donnell and Eggemeier, 1986; Kramer, 1991, for reviews). Similar techniques have also been explored in the context of 'traditional' human–computer interaction, although the literature is more limited. For example, Lin and Imamiya (2006) found HRV to be sensitive to a manipulation of computer game difficulty. Wilson and Sasse (2000) and Wilson (2001) report changes in HR, GSR and blood volume pulse (BVP) resulting from degradations of media quality (video frame rates and audio quality), and Iqbal et al. (2005) found pupil diameter to be sensitive to load manipulations for route planning and document editing tasks and also to sequence of task execution as defined by a hierarchical breakdown of task components.

Interest in the association between psychophysiological measures and 'traditional' methods of assessing usability continues (e.g. Lin et al., 2005). However, changing views on the importance of the user's affective state (as described above), and a substantial literature on the use of physiological measures to assess emotion (see, e.g., Cacioppo et al., 2000), have led to a wider application of these measures in relation to HCI. These methods are now being used in the context of affective computing to assess affective state of computer users. However, it is possible that there is a good deal of overlap in the psychophysiological findings from these two areas (workload and affective computing). Both are concerned with detecting a negative, stressful state in the user. Many of the manipulations used in affective computing studies to induce frustration could be considered to increase the workload of the user, and some explicitly change cognitive demand. There are, however, some interesting differences between the two areas of application. For example, HCI research has tended to investigate HR as a predictor of the state of the user, whereas HRV has proved the more successful index of operator workload (Kramer, 1991). Facial EMG has been used in many studies to assess emotion (see, e.g., Cacioppo et al., 1990) and has been applied with some success in the context of affective computing.

So far as we are aware, these measures were not used in human factors studies of workload. Conversely, EEG and ERPs have been used extensively to examine operator workload but have not been used in the context of affective computing.

A substantial advantage of psychophysiological measures is that they provide continuous monitoring of user state and, usually, are not disruptive of task performance (although recording of baseline periods can be an issue). In these respects, they provide a strong contrast with self-report methods. When coupled with a time-stamped record of users' activities, this can provide a powerful tool for assessing many aspects of user experience as they relate to specific features of the software. However, psychophysiological measures tend to be sensitive to uncontrolled environmental variations, such as changes in heating or lighting.

A further difficulty with several physiological methods of assessment (e.g. GSR) is that they provide information on arousal but not valence (Ward and Marsden, 2004). Given the earlier discussion on user motivation, this may not be sufficient and means that they need to be supplemented with other measurement techniques. Most physiological measures are also relatively intrusive, insofar as they require electrodes to be placed on the user. Although less intrusive methods of gathering, physiological data are being developed. These include, for example, the use of sensors embedded in an office chair to detect heart rate (Anttonen and Surakka, 2005); sensors in glasses to detect facial muscle activity (Scheierer et al., 1999); sensors in a computer mouse to collect measures of skin temperature, GSR and HR (Crosby et al., 2001); and the use of thermal imaging to detect difference in blood flow in the face that relate to muscle activity (Puri et al., 2005). Ethical issues must also be considered in this context as users have very little control over the responses that are being recorded by some of these measures (cf. Reynolds and Picard, 2004 and the final part of the handbook).

4 Evaluation Methods Inspired by Art and Design

Apart from these more usability-oriented methods, there are also methods that try to capture the qualitative, subjective experiences of users. In the previous chapters, we have mentioned a range of methods that may be as useful for the final evaluation of a system as they are in the context of on-going evaluation to improve early prototypes. The repertory grid technique (RGT), sensual evaluation instrument and in situ evaluation are all relevant methods for evaluation. Let us now turn to a couple of methods that have borrowed from the humanities and art in order to capture the meaning of systems as part of a larger social setting.

4.1 Cultural Commentaries

Any artefact we produce will be made part not only of a singular user's life, but it will affect the whole context in which it is used. It will be placed on the social, societal and political arena. If it is successful, it will have effects on "aesthetics, emotional effects, genre, social niche, and cultural connotations" (Gaver, 2007). In what we have named the third wave of human–computer interaction, these complex experiences, going beyond the work-related, desktop-oriented applications, have to be scrutinized in a new, more holistic perspective. Gaver's argument is that there is already a group of professionals whose task it is to place artefacts and phenomena into the whole picture of a society. He names them "cultural commentators, people whose profession it is to inform and shape public opinion, as resources for multi-layered assessments of designs for everyday life." In this group, he includes journalists, literature critics, documentary film makers and ethnographers. These groups are accustomed to providing complete pictures, whole narratives, of how artefacts are made part of our lives, reflect or alter aesthetics, enter the social or political arena. They provide a polyphonic assessment of artefacts. Gaver has therefore been experimenting with evaluation of some of his and his colleagues' research prototype systems. By first installing the research prototype in somebody's home and then allowing documentary filmmakers to make their own narrative of how that person makes sense of it, how it enters the home, its use, its importance and how it is related to the overall picture of what is going on in our society, Gaver is given an independent point of view on how to understand his own prototypes. The stories he gets are not only independent but also powerful dramatizations stories of the use.

The weakness of the method, its subjective nature, may also be seen as its strength. The meaning of the artefacts we produce in the field of affective interaction is not one simple story but of major importance in how we see machines and their roles in our society.

The cost of the method lies not only in finding and paying for those independent cultural commentators but also in taking the research prototypes to be assessed to such a stage that it can be experienced for a longer time period, as part of someone's life. Looking upon cultural commentators as a method to be used for evaluation is a novel approach in a sense; on the other hand, journalists, critics and documentary filmmakers have been part of our society for a long time and have a well-established presence at the university as both an academic method and discipline and part of the society.

4.2 Logbook Probes

As briefly mentioned above, one technique that appears to hold promise for characterizing users' experience is a variation on the cultural probe (Gaver et al., 1999) called the logbook (or, occasionally, 'diary') probe (Kaye et al., 2005; Kaye, 2006). The aim of the logbook probe is to provide a daily set of open-ended questions to be answered by those using a technology to provide a rich characterization of the users' experiences with the technology.

There is a tradition of using diary studies as an alternative to laboratory-based work in HCI (Rieman, 1993). Diary studies permit users to capture their thoughts and feelings in the course of using a technology over several days and as such have an ecological validity that cannot be matched by user studies within the confines of the laboratory. However, such studies are open to criticism due to the difficulty, particularly for mobile users, of subjects recording meaningful entries (Brandt et al., 2007), and for the 'thin' quality of reminiscences captured in traditional approaches (Carter and Mankoff, 2005).

However, there can also be other problems with diary studies. Researchers typically want to know details about interactions with the device under study, such as a mobile phone or some other novel technologies. But asking users to specifically discuss their uses of the technology can overemphasize the role of the technology itself at the expense of understanding the ways in which it fits into their everyday lives, interactions and cultural context. In addition, answers can frequently reflect a 'natural', unconsidered approach to the technology, rather than the reflective approach that researchers may find the most instructive.

In a typical study, a logbook probe is a booklet consisting of 7 days worth of questions. This booklet is given to a participant in a study, with instructions to complete one page a day. In situations where multiple people, such as a family, are interacting with a technology for multiple weeks, logbooks are given out to different members of the family in different weeks. Depending on the study, there may be two different varieties of logbooks: one type to capture initial impressions and one type to study reactions after users have become more familiar with the technology. Each page includes four or five questions, which, like the questions in cultural probes, are the following:

- *Open ended*: they allow for rich answers to simple questions
- *User interpreted*: they encourage the user to explain their answers
- *Defamiliarizing*: encourage the user to rethink their reactions in potentially novel ways
- *About the situation*: concentrate not just on the technology but also on the situation the technology is intended to impact or change, as well as the technology itself
- *Leveraging both users' and designers' skills in cultural interpretation*: taking advantage of shared bodies of cultural reference

For example, in Kaye (2006), a logbook probe to study a technology designed to help couples in long-distance relationships to maintain intimacy asks the following question:

- What colour best represents your relationship today? Why?

Subjects responded with answers including the following:

- Amber/yellow – do I proceed with caution or speed up to beat the red or slow down anticipating a step?
- Purple – we have a more matured, aged relationship rather than a new, boundless one which would best be described by red. Purple is the more aged, ripened form of red.
- Yellow! Like a sun, like a summer. I often laugh with Sven especially in those days. Using VIO [the technology under study] is really funny and interesting.

While these answers do not directly inform the next steps in technology design, they provide a richness and context for understanding the ways in which the technology is used within the relationship. Other questions included were the following:

- What TV family is most like your family? Why?
- What song best represents your relationship today? Why?
- If you had to rename [the technology], what name would you give it? Why?

Perhaps most importantly, these questions and the answers to them provide a rich source of material both for interpretation by the designer or the researcher themselves for future discussion with the users. It can be difficult to interview users about aspects of technology use without presupposing a framework for understanding that use. These questions and answers provide a way for researchers to build up a more complete picture of the experience of technology use based on the ways that the users make sense of the technology, rather than starting from the assumptions of the designer.

5 Challenges Related to Affective Interaction

While there are many more methods for understanding users, designing systems and evaluating them that we have left out of this discussion, there is still a lack of methods in this field where more research is needed.

One such area concerns designs that involve users not only cognitively but also physically. Such systems can have a strong impact on emotion as they address more than one aspect of emotional processes. In the previous chapters we have mentioned Laban analysis as one way of approaching emotional body language and the affective loop as a design concept by which we can approach such design. But when it comes to making a tight design loop and in particular evaluation, there is not much we can use. This concerns in particular games where devices such as the Nintendo Wii or special-made devices such as the guitar for Guitar Hero for the Sony Playstation 2 require a thoughtful integrated design. It is also relevant to the design of interactive characters who need to have not only facial expressions but also body language. It is also true for affective interactive systems that make use of body sensors picking up on emotional arousal, movement and other signals. A range of potential applications could make use of body sensors to create applications for health, games, mobile applications and more traditional desktop applications. But how do we design for these?

A second area that we have avoided in the previous chapters is that of considering the unique aspects of the material we use to build our applications. The use of mobiles, ubiquitous technology, body sensors, physical interaction devices, game consoles, advanced facial or voice recognition systems, etc. posits both possibilities and limitations. Most of the methods discussed above do not provide any explicit way to experiment with the materials as such.

Finally, as discussed in other parts, and in particular in the final part of this handbook, affective interaction will inevitably produce applications that touch upon important human values such as autonomy, privacy and bodily integrity. In a vision for HCI in the year 2020, Microsoft Research and a range of invited researchers have outlined ideas for the future of this field.[1] One of their important predictions concerns the importance of considering the value(s) of technology:

> HCI can no longer be the scientific investigation of what role technology might have, it will need to be part of the empirical, philosophical and moral investigation of why technology has a role. It will entail asking what the use of computing implies about our conceptions of society for example, just as it will entail asking new questions about how we ought to interact with technology in this new world. Even philosophical questions will be important, where our concepts of how the mind works will affect the way we design technologies to support memory, intelligence and much more besides. Above all, a new HCI will entail asking, what are the human values that might be designed for?

We find this to be particularly true for the field of affective interaction.

References

Anttonen J, Surakka V (2005) Emotions and heart rate while sitting on a chair. In: Proceedings of the SIGCHI conference on human factors in computing systems, ACM, Portland, OR, pp 491–499

Bailey JE, Pearson SW (1983) Development of a tool for measuring and analysing computer user satisfaction. Manage Sci 29:530–545

Banziger T, Scherer KR (2005) The role of intonation in emotional expressions. Speech Commun 46(3–4):252–267

Basso A, Goldberg D, Greenspan S, Weimer D (2001) First impressions: emotional and cognitive factors underlying judgments of trust e-commerce. In: Proceedings of EC'01, October 14–17, Tampa, FL, ACM Press, New York, NY, pp 137–143

Blythe M, Hassenzahl M (2003) The semantics of fun: differentiating enjoyable experiences. In: Blythe MA, Overbeeke K, Monk AF, Wright PC (eds) Funology: from usability to enjoyment. Kluwer, Dordrecht

Blythe M, Wright P (2003) From usability to enjoyment. In: Blythe MA, Overbeeke K, Monk AF, Wright PC (eds) Funology: from usability to enjoyment. Kluwer, Dordrecht

Blythe MA, Overbeeke K, Monk AF, Wright PC (eds) (2003) Funology: from usability to enjoyment. Kluwer, Dordrecht

[1] See http://research.microsoft.com/hci2020/default.html (visited 2010-05-30).

Boren MT, Ramey J (2000) Thinking aloud: reconciling theory and practice. IEEE Trans Prof Commun 43(3):261–278
Brandt J et al (2007) Txt 4 l8r: lowering the burden for diary studies under mobile conditions. In: CHI '07 extended abstracts on human factors in computing systems. ACM, San Jose, CA
Cacioppo JT, Tassinary LG (1990) Inferring psychological significance from physiological signals. Am Psychol 45:16–28
Cacioppo JT, Tassinary LG, Berntson GG (2000) Handbook of psychophysiology. Cambridge University Press, New York, NY
Carter S, Mankoff J (2005) When participants do the capturing: the role of media in diary studies. In: Proceedings of the SIGCHI conference on human factors in computing systems, ACM, Portland, OR
Chin JP, Diehl VA, Norman KL (1988) Development of an instrument measuring user satisfaction of the human-computer interface. In: Proceedings of SIGCHI '88, ACM/SIGCHI, New York, NY, pp 213–218
Crosby ME, Auernheimer B, Aschwanden C, Ikehara C (2001) Physiological data feedback for application in distance education. In: Proceedings of the 2001 workshop on perceptive user interfaces, ACM, Orlando, FL, pp 1–5
Dambrot FH, Watkins-Malek MA, Silling SM, Marshall RS, Garver J (1985) Correlates of sex differences in attitudes towards and involvement with computers. J Vocat Behav 27:71–86
Dillon A (2001) Beyond usability: process, outcome, and affect in human-computer interactions. Can J Inf Library Sci 26(4):57–69
Doll WJ, Torkzadeh G (1988) The measurement of end-user computing satisfaction. MIS Q 12:259–274
Edwardson M (1998) Measuring consumer emotions in service encounters: an exploratory analysis. Australas J Market Res 6:34–48
Ericsson KA, Simon HA (1984) Protocol analysis: verbal reports as data. MIT Press, Cambridge, MA
Flavian C, Guinaliu M, Gurrea R (2006) The role played by perceived usability, satisfaction, and consumer trust on website loyalty. Inf Manage 43:1–14
Gaver WW et al (1999) Design: cultural probes. Interactions 6(1):21–29
Gaver W (2001) The presence project. RCA CRD Research, London
Gaver WW (2007) Cultural commentators: non-native interpretations as resources for polyphonic assessment. Int J Hum Comput Stud 65(4):292–305
Gedgia G, Hamborg KC, Düntsch I (1999) The IsoMetrics inventory: an operationalization of ISO 9241-10 supporting summative and formative evaluation of software systems. Behav Inf Technol 18:151–164
Harper BD, Norman KL (1993) Improving user satisfaction: the questionnaire for user interaction satisfaction version 5.5. In: Proceedings of the 1st annual mid-Atlantic human factors conference, Virginia Beach, VA, pp 224–228
Hassenzahl M (2004) The interplay of beauty, goodness, and usability of interactive products. Hum Comput Interact 19:319–349
Hassenzahl M, Platz A, Burmester M, Lerner K (2000) Hedonic and ergonomic quality aspect determine a software's appeal. CHI Lett 2(1):201–208
Hassenzahl M, Tractinsky N (2006) User experience – a research agenda. Behav Inf Technol 25:91–97
Hodes R, Cook III, EW, Lang PJ (1985) Individual differences in autonomic response: conditioned association or conditioned fear?. Psychophysiology 22:545–560
Höök K et al (2003) FantasyA and SenToy. In: CHI '03 extended abstracts on human factors in computing systems. ACM, Ft. Lauderdale, FL
Höök K (2004) User-centred design and evaluation of affective interfaces. In: Ruttkay Z, Pelachaud C (eds) From brows to trust: evaluating embodied conversational agents. Kluwer's human–computer interaction series, vol 7. Kluwer, Dordrecht, pp 127–160

Hornbæk K (2006) Current practice in measuring usability: challenges to usability studies and research. Int J Hum Comput Stud 64(2):79–102

Huang M-H (2003) Designing website attributes to induce experiential encounters. Comput Hum Behav 19:425–442

Huang M-H (2005) Web performance scale. Inf Manage 42:841–852

Ioannou SV et al (2005) Emotion recognition through facial expression analysis based on a neurofuzzy network. Neural Netw 18(4):423–435

Iqbal ST, Adamczyk PD, Zheng XS, Bailey BP (2005) Towards an index of opportunity: understanding changes in mental workload during task execution. In: Proceedings of the SIGCHI conference on human factors in computing systems, ACM, Portland, OR, pp 311–320

Kay RH (1993) An exploration of theoretical and practical foundations for assessing attitudes towards computers: the computer attitude measure. Comput Hum Behav 9:371–386

Kaye JJ et al (2005) Communicating intimacy one bit at a time. In: CHI '05 extended abstracts on human factors in computing systems. ACM, Portland, OR

Kaye JJ (2006) I just clicked to say I love you: rich evaluations of minimal communication. In: CHI '06 extended abstracts on human factors in computing systems. ACM, Montréal, QC

Kim D, Benbasat I (2006) The effects of trust-assuring arguments on consumer trust in Internet stores: application of Tourmin's model of argumentation. Inf Syst Res 17:286–300

Kim J, Moon JY (1998) Designing towards emotional usability in customer interfaces – trustworthiness of cyber-banking system interfaces. Interact Comput 10:1–29

Kirakowski (1994) The use of questionnaire methods for usability assessment. Accessed from http://sumi.ucc.ie/sumipapp.html

Kirakowski J, Corbett M (1993) SUMI: the software usability measurement inventory. Br J Educ Technol 24:210–212

Kramer AF (1991) Physiological metrics of mental workload: a review of recent progress. In: Damos DL (ed) Multiple-task performance. Taylor & Francis, London, pp 279–328

Lang PJ (1995) The emotion probe. Am Psychol 50(5):372–385

Lavie T, Tractinsky N (2004) Assessing dimensions of perceived visual aesthetics of web sites. Int J Hum Comput Stud 60:269–298

Lazzaro N (2004) Why we play games: four keys to more emotion without story. Technical report

Lewis JR (1991) Psychometric evaluation of an after-scenario questionnaire for computer usability studies: the ASQ. SIGCHI Bull 23(1):78–81

Lin T, Imamiya A (2006) Evaluating usability based on multimodal information. In: Proceedings of the 8th international conference on multimodal interfaces – ICMI '06, Banff, AB, p 364

Lin T, Omata M, Hu W, Imamiya A (2005) Do physiological data relate to traditional usability indexes? In: Proceedings of the 17th Australia conference on Computer-Human Interaction: Citizens Online: Considerations for Today and the Future. Computer-Human Interaction Special Interest Group (CHISIG) of Australia, Canberra, ACT, pp 1–10

Lloyd BH, Gressard C (1984) Reliability and factorial validity of computer attitude scales. Educ Psychol Meas 44:501–505

Matthews G, Jones DM, Chamberlain AG (1990) Refining the measurement of mood: the UWIST mood adjective checklist. Br J Psychol 81:17–42

Mehrabian A, Russell JA (1974) An approach to environmental psychology. MIT Press, Cambridge, MA

O'Donnell RD, Eggemeier FT (1986) Workload assessment methodology. In: Boff K, Kaufman L, Thomas J (eds) Handbook of perception and human performance, vol II: cognitive processes and performance. Wiley Interscience, New York, NY, pp 42/1–42/9

Park S, Choi D, Kim J (2004) Critical factors for the aesthetic fidelity of web pages: empirical studies with professional web designers and users. Interact Comput 16:351–376

Puri C, Olson L, Pavlidis I, Levine J, Starren J (2005) StressCam: non-contact measurement of users' emotional states through thermal imaging. In: CHI '05 extended abstracts on human factors in computing systems, ACM, Portland, OR, pp 1725–1728

Reynolds C, Picard R (2004) Affective sensors, privacy, and ethical contracts. In: CHI '04 extended abstracts on human factors in computing systems, ACM, Vienna, Austria, pp 1103–1106
Rieman J (1993) The diary study: a workplace-oriented research tool to guide laboratory efforts. In: Proceedings of the INTERACT '93 and CHI '93 conference on human factors in computing systems. ACM, Amsterdam
Rubin J (1994). Handbook of usability testing how to plan, design, and conduct effective tests. Wiley, New York, NY
Russell JA (2003) Core affect and the psychological construction of emotion. Psychol Rev 110:145–172
Russell JA, Weiss, A, Mendelsohn GA (1989) Affect grid: a single item scale of pleasure and arousal. J Pers Soc Psychol 57(3):493–502
Scheirer J, Fernandez R, Picard RW (1999) Expression glasses: a wearable device for facial expression recognition. In: CHI '99 extended abstracts on human factors in computing systems, ACM, Pittsburgh, PA, pp 262–263
Scherer KR (2005) What are emotions? And how can they be measured? Soc Sci Inf 44(4):693–727
Sengers P, Boehner K, Warner S, Jenkins T (2005) Evaluating affector: co-interpreting what 'Works'. In: CHI 2005 workshop on innovative approaches to evaluating affective interfaces, Portland, Oregon, Mar 2005
Sillence E, Brigss P, Peter H, Fishwick L (2006) A framework for understanding trust factors in web-based health advice. Int J Hum Comput Stud 64:697–713
Tractinsky N, Katz AS, Ikar D (2000) What is beautiful is usable. Interact Comput 13:127–145
Van der Heijden H, Verhagen T, Creemers M (2003) Eur J Inf Syst 12:41–48
Vasalou A, Bänziger T (2006) Using the *think aloud* for affective evaluations in the lab: a work-in-progress. In: WP10 Humaine workshop on Ethics, Vienna
Voss KE, Spangenberg ER, Grohmann B (2003) Measuring the hedonic and utilitarian dimensions of consumer attitude. J Mark Res 40:310–320
Ward RD, Marsden PH (2004) Affective computing: problems, reactions and intentions. Interact Comput 16:707–713
Watson D, Clark LA, Tellegen A (1988) Development and validation of brief measures of positive and negative affect: the PANAS scales. J Pers Social Psychol 54:1063–1070
Webster J, Trevino LK, Ryan L (1993) The dimensionality and correlates of flow in human-computer interactions. Comput Hum Behav 9:411–426
Wilson GM (2001) Psychophysiological indicators of the impact of media quality on users. In: CHI '01 extended abstracts on human factors in computing systems, Seattle, Washington, pp 95–96ACM
Wilson GM, Sasse MA (2000) Do users always know what's good for them? Utilising physiological responses to assess media quality. In: The proceedings of HCI 2000: people and computers XIV—usability or else! HCI, pp 327–339
Wright PC et al (1989). Cooperative evaluation: the York manual. Technical report. Department of Psychology, University of York, York

Part VIII
Ethics and Good Practice

Editorial: 'Ethics and Good Practice' – Computers and Forbidden Places: Where Machines May and May Not Go

Roddy Cowie

Abstract Technology always faces questions about what it should and should not do, but they are unusually difficult to answer in the case of emotion-oriented computing – partly because they affect particularly sensitive areas of human life, and partly because of uncertainties surrounding the ways that machines might affect those areas. Those difficulties can lead to extraordinary extrapolations being treated as real reasons for concern and real concerns being overlooked. The chapters aim to provide a framework for addressing abstract concerns and dealing with concrete ones. They propose that a philosophical approach known as principalism is the natural framework for ethical debates. They examine the basis of abstract concerns, from generalised fear of the unknown to focused concerns about autonomy. They also look at the practical measures needed to ensure that reasonable ethical concerns are identified and not violated.

This handbook is about a branch of engineering. The characteristic question in engineering is: Can it be done? Increasingly, though, engineers have had to reckon with another question: Should it be done? Like it or not, that second kind of question looms large in emotion-oriented computing. This part of the handbook aims to encourage people to address it and to provide resources that allow them to address it effectively.

Some of the arguments around the 'should' question are thoroughly familiar. How do the costs and benefits balance financially? Will there be a disproportionate impact on certain social groups? How will the quality of the environment be affected? These are not new issues, but they are still difficult.

R. Cowie (✉)
Department of Psychology, Queen's University, Belfast, Ireland
e-mail: roddy.cowie@qub.ac.uk

P. Petta et al. (eds.), *Emotion-Oriented Systems*, Cognitive Technologies,
DOI 10.1007/978-3-642-15184-2_37,

The research described in this handbook is about machines whose functions are much less familiar. Generally, they are machines designed to do things that we associate with humans. Specifically, they are machines designed to do things that we associate with the emotional side of human life. It should be no surprise that extending into these new domains raises new questions in the domain of 'should'. The difficulty is that the arguments on all sides seem to be shrouded in mist.

Concerns often begin when popular writing throws up images of machines emerging from the laboratory fully equipped to carry out a function that we had thought of as uniquely human – very possibly with more than human accuracy. Closer inspection almost always shows that the machines reproduce at most a small part of the corresponding human ability – using the same word to refer to both is a matter of convenience rather than accuracy. Nevertheless, the mist remains: Who knows how much further the next generation of machines might go? Questions extend from there: If the machines acquire these new abilities, how will humans be affected? Of course there will be positive effects, otherwise the machines would not be taken up; but they may be a Trojan horse, and far-reaching effects on human society may be waiting to emerge. Compounding that, it is not clear what standards should be used to measure the good or the harm that a possible new system might do. In the extreme, might it not be better to replace the current version of humanity with a mechanised successor? The mist stretches on into the far distance.

Unfortunately, the mistiness of the arguments does not mean that there is nothing worrying beneath them. There are close links between emotion and morality. For example, saying that someone is jealous or sulking is to pass a negative moral judgement. It is disturbing to think that we might give machines licence to pass moral judgements under the cover of describing emotions. On the other side, there is an obvious case for saying that a machine showing signs of emotion is engaged in a kind of deception. There are very strong moral reasons for objecting to technologies that are designed specifically to achieve deception – particularly if the deception is likely to be targeted on the most vulnerable in society.

Even more concretely, contemporary technologies depend on databases of material that shows humans expressing relevant emotions. That raises unmistakable ethical questions. There may be good reasons to develop systems that can recognise people in fear of their lives, but it is absolutely not ethical to obtain the necessary recordings by inducing that kind of fear in randomly selected members of the public. Less obviously, there are ethical issues surrounding the dissemination of audio and visual recordings that are recognisable. It can be thoroughly disturbing for the subject of a recording to find him- or herself appearing in an undignified pose in conferences, journals or, worse still, press coverage. Those who are not personally moved by that line of thought should remember that the law is.

The chapters in this part of the handbook aim to give people working in the area the resources that they need to steer between unreasonable fear of misty possibilities and cavalier indifference to serious concerns. They range from pure principle to solid practicality.

1 The Chapters

The chapter by McGuinness deals head-on with a problem which, as many people in the area know from painful experience, can all too easily make it impossible to reach any kind of conclusion in an ethical debate. When people try to justify ethical conclusions in a particular case, they naturally try to ground their arguments in general ethical principles. The problem is that there are several quite different kinds of premise that can be used to ground ethical arguments. Most of them can be traced back in one form or the other for millennia and millennia of arguments have produced no way of deciding which is best. As a result, attempts to argue through specific ethical issues tend, with awful regularity, to slide back into arguments about the alternative types of ethical principles that might justify one choice or other and those are arguments that we know there is no prospect of resolving.

McGuinness points emotion-oriented computing towards a solution that has gained widespread acceptance in other areas which have real ethical concerns – above all, medical practice. It is called principalism. Its strength is that it refuses to begin the doomed search for moral axioms. It argues that particular moral judgements are essentially perceptual. We see that some things are acceptable and some are not. By and large, those that are acceptable are acceptable on any reasonable set of moral premises – naturally enough, because perceptual judgements provide the tests by which we measure the acceptability of a set of moral premises. What principalism adds to pure intuition is a set of guidelines that remind us what are the broad areas that should be considered in a moral judgement. Also, and not to be underestimated, it provides good and sophisticated authority for insisting that it is naive to think that ethical judgements should be derived from ethical first principles.

The two following articles move from the general basis of ethical judgement to particular ethical issues that affect emotion-oriented computing. Goldie, Döring and Cowie offer a broad view of the reasons why the area gives rise to particular moral and ethical concerns, while Baumann and Döring focus on one key concern, which is the impression that emotion-oriented systems might infringe human autonomy.

Goldie et al. describe their brief as dealing with 'the obstinate, vaguely defined concerns that seem troubling to the general public'. The argument is not that the concerns are objectively valid but that people working in the area need to understand them – otherwise, they will be permanently at cross purposes with the jury in the court of public opinion. Four problem areas are identified. The first is that people in general lack a framework that allows them to think clearly about even natural emotion, much less artificial analogues. Lacking a framework, they have no way to judge what is possible and therefore, they have no way to think through possible risks and benefits. The second problem area is what seems to be a deep-rooted human mistrust of things that behave or look like humans but that are actually not – in a word, impostors. Creatures that lure unwary people into trusting them, and then suddenly reveal their inhumanity, are part and parcel of the world's folklore. As the authors put it, that kind of feeling 'is an obstinately disturbing echo that disrupts attempts to hold cool, rational discussion about EOT'. Thirdly, the authors discuss the threat that ethics itself may be undermined by bringing people into contact with new kinds of agent. If we decide that we should treat machines as moral entities, we risk finding that humans slide alarmingly down the scale of moral value – for we would hardly build machines that reproduced the human race's uglier tendencies. If we deny them moral status, then we risk promoting systematic indifference to the fate of agents that are complex, intelligent and at least apparently emotional, and there is no guarantee that the indifference will stop with machines. Lastly, there are hugely complicated issues in ensuring proper legal control over a new type of device – particularly when it is unlikely that the law makers will have expert understanding of its potential. It will be no surprise to those who know Goldie's (2010) work that the chapter is intensely thought provoking.

In contrast, Baumann and Döring focus on one type of concern, which is at least philosophically relatively well defined. One of the areas highlighted by principalism is autonomy, implying 'a duty of non-interference, for example, respect for the decision-making capacity of an individual even if the consequences of these decisions are not in their best interests'. Emphasis on autonomy is not unique to principalism; it is a major theme in contemporary philosophy. Baumann and Döring show that many of the issues that worry the public, including those that are raised in the last part of the previous chapter, can be linked to that theme and that leads them to a clear prescription: 'developers have a *prima facie* duty to build machines that cannot (under normal conditions) infringe the autonomy of persons'. The chapter aims to equip people to fulfil that duty. The first part summarises contemporary thinking about autonomy. It then moves on to show why emotion-oriented technology carries a risk of infringing personal autonomy. It then considers principles that system developers might use to minimise the risk. It is worth noting, as the authors do themselves, that the chapter expands one part of the agenda that was identified in the previous two chapters. Autonomy is one of four themes in principalism, and present risks of violating clear ethical principles form one of Goldie et al.'s four sources of public concern. It is an expansion that may be particularly useful in practice, though.

The final chapter in the section focuses completely on practicalities. The established method of keeping research on track is monitored by an ethical committee. The first author of the chapter is Ian Sneddon, who has chaired ethical committees both in his own institution and for HUMAINE. The chapter is grounded in the experience of ensuring that committees are adequately informed and can reach decisions that are defensible and timely. The issues are set in the historical framework of the HUMAINE ethics committee, which was the first committee to be concerned specifically with emotion-oriented computing. The chapter sets many of the issues discussed in earlier chapters in the context of a working committee, such as the philosophical basis of ethical decision making. However, it also works through the very varied pragmatic issues, such as the codes that can be used, obtaining informed consent and storage of emotion data, which are the bread and butter of an ethics committee's work.

2 The Good We Hope to Do, the Harm We Should Avoid

Several of the chapters make the point that it would be wrong to focus exclusively on the problems that emotion-oriented technology might create if we were not vigilant. It is worth ending on that note.

It is a feature of the area that the research is idealistic. Our society requires people to interact with automatic systems. At best, the interactions currently depend on the humans adopting modes of interaction that suit the computers. That excludes people who find that kind of adjustment difficult or impossible, and it creates dangers in situations where, for instance, an emergency disrupts the human's ordinary ability to make the adjustment. It means that huge amounts of information that are held electronically are in practice highly unlikely to be absorbed by humans who could benefit from it (handbooks and manuals are a prime example). It leads to interactions that may be designed with the best will in the world but that most humans find aversive (automatic answering systems are a prime example). It would be easy to extend the list.

The point is not at all new, but it is worth repeating that in all these areas, pressures to develop systems with at least some emotional competence are not created by a community that has chosen to pursue its own research agenda. Rather:

> They arise willy-nilly when artefacts are inserted, into situations that are emotionally complex, to perform a task to which emotional significance is normally attached. There is a moral obligation to attend to them, so that if we intervene in such situations, we do it in ways that make things better, not worse (Cowie, 2010, p. 172).

Fulfilling moral obligations tends not to be easy, but ignoring them should not be an option.

References

Cowie R (2010) Companionship is an emotional business. In: Wilks Y (ed) Close engagements with artificial companions. John Benjamins, Philadelphia, PA, pp 169–172

Goldie P(ed) (2010) The oxford handbook of the philosophy of emotion. Oxford University Press, Oxford

Principalism: A Method for the Ethics of Emotion-Oriented Machines

Sabine Döring, Peter Goldie, and Sheelagh McGuinness

Abstract This chapter outlines the 'four principles' approach which is prevalent in medical ethics. Principalism was adopted as the ethical method of HUMAINE. This chapter introduces this method and also provides an account of the various criticisms of it. The chapter also includes some discussion of the relationship between ethics and scientific research. The purpose of this discussion is to show how ethics and good ethical research can be embedded in scientific practice. In conclusion, the chapter addresses the importance and the usefulness of considering fears that are embodied in works of science fiction when trying to deal with concerns and fears of the public. Consideration of the ethics of the possible is of massive practical importance and indeed should be a priority for those who are working within research groupings like HUMAINE. It is, however, often important to consider the ethics of what may never be possible – the science fiction if you like. When considering the impossible it will be important to stress the fact that these things are not and may never be possible. The role of this type of consideration lies in the importance of public engagement and showing that possible future scenarios are being taken into account by those who are pushing forward science and technology in this area.

1 Introduction

In this chapter I will give a brief outline of the 'four principles' approach that is prevalent in bioethical discussion. Principalism was adopted as the ethical method of HUMAINE. I will discuss some of the major criticisms of principalism – I do this believing that it is important to be aware of the weaknesses and the strengths. The chapter then goes on to raise some issues about the relationship between ethics and scientific research. In conclusion, the chapter addresses the importance and

S. Döring (✉)
Department of Philosophy, University of Tübingen, Tübingen, Germany
e-mail: mail@sabinedoering.de

P. Petta et al. (eds.), *Emotion-Oriented Systems*, Cognitive Technologies,
DOI 10.1007/978-3-642-15184-2_38, © Springer-Verlag Berlin Heidelberg 2011

the usefulness of considering fears that are embodied in works of science fiction when trying to deal with concerns and fears of the public. At a discussion held in Manchester we came to the conclusion that when dealing with ethics it may be useful to take a two-strand approach: consider on the one hand the ethics of the possible – this will be of massive practical importance and indeed should be a priority, but it was felt that it will often be important to also consider the ethics of what may never be possible – the science fiction if you like. When considering the impossible it will be important to stress the fact that these things are not and may never be possible. But it is important to consider them anyway as this is a way of dealing with public fears and engaging with those fears fully.

When considering how best to apply ethics to the science and technology which HUMAINE has been developing it is necessary to consider some theoretical issues. These are issues which engineers and scientists should be aware of as they carry out their research:

- Who or what are we considering ethics for?
- Are there more general obligations that we should be aware of, e.g. issues of social responsibility?
- What are the implications of this research?
- What are the aims of the research?
- Is the aim to make autonomous agents and if so what standards of autonomy would have to be reached in order to achieve this?
- If it is not possible to make actually autonomous agents but rather agents that appear to the user to be autonomous then different concerns will be at issue. These concerns will be heightened when these apparently autonomous agents are 'trying' to persuade the agent to act in a certain fashion.

This chapter will act as an introduction to the method of 'principalism', which it is hoped will help guide individuals through questions like these.

2 Principalism

The four principles approach was first championed by Beauchamp and Childress in the 1970s (Beauchamp and Childress, 1979). Also referred to as 'principalism' or the 'Georgetown Mantra', after the institute to which the pair belonged at the time, it has become a standard approach to ethical decision making among healthcare professionals.

> An examiner for the Royal College of General Practitioners membership exam says 'I . . . expect all candidates to not only be conversant with the four principles but also to be able to apply them appropriately'. (Gardiner, 2003)

The approach has been subject to a number of criticisms and its popularity, among ethicists, has seen a decline in the 1990s (Callahan, 2003). However, Beauchamp and Childress's book is now in its fifth edition (Beauchamp and

Childress, 2001). Beauchamp and Childress released the sixth edition of their book in 2008. Raanan Gillon has been the most influential supporter of principalism in England (Gillon, 1985). A trained practitioner as well as a philosopher, for Gillon, the beauty of the approach lies in the fact that he believes that it can provide an easily accessible approach to ethical decision making for those who need it most – medical practitioners. He says:

> As I began to teach using this framework I was impressed by the readiness with which doctors could agree that the four principles were indeed consistent with their own perspective on medico-moral issues. When I lectured more widely I found similar reactions from nurses, other healthcare workers, medical students .... (Gillon, 1994a)

Given that many of the individuals involved in emotion-oriented computing will also have a scientific background, it could be suggested that the method will have a similar attraction in this field as it does for clinicians in medicine. Later on in the chapter we will discuss some of the drawbacks of this approach when it is used by those who do not have an ethical background.

3 Beauchamp and Childress on the Four Principles

> Normative ethics is a form of inquiry that attempts to answer the question, 'which general moral norms for the guidance and evaluation of conduct should we use and why?' (Beauchamp and Childress, 2001a)

Beauchamp and Childress see the four principles of beneficence, non-maleficence, autonomy and justice as norms in a universal, common morality (Beauchamp and Childress, 2001b). By this they mean that the principles are so fundamental that they will appeal to 'all persons in all cultures who are serious about moral conduct' (Beauchamp and Childress, 2001c). The principles can be broadly defined in the following way:

- Beneficence implies an obligation to do good for your patient.
- Non-maleficence implies a duty to do no harm.
- Autonomy implies a duty of non-interference, for example, respect for the decision-making capacity of an individual even if the consequences of these decisions are not in their best interests.
- Justice is more problematic to define but at its most basic probably concerns access to health care and just distribution of healthcare resources.

The reason they choose the principles of beneficence and non-maleficence is that they have a long established place in healthcare ethics. The inclusion of autonomy and justice is due to the fact that they have in the past been neglected in medical ethics, but are now considered to be of great importance. The neglect of autonomy can probably be traced to the fact that traditionally paternalism was more acceptable and patients were more willing to defer to the 'better' judgement of their doctor. Justice is increasingly becoming a cause for concern in medical ethics as limits of available financial resources in health care have a direct effect on its distribution and accessibility.

The choice of these particular principles and their content has been widely criticized as being overly 'American'; I will look at these criticisms later. Beauchamp and Childress wished to create a method for use specifically in the field of bioethics and not a more general ethical theory (Beauchamp and Childress, 2001d). They see the principles as giving rise to prima facie rather than absolute obligations (Beauchamp and Childress, 2001e). Prima facie obligations can conflict with each other and a situation or decision can be both prima facie right and prima facie wrong at once. When this occurs these obligations must be further analysed and weighted against each other to ascertain what is the 'greatest balance' between right and wrong (Beauchamp and Childress, 2001e).

> What agents ought to do is in the end determined by what they ought to do all things considered. (Beauchamp and Childress, 2001f)

Throughout their account of principalism Beauchamp and Childress stress that the approach is not to be taken as a general moral theory. They recognize that alone the four principles are abstract and have insufficient content to deal with most problems in ethical reasoning. They see the principles as providing a framework and the processes of *specifying* and *balancing* build on this framework to move from principles to rules and ultimately decisions (Beauchamp and Childress, 2001f). Therefore when we consider principalism as a method for emotion-oriented machines, we must remember the principles are only the beginning of the method. It provides a framework within which to consider various ethical questions. It should not be seen as an algorithmic method in ethics through which definitive answers can be sought.

Specification plays a central role in principalism. While the principles provide a basis of values, specification removes a layer of abstraction and applies these values to specific issues that give rise to general rules. Specification provides 'action-guiding content' to principalism (Beauchamp and Childress, 2001g). However, it is not enough to create general rules; reasons and justification must also be provided in specification.

> Specification is an attractive strategy for hard cases of moral conflict as long as specification can be justified. Many already specified rules will need further specification to handle new circumstances of conflict. Progressive specification often must occur to handle the variety of problems that arise, gradually reducing the dilemmas and conflicts that abstract principles lack sufficient content to resolve .... All moral norms are, in principle, subject to such specification. They need this further content, because, as Henry Richardson puts it, 'the complexity of the moral phenomena always outruns our ability to capture them in general norms'. (Beauchamp and Childress, 2001h)

As can be seen by this passage, specification is a rigorous and dynamic process. However, the process of specification does not on its own solve conflicts that may arise between principles, as conflicts may also arise between specified rules. Indeed the process of specification requires the individual to make different judgements about different principles without always giving guidance for these judgements.

While specification deals with the 'range and scope' of each rule and principle, the process of balancing deals with the 'weight and strength' these rules should be given in ethical decision making.

> Moral progress is made through this work of specification, which often involves a balancing of considerations and interests, a stating of additional obligations, or the development of policy. (Beauchamp, 2001)

Balancing is an extension of specification and is not always necessary as some specified rules are taken as absolute. Balancing is important when rules come into conflict with each other and no over-riding interest is easily identifiable. When balancing one rule against the other, reasons must be given and these reasons cannot merely be subjective intuitions for believing one rule more important than the other. They must be morally justifiable reasons. Beauchamp and Childress lay down eight conditions which should be met in the process of balancing (Beauchamp and Childress, 2001i). The processes of specification and balancing will not always give one answer to an ethical problem, but they do not see this as being a weakness in the four principles approach. They say:

> The fact of unresolvable disagreement in some cases does not undermine an expectation that in most cases, the common morality ... affords us with adequate content to reach agreement or at least an acceptable compromise .... What one person may and should do may not be what other persons should do, even when faced with the same problem. (Beauchamp and Childress, 2001k)

Beauchamp and Childress also acknowledge that things other than principles should be taken into account in decision making, for example, 'cultural expectations' and 'precedent', all these factors combine to give weight to one particular outcome over another. In this final process of weighing one particular outcome against the other, Beauchamp and Childress use a method similar to that of Rawls' 'reflective equilibrium' (Beauchamp and Childress, 2001l). This process helps ensure that all the principles and rules used in reaching a decision are as coherent as possible. For Beauchamp and Childress ethical theories should be modified and revised in a manner similar to that of scientific hypotheses (Beauchamp and Childress, 2001m).

> This method requires, as does the associated method of reflective equilibrium, that we match and adjust all of our well substantiated moral judgments in order to render them coherent with the full range of our moral commitments. (Beauchamp, 2003)

4 Some Criticisms on Content and Method

Now that I have given an account of principalism I wish to examine some of the criticisms which have been levelled against it. Takala has questioned whether Beauchamp and Childress are justified in their claim of a common moral language.

She highlights that each principle will be subject to massively different interpretations depending on the context of the society in which they are being considered. This would mean that although the titles of the four principles would remain universal, their content would be hugely varied. Takala gives the following account of the difficulty in defining justice:

> In the spring of 1999, there was again in Finland the time for parliamentary elections. During the campaigns it became obvious that there was an over-whelming consensus among the rival parties that justice is important and that we should aim for a more just society. The only small difference between the parties was the understanding of what justice is and what measures should be taken that justice would prevail ... same word, but different interpretations of what justly belongs to whom. (Takala, 2001a)

She goes on to show the same difficulty with the other principles. This highlights the fact that only the titles of the principles are universal, their interpretation across cultures and within societies can be diverse. Holm raises similar criticisms about the difficulties of trying to create a universal ethic:

> The fact that common-morality theory necessarily uses the shared morality of a specific society as its basic premise is often overlooked by both proponents and opponents of the four principles. (Holm, 2001a)

This relativity seems to weaken Beauchamp and Childress' claim of a common morality. Holm states that this could be overcome by letting the four principles point to important aspects of morality across cultures but leaving the exact content to be decided within each culture. However, four chapters of *Principles of Biomedical Ethics* are devoted to a description to what the content of each principle should be (Holm, 2001a).

Clouser and Gert focus on this weakness in their criticisms of the four principles (Clouser and Gert, 1990a). They have three main objections to 'principalism'. The first of these is the criticism that the four principles are analogous to a checklist or a chapter heading but that they contain no moral substance. The four principles themselves seem to consist in nothing other than pointers, which may highlight considerations for ethical decision making but provide little guidance in coming to a decision. Beauchamp and Childress openly concede this. They say that it was never their intention that the four principles alone would be sufficient to deal with ethical dilemmas or problems.

> we agree that principles, order, classify and group moral norms that need additional content and specificity. Until we analyze and interpret the principles ... and then specify and connect them to other norms ... it is unreasonable to expect much more ... (Beauchamp and Childress, 2001n)

In light of the above criticisms it would seem that the principles are little more than chapter headings with assorted moral contents in different cultures and societies. Beauchamp and Childress may be overly optimistic in their claim to a global ethic. In their second criticism of principalism, Clouser and Gert state that the four principles provide no guidance to an individual when confronted with a moral problem (Clouser and Gert, 1990b). This lack of guidance leaves the individual free to give a subjective account of what is to be considered in the problem – they can focus on whatever principle they believe to be the most important and weigh the corresponding rules accordingly. This leads on to Clouser and Gert's third criticism: when two principles come into conflict, principalism is too indeterminate to give an account of ethical decision making (Clouser and Gert, 1990c). In response to this objection Beauchamp and Childress point to the processes of specification and balancing as guiding processes. According to Beauchamp and Childress these processes give structure to the process of ethical reasoning. Takala and Holm point to considerable problems with these processes. The obligations that arise from the four principles are, as I have mentioned earlier, prima facie and this means that they can be discarded if there is good moral reason to do so. However, as Takala points out, 'even this prohibition against breaking a promise turns out to be empty as a universal principle' (Takala, 2001b). What will be considered a morally good reason will be subject to the same relativism in interpretation as the principles are themselves. Holm criticizes the lack of technique involved in the processes of balancing and specification themselves. He claims that Beauchamp and Childress give little criteria as to what should be considered morally relevant in our ethical deliberation.

> Strangely enough the authors of PBE4 seem to see this as a strength of their theory: 'As with specification, the process of balancing cannot be rigidly dictated by some formulaic "method" in ethical theory. The model of balancing will satisfy neither those who seek clear-cut specific guidance about what one ought to do in particular cases nor those who believe in a lexical or serial ranking of principles with automatic overriding conditions'. (Holm, 2001b)

This statement seems to beg the question, what exactly is the purpose of the balancing then? This rejection of a formulaic method seems inconsistent with their previous descriptions of specification and balancing as dynamic processes analogous to scientific revision of hypotheses. Holm also highlights the fact that specification and balancing centre on subjective interpretations of what is important with regard to each of the four principles. The conditions, which are to be met in the process of balancing, are, on Holm's account, 'tautological' at best (Holm, 2001b).

5 What Is the Use of the Four Principles?

At this point I want to ask the questions, how useful is the four principles approach and to whom is it useful? On its own the approach appears to be of limited value; it is therefore not surprising that its supporters stress the fact that it is complementary to and can be used alongside other general moral theories. On Beauchamp and Childress' account, casuistry and accounts of virtues seem to be the favoured bedfellows for principalism. Considerable space is used to highlight the compatibility in *Principles of Biomedical Ethics*. Gillon even believes that principalism has the capacity to bridge the gap between utilitarianism and Kantianism: ethical theories that are generally thought to be in conflict with one another. He tells us:

> The elegance of the four principles approach is that it need say nothing about the deep and some claim untraversable, philosophical chasm separating these two types of philosophical theory – instead it offers each a meeting place in practical ethics. (Gillon, 1994b)

Even if principalism is useful in this respect it requires a basic knowledge of these theories that the majority of the healthcare professionals are not likely to have it would seem that supporters of principalism do not believe that they need it, as Gillon goes on to say:

> Nor, I believe, do all healthcare workers themselves need to try to acquire sufficient philosophical skills to come to a soundly based, well defended philosophical decision about which moral theory they accept as grounding those principles and why they reject all the others. If life long philosophers cannot succeed in this enterprise to the satisfaction of their philosophical opponents it would surely be ludicrous even to suggest that it is appropriate for healthcare workers spending a small part of their student and professional lives on the study of healthcare ethics to attempt it. (Gillon, 1994c)

As a method it may still have much to offer; it highlights four important principles which are worth considering in ethical analysis – however, it is a mistake to think that they are the only principles worth considering. It may provide a valuable method for ethical reasoning in some circumstances, if healthcare professionals, or those working with emotion-oriented computers, have sufficient philosophical knowledge to back up the approach. The biggest problem with principalism is what it does not do: it does not seem to motivate individuals to look beyond the four principles at more general ethical theories and this can only lead to underdeveloped ethical reasoning.

While it may seem at this juncture that I reject the four principles entirely, this is not true – they have many benefits. Time and again it is stated by those involved in clinical practice and ethics committees that the four principles provide a useful starting point. Also many of the criticisms that I have mentioned relate to the fact that medical treatment is often as cultural as it is scientific, and principalism will take on different forms in different cultures – whether these criticisms will be true of the work being carried out by HUMAINE is not yet clear to me. But it is important when considering any methodological approach to be aware of the perceived downfalls as well as the merits of the approach.

6 Some General Considerations

Science can move at astounding paces so it is important to try to pre-empt problems before they arise – this is so even if we sometimes have to delve into the realms of what is at present fiction. It will be better to have given prior consideration where possible. This is especially important when we are trying to gain public confidence in the work being carried out. Careful consideration should be given to how the possible results of research are presented. Scientists are often under huge pressure to get funding and the best way to do this is to present the best possible outcomes of their research. However, caution should be exercised. We have seen time and time again in the field of biotechnologies the presentation of the next 'panacea'. At present it is arguably stem cell therapies. While stem cell therapies may provide many astounding therapies the science is far from certain and this should be acknowledged. Again this will help with public perception of the sciences and allow for honesty and transparency. Also there should be on both sides a wariness to walk the road of either 'scientific imperatives' or 'precautionary principles'. That we can do something will not always imply that we should, and similarly epistemic uncertainty need not automatically give rise to the conclusion that we should not.

Another general point to consider is that when creating interactive machines, there must be an awareness of the ethical issues that will arise between the user and the machine. When machines are not themselves capable of ethical reasoning, many of the responsibilities will lie with the programmer. It is the programmer who must consider what it is they wish their machine to do and what possible effects this may have on users. It is also important to be aware of how the machines could be misused by others in the future, for example machines with the capacity to deceive the user may be able to cause the user much harm. If it is a persuasive agent, the machine may challenge the user's autonomy.

7 Separating Science from Fiction

Discussion of biotechnologies is often permeated with fears of these technologies; often these fears are illustrated by reference to works of fiction. In many of these, biotechnologies are often highlighted for the manner in which they can be abused and thus lead to a breakdown in social order. So if we look at one of the most famous books on the topic, *Brave New World* by Aldous Huxley, we see how society could be transformed by biotechnologies into a place where the norms and values we now have are alien (Huxley, 1994). Similarly the film *GATTACA* depicts a world in which a genetic underclass is created and although ultimately the system is effectively beaten by the 'defective' this is seen as triumph against biotechnologies (Niccol, 1997). Again in the film *The Island* we see how industry could exploit biotechnologies for profit and effectively create human beings in order to take their organs at some later stage (Bay, 2005). These are examples and there are many other works that could be cited. What can be taken from them is the fact that society does have concerns about technologies, although perhaps not as dramatic as these works would imply. These fears should therefore be acknowledged and accounted for and bioethical arguments should hope to answer/dissuade them. These works are the materials that fuel the myths surrounding biotechnologies.

We can learn a lot from fictional works if we accept the idea that they are a manifestation of the fears that people have with regard to biotechnologies. We can see similar fears of computer technologies. These fears are heightened when considering those technologies that aim to create smarter, more human like machines. Consider the films *I, Robot* (Proyas, 2004) and *Artificial Intelligence:*

A.I. (Spielberg, 2001) as examples. These works illustrate the fears that individuals have about 'clever' computers. However, a fine line must be walked between addressing concerns and the over-indulgence of the impossible.

8 Conclusion

So in conclusion when we use principalism as a method we are by no means confined by the approach championed by Beauchamp and Childress and indeed this may not be the most appropriate vehicle. However, there will be many things we can learn from the criticisms that have been levelled at their model. Their account can be amended and improved to better accommodate the needs of those working with emotion-oriented machines. Finally it will be in the interests of those working with emotion-oriented machines to identify and address the fears that the public may have about the technologies that are being developed – even though these fears may be based in fact but illustrated and perpetuated by fiction. In biotechnology there has often been unnecessary controversy and backlashes against certain breakthroughs – these could be avoided if careful consideration is given to the fears that fuel these reactions.

In spite of all the criticisms outlined above, principalism has many attractions, primarily the fact that as a method it is just intended to encapsulate our commonsense morality. Commonsense morality is not aiming to deliver up answers algorithmically and so in that sense the criticism of principalism is misplaced. The theoretical approach of HUMAINE is that of commonsense morality as embodied in principalism. Within the principles we find simply an account of everyday common sense that can be used to achieve the best outcome to any problems that may arise. This will help the researchers involved in this project to think sensibly in terms of ethics – something that it can be seen has been achieved so far. When it appears that the principles conflict with each other or are mutually incompatible it will be necessary to consider background conditions. Situations should always be considered in the appropriate context, which is not to say that we should give way to relativist accounts of 'anything goes'. These principles act merely as 'rules of thumb' and it may not always be necessary to consider all four in every situation. They do, however, provide a clear framework that can be used methodologically.

While it is important to be aware of the ethical issues involved in emotion-oriented computing one must not be *too* aware of them. Whether human–machine interaction can create genuinely novel situations is arguably unlikely, the principles of beneficence, non-maleficence, autonomy and justice will provide a solid framework within which different possible outcomes can be examined. When considering the uses of automated machines it is important to remember the general guidelines that apply in that field, for example when creating persuasive machines for use in the advertisement industry there may be advertising standards or regulation of advertising methods which must be abided by. There should not be a presumption that novel situations or technology necessarily lead to novel ethical issues.

References

Bay M (2005) The Island. DreamWorks SKG
Beauchamp T (2001) Principalism and it's alleged competitors. In: Harris J (ed) Bioethics. Oxford University Press, Oxford, p 82
Beauchamp T (2003) Methods and principles in biomedical ethics. J Med Ethics 29:269
Beauchamp TL, Childress JF (1979) Principles of biomedical ethics, 1st edn. Oxford University Press, New York, NY
Beauchamp TL, Childress JF (2001) Principles of biomedical ethics, 5th edn. Oxford University Press, New York, NY
Beauchamp TL, Childress JF (2001a) Principles of biomedical ethics, 5th edn. Oxford University Press, New York, NY, p 2
Beauchamp TL, Childress JF (2001b) Principles of biomedical ethics, 5th edn. Oxford University Press, New York, NY, p 3
Beauchamp TL, Childress JF (2001c) Principles of biomedical ethics, 5th edn. Oxford University Press, New York, NY, p 4
Beauchamp TL, Childress JF (2001d) Principles of biomedical ethics, 5th edn. Oxford University Press, New York, NY, p 23; Beauchamp T (2001) Principalism and it's alleged competitors. In: Harris J (ed) Bioethics. Oxford University Press, Oxford, p 480
Beauchamp TL, Childress JF (2001e) Principles of biomedical ethics, 5th edn. Oxford University Press, New York, NY, p 14
Beauchamp TL, Childress JF (2001f) Principles of biomedical ethics, 5th edn. Oxford University Press, New York, NY, p 15
Beauchamp TL, Childress JF (2001g) Principles of biomedical ethics, 5th edn. Oxford University Press, New York, NY, p 16
Beauchamp TL, Childress JF (2001h) Principles of biomedical ethics, 5th edn. Oxford University Press, New York, NY, p 17
Beauchamp TL, Childress JF (2001i) Principles of biomedical ethics, 5th edn. Oxford University Press, New York, NY, p 20
Beauchamp TL, Childress JF (2001j) Principles of biomedical ethics, 5th edn. Oxford University Press, New York, NY, p 385
Beauchamp TL, Childress JF (2001k) Principles of biomedical ethics, 5th edn. Oxford University Press, New York, NY, p 22
Beauchamp TL, Childress JF (2001l) Principles of biomedical ethics, 5th edn. Oxford University Press, New York, NY, p 397
Beauchamp TL, Childress JF (2001m) Principles of biomedical ethics, 5th edn. Oxford University Press, New York, NY, p 399
Beauchamp TL, Childress JF (2001n) Principles of biomedical ethics, 5th edn. Oxford University Press, New York, NY, p 19
Beauchamp TL, Childress JF (2008) Principles of biomedical ethics, 6th edn. Oxford University Press, New York, NY
Callahan D (2003) Principalism and communitarianism. J Med Ethics 29:287
Clouser KD, Gert B (1990a) A critique of principalism. J Med Philos 15:279–310
Clouser KD, Gert B (1990b) A critique of principalism. J Med Philos 15:383
Clouser KD, Gert B (1990c) A critique of principalism. J Med Philos 15:391
Gardiner P (2003) A virtue ethics approach to moral dilemmas in medicine. J Med Ethics 29:297
Gillon R (1985) Philosophical medical ethics. Wiley, Chichester; Gillon R (ed) (1994) Principles of healthcare ethics. Wiley, Chichester
Gillon R (1994a) The four principles revisited – a reappraisal. In: Principles of healthcare ethics. Wiley, Chichester, p 319
Gillon R (1994b) The four principles revisited – a reappraisal. In: Principles of healthcare ethics. Wiley, Chichester, p 321

Gillon R (1994c) The four principles revisited – a reappraisal. In: Principles of healthcare ethics. Wiley, Chichester, p 325
Holm S (2001a) Not just autonomy: the principles of american biomedical ethics. In: Harris J (ed) Bioethics. Oxford University Press, Oxford, pp 494–506
Holm S (2001b) Not just autonomy: the principles of american biomedical ethics. In: Harris J (ed) Bioethics. Oxford University Press, Oxford, pp 504
Huxley A (1994) Brave new world. Flamingo, London
Niccol A (1997) GATTACA. Columbia Pictures, Columbia, SC
Proyas A (2004) I, Robot. 20th Century Fox
Spielberg S (2001) Artificial intelligence: A.I. Warner Bros. Picture
Takala T (2001a) What is wrong with global bioethics? On the limitations of the four principles approach. Camb Q Healthc Ethics 10:72–77
Takala T (2001b) What is wrong with global bioethics? On the limitations of the four principles approach. Camb Q Healthc Ethics 10:76

The Ethical Distinctiveness of Emotion-Oriented Technology: Four Long-Term Issues

Peter Goldie, Sabine Döring, and Roddy Cowie

Abstract In this chapter we consider certain long-term ethical issues which are peculiar or special to emotion-oriented technology and which make the topic particularly charged ethically, both for the lay public and for those working in the area. We identify four such issues. First, it is far from clear whether technologies made by humans are conceivably capable of emotionality and, more generally, of phenomenal consciousness. Second, where we are dealing with a technology that simulates emotionality, we have responses that are often far from cool and rational. Third, as discussion of the first two issues will have illustrated, our ethical responses to emotion-oriented technology are often emotionally charged, lending a peculiar reflexivity to our ethical deliberations. This leads to a discussion of the kind of value that such technologies have, and of how they should be ethically treated by humans. Fourth, emotion-oriented technology impinges on many matters of law (such as laws of privacy); we discuss in particular the importance of technology that is used to filter raw emotional data on people for further use.

1 Introduction

The aim of this chapter is to consider what is special about the ethics of emotion-oriented technology.

One way of thinking about emotion-oriented technology (EOT) is that it is just another kind of technology, and that there is nothing special about it so far as ethics is concerned. On this view, we are just faced with the usual difficulties with technology concerning judgements under uncertainty about risk (Kahneman et al., 1982). We are certainly faced with these difficulties, and they should not be underestimated. But there is a good case for believing that EOT raises issues that are not covered in every text that deals with ethics and technology.

P. Goldie (✉)
Department of Philosophy, University of Manchester, Manchester, UK
e-mail: peter.goldie@manchester.ac.uk

P. Petta et al. (eds.), *Emotion-Oriented Systems*, Cognitive Technologies,
DOI 10.1007/978-3-642-15184-2_39,

If there are special ethical issues to be faced, then it makes sense for the community to understand them in some depth. This chapter is directed towards people who take that view seriously. Certainly for most computer scientists (or engineers or even psychologists) working in the area, it will be possible to cope most of the time without thinking more deeply about ethics than the minimum practical requirement ensuring proper scrutiny by an ethical committee. But there are also times when the community is challenged on ethical grounds, and then it needs people who can engage with the non-routine issues that this chapter raises.

We propose that what makes the area distinctive rests on four fundamental issues. The chapter will consider these four fundamental issues and various ways in which they manifest themselves. We should emphasise that it is not our concern here to consider the ethical implications connected with any particular existing research project or with the risks of ethically sound technologies being misused or abused through getting into the wrong hands. These are matters that can and should be dealt with by an ethics committee.

The issues that this chapter deals with are the obstinate, vaguely defined concerns that seem troubling to the general public, who are not generally deeply informed about the technical details of research that ethics committees deal with; and to the individuals and bodies involved in funding and monitoring such research. It is, one might say, a *tour d'horizon*.

2 Uncharted Conceptual Territory

EOT is distinctive because we lack a sound conceptual framework for understanding emotion itself, and this conceptual deficit makes it difficult to think clearly about risk. For example, there may be risks involved in mobile phone technology, and these risks may be hard to assess, but there is not also a lack of a firm conceptual grasp of what is involved, of what mobile phones are. And much the same applies to nuclear technology. In contrast to the nuclear example, it should be emphasised that the lack of a sound conceptual framework for the emotions is not confined to the lay public, but also affects philosophers and scientists working on the emotions – and, of course, on consciousness more generally, often these days called the last frontier of science. There is no settled consensus on many conceptual issues concerning consciousness and emotions, and there is no sign that one will emerge in the near future.

That conceptual uncertainty manifests itself in our attitude towards emotions in EOT – towards machines that, in some way, engage with emotionality. We are, these days, quite untroubled by computing machines. We perceive no threat to our humanity from, for example, supercomputers that can compute over highly complex material, often at speeds and with accuracy that ordinary mortals cannot aspire to. We are, however, troubled about two aspects of technology, which many people do intuitively feel threatened by. The first is where the technology has what might broadly be called the capacity for creativity, and in particular artistic creativity. The second is where the technology has the capacity for emotion and emotionality. (We will not address the creativity question, although we believe that there are important emotional aspects to creativity which may explain some of our difficulties here too.)

One of the manifestations of the lack of a framework for comprehension referred to above is that we are uncertain whether it is at all imaginable for technology to have the capacity for emotion. It is quite possible to argue that the concept of emotion has no place outside the human being and other animals. For example, it is disputed whether something made of metal and carbon fibre could ever be capable of emotion, or whether it is necessary for emotionality that whatever is to have it must be composed of the same kind of stuff as we are composed of Searle (1992). And, perhaps most fundamentally, it is disputed whether, if something is to have an emotion (to be afraid for example), that thing must also feel that emotion – to have (or at least be capable of having) certain feelings that are characteristic of that emotion.

The contrast can be put in terms of the more general contrast between two kinds of consciousness: what the philosopher Ned Block has called access consciousness and phenomenal consciousness (1995). Roughly, access consciousness is the kind of consciousness involved in mere cognition – information storage and processing for example. So, for example, the capacity of something to recognise a threat and to respond with evasive behaviour has access consciousness. And, still as part of access consciousness, a more complex organism might also be capable of recognising its own internal states, such as the state which represents that it is threatened and that a certain kind of evasive response is called for. Phenomenal consciousness, in contrast, is what is involved when there is something that it is like for the organism – in this case, where there is something that it is like to feel fear (Nagel, 1974). Clearly there is something that it is like to be a human, a dog, or a cow – they all have phenomenal consciousness, and they all can feel fear – but it seems obvious to most people that there is nothing it is like to be a stone, or a computer.

We do not know whether there could ever be something that it is like to be a robot – could a robot ever feel fear? However, as science fiction literature and film attest, people can easily be disturbed by the idea of non-animal things that are capable of emotional feelings: consider, for example, the Nexus-6 replicants in *Blade Runner* who are programmed with a fail-safe device to cease functioning after 4 years in case they start to develop empathy (Goldie, forthcoming) and Hal in *2001: A Space Odyssey*, who seems to be motivated emotionally, by revenge or envy perhaps, and who seems to suffer as his systems are shut down.

The point of these examples is to underline the slightness of the conceptual framework that human beings in general bring to understanding emotion itself. The examples are based on thought experiments: the real work in EOT is very far from all this. Even to say that it is tomorrow's problem would be misleading, as we have no idea whether it is even conceptually possible that such things might exist outside science fiction and philosophers' thought experiments. But the experiments are about issues that go the heart of our humanity, and they expose uncertainty about those issues. It is no wonder, then, that people tend to have a nervousness about the more practical and feasible aspects of EOT. They may not know the reasons, but it is reasonable to be nervous given the wider conceptual uncertainties that we have been discussing in this section.

Any research project in EOT must be properly sensitive to these issues. Science fictional characters should not simply be dismissed as mere science fiction: science fiction they are, but, in respect of their emotional resonance to current research and technology, they are not *mere* science fiction. The science fiction is a reflection of the fact that the capacity for emotion is in some sense special for us humans.

Quite what the sense is in which it is special is very hard to get a grip on, in large part because of the lack of a framework for comprehension just referred to. But we do not need to know exactly (or even roughly) what emotions are, to know that they are special. We have said that emotions (and consciousness more generally) are the last frontier for science. But unlike other frontiers of science – the unification of relativity and quantum theory, and so on – they are also personally significant: the last bastion of humanity.

3 The Last Bastion of Humanity

We now put science fiction aside to consider people's confused attitudes towards simulation of humanity and emotionality in EOT. Our concern shifts from the elusive notion of EOTs that have emotion, to the likely future reality of EOTs that give the impression of emotionality. This is likely to be possible in various manifestations in robots, in avatars and in embodied conversational agents (ECAs), through speech, appearance, behaviour and in other ways.

First, we want to mention an important and often neglected point. In human–machine interaction, we are, and have long been, very familiar with the idea that we interact with machines that simulate emotion and that are involved in bringing about emotional responses in the human user. Consider, for example, how emotionally charged is our reaction to the pre-recorded telephone message from the airline company, telling us they are 'sorry' to keep us waiting, and that our call is 'valuable' to them. We often have strong negative emotions in response – frustration, anger, feelings of inadequacy – at this blatant pretence of caring from a 'system' in which we are at best a number on a screen. This kind of interaction with technologies seems to be a near inevitability in our lives these days. The neglected point, then, is that, given this state of affairs, it is an excellent thing that there are people working in EOT aiming to develop better systems that will reduce or even eliminate such negative emotions in our interactions with technology. The following remarks should be understood in that overall context.

In the particular case of robots, it has been argued that our emotional responses take a curious shape, in what is called the uncanny valley, a term coined by Mori (1970), and now much discussed in robotics and computer science. The essential idea, captured in Fig. 1, is that our emotional attitude towards robots changes as they become more and more similar to human beings (in behaviour, in facial and verbal expression and so on). The claim is that, as the diagram shows, the relation is

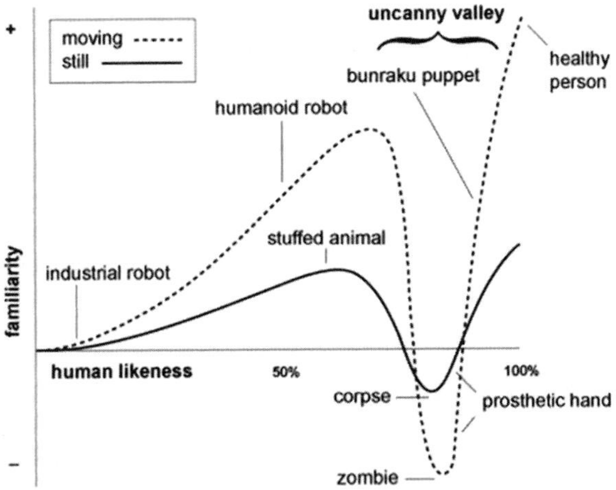

Fig. 1 'The uncanny valley' (Mori, 1970); adapted by MacDorman and Ishiguro (2006)

not a linear one. We are more comfortable with a humanoid robot than an industrial robot, but when the artefact becomes close to being like a healthy human but is still clearly not human, our feelings of comfort and familiarity decline: we are in the uncanny valley.

The concept of an uncanny valley has an immediate appeal, but it is not clear that it is accurate. The most systematic study of the topic suggests that it is not (Hanson et al., 2005). Creating progressively closer approximations to humanity may create artefacts that are 'bukimi' in Mori's sense – weird, ominous, eerie – but it is not clear that it usually does so. Not everything that comes close to human appearance is like a corpse, or a prosthetic hand, or a zombie. These very particular human-like things may signal potholes that a sensible technologist can and should steer round rather than a long and unavoidable valley.

Even if that is so, the potholes need to be understood. When people react in that way to particular artefacts, it may be because they give rise to disgust; because they deviate from the norms of physical beauty; because they frustrate our (largely unconscious) expectations; because they give rise to fear of death (MacDorman and Ishiguro, 2006). When these emotional responses occur, they are generally not of the kind that can be seen as rational, in the way that, for example, fear of a savage dog would be rational. They are, rather, more visceral, more primitive. Technology is both ethically and practically bound to deal carefully with these responses and to make sure that they are not roused inadvertently.

Furthermore, people's readiness to believe that there is an uncanny valley, on only the slightest evidence, suggests that visceral responses also penetrate what people believe is rational consideration of the issues. It seems likely that these responses are, at least in part, yet another expression of our confusions about EOT. They run through all discussions of the ethics of the issues.

There is a debate in robotics as to whether it should work towards developing a robot that we might, in our interactions with it, mistake for a human being, or at least treat in way that reflects confusion on our part about how it should be treated (perhaps as responsible autonomous moral agents). Recent research indicates that such confusion arises to some extent even in relation to avatars and robots that are known not to be human (Reeves and Nash, 1996; Slater et al., 2006; Rosalia et al., 2005; Bartneck et al., 2006). It may therefore be less difficult than one might assume to build a robot that a human might genuinely confuse with another human being. It is, as yet, far into the realm of science fiction, and so it does not impinge on practical ethics at the present time. But like uncertainty about emotion itself, it is an obstinately disturbing echo that disrupts attempts to hold cool, rational discussion about EOT.

4 The Reflexivity of Ethics and EOTs

Our ethical responses to EOTs are themselves emotional. The old idea that ethical judgements are cool and dispassionate has now been replaced, in philosophy and in psychology, with a picture where emotions are central to our ethical intuitions and judgements (Haidt, 2001, 2007). So there is a tricky reflexive aspect to ethics and EOT: the tool that we are using in our ethical deliberations is the very tool that is under examination in those deliberations.

An immediate response to the question of what ethical stance we should take towards technology that merely simulates emotionality might be a dismissive one: such machines are of mere instrumental value and should be treated no differently to the way we treat a can opener or a laptop computer. The idea that such machines could have rights, or that we could have duties towards them, might accordingly be thought insupportable, or even absurd. This may well the correct reaction, at least so far as concerns rights and duties. But still, there might be good reasons to treat such machines as non-instrumentally valuable (Goldie, unpublished; Bartneck et al., 2008). Our feelings towards, and the way we treat, EOTs, is expressive of our personality, and personality traits (of this kind) are largely a matter of habit. So there is a risk that we can become habituated in treating EOTs badly, and from this (especially bearing in mind the uncanny valley) there is a further risk that we will start to treat certain humans like this too, merely as means. This idea too is familiar from literature and film, and is often associated with a dystopia (Fritz Lang's *Metropolis* for example). On this view, then, we should cultivate our personality to make sure this does not happen; that we do not slide down the slippery slope to treating human beings in this way too.

There is another way in which reflexivity of ethics and EOT is manifested. EOTs are capable, to an increasing degree, of using emotions to persuade users. They can 'use' emotions in two senses. The first sense is the one that is familiar to us through our encounter with TV and other advertisements: the way in which they can appeal to our emotional sensitivities to persuade us to act in certain ways – to buy a product, or to take a holiday somewhere. The second is less familiar: the way in which EOTs could simulate emotionality in themselves in order to generate emotional responses in the user, and thus persuade us to act in certain ways (as discussed by Marco Guerini for HUMAINE). As evidenced by the use of Tagamochi toys with children, this can be highly effective, possibly in deleterious ways. A number of familiar ethical issues have application here. Let us mention just two. Issues arise concerning whether the end can justify the means. For example, if an EOT is more likely than a doctor to get true answers from patients to a medical questionnaire, would the end (better health for the patient) be justified by the means (rhetorical persuasion, perhaps subliminal, by the EOT)? And, second, issues arise concerning whether rhetorical emotional persuasive devices in EOTs undermine the autonomy of the user. For example, would a patient using a persuasive EOT justifiably consider his autonomy to have been undermined if he is not properly informed of the procedures (which might in itself eliminate their usefulness)? Again, as we saw earlier, perception of risk with regard to such issues itself involves emotion (Slovic, 2007) – another aspect of reflexivity.

5 EOT, Ethics and the Law

What is legal and what is ethical are not, of course, co-extensive. But in many cultures and in many circumstances, the law will often embody our intuitions about what is right and wrong. There is in western Europe little legislation that is specifically aimed at EOT, at least as far as we are aware (legislation on polygraphs or lie detectors being perhaps an exception). So any cases would have to be decided on existing legislation and case law, interpreted and applied as thought appropriate – for example, in employment legislation, human rights and privacy laws. Difficulties may well arise here, in part because of the uncharted conceptual waters which we have already discussed. For example, if there is an EOT which has the capacity to recognise someone's emotion from their facial expression in public, is it an invasion of privacy to record emotions in this way?

This example illustrates an important general issue concerning EOT. There are enormous practical complexities in deciding what is consistent with the law as it currently stands, and in seeing how the law could be adapted to allow applications that are benign, and rule out those that are not. Behind those complexities lie ethical questions about real devices rather than imaginary futures, which hinge as much on use as the device itself.

The issue is well illustrated in a class of applications that have been called SIIFs: semi-intelligent information filters. The hypothetical emotion recogniser mentioned above is a case in point. It is semi-intelligent in that it records more than just raw data – bodily posture and movements, facial expressions, eye saccades and so on; it also interprets this data in a meaningful way: for example, recording that the person is upset or feeling aggressive. SIIFs can be, and often are, enormously useful; for example, in car technology they can be used to determine whether a driver is safe to drive, and to advise him accordingly. As a filter, a SIIF will, as in this example, characteristically have the power to transmit data for further action. The same data, though, can be used for a high-impact judgement about the person being observed. For example, a SIIF of the kind envisaged might be useful in monitoring employees in a call centre, or for use by anti-terrorist police in monitoring people in crowded public areas, such as shopping malls or railway stations. In some of those situations, there is a real risk that a false positive could affect a person's life, or end it.

As these examples illustrate, even SIIFs with clear potential to do good raise complex issues on the borderline between ethics and the law, some of which are analogous to those that have arisen with polygraphs: for example, how to ensure proper understanding of a device's accuracy, and to ensure that questions of admissibility as evidence and invasion of privacy are properly addressed.

Those working in EOT, as researchers, or as funding bodies and monitors, are well aware of these issues. And it is right that they should be, because there is potential for abuse of EOT. There is always a risk that it will come into the public gaze at times when abuse has happened, or at least when it is being alleged; and this could affect the public attitude towards a research programme that aims to do good.

6 Conclusion

We are in uncharted conceptual waters with EOT, and emotion is seen as the last bastion of humanity. Accordingly, emotions run high about the ethics of EOT (the reflexivity point), and it is essential for us all to be sensitive to this. But it must not be forgotten that EOT is an enormous force for good, for example in reducing or eliminating the negative emotions that we so often feel in our interactions with complex technologies, from computers to car navigation systems to online questionnaires and booking services. It is part of the task for people working in the field to proselytise the benefits of EOT to humanity: to explain how it can make our lives easier and better. And at the same time, it is essential that there are in place adequate systems for ethical governance of the kind that were introduced in HUMAINE, able to draw on real depth and breadth of expertise in science, in emotion theory and in ethics.[1]

References

Bartneck C, et al. (2006) 'To Kill a Robot'. In: Proceedings of the workshop on misuse and abuse of interactive technologies in cooperation with the conference on Human Factors in Computing Systems (CHI2006), Montreal, QC

Bartneck C, Brahnam S, De Angeli A, Pelachaud C (2008) Misuse and abuse of interactive technologies. Interact Stud – Social Behav Commun Biolog Artif Syst 9(3):397–401

Block N (1995) On a confusion about the function of consciousness. Behav Brain Sci 18:227–287

Goldie P Forthcoming. What is it like to be a Nexus-6 replicant? In: Coplan A (ed) Blade runner. Routledge, London

[1] For those who are interested in an outline to some of the current philosophical issues in emotions, a good place to start is Ronald de Sousa's entry in the online Stanford Encyclopedia of Philosophy: http://plato.stanford.edu/entries/emotion (accessed 17 May, 2010)

Haidt J (2001) The emotional dog and its rational tail: a social intuitionist approach to moral judgement. Psychol Rev 108:814–834

Haidt J (2007) The new synthesis in moral psychology. Science 5827:998–1002

Hanson D, Olney A, Prilliman S, Mathews E, Zielke M, Hammons D, Fernandez R, Stephanou H (2005) Upending the uncanny valley. In Proceedings of the 20th national conference on Artificial Intelligence, AAAI Press, Menlo Park, CA, pp 1728–1729

Kahneman D, Slovic P, Tversky A (1982) Judgment under uncertainty: heuristics and biases. Cambridge University Press, Cambridge, MA

MacDorman KF, Ishiguro H (2006) The uncanny advantage of using androids in cognitive science research. Interact Stud 7(3):297–337

Mori M (1970) The uncanny valley. Energy 7(4):33–35

Nagel T (1974) What is it like to be a bat? Philos Rev 83:435–450

Reeves B, Nash C (1996) The media equation: how people treat computers. Cambridge University Press, Cambridge, MA

Rosalia C, Menges R, Deckers I, Bartneck C (2005) Cruelty towards robots. In: Robot workshop – Designing robot applications for everyday use, Göteborg

Searle J (1992) The rediscovery of the mind. MIT Press, Harvard, MA

Slater M, Antley A, Davison A, Swapp D, Guger C et al (2006) A virtual reprise of the Stanley Milgram obedience experiments. PLoS One 1(1):e39

Slovic P (2007) 'If 'I look at the mass I will never act': psychic numbing and genocide. Judgm Decis Mak 2:1–17

Emotion-Oriented Systems and the Autonomy of Persons

Holger Baumann and Sabine Döring

Abstract Many people fear that emotion-oriented technologies (EOT) – capable of registering, modelling, influencing and responding to emotions – can easily affect their decisions and lives in ways that effectively undermine their autonomy. In this chapter, we explain why these worries are at least partly founded: EOT are particularly susceptible to abuse of autonomy, and there are ways of respecting the autonomy of persons that EOT are unable to accomplish. We draw some general ethical conclusions concerning the design and further development of EOT, contrasting our approach with the "interactional design approach". This approach is often thought to avoid infringements of user autonomy. We argue, however, that it unduly restricts possible uses of EOT that are unproblematic from the perspective of autonomy, while at the same time it allows for uses of EOT that tend to compromise the autonomy of persons.

1 Introduction

Many people react with scepticism to the prospect of machines being able to engage emotionally with them. One important source of this scepticism is concerns about autonomy. People do not want machines to invade their emotional sphere, it seems, because they fear that by losing 'emotional privacy' they become vulnerable to 'emotional piracy'; it appears to them, less metaphorically, that emotion-oriented technologies (EOT) – capable of registering, modelling, influencing and responding to emotions[1] – can easily affect their decisions and lives in ways that effectively undermine autonomy. Consider, for example, an emotion-oriented system that detects how a person feels about all the gadgets in a computer store and starts to run

H. Baumann (✉)
Universitärer Forschungsschwerpunkt Ethik, Universität Zürich, Zürich, Switzerland
e-mail: baumann@ethik.uzh.ch

[1] Cf. http://emotion-research.net/aboutHUMAINE

P. Petta et al. (eds.), *Emotion-Oriented Systems*, Cognitive Technologies,
DOI 10.1007/978-3-642-15184-2_40,

advertisements for those the person is most excited about; or imagine a system that makes use of information about a user's emotional states in order to persuade him to take his medication (e.g. by purposely confronting him with those facts that increase his anxiety and make him feel guilty).

The primary aim of our chapter is to provide a framework that helps to articulate and to make sense of the uneasiness that is generated by such scenarios. In order to do so, several issues need to be raised. *First*, a thorough understanding of autonomy and its relationships to privacy and the emotions is called for: What is meant by autonomy, and how is the principle of respect for autonomy to be understood? (Sect. 2) In which ways are emotions significant to autonomy? Why is privacy of importance in connection with autonomy, and what is the status of emotional privacy in this regard? (Sect. 3) Addressing these far-reaching questions will allow us, *second,* to attend to the more specific questions: Why should it be considered more problematic that machines, and not human beings, emotionally interact with people? Is this due to the fact that there are ways of respecting autonomy that machines are (in principle) unable to accomplish? Are EOT particularly susceptible to abuse? (Sect. 4) And if there are any such problems, as we will suggest, what consequences should be drawn for the design and further development of EOT, or for possible fields of applications? (Sect. 5).

2 Outline of a Theory of the Autonomous Person

The autonomous person decides for herself and leads a life according to her own values and beliefs. She is neither a passive bystander of external forces that move her to act, nor does she always mindlessly accept and act on the commands and opinions of others. Instead she raises the questions of what to believe, what to do and what kind of person she wants to be and directs her actions and life correspondingly.[2]

The *condition* of autonomy must be distinguished from the *capacity* for autonomy.[3] Being autonomous presupposes that a person has several capacities, most prominently capacities for self-reflection and rationality (rational self-control). But merely having these capacities does not guarantee that a person is autonomous because other persons or the social environment she lives in might prevent her from *effectively exercising* them. In order to be autonomous, a person must also enjoy what is sometimes called 'procedural independence' – she must be free from distorting factors that subvert or inhibit her self-reflective and rational faculties. This condition is crucial to our understanding of the principle of respect for autonomy,

[2]This characterization is supposed to be compatible with a variety of views on autonomy. More elaborate characterizations and helpful introductions to the notion of autonomy can be found in Christman (1989, 2003), Dworkin (1988, Chap. 1), Friedman (2003, Chap. 1), Feinberg (1989) and Oshana (2006, chap. 1).

[3]The capacity/condition distinction is discussed, e.g., in Christman (1989, p 5), Feinberg (1989, 28 ff.) and Oshana (2006, pp 6–9).

and it is also important with regard to the relationships between autonomy, the emotions and privacy we want to emphasize. Before we consider this condition in some detail, however, the notions of self-reflection and rationality need to be introduced.

2.1 Self-Reflection

We propose to distinguish between three modes of self-reflection: self-awareness, self-understanding and self-evaluation. To begin with, a person must reflect on *who she is* in order to become aware of her standing desires, values, motivational dispositions, as well as her background beliefs and the position she occupies in her social environment.[4] A person who fails to attain self-awareness is unintelligible to herself and cannot control her life, partly because she is unable to engage in other modes of self-reflection. Consider Mary who is ignorant of her strong resentment against a colleague. She is often puzzled by her behaviour in his presence – that she becomes more aggressive and impatient – and resorts to rationalizing explanations until she finds out about or owns up to her hostile feelings. She is then able to make sense of her behaviour. Furthermore, she is in a position to ask herself why she has these feelings, whether she wants them to have influence on her actions and how she can possibly cope with them. In other words, she can try to achieve self-understanding and engage in self-evaluation.

A person who reflects on *why she has* certain desires, values, feelings, etc. can gain self-understanding in different ways. Mary might figure out that her resentment is a reasonable (or appropriate) response to the colleague's constant devaluation of herself and of others.[5] Sometimes, however, no such 'reason explanations' are available and one must attend to the causal processes or social influences under which certain desires or feelings have developed.[6] To stick to the example, Mary might initially find her resentment unintelligible because she regards her colleague to be a decent and sensitive person. But by attending to the genesis of her resentment she might realize that it has originated from an incident some time ago in which the colleague made her look silly and that she got somehow stuck with this feeling. In both cases, Mary gets a better grip on her feelings and can cope with them. In general, self-understanding is essential to autonomy because people become more intelligible to themselves; in addition, they gain access to another mode of self-reflection: self-evaluation.

In self-evaluation a person asks herself – from the background of her values, commitments, beliefs, etc. – what kind of person she wants to be, what desires or feelings she wants to be motivated by, and so on. In doing so the person is active rather than

---

[4] Cf. Christman (1991, 16 ff.), Velleman (1989, part I) and Meyers (1989) who speak of a competency for 'self-discovery'.

[5] Joseph Raz (1997) claims, e.g., that we are active rather than passive if we are responsive to reasons, i.e. if we can rationally make sense of our beliefs and desires (feelings, etc.).

[6] Cf. Christman (1991) and Young (1980).

passive. She is not a passive bystander but instead plays an active part in shaping her life. Many philosophers have concentrated on this mode of self-reflection since it seems to be the hallmark of autonomy.[7] But without self-awareness and self-understanding, a person would not be able to engage in self-evaluation in a meaningful way. There would be nothing she could (competently) evaluate, as the above example shows. Mary must first become aware of her feeling and understand its precise content before she can raise the question whether she wants her resentment to play a role in her life. Thereafter she can decide to express her feelings to the colleague and let him know that she finds his devaluing behaviour unacceptable; or she can resolve to let the matter rest and to get rid of her resentment, given her insight that it has derived from an isolated and uncommon incident. In order to shape one's life and to exercise control, one must engage in all three modes of self-reflection.

But there is more to autonomy, most importantly rationality; and as will become clear, capacities for self-reflection and for rationality are intimately connected.

2.2 Rationality

Again, we differentiate between three modes of rationality.[8] At first, an autonomous person must display instrumental rationality. If Paul intends to visit his father on an island, and if he believes that the only way to get there is by ferry, it is rational for him to buy a ticket and to take the ferry. If Paul is not instrumentally rational and finds himself sitting in a train to another place, or just doing nothing at all, he will be puzzled by his own behaviour, given his intention to visit the father. He will have contradicted his own will. In general, persons who are not instrumentally rational are most likely to get nowhere and are unable to direct their lives according to their own decisions and plans.

As to the second mode of rationality, rational coherence, a person must try to detect and resolve possible conflicts between her desires, beliefs, plans, etc.[9] Without some degree of rational coherence, a person will often be confronted with (practical) conflicts that are detrimental to autonomy. If Paul intends both to visit his father on the island and to visit a friend in Vienna at the same weekend, he

[7] Some philosophers have distinguished this mode of self-reflection from the other two modes, because at this point the notions of 'identification' and 'authenticity' come into play. In self-evaluation a person is said to 'identify' with certain desires and to 'make them her own'. The most prominent defender of this idea is Harry Frankfurt (1971). According to his 'hierarchical model' a person acts autonomously if she is moved to action by desires she identifies with (she desires to be effective in action). It is important to note that in order to be autonomous, a person must not only satisfy this so-called 'authenticity condition' but also have the general capacities or competencies we describe. Cf. Christman (2003).

[8] Cf. Oshana (2006, p 78); for more general discussions of the relationship between autonomy and rationality: Christman (1991, 13 ff.), Haworth (1986, Chap. 2) and Young (1980, 567 ff.).

[9] Cf. Ekstrom (1993).

entangles himself in two incompatible courses of action. Again, he will be unintelligible to himself and in some sense 'paralysed'. But rational coherence is not only a precondition for directing one's life in a self-controlled manner. It also fosters self-understanding because it allows people to make better sense of their beliefs, desires, feelings, etc.

Finally, a person must engage in critical reasoning. She must be able to reasonably respond to arguments by others and to process information about the world; when exposed to new or complex situations she must be able to distinguish the relevant features and norms; and so on. While it is disputed to what extent a person must engage in critical reasoning,[10] it seems obvious that without some degree of critical reasoning she is prone to unwarranted interferences from others and to 'shiftlessness'. Also critical reasoning plays some role in the process of self-evaluation – take Mary who evaluates her resentment against the background of shared norms of reasonableness.

A person who exhibits all three modes of rationality is more apt to achieve her goals and to be in control. She is more likely to govern her life in accordance with her own (considered) decisions and plans. Taken together, the capacities for self-reflection and rationality enable a person to achieve an important form of autonomy – provided that she can effectively exercise them. This leads us to the condition of procedural independence.

2.3 Procedural Independence

A person who enjoys procedural independence is not under the influence of factors that compromise her capacities for self-reflection and rationality.[11] This common formulation allows for two readings, both of which are important for our purposes: A person's abilities might be 'compromised' in the sense that (i) she is unable to engage in self-reflection and rational deliberation or that (ii) although she can engage in these activities her attempts are somehow in vain. The following paradigmatic cases are supposed to illustrate this difference.

First, consider Mary who is, unknown to her, given a drug that makes her extremely aggressive. She offends people all the time and is driven by the sheer force of her aggression. Neither can she control her behaviour nor can she adequately react to others' criticisms anymore, although she normally is a very self-controlled person and open to criticism. Mary is not autonomous because her capacities for self-reflection and rational deliberation are bypassed. She lacks procedural independence in the first sense.

---

[10]We want to avoid an 'external' or 'substantive' account of rationality according to which a person must be able to understand the correct reasons and/or have true beliefs in order to count as autonomous (cf. Benson, 1987). For a discussion of 'internal' vs. 'external' rationality, see Christman (1991, 13 ff.).

[11]Cf. Dworkin (1976, 25 ff.), Dworkin (1988, p 18) and Christman (1991, 18 f.).

Now, consider a different scenario in which a friend lies to Mary that a colleague denigrates her all the time. Mary trusts her friend and has no reason to believe that he wants to deceive her. She thus begins to feel strong resentment against her colleague and expresses, to the colleague's surprise, her feelings in distinct ways. In this case Mary fails to be autonomous because she does not, in the relevant sense, act on her *own* values and beliefs. She is acted upon. Nonetheless it is inadequate to say that she does not engage in self-reflection and rational deliberation. Her lack of autonomy is rather due to the fact that she lacks procedural independence in the second sense: Her capacities are compromised because another person has 'conquered' these.

The difference can now be formulated in a slightly different way: Under normal conditions a person can be said to achieve autonomy when she is able to reflect about herself and to rationally control her actions in accordance with her decisions. But under 'abnormal conditions' it is not true that exercising the capacities for self-reflection and rationality is sufficient for autonomy. In these cases the circumstances deprive a person from effectively using her capacities – the capacities are, so to speak, not her own and thus do not speak for herself.

Spelling out the conditions that constitute procedural independence (in both senses) is an intricate task. There are various ways in which a person might be prevented from engaging in self-reflection and rational control – drugs, hypnosis, certain forms of psychological pressure, etc. or in which the exercise of these capacities is, as we have put it, in vain: lying, delusion, brainwashing, coercion and so on. In each case it needs to be shown why and in which ways these conditions compromise a person's reflective capacities.[12] We do not attempt to give an exhaustive interpretation of this condition but will instead concentrate on those autonomy-undermining conditions that may become especially relevant in the context of EOT. Before, however, we need to link the notion of procedural independence to the notion of respect for autonomy.

2.4 Respect for Autonomy

The principle of respect for autonomy – one of the most important moral principles in modern liberal societies – can be given a negative and a positive reading. Put negatively, it says that we should not interfere with a person's autonomy; in its positive version the claim is that we should foster a person's autonomy.[13] The negative claim is commonly regarded as making stronger claims on people.

But what does it mean to interfere with/foster a person's autonomy? What does respect for autonomy amount to? Our suggestion is that this notion can be illuminated by the notion of procedural independence. More to the point, we contend that

---

[12]See Dworkin (1976) and the critical discussion by Oshana (2006, Chap. 2).

[13]The 'principle of respect for autonomy' figures prominently in Beauchamp and Childress (1994); see also Childress (1990).

(at least one important form of) *respect for autonomy can be understood as respect for a person's procedural independence*. This is tantamount to the negative claim that we should not prevent persons from effectively exercising their capacities for self-reflection and rationality; in its positive version the claim is that we should help persons to develop and to make effective use of their capacities.

This interpretation of the principle of respect for autonomy completes our outline of a 'theory of autonomy'. As will become clear in the next sections, things tend to be more complicated and the conditions of (respect for) autonomy have to be amended and refined. Nonetheless it is helpful to have a basic framework to build upon as the following discussion of emotional privacy and its connection to autonomy shows.

3 Autonomy, the Emotions and Privacy

It is often said that persons hide or repress their emotions because they want to 'protect themselves'; 'being overcome by emotion' at the wrong time and place is something many are afraid of; persons only allow for expression of emotions in the presence of 'those whom they trust'; they 'entrust' their emotions; and so on. These common expressions convey that (i) emotions make people specifically vulnerable, for which reason they (ii) try to keep their emotions private (in a sense to be explained), albeit (iii) these attempts are not always successful because emotions defy control to some extent. Emotional privacy, then, is a fragile state that people deeply care about because they want to protect ... their *autonomy*, as we will now illustrate starting from an everyday example.

3.1 Autonomy and Emotional Privacy

Consider John who falls in love with Laura, his best friend's wife. He takes pains to control his emotions in her presence and tries to keep distance because he does not want Laura or his friend to notice his affection before he has 'made up his mind'. After some time, John realizes that he is unable to cope with the situation on his own and decides to disclose his feelings to Paul, whom he trusts. But Paul is not as confidential as John believes and secretly tells the couple about the situation. He thereby changes the position of John in a drastic way: John has no longer control over the course of events and cannot decide for himself how to resolve the situation. Trying to overcome and not reveal his affection is no longer an option; the same holds for the possibility to determine when or to whom he reveals his feelings. The fact that he initially does not know about Paul's indiscretion makes the situation even worse. If he were attempting to find a solution, the friend and Laura would always see his actions in a different light against the background of their knowledge. It would be easy for them to play on the situation: The friend could expose John to compromising situations in which he has to lie or betray his own feelings; Laura could pretend affection and show him up in the end. Of course, both of them do not undertake any such actions. Instead the friend informs John about Paul's indiscretion as soon as possible and thus restores his position of authority over his life to some extent.

The story serves to highlight the followings aspects of the relationship between autonomy, the emotions and privacy: It becomes clear that privacy – understood as control of access to information about oneself or, more generally, to realms of one's life – is intimately connected to autonomy.[14] Breaches of privacy often undermine the procedural independence of persons: John is no longer in a position to effectively exercise his capacity for self-evaluation and to actively shape his life because his options are significantly narrowed down by Paul's indiscretion. His temporary ignorance about the information conveyed to the married couple also inhibits John's self-awareness – he does not know about his social position, so to speak – and thus compromises his capacity for self-reflection in general.[15] As a consequence, he becomes particularly vulnerable to manipulation since he can easily be acted upon.

These remarks about the relationships between privacy and autonomy do not apply solely to emotional privacy – dependent upon the kind of information many breaches of privacy change for the worse or irresponsibly put at risk the autonomy of persons. But there are two reasons, in our view, why access to emotions is of specific importance in this regard: On the one hand, emotions have a deep impact on persons' lives because they determine to a significant degree what matters to them.[16] On the other hand, persons can neither plainly decide to undergo or to get rid of an emotion, nor is it always possible for them to regulate the strong influence emotions exert on their thoughts and actions. Hence, by obtaining knowledge about a person's emotional states other persons often (i) find out something that is integral to that person's 'identity' and (ii) become acquainted with her potential 'weaknesses'. The case of John exemplifies that *in combination* these two aspects – personal significance of emotions and partial lack of control over their influence – *can* be especially damaging to a person's autonomy; John hides his affection from Laura and his friend because he wants to retain his position of authority in matters

---

[14]DeCew (2006) provides a helpful overview of the concept of 'privacy'; an elaborate discussion of the relationships between autonomy and privacy is to be found in Rössler (2001).

[15]One might argue that, on our account, the autonomy of Laura and her husband is also compromised, because John does not inform them about his feelings. But we can easily explain why our account of (respect for) autonomy does not yield this result: Most importantly, one reason why John *keeps distance* might exactly be that he does not want to interact with Laura and his friend in ways that would compromise their autonomy. John's autonomy is undermined because Paul compromises his ability to resolve the situation on his own, while Laura and her husband do not face a problem and thus do not have to resolve anything (yet).

[16]Cf., e.g., the work of Aaron Ben-Ze'ev, Ronald de Sousa, Sabine A. Döring, Peter Goldie, Patricia Greenspan, Bennett Helm, Martha Nussbaum, David Pugmire, Amelie Rorty, Robert Solomon, Holmer Steinfath, Michael Stocker, Christine Tappolet, Gabrielle Taylor, Bernard Williams and Richard Wollheim.

that are important to him: He knows about his own vulnerabilities. But Paul's indiscretion defeats these attempts. It undermines John's position of authority and makes him vulnerable to manipulation.

3.2 Emotional Privacy, Emotional Interaction and Respect for Autonomy

We have elaborated on some of the reasons why emotional privacy is important to the autonomy of persons. But what follows from these considerations for the question how persons should treat other persons in order to respect their autonomy? Our reflections on the story of John suggest the following two (very general) duties: *First*, persons should respect other persons' control of access to information about their emotional states. For example, without a person's consent other persons should not pass such information about the person to others.[17] Paul's indiscretion is a case in point – he clearly shows disrespect for John's autonomy by telling Laura and her husband about John's affection. *Second*, the fact that persons obtain or are entrusted with knowledge about emotional states of a person imposes special responsibilities upon them: They must not misuse the information and exploit the vulnerabilities of that person. Laura and her friend show respect for John's autonomy because they do not play on his affection, but instead foster his self-reflection by unveiling Paul's indiscretion.

Given that persons should comply with these duties, we claim that the same holds for emotion-oriented systems to which we now turn to. As we will show, there are specific problems that come to light when considering what it would mean for EOT to comply with duties of autonomy.

4 Duties of Developers of Emotion-Oriented Systems

The claim that machines should comply with certain duties is, of course, not to be taken literally. Unless machines are autonomous moral agents, persons cannot sensibly make demands on them and take up the 'second-person standpoint'.[18] The responsibility rather lies with developers. They have a *prima facie* duty to develop machines that *cannot* (under normal conditions)[19] infringe the autonomy of persons. In order to come up to this duty, they must first identify potential threats that are posed by the use of EOT and then translate their findings into design. In this section, we take up the first task and consider two problems: (i) Widespread and automatic collection of information about emotional states opens up a great

[17] This formulation might strike many as way too strong. But note that we always speak from the perspective of autonomy only.
[18] Cf. Darwall (2006).
[19] This condition is meant to exclude too far-fetched worst case scenarios.

many opportunities for abuse. (ii) The fact that machines isolate emotions and lack more fine-grained interpretive skills can lead to infringements of autonomy in human–machine interaction.[20]

We focus on these problems because we regard them to be *specific* – they arise from the fact that machines, and not other persons, interact emotionally with persons. Of course, there are many 'non-specific' problems or aspects related to the notion of respect for autonomy that must be taken into account by developers. To give just one example, consider an emotion-oriented system that attempts to persuade a person to take medication by confronting her with those facts that increase her anxiety and feelings of guilt (see Sect. 1). This system clearly shows disrespect for that person's autonomy. By appealing to feelings of guilt and anxiety it tends to undermine the person's procedural independence. Not surprisingly, there is no dispute that non-rational persuasion by, e.g., doctors is incompatible with the principle of respect of autonomy.[21] In general, what holds for persons holds for persuasive systems, too: They should not lie to persons, not hide their true intentions from them, not withhold important information and so on.[22,23] The framework in Sect. 2 provides a starting point to answer the question of what respect for autonomy demands in specific situations. However, as we will now indicate, there are specific problems with regard to (respect for) autonomy that developers of EOT must face in advance.

4.1 EOT and the Gathering of Information About Emotional States

Consider an emotion-oriented system that detects emotional states and conveys its findings to users. If a person shows signs of increasing frustration, for example,

---

[20]As should be clear, these problems relate to the two general duties we have set out in Sect. 3.

[21]Cf. Beauchamp and Childress (1994, Chap. 3).

[22]In passing, we want to mention an important qualification: Whether some action undermines a person's procedural independence cannot always be answered without reference to the person's actual capacities. For example, a father does not respect his child's procedural independence if he repeatedly tells her that the best thing to do in life is to become a check-out girl. The child is not in a position to evaluate and critically assess these claims from the background of a stable self-conception, and thus, her father interferes with her procedural independence. By contrast, a father does not disrespect autonomy if he tells his well-educated and self-reflective daughter that the best thing to do in life is to become a check-out girl. She will most probably laugh at him. This suggests a principle that could be labelled in a provocative way the 'low autonomy, high respect' principle. The less autonomous a person actually is, the more other persons should respect her autonomy – they are all the time in danger of undermining her procedural independence. This idea fits well intuitions concerning, for example, the treatment of children. In discussing the ethicality of persuasive systems, one needs to make use of something like 'normality conditions' and assume that a person fulfils to some degree the conditions of self-reflection and rationality.

[23]For further discussions of persuasive systems from the perspective of autonomy, see Baumann and Döring (2006). More on persuasive systems and the role of emotions in section WP8; for the ethicality of persuasive systems see Guerini and Stock (2006); see also the discussion in Goldie and Döring (2005a) (CyberDoc).

the system immediately notifies her about that – possibly long before she would otherwise become aware of it. The person can then relate the upcoming frustration to her current activities/environments in order to figure out effective means of counteracting it.

Obviously, this system is built 'with good intention'. It is supposed to foster persons' self-awareness and control (i.e. their autonomy). Conceding this, a sceptic might still oppose the use of such systems by bringing up general worries like the following: "What if the system passes information about my emotional states to others or collects much more information than I know? What if the data are correlated with additional information about myself, or are analysed in different ways? I do not want my supervisor to know that I often feel frustrated when I have talked to him. Nor do I have any intention to deliver information about my high blood pressure to the insurance company. Also, I dislike the idea of receiving advertisements or special offers from a computer store that, unknown to me, play on my excitement about a gadget I have seen there."

The sceptic knows that information about emotional states can be used in ways that undermine her position of authority and/or make her vulnerable to undue interferences and manipulation. She is worried that the supervisor will take action after he has come to know about her frustration, while she has found her own way of dealing with the situation and does not want to 'talk things out'. She is afraid that she will be manipulated by personalized advertisements she does not recognize as such. And so on. These possible threats to her autonomy are structurally similar to the ones we have already elaborated in Sect. 3 – breaches of privacy put at risk the autonomy of persons by compromising their procedural independence. However, the scenarios envisioned by the sceptic highlight that the use of machines generates *a great many* opportunities for abuse. While there are good reasons not to entrust one's emotions to other persons carelessly, there are even better reasons to be sceptical about using emotion-oriented systems and giving away information about one's emotional states.[24] This is due to several features that distinguish machines from persons with respect to the process of gathering and collecting information, some of which are the following.

In contrast to persons, machines can record every single emotional state of persons over long periods of time. They need no 'driving interest' and there are potentially no limitations on the amount of data that can be stored. Furthermore, machines are more reliable than persons in gathering certain kinds of information. In monitoring objective features like bodily expressions of emotions they are not 'disturbed' by other aspects of the situation. Finally, and maybe most importantly, these data can without much effort be related to other information or be analysed for different purposes by anybody who gains access to them – if a person's excitement is determined partly by monitoring her blood pressure, an insurance company might easily extract and use the relevant data for their purposes.

[24] Problems of privacy are also discussed by, e.g., Höök and Laaksolahti (2008) and Reynolds and Picard (2004).

Even if emotion-oriented systems are built with good intention, their use involves the risk of infringing a person's autonomy. Developers have a duty to avoid or at least to minimize these risks by taking emotional privacy seriously.

4.2 EOT and the Lack of Interpretive Skills

The problems we have pointed at in the last paragraph arise from the fact that emotion-oriented systems *gather* information about emotional states.[25] In this paragraph we turn to more intricate problems that crop up because such systems *model* and *respond* to emotional states. To begin with, compare the following two scenarios: *iCalm*™ is an emotion-oriented system that detects craving signals on part of abstinent drug addicts and, in response, sends 'personalized relapse-prevention messages' to them.[26] *iNerve* (as we call it) is an emotion-oriented system that detects frustration on part of persons and sends 'You make me feel frustrated' messages to those persons who are responsible for these feelings.

The scenarios obviously evoke diverse reactions. The former does not generate immediate concerns about autonomy, but the latter clearly does. We want to explore possible reasons for these intuitive reactions, by isolating factors that distinguish the scenarios and that seem to be of relevance to the question of autonomy.

A first explanation suggests itself: *iCalm* only provides the addicts with feedback about their emotional states, while *iNerve* undertakes further action. Hence, users of *iCalm* can reflect on their situation and decide what to do, while users of *iNerve* cannot engage in self-reflection and are acted upon. The latter system thus infringes the autonomy of persons because it brings about action that is not proximately caused by autonomous decisions of theirs. The following principle emerges:

> (OneAction) Emotion-oriented systems should not undertake any action in response to emotional states, except of providing feedback to users.

This principle, however, seems to make an overly restrictive claim on the use of EOT. It seems possible to imagine scenarios in which a person's autonomy is not infringed, albeit emotion-oriented systems bring about further action. Consider a panic detection system (*iPanic*) that triggers the emergency break in response to a panicking engine driver. Or consider a slightly modified *iCalm* system that administers substitutes to addicts instead of sending messages. *Provided that* persons consent to using these systems, neither of them seems to infringe their users' autonomy.

Are concerns about autonomy unfounded then, if a person consents to the use of an emotion-oriented system, no matter what action is taken by the system? How

---

[25]See also the discussion of 'Semi-Intelligent Information Filters' (SIIF) in chapter "The Ethical Distinctiveness of Emotion-Oriented Technology" by Sabine Döring et al.

[26]This system has been developed by Picard et al., http://affect.media.mit.edu/projects.php?id=2145 (accessed January 28, 2008). The other two systems – 'iNerve' and 'iPanic' – are only invented for purposes of discussion.

could the misgivings about *iNerve* be accounted for in this case? Here is an explanation: Everybody understands the reasons why addicts avail themselves of *iCalm* – it fosters their self-awareness and helps them to stay in control. But it is hard to imagine why anybody should decide to use *iNerve*. Hence, people implicitly presuppose that *iNerve* is not consented to by persons. The system is thus said to infringe the autonomy of persons because it undertakes action behind their backs. A second, competing principle comes into play:

> (AnyAction) It is permissible that emotion-oriented systems undertake any action, provided that their use is consented to.

In contrast to the first, overly restrictive principle, this one seems to be too moderate. To see why, consider Anna who often feels frustrated because of other people, but is unable to tell them. She strongly suffers from this inability, until one day she stumbles upon *iNerve*. Anna accedes to the use of *iNerve* in order to overcome her silence and to give expression to her feelings. Now, if the above principle were correct, there would be nothing to worry about, at least from the perspective of autonomy. After all, the use of *iNerve* is consented to. But Anna herself might stop using *iNerve* after a short while because she feels acted upon: "I have acceded to the use of *iNerve* because I wanted to give expression to my feelings of frustration. But the system undertakes actions I do not endorse. One evening it told my friend that he makes me feel frustrated, although I was frustrated because of the news on TV. Another time, *iNerve* sent one of its messages to my supervisor during a meeting, although my frustration was about my own lack of preparation."

These complaints are illuminating because they highlight two additional, hitherto neglected differences between systems like *iCalm/iPanic* and *iNerve*. The former systems (i) only process emotions that do not call for interpretation and (ii) respond to the occurrence of these emotions with one determinate action. In contrast, *iNerve* responds to an emotion that cannot be understood without interpretation, and it can initiate different courses of actions. As we have said, the aim of *iNerve* is to send messages to those persons who are responsible for the user's frustration. But in order to decide who should receive the 'You make me feel frustrated' messages, *iNerve* must know what the frustration is all about. Is Anna frustrated about her lack of preparation, or because her supervisor always criticizes her? Does her friend's inattentiveness make her feel frustrated, or is it the news about poverty in poor countries? And so on.

The crucial problem is that *iNerve* cannot accomplish this task. It monitors bodily expressions of persons that are indicative of frustration, looks for possible causes and sends along its messages. The content of the emotion – what Anna is frustrated about – is not taken into account.[27] Since *iNerve* does not adequately understand Anna's frustration, then, it cannot do what she wants. Put slightly differently, even if *iNerve* reliably detects frustration on the part of Anna, it does not undertake

[27] Cf. Döring and Goldie (2005b).

actions that are expressive of her frustration – it does not send out messages to people only when Anna feels frustrated about them/because of them. In consequence, she feels acted upon and decides, for reasons of autonomy, not to use it anymore. This motivates a third principle:

> (SomeAction) Emotion-oriented systems should not undertake any actions that users – as autonomous persons – do not or cannot endorse.

The rationale behind this principle is that persons can use emotion-oriented systems without forfeiting their autonomy if it is possible for them to transfer their autonomy to these systems. By this we mean that in some cases a person retains control because the actions undertaken by a system are expressive of her autonomous (self-reflective and rational) decisions and plans. An engine driver who uses *iPanic* delegates his authority to the system because he knows that if he panics, he will no longer be able to deal with the situation autonomously. An addict who avails himself of *iCalm* allows the system to send feedback because he believes that the messages help him to get a grip on his addiction. In both cases, it is very unlikely that any user will ever feel acted upon. The systems only undertake such actions that users do and can endorse as autonomous persons. In contrast, systems like *iNerve* are likely to infringe the autonomy of persons because the latter cannot stay in control while using them, due to the systems' lack of understanding of the emotion at stake. Such an understanding would be necessary if a person were to delegate her authority to, e.g., *iNerve*. This explains, finally, why most people find it hard to imagine that anybody could decide to use *iNerve* – the user would thereby confer responsibilities on a system to which it cannot come up.

In Sect. 3 we have claimed that in emotional interaction, people who are acquainted with a person's emotional states have special responsibilities. In particular, they must not exploit the vulnerabilities of that person. The foregoing discussion suggests that emotion-oriented systems tend to systematically fall short of this duty: At least more complex emotions call for interpretation and cannot be understood without attending to their content – what is a person frustrated *about* or afraid *of*? Since emotion-oriented systems only monitor bodily expressions of emotions, however, they will often remain ignorant about the vulnerabilities of persons. Consequently, they are likely to infringe the autonomy of persons. Developers have a duty not to build systems that can 'run amok' due to this lack of understanding.

5 The Interactional Perspective on Developing Emotion-Oriented Technology

Given that developers have a *prima facie* duty to build machines that cannot (under normal conditions) infringe the autonomy of persons, what follows from the foregoing considerations for the design of EOT, or for possible fields of application? How can the problems we have pointed at in the last section possibly be avoided? In this section we want to discuss very briefly the 'interactional view' which has been put forward by some developers, partly because of concerns about autonomy. We first explain why this approach actually meets some of the problems we have alluded to in the last section. In a second step, we then outline two possible criticisms.

5.1 The Interactional View and Respect for Autonomy

Designs that are built from the interactional perspective "assume that the *meaning* of an emotional process is created by people and that affective interactive systems should be such that users are encouraged to negotiate these meanings themselves".[28] Neither are bodily expressions of persons related to specific emotions by the systems, nor do they undertake any action in response to the detection of bodily states. Instead, everything is up to users. They are provided with feedback about their bodily expressions of emotions in ways that are open to or even call for interpretation. Users can then give meaning to this feedback by relating it to their current situation, etc. They can thus determine what is done with the feedback provided by the emotion-oriented system. Furthermore, it is up to them whether action is undertaken in response to the emotional state.

While we abstain from giving detailed examples of systems that are built from the interactional perspective,[29] this outline already indicates why problems with regard to the autonomy of persons in using emotion-oriented systems seem to be avoided. *First,* the way/form in which data are gathered ensures that (i) data cannot easily be used for other purposes and that (ii) it is always transparent to users which data are collected and for what purposes. Bodily expressions are not related to specific emotions but are collected in a form that is open to interpretation. Hence, to take up one of the examples mentioned in Sect. 4, the supervisor cannot figure out that his employee is frustrated every time she has talked to him. If he gains access to the data at all, he will only come across data that map the bodily expressions of emotions in abstract ways. Furthermore, users can give meaning to the data and determine what is done with these. They are thus put in a position of power that provides them control of access to information about their emotional states.[30]

Second, threats to autonomy that arise from the inability of emotion-oriented systems to adequately understand/interpret complex emotions, as described in Sect. 4, are avoided. It is the person who gives meaning to the feedback that an emotion-oriented system provides. She can, for example, relate the state of arousal that is displayed to her either to a conversation with her supervisor or to the spider at the window. Accordingly, she can either figure out that she is frustrated because of her advisor or that she fears the spider. If she figures out that she is frustrated, she might command a system to send one of those messages to her supervisor that *iNerve* is supposed to send on its own. But in this case, the action is brought about by the

[28] Höök and Laaksolahti (2008); the interactional approach has been formulated by Boehner et al. (2005).

[29] See Lindström et al. (2006) and Höök and Laaksolahti (2008).

[30] A different approach to come up to problems of privacy is to be found in Picard (2004).

person's autonomous decision and really gives expression to her frustration. The possibility that an emotion-oriented system runs amok is excluded, then, because the user is in control of everything.

5.2 Two Possible Criticisms of the Interactional Approach

While the interactional approach obviously helps to avoid some important threats to autonomy, it is vulnerable to criticisms from two perspectives. On the one hand, developers might argue that the interactional approach unduly restricts the use of EOT.[31] Even from the perspective of autonomy, there seem to be good reasons to extend the limits set by the interactional approach. Take, for example, *iPanic*: It automatically relates bodily states to a specific emotion (panic) and then undertakes action on its own (trigger the emergency break). It thus does not conform to the interactional design approach at all. But in contrast to what defenders of the interactional view seem to believe, it is neither true that every emotion calls for an interpretation,[32] nor true that the autonomy of persons is always infringed if further action is undertaken by the system. If a person can transfer her autonomy to an emotion-oriented system, as we have put it, there is no problem about autonomy. However, the transfer of autonomy is only possible in those cases in which the emotional state in question does not call for an interpretation, and in which the course of action a system can undertake is determinate. In every other case, the interactional approach is to be preferred.

On the other hand, sceptics about EOT might argue that the use of affective interactive systems inhibits persons' ability for introspection and compromises their capacity for self-reflection. The idea is that, in the long run, people lose other modes of access to their emotions – most notably introspection – if they rely on feedback from machines about their bodily states. Furthermore, it is likely that persons will always try to make sense of the feedback provided by machines because it is somehow objective. They might thus be forced to reflect on their emotional states in ways that actually distort their self-awareness or self-understanding.[33] This last point relates to the problem of interpretation: It might sometimes be the case that bodily states are not related to any emotion on the part of the user, but are brought about by other means. By getting feedback about their arousal, for example, users might nonetheless try to figure out the emotional state they are (not) in.

---

[31] An overview of the fields of application is given by Schröder et al. (2006).

[32] It is important to note that some might want to distinguish between emotions and mere affective states, the former being more complex intentional states. Although we welcome such attempts, this move is not helpful in discussions of EOT, because 'emotion' or 'affect' is generally used as umbrella terms in this connection.

[33] Such worries are also mentioned in Picard and Klein (2002).

6 Conclusions

It is time to sum up. In the first part of this chapter we have outlined a theory of autonomy in order to provide a general framework of discussion. We have argued that capacities for self-reflection and rationality are essential to the autonomy of persons, and that respect for autonomy can be understood as respect for the procedural independence of persons: Persons should (negatively) not prevent or (positively) make possible that other persons can effectively exercise their capacities for self-reflection and rationality. In Sect. 3 we have considered some of the reasons why emotional privacy is important to autonomy, and we have indicated what respect for autonomy in emotional interaction with other persons might amount to. In the second part of the chapter, we have then presented two problems that crop up when EOT are supposed to show respect for autonomy. On the one hand, the fact that emotion-oriented systems collect information about emotional states generates a great many opportunities for abuse; on the other hand, emotion-oriented systems tend to infringe the autonomy of persons because they do not 'understand' what emotions are about and thus remain ignorant about persons' vulnerabilities. In the last section we have discussed one way to deal with these problems. The interactional approach minimizes opportunities for abuse and avoids the emotion-oriented systems 'running amok' by 'empowering' the user. However, this approach unduly restricts the use of EOT and at the same time runs the risk of compromising a person's capacities for introspection and self-reflection. We must leave open how these issues can be resolved. Also, we cannot even begin to discuss the question of how reasons of autonomy are to be weighed against other reasons, e.g., of beneficence or justice.[34] In this chapter we have only viewed emotion-oriented technologies from the perspective of autonomy, and we hope to have done so in a way that facilitates and initiates further discussion.

References

Baumann H, Döring S (2006) Emotion-oriented systems – threats to user autonomy. http://emotion-research.net/ws/wp10/presentation-materials/HolgerBaumann-SabineDoering-wp10ws-EmotionOrientedSystems-ThreatsToUserAutonomy-final.pdf. Accessed 17 May 2010

Beauchamp TL, Childress JF (1994) Principles of biomedical ethics, 4th edn. OUP, Oxford

Benson P (1987) Freedom and value. J Philos 84:465–486

Boehner K et al (2005) Affect: from information to interaction. In: Proceedings of the 4th decennial conference on critical computing: between sense and sensibility, Aarhus, Denmark

Childress JF (1990) The place of autonomy in bioethics. Hastings Cent Rep 20:12–17

Christman J (1989) Introduction. In: Christman J (ed) The inner citadel. OUP, New York, NY, pp 3–23

Christman J (1991) Autonomy and personal history. Can J Philos 21:1–24

Christman J (2003) Autonomy in Moral and Political Philosophy. In: Stanford encyclopedia of philosophy. http://plato.stanford.edu/entries/autonomy-moral. Accessed 17 May 2010

[34]See also chapter "Principalism: A Method for the Ethics of Emotion- Oriented Machines" by Sheelagh McGuinness.

Darwall S (2006) The value of autonomy and autonomy of the will. Ethics 116:263–284
DeCew JW (2006) Privacy. In: Stanford encyclopedia of philosophy. http://plato.stanford.edu/entries/privacy/
Döring S, Goldie P (2005a) Interim report to plenary meeting on ethical frameworks for emotion-oriented systems. HUMAINE deliverable D10b in: http://emotion-research.net/deliverables/D10b.pdf. Accessed 17 May 2010
Döring S, Goldie P (2005b) Categories of emotion: everyday psychology and HUMAINE. http://emotion-research.net/ws/wp3/ExtraMaterial/HUMAINE-Goldie.pdf. Last visited 7 November 2010
Dworkin G (1976) Autonomy and behavior control. Hastings Cent Rep 6(1):23–28
Dworkin G (1988) The theory and practice of autonomy. CUP, Cambridge, MA
Ekstrom L (1993) A coherence theory of autonomy. Philos Phenomenol Res 53:599–616
Frankfurt H (1971) Freedom of the will and the concept of a person. J Philos 86:5–20
Feinberg J (1989) Autonomy. In: Christman J (ed) The inner citadel. OUP, New York, NY, pp 27–53
Friedman M (2003) Autonomy, gender, politics. OUP, Oxford
Guerini M, Stock O (2006) Ethical guidelines for persuasive systems. In: Proceedings of the HUMAINE WP10 workshop, Nov 2006, Vienna, Austria (EU). http://emotion-research.net/ws/wp10/presentation-materials/MarcoGuerini-OlivieroStock-wp10ws-EthicalGuidelinesForPersuasiveSystems-final.pdf/. Accessed 17 May 2010
Haworth L (1986) Autonomy. An essay in philosophical psychology and ethics. Yale UP, Yale
Höök K, Laaksolahti J (2008) Empowerment: a strategy to dealing with human values in affective interactive systems. http://citeseerx.ist.psu.edu/viewdoc/download?doi=10.1.1.96.5891&rep=rep1&type=pdf. Last visited 7 November 2010
Lindström M et al (2006) Affective diary – designing for bodily expressiveness and self-reflection. http://www.sics.se/~petra/affd.pdf. Last visited 17 May 2010
Meyers D (1989) Self, society, and personal choice. Columbia UP, New York, NY
Oshana M (2006) Personal autonomy in society. Ashgate, London
Picard RW, Klein J (2002) Computers that recognise and respond to user emotion: theoretical and practical implications. Interact Comput 14:141–169
Raz J (1997) When we are ourselves: the active and the passive. Reprint In: – (1999) Engaging reason. On the theory of value and action. OUP, Oxford, pp 5–21
Reynolds C, Picard RW (2004) Affective sensors, privacy, and ethical contracts. http://affect.media.mit.edu/pdfs/04.reynolds-picard-chi.pdf. Last visited 17 May 2010
Rössler B (2001) Der Wert des Privaten. Suhrkamp, Frankfurt
Schröder M, Cowie R, Kollias S (2006) The future of emotion-oriented computing. HUMAINE Plenary presentation, Paris, Jun 2007. http://emotion-research.net/ws/plenary-2007/2007-FutureOfEmotionOrientedComputing.pdf. Last visited 17 May 2010
Velleman D (1989) Practical reflection. Princeton UP, Princeton, NJ
Young R (1980) Autonomy and socialization. Mind 89:565–576

Ethics in Emotion-Oriented Systems: The Challenges for an Ethics Committee

Ian Sneddon, Peter Goldie, and Paolo Petta

Abstract The development of emotion-oriented systems has the potential to raise a range of ethical issues, not all of which were clearly understood at the inception of the project. This chapter discusses these issues and details the practical measures taken and the challenges faced in addressing them. An ethical audit revealed a lack of consistency in ethical procedures across the institutions and disciplines involved in the network, and HUMAINE established its own ethics committee to offer ethical advice and scrutiny when required. In addition, space was provided within the project for discussion of ethical issues – a process that allowed the emergence of a wider understanding of the issues themselves and of the sensitivities of different disciplines and users to them.

1 Introduction

It was a distinctive feature of HUMAINE that it established its own ethics committee. The intention of this chapter is first to explain why that was done; second, to examine some of the issues that should be addressed when establishing a similar ethics committee in the context of an interdisciplinary research programme; third, to chart the various initiatives and approaches to ethics adopted during the life of HUMAINE; and finally, to touch on some of the key conceptual issues that an ethics committee in the area has to deal with. There are two types of issues to address. One arises simply because some research on emotion-oriented computing involves human participants, and the ethical issues surrounding their participation have to be addressed, just as they do in any other field that uses human participants in experiments. The other is unique to emotion-oriented computing, and it arises because of the particular type of application that the area aims to develop.

To say that there are specific ethical issues is not to imply that the area is fraught with ethical problems. The HUMAINE project was at pains throughout to emphasise

I. Sneddon (✉)
Department of Psychology, Queen's University, Belfast, Northern Ireland, UK
e-mail: I.Sneddon@qub.ac.uk

P. Petta et al. (eds.), *Emotion-Oriented Systems*, Cognitive Technologies,
DOI 10.1007/978-3-642-15184-2_41,

that the aim of the research is benign. It is self-evident that many of our current interactions with machines are limited and often at best unsatisfying, at least in part due to the very limited communication abilities of current systems. There are of course human–machine interactions that would not benefit from the machine having any sensitivity to the emotional state of the human. However, just as with human to human communication, there are many situations where success in communication is improved by the ability of one party to interpret the emotional status of the other. It is easy to see that many machine systems involved in, for instance, education or entertainment would simply work better if they had a capacity to interpret changes in the emotion of the person and to respond in an appropriate emotional manner.

However, even with the best of intentions, there is a potential for unintentional harm or intentional abuse of emotion-oriented systems. There is also a problem of perception. Dystopian futures inhabited by machines with just such abilities are a common theme in science fiction and such imaginary worlds do have an impact on public ideas about this kind of research; and allaying public fears is a serious issue in itself.

2 The HUMAINE Ethics Committee: A Brief History

Increasingly, research programmes involving human participants have to make detailed provision for the treatment of ethical issues that might arise either during or as a result of their research. Five years ago, at the planning stages of the HUMAINE programme, it was not at all clear what provision the project would need in that area. At that time, we had a clear idea neither of the ethical climate existing in each of the partner countries nor of ethical practice and awareness in the various disciplines from which partners came. It seemed likely that their backgrounds would range from areas where the ethical issues were thoroughly familiar, to areas where they were never considered.

Against that background, the project adopted a proactive strategy. From the first plenary meeting of the network, researchers were alerted to some of the potential ethical problems that could arise from the development of emotion-oriented systems and asked to think about any possible ethical implications in their own research. Philosophers were included in the research programme to ensure that the response had a sound intellectual basis. Their input has proved invaluable throughout the lifetime of the project, particularly in dealing with the novel issues that the research raises.

To clarify the starting position, an ethical audit was conducted during the first year of the project (Goldie et al., 2004). This revealed that the majority of partners were conducting research that involved human participants in some way, but that they were operating under a very wide range of regulatory regimes. In the UK at the time, all research institutions were at various stages of implementing the policy that all research on human participants must be approved by a research ethics committee and the pace of demand for such change had quickened over the previous few years.

In partner institutions in other countries the picture was more varied. In some countries researchers did not yet have access to research ethics committees and in others, although access existed, submission of research for scrutiny was only required if the research would threaten the participant's health or psychological well-being – effectively introducing a self-selection element into the process. There was also variation across disciplines. Most social scientists were used to submitting their work to ethical scrutiny by others, but the concept was alien to most engineers and others from an IT background. It was the lack of consistency revealed by the ethical audit that suggested that HUMAINE itself might usefully set up a standing ethical committee.

There were mixed feelings about the proposal, and it was not at all clear to some members of the network what the function of such a committee might be. There was a strong view that a heavy-handed or over-regulatory approach would be counterproductive. Rather than encouraging researchers to reflect on the ethical dimension of their research, any attempt to impose regulation from above (without the legal basis for enforcement) would clearly have resulted in many researchers disengaging from any discussion of ethics altogether. There was also debate about the proper composition of a committee.

With this in mind, a group from diverse countries, disciplines and backgrounds was identified and approached to form an initial committee that held its first face-to-face meeting in Athens in June 2006. In the light of the concerns expressed above, the first matter for discussion was obviously the role of the committee. After much discussion, both between the members of the committee and between the committee and the other members of the network attending this 'cross-currents' meeting, it was decided that it could probably be of most benefit to HUMAINE researchers by taking on two main roles.

The first role was to answer questions on matters of research ethics and to offer advice on procedure and best practice for those involved in research on human participants. It was strongly hoped that the network would see the committee as a useful source of help and advice rather than as an unnecessary piece of bureaucracy. Informal requests for information or advice were, in the first instance, channelled through the chair of the committee. This role of the committee has been reasonably successful, with a number of requests for mainly practical advice on ethical matter being submitted to the committee.

The second possible role of the committee was to review any HUMAINE research involving human participants that had not already been scrutinised by an institutional research ethics committee. Although the committee had no legally constituted role for the formal approval or rejection of research proposals, it could nonetheless act as an advisory body and help researchers to ensure that their practice and procedures conformed to appropriate ethical guidelines.

Although the committee dealt with only a small number of full research submissions, a larger number of researchers asked for advice on specific aspects of a project. As with almost all administrative structures in large international projects, most business was conducted via e-mail or by telephone and after the committee had been established there was felt to be little need for face-to-face meetings.

It is revealing that the committee's last activity involved a request from another project looking for guidance that was not available in some of its participant institution. There does appear to be a useful function in the area for a properly constituted ethics committee that can cover gaps in local provision. It remains to be seen how the need will be met after the end of the HUMAINE project.

3 General Principles Underpinning an Ethics Committee

This section covers issues that would be faced by any project or group considering establishing an ethics committee. These range from highly abstract to extremely practical.

3.1 The Intellectual Framework of the Committee

Early debates about ethics in HUMAINE were constantly sidetracked into the maze of meta-ethical theory. Discussion visited and revisited questions about the possibility of justifying any ethical stance at all and about the relative merits of ideas derived (not always expertly) from Aristotle, Hobbes, Kant, the Utilitarians, Rawls and others. Fascinating as these debates are, they do not offer a basis for practical responses in a finite time.

HUMAINE addressed the difficulty by agreeing to base its practical analysis on the framework known as principalism (see Beauchamp and Childress, 2001, Principalism: A Method for the Ethics of Emotion Oriented Machines by McGuinness, this volume). Beauchamp and Childress reconstruct the core of morality in terms of a four-level model where each level is characterised by one of the following principles: *nonmaleficence*, *autonomy*, *beneficence* and *justice*.

Principalism was first designed for biomedical ethics, but it can easily be applied to other areas of ethical research, including research on the ethics of emotion-oriented systems. In fact, it has become one of the most influential approaches in the so-called *applied ethics*. At first glance it may seem as if 'application' means to apply a general ethical principle like the Utility Principle of Utilitarianism or Immanuel Kant's Categorical Imperative to a particular situation. Things are much more complicated, however. What familiar problem cases show is that there are moral intuitions that we all share, and against which we measure general ethical principles. By that measure, all of the familiar accounts can be found wanting. Against the Utility Principle, for instance, Bernard Williams offered the example of a botanist who wanders into a village in the jungle where 10 innocent people are about to be shot. He is told that nine of them will be spared, if only he will himself shoot the tenth (see Williams, 1973). In a similar way, many have argued against Kant's claim that we have a 'perfect duty' not to lie (or, a perfect duty to be truthful) that sometimes lying can be the only way to save an innocent life from death.

In contrast, principalism is a form of perception model (see Quante and Vieth, 2002). The principles that Beauchamp and Childress employ are not first justified through a specific ethical theory. They enable us to articulate what we perceive as being ethically sound or unsound in a particular situation. They are rules of thumb, or prima facie duties, which reflect the core stock of moral beliefs held in common in a modern pluralistic world: a framework for expressing the moral intuitions we all share.

Naturally, principalism has its critics (e.g. Clouser and Gert, 1990; Gert et al. 1997). Nevertheless, it provides a very valuable service by providing a rationale for closing down debates that are intriguing, but fruitless.

3.2 The Membership of a Committee

The guiding principle for membership is that a committee should reflect as wide a range of expertise as possible. That allows members to offer expert advice on as many research areas as possible and also makes the committee less likely to be unduly swayed by the interests of any one group. Perhaps less obvious, but common practice on medical ethics committees and increasingly on social science committees is the inclusion of a 'lay' person. On medical committees it is felt to be important that the lay representative is from outside the medical profession altogether. In the case of interdisciplinary projects such as HUMAINE, where members come from many diverse research cultures, there is less risk of developing a single 'establishment' view that needs to be challenged by someone from outside the community altogether. Nevertheless, it is important to have a lay representative who has no direct stake in the project; and preferable, even in intellectually diverse projects, one who is independent of the disciplines involved.

3.3 Choosing a Code

The most significant decision was to adopt a code derived from psychology rather than medicine. Medical codes are geared to situations where success may be life-preserving, unforeseen effects may be life-threatening, and there is an extreme disparity in power between clinicians and patients. As such, they tend to be more restrictive than codes directed to the less extreme situations that are the norm in psychology.

There are various psychological codes for research involving human participants. HUMAINE decided to adopt the guidelines of the British Psychological Society (BPS) as described in the 'Code of Ethics and Conduct' (2006), the 'Guidelines for Minimum Standards of Ethical Approval in Psychological Research' (2004) and the Code of Conduct, Ethical Principles and Guidelines (2000). There is a great deal of common ground between the codes of practice available in several countries and even in different disciplines. Other codes are mentioned later in this chapter.

3.4 The Value of Scrutiny

It is useful to sketch what, in practice, research ethics committees are likely to achieve. In an area where people are new to ethical scrutiny, they will clearly have an educational function. However, experience of reviewing research applications as part of a psychology research ethics committee suggests that research psychologists actually have rather a good grasp of most of the ethical issues in principles, and yet they repeatedly fall down in turning these abstract guidelines or codes of conduct into good research practice. It is as if having grasped the big picture, they assume that the details can look after themselves. Unfortunately it is the details that can make the difference between participants leaving an experiment with a positive view of science and scientists – or leaving upset, vowing never to return – and perhaps transmitting a negative image of the discipline.

It seems to be increasingly accepted that it is unwise to leave decisions about the conduct of experiments with human participants to the researcher alone. Researchers tend to be enthusiasts. Nobody advances as an academic researcher without some degree of passion for the research. That enthusiasm drives research forward; but on the other hand, it can make the researcher (with a clear focus on the research itself rather than on the participant) a poor judge of some of these issues.

For example, the principle of informed consent implies that the participant gives consent to take part on the basis of full knowledge of what they are likely to experience. But often the experimenter does not think of the situation from the viewpoint of the participant at all, leaving the participant to decide whether to continue based on very patchy and incomplete information. The ability to view the experimental procedure from the outside is one of the most valuable contributions that an ethics committee can make to the process of designing research that works.

3.5 Due Care and Indemnity

University ethics committees typically have two functions: to prevent undesirable things from happening and to provide a kind of indemnity if they do (so that an institution, or preferably its insurers, will bear the cost if anything does go wrong).

Unless there are radical changes, groups like the HUMAINE ethics committee will have no role in providing indemnity. Linked to that, approval from them will not satisfy institutions that require employees to go through their own ethical approval procedure. For that reason, some people have suggested that the committee was useless or worse than useless.

That ignores the function of preventing undesirable things from happening. What an ethical committee like HUMAINE's can do is to identify potentially unethical things that are entailed in an original proposal and to direct the research onto sounder tracks. In other words, that kind of committee is of use to people who do not want to do unethical things or who want to minimise the risk of doing things that would be unethical because they would do harm. It is of no use to people who simply want a fireproof wall if something does go wrong.

Even that may not be quite true. It remains to be tested who is legally responsible if a partner in a large project causes harm to participants. If the project has not taken due care, including appropriate ethical scrutiny, the whole group may be considered liable.

4 Specific Issues in Studying Human Emotion

HUMAINE aims to develop emotionally oriented technology, and so there is clearly a requirement for suitable examples of human emotion. Systems capable of interpreting human emotion or of sending emotionally appropriate signals must have examples from which to learn. Such examples have to be gathered, stored and labelled or annotated in a form that is comprehensible to the artificial system. For a number of theoretical reasons, an explicit decision was taken at the start of the project that the examples used would not be based on acted emotion but would, rather, be as natural as possible. This poses a number of practical difficulties for researchers, as natural episodes of emotion are neither predictable nor particularly common. The practical challenge of capturing, storing and annotating natural emotion also poses a number of particular ethical problems.

4.1 The Ethics of Inducing Emotion

The first approach to solving the problem of the rarity of natural emotional episodes might be to find some method of inducing the emotion. Of course the consequences of inducing emotion in a participant can vary from the trivial to the profound, depending on the nature and intensity of the emotional state generated. However, induction of emotion, particularly negative emotion, is generally not a process to be undertaken lightly. The BPS' revised Ethical Principles for Conducting Research with Human Participants (2000) indicate that on completion of an experiment that has involved induction of a negative emotional state, simply offering a descriptive information-based debriefing is not sufficient. The researcher should ensure that the participant is returned to at least a neutral emotional state before leaving. Thus, if a negative mood has been induced then it should be dispelled or a positive emotional state induced before the participant leaves. However, experience would suggest that this is a process that is easier to achieve with some emotions than with others. Some emotions such as anger seem to be relatively easy to dispel by using a short film or story to invoke amusement or pleasure. Other emotions can be much harder to dispel easily. For instance, the common emotion induction technique of asking someone to recall an emotional event from their past could, if focussing on sadness, easily unlock quite profound memories from their past that neither the participant nor the experimenter is prepared to deal with. Such eventualities may be rare, but when dealing with human participants with all of their frailties, there is a real responsibility on researchers to try to anticipate such possible problems and put safeguards in place for their protection. It is the sharing of such infrequent but important experiences and insights among the researchers in HUMAINE that has been one of the most significant roles of the ethics strand within HUMAINE and of the ethics committee in particular.

4.2 The Ethics of Observing Emotion

If our aim is to collect examples of emotion that are as natural as possible, there is a second problem with inducing emotion: it may be genuine, but it is not natural. For that reason, some teams opt for observing people as they go about their daily business. In adopting this approach you might or might not choose to inform the participant in advance that they are being observed. The former method is open to the criticism that the participant's behaviour could be affected by the knowledge that they are being observed as part of a scientific study. The latter method, of course, raises the ethical issue of an individual's right to privacy.

Indeed, many codes of conduct recognise the particular difficulties of observational research, and the BPS principles state that researchers should normally only make recordings of participants with their full consent both to the recording itself and to any subsequent use(s) of the recordings. However, they do suggest that observational research may be acceptable in situations "where those observed would expect to be observed by strangers". It is suggested that in a public context it may be permissible to record their behaviour and to then ask for permission afterwards – a process that clearly involves 'incomplete disclosure'. However, in the introduction to their revised ethical principles (2000) the BPS make it clear that in the case of deception "the central principle was the reaction of participants when deception was revealed. If this led to discomfort, anger or objections from the participants then the deception was inappropriate."

Thus the guidance seems to be that if you are going to record people's public behaviour without their knowledge, they need to be involved as soon as possible and debriefed. If at that point you encounter negative reactions, then you need to rethink.

Realistically, someone who has been recorded like this should be debriefed as soon as is practical and should be informed about the research. The process naturally goes through many of the steps that would be involved in looking for informed consent in advance of a recording. Of course the participant's real safeguard here is that at this point the investigator must offer them the possibility of withdrawing and of having their data destroyed.

Best practice for a debriefing like this suggests that the researcher should try to involve the participant, to give a sense of having contributed something to the advancement of our knowledge about emotions. There is also an onus on the researcher to find out how the participant feels about the study and their place in it. It would be only too easy to ignore the fact that the exercise has made someone feel uncomfortable or upset in these circumstances – so rather than wait for the participant to express negative feelings, the researcher should ask.

4.3 Alternatives to Induction and Covert Observation

There is an increasingly vocal body of opinion that deception or incomplete disclosure should be avoided if any viable alternatives exist, and it does seem that there is a wide range of situations where researchers could use naturalistic observation after getting informed consent. People regularly take part in events or experiences where both the researcher and the participant know in advance that the participant is likely to experience and express a range of emotions. There is a wide range of situations such as weddings, sporting events (both participating and spectating), parachute jumps, public speaking, emerging after important exams and so on where observation would be possible.

In all of these cases, the participant could be asked in advance to consent to being filmed. It is not yet known whether any of these situations could provide sufficient diversion for the participant to become unaware of the observation and to behave in a natural manner but this is an empirical question and such possibilities should certainly be explored.

4.4 Privacy, Confidentiality and Anonymity

Even once the data are collected, there are further ethical issues raised by the particular nature of the recordings of emotion and the fact that these must be stored and then used in the development of emotionally sensitive systems. The particular problems here are privacy and confidentiality. For most psychological research these should not be major concerns. Usually the individual is included in the research as a representative of a larger group – either as a member of an experimental treatment group or as a representative of an age or gender group or social class and so on. The point is that the data are usually aggregated and there is no need or desire on the part of the researcher to ever reveal the performance of an individual. Even studies of patients with particular psychological problems are usually interesting because of the problem and not the individual, and their identity can be kept confidential with no difficulty. However, in the case of HUMAINE research that simply is not true – the point of the observing and recording emotional behaviour is to allow us to identify and label occurrences of emotion. In most cases the recordings will of course reveal the face or the voice and allow the individual (at least potentially) to be identified. In addition, the BPS ethical principles (2000) state that "in the event that confidentiality/anonymity cannot be guaranteed, the participant MUST be warned of this in advance of agreeing to participate". This brings us to probably the most important of the ethical principles that we will need to consider when we begin collecting and handling emotional data – the principle of informed consent.

4.5 Informed Consent

Informed consent is seen as an essential tool for safeguarding the rights of the individual. Most modern codes of conduct place great emphasis on the need for psychologists to avoid deception of participants whenever possible and the Canadian Code of Ethics for Psychologists (Canadian Psychological Association, 2000) explicitly equates incomplete disclosure with deception – emphasising the need to get fully informed consent before starting. The Canadian code also makes the point that we should approach this exercise as more than just getting the form filled in – we should see informed consent as "the result of a process of reaching an agreement to work collaboratively". Unless there are very good reasons for not doing so, obtaining fully informed consent in writing is seen as an essential prerequisite to most human research.

A typical consent form takes the participant through the essential elements that need to be addressed before asking someone to take part in a study. First they agree that they consent to take part in the study and that they are not being coerced or induced to consent. There is then usually some mention of the consent being genuinely informed. A common way to achieve this is to give participants a short written summary of what the participant should expect and to reassure them that no physical or psychological harm will occur to them. This summary might also briefly describe the aims and nature of the research but, in some cases where this knowledge is thought likely to influence the behaviour of the participant, that information might be withheld until the debriefing at the end of the experiment. There is usually an item that draws the participant's attention to the principle that they can withdraw from the research at any time with no adverse consequences – in effect the participant does not just have to consent at the beginning of the study, they must continue to consent throughout. They can withdraw their consent at any point and this also means that they can withdraw their data at any time. Finally on a typical consent form the issue of confidentiality is commonly addressed. If it will not be possible to guarantee the anonymity or confidentiality of the data then this would be an ideal opportunity to make the participant aware of that fact in advance of them agreeing to take part in the study.

4.6 Storage of Emotion Data

If it is really not possible to guarantee the anonymity or confidentiality of the data then there is an added responsibility to anticipate the uses to which the data are likely to be put. Will other research teams have access to it? Might it be used in public dissemination of research results? There are some notable picture sets of facial expression of emotion that have been very widely used indeed and it is not difficult to imagine that some participants might find that sort of exposure disturbing. We have found that while most participants have no problem consenting to a wide range of possible uses for such data a substantial minority are uncomfortable, particularly with the idea of possible public dissemination of their image or voice. Even if participants are willing to have their faces or voices appear in a variety of circumstances, it is conceivable they may have an objection to being portrayed in a negative way (for instance through the researcher labelling them or their behaviour as angry or ashamed or fearful).

A final issue about the storage of data raised in the Canadian Code of Ethics for Psychologists (2000) is about the security of data. During the process of giving consent to participate in the study and for any uses of the data specified, it is assumed that researchers will safeguard that data. However, the Canadian code asks if the researcher's responsibility ends when anything happens to them (e.g. if they fall ill or lose their job or, in extreme circumstances, if they die). The implication in the Canadian code is that researchers have a responsibility to anticipate such events and should make provision for these eventualities – ensuring, for instance, that there is another individual who can take over responsibility for the data or have it destroyed.

4.7 Who Are the Examples?

When research is compiling databases of emotional behaviour to be used as examples, it begs the question 'Who are to be used as examples?' HUMAINE was sensitive to this question from the beginning and concerned that examples of emotional behaviour should represent people of different age, gender, culture and so on. A broad collection is crucial for many reasons, not least that the developed systems will work only poorly if provided with a narrow range of emotional examples. The inclusion of examples of people of different ages implies the inclusion of children and, of course, the emotional development of children is surely one of the most theoretically interesting areas of study. However, in the UK at least, there has been a marked increase in sensitivity about anyone having contact in a professional capacity with children. Anyone now wishing to conduct research with children must go through a police vetting procedure to make sure they do not have any previous offences against children. In terms of ethical procedures, it is obviously necessary to get the informed consent of the parents as well as the consent of the child, and researchers would need to be much more sensitive to the dangers posed by inducing emotions.

Where researchers again stray beyond the usual experience of collecting data from child participants is in the recording and storage of data by which the child can be identified. In naturalistic studies it is difficult to imagine a research ethics committee granting permission for a study where the researcher planned to film children without their knowledge and only then to ask for consent from the child and the parents. Again, as with natural observation of adults, researchers may have to be much more creative in anticipating suitable emotional situations and gaining prior consent.

5 Predicting Future Ethical Issues

Of course the underlying rationale for the ethical scrutiny of any research is to protect people from any harm that might arise from the research. In most of the cases we have been considering, the focus of such scrutiny is on the research environment itself and on the protection of the participant while the research is being conducted. However, in the case of research on emotion-oriented systems, it was recognised early on in the HUMAINE project that the research might generate applications or products that could potentially have ethical implications far beyond the research environment itself.

Throughout the life of HUMAINE the researchers have tried to predict the ethical issues related to their research and have openly discussed the potential pitfalls and possible abuses of any projected applications (see also "The Ethical Distinctiveness of Emotion-Oriented Technology: Four Long-Term Issues" by Goldie et al., this volume). The issues are relevant in principle to an ethical committee: it would not be ethical to approve a project designed to produce technology that would clearly be damaging to substantial numbers of people. That issue has not arisen in HUMAINE because there is a benign motivation behind all the technology, as the introduction to this chapter pointed out. However, in an area of newly developing technology like this it can be difficult to predict how people will actually use any systems that are developed.

It is not surprising, therefore, that some of the discussions that have taken place over the last 5 years have spilled over into 'science fiction' rather than science. Such speculation is a key part of trying to foresee future developments of technology and its uses. It is increasingly expected that when research involves technological innovation that may have ethical implications, the scientists involved in its development take some responsibility for trying to foresee what the problems might be.

On the other hand, ethical committees should certainly not block research because of speculation that belongs in the realm of science fiction. The sections that follow try to give a sense of what the real issues are in core areas of activity.

5.1 Systems That Can Recognise Emotion

Several types of issues arise in the context of systems that are designed to be able to recognise some features of human emotion. Some of the most pressing hinge on overestimating what they can do (or will soon be able to). There has been a common historical tendency for us to attribute 'objectivity' to machine systems such as the polygraph that can recognise some aspect of human behaviour. Of course the arguments about the abilities and limitations of the polygraph have dragged on for over half a century and it is crucial that, for any new technology in this area, we recognise and spell out the limitations as clearly as possible. In fact the current ability of emotion-oriented systems to accurately and reliably read human emotion falls very far short of the human ability to do the same thing. This seems unlikely to change in the near/medium term and it is important that researchers ensure that their systems are accurately portrayed. Overselling the abilities of such systems would be ethically wrong and could lead to quite inappropriate decisions being taken and harm done.

A related issue is the extent to which people will regard the monitoring of their emotions as an intrusion into their private lives. In an extension of a point made above in relation to the observational study of human emotion, it is not clear if the use of machines to monitor people in public places without their knowledge should be regarded as a similar invasion of privacy. In addition, the use of technology such as semi-intelligent information filters (SIIFs) means that such monitoring is unlikely to remain purely descriptive, but will allow judgements to be made about emotional changes. This of course begs the question about the uses to which such technology might be put to. The monitoring of call centre employees might allow workers to be pulled off the job as their anger levels rise. This is a rather straightforward example of a much wider reaching ethical problem. After monitoring a large sample of workers we may be able to make some sort of probabilistic statement about the relationship between the emotion (e.g. anger) and problem behaviour (e.g. rudeness). However, the point is that the emotion does not *inevitably* lead to a change in behaviour. A worker may become very angry indeed with a caller, but manage to keep her feelings under control and not let the anger alter her behaviour towards the caller. So, even if it does become possible to identify emotional states reliably, then it is important that the information is not used unreasonably.

In the first report (D10a) on ethics for HUMAINE (Goldie et al., 2004) the point was also made that the very existence of machines that can monitor or probe inner metal states in ways that may be poorly understood could in itself be frightening to many people, The research community has a responsibility to reduce the risk that they will be used to create fear.

5.2 Systems That Can Manipulate Emotion

Next we come to systems that are designed to manipulate human emotions. Indeed, one area that has been the focus of an entire work-package within HUMAINE is persuasion. Of course this area raises many ethical issues – some of which were raised at the start of the project. At the first plenary meeting of the network members, in a talk intended to introduce many of the ethical issues that might lie ahead, the network coordinator warned of a 'nightmare possibility' – a system that could read user emotions and selectively reinforce or attenuate them without conscience or empathy.

However, although it is possible to conceive of such an extreme system in theory, most of the problems associated with persuasive systems seem to be more mundane. In fact the existing legal controls and guidelines applicable in most western countries already cover the most likely abuses. The replacement of a person with a machine in the role of the persuader does not alter the underlying ethical issues.

5.3 Systems That Can Express Emotion

Finally we come to systems that are capable of sending emotional signals (ultimately with the aim of simulating human emotion). Although some of the discussions regarding ethical issues in this area are definitely 'crystal-ball gazing' the ethical problems posed by machines that can simulate human emotion and one day might be mistaken for human should probably begin to be discussed now.

An issue that was raised at the first HUMAINE plenary meeting was that when people's expectations of a machine include emotional responsiveness, then you immediately enter an ethically sensitive situation – if that expectation exists then it becomes possible for the machine to respond inappropriately (either by failing to respond, or by responding with the wrong emotion) – for instance if someone is experiencing and expressing sadness or grief and receives a flat, unemotional response or even worse, a response of laughter, the result could be distressing.

Of course, the way that *we* respond to a machine that is expressing emotion also raises tricky questions. While it is true that the machine is only simulating rather than experiencing emotion and cannot 'feel' emotional distress if we react inappropriately, it has been argued that such behaviour may, in the long run, be harmful to us – by getting us into the bad habit of treating emotional beings as mere things.

These questions, related to the possible uses that people might eventually find for emotion-sensitive technology and how they might interact with such systems, leave us gazing into an uncertain future. Predictions about how the world might be influenced by developing technologies are notoriously unreliable. Such speculations definitely go beyond the activity of a traditional research ethics committee.

6 Conclusion

HUMAINE followed a dual-track approach: establishing an ethics committee to deal with immediate issues raised by ongoing research on one side and, on the other, allowing sufficient time and resources for discussion of the ethical implications of the systems that might develop out of the current research. The approach seems to have been very productive.

Over the lifetime of the project, attitudes towards the process of ethical scrutiny have changed considerably as the network members have become more familiar with the issues involved. The point was made at the beginning of this chapter that the aims of HUMAINE have been benign, but it is undoubtedly true that research on emotion-oriented systems has the potential for controversy. The community is far more likely to maintain public support for this research and the technology that emerges from it, if it is open about discussing the ethical issues and it communicates these discussions as widely as possible.

References

Beauchamp TL, Childress JF (2001) Principles of biomedical ethics, 5th edn. Oxford University Press, Oxford

Canadian Psychological Association (2000) Canadian code of ethics for psychologists, 3rd edn. Canadian Psychological Association, Ottawa, ON

Clouser KD, Gert B (1990) A critique of principalism. J Med Philos 15:219–236
Gert B et al (1997) Bioethics: a return to fundamentals. Oxford University Press, New York, NY
Goldie P et al (2004) Report on ethical prospects and pitfalls for emotion-oriented systems. HUMAINE report D10a. http://emotion-research.net/projects/humaine/deliverables/D10a.pdf. Last visited 17 May 2010
Quante M, Vieth A (2002) Defending principalism well understood. J Med Philos 27(6):621–649
The British Psychological Society (2000) Code of conduct, ethical principles and guidelines. The British Psychological Society, Leicester, UK
The British Psychological Society (2004) Guidelines for minimum standards of ethical approval in psychological research. The British Psychological Society, Leicester, UK
The British Psychological Society (2006) Code of ethics and conduct. The British Psychological Society, Leicester, UK
Williams B (1973) A critique of utilitarianism. In: Smart JJC, Williams B (eds) Utilitarianism for and against. Cambridge University Press, Cambridge, MA, p 98

Additional References (on Principalism)

Bryson J, Kime P (2003) Just another artifact: ethics and the empirical experience of AI. University of Edinburgh, Edinburgh
Dancy J (1993) Moral reasons. Blackwell, Oxford
Döring SA (2004) Gründe und Gefühle. Rationale Motivation durch emotionale Vernunft. Habilitationsschrift, Universität Duisburg-Essen. (in German)
Goldie P, Döring S (2004) Categories of emotion: everyday psychology and scientific psychology, with Peter Goldie. In: Proceedings of the first HUMAINE workshop, Geneva 17–19 Jun. http://emotion-research.net/ws/wp3/ExtraMaterial/HUMAINE-Goldie.pdf. Last visited 17 May 2010
Goldie P, Döring S (2005) Emotions as evaluations. In: Proceedings of the AISB'05 symposium on *Agents that Want and Like: Motivational and Emotional Roots of Cognition and Action*, Hertfordshire University, 12–13 April 2005, SSAIB, pp 45–50
Goldie P, Döring SA, WP10 members (2005) D10b Interim report to plenary meeting on ethical frameworks for emotion-oriented system. http://emotion-research.net/project/humaine/deliverables/D10b.pdf. Accessed 17 May 2010
McDowell J (1998) Virtue and reason. Reprinted in his mind, value, and reality. Harvard University Press, Cambridge, MA, pp 131–166
Quante M, Vieth A (2001) Wahrnehmung oder Rechtfertigung? Zum Verhältnis inferenzieller und nicht-inferenzieller Erkenntnis in der partikularistischen Ethik. Jahrbuch für Wissenschaft und Ethik 6:203–234. (in German)
Wiggins D (1998) Deliberation and practical reason. Reprinted in his Needs, values, truth, 3rd edn. Blackwell, Oxford, pp 215–237

Glossary

One of the problems with interdisciplinary fields is that they draw their vocabulary from very diverse sources. In addition, the original terminology is often adapted. The final writing can therefore be difficult to follow not only for readers lacking background in the underlying areas, but also for generalists who have to learn the new meanings attributed to terminology they already know. This glossary was compiled to help people deal with this problem. It does not claim to give exact definitions of the terms. The entries needed to do that would have to be long and would often have to describe many variant uses. The aim is simply to enable people to read the chapters without being confused by totally unfamiliar words or phrases.

Some terms are closely linked to particular individuals. In those cases, the names are given, without references. A web search will identify the relevant sources easily.

The entries are given in alphabetical order, prefixed by a specification of the part of the Handbook they primarily apply to. Many terms are inter-related. That is signalled by mentioning the relevant term, followed by 'qv' (meaning 'quod vide' – 'which see').

Action tendency: The tendency to act or react in a particular biologically significant way that is associated with an emotion. The term is due to Frijda, who argued that action tendencies form the kernel of emotions.

Action units (AU): visible results from the contraction or relaxation of one or more muscles, used also to describe higher level concepts in the Facial Action Coding System (qv).

Activation level: A measure of the organism's overall disposition to engage in action, corresponding to how active or lethargic it feels. Used in dimensional theories to describe one of the two most widely agreed dimensions, the other being valence (qv). Broadly equivalent to arousal level (qv).

Adaptivity: The process of altering the defining parameters of a system or context of interaction to match emerging information from the environment or the user.

Affect: A term which has no very precise meaning in everyday English and which various modern scholars have recruited to express theoretically motivated ideas. For instance, Hilgard revived the old scholastic idea

P. Petta et al. (eds.), *Emotion-Oriented Systems*, Cognitive Technologies,
DOI 10.1007/978-3-642-15184-2,

that mental life involves three divisions: affect, cognition, and will ('conation'); Davidson's group use 'affective' as a blanket term for emotion, mood, and other related states; Russell's group use 'affect' to denote a kind of self-evaluation that enters into emotions, moods, and other states; and Panksepp uses it to denote a form of consciousness that arises from particular physiological systems. There are many other uses.

Affect burst: Brief, discrete, and nonverbal expressions of affect in the voice, often accompanied by a facial gesture. Their vocal form ranges from non-phonemic vocalisations, such as laughter or a rapid intake of breath, to quasi-verbal interjections such as English 'yuck' or 'yippee' for which the segmental form transports the emotional meaning independently of the prosody.

Annotation: The manual or automatic attribution of information describing behaviour. For example, a word by word transcription including punctuation, following the LDC (Linguistic Data Consortium) norms for hesitations, breath, etc., at several temporal levels (whole video, segments of the video, behaviours observed at specific moments) and at several levels of abstraction. Global behaviour observed during the whole video is annotated with communicative act, emotions, and multimodal cues. See also: Quantised labelling, Trace labelling.

Appraisal: Evaluation made by the organism about the way events, things, and people in its environment relate to its 'weal or woe'. Appraisals are defined as rapid, automatic, unconscious, and ballistic evaluations involved in the recurrent process of emotion.

Appraisal theory: Any of several theories that propose the most important distinctions between emotions are the distinctive patterns of appraisal (qv) that they involve.

Arousal level: A measure of the organism's overall disposition to engage in action. It is broadly equivalent to activation level (qv), but the concept of arousal is rooted in a global theory of sleep, wakefulness, attention, etc., which has been superseded.

Artificial vision: Conventional computer vision schemes, where imagery representing the environment is processed in order to segment, detect, and recognise or conduct interpretations of constituent scene objects.

Audio-visual speech recognition: Speech recognition algorithms which also utilise visual information from the mouth/lip area, besides audio features.

Backchannel: Feedback and comments provided by listeners during face-to-face conversation, through short verbalisations and nonverbal signals, showing how they are engaged in the speakers' dialogue.

Bag-of-words: Representation of a document or sentence as an unordered collection of words, disregarding grammar and word order.

Basic emotion: According to many theorists, the natural unit of emotional life. Classical approaches list five to seven basic emotions, frequent members being: anger, contempt, disgust, fear, interest, joy/happiness, and sadness. Each is supposed to be a qualitatively different, transient affective reaction, evolutionarily justified and rooted in discrete physiological systems. See also: Big six.

BDI: See Belief-desire-intention (BDI) architecture.

Behaviour-based robotics: A subdiscipline of (embodied) AI and autonomous robotics that conceives robot architectures in terms of 'behaviours' or competence modules implementing the various activities that a robot can perform in the particular environment that it inhabits. A behaviour-based robot has a set of behaviour modules that compete with one another in order to gain control of the robot's actuators.

Behaviour Markup Language (BML): Behaviour Markup Language is a representation language (qv) comprising all those representations that are necessary for the realisation of multimodal behaviour. Thus BML stands between behaviour planning (qv) and behaviour realisation (qv). It includes directives for the realisation of textual and prosodic information, facial display, gestures and postures, eye gaze, and, very importantly, directives for the temporal synchronisation of behaviours (see Synch point and Temporal alignment). BML is developed under the frames of the SAIBA (qv) initiative (http://wiki.mindmakers.org/projects:bml:main; accessed 31 May 2010).

Behaviour planning: In the context of multimodal communication (qv), the process that addresses all aspects that are necessary for the realisation (how the communicative intent is conveyed) of multimodal communicative behaviour. This includes the specification of the communication channels involved, such as face, gesture, posture, speech, and their relative temporal alignment (see also SAIBA, Synchronisation point and Temporal alignment).

Behaviour realisation: Process that concretises the planned behaviour. In the context of multimodal behaviour this includes speech synthesis, the realisation of concrete facial expressions and gestures, and the transformation of relative timings to absolute times in milliseconds. At this stage of behaviour generation the planned behaviour is transformed into low-level representations that are input to an animation player. See also: SAIBA.

Belief-desire-intention (BDI) architecture: Within the research community concerned with software agents, the term belief-desire-intention (BDI) has been used variously to denote a position on theoretically useful mental state distinctions, particular models of how these mental states affect reasoning, and a genre of architectures or frameworks for developing software agents based on philosophical theories of practical reasoning.

Big five model: A widely used model of personality which identifies five main factors that describe human personality: openness to experience, conscientiousness, extraversion, agreeableness, and neuroticism. Each corresponds to a cluster of personality traits. For example, shyness is a personality trait in the extraversion factor.

Big six: An influential list of basic emotions (qv) proposed by Ekman which includes anger, disgust, fear, surprise, joy/happiness, and sadness.

BML: See Behaviour Markup Language (BML).

Body language: Popular term for visible movements and stances that convey information about emotion, intention, social relationships, etc.

Bottom-up attention: Selective mechanisms operating on raw sensory information associated with involuntary and rapid shifting of attention to visual features of potential importance. Contrasts with top-down attention (qv).

Central nervous system (CNS): The concentrations of nervous tissue (brain and spinal cord) that coordinate the activity of all parts of the body.

Cepstrum: The result of applying the Fourier transform (FT) to the decibel spectrum.

Cognitive dissonance: Term used in psychology to describe situations where a person simultaneously holds two cognitions (i.e. beliefs, attitudes, values, or feelings), that contradict or are inconsistent with one another.

Complex emotion: In this volume, informal term describing situations where several emotions are either in play simultaneously or blend into a quick succession of related emotions.

Component of emotion: Aspects of life that are not themselves emotions but which come together in an archetypal emotion. Componential theories argue that the emotion is simply the combination of components. Typical lists include appraisal, physiological reaction, expression, action tendency, and subjective feeling. Each component has been the focus of dedicated research threads.

Context: The environment and location where interaction takes place, including the particular characteristics of the users, their personality, mood, or other information used as a priori knowledge.

Contextual novelty: Contextual novelty refers to the situations in which an object is perceived in a new context or emerged in a given stable context. For example, a ball on the table, while such objects are usually on the floor, would induce a novelty detection phenomenon. See also: Partial novelty, Perceptual novelty, Real novelty, Semantic novelty.

Cover classes: Broad groupings of emotion-related terms, often used in labelling emotional databases.

Curse of dimensionality: The rapid increase in the amount of data needed to establish empirically how a set of phenomena behaves as the number of relevant dimensions is increased. To study a space of N dimensions with l levels each, a minimum of l^N data points are needed. A particular implication is that in a complex field such as emotion, valid conclusions require very large amounts of empirical data. See: Validity.

Dimensional theories of emotion: Theories that propose emotions are not qualitatively distinct: the differences between them are rather taken to be matters of degree, corresponding to differing positions on an underlying set of dimensions of emotion (qv).

Dimensions of emotion: A dimension is a meaningful continuum along which different emotional states can be placed. Core examples are valence (how positive or negative the person feels) and arousal (how active or lethargic they feel). For other dimensions, see the GRID study.

Display rules: According to Ekman, social and cultural customs and norms that lead people not to give emotions their natural expression under particular circumstances or situation.

Drives: Inbuilt pressures to do certain things or acquire certain resources that are fundamental to survival rather than others – for instance, to eat, to drink, to reproduce.

Dual process theory of emotion: A theory suggesting that emotions are processed along two routes: (1) a fast intuitive route and (2) a slower reflective route. The slower route can serve as a check or pro-active management device for the faster route.

ECA: see Embodied conversational agent.

Ecological validity: The extent to which the situation involved in an experiment or study corresponds to something that would occur in a natural setting. A goal of many experiments is to control the environment but this can result in low ecological validity, which in turn may mean the experiment casts very little light on the way the relevant systems normally function. See also: Validity.

Electroencephalograph (EEG): A record of electrical activity in the brain (e.g. firing of neurons) obtained from electrodes placed on the scalp.

Embodied artificial intelligence: Embodied artificial intelligence is researched in the context of 'complete' (embodied, situated) autonomous agents. It considers the richness of behaviour shown by an embodied agent that acts in the real world (as complex as it is) obtaining its (partial) information about the environment through its sensors in continuous interaction with the real world (situated agent). Special attention is paid to the boundedness of resources (size of memory, limitations of sensors and effectors, timeliness of decisions and actions, ...) and the kinds and relations of internal (i.e., the agent's) and external (i.e., environmental) contributions to overall performance. See also: Sensory honesty.

Embodied conversational agent (ECA): An embodied conversational agent (ECA) is a human-like conversational character able to engage with the user in multimodal communication. The usual modalities include speech, facial expression, eye gaze, head movement, body posture, and hand-arm gesture.

Emergent emotion: A term coined to describe the kind of delimited, relatively intense emotional episode that theories of emotion often focus on. Contrasts with pervasive emotion (qv).

Emotional corpora: Collections of video and audio materials used in experimental studies or to develop, validate and compare computational models. Emotional corpora have been gathered in a variety of contexts (laboratory, meeting, TV material, field studies).

Engagement: The extent of the user's attention towards a particular object or person, usually described in measurable terms by a theory of mind (qv).

Epistemic states: Mental states that have an emotional dimension, for example, doubt, questioning, rejection. Attention was called to these by the work of Simon Baron-Cohen.

eXtensible Markup Language (XML): A tightly defined format for representing machine-readable information. XML documents have a closely enforced tree structure consisting of elements which can have attributes and children, which again can be elements or free text. Names of valid elements and attributes for a given markup language are defined using a validation mechanism such as a Document Type Definition (DTD) or an XML Schema.

F0: See Fundamental frequency (F0)

Face-threatening acts: Behaviour with the potential to undermine another individual's sense of freedom or sense of self-worth.

Facial action coding system: A categorisation system for facial behaviours based on the underlying musculature. Facial behaviours are coded in terms of action units (qv) involved in a change in appearance as well as duration, intensity, and asymmetry. Originally developed by Paul Ekman and Wallace Friesen.

Feature extraction: The process by which continuous and/or multi-dimensional information is described using fewer and simpler variables.

Feedback: Broadly speaking, a signal that informs an agent about its own states or actions; in a more specific sense, implies that the signal originated in the agent's own action.

FML: See Functional Markup Language (FML).

Functional Markup Language (FML): The Functional Markup Language is a representation language (qv) designed to include all information regarding an Embodied conversational agent (ECA)'s (qv) mental, communicative, and affective state that is necessary to create a link between intent and behaviour planning. It needs to provide a large spectrum of information including semantic, communicative, discursive, pragmatic, and epistemic information. FML is developed under the frames of the SAIBA (qv) initiative (http://wiki.mindmakers.org/projects:fml:main (visited November14, 2010)).

Fundamental Frequency (F0): The lowest frequency in a harmonic series – the others are whole number multiples of it.

Fusion: The process of combining information from different sources into a single decision.

Galvanic skin response (GSR): A measure of the electrical resistance of the skin. It directly reflects the release of sweat (which is conductive) and is indirectly related to emotional or physiological arousal, which selectively affects some sweat glands.

GRID study: Research by Fontaine et al. on the dimensions needed to distinguish major emotions. Four dimensions were identified: valence (qv), potency (qv), arousal (qv), and unpredictability (qv).

Ground truth: A term, with origins in cartography and aerial imaging, used to describe data that can be taken as definitive and against which systems can be measured. Its application to emotion is controversial, since it is highly debatable whether emotions as they normally occur are things about which we can have definitive knowledge.

Grounding: The mutual belief shared by the partners in a conversation that they have understood what the exchanges meant.

Hidden Markov model: A statistical model where a system is assumed to be a Markov process involving both observed and unobserved states.

Individual differences: The branch of psychology that seeks to measure the ways in which people differ from one another in the long term. Areas of study include personality, intelligence, ability, emotional intelligence, and motivation.

Induction of emotion: Procedure that deliberately creates an emotion in an individual, ranging from self-induction (e.g. by imagining a situation) to experimental manipulations designed to amuse, irritate, etc.

Intelligent user interfaces: A subset of human–computer interfaces that is informed by research from artificial intelligence and statistical learning techniques.

Intent planning: Addresses all aspects of communicative intent (what to convey) including semantic, communicative, discursive, pragmatic, and epistemic information. See also: SAIBA.

Interaction analysis: A widely used method of describing the social meaning expressed by interactions between human beings and objects in their environment, using standard categories such as: request information, express solidarity, etc.

Interactional synchrony: Phenomenon in which the flow of movements of the listener is rhythmically coordinated with those of the speaker.

Markup language: A formally specified, machine-readable language that encodes information according to a pre-defined syntax, often represented in XML (eXtensible Markup Language qv). Markup languages exist for many different application areas and can either be written by hand, generated by humans with the help of authoring tools, or generated by software programs. 'Markup language' is often used synonymously to 'representation language' (qv). Where the two terms are distinguished, markup languages are understood as adding at a high level of abstraction nonverbal information to an existing text.

Mel-frequency cepstral coefficients (MFCC): Coefficients of a representation of the short-term power spectrum of a sound, derived from the cepstrum (qv).

Misleading emotions: Used in the context of Part V of the Handbook in particular, to describe emotions that are not useful in picking up saliences in the environment and enabling quick and effective action.

Mood: An emotion-related state that is routinely distinguished from emotion in a strong sense (see emergent emotion). It involves emotion-like feelings, but they are not about anything in particular. Moods tend to be relatively long lasting.

Multimodal communication: Communication that uses more than one of the human communicative modalities – speech, hand gestures, other bodily movements, facial expressions, gaze (and pupil dilation), posture, spatial behaviour, bodily contact, clothes (and other aspects of appearance), nonverbal vocalisations, smell.

Multimodality: The fact that relevant information is distributed across inputs from multiple, qualitatively diverse sensors (e.g. visual, aural, physiological data).

Natural language generation: The branch of natural language processing that deals with the automatic production of text (producing speech is generally regarded as a separate problem).

Neural networks: Computational models which try to simulate the structure and/or functional aspects of biological neural networks. In a more technical sense, a branch of nonlinear statistics (see: Statistical learning).

Neuromodulation: Neuromodulation refers to the action on neurons of a large family of chemicals called neuromodulators, e.g. dopamine, serotonin, and norepinephrine. Each neuromodulator activates specific receptors on the neural membrane, having specific effects on the functioning of the neuron.

Ontology: In computer science, a structured representation of knowledge shared by a community of practice. An ontology describes certain aspects of 'the world' by describing relations among classes (e.g. that a 'table' is a specific type of 'furniture') and instances of these classes.

Paralanguage: Nonverbal elements of oral/aural communication. These can include variations of pitch, volume, and intonation, as well as laughter, sobbing, breaks in voice, tremulous voice, gasp, sigh, exhalation, screams.

Partial novelty: Partial novelty would be involved when an organism perceives an object looking like an already perceived object in the past but presenting some perceptual differences on one or several of these characteristics. For example, if a baby used to play with a red ball, a new violet ball will not be totally new; it is only one or several of these characteristics which are new. The categorisation processes are very important in this concept of partial novelty. See also: Contextual novelty, Perceptual novelty, Real novelty, Semantic novelty.

Perceptual novelty: Perceptual novelty is the assembly of a new representation of an object never perceived in the past by the organism and requiring a new encoding in short-term and long-term memory. See also: Contextual novelty, Partial novelty, Real novelty, Semantic novelty.

Persuasion: Influencing people, often through communication, in order to make them perform certain actions, pursue goals, or collaborate in various activities that they would have not otherwise engaged in.

Pervasive emotion: Term coined to capture the broad sense of the everyday term 'emotion'. In contrast with emergent emotion (qv), which is concentrated in brief episodes, pervasive emotion is only absent on the relatively rare occasions when the person is completely unemotional.

Phoneme: Smallest segment in the sound system of a language that can be used to create a meaningful contrast between utterances (qv, e.g. bob/cob, bob/bib, bob/bog). See also: Viseme.

Physiological signal: Output of a sensor (usually electrical) that measures simple physical changes in humans.

Potency: Widely regarded as one of the key dimensions of emotion (qv). It refers to the subject's sense that he/she does or does not have the power to influence emotionally significant people, events, or situations.

Quantised labelling: Annotation (qv) strategy where labels are attached to discrete chunks of time. Beginning and end periods are identified and the label applies to the encapsulated behaviour. See also: Annotation.

Rapport: Mutual understanding signalled by behaviours such as head nods or smiles, mutual attentiveness (e.g. mutual gaze), and coordination (e.g. postural mimicry or synchronised movements). See also: Grounding.

Real novelty: Refers to a new object instance that has never been or rarely been encountered before, although it may fall into a known object category. For example, a car may be deemed novel as this particular instance has never been observed before. See also: Contextual novelty, Partial novelty, Perceptual novelty, Semantic novelty.

Recalcitrant emotions: Used in the context of Part V of the Handbook in particular to describe emotions involving a conflict between the emotional perception and an evaluative belief.

Reliability: One of two fundamental considerations relevant to evaluating a measure: the other is validity (qv). It is the degree to which a measure, experiment, or study will provide the same results on repeated trials or assessments. It can be quantified using various indices of inter-rater reliability. High reliability does not guarantee validity (the raters may agree on the wrong conclusion), and low reliability does not rule it out (measuring reliability does not clarify the validity of answers to the question 'What were you thinking a minute ago?').

Representation language: A formally specified, machine-readable language that provides information according to a pre-defined syntax, often represented in XML (qv). Representation languages exist for many different application areas and can either be written by hand, generated by humans with the help of authoring tools, or generated by software programs. Due to the complexity of their representations they are best generated with the help of software tools. Representation language is often used synonymously to Markup language (qv). Where the two terms are distinguished, markup languages are understood as annotating text with high-level nonverbal information, whereas representation languages (declaratively) represent information relevant at different stages of the multimodal generation process at different levels of granularity.

SAIBA (situtation, agent, intention, behaviour, animation): The generation of natural multimodal (qv) output for embodied conversational agents (qv) requires a time-critical production process with high flexibility. To scaffold this production process and encourage sharing and collaboration, a working group of ECA (qv) researchers has introduced the SAIBA framework. The framework specifies multimodal generation at a macro-scale, consisting of processing stages on three different levels: (1) planning of a communicative intent (Intent planning, qv), (2) planning of a multimodal realisation of this intent (Behaviour planning, qv), and (3) realisation of the planned behaviours (Behaviour realisation, qv).

Saliency map: A two-dimensional grey-scale representation of the most likely areas of a scene to pop out at a viewer, based on the low-level contrast of constituent features such as colour, intensity, depth, and motion.

Scripting language: A formally specified, machine-readable language that provides information according to a pre-defined syntax, often represented in XML (qv). Basically it is a representation language (qv) with procedures of the kind 'if event X happened, then trigger Y' added. Thus a scripting language is more like a high-level programming language, whereas a representation language is a language for describing/declaring aspects of multimodal behaviour at a higher level of detail.

Self-organising map: A type of neural network (qv) which produces a low-dimensional, discretised representation of the input space (called the 'map').

Semantic novelty: Semantic novelty refers to a situation in which the relationships between the objects or the concepts are organised in a new manner and have never been perceived such as in the past. The fact that individuals are able to create a new tool or a new concept from a series of well-known objects or ideas is characteristic of this type of novelty. See also: Contextual novelty, Partial novelty, Perceptual novelty, Real novelty.

Sensory honesty: Constraint on the modelling of agents in a virtual environment. They can potentially be made aware of everything happening in it with perfect accuracy, but if they are to simulate a human-like character, perceptual constraints should be imposed so as to limit their capabilities to those of the character being simulated. See also: Embodied artificial intelligence, Synthetic perception.

Sign language: A language which uses visually transmitted sign patterns (hand shapes; orientation and movement of the hands, arms, or body; and facial expressions) to communicate meanings; native language of deaf communities.

Signal processing: An area of computing that uses mathematical techniques to transform, compress, and partition information contained in signals generated by sensors (such as cameras, microphones, electrodes). It is generally distinguished from 'high-level' perceptual techniques, which deal with problems such as interpreting the signal.

Social influence: Affecting or changing the way someone behaves or thinks, by using social processes to change their mental state.

Speech act: Concept from linguistics, highlighting the fact that utterances (qv) do not simply describe situations, they change them. From a speech act perspective, any utterance is some kind of invitation to the addressees to participate in a particular configuration of actions: attend to what is being said, try to figure out what is meant, and carry out what was intended by the speaker, which could range from updating a belief state, to feeling offended, or closing the window.

Statistical learning: Formal techniques for inferring probabilistic relationships from large bodies of data. Key examples include regression techniques, principal component analysis, neural networks (qv), support vector machines, linear discriminant analysis, Bayesian networks, and hidden Markov models (qv).

Stimulus evaluation checks (SECs): According to Scherer's component process model, the appraisals (qv) made by the organism take the form of distinct and sequential 'stimulus evaluation checks', each concerned with a different aspect of potentially emotion-laden stimuli in the environment.

Synchronisation point: A synch or synchronisation point is a point on the time line of the signal of a verbal or nonverbal gesture where it may be aligned with signals from other modalities in order to lead to integrated multimodal behaviours. Take for instance the multimodal behaviour where the utterance (qv) 'give me this cake' is accompanied by two nonverbal gestures, an eye gaze and a pointing gesture at the cake, and the nonverbal gestures start immediately before this is uttered. Such a fine-grained timing is only possible with the a priori stipulation of synch points in the signals. See also Temporal alignment.

Synthesis: In the context of Part III of the Handbook in particular, generally it refers to the technology of generating stimuli that convey emotion, which may take the form of speech, facial expression, gesture, or signals in other modalities.

Synthetic perception: Consideration of ways in which scene data from a virtual environment should be filtered, restricted, or integrated for use by agents in order to limit their sensory capabilities (see also Embodied artificial intelligence, Sensory honesty, Synthetic Vision).

Synthetic perceptual maps: Grey scale topographic retinotopic maps that represent the visual world as seen through the eyes of a viewer, where the value of a location in the map is the strength of a feature or resultant operation based on the corresponding spatial location.

Synthetic vision: Specific case of synthetic perception (qv) for the visual modality, imposing constraints relating to field-of-view, distance, and occlusion on data sensed by agent. A common technique for implementing synthetic vision involves rendering the scene from the agent's perspective. See: Synthetic perceptual maps.

Temperament: In the psychology of individual differences (qv), long-term predisposition to behave, react, and feel in particular ways. Temperament helps to form personality.

Temporal alignment: It is the synchronisation of signals from different communication channels or modalities such as face, gaze, hand-arm gesture, speech. See also Synch point.

Theory of mind: Term used in psychology and cognitive science to describe people's intuitive ability to understand their own minds in terms of concepts such as thoughts, beliefs, and desires and to understand that other people have thoughts, beliefs, and desires which are different from one's own.

Top-down attention: Voluntary, goal-oriented strategies biasing attention towards locations of relevance to an entity. Contrasts with bottom-up attention (qv).

Trace labelling: A method of annotation (qv) that results in a continuous, time-varying trace of the way some attribute fluctuates from moment to moment. Typically the attribute is a dimension, such as valence (qv) or arousal (qv), but the technique can also be used to record the strength of a specified emotion (e.g. anger) or the salience of an appraisal (qv, e.g. goal obstructiveness).

Turn-taking: The process that ensures an orderly transition from communicative action taken by one interactant (a 'turn') to a communicative act by another.

Two-resource problem: In the context of Part V of the Handbook in particular, a task where a robot must maintain appropriate levels of two internal variables by consuming two external resources.

Unpredictability: A feature of situations which has a major bearing on the agent's emotional response to them (for instance, it is a fundamental trigger for surprise). Highlighted theoretically by appraisal theories (qv) and empirically by the GRID study (qv).

User model: A formal representation of the main characteristics of a user that may affect his/her interaction with software products or, more in general, with technology.

User modelling: The process of developing user models (qv) for the purposes of tailoring messages and interactions to the individual concerned. Models may be static or dynamic and universal, broad, or specific.

Utterance: A complete unit of speech, usually bounded by silence.

Valence: The most widely acknowledged dimension of emotion (qv). Valence measures the strength of positive or negative feeling that the agent has. In the case of mood (qv), the feeling is a 'barometer' of the agent's overall state. In the case of emotion, it is about the people, things, etc. on which the emotion is focused.

Validity: One of two fundamental considerations relevant to evaluating a measure: the other is reliability (qv). The degree to which a measure, experiment, or study assesses the actual object or concept that is the focus of a study.

Viseme: The visible pattern of lip movements that corresponds to a phoneme (qv).

Wizard-of-Oz: An experiment where users interact with a system that they believe is autonomous, but that is actually being operated (wholly or partially) by a human operator.

Index

A

ABL (representation language), 577
Accompaniment signals, 594
Acoustic feature, 77–78, 80–81, 83, 91
Acronym generation, 621, 623
Acted data, 164, 172, 198–203, 245, 258
Action
  inducement, 561
  tendencies, 377, 380
  tendency, 19
Action Unit (AU), 102–103, 352
Activation, 173, 180, 185, 198, 223–224, 232, 236, 255–257, 259–260, 268
Active appearance model, 105
Active goal, 533
Activity
  Data, 246, 248–249, 252, 255, 268, 277, 282–283
  space, 672
Adaptation
  algorithm, 127–128
  phase, 127, 130
Adaptive neural network, 128–130
Adult slang (in humour), 617, 628–629, 631
Advertising, 527, 535
Aesthetic experiences, 658
Aesthetic response, 692
Affect, 9, 11–18, 20–21, 24–28
  affective disposition, 24
  bursts, 334–338
  dimensions, 177, 223–224, 256, 259
  Grid, 691
Affective body behaviours, 663
Affective computing, 177, 247
Affective contagion, 592
Affective evaluation methods, 688
Affective induction through verbal language, 610
Affective interaction, 638–639, 689, 694, 697, 699–700
Affective involvement, 643
Affective language generation, 576
Affective loop, 667–669
Affective Presentation Markup Language (APML), 371–372, 396
Affective reasoner, 576
Affector (system), 689
Agent perception, 293–315
Agreement, 589, 594–595, 597–598, 603
AIBO (Sony Artificial Intelligence Robot), 178
ALICE, 510
Alliteration (in humour), 617, 624, 627–631
Ambiguity in humor, 614
Ambiguous design, 668, 681
AMI project, 186
Amygdala, 50–58
Anger, 10–11, 13, 19, 22, 25–27, 34, 36, 39, 50, 58, 664–666
Animated text, 578
Annotated *corpora*, 184, 514
Annotation, 175, 202, 222, 234, 236, 265, 353–354, 357, 361
  of dialogue features, 234
Anterior cingulate cortex (ACC), 54
Anticipated emotions, 538–539
Anticipation, 483–497
  and emotion, 483–497
  and goal-directed systems, 484
Anticipatory function of emotion, 485
Antonymy (in humour), 617–621, 627–631
ANVIL platform, 252, 254–255
Apathy, 12
Appearance feature, 106
Applied ethics, 756
Appraisal, 18–20, 24, 27–28, 51–53, 56, 58
  categories, 205, 230–231, 256
  framework, 517

Appraisal (*Cont.*)
processing, 442
theory, 176, 216, 230
Approach, 47–59
ARAUCARIA (argument analyzer), 575
AR face database, 182
ARGUER (persuasive system), 575
Argumentation, 528, 539, 551–554
scheme dialogue, 574
schemes, 553
-specific persuasion, 561
strategies, 570
Argumentative discourse, 593
Argumentative persuasion, 538, 550
Argument from popular practice, 550
Arousal, 13, 25, 51–54
Arousal measurement, 690–691, 696
Artificial intelligence, 39, 43
Artificial Neural Network (ANN), 87
Artificial vision, 301–302
Assenting signals, 594
Assessments (in persuasive nonverbal communication), 595
AT&T database, 184
Attention, 50, 54, 56
Attentional Blink (AB), 56–57
Attention signals, 594
Attitudes, 24–25, 528, 533, 535–536, 548
-incongruent information, 534
inducement, 561
towards emotion-oriented technology, 725
Attitudinal reactions, 595, 597
Attribution, 506–507
Audience specific persuasion, 561, 570
Authenticity, 164, 197, 231–232, 256, 259
Automatic humor generation, 525
Automatic humor recognition, 610, 616
Automatic persuasion, 524
Automatic Speech Recognition (ASR), 74
Autonomic Nervous System (ANS), 137–139, 145, 695

B

Backchannels, 324–325, 327–330, 332–336, 338, 341–342
in persuasive communication, 593–595
Bag-of-Words, 82–83
Balancing (Principalism), 716–717, 719
Basic emotions, 22, 28
Bayesian Interactive Argumentation System (BIAS), 570
Bayesian networks, 574
Becker, C., 504
Behavioural anticipation, 483–497
Behaviour
expressivity, 596
inducement, 561, 563, 572
planning, 390–391, 393, 397–398, 400–401, 406, 409
realisation, 390–391, 393, 397–398, 400–405, 408
Behaviour Markup Language (BML), 397
Being in love, 664–665
Belfast adventure, 186
Belfast boredom database, 179
Belfast Driving Simulator Data, 246, 251, 269
Belfast Green Persuasive Data Set, 207
Belfast Naturalistic, 177–179, 181, 185–186, 189, 201–202, 207, 235, 246, 252, 254, 258, 261, 266, 272–273, 275–277
Belfast Structured database, 181, 201
Belief, Desire, Intention (BDI), 397, 504, 513
Belief, Desire, Intention and Emotions (BDI&E), 567, 571
Benevolence
in persuasive nonverbal communication, 587, 590–592, 604
strategy (in persuasion), 590, 604
Berlin database, 176, 182
Big-five, 506
Big six, 174, 177, 228
Biological plausibility, 49
Blood pressure, 237
Blood Volume Pulse (BVP), 141, 155, 695
Bodily and physical practices, 660
Body
contraction, 300
expansion, 300
language, 586, 589–590
posture, 244
Bodystorming, 639, 652, 660–661, 674–675
Bottom-up attention, 306–311
Box-and-arrow approach, 48
Brainstorming methods, 639, 673–674
Brain waves, 148
Brevity in assessment, 691
British National Corpus (BNC), 625, 627
Brown & Levinson, 503, 512

C

Call centres, 181, 187, 202–203
Caltech Frontal Face Database, 183
Camshift algorithm, 298
Capability principle (of cognition), 535
Captology, 527, 530
Cardiovascular system, 140

Castaway Reality Television Data Set, 246–247, 252, 264
Categorical labels, 217
CEICES, 92, 178, 191, 206, 243
Central Nervous System (CNS), 135, 146
Central (systematic) mode of thinking, 550–551
Cepstrum, 81
Certainty (in persuasive nonverbal communication), 591, 595
Chameleon effect, 38–39
Character Markup Language (CML), 395
Circumplex model of affect, 691
Classification, 73–78, 80–81, 83–85, 91
CMU facial expression database, 183
CMU PIE database, 183
CMU Pronouncing Dictionary, 619
Coding scheme, 214, 234, 258
Coercion, 507, 530, 532, 563–564
Cognitive anticipation, 497
Cognitive aspects of humour, 615
Cognitive components of envy, 543
Cognitive consistency, 534
Cognitive inconsistency, 460, 476
Cognitive load variation (assessment of), 343
Cognitive models of emotion, 471
Cognitive neuroscience, 47–53
Cognitive revolution, 47
Cognitive science, 48
Cohn-Kanade database, 104
Cold persuasion, 531
Commercial design community, 673
Commercial environments, 639
Commitment, 533
Communicative action, 322–325, 327, 329
Communicative function, 351, 357
Communicative goal, 529, 564–565
Communicator corpus, 512, 514
Competence (in persuasive nonverbal communication), 587, 590–592, 604
Complementarity, 355, 357
Complex emotion expressions, 352–353, 359–361
Component process model, 442
Computational model of emotions, 460
Computational verbal humor, 525, 611–615
Computer anxiety, 691
Computer game difficulty (assessment of), 695
Computer games, 643
Computer-induced human change, 524
Computer Supported Cooperative Work (CSCW), 647–648
Concealment of emotion, 232
Conceptual level of processing, 446
Conceptual uncertainty of Emotion, 726, 728
Conclusion explicitness (in persuasion), 569
Confidentiality, 208–209
Connectedness, 20–21
Consciousness and emotion, 726
Constraints of commerce, 650
Context-generic psychometric measures, 690–691
Context labelling, 215, 219–220, 239
Contextual goals, 523, 559
Contextual novelty, 443, 450, 452
Context of use, 658
Continuous, 164, 170, 217, 232
Control, 10, 18, 20–21, 23–25
Convergent validity, 692
Conversational interactions, 207–208
Conversation analysis, 514–517
Coordination of modalities, 350
Copyright, 178, 185, 202, 258, 266
Co-regulation, 36
COrollary Discharge of Attention Movement (CODAM), 56
Corpora, 504, 514
Corpus analysis, 603
Corpus/*corpora*, 171, 179, 186, 236
Corpus engineering, 73–78
Correctness *vs.* effectiveness of argumentation, 563
COSMO (virtual tutor agent), 356, 599
Co-verbal gestures, 589
Cover classes, 177, 216, 225, 228, 263
Creativity, 610, 626–627
Credibility, 586, 598
CREST database, 180, 187
Critical design practice, 683–684
Critical Discourse Analysis, 516–517
Critical Reasoning in Human Autonomy, 709
Criticism of Principalism, 719, 712
Cronbach's alpha, 232
Cross-cultural difference, 265
Cultural commentaries, 697
Cultural commentators, 697
Cultural perspective, 666
Cultural probes, 651, 661–662, 671–672, 697
Cultural situations, 649
Curse of dimensionality, 84–87, 170, 190
CVL face database, 182

D

DARPA Communicator Corpus, 181, 187–188
Database development, 245
Data-driven approach, 75, 82

Data protection, 759–760
Deception, 530, 576, 598–602
Decision-level fusion, 120, 123–124
Defensive avoidance, 541
Definition of emotion, 51
Degradation in media quality (assessment of), 695
Dependable generalizations, 569
Description of the situation, 660
Design
 challenges, 659–660
 of emotional interaction, 637–639
 for intimacy, 661
 methods, 658, 662
 principles, 569
 problem framing, 665
 prototype, 642
 values, 639
Dialogical persuasion, 566–567, 573–575
Diary studies, 698
Differential amplification, 137
Dimensional encodings, 381
Dimensional representations of affect, 690–691
Dimensions, 4–5, 17, 19, 21, 24, 26, 50–51, 53
 of emotional behaviour, 354
 of verbal persuasion, 559–578
DIPLOMAT (negotiating agent), 573
Direct generation of goals, 540
Disappointment, 489–491, 494–495
Discouragement, 490–491, 497
Discourse
 delivery, 588
 plan, 552–553
Disgust, 50, 52, 54
Display rules, 219
Dissynchrony, 37–38
Distinctiveness of emotion-oriented technology, 725–732
Divergent validity, 692
Domain specific persuasion, 559, 572
Dramatic curve (of user's experience of an interactive story), 682
DRIVAWORK corpus, 250–251, 270
DRIVAWORK database, 185
Drives, 11, 19
Driving simulator, 185, 204, 246, 251, 269
Dual-process theories, 531, 535–536, 550
Durkheim, 503
Dynamic advertising, 559
Dynamic Bayesian Network (DBN), 87
Dynamic Belief Network, 576
Dynamic Time Warp (DTW), 87

E

Ecological validity, 76, 174, 201, 698
Educational role-play, 568
Edutainment, 559
Effective arguments, 560
Effectiveness assessment, 677–678
Effects of body behaviour on persuasion, 592
Efficiency, 638
Elaboration likelihood model of persuasion, 531, 535, 550
Electrocardiography (ECG), 140–141, 144–145, 147, 150, 192, 204–205, 252, 269–270
Electrodermal Activity (EDA), 137–138
Electroencephalography (EEG), 147–148, 150–151, 154, 696
Electromyography (EMG), 145–146, 148, 150–151, 153–154, 205, 237, 270
Eloquence generation, 573
EMA (Gratch & Marsella), 506–507
Embedded physiological sensors, 696
Emblematic messages, 351
Embodied Conversational Agent (ECA), 33, 42, 44, 288, 354–357, 359–360, 362, 389–390, 392–394, 396–398, 400, 402–408, 571, 577, 595–605, 678
Emergent emotion, 214, 217, 228–229, 256, 259
EMoTaboo, 178, 185–186, 205, 236, 246, 249–250, 252, 259, 262, 270
EMOTE, 577, 596
Emotion, 9–29
 basic, 4, 22, 28, 48, 50–52, 55–56, 58
 and biological evolution, 485
 categories, 372, 374, 383–384
 category labels, 22–23
 components of, 3, 5, 32–36
 and decision making, 487, 494–497
 design features of, 24
 emergent, 14–15, 18, 23–26
 explicit response to, 32–33
 and goals, 484–486, 495
 implicit response to, 32–33
 impressions of, 21–22
 labeling, 694
 and language, 507–514
 levels of processing of, 3
 in persuasive nonverbal communication, 586, 598–603
 pervasive, 14–16, 23
 semantic categories of, 4–5, 10
 in ethics, 732
 and goals, 531, 536–538, 541

states, 23–27
suppressed, 25
theories of, 3–6, 50–53, 56
units of, 16–17
Emotional contagion, 37–41
Emotional *corpora*, 350, 353–354, 357
Emotional expressions, 663, 665
Emotional persuasion, 531, 535–542
Emotional privacy and autonomy, 741–743
Emotional response, 536
Emotional space, 174, 253
Emotional strategies of persuasion, 531, 550
Emotion Annotation and Representation Language (EARL), 372–373, 383–385, 393
EmotionML, 76
EMoto system, 662, 668
EmoTV, 177–178, 185, 202, 236, 246, 254, 258, 266
Empathic design methods, 639
Empathic tour guide system, 573
Empathy induction, 679
Emulation (emotion), 531, 542–543, 545, 547, 549, 552
End users, 642–643, 645, 647, 651, 653
Energy, 78, 80–81, 88
Enjoyability, 638
Envy, 531, 537, 540, 542–549
EPA, 503
Epistemic mental states, 230
Erlangen AIBO database, 187
Erlangen driving database, 186
Ethical audit, 753–755
Ethical decision making, 710, 714–715, 717–719
Ethical duties and autonomy, 735–750
Ethical duties of developers of emotion-oriented technology, 743–748
Ethical issues, 696
Ethical reasoning, 571
Ethics, 191, 208
committee, 726, 753–766
Ethnography, 643, 648
Ethnomethodology, 515–516
Ethos strategy (in persuasion), 590, 592
Evaluation, 641–654
challenges, 654
of experience, 648–649, 658
methods, 640, 687–688, 690–699
in persuasive nonverbal communication, 586, 591–592, 595–596
Event Related Potentials (ERP), 444
Evolution, 485, 503
Excitement, 664–665
Expectation, 766
-based emotions, 460, 469–473, 488–489, 496
invalidation and emotions, 492
Expected and non pre-felt emotion, 494–496
Expected and pre-felt emotion, 494, 496–497
Experience
clips, 662
design, 654
Experimental psychology, 645–646
Exploitation, 530
Explorative behaviour, 454
Expressions, 11–15, 19–20, 372–373, 376–377, 381, 762
Expressive gestures, 595–596
Expressivity, 236, 354, 357, 360
eXtensible Markup Language (XML), 383
Extensible Multimodal Annotation language (EMMA), 384–385
Eye contact, 604

F

Face
detection, 105, 298
recognition, 168
threat mitigation by non-verbal means, 603
threats, 602–603
tracking, 106, 109
Facial Action Coding System (FACS), 102–104, 106, 109, 236, 352
Facial Action Packages, 236
Facial Action Units (FACS), 408
Facial animation, 189
Facial Animation Parameters (FAPs), 359, 405
Facial clues of deceit, 602
Facial expression
asymmetry, 599–600
of deception, 598–599, 602
masks, 599
masking, 694, 699
modeling, 588, 593, 595–596, 599–600, 602–605
timing, 599
Facial feature extraction, 105
Facial micro-expressions, 599
FACs coding, 192
Factor analysis, 233
Fake emotions, 576
False-coloured rendering, 301–302
False emotional expressions, 598
Familiarity, 443, 444, 449–450
F0 contours, 235

Fear, 34, 40, 50–55, 58
appeal, 539, 541, 561, 569
conditioning, 52, 54
module, 50, 52, 55
and ethics, 713–722, 744
Feature
extraction, 65–68
-level fusion, 122–124, 126
recognition, 65–68
reduction, 84–85
selection, 84–86
Feedback, 324, 326–327, 331, 335–343
expressions, 594
Feeling, 11–15, 17–18, 27–28, 51, 54
FEELtrace, 223–224, 232
Felt emotions, 562
Flexibly persuasive ECAs, 577
Flow, 692, 696
Focus group, 651–652
Formal user studies, 689
Fun, 638, 688, 692
Functional Magnetic Resonance Imaging (fMRI), 147
Functional Markup Language (FML), 397
Functional value of emotion, 484
Fusion, 65–68, 119–120, 122–124, 126
Futile goals, 531
Fuzzy rules, 600

G
Gain-framed and loss-framed appeals (in persuasion), 569
Galvanic Skin Response (GSR), 179, 204–205, 252, 269–270, 695
Games, 200, 204–205, 207
Gaze persuasive strategies, 590
GEMEP Data Set, 245–246, 252
General Theory of Verbal Humor (GTVH), 611–614
Generated goal, 533, 537, 540, 542, 551
Generic emotion labels, 228–230
Geneva Appraisal Wheel, 691
Geneva Multimodal Emotion, 201, 246
Geometric feature, 102, 106–107
Gesture, 171–172, 177–179, 184–185, 190, 201, 203, 205, 215, 234, 236, 244, 249–250, 253–255, 258–259, 262, 270–278, 280–283
recognition, 298–300
repositories, 350
in persuasion, 588–589
GESTYLE representation language, 396
Global label, 220, 222, 254, 258
Goal hooking, 534, 540
Goffman, 503, 507
Granularity, 174
Gratch & Marsella, 506
Greta (Embodied Conversational Agent), 357, 360, 577, 596, 600, 602–603
GRID study, 4–5
Groningen ELRA corpus, 182
Grooming, 502
Grounding, 327, 331
Ground truth, 175, 204, 218, 249, 268
Guilt appeal, 569
Gullibility, 535

H
HAHAcronym (computational humour system), 610, 616–624
Happiness, 50
Heart Rate (HR), 179, 237, 695–696
Heart Rate Variability (HRV), 695
Hedonic responses to software, 692
Heise, 503
Heuristics, 560, 573, 576
-systematic model of persuasion, 531, 535, 550
Historical perspective on evaluation, 644–649
Hot anger, 353
HUMAINE, 11, 14, 23–24
database, 164, 171, 185–186, 214, 220, 224, 228, 235–239, 243–283
portal, 165
Human autonomy, 735–750
Human–computer interaction, 638
Human–machine interaction, 728
Human Markup Language (HumanML), 393–394
Human values, 700
Human visual system, 301, 305
Humour
evaluation, 623–624
and culture, 615–616
and gender, 614–615, 617, 623
and personality, 615
in sociology, 611, 614–615
-specific stylistic features, 628–629
Hunger, 11, 13, 19

I
Identity, 504–505, 514
Image processing, 66
Imitation, 38–39, 41
Impedance plethysmography, 142, 144
Importance (in persuasive nonverbal communication), 585–605

Incidental emotions, 539
Incongruity theory (of humor), 611, 614
Independent Component Analysis (ICA), 85
Indian Institute of Technology Kanpur database, 183
Individual differences, 189, 217
Induced emotional state, 562
Induction
of emotion (ethics), 759, 775
techniques, 184, 188, 245, 251, 268
Inferiority, 542–549
Informatics of affect, 666
Informational view of emotion, 639, 666–667
Information
flow, 48–49
gathering (ethics), 744–746
Informed consent, 191, 208–209, 677, 758, 760–763
Inner dialogues (in persuasion), 574
Input, 48–49, 53, 55–56
In situ exploration, 662
Instrumental rationality in autonomy, 738–739
Insula, 50–52, 54
Intelligent virtual agent, 288
Intensity, 24
Intention, 528, 530–536, 538–542, 545, 548, 551
Intentional change of other's attitudes, 528
Intentionality, 20
Intentional stance, 509
Intent planning, 390, 397–399
Interaction, 501–502, 504–505, 507–511, 514–517
analysis, 517
design, 663
Interactional perspective on emotion-oriented technology, 748–750
Interactional Sociolinguistics, 517
Interactional synchrony, 324, 340
Interactional view of emotion, 639, 666–667
Interactive story, 682–683
Interactive virtual agents, 342
Interface paradigm, 645
Intermediary agent, 505–506
International Affective Picture Set, 203
Interpersonal sensitivity, 340
Interpersonal stance, 24
Interpretation loop, 688
Inter-reliability, 76
Intra-reliability, 76
Invalidation-based emotion, 488–494
Irony, 564, 571
ISLE project *corpora*, 179
ISO 9241 (Ergonomics of Human–System Interaction), 689, 691–692
ISO 13407 (User-Centred Design standard), 638, 642–643, 657

J
Japanese Female Facial Expression (JAFFE) database, 182
JAPE (computational humor system), 616
Jokes, 612–614, 616–617, 624, 626–628, 631
Joy, 34
JST/CREST Expressive Speech Corpus, 203

K
Kemper, 503
Kinetic typography, 571, 577–578

L
Laban annotation scheme, 596
Laban movement analysis, 662, 699
Labelling schemes (annotation), 190, 214, 222–223
Lack of interpretive skills (emotion), 746–748
Language & emotion, 507–514
Least effort principle (of cognition), 535
Leeds–Reading database, 175, 176–178, 180, 187, 202, 230
Legislation of emotion-oriented technology, 731
Lexeme, 82
Lexical devices in humorous text, 614
Limitations (of systems), 642, 644
Limiting factor (of systems), 644–647
Linea Discriminant Analysis (LDA), 84–86
Listener, 321–343
Listening behaviours, 321–343
Loebner Prize, 509
Logbook probes, 697–699
LOLITA (natural language system), 573
Low fidelity prototypes, 652
Low Level Descriptor (LLD), 79
Low-level feature, 65, 68
Ludology, 654

M
Machine learning, 84
Magtalo (multi-agent argumentation, logic, and opinion), 574
Manipulation, 530, 532, 534–535, 541, 553
Manipulative persuasion, 530, 532
Markup languages, 370–371, 376, 381–385
Max, 504–505
Meaning-text theory, 507
Means–end reasoning, 533–534, 537–538, 540

Mean-shift algorithm, 298
Measures of performance, 645
Media Equation, 570
Mehrabian, 503
Memorization, 571
Memory based learning, 631
Meta-analytic reviews of persuasion effects, 569
Metaphor (in persuasion), 569
Method(s)
  cards, 651
  of user study, 638
Migraine (persuasive system), 572
Mimicry, 37–40, 324, 340–341
Mirroring, 324, 341
Mixed emotions display, 599
Mixed initiative, 511–513
Mock-ups, 642
Modalities, 165, 172–173, 177–179, 184–185, 189, 190, 205, 244–245, 253, 262, 264, 271–283
Modelling, 47–59
Modifying other's beliefs, 529
Modifying other's goals, 529
Monological persuasion, 565–566, 572
Mood, 20, 22, 24–25
  boards, 672
  effects, 567
  induction, 203–204
Multicultural (cross-cultural), 185
Multimodal, 164, 172, 178–179, 184–186, 189–190, 192, 200–201, 205, 236–237, 243–244, 246, 270
Multimodal communication, 351–355
Multimodal expressions, 350–351
Multimodal fusion, 65–68
Multimodal fusion technique, 119
Multi-modal interfaces, 770, 777–778
Multimodal natural language generation, 390, 395, 400
Multimodal Presentation Markup Language (MPML), 395
Multimodal Utterance Representation Markup Language (MURML), 395, 404, 407
Mutuality, 340

N
Natural argumentation, 564, 570
Naturalistic, 164, 172, 177–179, 181, 184–189, 198–203, 207–209, 228, 230, 235, 245–246, 252, 254, 258, 261, 264, 266, 272–273, 275–277
Naturalistic studies and ethics, 763
Natural language
  applications, 677
  generation, 390, 395, 400–401, 406, 559, 564–571
Need for prediction, 493
Negative expectation, 487–489, 492–493, 495
Negotiation, 563–564, 566, 573
Neural network, 48, 52–59
Neuron, 53–56
N-grams, 82–83
Nice Argument Generator (NAG), 574
Non-linguistic vocalization, 81–82
Non manipulative persuasion, 530, 532
Nonverbal behaviour in face-to-face communication, 525, 586
Non-verbal communication, 351
Nonverbal persuasion, 585–605
Normative behaviour, 509
Normative component of positive expectation, 491–492
Norms (ethics), 715–716, 718, 721
Novelty processing, 441–455
Nucleus accumbens (Nacc), 54–56

O
Obligations, 507
  principalism, 714, 719
Observing of emotion (ethics), 760
Obtrusiveness, 136
OCEAN (five-factor model of personality), 398
Olivetti research database, 184
One-sided and two-sided messages, 569
Online Dispute Resolution, 566
Ontological reasoning, 619
Ontologies, 383, 385–386
Open design, 668
Open Mind Common Sense (OMCS) corpus, 625–627, 630–631
Optimal innovation, 632
Orbitofrontal Cortex (OFC), 51–52, 54, 58
ORESTEIA database, 179, 185
Organisational requirements, 657
Orientation response, 444, 446
Orthographic transcription, 234
Ortony, Clore, and Collins (OCC) model of emotions, 380, 398, 666
Oulu University physics-based face database, 183
Output, 48–49, 53, 55–56

P
Pain, 13–14, 19
PANAS (self-report psychometric test), 690

Paper prototyping, 675–676
Paralanguage, 235, 258
Paralinguistic, 72
Parasympathetic response, 137, 140
Partial novelty, 443, 451
Participatory design, 642, 651
Passion, 11–12, 14, 16, 27
Pathos strategy (in persuasion), 590, 592
Pattern recognition, 84
Perceptual novelty, 443, 450–452
Performative displays (in persuasive nonverbal communication), 597
Peripheral (heuristic) mode of thinking, 550
Personal construct theory, 682
Personality, 534, 539, 541–542, 546, 548–549
  and ethics, 730
Personas, 672–673
Persuader's reasoning and planning, 531
Persuasion, 527–542, 548–551
  and ethics, 731, 765
  strategies, 527, 530, 535–536, 549, 552
  task, 552
  text, 553
  through appeal to expected emotions, 531, 536, 538–539, 550
  through arousal of emotions, 531, 536–542, 551
Persuasive communication, 560–564, 569
Persuasive discourse, 588–589, 592, 594
Persuasive force, 562
Persuasive gaze, 591
Persuasive gesture, 590, 592–593
Persuasive humor, 610
Persuasive intention, 532, 536
Persuasive messages, 559–560, 563, 565, 568–571, 574, 576, 578
Persuasive plan, 530, 532, 541
Pervasive emotion, 244
Petit Mal robot, 683
P300 (evoked potential component), 444
Phonetic transcription, 234
Photoplethysmography, 141–144
Physiological data, 122
Physiological measures, 134–136, 149, 152–156
  for human factors assessment, 696
Physiological signal, 135, 137
Physiology, 13
Pitch, 77, 79–81, 88
Pleasure arousal dominance (PAD), 503–505
PML (representation language), 577–578
Politeness, 228, 258, 271–275, 278–281, 513–514
  behaviour, 603
  strategies, 603
Political discourse (non-verbal behaviour), 589
PORTIA (persuasive system), 574
Positive emotions, 610
Positron Emission Tomography (PET), 147
Potency, 17, 19, 24
Power, 503, 505–507, 515–518
PPP Persona, 577
Practical argumentation, 552
Pragmatist philosophy, 649
Predictability, 444
Prediction invalidation and emotions, 493–494
Preparatory function of emotion, 485
Preventive medicine, 559
Pride, 664
Primates, 501–503
Principal Component Analysis (PCA), 84–85
Principalism, 713–722, 756–757
  as common sense morality, 722
  as method, 713
  vs. general guidelines, 722
Prior considerations and ethics, 720
Privacy, 669
Procedural independence in autonomy, 736, 739–742, 744–745, 750
Product concept, 651
Promoter (persuasive system), 572–573
Prosody features (in humour), 617
Prospect theory, 570
Prototype idea generation, 673
Prototypical visual expressions (in persuasive nonverbal communication), 597
Prototyping, 638–639, 652–654, 675–676, 684
Psychological Image Collection, 182
Psychopathology, 51
Psychophysiology, 133, 135, 151, 153, 156
Pupil diameter, 695

Q

Qualitative and quantitative methods, 689
Qualitative *vs.* quantitative evaluation, 688–690
Quantised labelling, 220–221, 228
Quantity of motion, 300
Questionnaire measure, 690, 692

R

Random Words (brainstorming method), 673–674
Rapport, 237, 253, 256, 277, 322, 330, 338–343
Rapport Agent, 338, 340–343
Rational coherence in autonomy, 738

Rationality in autonomy, 736, 738–740, 744, 751
Rational persuasion, 531
Reactance, 535, 541
Reality Castaway TV database, 186
Redundancy, 355, 357
Referral-backfire effect, 567
Reflexive engineering, 683
Reflex level of processing, 358
Regulation, 32, 34, 36, 38, 41
Relationship establishment, 529
Reliability, 175, 218, 222–225, 231–233, 238, 263, 644–645, 654
Relief, 489–490, 492, 497
  theory (of humor), 612
Repertory grid technique (subjective experience report), 682
Representation languages, 370, 378–383, 385, 389, 392–396, 398, 406–407
Research prototypes, 652–653
Resistance to persuasion, 534, 553
Respect for autonomy, 736–737, 740, 742–744, 749–750
Respiratory effort, 237
Respiratory system, 143–144
Responsibility, 505–507
Rhetoric, 527, 530, 552–553, 561, 563–565, 573–574, 576
Rhetorical Structure Theory (RST), 565
Rich Representation Language (RRL), 372, 393, 395, 398, 577
Ritual, 509
Role(s)
  of emotions in systems, 637–639
  management, 514
  of ethics committees, 753–759, 760–766

S

Sadness, 50, 52, 54, 59, 664–665
Safety, 638
SAL Hebrew database, 186
Saliency maps, 305–309, 311–312
Sampling
  rate, 137
  resolution, 137
Scene analysis, 66
Schematic level of processing, 446
Science fiction, 714, 727–728, 730
SCREAM (Prendinger & Ishizuka), 505
Scripting languages, 393–395, 404
Script opposition (in humor theories), 612–613, 616
Second-order
  beliefs, 548
  reasoning, 574
Segmental feature, 78–79
Self
  -awareness in autonomy, 737, 742
  -determination, 535
  -esteem, 58, 534, 541, 544, 547–549, 552
  -evaluation in autonomy, 737
  -reassurance gestures, 605
  -reflection in autonomy, 736
  report affect, 680
  -report psychometric tests, 690
  -understanding in autonomy, 737
Self-assessment Mannikin, 224, 690
Semantic fields, 619, 621–623
Semantic novelty, 443–444, 446, 452
Semantic Script-based Theory of Humor (SSTH), 611–612
Sememe, 82
Semi-intelligent information filters, 731, 765
Sense of injustice, 490–492, 497
Sensitive Artificial Listener (SAL), 178–179, 185–186, 208, 246–248, 252, 258, 261, 267, 277–282, 354
Sensory honesty, 295
Sensual evaluation instrument, 651, 679–682
SenToy (tangible interaction interface), 678
Shades of emotion experiences, 639
Shielding, 137
Signs, 65
  of emotion, 175, 177, 186, 197, 217, 219, 234–237, 252, 254–255, 258–259
Signals, 3, 54, 56, 65, 67–68, 165, 171, 184, 197, 200, 218, 234–237, 249, 264, 270
Signal isolation, 136
Signing Gesture Markup Language (SiGML), 395
Sincere and deceitful persuasion, 532
Situation, Agent, Intention, Behaviour, and Animation (SAIBA), 397, 406–408
Six Thinking Hats (brainstorming method), 673–674
Skin conductance response (SCR), 138
Skin temperature, 204–205, 237, 252, 269–270
SMARTKOM database, 185
SMIL-Agent, 577
Social action, 559
Social comparison, 543–545, 547, 549
Social glue, 340
Social influence, 527–529, 591
Socialization, 502
Social lies, 576
Social relations, 560

Social signal processing, 68
Social virtual agents, 586
Society, 502–503, 507, 516
Sociological models, 503–504
Socratic effect, 534
Somatic Nervous System (SNS), 137–146
Spaghetti databases, 186
Specification (principalism), 716–717, 719
Spectrum, 80–81
Speech act, 324–326
 theory, 325
"Speech-genre" think aloud protocol, 694
Speech processing, 67
Staged lived experiences, 676–677
State transition, 50–53
Statistical learning, 169
Status & power, 503
Status, 503
Stemming, 82–83
Stimulus evaluation checks (SECs), 18, 34, 295
STOP (behaviour inducement system), 572
Stopping, 82–83
Storytelling, 568–569, 573
Strength of persuasive strategies, 570
Structure, 48–52, 54–55, 57, 59
Subjective experiences, 659–660, 682–683
Suddenness, 444, 450–451
Sulkiness, 664
Superiority theory (of humor), 611
Superposition of felt emotions, 600–601
Support Vector Machine (SVM), 83, 87–88
Supra-segmental feature, 79
Surprise, 442, 492–494, 497, 664
SUSAS database, 187
Suspicion of manipulation, 534–535
Switchboard corpus, 511
SYMPAFLY, 187, 206
Sympathetic response, 137, 140
Synchrony, 37–41
Synthesis, 163, 169–171, 185, 243, 249
Synthetic perception, 294–295, 300, 303–304
Synthetic perceptual maps, 453–454
Synthetic sensing, 448–449
Synthetic vision, 301–302, 307, 309–310
Systematic Evaluation for Stakeholder Learning (SESL), 648

T

TALKAPILLAR, 180
Tangible emotional interaction interfaces, 678
Tangible interfaces, 659
Target group, 658
Taxonomies, 174
Technology probes, 661–662
Text-to-speech, 395, 403
Thalamus, 54
Theoretical issues (ethics), 714
Theory of mind, 509, 533
Think aloud protocol, 693–694
Third wave of human–computer interaction, 697
Threatening argumentation, 539
Time course, 198, 220, 230
ToM, 509
Top-down approach, 309
Top-down attention, 306, 308
Trace labelling, 221, 260
Transfer
 of autonomy, 750
 function, 49
Transformative design, 650
Transitional emotion, 329
Transliteration, 73–74
Trust, 505–506, 692
Turn-taking, 329–330, 508
Two levels of evaluation, 688

U

UMIST database, 183
Uncanny valley, 728–730
Uncertainty, 450–451, 454, 460, 467, 472, 474, 479, 501
 in persuasive strategies, 570
Undercutting, 548, 567
Unfairness, 530
Units of analysis, 171
Universal persuasion, 570
Unpredictability, 51
Unselfish concern, 534
Usability, 638
 assessment, 689
 goals, 638
 metrics, 649
 questionnaires, 693
 testing, 650–653, 693–694
Usefulness (principalism), 714
Users, 641–646, 650–654
 agency (in interactive stories), 682
 autonomy, 700
 data, 672
 experience, 638, 654, 658–659, 676
 experience goals, 638
 immersion (into interactive stories), 682
 involvement, 642, 650–651
 model, 570

Users (*Cont.*)
needs, 642
practices, 642, 647–653
requirements, 657
satisfaction (with computer based systems), 692
scenarios, 672–673
transformation (by interactive stories), 681
User-centred design, 638, 639, 641–644
User-centred design activities, 641–644
User-centred design methods, 641, 657, 662
User-centred design processes, 638
UWIST MACL (self-report psychometric test), 690

V
Valence, 17, 24, 51–53, 219, 223–224, 232–233, 254, 256, 259
Valentino (affective text generation system), 576
Validation and emotions, 492–494
Validity, 174, 201, 218
Values, 524, 530
Variety of emotions, 681–682
Velten technique, 203
Ventral tegmental area (VTA), 54–56
Verbal persuasion, 559–579
Virtual Human Markup Language (VHML), 371–372, 393–394, 406
Visemes, 392, 403
Voice-based systems, 507
Voice quality, 79, 81
VoiceXML, 510

W
Warm persuasion, 531
Wavelet, 78, 81
Withdrawal, 51
Wizard of Oz, 72, 185, 677–679
WordNet (English language thesaurus), 618–621, 628–629

X
Xface, 577
XM2VTS multimodal face database, 189
XSTEP, 577

Y
The Yale face database, 183–184

Z
Zoom-lens model of attention, 308